June 11–14, 2012
Hilton Head, SC, USA

I0054757

Association for Computing Machinery

Advancing Computing as a Science & Profession

MobiHoc'12

Proceedings of the 13th ACM International Symposium on
Mobile Ad Hoc Networking and Computing

Sponsored by:
ACM SIGMOBILE

Supported by:
NEC Laboratories America

**Association for
Computing Machinery**

Advancing Computing as a Science & Profession

The Association for Computing Machinery
2 Penn Plaza, Suite 701
New York, New York 10121-0701

Notice to Past Authors of ACM-Published Articles

ISBN: 978-1-4503-1281-3 (Digital)

ISBN: 978-1-4503-1744-3 (Print)

Additional copies may be ordered prepaid from:

ACM Order Department
PO Box 30777
New York, NY 10087-0777, USA

Phone: 1-800-342-6626 (USA and Canada)
+1-212-626-0500 (Global)
Fax: +1-212-944-1318
E-mail: acmhelp@acm.org
Hours of Operation: 8:30 am – 4:30 pm ET

Printed in the USA

Message from the MobiHoc 2012 General Chair

On behalf of the organizing committee, it is my pleasure to welcome you to *MobiHoc 2012, the Thirteenth ACM International Symposium on Mobile Ad Hoc Networking and Computing.* *MobiHoc* continues its tradition of serving as the premier forum for bringing together researchers and practitioners on a broad spectrum of wireless networking and mobile computing topics that include mobile ad hoc networks, wireless sensor networks, wireless mesh networks, cognitive radio networks, and other emerging areas. *MobiHoc 2012* is being held on the beautiful Hilton Head Island, SC, which we hope will provide an enjoyable environment for presenting the most recent research achievements and connecting with the research community in these areas.

This year's *MobiHoc* includes seven exciting technical sessions that include 24 exceptional papers, four workshops, and a poster session. The technical program committee led by the chairs Yih-Chun Hu and Sanjay Shakkottai did an outstanding job in reviewing and selecting the *MobiHoc* papers, while maintaining the *MobiHoc* tradition of high selectivity and fairness. The Workshop chairs Ryuji Wakikawa and Wenye Wang did an excellent job in getting together the *MobiHoc* workshops: the *Workshop on Sensor-Enhanced Safety and Privacy in Public Spaces*; the *Workshop on Airborne Networks and Communications*; the *Workshop on Name-Oriented Mobile Networking Design - Architecture, Algorithms, and Applications*; and the *Workshop on Pervasive Wireless Healthcare*. These workshops are focused on areas of emerging interest and provide significant diversity to the *MobiHoc* program. The poster chair Karthik Sundaresan did substantial work to have poster papers reviewed and accepted in time for the proceedings. I am deeply thankful to all of these committee members.

I am delighted to announce keynote presentations from two renowned contributors to *MobiHoc* (as well as their respective research areas): Dr. Edward Knightly from Rice University, who will be presenting on *Urban-scale Wireless Networks in Unlicensed sub-GHz bands*; and Dr. Nitin Vaidya from the University of Illinois at Urbana-Champaign, who will be presenting on *Resilient Distributed Consensus*.

The success of *MobiHoc 2012* is due to the tremendous efforts of the organizing committee members. In addition to the technical program committee chairs, the workshop chairs, and the poster chair, I would like to gratefully acknowledge the efforts and contributions of the finance chair Yu Wang, the local arrangements chair Yiming Ji, the registration chair Srihari Nelakuditi, the publications chair Jiang (Linda) Xie, the TPC meetings chair Sujay Sanghavi, the publicity co-chairs Kaushik Chowdhury and Shweta Jain, the web chair Justin Manweiler, and the student travel grant chair Mainak Chatterjee. Special thanks go to the steering committee, especial the chair P. R. Kumar, and Nitin Vaidya for providing valuable help in shaping *MobiHoc*.

Finally, I would like to extend my sincere thanks to NEC Laboratories for providing financial support for the event, and thank our sponsors ACM and SIGMOBILE and all ACM staff who helped in organizing this event.

We hope that you will enjoy *MobiHoc 2012*.

Asis Nasipuri
The University of North Carolina at Charlotte
MobiHoc 2012 General Chair

MobiHoc'12 Program Chairs' Welcome

Welcome to the *Thirteenth International Symposium on Mobile Ad Hoc Networking and Computing (MobiHoc 2012)*. It was our privilege to coordinate the Technical Program process, in which 24 papers were selected from 120 submissions. Each submission received at least three reviews from Program Committee members, and the top rated papers were discussed at the Technical Program Committee meeting, where the committee reached consensus on the 24 accepted papers.

We are grateful to all of those who have helped put this program together: the authors who submitted their work, without whom we could not have a program, to the TPC members, who wrote prompt expert reviews to help with the selection process, the TPC members who made the trip to Austin to attend the TPC meeting and to discuss these papers, our TPC meeting chair, Sujay R. Sanghavi, and the student volunteers, Siddhartha Banerjee, Praneeth Netrapalli, and Sharayu Moharir, who made our meeting run so smoothly. We're also grateful that you, our attendees, have come to hear about some of the latest work in mobile ad hoc networking and computing. We hope you'll enjoy the technical program we've put together.

Yih-Chun Hu and Sanjay Shakkottai
MobiHoc 2012 Technical Program Co-Chairs

Table of Contents

Message from the MobiHoc 2012 General Chair iii
Asis Nasipuri *(The University of North Carolina at Charlotte)*

MobiHoc'12 Program Chairs' Welcome iv
Yih-Chun Hu *(University of Illinois at Urbana-Champaign)*, Sanjay Shakkottai *(University of Texas, Austin)*

MobiHoc'12 Thirteenth ACM International Symposium on Mobile Ad Hoc Networking and Computing (MobiHoc) Organization viii

MobiHoc'12 Sponsor & Supporter xi

Keynote Talks

- **Urban-Scale Wireless Networks in Unlicensed sub-GHz Bands** 1
  Edward W. Knightly *(Rice University)*
- **Resilient Distributed Consensus** 3
  Nitin Vaidya *(University of Illinois at Urbana-Champaign)*

Session 1: Applications

- **Your Friends Have More Friends Than You Do: Identifying Influential Mobile Users Through Random Walks** 5
  Bo Han, Aravind Srinivasan *(University of Maryland)*
- **Encounter Based Sensor Tracking** 15
  Andrew Symington, Niki Trigoni *(University of Oxford)*
- **EV-Loc: Integrating Electronic and Visual Signals for Accurate Localization** 25
  Boying Zhang, Jin Teng, Junda Zhu, Xinfeng Li, Dong Xuan, Yuan F. Zheng *(The Ohio State University)*

Session 2: Scheduling and Interference

- **Low-Complexity Scheduling for Wireless Networks** 35
  Guanhong Pei, V.S. Anil Kumar *(Virginia Tech)*
- **Stability Analyses of Longest-Queue-First Link Scheduling in MC-MR Wireless Networks** 45
  Peng-Jun Wan, Xiaohua Xu, Zhu Wang, Shaojie Tang *(Illinois Institute of Technology)*,
  Zhiguo Wan *(Tsinghua University)*
- **Enabling Real-Time Interference Alignment: Promises and Challenges** 55
  Kyle Miller, Atresh Sanne *(University of Texas, Austin)*, Kannan Srinivasan *(The Ohio State University)*,
  Sriram Vishwanath *(University of Texas, Austin)*
- **Optimization Schemes for Protective Jamming** 65
  Swaminathan Sankararaman *(Duke University)*, Karim Abu-Affash *(Ben Gurion University)*,
  Alon Efrat *(The University of Arizona)*, Sylvester David Eriksson-Bique, Valentin Polishchuk *(University of Helsinki)*,
  Srinivasan Ramasubramanian *(The University of Arizona)*, Michael Segal *(Ben Gurion University)*

Session 3: Sensors and Sensing

- **Taming Uncertainties in Real-Time Routing for Wireless Networked Sensing and Control** 75
  Xiaohui Liu, Hongwei Zhang, Qiao Xiang, Xin Che, Xi Ju *(Wayne State University)*

- **Minimum-Energy Connected Coverage in Wireless Sensor Networks with Omni-Directional and Directional Features** 85
  Kai Han *(Zhongyuan University of Technology & Nanyang Technological University)*,
  Liu Xiang, Jun Luo *(Nanyang Technological University)*, Yang Liu *(Henan University of Technology)*

- **Probabilistic Missing-Tag Detection and Energy-Time Tradeoff in Large-Scale RFID Systems** 95
  Wen Luo, Shigang Chen, Tao Li, Yan Qiao *(University of Florida)*

Session 4: Coding and Capacity

- **Adaptive Network Coding for Scheduling Real-Time Traffic with Hard Deadlines** 105
  Lei Yang *(Arizona State University)*, Yalin Evren Sagduyu, Jason Hongjun Li *(Intelligent Automation, Inc.)*

- **Distributed Network Coding-Based Opportunistic Routing for Multicast** 115
  Abdallah Khreishah *(Temple University)*, Issa M. Khalil *(United Arab Emirates University)*,
  Jie Wu *(Temple University)*

- **Throughput of Rateless Codes Over Broadcast Erasure Channels** 125
  Yang Yang, Ness B. Shroff *(The Ohio State University)*

- **Closing the Gap in the Multicast Capacity of Hybrid Wireless Networks** 135
  Shaojie Tang *(Illinois Institute of Technology)*, Xufei Mao *(Tsinghua University)*,
  Taeho Jung, Junze Han *(Illinois Institute of Technology)*,
  Xiang-Yang Li *(Illinois Institute of Technology & Dalian University of Technology)*,
  Boliu Xu, Chao Ma *(Illinois Institute of Technology)*

Session 5: Routing

- **Serendipity: Enabling Remote Computing Among Intermittently Connected Mobile Devices** 145
  Cong Shi, Vasileios Lakafosis, Mostafa H. Ammar, Ellen W. Zegura *(Georgia Institute of Technology)*

- **Oblivious Low-Congestion Multicast Routing in Wireless Networks** 155
  Antonio Carzaniga, Koorosh Khazaei *(University of Lugano)*,
  Fabian Kuhn *(University of Lugano & University of Freiburg)*

- **Dissemination in Opportunistic Social Networks: The Role of Temporal Communities** ... 165
  Anna-Kaisa Pietiläinen, Christophe Diot *(Technicolor)*

- **Optimal Energy-Aware Epidemic Routing in DTNs** 175
  MHR. Khouzani *(The Ohio State University)*, Soheil Eshghi, Saswati Sarkar *(University of Pennsylvania)*,
  Ness B. Shroff *(The Ohio State University)*, Santosh S. Venkatesh *(University of Pennsylvania)*

Session 6: Cognitive

- **Transmission Delay in Large Scale Ad Hoc Cognitive Radio Networks** 185
  Zhuotao Liu, Xinbing Wang, Wentao Luan *(Shanghai JiaoTong University)*,
  Songwu Lu *(University of California, Los Angeles)*

- **Enforcing Dynamic Spectrum Access with Spectrum Permits** 195
  Lei Yang *(Intel Labs)*, Zengbin Zhang, Ben Y. Zhao, Christopher Kruegel,
  Haitao Zheng *(University of California, Santa Barbara)*

- **Spatial Spectrum Access Game: Nash Equilibria and Distributed Learning** 205
  Xu Chen, Jianwei Huang *(The Chinese University of Hong Kong)*

Session 7: Cellular

- **Design and Implementation of an Integrated Beamformer and Uplink Scheduler for OFDMA Femtocells** 215
  Mustafa Y. Arslan *(University of California, Riverside)*, Karthikeyan Sundaresan *(NEC Laboratories America, Inc.)*,
  Srikanth V. Krishnamurthy *(University of California, Riverside)*,
  Sampath Rangarajan *(NEC Laboratories America, Inc.)*

- **A Case for Adaptive Sub-Carrier Level Power Allocation in OFDMA Networks** 225
  Shailendra Singh, Moloud Shahbazi *(University of California, Riverside)*,
  Konstantinos Pelechrinis *(University of Pittsburgh)*, Karthikeyan Sundaresan *(NEC Labs America Inc.)*,
  Srikanth V. Krishnamurthy *(University of California, Riverside)*, Sateesh Addepalli *(Cisco Systems, Inc.)*

- **A Distributed Resource Management Framework for Interference Mitigation
  in OFDMA Femtocell Networks** 235
  Jongwon Yoon *(University of Wisconsin-Madison)*, Mustafa Y. Arslan *(University of California, Riverside)*,
  Karthikeyan Sundaresan *(NEC Labs America, Inc.)*, Srikanth V. Krishnamurthy *(University of California, Riverside)*,
  Suman Banerjee *(University of Wisconsin-Madison)*

Posters

- **Towards Intelligent Antenna Selection in IEEE 802.15.4 Wireless Sensor Networks** 245
  Mubashir Husain Rehmani, Stéphane Lohier, Abderrezak Rachedi, Thierry Alves,
  Benoit Poussot *(Université Paris-Est Marne-la-Vallée)*

- **Localization Using Bluetooth Device Names** 247
  Troy A. Johnson, Patrick Seeling *(Central Michigan University)*

- **Channel Width Assignment Using Relative Backlog: Extending Back-Pressure
  to Physical Layer** 249
  Parth H. Pathak, Sankalp Nimbhorkar, Rudra Dutta *(North Carolina State University)*

- **Reliable Link Lifetime-Based Cluster-Head Election in Wireless Ad Hoc Networks** 251
  Dongsheng Chen, Alireza Babaei, Prathima Agrawal *(Auburn University)*

- **Price-Reward for Data Relaying and Handover Management in Wireless Networks** 253
  Muhammad Shoaib Saleem, Éric Renault *(Institut Télécom - Télécom SudParis)*

- **JIM-Beam: Using Spatial Randomness to Build Jamming-Resilient Wireless
  Flooding Networks** 255
  Jerry T. Chiang *(Advanced Digital Sciences Center, Singapore)*,
  Dongho Kim, Yih-Chun Hu *(University of Illinois at Urbana-Champaign)*

- **Extracting Jamming Signals to Locate Radio Interferers and Jammers** 257
  Zhenhua Liu *(University of South Carolina)*, Hongbo Liu *(Stevens Institute of Technology)*,
  Wenyuan Xu *(University of South Carolina)*, Yingying Chen *(Stevens Institute of Technology)*

- **A Novel Misbehavior Evaluation with Dempster-Shafer Theory
  in Wireless Sensor Networks** 259
  Muhammad R. Ahmed, Xu Huang, Dharmendra Sharma *(University of Canberra)*

- **Your Smartphone Can Watch the Road and You: Mobile Assistant
  for Inattentive Drivers** 261
  Sanjeev Singh, Srihari Nelakuditi, Yan Tong *(University of South Carolina)*,
  Romit Roy Choudhury *(Duke University)*

- **A Distributed Channel Selection Scheme for Multi-Channel
  Wireless Sensor Networks** 263
  Amitangshu Pal, Asis Nasipuri *(The University of North Carolina at Charlotte)*

- **Can Smartphone Sensors Enhance Kinect Experience?** 265
  Rufeng Meng, Jason Isenhower, Chuan Qin, Srihari Nelakuditi *(University of South Carolina)*

Author Index 267

Thirteenth ACM International Symposium on Mobile Ad Hoc Networking and Computing (MobiHoc) Organization

General Chair: Asis Nasipuri *(University of North Carolina at Charlotte, USA)*

Program Co-Chairs: Yih-Chun Hu *(University of Illinois at Urbana-Champaign, USA)*
Sanjay Shakkottai *(University of Texas at Austin, USA)*

Workshop Co-Chairs: Ryuji Wakikawa *(Toyota InfoTechnology Center, USA)*
Wenye Wang *(North Carolina State University, Raleigh, USA)*

Poster Chair: Karthik Sundaresan *(NEC Labs, USA)*

Finance Chair: Yu Wang *(University of North Carolina at Charlotte, USA)*

Publications Chair: Jiang (Linda) Xie *(University of North Carolina at Charlotte, USA)*

Registration Chair Srihari Nelakuditi *(University of South Carolina, Columbia, USA)*

Local Arrangements Chair: Yiming Ji *(University of South Carolina at Beaufort, USA)*

TPC Meetings Chair: Sujay Sanghavi *(University of Texas at Austin, USA)*

Publicity Co-Chairs: Kaushik Chowdhury *(Northeastern University, USA)*
Shweta Jain *(York College/CUNY, USA)*

Web Chair: Justin Manweiler *(IBM T. J. Watson Research Center, USA)*

Student Travel Grants Chair: Mainak Chatterjee *(University of Central Florida, USA)*

Steering Committee Chair: P. R. Kumar *(Texas A&M University, USA)*

Steering Committee: Mario Gerla *(University of California, Los Angeles, USA)*
Jean-Pierre Hubaux *(EPFL, Switzerland)*
Joseph Macker *(Naval Research Lab, USA)*
Sergio Palazzo *(University of Catania, Italy)*
Charles Perkins *(WiChorus, USA)*
Martha Steenstrup *(Clemson University and Stow Research L.L.C, USA)*

Program Committee: Anima Anandkumar *(University of California, Irvine, USA)*
Fan Bai *(General Motors, USA)*
Rajesh Balan *(Singapore Management University, Singapore)*
Randall Berry *(Northwestern University, USA)*
Sem Borst *(Technische Universiteit Eindhoven, Nederlands)*
Joseph Camp *(Southern Methodist University, USA)*
Jerry T. Chiang *(Advanced Digital Sciences Center, Singapore)*

MobiHoc 2012 Sponsor & Supporter

Sponsor:

Supporter:

NEC Laboratories
America
Relentless passion for innovation

Urban-Scale Wireless Networks in Unlicensed sub-GHz Bands

[Keynote Address]

Edward Knightly
Department of Electrical and Computer Engineering
Rice University
6100 South Main
Houston, TX 77005
knightly@rice.edu

ABSTRACT

The FCC ruled in September 2010 that unused UHF TV channels could be repurposed for unlicensed wireless Internet, a capability often termed "Super WiFi" because UHF bands have much greater range compared to today's GHz WiFi. Yet with so many channels used for TV in populated areas, few and sometimes none are left over in urban areas, severely hindering the capacity and market possibilities of Super WiFi in U.S. cities. Moreover, with a limited urban market hindering the cost savings of mass production, rural deployments are likewise hindered. Nonetheless, many urban areas world-wide have far greater unlicensed spectrum availability across many sub-GHz bands. In this talk, I will describe the global possibilities for UHF-band wireless networks, including specific deployment scenarios. Moreover, I will describe research, standardization, and policy challenges that must be overcome to realize urban-scale wireless networks in sub-GHz bands.

Categories and Subject Descriptors

C.2.1 [**Network Architecture and Design**]: [Wireless communication]; C.2.3 [**Network Operations**]: [Network management]

Keywords

UHF-band networks, urban-scale wireless networks

Author Biography

Edward Knightly is a Professor of Electrical and Computer Engineering at Rice University in Houston, Texas. He received his Ph.D. and M.S. from the University of California at Berkeley and his B.S. from Auburn University. He is an IEEE Fellow, a Sloan Fellow, and a recipient of the National Science Foundation CAREER Award. He received the best paper award from ACM MobiCom 2008 and serves on the IMDEA Networks Scientific Council.

Professor Knightly's research interests are in the areas of mobile and wireless networks with a focus on protocol design, performance evaluation, and at-scale field trials. He leads the Rice Networks Group. The group's current projects include deployment, operation, and management of a large-scale urban wireless network in a Houston under-resourced community. This network, Technology For All (TFA) Wireless, is serving over 4,000 users in several square kilometers and employs custom-built programmable and observable access points. The network is the first to provide residential access in frequencies spanning from unused UHF DTV bands to WiFi bands. The group is also co-developing a clean-slate-design hardware platform for high-performance wireless networks. The WARP platform is now operational, and ongoing research includes development of the first MUBF WLAN system.

MobiHoc'12, June 11–14, 2012, Hilton Head Island, SC, USA.
ACM 978-1-4503-1281-3/12/06.

Resilient Distributed Consensus

[Keynote Address]

Nitin Vaidya
Department of Electrical and Computer Engineering,
and Coordinated Science Laboratory
University of Illinois at Urbana-Champaign
1308 West Main Street
Urbana, IL 61801
nhv@illinois.edu

ABSTRACT

Consensus algorithms allow a set of nodes to reach an agreement on a quantity of interest. For instance, a consensus algorithm may be used to allow a network of sensors to determine the average value of samples collected by the different sensors. Similarly, a consensus algorithm may also be used by the nodes to synchronize their clocks. Research on consensus algorithms has a long history, with contributions from different research communities, including distributed computing, control systems, and social networks.

In this talk, we will discuss two resilient consensus algorithms that can perform correctly despite the following two types of adversities: (i) In wireless networks, transmissions are subject to transmission errors, resulting in packet losses. We will discuss how average consensus can be achieved over such lossy links, without explicitly making the links reliable, for instance, via retransmissions. (ii) In a distributed setting, some of the nodes in the network may fail or may be compromised. We will discuss a consensus algorithm that can tolerate Byzantine failures in partially connected networks.

Categories and Subject Descriptors

C.2.4 [**Distributed Systems**]: Distributed applications

General Terms

Algorithms

Keywords

Consensus, Byzantine faults, iterative algorithms

Author Biography

Nitin Vaidya is a Professor of Electrical and Computer Engineering at the University of Illinois at Urbana-Champaign. His research interests are in distributed algorithms, fault-tolerant computing, and wireless networks. Nitin has held visiting positions at Technicolor Paris Lab, TU-Berlin, IIT-Bombay, Microsoft Research-Redmond, and Sun Microsystems, as well as a faculty position at the Texas A&M University. He has co-authored papers that received awards at several conferences, including 2007 ACM MobiHoc and 1998 ACM MobiCom. He has served as the Editor-in-Chief for the IEEE Transactions on Mobile Computing. Nitin is a Fellow of the IEEE. For more information, please visit http://users.crhc.illinois.edu/nhv.

Your Friends Have More Friends Than You Do: Identifying Influential Mobile Users Through Random Walks

Bo Han
Department of Computer Science
University of Maryland
College Park, MD 20742, USA
bohan@cs.umd.edu

Aravind Srinivasan
Department of Computer Science and
Institute for Advanced Computer Studies
University of Maryland
College Park, MD 20742, USA
srin@cs.umd.edu

ABSTRACT

In this paper, we study the problem of identifying influential users in mobile social networks. Traditional approaches find these users through centralized algorithms on either friendship or social-contact graphs of all users. However, the computational complexity of these algorithms is known to be very high, making them unsuitable for large-scale networks. We propose a *lightweight* and *distributed* protocol, iWander, to identify influential users through fixed-length *random walks*. To the best of our knowledge, we are the first to design a distributed protocol on smartphones that leverages random walks for identifying influential mobile users, although this technique has been used in other areas.

The most attractive feature of iWander is its *extremely low* message overhead, which lends itself well to mobile applications. We evaluate the performance of iWander for two applications, targeted immunization of infectious diseases and target-set selection for information dissemination. Through extensive simulation studies using a real-world mobility trace, we demonstrate that targeted immunization using iWander achieves a comparable performance with a degree-based immunization policy that vaccinates users with large number of contacts first, while consuming only less than 1% of this policy's message overhead. We also show that target-set selection based on iWander outperforms the random and degree-based target-set selections for information dissemination in several scenarios.

Categories and Subject Descriptors

C.2.1 [**Computer Communication Network**]: Network Architecture and Design—*Wireless communication*

General Terms

Design, Performance

Keywords

Influential mobile users, random walks, distributed protocol, disease control, information dissemination.

1. INTRODUCTION

Mobile social networks, under the merging of social networks that link humans and Internet that connects computers [18], have emerged as a new frontier in the mobile computing research community. Mobile social networking is social networking where mobile users interact, communicate and connect with each other using their wireless devices. There have been several novel mobile social applications developed recently (e.g., Micro-Blog [14], PeopleNet [21], and SociableSense [29]). Meanwhile, more and more native mobile social networking services, such as Loopt and Foursquare, have been created.

Not all mobile users are equal in terms of mobility. Some of them, such as salespeople, may travel to many places during a day, while others, such as graduate students, may stay in their office for most of the working time. When considering the problem of information dissemination in mobile networks, if we employ these active salespeople as the initial physical carriers, they may be able to further propagate information to a much larger fraction of mobile users, compared with selecting initial carriers randomly. This is exactly the rationale behind the influence maximization problem of information diffusion in traditional social networks [10, 17]. Similarly, if we monitor these critical individuals, we may be able to detect the outbreaks of infectious diseases much earlier, for example, during the flu season [4].

In this paper, we address the following question: *how do we identify influential users in mobile social networks through distributed solutions with low message overhead?* Influential users are individuals with high centrality in their social-contact graphs. In our previous work [16], we propose a heuristic algorithm to select influential mobile users for information dissemination, which is an extension of the greedy algorithm by Kempe, Kleinberg, and Tardos [17]. Nguyen et al. [23] propose to find these users through the detection of overlapping community structures in dynamic networks. However, these solutions are all *centralized* and require the complete social-contact graphs.

There are two major challenges when finding these critical mobile users. First, given the large size of mobile social networks, the proposed solutions must be distributed. Besides the drawback of requiring complete contact graphs, centralized schemes are known to have high computational complexity, especially on large social graphs. For example, as reported by Chen et al. [3], finding a small set of nodes with high centrality in a graph with 15,000 vertices could take days on a modern server machine. Second, because these distributed protocols usually run on battery-supported mobile devices, such as smartphones[1], we need to control their communication overhead, as data transmission is the major source of smartphone energy consumption.

Our approach is motivated by the "friendship paradox" [13] that *"your friends have more friends than you do"* and leverages random walks to identify critical users. The reason behind this paradox is that people with larger numbers of friends may have a high probability of being observed among one's friend circle. Thus, the friends of randomly selected individuals may have higher centrality in friendship graphs than average. Although the original proof in Feld [13] is for the static friendship graph of traditional social networks, we can easily extend it for the dynamic contact graph of mobile social networks.

This paper makes the following contributions.

- We design a distributed and lightweight protocol, called iWander, to identify critical individuals in mobile social networks (Section 3). With the design principle of decreasing message overhead, iWander can significantly reduce the energy consumption on smartphones. We assume that everyone in the examined network has a smartphone that runs iWander in the background. The key idea behind iWander is to perform fixed-length random walks periodically by a small group of smartphones and estimate the centrality of individuals through their random-walk counters (i.e., the number of times their smartphones are visited by random walks). To verify the feasibility of iWander, we also implement a proof-of-concept prototype on Nokia N900 smartphones.

- We present a targeted immunization policy based on the centrality information provided by iWander to contain the spread of infectious diseases (Section 4). We evaluate the performance of our proposed random-walk based immunization through extensive simulation studies. The simulation results from a real-world mobility trace show that random-walk based immunization always outperforms random immunization and performs very close to degree-based immunization with less than 1% of its message overhead. The results also demonstrate that selecting monitors based on iWander can offer early outbreak detection of infectious diseases.

- We show how to benefit from iWander for information dissemination in mobile social networks (Section 5). Specifically, we study the target-set selection problem which chooses target users based on the random-walk counters of mobile users provided by iWander. Surprisingly, we find that differently from targeted immunization, if we choose all target users with high centrality, the resultant scheme only outperforms random selection for small target sets. We also propose another enhanced scheme that chooses both influential and non-influential users into the target set. Our simulation results verify that this enhanced scheme outperforms random selection for large target sets.

---

[1] We focus on smartphones, the most popular mobile devices, in this paper.

2. RELATED WORK

In this section, we review related work about identifying influential users in social networks, applications of random walks, infectious disease control for public health and information dissemination in mobile networks.

2.1 Identifying Influential Users

2.1.1 Traditional Social Networks

Identifying influential users has been extensively studied for information diffusion in traditional social networks [10, 17, 30]. Domingos and Richardson [10, 30] were the first to introduce a fundamental algorithmic problem of information diffusion: what is the initial target set of k users, if we want to maximize the propagation of information in a social network? Kempe et al. [17] prove that the information dissemination function of this influence maximization problem is submodular for several classes of models. They also propose a greedy algorithm that outperforms heuristics based on node centrality and distance centrality. Although the greedy algorithm of Kempe et al. [17] achieves the best known result so far with a provable approximation ratio of $(1 - 1/e)$, it is computationally expensive [3]. To solve this problem, Chen et al. [3] propose an improvement to reduce the algorithm's running time.

2.1.2 Mobile Networks

The problem of influence maximization has also been extended to mobile networks. Previously, we have studied the target-set selection problem for information delivery as the first step toward bootstrapping mobile data offloading [16]. In particular, we investigate how to select a target set with only k users among all subscribed users, such that we can maximize the number of users that receive the delivered information through mobile-to-mobile opportunistic communications. We propose a heuristic algorithm to select the target set by exploring the regularity of human mobility. Nguyen et al. [23] propose to select critical nodes through overlapping community detection in dynamic networks. They present a framework to adaptively update the community structure based on history information. They also show that this framework can improve the performance of forwarding protocols in delay-tolerant networks and schemes to contain worms in online social networks.

2.1.3 Targeted immunization

Targeted immunization has been proposed to eradicate infections for scale-free complex networks [8], by considering the heterogeneous connectivity properties of these networks. The idea is to cure the highly connected nodes, the hubs, to restore the epidemic threshold in the diffusion process. Christakis and Fowler [4] propose a mechanism for detecting contagious outbreaks. They recruited 390 Harvard College students to participate in an experiment and asked them to nominate up to three friends. Their work demonstrates that by monitoring only the friends of these randomly selected students they can provide an early detection of flu by up to 13.9 days. Christley et al. [5] evaluate the performance of several network centrality measures for identifying high-risk individuals, including degree, shortest-path betweenness and random-walk betweenness. Their results show that degree performs very close to other network measures in predicting risk of infection.

Remark: All the above approaches for various problems, ranging from influence maximization to targeted immunization, are based on *centralized* solutions. Motivated by the friendship paradox, we leverage smartphones to perform random walks among mobile users during their contacts and design a distributed lightweight protocol to identify the most influential individuals.

2.2 Random Walks

The term random walk was first introduced by Karl Pearson [27]. We are interested in random walks on graphs, where a walker starts from a source node to a destination node and for each step of this travel, the next node to visit is selected uniformly at random from the neighbor-set of the current node.

Random walks have been integrated into centrality measurement of social science. For instance, Newman [22] proposes the random-walk betweenness centrality, a relaxation of the shortest-path betweenness. This measure defines how often a node in a graph is visited by a random walker between *all* possible node pairs. Noh and Rieger [24] introduce the random-walk closeness centrality metric, which measures how fast a node can receive a random-walk message from other nodes in the network.

Random walks have also been widely explored in other fields, such as computer science, economics, biology and psychology, for various purposes. For example, Braginsky and Estrin [1] route queries on a random walk to sensor nodes around which a particular event occurs. Yu et al. [35] propose SybilGuard which uses a special kind of random walk, where every node chooses the next hop based on a pre-computed random permutation, to limit the bad effect of sybil attacks on peer-to-peer systems.

2.3 Infectious Disease Control

Public-health researchers have developed tools to study the spreading patterns of infectious diseases and mitigate the effects of epidemics. Eubank et al. [12] model the outbreaks of infectious disease in urban social networks. They find that the contact network is a small-world graph, but the locations graph is scale-free which enables efficient outbreak detection that places sensors in the hubs of the graph. They evaluate the performance of several vaccination strategies for smallpox using a realistic large-scale simulation framework. The results show that it is possible to contain outbreaks through a combination of targeted vaccination and early detection.

Recently, researchers started to measure human contact networks for infectious disease transmission using mobile devices, such as sensor motes or RFID badges. Salathé et al. [31] measure the close proximity interactions among 788 individuals at an American high school during a typical day. Through trace-driven simulation studies, they show that targeted immunization using the contact-network data is more effective than random immunization. Stehlé et al. [33] report a similar study in a primary school in French which measured face-to-face proximity of 6-12 years children and teachers. Based on the measurement results, they provide several public-health implications of infectious diseases, for example, closing selected classes instead of the whole school.

In this paper, we propose iWander to make it feasible to identify influential individuals for targeted immunization in a distributed way, through fixed-length random walks.

2.4 Mobile Information Dissemination

Information dissemination is an important application of mobile networks. Papadopouli and Schulzrinne [25] propose 7DS, a peer-to-peer information dissemination system, to increase data availability for mobile users with intermittent connectivity. With 7DS, mobile devices query data from neighboring peers when they fail to access Internet with their own connections. Small and Haas [32] propose a networking model, called the Shared Wireless Infostation Model (SWIM), which allows information to travel within a network using mobile users as physical carriers. They demonstrate the effectiveness of SWIM using a practical information system of radio-tagged whales.

McNamara et al. [20] propose a scheme to choose the best sources (peers who can remain co-located long enough to complete data transfer) for content sharing among co-located mobile users in urban transport. Using three different experimental traces, Zyba et al. [36] study fundamental properties of human interactions that may affect the performance of information dissemination in mobile ad-hoc networks. They find that the efficiency of content distribution depends on not only the devices' social status, but also the number and density of devices.

In this paper, we propose to identify critical mobile users for facilitating information dissemination through distributed random walks.

3. THE RANDOM WALKS PROTOCOL

In this section, we present the detail of iWander design and its proof-of-concept prototype implementation.

3.1 The Protocol

We propose to leverage *random walks* to design a distributed protocol, iWander, for identifying influential users in mobile social networks. The intuition is that if we periodically initialize random walks from a small group of smartphones, influential mobile users may be visited by these random walks more frequently than average.

The proposed iWander protocol works as follows. Every ΔT hours, iWander generates a tiny probing message with a given probability q on each smartphone. The message contains *only* a pre-configured time-to-live (TTL) field L. During the contacts of a smartphone with its peers, if it has a probing message in its local queue, it sends this message to another randomly selected peer. When a smartphone receives a probing message, it decreases L in the message by 1, and then stores it in its local queue, waiting for the opportunity to forward the message to other peers. A probing message with $T = 0$ will be finally discarded. iWander maintains a random-walk counter on each smartphone, initialized to zero, to record how many times it has received the probing messages (i.e., visited by these random walks).

After collecting the random-walk counters from all users recorded by their smartphones, we can determine the set of k critical users from the head of the user list sorted by these counters. The reason is that based on the friendship paradox, influential users have high probabilities to be visited by random walks and thus own large random-walk counters. Differently from the random-walk betweenness metric proposed by Newman [22], iWander applies *fixed-length* instead of *all-pairs* random walks for two reasons. First, in practice, it is difficult for a mobile user to know every other user and thus specify random-walk destinations. Second, the message

	discovery	idle
Bluetooth	253.05 (5.51)	16.54 (1.11)
WiFi	836.65 (8.98)	791.02 (5.23)

Table 1: The power level of Bluetooth and WiFi on Nokia N900 during discovery and idle modes (in mW).

overhead of all-pairs random walks may be much higher than fixed-length random walks, which makes them unsuitable for battery-powered smartphones.

The update and reset of random-walk counters are determined by the upper layer applications. In practice, they may reset these counters periodically, for example, at midnight (12:00 AM) of every day. They can also apply an exponential moving average to update these counters by assigning a higher weight to recent counters.

In summary, the performance of iWander relies on three parameters: q – the probability that a smartphone generates a probing message (i.e., the fraction of mobile users that initialize random walks), L – the length of random walks (i.e., the number of mobile users visited by a single random walk), and ΔT – the frequency of generating new random-walk probing messages. It is important to understand their impact on the performance of iWander, because they determine both the quality of identified influential users and the number of probing messages spreading over the network.

To reduce energy consumption on smartphones, we prefer short random walks with only a few steps. "Static" versions of social-contact networks are often very dense and "expander-like" (see, e.g., Eubank et al. [12]). In such highly-mixing networks, it is well-known that a random walk of length $O(\log n)$ or less, where n is the number of nodes in the network, suffices to come very close to the stationary distribution of the random walk (in which each vertex has a probability proportional to its degree). Our networks are inherently mobile and thus not static, but their static snapshots will likely be expander-like. The mobile networks will also likely mix well, serving to explain intriguing results such as those of Grossglauser and Tse [15]. Thus, the short random walks that we take will likely come quite close to sampling vertices approximately according to their degrees.

In Section 4.2, we show how the length of random walks L affects the performance of iWander through trace-driven simulation studies. We also evaluate the performance of iWander with different probabilities (q) and frequencies (ΔT) of the generation of random-walk probing messages. We leave the theoretical analysis of the optimal values for these parameters as our future work.

3.2 Proof of Concept

To demonstrate the feasibility of iWander, we implement a prototype in C language on Nokia N900 smartphones. We choose Bluetooth as the underlying communication protocol for iWander due to its low energy consumption. We measured the power of discovery and idle modes of Bluetooth and WiFi devices and summarize the average results and standard deviations for 10 runs in Table 1, which shows that in Bluetooth discovery mode the power of N900 is less than 1/3 of WiFi discovery. Moreover, when the Bluetooth device is in idle mode, the power of N900 is negligible. The reason for high power of WiFi idle mode is that to enable device discovery, a WiFi device needs to run in ad-hoc mode and sends out Beacon messages periodically. Given that the power of WiFi idle mode is also higher than that of Blue-

tooth discovery mode, no matter what the duration of device discovery is, the energy consumption of WiFi discovery will be higher than that of Bluetooth discovery.

Due to the simplicity of iWander design, its prototype implementation using the BlueZ[2] protocol stack has less than 300 lines of code and the size of the compiled file is only around 32 kB, which means that we can easily deploy it on a variety of mobile devices. Unfortunately, it is hard to evaluate the performance of iWander in practice because it is difficult to recruit a large number of participants. In the next two sections, we present two applications of iWander, targeted immunization of infectious diseases and target-set selection for information dissemination, and evaluate their performance through trace-driven simulation studies using a real-world mobility trace.

4. CONTROLLING INFECTIOUS DISEASES

In this section, we demonstrate how to leverage the critical individuals identified by iWander to control infectious diseases and perform early outbreak detection.

4.1 Random-Walk Based Immunization

Smartphones and Internet-related technologies have recently been used to collect data pertaining to the behavior of individuals for various purposes, including disease control and health care. For example, the FluPhone[3] study collects information on social encounters in Cambridge, UK using mobile phones, with the goal of helping medical researchers to better understand the propagation of close-contact infections. Pollak et al. [28] design a mobile phone based game to motivate children to practice healthy eating habits. Moreover, Cook et al. [7] propose Google Flu Trends which uses aggregated Google search queries to provide near-real time estimates of the level of flu in 121 cities of the US.

We propose to perform targeted immunization of infectious diseases based on the random-walk counters maintained by iWander. For example, during the flu season, iWander can periodically report these counters on the smartphones of college students to the university health center. The medical staff can then vaccinate students with high random-walk counters first to contain the spread of flu. We can also use these counters to detect the outbreaks of infectious diseases, where the medical staff monitor the health condition of students with high counters instead of randomly selected students.

The centralized collection of random-walk counters is required by this specific application and the target-set selection for mobile information dissemination in Section 5. For other applications, such as distribution of self-generated content among users, it is possible to extend iWander and design a fully distributed protocol to compute and disseminate these counters among mobile users, for example, by leveraging diffusing computations [9].

There are several differences between our proposed targeted immunization scheme and those in the literature, for example, by Christakis and Fowler [4] and Christley et al. [5]. First, our scheme can benefit from the social contacts detected directly by smartphones, instead of using the estimation through friendship graphs generated from surveys [4].

---

[2]The default Bluetooth protocol stack of most Linux distributions, http://www.bluez.org/

[3]https://www.fluphone.org/

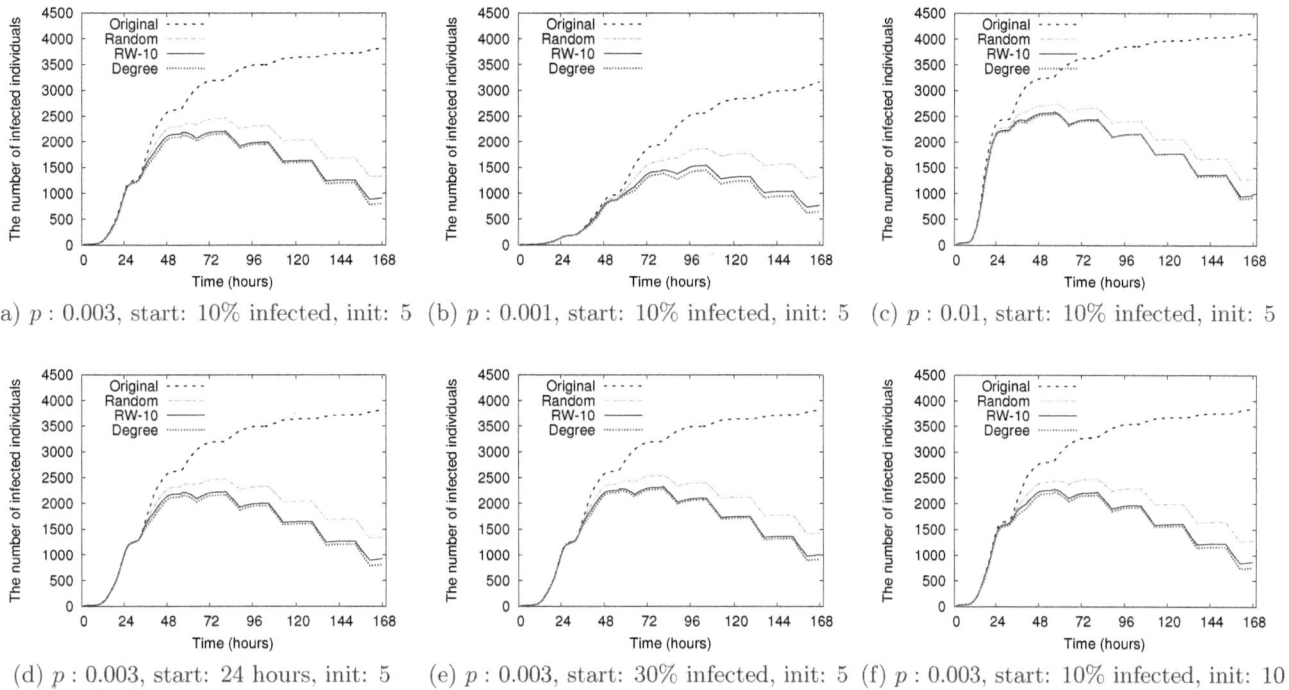

(a) $p : 0.003$, start: 10% infected, init: 5   (b) $p : 0.001$, start: 10% infected, init: 5   (c) $p : 0.01$, start: 10% infected, init: 5

(d) $p : 0.003$, start: 24 hours, init: 5   (e) $p : 0.003$, start: 30% infected, init: 5   (f) $p : 0.003$, start: 10% infected, init: 10

Figure 1: Comparison of the evolution of infected individuals for three immunization policies, random, degree-based, and random-walk-based, with different infection probabilities, immunization start conditions, and initial infections.

Second, our scheme can reflect the dynamics of social contacts in a timely way and avoid the computation-extensive centralized data analysis. Finally, our fixed-length random-walk metric is an extension of the general all-pairs random-walk betweenness centrality [22] and the one-step diffusion-style estimation of node centrality [4], and its low message overhead makes it amenable to run on smartphones.

4.2 Performance Evaluation

We evaluate the performance of iWander for infectious disease control through extensive trace-driven simulations.

4.2.1 Simulation Setup

We implement a simulator in C based on the SIR model, to simulate the spread of infectious diseases. Each individual can be in one of three states: susceptible, infectious, and recovered. Initially, all individuals are in the susceptible state. At the beginning of the simulation, we randomly select a small group of individuals and set their status to be infectious. Transmission of disease occurs from an infectious to a susceptible individual with a probability of p per 60-second contact. Thus, the probability of disease transmission from an infectious individual to a susceptible individual, co-located for t seconds, is $1 - (1 - p)^{\lfloor t/60 \rfloor}$. Finally, an infectious individual is recovered from the disease if he or she is vaccinated.

To simulate the social contacts of individuals, we use a real-world mobility trace, the Dartmouth data set [19], which records at WiFi access points the association and disassociation events of wireless devices. We use a one-week trace of this data set, from 2004-03-01 to 2004-03-07, which includes 4522 devices. As in many previous studies that use this kind of data set, for example in Zyba et al. [36], we consider that the owners of wireless devices are in "social contacts" if their devices are associated with the same access point. We note that although the Dartmouth data set is based on WiFi association data, the user mobility derived from it is for general purpose and has been widely used in the literature [2, 6, 34, 36].

The main reason we chose the Dartmouth data set is that it involves a large number of mobile users, although this data set has its own limitations. For example, the user mobility derived from WiFi association events may not be complete (only around WiFi APs). There are some other publicly available data sets, such as the Reality Mining data set of mobile phone users [11] and the Cabspotting traces of San Francisco's taxi cabs[4]. However, compared to them which either is too small (e.g., the Reality Mining data set with only less than 100 users) or cannot represent the human mobility (e.g., the traces of cabs), we believe the Dartmouth data set is more suitable for our purpose.

For all figures presented in this section, we run the simulation 1,000 times to get average values and standard deviations. For the sake of clarity, we plot standard deviations only in Figure 3 of message overhead.

4.2.2 Targeted Immunization

We compare the performance of random-walk based immunization with random immunization, Random, and degree-based immunization, Degree. With Random, the medical staff vaccinate college students randomly. Using Degree, the smartphone attached with a student performs device discovery every 60 seconds to record the number of smartphones it has contacts with (i.e., node degree in the aggregated social-contact graphs). Then the medical staff vaccinate students with large number of contacts first. During random-walk

---

[4] http://cabspotting.org/

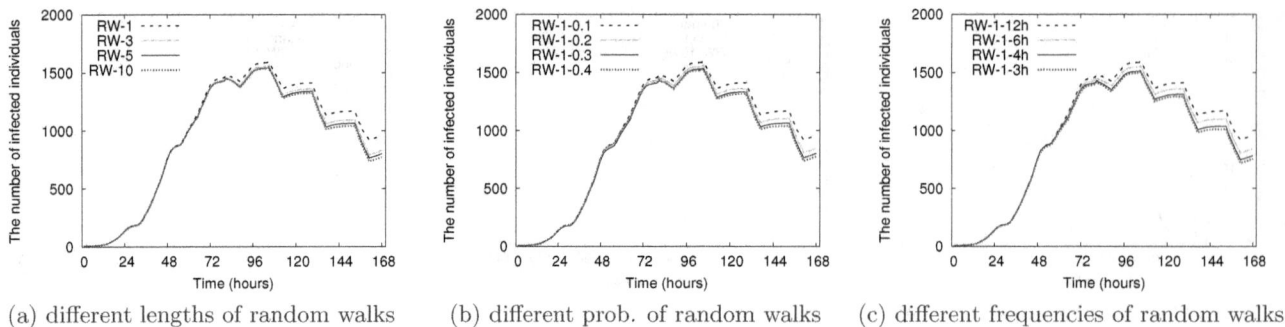

(a) different lengths of random walks   (b) different prob. of random walks   (c) different frequencies of random walks

Figure 2: Comparison of random-walk based immunizations with different lengths, probabilities and frequencies.

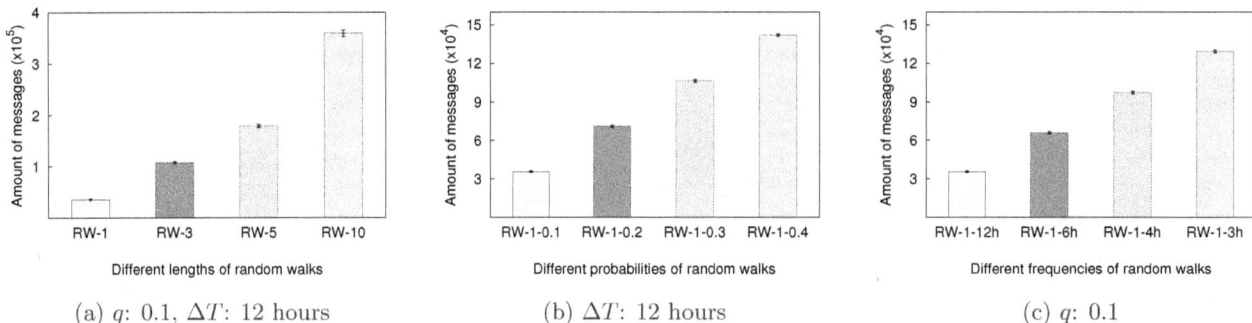

(a) q: 0.1, ΔT: 12 hours   (b) ΔT: 12 hours   (c) q: 0.1

Figure 3: Comparison of the amount of messages for different lengths, probabilities and frequencies of random walks. The number of messages for the degree-based scheme is 1.26×10^8.

based immunization, `iWander` also performs device discovery every 60 seconds only when the message queues on smartphones are not empty. Finally, we assume that vaccinations happen only during the day time, from 9:00AM to 5:00PM, and that on average 60 students are vaccinated every hour.

There are two reasons why we chose degree-based immunization for comparison. First, Christley et al. [5] report that for the networks they examined, degree performs at least as good as other network centrality metrics, such as shortest-path or random-walk betweenness, in predicting risk of infection. Second, it can be easily implemented in a distributed way. For example, Pásztor et al. [26] propose a selective reprogramming mechanism for sensor networks, which determines target sensor nodes using the results of distributed community detection based on node degrees.

For the random-walk based and degree-based immunizations, we update the medical staff with the latest random-walk counters and the number of contacts of all students every 12 hours. Smartphones can send this information to a centralized server through cellular networks. This message overhead should be low, because it contains only a number and two bytes should be enough for the most of the cases. During the immunizations, the medical staff use the most recent information to get a sorted list of all students and then select from this list the student to be vaccinated for the next minute.

We plot the evolution of the number of infected individuals during the one-week simulated period in Figure 1 for various immunization policies, with different infection probabilities, immunization start conditions, and initial infections. During the outbreak of an infectious disease, we assume that

the medical staff start immunizations under two conditions: (1). they have an estimation of the percent of infected individuals and start immunizations after a certain percentage of students are infected; (2). the medical staff start immunizations after a certain amount of time, say 24 hours.

In Figure 1, `Original` plots the curves without immunization as the baseline. As we can see from these subfigures, the number of infected individuals increases much more slowly from the midnight till the morning, compared with other periods in a day, mainly because college students move less frequently during that time period. It is true especially for the first 2 or 3 days, when a large number of students get infected. In all figures of this paper, `RW-n` plots the curves for random walks with n steps.

Among these 6 subfigures, Figures 1a, 1b, and 1c plot the number of infected individuals with different infection probabilities, 0.003, 0.001 and 0.01, 5 initial infections and immunizations after 10% of students are infected. Figures 1d and 1e plot the cases for immunizations after 24 hours and 30% of infections with 0.003 infection probability and 5 initial infections. Finally, Figure 1f plots the case with 0.003 infection probability, 10 initial infections and immunizations after 10% infections. In all these 6 subfigures, `RW-10` performs very close to `Degree` and they all outperform `Random`. Compared to `Random`, the improvement of `RW-10` ranges from 14.10% (Figure 1c) to 25.36% (Figure 1b).

4.2.3 Effects of Various Random-Walk Parameters

We also evaluate the performance of random-walk based immunization with different lengths, probabilities and frequencies of random walks, and plot the simulation results

10

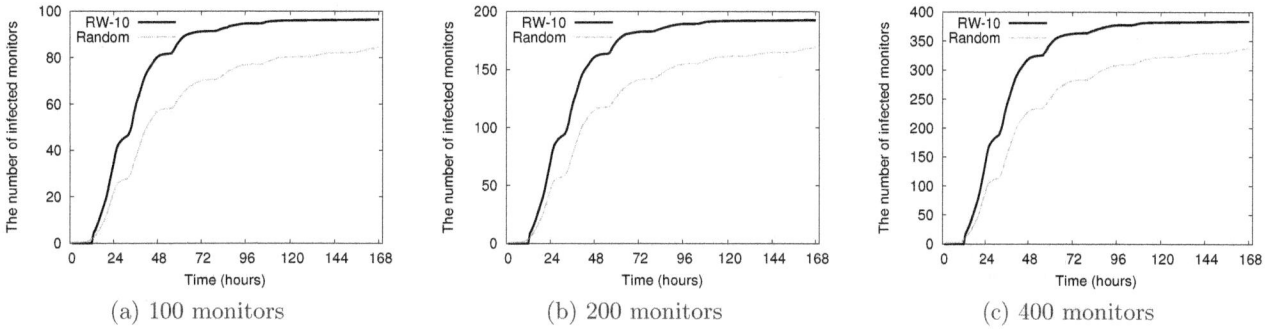

(a) 100 monitors	(b) 200 monitors	(c) 400 monitors

Figure 4: Comparison of early detection of outbreaks with randomly selected monitors and those selected using RW-10.

in Figures 2a, 2b and 2c. All the curves in Figure 2 show the number of infected individuals under random-walk based immunization with 0.001 infection probability, 5 initial infections and immunizations after 10% infections. As we can see from these 3 subfigures, we can improve the performance of random-walk based immunization when increasing the length of random walks from 1 to 10, increasing the probability from 0.1 to 0.4, or increasing the frequency from once every 12 hours to 3 hours. However, we achieve these improvements at the expense of higher message overhead.

We plot the message overhead of iWander with different lengths, probabilities and frequencies of random walks in Figures 3a, 3b and 3c. There are three types of messages, probing request and probing response messages for device discovery, and random-walk probing messages for iWander. In all these subfigures, the baseline is iWander with 1-step random walks and smartphones generate random-walk messages with probability 0.1 every 12 hours. For Degree, all messages are generated during device discovery and the total number of messages is 1.26×10^8 for the simulated period. The amount of messages generated by iWander is extremely low, less than 1% of Degree (3.6×10^5 for RW-10 in Figure 3a and less than 1.5×10^5 for RW-1-0.4 in Figure 3b and RW-1-3h in Figure 3c).

4.2.4 Early Detection of Outbreaks

We can also benefit from iWander for early outbreak detection, which is important to control the spread of infectious diseases [4, 12]. We investigate how to choose a subset of students whose health conditions are monitored to provide early detection, similar to the approach in Christakis and Fowler [4]. Motivated by the observation that monitoring a sample of individuals with high centrality in social-contact networks could allow early detection of contagious outbreaks before they happen in the whole population [4], we propose to choose monitors based on the random-walk counters maintained by iWander.

We plot the evolution of the number of infected monitors chosen randomly and based on iWander in Figures 4a, 4b and 4c with 100, 200, and 400 monitors. In this scenario, the infection probability is 0.003 and there are 5 initial infections. Smartphones generate random-walk messages with probability 0.1 every hour. The medical staff choose a group of monitors based on the random-walk counters reported at the noon of 2004-03-01. These subfigures confirm that iWander does offer early outbreak detection, compared with the random selection scheme. For example, if we draw the

conclusion that an outbreak is occurring when 60% of the monitors are infected, we can detect the outbreak around 21 hours earlier.

5. FACILITATING INFORMATION DISSEMINATION

In this section, we illustrate how to benefit from iWander for target set selection of information dissemination.

5.1 Target-Set Selection Using Random Walks

Motivated by the importance of influence maximization in traditional social networks, in our previous work we study the target-set selection problem for information dissemination in mobile social networks [16]. We leverage opportunistic communications and social participation to facilitate information dissemination and thus reduce the amount of data traffic over 3G networks. We also propose a centralized heuristic algorithm based on the regularity of human mobility, which requires the complete social-contact graph of a given time period and shares the same computational inefficiency as the original greedy algorithm by Kempe, Kleinberg, and Tardos [17].

In this paper, we leverage the random-walk counters of iWander to select target users without requiring global network structure and thus design a distributed solution for the target-set selection problem. Smartphones attached with mobile users run iWander in the background and periodically report their random-walk counters to a centralized server of information service providers. The providers then sort all users based on these counters and then choose the top-k users into the target set. In this scenario mobile users not in the target set can also help to propagate information once they receive it from either target users or others.

The process of information dissemination in mobile social networks is mainly determined by user behaviors. Usually, smartphones can start the exchange of information after they know each other through periodic device discovery. A key concept in the target-set selection problem is the *information dissemination probability* and it is defined as the probability p that information propagates among mobile users after each device discovery. The value of p may be affected by several factors, including status of mobile users and their privacy concerns. Mobile users with high levels of privacy concerns or those who are very busy with their work may have a low probability to involve in information dissemination process. Similar to the transmission of in-

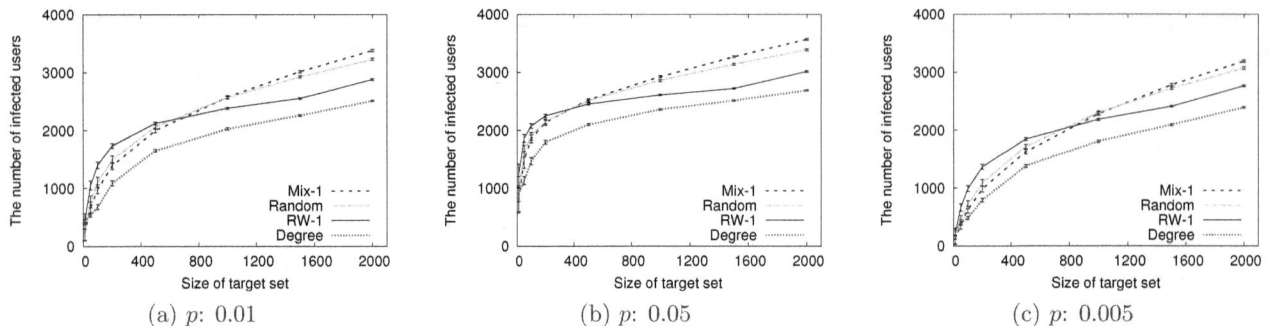

| (a) p: 0.01 | (b) p: 0.05 | (c) p: 0.005 |

Figure 5: Comparison of the number of infected users for four target-set selection schemes with different values of p.

fectious diseases, given the value of p, the probability that two mobile users with a 60-second device discovery interval can exchange information during a t-second contact is $1 - (1 - p)^{\lfloor t/60 \rfloor}$.

We note that the purpose of target-set selection for mobile information dissemination is different from targeted immunization, although the usage of random-walk counters is similar in these two applications. For targeted immunization, we want to vaccinate all influential individuals as early as possible. For target-set selection, if two influential mobile users spend most of their time together, a good choice may be selecting only one of them into the target set. Moreover, as we will show in Section 5.2.2, adding non-influential users into the target set can increase the number of infected users for large target sets.

5.2 Performance Evaluation

We develop another trace-driven simulator also in C, using the same Dartmouth data set [19], to evaluate the performance of random-walk based target-set selection. In this simulator, we assume that the underlying wireless communication is reliable. We have measured the performance of Bluetooth-based opportunistic communications on Nokia N900 smartphones, such as the device discovery probability [16]. We are currently working on a packet-level simulator to take into account the low layer issues, including the failure of random-walk probing messages and the transmission of data packets in information dissemination.

5.2.1 Simulation Setup

The simulator first generates the contacts trace of mobile users under the same assumption that they are in contacts if their wireless devices are associated with the same access point. It then replays the contact events for the given information dissemination period, from 6:00PM to 10:00PM on 2004-03-01.[5] Based on the pre-configured information dissemination probability, the simulator determines randomly whether a user can receive information from peers after each device discovery. We also call the users that can receive information before delivery deadline *infected users*. Usually, information providers will send information to uninfected users at the end of dissemination period, to guarantee that every user can finally receive the delivered information [16].

We compare the performance of random-walk based target-set selection, RW-1, with random selection, Random, and the degree-based selection, Degree. The interval of device discovery is 60 seconds, which means that smartphones have the chance to start the exchange of information every 60 seconds. Similar to degree-based immunization, Degree also uses the number of other smartphones that a smartphone has contacts with as the metric to select target users. For RW-1, smartphones generate 1-step random-walk messages of iWander with probability 0.1 every hour. RW-1 and Degree choose target users based on the updated random-walk counters and the number of contacts of smartphones at the beginning of information dissemination period.

5.2.2 The Number of Infected Users

We plot the number of infected users I for RW-1, Random and Degree in Figure 5. Suppose the number of subscribed users is n, the amount of reduced mobile data traffic will be $n - (k + (n - I)) = I - k$ [16]. We run the simulation 1,000 times and report the average values with standard deviations. The information dissemination probability p is 0.01, 0.05 and 0.005 for Figures 5a, 5b and 5c. We vary the size of target set from 10 to 2,000. As we can see from these subfigures, RW-1 and Random outperform Degree when the size of target set is larger than 10. RW-1 performs better than Random for small target sets. For example, for a target set with 50 users, RW-1 can deliver information to 51% more users than Random (667 vs. 441) when p is 0.005. The improvement is 37% when p is 0.01 (1054 vs. 772) and 14% when p is 0.05 (1863 vs. 1639).

The performance of RW-1 becomes worse than Random when the size of target set is larger than 1,000. One of the possible reasons is that non-influential users (i.e., users with low centrality in social-contact networks) also play an important role in information dissemination. These users are called vagabonds in Zyba et al. [36], which demonstrates that under certain circumstances the effectiveness of information dissemination in mobile social networks predominantly depends on the number of vagabonds. When the size of target set is large, Random has a higher probability to select more vagabonds into the target set, who may have very little chance to receive information before delivery deadline. However, Degree and RW-1 select only mobile users with high centrality into the target set and ignore these vagabonds.

To verify this possible reason, we modify RW-1 by selecting 90% of target users with low centrality from the end of the user list sorted by random-walk counters. We call this en-

---

[5]We have also evaluated other information dissemination periods with different durations and got similar results with those presented in this paper.

12

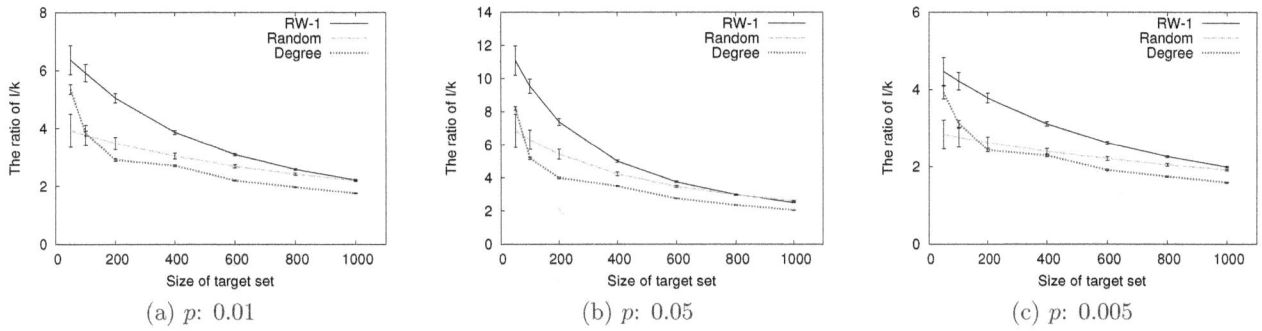

(a) p: 0.01   (b) p: 0.05   (c) p: 0.005

Figure 6: Comparison of the ratio between the number of infected users I and the size of target set k for three target-set selection schemes with different values of p. Only target users can propagate information to others.

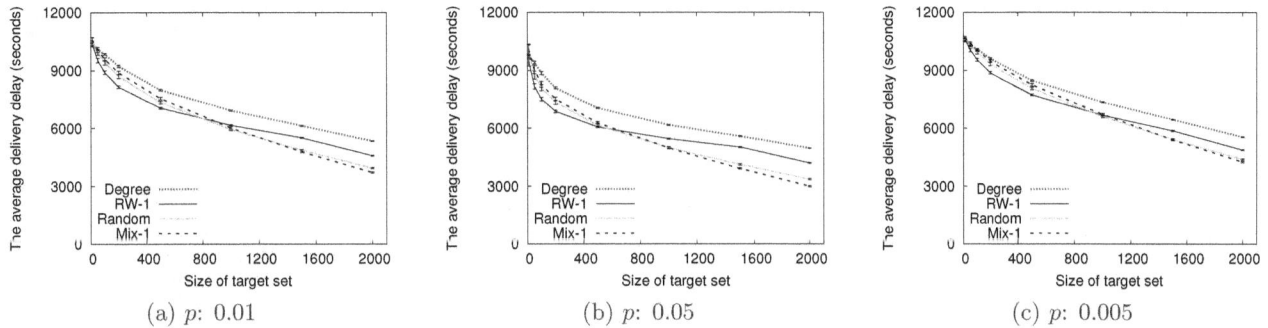

(a) p: 0.01   (b) p: 0.05   (c) p: 0.005

Figure 7: Comparison of delivery delay for 4 target-set selection schemes with different values of p.

hanced scheme `Mix-1`, which also uses 1-step random walks. The three subfigures in Figure 5 show clearly that `Mix-1` outperforms `Random` for large target sets. We tried other different percentages of non-influential target users and these variations perform very close to each other.

We also evaluate the performance of these schemes for another scenario where only target users are willing to propagate information to others. We show the results of only `RW-1`, `Random`, and `Degree` with k ranging from 50 to 1,000 in Figure 6 for clarity. These subfigures plot the ratio between the number of infected users, I and k for different target-set sizes. In this uncooperative scenario, `RW-1` performs much better than `Random` and `Degree` for small target sets. For example, when $p = 0.05$ and $k = 100$, the improvement of this ratio is 51% and 85% compared with `Random`, and `Degree`. However, for the cooperative scenario, the improvement under the same condition in Figure 5 is only 9% (`Random`) and 42% (`Degree`). For large target sets, `Random` performs very close to `RW-1` because in these cases `Random` has more chances to select influential mobile users into a target set.

Differently from targeted immunization, increasing the values of q, L, or ΔT has limited impact on the performance of random-walk based target-set selection. We omit these results due to the limited space.

5.2.3 Delivery Delay

We finally compare the delivery delay of these four target-set selection schemes for the cooperative scenario. We set the delivery delay of target users to be 0 and the users who cannot receive information before delivery deadline to be 10,800 seconds, the same as the duration of information dissemination period. We plot the delivery delay for different information dissemination probabilities in Figure 7. Similarly to the observation from Figure 5, `RW-1` outperforms `Random` for small target sets and `Mix-1` outperforms `Random` for large target sets, in terms of delivery delay. Moreover, they all perform better than `Degree` when the size of target set is larger than 50.

In summary, when information service providers can deliver information directly to only a small number of users, we should use the pure random-walk based target-set selection policy. However, the enhanced scheme that mixes both influential and non-influential users into the target set is preferable when it is possible to deliver information to a large number of users directly.

6. CONCLUSION

In this paper, we propose a lightweight and distributed protocol, named `iWander`, to identify influential mobile users who have high centrality in their social-contact networks. `iWander` leverages fixed-length random walks and runs in the background of smartphones attached to mobile users. It estimates the centrality of individuals based on the number of times their smartphones are visited by random walks. We evaluate the performance of `iWander` using trace-driven simulations for two applications, targeted immunization of infectious diseases and target-set selection for information dissemination.

Our simulation results show that the proposed random-walk based immunization outperforms random immunization and performs very close to degree-based immunization, but generating only less than 1% of its message overhead. For the information dissemination application, the proposed random-walk based target-set selection performs better than random selection for small size of target set and another proposed scheme that chooses also users with low centrality into the target set outperforms random selection when the size of target set is large.

We are exploring the design space of device discovery to further reduce the message overhead of iWander. We also plan to evaluate its performance using other real-world human-contact traces [31].

7. ACKNOWLEDGEMENT

We thank our shepherd Konstantinos Pelechrinis and the anonymous reviewers for their insightful comments. We thank Madhav V. Marathe for valuable inputs. We also thank Pan Hui for the equipment support of several experiments. Aravind Srinivasan and Bo Han were supported in part by US National Science Foundation (NSF) ITR Award CNS-0426683, NSF Award CNS-0626636, and NSF Award CNS 1010789.

8. REFERENCES

[1] D. Braginsky and D. Estrin. Rumor Routing Algorithm For Sensor Networks. In *Proceedings of MobiCom 2002*, pages 22–31, Sept. 2002.

[2] A. Chaintreau, P. Hui, J. Crowcroft, C. Diot, R. Gass, and J. Scott. Impact of Human Mobility on Opportunistic Forwarding Algorithms. *IEEE Transactions on Mobile Computing*. 6(6):606–620, June 2007.

[3] W. Chen, Y. Wang, and S. Yang. Efficient Influence Maximization in Social Networks. In *Proceedings of SIGKDD 2009*, pages 199–207, June-July 2009.

[4] N. A. Christakis and J. H. Fowler. Social Network Sensors for Early Detection of Contagious Outbreaks. *PLoS ONE*, 5(9):e12948, Sept. 2010.

[5] R. M. Christley, G. L. Pinchbeck, R. G. Bowers, D. Clancy, N. P. French, R. Bennett, and J. Turner. Infection in Social Networks: Using Network Analysis to Identify High-Risk Individuals. *American Journal of Epidemiology*, 162(10):1024–1031, Nov. 2005.

[6] V. Conan, J. Leguay, and T. Friedman. Fixed Point Opportunistic Routing in Delay Tolerant Networks. *IEEE Journal on Selected Areas in Communications*, 26(5):773–782, June 2008.

[7] S. Cook, C. Conrad, A. L. Fowlkes, and M. H. Mohebbi. Assessing Google Flu Trends Performance in the United States during the 2009 Influenza Virus A (H1N1) Pandemic. *PLoS ONE*, 6(8):e23610, Aug. 2011.

[8] Z. Dezső and A.-L. Barabási. Halting viruses in scale-free networks. *Physical Review E*, 65(5):055103, May 2002.

[9] E. W. Dijkstra and C. S. Scholten. Termination Detection for Diffusing Computations. *Information Processing Letters*, 11(1):1–4, Aug. 1980.

[10] P. Domingos and M. Richardson. Mining the Network Value of Customers. In *Proceedings of SIGKDD 2001*, pages 57–66, Aug. 2001.

[11] N. Eagle, A. S. Pentland, and D. Lazer. Inferring friendship network structure by using mobile phone data. *Proceedings of the National Academy of Sciences*, 106(36):15274–15278, Sept. 2009.

[12] S. Eubank, H. Guclu, V. S. A. Kumar, M. V. Marathe, A. Srinivasan, Z. Toroczkai, and N. Wang. Modelling Disease Outbreaks in Realistic Urban Social Networks. *Nature*, 429(6988):180–184, May 2004.

[13] S. L. Feld. Why Your Friends Have More Friends Than You Do. *American Journal of Sociology*, 96(6):1464–1477, May 1991.

[14] S. Gaonkar, J. Li, R. R. Choudhury, L. Cox, and A. Schmidt. Micro-Blog: Sharing and Querying Content Through Mobile Phones and Social Participation. In *Proceedings of MobiSys 2008*, pages 174–186, June 2008.

[15] M. Grossglauser and D. N. C. Tse. Mobility Increases the Capacity of Ad Hoc Wireless Networks. *IEEE/ACM Transactions on Networking*, 10(4):477–486, Aug. 2002.

[16] B. Han, P. Hui, V. S. A. Kumar, M. V. Marathe, J. Shao, and A. Srinivasan. Mobile Data Offloading through Opportunistic Communications and Social Participation. *IEEE Transactions on Mobile Computing*, 11(5):821–834, May 2012.

[17] D. Kempe, J. Kleinberg, and Éva Tardos. Maximizing the Spread of Influence through a Social Network. In *Proceedings of SIGKDD 2003*, pages 137–146, Aug. 2003.

[18] J. Kleinberg. The Convergence of Social and Technological Networks. *Communications of the ACM*, 51(11):66–72, Nov. 2008.

[19] D. Kotz, T. Henderson, I. Abyzov, and J. Yeo. CRAWDAD trace dartmouth/campus/movement/01_04 (v. 2005-03-08). Downloaded from http://crawdad.cs.dartmouth.edu/dartmouth/campus/movement/01_04, Mar. 2005.

[20] L. McNamara, C. Mascolo, and L. Capra. Media Sharing based on Colocation Prediction in Urban Transport. In *Proceedings of MobiCom 2008*, pages 58–69, Sept. 2008.

[21] M. Motani, V. Srinivasan, and P. S. Nuggehalli. PeopleNet: Engineering A Wireless Virtual Social Network. In *Proceedings of MobiCom 2005*, pages 243–257, Aug.-Sept. 2005.

[22] M. E. Newman. A measure of betweenness centrality based on random walks. *Social Networks*, 27(1):39–54, Jan. 2005.

[23] N. P. Nguyen, T. N. Dinh, S. Tokala, and M. T. Thai. Overlapping Communities in Dynamic Networks: Their Detection and Mobile Applications. In *Proceedings of MobiCom 2011*, pages 85–95, Sept. 2011.

[24] J. D. Noh and H. Rieger. Random Walks on Complex Networks. *Physical Review Letters*, 92(11):118701, Mar. 2004.

[25] M. Papadopouli and H. Schulzrinne. Effects of Power Conservation, Wireless Coverage and Cooperation on Data Dissemination among Mobile Devices. In *Proceedings of MobiHoc 2001*, pages 117–127, Oct. 2001.

[26] B. Pásztor, L. Mottola, C. Mascolo, G. P. Picco, S. Ellwood, and D. Macdonald. Selective Reprogramming of Mobile Sensor Networks through Social Community Detection. In *Proceedings of EWSN 2010*, pages 178–193, Feb. 2010.

[27] K. Pearson. The Problem of the Random Walk. *Nature*, 72(1865):294, July 1905.

[28] J. Pollak, G. Gay, S. Byrne, E. Wagner, D. Retelny, and L. Humphreys. It's Time to Eat! Using Mobile Games to Promote Healthy Eating. *IEEE Pervasive Computing*, 9(3):21–27, July-Sept. 2010.

[29] K. K. Rachuri, C. Mascolo, M. Musolesi, and P. J. Rentfrow. SociableSense: Exploring the Trade-offs of Adaptive Sampling and Computation Offloading for Social Sensing. In *Proceedings of MobiCom 2011*, pages 73–84, Sept. 2011.

[30] M. Richardson and P. Domingos. Mining Knowledge-Sharing Sites for Viral Marketing. In *Proceedings of SIGKDD 2002*, pages 61–70, July 2002.

[31] M. Salathé, M. Kazandjieva, J. W. Lee, P. Levis, M. W. Feldman, and J. H. Jones. A high-resolution human contact network for infectious disease transmission. *Proceedings of the National Academy of Sciences*, 107(51):22020–22025, Dec. 2010.

[32] T. Small and Z. J. Haas. The Shared Wireless Infostation Model - A New Ad Hoc Networking Paradigm (or Where there is a Whale, there is a Way). In *Proceedings of MobiHoc 2003*, pages 233–244, June 2003.

[33] J. Stehlé, N. Voirin, A. Barrat, C. Cattuto, L. Isella, J.-F. Pinton, M. Quaggiotto, W. V. den Broeck, C. Régis, B. Lina, and P. Vanhems. High-Resolution Measurements of Face-to-Face Contact Patterns in a Primary School. *PLoS ONE*, 6(8):e23176, Aug. 2011.

[34] J. Yoon, B. D. Noble, M. Liu, and M. Kim. Building Realistic Mobility Models from Coarse-Grained Traces. In *Proceedings of MobiSys 2006*, pages 177–190, June 2006.

[35] H. Yu, M. Kaminsky, P. B. Gibbons, and A. Flaxman. SybilGuard: Defending Against Sybil Attacks via Social Networks. In *Proceedings of SIGCOMM 2006*, pages 267–278, Sept. 2006.

[36] G. Zyba, G. M. Voelker, S. Ioannidis, and C. Diot. Dissemination in Opportunistic Mobile Ad-hoc Networks: the Power of the Crowd. In *Proceedings of INFOCOM 2011*, pages 1179–1187, Apr. 2011.

Encounter Based Sensor Tracking

Andrew Symington
Department of Computer Science
University of Oxford
Oxford, United Kingdom
andrew.symington@cs.ox.ac.uk

Niki Trigoni
Department of Computer Science
University of Oxford
Oxford, United Kingdom
niki.trigoni@cs.ox.ac.uk

ABSTRACT

This paper addresses the problem of tracking a group of mobile sensors in an environment where there is intermittent or no access to a localization service, such as the Global Positioning System. Example applications include tracking personnel underground or animals under dense tree canopies. We assume that each sensor uses inertial, visual or mechanical odometry to measure its relative movement as a series of displacement vectors. Each displacement vector suffers a small quantity of error which compounds, causing the overall accuracy of the positional estimate to decrease with time. The primary contribution of this paper is a novel offline method of counteracting this error by exploiting opportunistic radio encounters between sensors. We fuse encounter information with the displacement vectors to build a graph that models sensor mobility. We show that two dimensional sensor tracking is equivalent to finding an embedding of this graph in the plane. Finally, using radio, inertial and ground truth trace data, we conduct simulations to observe how the number of anchors, transmission range and radio noise affect the performance of the proposed model. We compare these results to those from a competing model in the literature.

Categories and Subject Descriptors

C.2.1 [**Network Architecture and Design**]: Wireless communication

Keywords

Localization, Tracking, Encounters, Rigidity

1. INTRODUCTION

A Wireless Sensor Network (WSN) is a group of interconnected, resource-constrained devices ("sensors") that instrument the environment, store data, perform processing tasks and forward information. There are many applications in which sensor location awareness is advantageous, or even essential. Example applications include tracking mineworkers underground, or groups of animals over long periods.

One widely-used system for localization is the Global Positioning System (GPS). However, many sensor deployments preclude the use of GPS for one or more of the following reasons. Firstly, the service may not be available in satellite-denied environments. Secondly, GPS consumes a significant amount of energy, which reduces the sensors' lifetime. Thirdly, the technology requires expensive hardware, making it infeasible for use in large-scale sensor networks that feature many nodes. GPS is one example of anchor-based localization, which assumes the availability of a set of fixed sensors. Such an infrastructure is often difficult – or impossible – to set up, for example in disaster scenarios.

When a fixed localization infrastructure is unavailable, one may measure relative changes in position using an inertial navigation system [9]. The micro electro-mechanical sensing (MEMS) devices used to instrument inertial changes are becoming increasingly energy efficient and cheaper to manufacture, making them feasible for use in WSNs. However, measurements from such devices are typically corrupted by a small quantity of error, arising from bias, noise or information quantization. Since the measured changes are relative, these small errors compound over time. This causes the positional estimate to *drift*, which means that one cannot accurately track a sensor by inertial measurement alone.

In this paper we propose a centralised model that counteracts sensor drift by exploiting opportunistic radio encounters between pairs of sensors, as well as between sensors and anchors. In so doing we provide a method by which sensors may be tracked for longer periods in the absence of a localization infrastructure. We use encounters and displacement estimates to build a graph that models sensor mobility. A vertex represents either the fixed position of an anchor, or the position of a sensor during an encounter. Edges represent the physical distances between vertices. Our proposed tracking model then finds an embedding of this graph in the plane, thereby *localizing* the encounter points. Finally, the original positional estimates are threaded through the embedding to obtain a drift-corrected trajectory. To summarize, our work makes the following contributions:

1. We propose a novel *Encounter Based Tracking* (EBT) model that converts encounters and displacement vectors to a graph, solving the tracking problem with a single application of graph realization.

2. We propose a novel drift correction algorithm, entitled *Radial Drift Correction*, that outperforms *Linear Drift Correction* when used in conjunction with EBT.

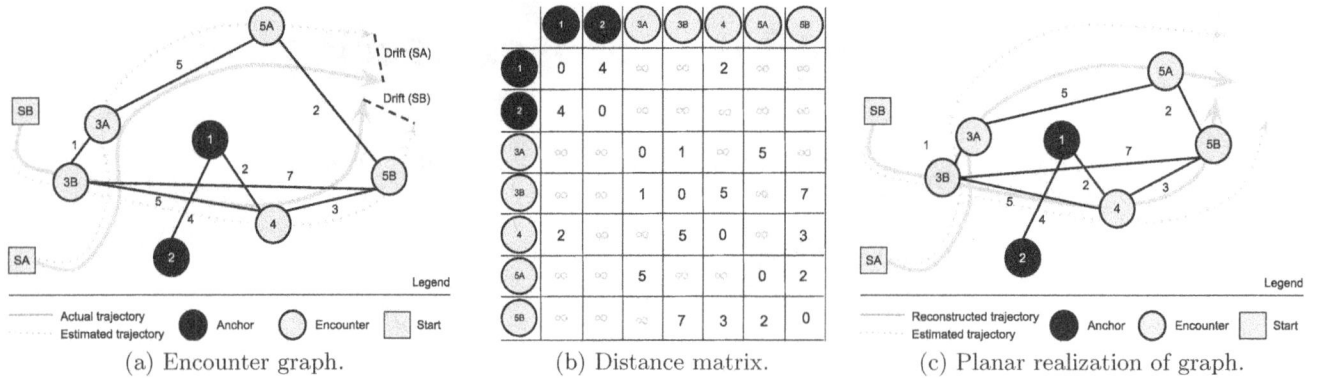

| (a) Encounter graph. | (b) Distance matrix. | (c) Planar realization of graph. |

Figure 1: Overview of Encounter Based Tracking: Diagram 1a shows the actual (solid) and estimated (dotted) trajectories for two sensors, starting at SA and SB. The vertices and edges for the corresponding graph are shown by circles and solid lines respectively. We use displacement vectors drawn from estimated trajectories, as well as distance estimates from radio encounters between sensors and anchors to insert edges into the graph. The positions of the vertices are unknown, and so the graph is described by the distance matrix in 1b. We perform graph realization on this matrix to obtain the planar embedding in 1c. The estimated trajectories are threaded through this embedding to obtain a drift-corrected trajectory for each sensor.

3. We evaluate the performance of EBT using a realistic radio model and trajectories drawn from real trace data. We conduct a sensitivity analysis of EBT and show that it performs well relative to *Directed Diffusion Tracking* [6], a related model from the literature.

The remainder of this paper is organized as follows. Sec. 2 positions our work relative to existing localization literature. Sec. 3 provides an overview of our proposed model. Secs 4 to 6 describe each of our model's three main steps in detail. Sec. 7 discusses the experimental setup used to calibrate our model, and measure its performance. Sec. 8 discusses the results of our experimental study. Finally, Sec. 9 concludes this paper and highlights directions for future work.

2. RELATED WORK

From a sensor networking perspective, localization is typically seen as the process of converting low-level measurements (received signal strength, time of arrival, time difference of arrival) to higher-level information (proximity, range or angle between pairs of sensors) that is then fused together to obtain an estimate of a sensor's location (in relative or global coordinates). Traditional methods localize each sensor in isolation, by estimating multi-hop distances to nearby anchors. Patwari et al [18] define *cooperative localization* as a method whereby the pairwise distances between neighbouring sensors are used to simultaneously localize all sensors in the network. Graph realization is a well-studied [3, 4, 15, 20] family of solutions to the cooperative localization problem, which has seen little application to tracking mobile sensors. Macagnano and de Abreu [17] propose a method of using MDS to track sensors relative to a set of anchor points. However, their representation does not take into account peer to peer encounters, dynamic or kinematic constraints. Cabero et al. [5] propose a Dynamic Weighted Multidimensional Scaling (DWMDS) model for pedestrian tracking that fuses encounters with ranging information. In this model dynamic constraints are encoded directly into a particular graph realization technique. In contrast, our proposed model encodes *measured* displacements directly into

the graph itself, allowing a wide variety of graph realization algorithms to be used. The key feature that distinguishes our model is that we integrate mobility information with ranging information to localize sensors simultaneously over space and time. Constandache et al. [6] propose a similar model that uses directed diffusion to propagate position corrections amongst a set of mobile sensors. Their model considers nodes to be collocated during encounters, and thus ranging information is essentially ignored, which we show to significantly affect tracking accuracy.

From a robotics perspective, localization and mapping are dealt with together in the *Simultaneous Localization and Mapping* (SLAM) problem. The objective of SLAM is to fuse process updates (relative changes in the robots pose) with measurement updates (distances to features in the world) in order to find a maximum likelihood estimate (MLE) of the robot's state over time (localization) and the fixed positions of features (mapping). This is closely related to mobile sensor localization, the key exception being that in sensor networks *feature positions and correspondences* are typically known *a priori*. In Thrun and Montemerlo's GraphSLAM [22] a single robot's trajectory is modeled as a graph, which is similar to our model. Subsequently, Kim et al [13] extended GraphSLAM to leverage encounters between robots. Wymeersch et al [23] propose a distributed model called SPAWN that uses a statistical technique, in conjunction with process and measurement models, to infer the state of the sensors over time. Importantly, this model makes the restrictive assumption that sensors move independently according to a memoryless walk. Our approach differs from this family of approaches because we consider the sensor's state as a position only. The 'dissimilarity' between encounters is therefore simply the Euclidean distance, which means that we are able to exploit existing, fast dimensionality reduction algorithms to obtain a solution.

3. ENCOUNTER BASED TRACKING

We begin with the assumption that mobile sensors are equipped with radios, and that they exchange beacons at regular intervals. When a sensor receives a beacon an *encounter* is recorded, which comprises of a time stamp, two sensor identification numbers and an estimated distance. This distance is obtained by passing the signal strength of a radio beacon through a path loss model. We assume further that the sensors use instrumentation in conjunction with a navigation system to determine an *estimated trajectory*. This is essentially a series of two-dimensional displacement vectors, which model the position of the sensor relative to its starting point $(0,0)$. We assume further that this estimated trajectory suffers drift, and so these positional estimates become less accurate with time. At the core of our model lies the intuition that whenever two sensors encounter one another, it relates them in space and time. We model these encounters, and hence sensors' mobility, with a connected graph. The overall goal of EBT is to firstly find an assignment of two dimensional coordinates to the graph vertices (a planar embedding) that satisfies the observed distances, and to then use this embedding to subtract the drift out from the estimated trajectories. EBT does not require anchors to drift-correct the estimate trajectories. However, several anchors (these can be known start or end points on the estimated trajectory) are required to project the final embedding into a usable coordinate frame, such as meters north or east. Fig. 1 provides an overview of EBT, which may be broken down in to the following three steps:

1. *Graph construction* – A graph is built using the encounters and estimated trajectories. Fig. 1a shows the sensors' positions during an encounter as vertices in this graph; edges represent the observed pairwise distances between the vertices. In general, the relative positions (coordinates) of the vertices are not known, and so the problem can be entirely described by the distance matrix given in Fig. 1b.

2. *Graph realization* – This is the process of assigning a two dimensional coordinate to each vertex of the graph, such that the resultant pairwise distances agree with the observed distance matrix in Fig. 1b.

3. *Drift correction* – Each sensor's estimated trajectory is threaded through the embedding. The drift-corrected trajectories are shown as solid lines in Fig. 1c.

The following sections examine each step in closer detail.

4. GRAPH CONSTRUCTION

Fig. 2 shows a toy example that illustrates how the graph is constructed. In this example there are two anchors, shown as dark circles with the numbers 1 and 2. There are two sensors labeled A and B, which follow the path indicated by the corresponding dotted arrows. When these two sensors come within communication range they exchange a radio beacon, and an encounter is recorded. In the example there are two such encounters that occur, depicted as light circles marked 3A/3B and 5A/5B. Two vertices are added for each encounter – one for each of the two sensor positions at that point in time. When encounter is between a sensor and an anchor, only one vertex is added (Circle 4 in Fig. 2).

With the exception of the anchors, the coordinates of the graph vertices are unknown. However, the estimated trajectories, encounters, and anchor positions (if available) provide us with information about the pairwise distances be-

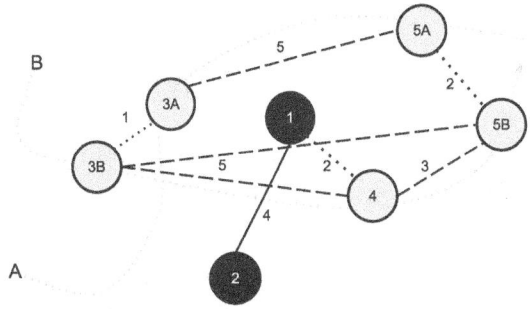

Figure 2: Sensor and anchor positions are shown by light and dark circles respectively. The dotted arrows show the trajectories of sensors A and B. Circles and straight lines correspond to graph vertices and edges respectively. Dashed, dotted and solid edge lengths are obtained respectively from estimated trajectories, the signal strength of a radio beacon, and the known distances between anchors.

tween vertices. In our model this information is captured by graph edges. We distinguish between three edge types:

1. *Mobility edges* – Recall that encounters occur at discrete points in time. The distance between two encounters on a single sensor's trajectory is given by sampling that sensor's estimated trajectory over the appropriate time interval. Mobility edges are shown by dashed lines in Fig. 2.

2. *Anchor edges* – These are included only when anchor information is available. Since the position of each anchor is known *a priori*, it follows that the subgraph for the anchors is fully-connected. Anchor edges are shown by solid lines in Fig. 2.

3. *Radio edges* - The distance separating two sensors, or sensor and an anchor, is found using a path loss model, which estimates a distance between the two vertices from the signal strength of the received beacon frame. Radio edges are shown by dotted lines in Fig. 2.

4.1 Vertex merging

In Fig. 2 the layout of the graph is relatively simple, as there are very few encounters. Depending on the radio communication range and beaconing rate, a series of duplicate encounters may arise from sensors being spatially and temporally collocated for short periods. Since the complexity of EBT scales with the number of graph vertices, this number should be kept to a minimum. To this end, we propose the following vertex merging algorithm. We begin by selecting some constant time threshold T_m, which defines the time granularity of our graph. The merging algorithm searches for a pair of duplicate encounters with the smallest time difference. If this time difference is less than T_m, both encounters are *merged* into a single encounter, with a time and distance equal to the mean of the merged pair. The algorithm then repeats until no matching pair is found.

4.2 Edge selection

Consider the subgraph that contains only the vertices for a single sensor (eg. vertices 3B, 4 and 5B form an example subgraph for Sensor B in Fig. 2). Such a subgraph

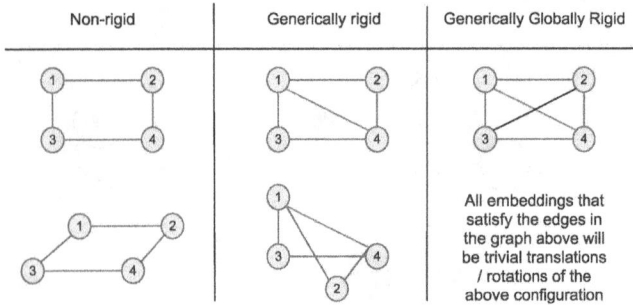

Non-rigid	Generically rigid	Generically Globally Rigid

All embeddings that satisfy the edges in the graph above will be trivial translations / rotations of the above configuration

Figure 3: Left, a flexible graph that fails generic local rigidity: a continuous force applied to vertices 1 and 2 shifts them relative to vertices 3 and 4. Middle, a locally rigid graph structure that is not globally rigid: vertices are rigid against a small continuous force, there exists a reflection (edge $e_{1,4}$) that satisfies the observed edges. Right, a globally rigid graph with a unique planar embedding.

forms a crude, discrete time trace of the given sensor's trajectory. We are able to measure the distance between any pair of vertices in this subgraph by sampling the estimated trajectory. We therefore have sufficient information to fully-connect this subgraph. We refer to this as the *complete edge selection rule*. This rule always yields a subgraph that can be uniquely embedded in the plane.

As a result of drift, a displacement vector measured between two points on an estimated trajectory contains a quantity of error, which increases proportionally to the time separating the two points. An alternate approach to the *complete edge selection rule* is to remove those edges connecting vertices with a large time differences. However, we should only remove those vertices that retain the subgraph's unique embeddability. To achieve this, we must first introduce the notion of graph rigidity [8]. We will use this theory to derive a *conservative edge selection rule*, that seeks to connect the graph with a minimum number of short edges in a way that preserves unique planar embeddability.

We begin by considering only *generic graphs* – those having no $d + 1$ vertices that are collinear in $\mathbb{R}^d$. Such graphs are said to be *rigid* if they do not bend or flex under a small force. Generally speaking, the more edges one adds to a graph the more rigid the graph becomes, as a result of more constraints being placed on motion. A graph containing n vertices has $2n$ possible independent motions in two dimensions, corresponding to a horizontal and vertical translation of each node. It is impossible to constrain the three global graph motions, namely a horizontal translation, vertical translation and rotation. This intuitive notion of constrained movement lies at the core of Theorem 1 below, which provides a test for two dimensional rigidity [16].

THEOREM 1. *The edges of a graph $G = (V, E)$ are independent in $\mathbb{R}^2$ iff. no subgraph $G' = (V', E')$ has more than $2n' - 3$ edges, where n' is the number of vertices in G'.*

Graphs that satisfy the Laman condition are referred to as being *locally rigid*. Furthermore, a graph is said to be *redundantly rigid* if and only if the removal of any single edge results in a graph that is locally rigid. Local and redundant rigidity do not necessarily imply unique graph realizability.

Graphs that satisfy a stronger condition, *global rigidity*, are rigid against both continuous and discontinuous forces. Refer to Fig. 3 for examples of flexible and rigid graphs.

Theorem 2 frames global rigidity certification as composition of 3-connectivity[1] and redundant rigidity testing, both having polynomial time algorithms. Refer to Jackson and Jòrdan [10] for a proof of this theorem. No equivalent result exists for three dimensions or greater.

THEOREM 2. *A graph $G = (V, E)$ with $n \geq 4$ vertices is generically globally rigid in $\mathbb{R}^2$ if and only if it is 3-connected and redundantly rigid in $\mathbb{R}^2$.*

Having outlined conditions for unique localizability, we may now define the *conservative edge selection rule*. Consider the case where we connect every vertex i to its three neighbours, as shown in Fig. 4. The resulting graph is an instance of a trilateration graph [24] and is therefore globally rigid. Applying this rule over the entire trajectory results in the subgraph being globally rigid. However, this rule does not guarantee global rigidity of the entire graph.

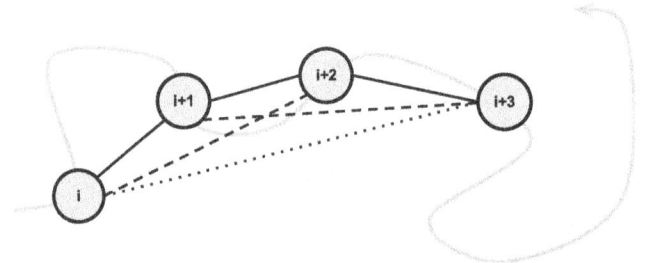

Figure 4: Dotted lines show a globally rigid subgraph formed by connecting each vertex to its three adjacent neighbours. If all encounters are connected in this way, the resultant graph is globally rigid.

5. GRAPH REALIZATION

Consider a graph $G = (V, D)$ with V being a set of n vertices, and D being a $n \times n$ symmetric matrix of pairwise distances between these vertices. The *graph realization problem* seeks to find a d-dimensional embedding – the coordinates of V in $\mathbb{R}^d$ – that preserves D. The d-dimensional graph realization problem is an instance of dimensionality reduction that is provably NP-Hard in the general case [19], and also when radio propagation is modelled as a unit disc graph [2]. We showed in the previous section that if one assumes that the graph vertices are algebraically independent, then it is possible to certify the uniqueness of a planar embedding. The embedding itself may be found using one of many graph realization algorithms.

In this section we survey three types of graph realization that are commonly used for static sensor localization. We conclude this section with an analysis of their applicability for use in our proposed Encounter Based Tracking model.

---

[1] A graph $G = (V, E)$ with n vertices is *k-connected* if the removal of any $i < k$ vertices results in a connected subgraph $G' = (V', E')$ with $n - i$ vertices.

18

5.1 Multidimensional Scaling (MDS)

The objective of *multidimensional scaling (MDS)* is to find an embedding $X = \begin{bmatrix} \mathbf{x}_1 & \cdots & \mathbf{x}_n \end{bmatrix}^T$ of all n vertices in d-dimensional space, given some distance matrix D, through the minimization of the stress function shown in Eqn 1.

$$X = \arg\min_X \sum_{<i,j> \in D} \left(\left\| \mathbf{x}_i - \mathbf{x}_j \right\| - d_{ij} \right)^2 \quad (1)$$

A shortcoming of MDS is that a complete distance matrix is required to find a solution, which is typically unavailable. Typically, unknown edges are calculated using a shortest path method, as in MDS-MAP by Shang et al. [21].

MDS has the side-effect of pulling nodes towards the centroid of the embedding, meaning that it tends to perform poorly on irregularly-shaped graphs. Shang and Ruml [20] proposed an extension, called MDS-MAP(P), in which MDS-MAP is applied in patches across the network. A patch is a subgraph produced for every node in the network, including all neighbours within some hop count (typically 2-hops). For each patch an MDS solution is first obtained, which is then followed by an iterative least squares minimization of stress between the measured and MDS distance values. The maps are then merged together to form a global map.

5.2 Spectral Graph Drawing (SGD)

Like MDS, spectral graph drawing (SGD) may be expressed as a minimization problem [15]. The objective is to find a planar realization of n vertices. The *neighbourhood* $N(i)$ is a set of vertices connected by an edge to vertex i. Eqn 2 defines W as a *weighting matrix* derived from distances, K as the *degree matrix* and L as the *Laplacian matrix*.

$$[W]_{ij} = w_{ij}, \quad [K]_{ii} = \sum_{j=N(i)} w_{ij}, \quad L = K - W \quad (2)$$

SGD finds an optimal embedding $\tilde{\mathbf{x}}$ that satisfies the minimization problem shown in Eqn 3. It therefore favours solutions that push together those vertices connected by high weighted edges. Clearly, the edge weights should be inversely proportional[2] to distance. One trivial solution that satisfies this minimization problem is to select all nodes to be at the same position. To prevent this, a constraint $\mathbf{x}^T\mathbf{x} = 1$ is added, which forces the solution to have a non-zero variance. The selection of the variance to be $1/n$ is arbitrary and simply governs the spread of the solution. The second constraint $\mathbf{x}^T\mathbf{1}_n$ sets the mean of the solution to zero, making the solution translation-invariant.

$$\begin{aligned} \tilde{\mathbf{x}} &= \arg\min_{\mathbf{x}} \left[\mathbf{x}^T L \mathbf{x} \right] \\ s.t \quad & \mathbf{x}^T \mathbf{x} = 1 \\ & \mathbf{x}^T \mathbf{1}_n = 0 \end{aligned} \quad (3)$$

Koren [15] argues that SGD performs poorly on irregular graphs, as it tends to draw nodes into the graph centroid. Koren goes on to propose a variant of SGD, called *degree normalized spectral graph drawing* (DN-SGD). This model updates the variance constraint of SGD to weight the position of the graph vertices by their degree. That is, the constraint $\mathbf{x}^T\mathbf{x} = 1$ in Eqn 3 is replaced by $\mathbf{x}^T K \mathbf{x} = 1$. For regular graphs, the degree matrix is close to a scaled

---

[2]In spectral graph drawing the edge weighting $w_{ij} \in W$ is typically calculated as either $\frac{1}{1+d_{ij}}$ or $e^{-d_{ij}}$ for $d_{ij} \in D$.

identity matrix, which makes the constraint optimization problem equivalent to that of regular SGD. One of Koren's primary contributions is a proof showing that the solution to this constrained optimization problem is equivalent to finding the generalized eigenvectors of (L, K). Broxton et al. [4] applied DN-SGD to static localization and found that it required a refinement step to obtain a reasonable embedding.

5.3 Semidefinite Programming (SDP)

Assume that a graph contains n nodes and m anchors, with known positions $A = \{a_1, \ldots, a_m\}$. Let $D \in \mathbb{R}^{n \times n}$ represent the pairwise distances between all nodes. Similarly, let $\bar{D} \in \mathbb{R}^{m \times n}$ represent the pairwise distances between all nodes and anchors. A value of zero in either $d_{ij} \in D$ or $\bar{d}_{ij} \in \bar{D}$ indicates that no distance information is available. The sets $N_x = \{(i,j) : \forall_{i<j} d_{ij} \neq 0\}$ and $N_x = \{(k,j) : \bar{d}_{kj} \neq 0\}$ therefore contain all unique node-node and node-anchor edges respectively. Semidefinite programming (SDP) is a method of solving a minimization problem with a specific form, using a interior point algorithm. The sensor localization problem may be expressed as the minimization problem Eqn 4, with $X = \{x_1, \ldots, x_n\}$ being a candidate embedding, and $\tilde{X}$ being the optimal embedding. The α and β terms are slack variables that seek to minimize the error between the observed and the predicted distances. Although we have not included it in this paper for space reasons, this formulation of the sensor localization problem may be relaxed into SDP form.

$$\begin{aligned} \tilde{X} &= \arg\min_X \left[\sum_{(i,j) \in N_x} (\alpha_{ij}^+ + \alpha_{ij}^-) + \sum_{(k,j) \in N_a} (\beta_{kj}^+ + \beta_{kj}^-) \right] \\ s.t \quad & \|x_i - x_j\|^2 + \alpha_{ij}^+ - \alpha_{ij}^- = d_{ij}^2 \quad \forall_{(i,j) \in N_x} \\ & \|x_i - a_j\|^2 + \beta_{kj}^+ - \beta_{kj}^- = \bar{d}_{kj}^2 \quad \forall_{(k,j) \in N_a} \\ & \alpha_{ij}^+ \geq 0, \quad \alpha_{ij}^- \geq 0, \quad \forall_{(i,j) \in N_x} \quad (4) \\ & \beta_{kj}^+ \geq 0, \quad \beta_{kj}^- \geq 0, \quad \forall_{(k,j) \in N_a} \quad (5) \end{aligned}$$

Biswas and Ye [3] applied SDP to the problem of sensor localization with noisy range estimates. Subsequently, Kim et al. [14] expanded on this work, exploiting matrix sparsity to reduce the amount of time required to obtain a solution. Recent results by Javanmard and Montanari [12] provide analytical performance bounds for SDP localization where nodes are distributed randomly according to a uniform distribution within a bounded d-dimensional hypercube.

5.4 Graph realization and EBT

In our particular sensor tracking problem, there are two aspects that affect the performance of the realization algorithms. Firstly, the graph is likely to be regularly shaped, as the sensors' movement is sampled from an overlapping trajectory. Secondly, the graph is comprised of several clusters of densely connected vertices (see Fig. 6a). Each cluster corresponds to a single sensor's trajectory, and it is densely connected because is possible to sample an inertial trajectory between any two points. These clusters are loosely interconnected by comparatively fewer edges, which arise from opportunistic contacts. Our first expectation is that DN-SGD will 'push' these clusters away from one other. This will result in there being little spatial overlap between the sensors' trajectories, which will yield an incorrect embedding. Although an iterative refinement step will reduce this error, it will invariably converge to a local minimum. Our second expectation is that MDS will outperform SGD, which we infer directly from the minimization functions. MDS seeks to directly minimize the sum squared error between the observed and predicted distances, whereas SGD seeks to push together those vertices connected by short edges. They key problem is that SGD places no constraint on how close any pair of vertices may be to one another, but only on the spread of the entire embedding. We therefore expect that SGD will yield an acceptable overall distribution of vertices, but on a local level vertices will deviate significantly from their correct position. We expect SDP to perform similarly to MDS, but at a higher computational cost. Finally, we have chosen not to test MDS-MAP(P) for two reasons. Firstly, we expect the graph to be regular, implying that MDS-MAP(P) will perform approximately just as well as straight MDS. Secondly, the computational complexity required to obtain a solution scales poorly with the number of vertices: preliminary experiments showed a run time of 200 seconds for a graph containing fewer than 100 vertices.

6. DRIFT CORRECTION

In the final step of EBT we use the projected graph embedding to drift-correct the sensors' estimated trajectories. Piecewise drift correction is applied independently to each sensors' estimated trajectory between encounter points. We begin with a description of an existing approach from the literature, *Linear Drift Correction*. We then propose a novel alternate method, which we call *Radial Drift Correction*.

Linear Drift Correction was proposed by Constandache et al. [6], and it assumes that the positional estimate drifts linearly with time. Let $\tilde{P}(t)$ be the trajectory estimate for a given node, at some time t. Let t_i and t_{i+1} be the time associated with two adjacent graph vertices representing encounters along a given sensor's path. Let $X(t_i)$ and $X(t_{i+1})$ be the coordinates of the first and second encounter respectively, obtained directly from the graph embedding.

Linear Drift Correction begins by shifting the sensor's estimated trajectory to begin at the position of the first encounter $X(t_i)$. The approach then adds the difference between the estimated and known end points linearly over the period $(t_{i+1} - t_i)$. This is done by first calculating a shift vector $\vec{v} = X(t_i) - \tilde{P}(t_i)$ and then a correction vector $\vec{c} = X(t_{i+1}) + \vec{v} - \tilde{P}(t_{i+1})$. Eqn 6 shows how the final drift-corrected trajectory segment $P(t)$ is calculated.

$$P(t) = \tilde{P}(t) + \vec{v} + \frac{t - t_i}{t_{i+1} - t_i}\vec{c}, \quad t_i \leq t \leq t_{i+1} \quad (6)$$

We observed that Linear Drift Correction distorted the shape of the curve, most notably in cases where there was a large angular difference in the direction vector between the two points on the estimated trajectory and the two encounter points. To counteract this, we propose an alternate drift-correction method called Radial Drift Correction. In this method the starting point of the estimated trajectory is first shifted to the first encounter. Rather than adding the drift correction linearly along the curve, we rotate the curve about the starting encounter by some angle α, until the estimated displacement vector and actual displacement vector are aligned. Finally, we scale the estimate trajectory by β to align its end point with the second encounter. We denote $\vec{a} = X(t_i) - X(t_{i+1})$ and $\vec{m} = \tilde{P}(t_i) - \tilde{P}(t_{i+1})$ to be the displacement vectors measured between the two given

encounter points, according to the embedded graph vertices and the trajectory estimate. Eqn 7 and Eqn 8 show how $\vec{a}$ and $\vec{m}$ are used to find the rotation angle α and scale β.

$$\alpha = atan2(\vec{a}_y, \vec{a}_x) - atan2(\vec{m}_y, \vec{m}_x) \quad (7)$$

$$\beta = \frac{\|\vec{a}\|}{\|\vec{m}\|} \quad (8)$$

The updated trajectory segment $P(t)$ is given by Eqn 9 using the rotation angle α and scale factor β defined above.

$$P(t) = X(t_i) + \beta \begin{bmatrix} cos\alpha & -sin\alpha \\ sin\alpha & cos\alpha \end{bmatrix} \left(\tilde{P}(t) - \tilde{P}(t_i) \right) \quad (9)$$

7. EXPERIMENTAL SETUP

To evaluate the performance of our model we synthetically created five sensors by randomly sampling the source data, discussed in Sec. 7.1.1, at different 30 second intervals. We shifted the all trajectories to begin at time zero, and calculated the encounter points between sensors and anchors. At these encounter points the pairwise distances between sensors were convoluted by additive white noise, with parameters drawn from experiments, which are discussed in Sec. 7.1.2. In Sec. 7.2 we discuss the parameters that were varied, models that were tested and metrics for analysis, while Sec. 7.3 shows how our EBT was calibrated. Each combination of the simulation parameters was repeated 100 times in order to obtain 99% confidence intervals. The simulations were conducted over two days using three parallel Matlab processes on a 2.4 GHz Core 2 Quad with 4GB RAM.

7.1 Source data

7.1.1 Inertial and ground truth traces

Our model was evaluated on trace #13 from Angermann et al. [1], which is a recording of a random walk in a 6 x 8 meter room. The trace offers two time-synchronized data streams: a *ground truth* trajectory recorded from an optical localization system, and an *inertial trajectory*, which the authors obtain by passing raw measurements from a shoe-mounted inertial sensor through Foxlin's *Inertial Pedestrian Dead Reckoning* algorithm [9].

7.1.2 Radio noise model

To test the performance of our tracking model under more realistic conditions, we used the radio model proposed by Patwari et al [18]. We staggered 10 wireless stations at 1.8m intervals from a given receiver, and measured their signal strength for a 10 minute period. We assumed that the distance d in meters relates to the RSSI r in dBm according to the free space path loss model shown in Eqn. 10. We used a non-linear fitting technique to find p_1 and p_2 for our data set. We assumed further that the received signal strength was corrupted by Additive White Gaussian Noise (AWGN) with a constant standard deviation p_3, measured directly from the trace data. Our radio data and path loss model is summarized in Fig. 10.

$$r = p_1 - p_2 10log_{10}(d) \quad (10)$$

We used the following method to model the effect of radio noise on our tracking algorithms. Each simulation run was assigned a radio noise multiplication factor n_f, which weights the standard deviation of the additive white Gaussian noise. Assuming an observed distance of d_{ij}^t between

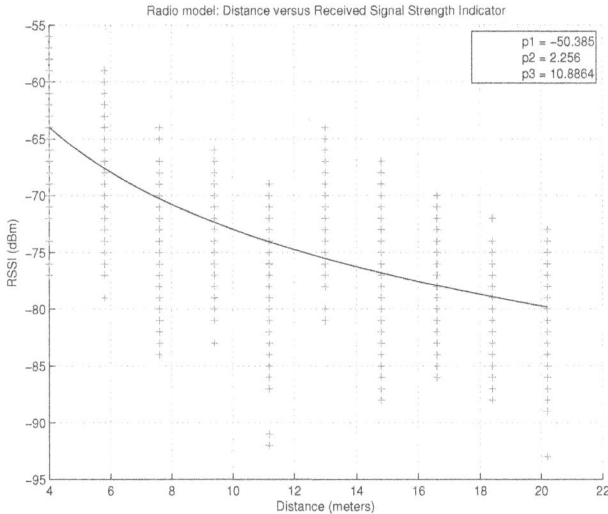

Figure 5: This graph shows the received signal strength (in dBm) as a function of the distance between a sender and receiver. Small crosses represent beacon frame measurements, while the solid line is the path loss model fitted to the measured data.

two sensors i and j at time t, then the RSSI for the beacon frame is given by $\tilde{r}_{ij}^t = p_1 - p_2 10 log_{10}(d) + e$, where $e \sim N(0, n_f p_3)$. This RSSI was fed back through the path loss model to obtain a noise-corrupted distance $\tilde{d}_{ij}^t$.

7.2 Models and Performance Metric

The objective of our experiments was to measure how the following two tracking algorithms perform as a function of the number of anchors (4 to 12), communication range (0.25m to 1.5m) and radio noise factor n_f (0 to 2):

- *Drift Correction Tracking* (DDT) - This model was taken from Constandache et al. [6] and operates in the following way: whenever a sensor encounters another sensor, or anchor, with a fresher positional estimate, it assumes the position of its peer, and corrects for drift between the previous and current encounter.

- *Encounter Based Tracking* (EBT) - This is the proposed tracking algorithm, which was described in Secs 4-6. The model was calibrated as per Sec. 7.3.

We used the Procrustes Transform [7] to express the output trajectories in a common coordinate frame, relative to a minimum of four non-collinear anchor points. The performance of both tracking models was measured by the *root mean square error* (RMSE), over the errors between points along the output trajectory P_i and the ground truth trajectory G_i, for every sensor i. Assuming that we have S sensors and T discrete time ticks, then the RMSE e is given by Eqn 11, where $\|\cdot\|$ represents the standard Euclidean norm.

$$e = \sqrt{\frac{\sum_{s=1}^{S} \sum_{t=1}^{T} \|P_s(t) - G_s(t)\|^2}{ST}} \quad (11)$$

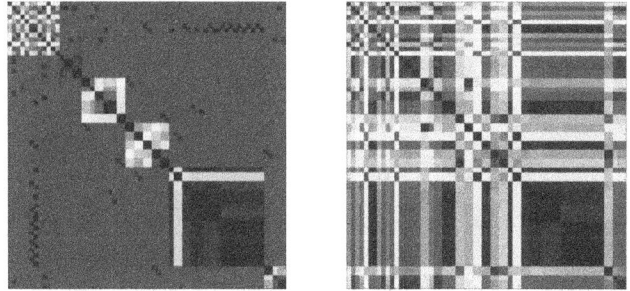

(a) Before Floyd-Warshall   (b) After Floyd-Warshall

Figure 6: The heat map representation of a sample EBT distance matrix. The upper-left block on the diagonal corresponds to the known anchor distances, while the remaining five blocks correspond to pairwise distances between points on each trajectory. Off-diagonal elements correspond to encounters.

7.3 Calibration of Encounter Based Tracking

7.3.1 Vertex merging

If the vertex merging threshold is set to a value that is too high, then EBT may merge encounters that take place in very different locations. In this case the average distance does not accurately represent the samples that are merged, and the end result is a lower tracking resolution. Conversely, if the value is set too low then the graph contains a large number of vertices, increasing the complexity of graph realization. We determined empirically that five seconds is a sufficient threshold for vertex merging on our trace data set. With this threshold set, in the worst case all graph realization algorithms obtained a solution within one minute.

7.3.2 Edge selection

In Sec. 4 we introduced the idea of graph rigidity, and we showed how it relates to unique localizability. We also proposed two edge selection rules for connecting encounters along a single sensor's path. The *complete edge selection rule* fully-connected all encounters on a single sensor's trajectory. The *conservative edge selection rule* sought to reduce the amount of inertial error in the representation, while preserving generic global rigidity. Given perfect distance measurements, it should be possible to find a unique embedding in both cases. However, because of the NP-Hardness of graph realization, in practice this cannot be achieved. Through a series of experiments, we observed that all of the tested approximation algorithms performed favourably using the complete edge selection rule. We attribute this behaviour to the density of the distance matrix and its effect on approximation algorithms for graph realization. Although the remaining edges introduce larger quantities of measurement error, this error has less impact on the final solution that the loss of information about vertex connectivity.

7.3.3 Graph realization

Here, our objective was to determine which graph realization algorithm performs well with EBT. To do this we implemented MDS, SGD and DN-SGD, and used the SDP localization implementation by Kim et al. [14]. Fig. 7a and Fig. 7b show how, for these four graph realization al-

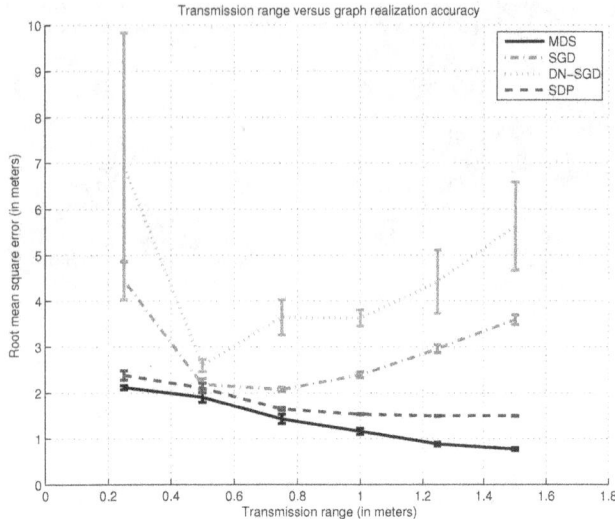

(a) Transmission range versus graph realization accuracy.

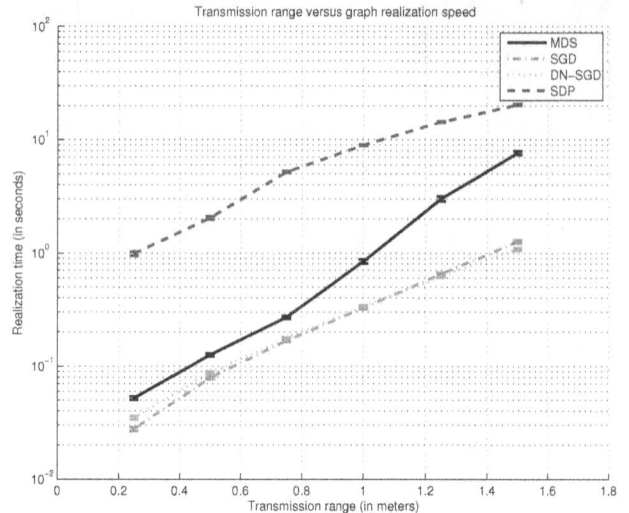

(b) Transmission range versus graph realization speed.

Figure 7: Comparing the accuracy and speed of EBT using four different graph realization algorithms.

gorithms, EBT accuracy and speed change as a result of an increase in radio transmission range. The RMSE metric used to assess accuracy is given by Eqn 11. An increase in transmission range results in a greater number of encounters, increasing the overall number of vertices in the graph, and thus the speed of graph realization (in our experiments the number of vertices is bounded to approximately 500 by the vertex merging mechanism). As predicted in Sec. 5.4 the accuracy of the two SGD algorithms was significantly lower than the others. SGD, DN-SGD and MDS implementations all rely on eigendecomposition of a square matrix, which means that they should exhibit similar time performances. However, a MDS requires preliminary step to determine unknown elements of the distance matrix (see Fig. 6). Our implementation uses a Floyd-Warshall shortest path algorithm, which has a complexity of $O(n^3)$ for a matrix with $n \times n$ vertices. This is why the speed of MDS is lower than the other two methods. Our results suggest that for problems with over 500 encounters, switching from MDS to SDP may provide a good trade-off between speed and accuracy.

8. RESULTS AND DISCUSSION

Fig 8 provides a summary of our sensitivity analysis. We have included an *Inertial* curve, which shows the RMSE of the inertial data without any drift-correction. It serves as a worst-case benchmark for the tracking algorithms. Take note that only noise-free data is used in Figs 8a and 8b. In EBT, Radial Drift Correction showed a consistent 10cm improvement over Linear Drift Correction.

8.1 Sensitivity analysis for EBT and DDT

8.1.1 Number of anchors

For DDT, increasing the number of anchors is equivalent to sampling the search space at a greater resolution. As Fig. 8a illustrates, the DDT error drops at a decreasing rate, as more anchors are added to the system. Presumably, this is because a doubling in resolution requires a square

increase in the number of anchors. For EBT, we observe that an increase in the number of anchors greatly improves the tracking performance. Increasing the number of anchors yields more encounters, thereby increasing the total number of sensor to anchor encounters in the experiment. In addition to adding more information to the model, such encounters help 'pin' the sensors' trajectories to the rigid cluster formed by the anchors, thereby improving graph rigidity.

8.1.2 Transmission range

Fig. 8b shows how the performance of DDT and EBT change as the radio transmission range increases, equally for both sensors and anchors. In general, an increase in the transmission range yields a greater number of encounters. This adds more information to both EBT and DDT, thus improving performance. A transmission range of less than 0.5m yields very few encounters. For DDT this means that positional updates are less frequent, yielding a poor tracking performance. For EBT this poses a bigger problem – a small number of encounters causes a flexible encounter graph, resulting in a bad realization (see Sec. 8.2). This is why the EBT error for small transmission ranges is greater than DDT. However, once the graph is rigid, EBT performs significantly better than DDT. We attribute this to two things. Firstly, DDT assumes that when an encounter takes place, the two devices are collocated. As the transmission range increases, so this assumption becomes less valid. Secondly, EBT performs a global correction of encounter points, whereas DDT perform a local correction that only involves the two sensors at the point of encounter.

8.1.3 Transmission noise

Fig. 8c shows how the performance of EBT and DDT scale as more white noise is added to radio distance estimates. Since DDT does not take into account radio distance, it follows that its performance remains consistent across all experiments. EBT showed an exponential drop in performance as a function of noise. This relationship arises directly a result of the path loss model: a linear increase in RSSI error

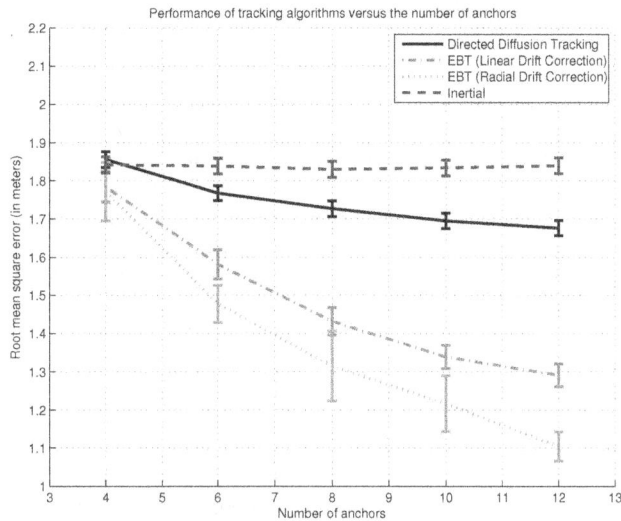

(a) Increasing the number of fixed anchors.

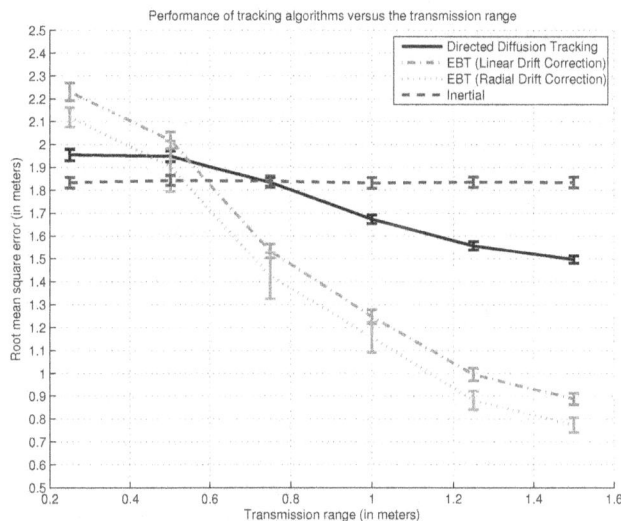

(b) Increasing the transmission range.

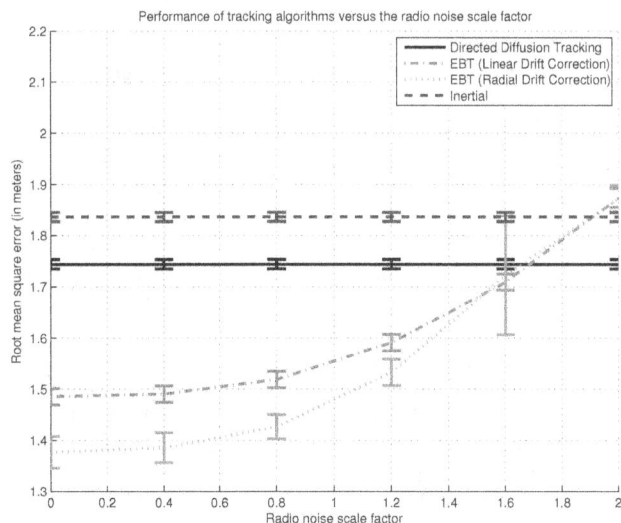

(c) Increasing the RSSI additive white noise variance.

Figure 8: Sensitivity analysis results for Encounter Based Tracking and Directed Diffusion Tracking.

yields an exponential error in distance. Extremely noisy distance data results in EBT pushing apart sensor's subgraphs in the plane, in order to satisfy artificially-long edges.

8.2 The effect of graph rigidity on EBT

One limitation of EBT is that it requires that the underlying graph is globally rigid. Although we can construct globally rigid subgraphs for each sensor, the global rigidity of the joint graph is a function of the opportunistic encounters made by each sensor. If a graph does not satisfy global rigidity, multiple embeddings will exist that satisfy the observed pairwise distances. Fig. 9a shows a graph from an example simulation run that exhibits such behaviour. We analysed the graph with the 2D Pebble Game [11] in order to find the minimum number of locally rigid clusters. Local flexibility results in unconstrained vertices, which causes an indeterminacy in the realization solution space. This manifests in a poor embedding, resulting in the trajectory being reconstructed incorrectly, which we highlight in Fig. 9b.

9. CONCLUSION AND FUTURE WORK

In this work we proposed a novel offline Encounter Based Tracking (EBT) algorithm, which uses opportunistic peer-to-peer and sensor-to-anchor radio encounters to correct drift-corrupted trajectories. Although EBT may also be used for three dimensional tracking, polynomial time rigidity certification for three dimensions and higher remains an open problem. Through experimentation, we have shown that 2D EBT tracking outperforms a competing model in the literature, over a wide range of network parameters. Our results also show that EBT scales predictably with additive white Gaussian noise corrupted radio distances. In future work we plan to investigate more sophisticated drift-correction algorithms, how to include measurement uncertainty into EBT, and how to calibrate EBT for scenarios containing irregular graph formations – those in which the encounter points are not uniformly distributed on the plane.

Acknowledgements

This research was supported by the Sensing Unmanned Autonomous Aerial Vehicles (SUAAVE) research project under EPSRC grant EP/F064217/1. We would like to thank Hongkai Wen for his help collecting RSSI data, and Simon Julier, Mihai Cucuringu, Andrew Markham, Sarfraz Nawaz and Nadine Levin for reviewing earlier versions of this work.

10. REFERENCES

[1] ANGERMANN, M., ROBERTSON, P., KEMPTNER, T., AND KHIDER, M. A high precision reference data set for pedestrian navigation using foot-mounted inertial sensors. In *Indoor Positioning and Indoor Navigation (IPIN), 2010 Intl Conference on* (2010), pp. 1–6.

[2] ASPNES, J., GOLDENBERG, D., AND YANG, Y. R. On the computational complexity of sensor network localization. *Algorithmic Aspects of Wireless Sensor Networks* (2004), 32–44.

[3] BISWAS, P., AND YE, Y. Semidefinite programming for ad hoc wireless sensor network localization. In *Proc. of the 3rd Intl Symposium on Information Processing in Sensor Networks* (2004), pp. 46–54.

[4] BROXTON, M., LIFTON, J., AND PARADISO, J. A. Localization on the pushpin computing sensor network

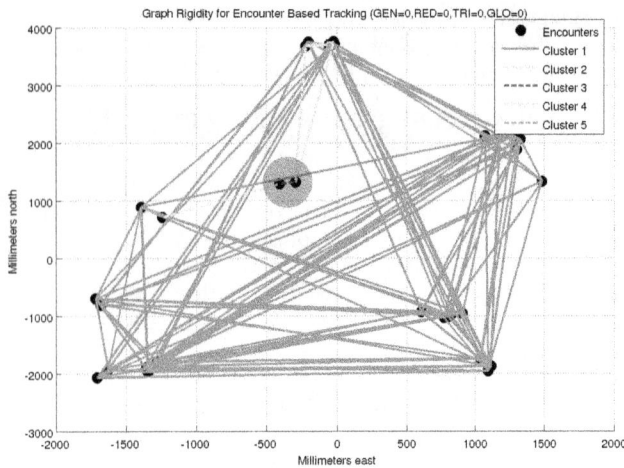

(a) Rigidity graph highlighting two non-rigid vertices.

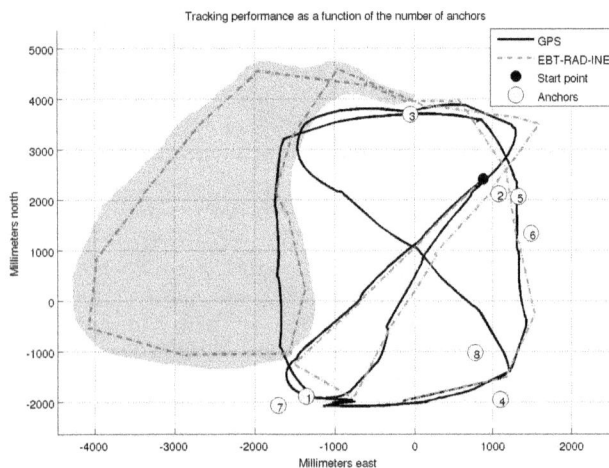

(b) The net effect of non-rigid vertices on sensor tracking.

Figure 9: The effect of non-rigid graph vertices on the accuracy of Encounter Based Tracking.

using spectral graph drawing and mesh relaxation. *ACM SIGMOBILE Mobile Computing and Communications Review 10*, 1 (2006), 1–12.

[5] CABERO, J. M., TORRE, F. D. L., SANCHEZ, A., AND ARIZAGA, I. Indoor people tracking based on dynamic weighted multidimensional scaling. In *MSWiM '07. Proc. of the 10th ACM Symposium on Modeling, Analysis, and Simulation of Wireless and Mobile Systems* (2007), pp. 328–335.

[6] CONSTANDACHE, I., BAO, X., AZIZYAN, M., AND CHOUDHURY, R. R. Did you see Bob?: human localization using mobile phones. In *Proceedings of the sixteenth annual international conference on Mobile computing and networking* (2010), pp. 149–160.

[7] COX, M., AND COX, T. Multidimensional scaling. *Handbook of data visualization* (2008), 315–347.

[8] EREN, T., GOLDENBERG, O. K., WHITELEY, W., YANG, Y. R., MORSE, A. S., ANDERSON, B. D. O., AND BELHUMEUR, P. N. Rigidity, computation, and randomization in network localization. In *INFOCOM 2004. 23rd Conf. of the IEEE Computer and Comms Societies* (2004), vol. 4, pp. 2673–2684.

[9] FOXLIN, E. Pedestrian tracking with shoe-mounted inertial sensors. *IEEE Computer Graphics and Applications* (2005), 38–46.

[10] JACKSON, B., AND JORDÁN, T. Connected rigidity matroids and unique realizations of graphs. *Journal of Combinatorial Theory, Series B 94*, 1 (2005), 1–29.

[11] JACOBS, D. J., THORPE, M. F., AND CHUBYNSKY, M. 2D Pebble Game with Central Forces, November 2011.

[12] JAVANMARD, A., AND MONTANARI, A. Localization from incomplete noisy distance measurements. In *Information Theory Proceedings (ISIT), 2011 IEEE International Symposium on* (31 2011-aug. 5 2011), pp. 1584–1588.

[13] KIM, B., KAESS, M., FLETCHER, L., LEONARD, J., BACHRACH, A., ROY, N., AND TELLER, S. Multiple relative pose graphs for robust cooperative mapping. In *Robotics and Automation (ICRA), 2010 IEEE Intl Conf. on* (may 2010), pp. 3185–3192.

[14] KIM, S., KOJIMA, M., AND WAKI, H. Exploiting sparsity in SDP relaxation for sensor network localization. *Tokyo, Japan: Department of Mathematical and Computing Sciences, Tokyo Institute of Technology* (2008).

[15] KOREN, Y. On spectral graph drawing. *Computing and Combinatorics* (2003), 496–508.

[16] LAMAN, G. On graphs and rigidity of plane skeletal structures. *Journal of Engineering mathematics 4*, 4 (1970), 331–340.

[17] MACAGNANO, D., AND DE ABREU, G. Tracking multiple targets with multidimensional scaling. In *Wireless Personal Multimedia Comms* (2006).

[18] PATWARI, N., ASH, J. N., KYPEROUNTAS, S., HERO, A. O., . I. I. I., MOSES, R. L., AND CORREAL, N. S. Locating the nodes: cooperative localization in wireless sensor networks. *Signal Processing Magazine, IEEE 22*, 4 (july 2005), 54–69.

[19] SAXE, J. B. *Embeddability of weighted graphs in k-space is strongly NP-hard.* Carnegie-Mellon University, Dept. of Computer Science, 1980.

[20] SHANG, Y., AND RUML, W. Improved MDS-based localization. In *INFOCOM 2004. 23rd Conf. of the IEEE Computer and Communications Societies* (2004), vol. 4, pp. 2640–2651.

[21] SHANG, Y., RUML, W., ZHANG, Y., AND FROMHERZ, M. P. J. Localization from mere connectivity. In *Proc. of the 4th ACM international symposium on Mobile ad hoc networking & computing* (2003), pp. 201–212.

[22] THRUN, S., AND MONTEMERLO, M. The graph SLAM algorithm with applications to large-scale mapping of urban structures. *The International Journal of Robotics Research 25*, 5-6 (2006), 403–429.

[23] WYMEERSCH, H., LIEN, J., AND WIN, M. Z. Cooperative Localization in Wireless Networks. *Proceedings of the IEEE 97*, 2 (feb. 2009), 427–450.

[24] ZHU, Z., SO, A. M.-C., AND YE, Y. Universal Rigidity and Edge Sparsification for Sensor Network Localization. *SIAM J. on Optimization 20*, 6 (oct 2010), 3059–3081.

EV-Loc: Integrating Electronic and Visual Signals for Accurate Localization

Boying Zhang [†*], Jin Teng [†], Junda Zhu [‡], Xinfeng Li [†], Dong Xuan [†] and Yuan F. Zheng [‡]

[†]Department of Computer Science and Engineering, [‡]Department of Electrical and Computer Engineering
The Ohio State University, Columbus, Ohio, USA, 43210
{zhangboy, tengj, lixinf, xuan}@cse.ohio-state.edu, {zhuj, zheng}@ece.osu.edu

ABSTRACT

Nowadays, an increasing number of objects can be represented by their wireless electronic identifiers. For example, people can be recognized by their phone numbers or their phones' WiFi MAC addresses and products can be identified by their RFID numbers. Localizing objects with electronic identifiers is increasingly important as our lives become increasingly "digitalized". However, traditional wireless localization techniques cannot meet the fast growing needs of accurate and cost efficient localization. Some of these techniques require expensive hardware to achieve high accuracy, which is impractical for massive deployment. Others, such as WiFi RSSI based localization, are inaccurate and not robust to environmental noise. In this paper, we propose a new localization technique called *EV-Loc*. In EV-Loc, we use visual signals to help improve the accuracy of wireless localization. Our technique fully leverages visual signals' high accuracy and electronic signals' pervasiveness. To effectively couple these two signals, we design an E-V match engine to find the correspondence between an object's electronic identifier and its visual appearance. We implement our technique on mobile devices and evaluate it in real-world scenarios. The localization error is less than 1 m. We also evaluate our approach using large scale simulations. The results show that our approach is accurate and robust.

Categories and Subject Descriptors

C.2.1 [**Network Architecture and Design**]: Wireless Communication

General Terms

Algorithm, Design, Experimentation

Keywords

Localization, Wireless Devices, Visual, Matching

---

*The two authors are co-primary authors.

1. INTRODUCTION

1.1 Motivation

Mobile devices have proliferated in recent decades. Almost everyone carries one or more such devices, e.g., smartphones and laptops. RFID technology is also developing rapidly. RFID applications like product tags are becoming an integral part of our daily lives. Following this trend, our lives are becoming increasingly "digitalized". We are living in and connecting with a new interactive "web" that involves almost everything around us. It is imaginable in the near future, our intelligent house will respond to our needs automatically. For example, the door will open and the lights will turn on when and only when the house owner approaches and wants to enter. In supermarkets, if we want to find something, a robotic assistant will come up to help localize or even retrieve the product (with an RFID tag) amidst racks of merchandise.

One key enabling technique for the digital life described above is accurate localization of an object, such as a human, a product, a robotic assistant or even a phone. In a digitalized world, every object can be assigned an electronic identifier, e.g., we can represent a person by his phone number or a product by its RFID number. Therefore, we are interested in the following localization problem: given an electronic identifier, how can we localize the object bearing this electronic identifier accurately? Here accuracy is critical. In the above examples, with large localization errors, the back door may be opened when the house owner is at the front door, or the robotic assistant may bring back the wrong product. As we consider the electronic identifier and its bearer as one logical point in the space, we will use 'localizing the electronic identifier' and 'localizing the object with the electronic identifier' interchangeably hereafter.

To localize an object with an electronic identifier, people naturally use wireless technologies to capture and measure wireless signals. Different wireless localization techniques have been proposed in recent years [19]. However, the performance of these localization techniques (shown in Table 1) is still not satisfactory. Some techniques like Cricket [25] or Pinpoint [32] provide accurate localization results, but their accuracy relies on special hardware, which is often costly. These techniques are impractical for civilian use. Other techniques, e.g., Virtual Compass [5], do not rely on special hardware, but their localization results have relatively low accuracy, because these techniques rely on such measurements as signal strength, which are very vulnerable to the noise in the electronic environment.

System	Technologies	Accuracy
Pinpoint [32]	RF TOA	1.3 m
Cricket [25]	TDOA(Ultrasound+RF)	5 cm
RADAR [3]	WiFi RSSI	5.9 m
Horus [31]	WiFi RSSI	2 m
TIX [15]	WiFi RSSI	5.4 m
Virtual Compass [5]	WiFi RSSI	3.2 m

Table 1: Accuracy of representative localization techniques

1.2 Our Contributions

Given the limitations of wireless localization, we consider using visual signal as an auxiliary tool for accurate and practical localization. Compared with electronic signals, visual signals are relatively accurate in localization (less than 1 m) and less affected by noise [30]. The proliferation of camera phones and commercial cameras make visual localization affordable. Therefore, visual signals are a good candidate to help traditional wireless localization.

However, visually localizing an object does not mean knowing what the object is, hence is not enough. For example, we can visually localize a human, but we may not find out who he or she is. It is well known that *recognizing* a human is very hard just based on his or her visual appearance [14]. However, if we know the correspondence (or mapping) between a human's visual appearance and electronic identifier, we can localize the human with the electronic identifier (as well as recognizing him or her) based on the visual localization result and the correspondence. Hence, with the visual localization result, the key problem is to find out such correspondence or mapping. In some cases, it is possible to build up the correspondence library based on *a priori* information. But inputting the precise visual appearances of every object to be localized is cumbersome and undesirable. Moreover, even with the precise visual appearances, as mentioned above, recognizing a very specific human or object among others is still very difficult and time consuming.

In this paper, we propose EV-Loc, a technique integrating electronic and visual signals for accurate localization. In EV-Loc, we assume that each object has an electronic identifier, such as a smartphone's WiFi MAC address, is given. We want to localize an object associated with the electronic identifier. Specifically, EV-Loc takes electronic identifiers as the input and automatically corresponds/maps them with their visual signal counterparts extracted from the images. Then we can leverage visual localization for more accurate localization results. We claim the following contributions:

- We propose a methodology that leverages the accuracy of visual localization to help wireless localization.

- We propose an effective approach for E-V matching, i.e., corresponding an object's electronic identifier with its visual appearance. We also propose a novel method using a distance based location descriptor to model the uncertainty in E-V matching. We derive appropriate thresholds to produce correct matching results.

- We prototype the EV-Loc system and conduct real-world experiments. We can achieve high accuracy for cellular phone localization with median error at ~0.5 m and 90 percentile error at ~1 m. We further perform large scale simulations to evaluate EV-Loc's performance. The simulation results show that our approach is efficient and robust.

The rest of this paper is organized as follows. Section 2 reviews related work. Section 3 details the proposed localization technique, including cases where electronic and visual signals are indistinct and missing. Section 4 presents our experimental and simulation results. Section 5 discusses practical issues related to EV-Loc. Finally, Section 6 concludes the paper.

2. RELATED WORK

Wireless localization has been an active research area in recent years. Many researchers have conducted extensive work to advance core localization technologies and systems. In this section, we summarize representative work in this area. Generally, these works can be classified into three classes: range-based, range-free, and fingerprinting based.

Range-based localization uses electronic signals to measure the distance or angle between neighbor nodes and performs trilateration/triangulation to estimate an object's position. Based on ways for measuring electronic signals, different localization techniques are proposed. For example, Pinpoint [32] relies on time of arrival (TOA) for localization. Cricket [25] implements time difference of arrival (TDOA) using ultrasound and RF. Virtual Compass [5] uses RSSI (received signal strength indicator) for relative positioning by combining the Bluetooth and WiFi RSSI readings.

Range-free localization does not rely on measurement of distance or angles. Instead, it assumes nodes can estimate distances between each other and the anchor nodes' positions are known. For example, the Centroid algorithm [6] and APIT [16] use area estimation to estimate an object's position. DV-Hop [23] and Amorphous [22] estimate the minimum hop count between the unknown target node and the anchor node as well as the average hop distance, then calculate distance based on these estimations.

Fingerprinting based localization differs from the previous techniques. It fingerprints each location in a scene with a vector of RSSIs from various transmitters, e.g., WiFi APs and GSM towers. Then an object's position is estimated by matching the observed RSSI readings with the closest *a priori* location fingerprints. RADAR [3], Horus [31], and TIX [15] are representative works in this category.

Our work is also closely related to visual tracking and sensor fusion techniques. Yilmaz et al. overview this field in [30]. It has many real applications in a variety of areas such as robotics or bioinformatics. Smith et al. survey the use of multisensor fusion for tracking [27]. SurroundSense [2] uses WiFi RSSI, sound, light, etc. for indoor localization. There is a recent work [28] using combinations of electronic and visual signals for object identification. This work focuses on object filtering over a long period of time in which objects' locations change. Thus, it is not a localization problem.

3. EV-LOC DESIGN

In this section, we present the design of EV-Loc. We start with an overview of EV-Loc, then introduce its workflow, followed by its core component, the E-V match engine.

3.1 Design Overview

As discussed in Section 1, our goal is to improve the localization accuracy of an electronic identifier with the help of video cameras. This can be achieved by corresponding a wireless device, i.e., its electronic identifier, with its shape or owner in the video and then incorporating the visual localization results. In reality, the electronic identifiers and their visual signal counterparts can take any form. In this paper, we take smartphone localization as an example; the visual object corresponding to a smartphone is its owner.

In order to find the correspondence, we estimate the wireless device's location with traditional wireless localization, and match this location to a human's location. If the match is successful, i.e., we find the smartphone's owner, we can fuse the electronic and visual location results to get a more accurate result than can be obtained from purely electronic or visual localization.

However, the wireless localization result can be highly inaccurate. The circles in Fig. 1a give the possible localization result range. The circle in the wireless "vision" is much bigger than that in the camera vision.Consider Fig. 1 as an example. If there is only one smartphone in the wireless vision and one person in the camera vision (Fig. 1a), corresponding the smartphone and person is straightforward. However, if there are several people close to each other (Fig. 1b), corresponding smartphones with people is very hard. In this case, we cannot directly fuse the location results, as a mismatch will result in a large localization error. The correspondence difficulty increases as the number of smartphones and people in our vision increases.

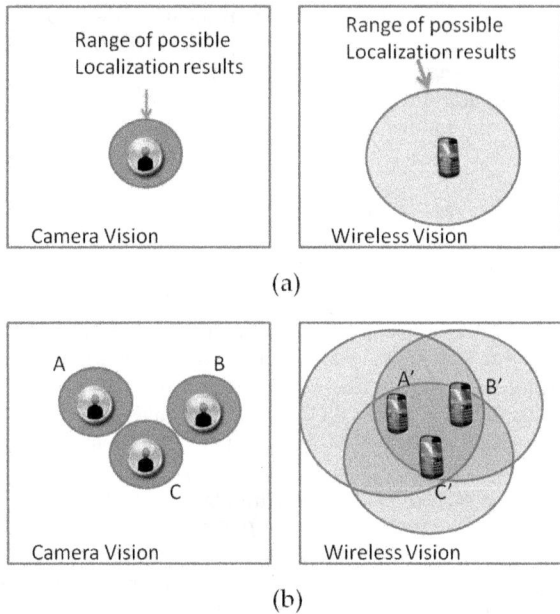

(a)

(b)

Figure 1: Associating wireless devices with their owners

Though inaccurate wireless localization makes correspondence difficult, we can still infer the correspondence statistically. In Fig. 1, if person A really corresponds to smartphone A', the average geometric distance between their location estimates from wireless and visual localization should be the smallest in the long run, e.g., smaller than the average distance between A and B'. Given enough time for location estimation, we can have a certain degree of confidence in determining the correspondence between people and smartphones.

We have two ways to describe the distance between two objects. The traditional way is to use the Euclidean distance between the location estimate points. However, we introduce here the concept of a *location descriptor*. A location descriptor of an object is a tuple of distances between this object and other objects, including the wireless detectors, i.e., access points (APs). We can define the distance between two tuples using any reasonable measure. The advantage of using location descriptors is two-fold. First, location descriptors capture the geometric topology among the objects, and carry more information than an estimated location point. In fact, the location estimate point is calculated from the topology among the object and several APs, so it cannot carry more information than raw distances. Furthermore, we can add the distance between any pair of objects into the descriptor to increase its power and flexibility. Second, the descriptor's error distribution is easier to model than pure estimated location points. The elements in each tuple are distance measurements, whose distribution can be analyzed or at least approximated in closed form. On the other hand, the error distribution of the final point localization result is almost impossible to derive. Though we can use the Cramér-Rao lower bound to estimate the variance [9], this bound is not typically useful in our case. As we will compare many pairs of locations, we need a finer distribution of the error than mere variances.

Note that using the descriptor brings extra benefits if the object to be localized is moving, e.g., humans with smartphones or robotic assistants. Normally, mobility is thought of as a curse for localization, as we cannot average away the noises from multiple measurements. However, if we consider the topological information of all objects to be localized, mobility can be useful to distinguish one object from the others. Specifically, a local position change of one object will change all the distances, i.e., the whole topological information. This more significant change in the topological information can help us uniquely identify and localize this object. For example, when three objects are very close, they cannot be distinguished. But if one of them moves farther, it can be easily identified from the topological change. The faster the topology changes, the faster the objects can be distinguished. Nevertheless, it should be pointed out that our scheme works with both mobile and static objects. While object mobility helps the matching converge faster, multiple readings of a static object help to cancel the noise, which also leads to faster convergence of the matching. Our EV-Loc scheme fully exploits any statistical opportunity for both mobile and static objects.

3.2 Workflow

Fig. 2 describes the main components and workflow for realizing our proposed localization technique.

The data collector has two parts: the electronic signal measuring unit and the visual signal measuring unit. The electronic signal measuring unit is installed on APs or mobile devices used as reference points. It periodically measures RSSI readings from smartphones. The visual signal measuring unit is installed on a central camera, which continuously records time-sequenced visual snapshots containing all objects and their background scene information.

The signal processor conducts two following tasks: signal error modeling and signal to distance conversion. The signal error modeling module is to find a statistical model for

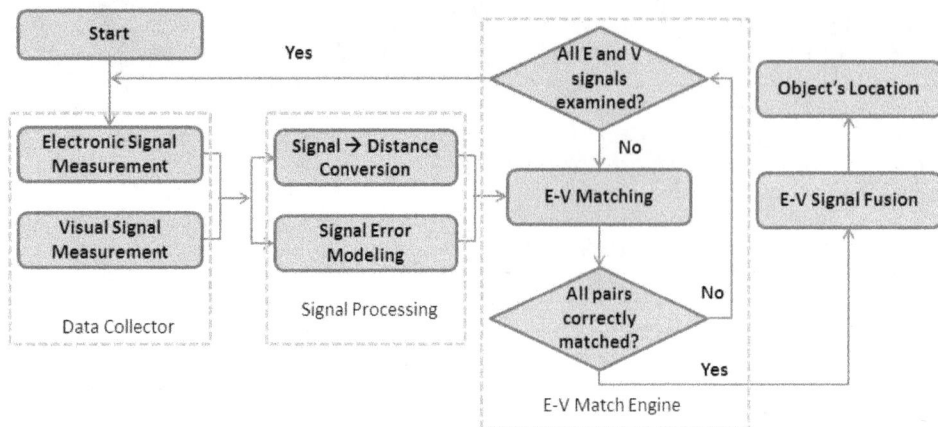

Figure 2: Workflow

the random observation of the electronic and visual signals collected from each scene. The signal to distance conversion uses site-specific environmental information to calibrate the parameters of the log-distance path loss model and the visual coordinate system. Then it converts each object's electronic and visual signals to estimated distances from the object to all surrounding APs.

The E-V match engine corresponds an object's electronic signals to its visual signals. Its inputs are the estimated distance (converted from electronic and visual signals) and the signal error model. Since the aforementioned E-V correspondence is not known *a priori*, we need to compare each pair of electronic and visual signals. To facilitate this comparison, we use a cost matrix to represent the similarity between each pair of converted distances from the electronic and visual signals. Based on this matrix, the match engine generates a "best match" result with highest similarity between each pair.

The workflow of our approach is as follows. Given an object's electronic identity, e.g., its WiFi MAC address, we first use the data collector to gather RSSI readings from different APs and visual snapshots from the central camera. All collected data are sent to a back-end server for further processing. After processing, the statistical characteristics of the collected signals are generated with the signal error model; the RSSI readings and visual appearances are converted to estimated distances between the object and different APs. Using this information as inputs, the E-V match engine can generate a best match between an object's electronic and visual location descriptors. To ensure the matching result is correct with high confidence, we repeat the matching process using the processed electronic and visual signals at other time points until a certain threshold is reached. Finally, when the match result is stable, we leverage visual localization to estimate the object's location.

It is worth noting that we use highly modular function blocks for adaptability. For example, instead of using a range-based location descriptor, we may also use range-free localization results for E-V matching. However, we need a new error model for each different localization method.

3.3 E-V Match Engine

Now we introduce the core of the above workflow, the E-V match engine. For simplicity, we assume here that the elec-

tronic signals and visual signals are complete with no false negatives or positives, i.e., there are no "ghost" or missing objects. Also, we assume the visual signals are distinct, i.e., we can distinguish people in different frames. More practical considerations, e.g., indistinct visual signals or incomplete signals, will be discussed in Section 3.5.

Suppose we have n wireless devices and n people. Each person carries a wireless device, which is uniquely identifiable. The set of electronic location descriptors is $\mathbf{x} = (x_1, \ldots, x_n)^T$, and the set of visual location descriptors is $\mathbf{y} = (y_1, \ldots, y_n)^T$. π_i is a permutation of the sequence $(1, 2, \ldots, n)$, and $\mathbf{y}_{\pi_i} = (y_{\pi_1}, y_{\pi_2}, \ldots, y_{\pi_n})^T$ is a permutation of the original vector $\mathbf{y} = (y_1, \ldots, y_n)^T$. Then we can formulate the following best match problem:

$$\arg\min_{\pi_i} \sum_{i=1}^{n} \|x_i - y_{\pi_i}\| \quad (1)$$

$$z_i = \alpha x_i + \beta y_{\pi_i} \quad (2)$$

The problem defined in (1) can be understood as we first find a permutation of $\mathbf{y}$ to match $\mathbf{x}$. (1) can be solved with the standard Hungarian algorithm [18]. After finding such a permutation π_i, we fuse the locations acquired wirelessly and visually into $\mathbf{z} = (z_1, \ldots, z_n)^T$. α and β are the coefficients that reflect the measurement confidence. If the measurement is inaccurate, i.e., the standard deviation is large, we give the location estimate less weight, and vice versa. Suppose every $x_i \in \mathbf{x}$ and $y_i \in \mathbf{y}$ have standard deviations σ_1 and σ_2, respectively. σ_1 and σ_2 are determined by the equipment used and the experiment environment. They remain relatively stable throughout the time. Then we can let $\alpha = \sigma_1^{-2}/(\sigma_1^{-2} + \sigma_2^{-2})$ and $\beta = \sigma_2^{-2}/(\sigma_1^{-2} + \sigma_2^{-2})$. These are the two optimal coefficients in the maximum likelihood sense given the two standard deviations σ_1 and σ_2. $\|\cdot\|$ is the norm operation. If we take x_i and y_i as coordinates, $\|\cdot\|$ can be the Euclidean distances or squared Euclidean distances. As we note that it is often far more convenient to deal with the squared Euclidean distance, we will stick to squared Euclidean distance in the following part, especially in Section 3.5.

So far, we have assumed that x_i, y_i, and z_i are static coordinates. We can extend them to a function of time, i.e., $\mathbf{x}(t) = (x_1(t), \ldots, x_n(t))^T$, $\mathbf{y}(t) = (y_1(t), \ldots, y_n(t))^T$ and $\mathbf{z}(t) = (z_1(t), \ldots, z_n(t))^T$. We correspondingly define the norm $\|x_i(t) - y_j(t)\| = \int |x_i(t) - y_j(t)| dt$, where $|\cdot|$ is the Euclidean distance (or squared Euclidean distance). With these adaptations, we can use (1) and (2) to solve the dynamic version of the problem. Note that we need to adapt the Hungarian algorithm to this dynamic inflow of information. After a new frame arrives at time t_i, we need to recompute the distance and re-run the Hungarian algorithm. However, we notice that we can keep the final matrix obtained in the last round of the Hungarian algorithm and increase the distance based on that matrix. We can consider this an incremental version of the Hungarian algorithm.

In practice, the accuracy of visual localization is far better than that of wireless localization, so it is possible that α is close to 1 and β close to 0. Under this circumstance, it is reasonable to take the visual localization result alone as the final fusion result.

3.4 Derivation of Matching Threshold

The E-V match engine provides a best match and a potential correspondence between each object's electronic identifier and visual appearance. A remaining problem is guaranteeing the correctness of the matching and the ensuing localization results. To address this issue, we derive a matching threshold based on the deviation of the estimated distances.

In general, the location descriptors for the same object from the electronic and visual sides should be the same. (In our cases, they are sets of distances.) But the electronic and visual descriptors actually differ because of noise. We model the noise in the visual distance reading as Gaussian. We also model the signal strength readings as Gaussian [26]. This means the converted distance from signal strength has a log-normal distribution, as signals attenuate exponentially with distance. The above measurements fluctuate around the mean values. If we average the measurements, the result is unlikely to deviate far from the mean values by the central limit theorem. Thus we can bound the deviation and determine with a certain confidence that the electronic descriptor and the visual descriptor do not belong to the same object if their deviations are too large. In the following, we give a rigorous mathematical description of this process.

First, we define the electronic and visual location descriptor as $x_i = (x_i^{AP_1}, x_i^{AP_2}, x_i^{AP_3})$ and $y_i = (y_i^{AP_1}, y_i^{AP_2}, y_i^{AP_3})$, where $x_i^{AP_j}$ or $y_i^{AP_j}$ is the measured distance between the i-th object and the j-th AP. For brevity, we will write x_{ij} and y_{ij} hereafter. In the above descriptor, we only consider the topological relation of objects with APs. It can be easily extended to cases where more comprehensive topology information is used. We can simply append the tuple with the distances to reference objects other than APs in a consistent manner, i.e., all the tuples should include the distance to these objects. Also, we use three APs here for illustration purposes, but we could easily extend the tuple to accommodate the additional AP measurements.

With the above transformation, we can model the variance of each x_i and y_i. We model x_{ij} as a log normal variable. According to [26], we can model $P(d)$, the RSSI reading at distance d, as

$$P(d) = P_0 - a \log_{10} \frac{d}{d_0} + P_n, \qquad (3)$$

where P_0 is the original transmission power (known), a is the attenuation coefficient (known), d_0 is a reference distance (known), and P_n is the noise, which can be modeled as a

normal random variable, $N(0, \sigma'_x)$. So $x_{ij} = d_0 \cdot \exp\{\ln 10 \cdot (RSSI - P_0 - P_n)/a\}$. We can choose the reference distance d_0 to have unit length and define the exponent as a random variable $N(\mu'_{x_{ij}}, \sigma'_x)$. Here, $\mu'_{x_{ij}}$ can be considered the accurate path loss reading $RSSI - P_0$ for object x_i by AP_j; σ_x^2 is the measurement variance of the AP in that wireless setting, which remains invariant. Let $\mu_{x_{ij}} = \ln 10 \cdot \mu'_{x_{ij}}/a$ and $\sigma_x = \ln 10 \cdot \sigma'_x/a$. Then $x_{ij} \sim \log N(\mu_{x_{ij}}, \sigma_x)$. It is worth noting that, if $x \sim \log N(\mu, \sigma)$, then $E[x] = \exp\{\mu + \sigma^2/2\}$, and $Var(x) = \{\exp(\sigma^2/2) - 1\} \exp\{2\mu + \sigma^2\}$.

On the other hand, y_{ij} can also be modeled as a normal variable, $y_{ij} = y \sim N(\mu_{y_{ij}}, \sigma_y)$. For the same object i, the distances measured wirelessly and visually without noise should be the same, i.e., $\exp\{E[x]\} = E[y]$.

Suppose x_i and y_i represent the same object. Then we can have $\exp\{\mu_{x_{ij}}\} = \mu_{y_{ij}}$. The squared Euclidean distance between $x_i = (x_{i1}, x_{i2}, x_{i3})^T$ and $y_i = (y_{i1}, y_{i2}, y_{i3})^T$ is written as:

$$\Delta_i = \sum_{j=1}^{3} (x_{ij} - y_{ij})^2. \qquad (4)$$

If x_i and y_i represent the same object, Δ_i, the distance between x_i and y_i should be lower-bounded by a threshold. If Δ_i is larger than the threshold, we can say with high confidence that the matching is wrong. Now we will calculate the threshold and the confidence.

Let us fix j and look at a single term $M_{ij} = (x_{ij} - y_{ij})^2$. We can bound M_{ij} within $[0, (\mu_{y_{ij}} \exp\{3\sigma_x\} + 3\sigma_y)^2]$ with the 3-σ rule. Though the 3-σ rule is just an approximation, it accurately reflects the fluctuation range of RSSI readings. From our large amount of empirical data and many previous research data, e.g., [20, 3], we find that the 3-σ rule holds in general. Then we apply the Hoeffding inequality, which states that if we have n independent variables $X_1, \ldots, X_n$, $\Pr(X_i \in [a_i, b_i]) = 1$ and $\bar{X} = \sum_i X_i/n$, then

$$\Pr(\bar{X} - E[\bar{X}] \geq t) \leq \exp\left\{-\frac{2t^2 n^2}{\sum_{i=1}^{n} (b_i - a_i)^2}\right\}. \qquad (5)$$

Let each M_{ij} be a random variable and substitute it in (5). We find that

$$\Pr(\Delta_i - E[\Delta_i] \geq t) \leq \exp\left\{-\frac{2t^2}{\sum_j (\mu_{y_{ij}}(\exp\{3\sigma_x\} - 1) + 3\sigma_y)^4}\right\}. \qquad (6)$$

It can be seen that $E[\Delta_i] = (\mu_y^2 \cdot (\exp\{2\sigma_x^2\} - 2\exp\{\sigma_x^2/2\} + 1) + \sigma_y^2)$. From (6), we know that if we have $\Delta_i > t + E[\Delta_i]$, then the possibility that x_i is not y_i is at least $1 - \exp\{-2t^2/\sum_j \mu_{y_{ij}}(\exp(3\sigma_x) - 1) + 3\sigma_y^4\}$.

The above inequalities form the basis for evaluating the quality of the matching. A good matching must satisfy the following conditions. If x_i is matched with y_j, Δ_{ij}, which is the squared Euclidean distance between the x_i and y_j tuples, must be below a certain threshold with a certain confidence, and Δ_{ik} ($k \neq j$), the distance between x_i and any other y_k tuples than y_j, should be larger than a threshold with a certain confidence. Specifically, $\Pr(\Delta_{ij} - E[\Delta_{ij}] \leq th_1) \geq c_1$, and $\forall k \neq j$, we have $\Pr(\Delta_{ik} - mE[\Delta_{ik}] \geq th_2) \geq c_2$. Empirically, we can set c_1 very large, e.g., 99.9%, and let $c_2 \in [90\%, 99\%]$. We can then compute thresholds th_1 and

th_2 and use them to decide the appropriateness of the matching.

We note that we do not actually know $\mu_{x_{ij}}$ and $\mu_{y_{ij}}$ in our calculation. In practice, we may take the visual measurements as approximate $\mu_{y_{ij}}$s as they are relatively accurate. With $\mu_{y_{ij}}$, a simple logarithmic operation can give us $\mu_{x_{ij}}$.

3.5 Extensions to Practical Settings

In this subsection, we discuss practical issues beyond our baseline cases given above. We examine the cases where some objects' visual signals are indistinct, or some objects' electronic or visual signals are missing, i.e., false negatives in the detection. It is worth noting that false positives with respect to electronic sensing can be considered as false negatives with respect to visual sensing, and vice versa. For example, if we detect an irrelevant visual object without wireless devices or a post is mistakenly identified as a person, a false positive takes place. We can view this false positive object "missing" an electronic identifier. If our E-V match engine is robust enough, we will not associate any electronic object with this irrelevant visual object.

To handle the missing object cases, we look deeper into the Hungarian algorithm. There are several implementations of the Hungarian algorithm. One of them is based on maximum flow. It finds a maximum flow in a bipartite graph, regardless of the number of nodes in each graph bipartition. This means that such an implementation can have different numbers of x_is and y_is and still finds a best match. We can use this Hungarian algorithm implementation to handle the missing objects cases. We will evaluate the robustness of our proposed approach when facing the indistinct and missing visual signal cases in Section 4.

With indistinct visual objects, we cannot distinguish among people in different visual frames. We need to determine this correspondence. We first define some terminology before formulating the problem. Suppose we have m frames and n wireless and visual objects in each frame. Then we have $\mathbf{x}(t) = (x_1(t), \ldots, x_n(t))^T$ as wireless location descriptors and $\mathbf{y}(t) = (y_1(t), \ldots, y_n(t))^T$ as visual location descriptors, where $t = t_1, \ldots, t_m$. Here we notice that $x_i(t_1)$ is the same object as $x_i(t_2)$, but $y_i(t_1)$ is not necessarily the same object as $y_i(t_2)$. We only know that there are n visual objects at times t_j and t_k. We randomly place the visual location descriptors in the descriptor tuple. Thus, we can only say that $y_i(t_1)$ is some $y_j(t_2)$, but we do not know which y_j. However, we have a distance matrix containing the distances between each pair of visual objects in different images. We want to find a permutation π_i for each time point, i.e., $\pi_i(t)$, that minimizes the total sum of location differences and visual object distances.

We can visualize the problem with the help of Fig. 3. We want to match one x_i with one y_j at each time point $t_1, \ldots, t_m$. We have a location distance matrix XY_i between $\mathbf{x}$ and any one of the m $\mathbf{y}(t_i)$s and a visual distance matrix Y_{ij} between each pair of $\mathbf{y}(t_i)$ and $\mathbf{y}(t_j)$. Y_{ij} details the visual dissimilarity between an object in $\mathbf{y}(t_i)$ and an object in $\mathbf{y}(t_j)$. By associating an E signal with one or two V signals from two different visual frames, we actually pick a cost in the distance matrix. A natural way to formulate the problem is to find $\pi_i(t)$ to minimize the total association cost from each XY_i and Y_{ij}. However, we face two difficulties here.

First, for comparisons of every pair of visual objects in different frames (m frames in total), the visual distance itself

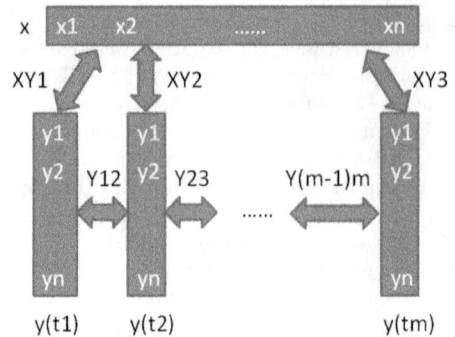

Figure 3: Problem formulation

has space complexity $O(n^m)$, which is exponential in m. Then the time complexity is at least exponential in m. If m is relatively small, e.g., $m \leq 10$, we may try enumeration. But as m increases, brute force enumeration will soon fail.

Second, the above problem is in fact a multi-dimensional assignment problem [7]. It is NP-hard and even inapproximable when the dimension is at least 3. Even if we only consider visual tuples at three time points, finding a best match to minimize the cost is infeasible for large n.

In order to cope with the above two difficulties, we take the following strategies:

First, we only consider the distance between two frames neighboring in time. This practice makes sense, because two frames far away in time may have low correlation, and the distance of objects therein can be heavily distorted. We may extend the processing scope to three or more frames neighboring in time. In essence, we only consider a finite number of frames close in time, not the entirety of frames. Then, we need to solve the following problem:

$$\arg\min_{\pi_j(t)} \sum_{i=1}^{m} \sum_{j=1}^{n} |y_{\pi_j}(t_i) - y_{\pi_j}(t_{i+1})|, \qquad (7)$$

where, for convenience, we write $y_{\pi_i}(t)$ to denote $y_{\pi_i(t)}(t)$, and define $|y_{\pi_j}(t_m) - y_{\pi_j}(t_{m+1})| := 0$.

Second, we perform visual object matching first. After getting the matching, the objects' visual appearances can be considered distinct, and then we run the E-V match engine discussed in Section 3.

In this paper, we leverage visual tracking techniques to generate a similarity matrix between every pair of consecutive visual frames. Then we perform $m - 1$ rounds of the Hungarian algorithm to find the best match between every pair of neighboring visual frames. After this step, we can use the E-V match engine to localize an object as discussed in previous section. The performance of this approach is evaluated in Section 4.

4. EVALUATION

In this section, we evaluate the performance of our proposed approach. To evaluate the localization accuracy in real-world settings, we conduct both indoor and outdoor experiments using mobile phones and laptops. To evaluate the efficiency and robustness of our proposed approach at large scale, we perform simulations with different environmental settings and population sizes.

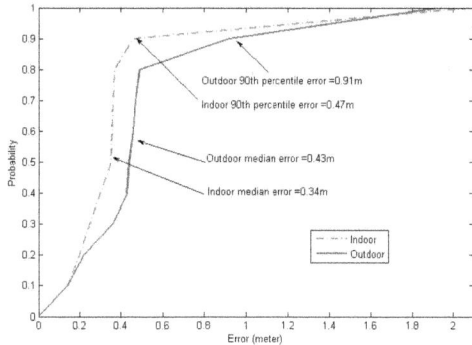

Figure 4: Cumulative distribution function (CDF) of the error distance

4.1 Real World Experiments

We conduct experiments to localize mobile phones in both indoor and outdoor settings. The indoor experiments are conducted in a 10 m × 10 m area inside a research building. The outdoor experiments are conducted in a 20 m × 20 m area outside the research building. As mobile phones are too small to see in the image, we associate a discernible object with each phone. In this case, we choose its owner as the visual object, and we view the identifier and the person as an integral entity, i.e., we do not consider the impact of human body on the phone signals. So we let the phones lie on the ground with the owners. We will discuss the impact of human bodies in the next section. In all experiments, we have five colleagues with WiFi-equipped mobile phones as target objects. We performed 10 indoor experiments and 10 outdoor experiments. In each experiment, we recorded 15 electronic frames and 15 visual frames. The RSSI readings recorded in each frame are averaged over 100 measurements. The time for APs to conduct 100 measurements varies between 5 to 20 seconds.

To capture target objects' visual information, we set up a camera shooting from above and covering the entire area. The surveillance camera is calibrated with known intrinsic parameters including focal length, lens distortion, and the relative rotation and translation with respect to the scene. The target objects' visual appearances are detected using the HoG pedestrian detector [12] and the correspondences across frames are established using the detection scheme in [1]. A person's planar coordinates in the scene are calculated by the pixel location of the bottom center of each bounding box. To capture target objects' electronic information, we use three laptops on the ground as APs to detect nearby mobile devices' WiFi MAC addresses. The WiFi on target objects' mobile phones is turned on, enabling them to be continuously detected.

The result of all experiments is shown in Fig. 4. For the indoor experiments, we achieve a median error of 0.34 m and 90 percentile error for 0.47 m. For the outdoor experiments, the median error is 0.43 m and the 90 percentile error is 0.91 m. Specifically, our E-V match engine can achieve a 90% (indoor) and an 80% (outdoor) success rate for correctly pairing all target objects' WiFi MACs and visual appearances. On average, the E-V match engine uses 6 per 15 frames to converge indoors and 8 per 15 frames to converge

outdoors. The parameter settings of our matching threshold are $\sigma_x = P \cdot 5/\sqrt{K}$ and $\sigma_y = 0.3$, where P is the tuning parameter for RSSI to distance conversion and is learned from each scenario. K records the number of RSSI measurements (100) for each frame.

4.2 Large-Scale Simulations

Since it is difficult to conduct larger scale real experiments, we conduct extensive simulations to further validate the performance of our proposed approach in different environment settings. In our simulations, we want to localize electronic identifiers, which may be static or moving, e.g., humans with mobile phones or the robotic assistants. All objects are randomly distributed in an area of 10 m × 10 m, and, if they are mobile, they move with a constant speed under the random waypoint model [8]. Depending on the coordinates of a given object within the area, its RSSI as received by the APs is simulated, and a perspective distortion determined from its distance to the camera is applied. For each location, we obtain 10 RSSI readings, and the standard deviation of RSSI is set to $5/\sqrt{10}$.

The visual appearances of objects are simulated by taking human-shape samples from the INRIA database [11] and scaling the sample in each frame according to its relative location with respect to the camera. (Whether the objects are human or not is not important, so long as the visual objects can be extracted from the images.) Samples of different poses from the same objects are randomly selected, and image noise is added to the visual appearances to simulate appearance changes in different frames. The HSV histogram feature with 8×8×4 bins is extracted for each visual object, and the similarity between two visual objects is determined using the Bhattacharya distance between their histograms [12, 24].

We measure the performance of our proposed scheme using three aspects: (1) The accuracy of our proposed approach as compared to localization using only electronic signals. We measure it by examining the variation of the localization accuracy when a number of frames are captured in a single scenario; (2) The efficiency of our proposed approach. We evaluate it by examining the number of frames needed to reach a given localization accuracy by changing the number of APs, the number of localized objects, and the motion speed of localized objects; (3) The robustness of our proposed approach when there are missing and indistinct

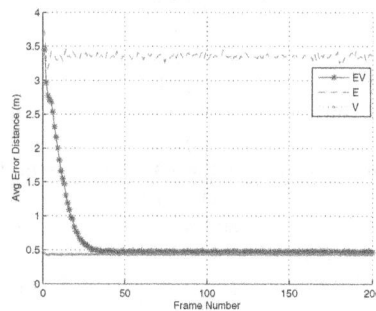

Figure 5: Localization performance comparison among different approaches

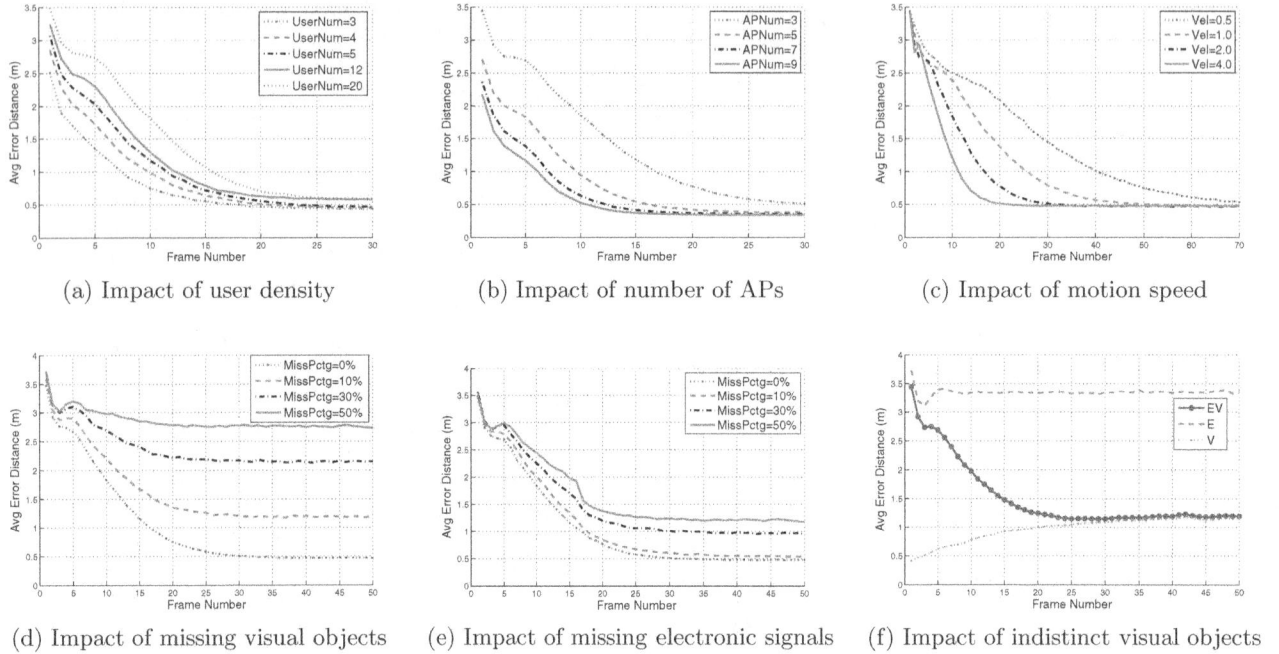

(a) Impact of user density

(b) Impact of number of APs

(c) Impact of motion speed

(d) Impact of missing visual objects

(e) Impact of missing electronic signals

(f) Impact of indistinct visual objects

Figure 6: Efficiency and robustness of our approach

visual signals. All simulation results are reported in Fig. 5 and 6. All results are averaged over 5000 runs.

Fig. 5 shows the accuracy of our proposed approach with respect to the number of input electronic and visual frames. We compare our approach with pure wireless and visual localization. For wireless localization, we use RSSI-based distance estimation plus multilateration to estimate the position of localized objects, as this technique is more practical and more widely available than other wireless localization techniques. We let 10 objects move with a speed of 2 m/s and deploy 3 APs on the boundary to form an isosceles triangle. There are three curves in the figure. The top curve shows the error distance using RSSI localization, the middle curve is the error distance of our approach, and the bottom curve is the error distance of visual localization. We can see that the wireless localization consistently has an error of around 3.4 m, while our approach has a decreasing error distance as the frame number increases. This is because more and more objects are corresponded between their electronic identifiers and visual appearances. After around 40 frames, our approach can fully leverage the high accuracy of visual localization to achieve an average error about 0.5 m.

In Fig. 6a, we evaluate the impact of the number of localized objects on the localization error distance. The initial position of each object is random, and their movements are independent. The figure shows that we need more frames to reach a certain error distance when there are more objects. However, according to the trend of the displayed curves, we can also discern that the number of frames needed to reach a certain accuracy does not remarkably increase with more objects. Eventually, they all reach a very high accuracy. This result shows that our approach is efficient even if the density of localized objects is high.

In Fig. 6b, we evaluate the impact of the number of APs on the localization error distance. All the APs are deployed

along a regular polygon on the boundary. When the number of APs exceeds 8, we deploy them in a grid shape. As more APs are deployed, the error distance of our approach decreases rapidly. This shows that additional APs improve the accuracy of EV-Loc. Eventually, the error distance becomes stable at around 0.5 m.

In Fig. 6c, we evaluate the impact of the motion speed of the localized objects on EV-Loc's error distance. The trend of the curve shows that we need less frames to reach a certain accuracy when the localized objects move faster. This result confirms that the combination of objects' movement and their topological relationship can effectively reduce the convergence time of the E-V match engine.

In Fig. 6d, we evaluate the error distance of EV-Loc in the case where objects are not all visually detected (misses). Generally, the localization accuracy decreases when the objects' visual miss rate increases. We also observe that the localization error distance is decent at around 1.2 m even if the miss rate is as high as 10%.

In Fig. 6e, we evaluate the error distance of EV-Loc in the case where objects are not all electronically detected (misses). In this situation, some objects' visual appearances simply do not have corresponding electronic identifiers. According to the simulation result, we find the localization accuracy decreases when the objects' electronic miss rate increases. However, such decline does not significantly impact the accuracy of EV-Loc. Its error distance remains around 1 m even when 30% of objects are missing.

In Fig. 6f, we evaluate the localization error distance in the case where the objects' visual appearances are indistinct. To achieve this effect, we randomly permute the order of visual appearances in different frames. Generally, the localization accuracy decreases when the objects' visual appearances become indistinct. This is because the object that fails to find a match will simply use wireless localization results. As is shown in the figure, our solution for handling the object's visual indistinction can achieve similar performance to the case with distinct visual appearances.

5. DISCUSSIONS

Though our paper focuses on introducing a new idea and enabling it, this section discusses some problems that the conceptual EV-Loc system may encounter in practice.

The EV-Loc system requires two sensing systems working at the same time, one wireless network, and one camera network. Compared with traditional wireless localization, there will be additional costs and coordination efforts in deployment. However, we believe these issues do not pose a major hurdle to deployment. Cameras are becoming increasingly affordable: common webcams normally cost under 100 dollars. Moreover, we can use existing surveillance networks, as they are increasingly pervasive for public or private usage. The camera network needs to be calibrated for localization, i.e., the transformation matrix from an image point to a real location should be known. This can be done during installation. The cameras also need to be synchronized with the wireless network, as we need to match objects from both sensors. Here a small temporal drift of up to 1 s for synchronization is usually tolerable. We can achieve it via the NTP protocol. Other relevant infrastructure issues for EV-Loc include sensor deployment and data management. For cost efficiency, we want to cover the whole monitored area with as few sensors as possible. This is the optimal sensor deployment problem. We plan to follow the research of [4, 33] and design optimal AP and camera deployment patterns as a part of our future work. On the other hand, in real implementation, we may need a database and a mechanism, such as [29], for recording, managing and searching through electronic and visual signals and the final location results.

In the paper, we have implicitly assumed that there is little visual occlusion. We have shown that EV-Loc works under this assumption. However, in reality, there might be serious occlusions when there are many objects in the monitored area. One way to deal with occlusion is to follow the method proposed in Section 3.5 and treat erroneous detections due to visual occlusion as missing IDs. Another more advanced method is to enable visual tracking. Currently, visual tracking technologies allow continuous tracking when occlusion occurs in some circumstances. Using multiple cameras and tracklets are two major approaches [1, 21]. Though it is well beyond the scope of this paper, we plan to incorporate these technologies in our future work.

Environmental interference can also impact EV-Loc's accuracy. For example, human body can cause a blocking effect on transmitted electronic signals [34]. To examine the impacts of such interference, we conduct a preliminary experimental study in an outdoor environment. There are five experiment participants, all holding mobile phones in their hands and standing in random positions. The purpose of such settings is to consider the human body's blocking effects on electronic signals. Based on our experimental result, we find that the median error of EV-Loc is around 0.5 m, while the 90 percentile error is around 2.5 m. The elevated 90 percentile error distance indicates a simple error distribution model of electronic signals may not suffice to describe human body interference. Based on this observation, we plan to investigate the impact of the human interference on EV-Loc's localization accuracy in our future work.

The EV-Loc technique can easily work with other electronic and visual localization methods. We can simply input the location estimation results obtained via other localization methods as location descriptors into our workflow (with a well-defined distance measure), model the error distribution of each localization method separately, and then run our E-V match engine. For example, to accommodate range-free localization, we take coordinates-based location descriptors as input to our E-V match engine and set the cost function of associating electronic and visual objects appropriately. By properly selecting the error distribution model (e.g. [17]), we can match an object's electronic and visual signals and then apply the proposed technique to localize it. With more accurate localization estimates on the electronic side, the association time is expected to be shorter.

EV-Loc is largely centralized, like most current existing localization algorithms [10, 3, 2, 5, 13]. In some circumstances, a distributed algorithm is desirable for efficiency and privacy reasons. We can achieve different levels of distributed processing under the EV-Loc framework. In the simplest distributed version, electronic and visual sensors perform part of the computations, e.g., the cameras complete human detection and location estimation on their own rather than delegating the tasks to the central server. In a more complicated version, a hierarchy can be established in the EV-Loc system. We partition the EV-Loc sensors, both electronic and visual, into clusters and assign cluster headers. This is useful for large area deployment as global information is rarely useful for localizing an object that is unlikely to move much. Finally, in a fully distributed version of EV-Loc, each sensor performs its own object localization, but sufficient information exchange needs to happen in a properly large geographical area, as is required for E-V matching. The fully distributed version introduces further scheduling and error tolerance issues, which we aim to explore in future work.

6. FINAL REMARKS

In this paper, we presented a technique called EV-Loc for accurate localization based on electronic and visual signals. Given an object's electronic identifier, we aimed to accurately localize the object with the help of visual signals. In order to achieve this, we proposed the E-V match engine that can accurately and efficiently correspond an object's electronic and visual signals. Following the matching results, we used visual localization to precisely estimate the given object's position. We also considered practical situations, e.g., missing or indistinct electronic and visual signals, and devised schemes to eliminate their impacts. We implemented EV-Loc on mobile devices and conducted real world experiments and large scale simulations to evaluate our proposed approach. The results showed our approach can achieve high localization accuracy and the underlying matching algorithm is efficient and robust.

Acknowledgments

This work was supported in part by China 973 Project No. 2011CB302800, the National Science Foundation of China (NSFC) under grant No. 61070221, and the US National Science Foundation (NSF) under Grant No. CNS0916584, CNS1065136. Any opinions, findings, conclusions, and recommendations in this paper are those of the authors and do not necessarily reflect the views of the funding agencies.

We would like to thank Adam C. Champion, Gang Li and Qiang Zhai for their help in the discussions and experiments. We would also like to thank the anonymous reviewers and our shepherd for their constructive comments which help to bring the paper to its current state.

7. REFERENCES

[1] M. Andriluka, S. Roth, and B. Schiele. People-tracking -by-detection and people-detection -by-tracking. In *Proc. of IEEE CVPR*, 2008.

[2] M. Azizyan, I. Constandache, and R. R. Choudhury. Surroundsense: mobile phone localization via ambience fingerprinting. In *Proc. of ACM MobiCom*, 2009.

[3] P. Bahl and V. N. Padmanabhan. RADAR: an in-building rf-based user location and tracking system. In *Proc. of IEEE INFOCOM*, March 2000.

[4] X. Bai, C. Zhang, D. Xuan, J. Teng, and W. Jia. Low-connectivity and full-coverage three dimensional networks. In *Proc. of ACM Mobihoc*, 2009.

[5] N. Banerjee, S. Agarwal, P. Bahl, R. Chandra, A. Wolman, and M. Corner. Virtual compass: relative positioning to sense mobile social interactions. In *Pervasive*, 2010.

[6] N. Bulusu, J. Heidemann, and D. Estrin. Gps-less low cost outdoor localization for very small devices. *IEEE Personal Communications Magazine*, 7(5):28–34, October 2000.

[7] R. E. Burkard and E. Cela. *Handbook of Combinatorial Optimization, Volume A*. Kluwer Academic Publishers, 1999.

[8] T. Camp, J. Boleng, and V. Davies. A survey of mobility models for ad hoc network research. *Wireless Communications and Mobile Computing*, 2(5):483–502, 2002.

[9] C. Chang and A. Sahai. Cramér-rao-type bounds for localization. *EURASIP Journal on Applied Signal Processing*, pages 1–13, 2006.

[10] K. Chintalapudi, A. Padmanabha Iyer, and V. N. Padmanabhan. Indoor localization without the pain. In *Proc. of ACM MobiCom*, pages 173–184, 2010.

[11] N. Dalal. INRIA person dataset. http://pascal.inrialpes.fr/data/human/, 2005.

[12] N. Dalal and B. Triggs. Histograms of oriented gradients for human detection. In *Proc. of IEEE CVPR*, pages 886–893, June 2005.

[13] M. Ding, F. Liu, A. Thaeler, D. Chen, and X. Cheng. Fault-tolerant target localization in sensor networks. *EURASIP J. Wirel. Commun. Netw.*, 2007(1):19–19, Jan. 2007.

[14] A. Ferencz, E. Learned-Miller, and J. Malik. Learning hyper-features for visual identification. In *Proc. of NIPS*, pages 425–432, 2005.

[15] Y. Gwon and R. Jain. Error characteristics and calibration-free techniques for wireless lan-based location estimation. In *Proc. of ACM MobiWac*, September 2004.

[16] T. He, C. Huang, B. Blum, J. Stankovic, and T. Abdelzaher. Range-free localization schemes for large scale sensor networks. In *Proc. of ACM MobiCom*, pages 81–95, 2003.

[17] A. Karbasi and S. Oh. Distributed sensor network localization from local connectivity: performance analysis for the hop-terrain algorithm. *SIGMETRICS Perform. Eval. Rev*, 38(1):61–70, June 2010.

[18] H. W. Kuhn. The hungarian method for the assignment problem. *Naval Research Logistics Quarterly*, 2:83–97, 1955.

[19] H. Liu, H. Darabi, P. Banerjee, and J. Liu. Survey of wireless indoor positioning techniques and systems. *IEEE Trans. Syst., Man, Cybern. C: Applications and Reviews*, 37(8):1067–1080, November 2007.

[20] Y. Liu, Z. Yang, X. Wang, and L. Jian. Location, localization, and localizability. *Journal of Computer Science and Technology (JCST)*, 25(2):274–297, 2010.

[21] C. C. Loy, T. Xiang, and S. Gong. Multi-camera activity correlation analysis. In *Proc. of IEEE CVPR*, pages 1988–1995, june 2009.

[22] R. Nagpal. Organizing a global coordinate system from local information on an amorphous computer. In *A.I. Memo 1666*. MIT A.I. Laboratory, August 1999.

[23] D. Niculescu and B. Nath. DV based positioning in ad hoc networks. *Journal of Telecom. Systems*, 2003.

[24] K. Nummiaro, E. Koller-Meier, and L. V. Gool. An adaptive color-based particle filter. *Image and Vision Computing*, 21(1):99–110, January 2003.

[25] N. B. Priyantha, A. Chakraborty, and H. Balakrishnan. The cricket location-support system. In *Proc. of ACM MobiCom*, pages 32–43, 2000.

[26] S. Y. Seidel and T. S. Rappaport. 914MHz path loss prediction models for indoor wireless communications in multifloored buildings. *IEEE Transactions on Antennas and Propagation*, 40(2):209–217, 1992.

[27] D. Smith and S. Singh. Approaches to multisensor data fusion in target tracking: a survey. *IEEE Transactions on Knowledge and Data Engineering*, 18(12):1696–1710, December 2006.

[28] J. Teng, J. Zhu, B. Zhang, D. Xuan, and Y. Zheng. E-V: efficient visual surveillance with electronic footprints. In *Proc. of IEEE INFOCOM*, March 2012.

[29] H. Wang, C. Tan, and Q. Li. Snoogle: A search engine for the physical world. In *Proc. of IEEE INFOCOM*, April 2008.

[30] A. Yilmaz, O. Javed, and M. Shah. Object tracking: a survey. *ACM Comput. Surv*, 38(4), December 2006.

[31] M. Youssef and A. Agrawala. The Horus WLAN location determination system. In *Proc. of ACM MobiSys*, June 2005.

[32] M. Youssef, A. Youssef, C. Reiger, A. Shankar, and A. Agrawala. Pinpoint: an asynchronous time-based location determination system. In *Proc. of ACM MobiSys*, pages 165–176, 2006.

[33] Z. Yu, J. Teng, X. Bai, D. Xuan, and W. Jia. Connected coverage in wireless networks with directional antennas. In *Proc. of IEEE INFOCOM*, 2011.

[34] Z. Zhang, X. Zhou, W. Zhang, Y. Zhang, G. Wang, B. Y. Zhao, and H. Zheng. I am the antenna: accurate outdoor AP location using smartphones. In *Proc. of ACM MobiCom*, pages 109–120, 2011.

Low-complexity Scheduling for Wireless Networks

Guanhong Pei
Department of Electrical and Computer
Engineering, and Virginia Bioinformatics Institute
Virginia Tech, Blacksburg, VA, USA
guanhong@vt.edu

V.S. Anil Kumar
Department of Computer Science, and Virginia
Bioinformatics Institute
Virginia Tech, Blacksburg, VA, USA
akumar@vbi.vt.edu

ABSTRACT

Designing efficient scheduling and power control algorithms for distributed wireless communication has been a challenging issue, especially in the physical interference model based on SINR constraints. In this paper, we discuss the first local distributed scheduling and power control algorithm in the SINR model that achieves an $O(g(L))$ approximation factor of the rate region, where $O(g(L))$ denotes the link diversity. As an intermediate step, we develop a scheduling algorithm in a k-hop interference model, which is used in the analysis of the more general model. Our algorithms are based on random-access and use local queue size information. Exchanging queue size information is challenging, and the complexity of this step is often ignored in prior local distributed algorithms. A novel aspect of our paper is the use of stale and infrequently updated queue size info, which significantly improves the complexity of our algorithms.

Categories and Subject Descriptors

C.2.1 [**Computer-Communication Networks**]: Network Architecture and Design—*wireless communication, distributed networks*; G.1.6 [**Numerical Analysis**]: Optimization—*stochastic programming*; G.3 [**Mathematics of Computing**]: Probability and Statistics—*queueing theory, probabilistic algorithms*

Keywords

Wireless Networks, Physical Interference, SINR, Distributed Algorithms, Scheduling, Throughput Optimization, Cross Layer Design

1. INTRODUCTION

A fundamental problem in wireless networks is to maximize the network throughput capacity through proper scheduling and resource allocation — given a set of wireless nodes, what network throughput comparing to the optimum we can obtain by appropriate link scheduling and power management. Tassiulas and Ephremides in their seminal work [33], introduce *stability region* to characterize the maximum attainable network throughput and also provide a stable scheduling strategy that attains optimal throughput region in

an arbitrary wireless network. In this paper, we extend the concepts of throughput region and capacity region to the SINR setting w.r.t. both scheduling and power control. Stability region or *throughput region*, denoted by Λ, is referred to as the convex hull of all arrival rate vectors under which the network can be stably scheduled and power-controlled by a specific algorithm; and *capacity region*, denoted by Λ^{OPT}, refers to the maximum throughput region under any scheduling and power control scheme. The fraction of the capacity region that an algorithm achieves is called *efficiency ratio*, which is the reciprocal of the approximation ratio. Solving the throughput maximization problem generally involves solving an NP-hard problem at every time slot with global information [22], *e.g.*, the back-pressure algorithm [33]. Since then, multitudes of research efforts have been made on achieving the following three goals at the same time: (1) lowering the scheduling complexity; (2) ensuring a certain throughput region; and (3) requiring as little information as possible. The amount of information required also implies the overhead or complexity for scheduling. Based on the idea of maximal matching, a large number of algorithms are proposed, *e.g.*, variants of max-weight scheduling [22], greedy maximal scheduling [15], maximal scheduling [6, 27]. As the scheduling complexity gets improved, the achievable throughput region usually gets traded-off.

One of the key challenges in wireless scheduling is interference, and a number of abstract models have been proposed for this, a majority of which involve "graph based constraints" in the form of interference sets for each link (e.g., the k-hop interference model and protocol model). These models are simpler to analyze, but are too simplistic [3,24], and the "physical interference model based on SINR constraints" (referred to as the SINR model) [12], is considered much more realistic. The SINR model is much more challenging, and the stability region for this model is not well understood. In this paper, we develop algorithms for scheduling and power control and for characterizing the rate region in the SINR model.

Many of the algorithms for link scheduling and power control (e.g., the back-pressure algorithm [33]) are centralized, and cannot be implemented in a practical manner in wireless networks which are inherently distributed. Developing low-complexity distributed algorithms for scheduling is a fundamental problem, and has been an active area of research. Several distributed scheduling schemes have been developed (e.g., Q-SCHED [11]) for graph based interference models, and these involve scheduling decisions based on information about queue sizes from interfering links. One limitation in many of these approaches is that the overhead of control communication is not explicitly taken into account (a common strategy is to assume a "control channel" for exchanging backlog information). As we discuss later, that may incur a large overhead if it takes place frequently (*e.g.*, in every time slot), when the complexity is rigorously analyzed.

In this paper, we focus on developing and analyzing low-complexity distributed scheduling and power control algorithms for approximating the throughput region, while addressing the communication complexity rigorously. Our algorithms are based on random-access maximal scheduling, which uses queue information from "close-by" links. We reduce the overhead of information exchange by using *infrequently* updated stale neighborhood information — this is intuitively reasonable, but makes the analysis much more involved. Our main contributions are summarized below.

Main Contributions. We present the first low-complexity distributed scheduling and power assignment algorithm (called RA-SCHED-SINR) in the SINR model which ensures a throughput region of $\Omega(\frac{1}{g(L)})\Lambda^{OPT}$, where $g(L)$ is the "link diversity" [10], which is the number of classes into which the links can be partitioned, so that links within each class have similar lengths. RA-SCHED-SINR uses random-access based on local queue information. Our analysis rigorously takes the communication complexity of all the steps of the protocol, including the information and data exchange steps. A novel aspect of our analysis is the use of out-of-date information to significantly improve the efficiency.

As an intermediate step, we first design a low-complexity distributed scheduling algorithm RA-SCHED in a graph-based interference model, which is of independent interest and helps in the analysis of the more complex RA-SCHED-SINR. Our algorithm builds on the Q-SCHED protocol of [11] and utilizes neighborhood queue size information for scheduling, but is simpler because of its explicit use of random-access. Other distinguishing features of RA-SCHED and its analysis include: (1) Links access the channel in a slotted Aloha fashion, which do not involve per-slot control phase; (2) Links make only infrequent local queue size information update, and use stale information for scheduling; rather than assuming a "magically" done per-slot queue size information exchange, we prove that the throughput loss due to the info-exchange process can be arbitrarily low. The achievable throughput region of RA-SCHED is close to $\frac{\Lambda^{OPT}}{e\Delta}$, where Δ is the interference degree defined in Section 3.

RA-SCHED-SINR is based on RA-SCHED, with a different notion of "interference sets," which leads to slightly simpler sufficiency conditions for the throughput region than [3]. Our power control policy is simple and flexible: we only require links of similar lengths use similar power values. This policy incorporates many oblivious power assignments as special cases. The scheduling is implemented in a framed manner, and achieves an order optimal throughput region for networks with links of similar lengths. Further, our results take the background noise fully into account.

Organization. We survey and discuss related work in Section 2. In Section 3, we present the system model for the throughput optimization problem. Section 4 presents RA-SCHED which solves the problem on a graph-based interference model; Section 5 presents RA-SCHED-SINR for scheduling and power control under the geometric SINR model. Section 6 provides conclusion and discussion. We defer some of the proofs to the appendix.

2. RELATED WORK

2.1 Graph-based Interference Model

In the past few years, much progress has been made in designing low-complexity distributed throughput-optimal scheduling schemes, most of which employ an *adaptive* approach, where the channel

access parameters are dynamically adjusted from time to time (by referring to queue sizes).

A line of work uses adaptive random-access maximal scheduling coupled with a per-slot control phase in each slot, as in [11, 16, 21], where knowledge of interference (or the conflict graph) is assumed. The control phase consists of multi-round message passing, during which contention resolution is realized by probabilistic channel requests and sensing. In [11], the algorithm Q-SCHED achieves a throughput region close to $\frac{\Lambda^{OPT}}{\Delta}$, comparable to that of maximal scheduling. Follow-up works include [36], which provides provable bounds on queue-overflow probability, and [19] which proposes a distributed algorithm with constant-time control phase under k-hop interference model. Many of these approaches make simplifying assumptions which might entail high overheads, such as: (i) the sender of a link can detect collision or the transmission of an interfering link immediately, and (ii) the backlog info-exchange among interfering links is assumed to be done instantaneously in each slot— this can be a problem if the interference set is large.

A second line of work employs randomized policies based on a priori knowledge of traffic rates [5, 30]. These algorithms does not involve any message passing or queueing information. However, traffic information is not often known. It is possible to estimate traffic rates with high accuracy for some stochastic arrival processes, while the estimation process may take a long time (*e.g.*, $O(e^n)$) for small rate values.

Another line of work is based on *adaptive CSMA*, where the channel occupation and back-off time parameters are adjusted according to local information (*i.e.*, queue size and transmission record). With elegant analysis, adaptive CSMA based schemes are shown to be optimal in their settings [14, 28]. Nardelli et al. [26] have shown that despite the promising properties, adaptive CSMA schemes may lead systems to a negative extreme in some cases, *e.g.*, with hidden terminals and symmetric channels, the high aggressiveness of adaptive CSMA "further increases collisions yielding a self-sustaining loop" and "eventually, the protocol enters a state where no successful transmissions occur."

This also suggests that interference information can be crucial in the performance of a scheduling scheme. For example, as discussed above, there could be some extreme cases that make it hard for adaptive CSMA to provide a lower bound on system performance. Our approach can make use of the knowledge of interference, namely, the conflict graph, which may be necessary for obtaining a performance lower bound in practice; it can also work without any the knowledge of interference, with added control time (which affects the network delay but not the throughput region). It is also worth noting that even if interference information is known, without knowing and sharing either the traffic rate information or the interference-neighborhood queueing information, adaptive CSMA may still suffer from the same issues under a stochastic traffic model. Therefore, neighborhood queueing information serves also as a crucial factor in algorithm design.

2.2 Physical Interference Model

Most of the work in the SINR model is focused on the link scheduling problem, and there is limited work addressing the throughput optimization problem, largely due to the complexity of SINR model and the difficulty of applying a Lyapunov function or fluid-limit approach. Goussevskaia et al. [10] develop a centralized algorithm for the weighted one-shot scheduling algorithm, which can be adapted and achieves a throughput region of $1/O(g(L))$ factor of Λ. Later, Goussevskaia, et al. [8] proposed an algorithm that solves the non-weighted version of the one-shot scheduling problem with approximation ratio of $O(1)$ with a centralized algorithm [8]. Kesselheim [17] developed the first $O(1)$-approximate algorithm for the non-weighted version of the one-shot scheduling problem with power control. It is not clear how to implement these algorithms in a decentralized manner. We also note that the results of [8,18] do not imply comparable results as ours, since our setting is dynamic and involves stochastic packet arrivals. Xi and Yeh [34] study distributed power control to achieve optimal throughput region based on "high-SINR approximation" (defined in [34]) for a CDMA network. Lee et al. [20] propose a power-control algorithm, which maximizes the throughput region but involves network-wise gossip and m-polynomial convergence time for scheduling at every slot. In this paper, we present a low-complexity distributed algorithm in a geometric SINR setting, and our approximation results refer to an optimality that is achieved by optimum scheduling and power assignment scheme.

2.3 On Using Delayed Network-State Info

Our use of stale backlog information features infrequent control information updating, and differs from previous techniques of using delayed information [31, 35]. Ying and Shakkottai [35] provide important results on routing and scheduling with homogeneously delayed queue-and-channel-state information for graph-based models. Interestingly their simulation show that information delay may cause throughput region to shrink. It is not clear if the algorithms of [31,35] can be implemented with low algorithmic complexity and if the overheads due to information (although delayed) that needs to be exchanged and updated in every step is bounded. We try to address some of these issues in this paper, by reducing the frequency of the distributed information exchange among links and rigorously analyzing the overheads of it under SINR constraints, without assuming instant detection of collision or transmission on interfering links.

3. SYSTEM MODEL

Table 1: Notation.

G	network graph	n	#nodes
V	set of nodes	L	set of links
$\mathcal{I}(l)$	interference set of l	$\hat{\mathcal{I}}(l)$	$\mathcal{I}(l) \cap \{l\}$
$x(l)$	transmitter of link l	$r(l)$	receiver of link l
$q_l(t)$	indicator: $Q_l(t) > 0$	$Q_l(t)$	backlog on l
$d(l)$	length of link l	$d(u,v)$	dist. of u and v
$s_l(t)$	indicator: l transmits	$u_l(t)$	departure of Q_l
$\lambda(l)$	mean arrival rate on l	$A_l(t)$	exogenous arrival
Λ	throughput region	Λ^{OPT}	capacity region
α	path-loss exponent	β	SINR threshold
N	background noise	$g(L)$	link diversity

Network Model. The wireless network is modeled as a directed graph $G = (V, L)$ in a Euclidean plane, where V is the set of nodes (or transceivers) and L the set of communication links. The network is synchronized and for simplicity we assume all slots have the same length. Links are directed, and for link $l = (x(l), r(l))$, $x(l)$ and $r(l)$ denote the sender and the receiver, respectively. For nodes u and v, let $d(u, v)$ denote the Euclidean distance. For link l, let $d(l) \triangleq d(x(l), r(l))$ denote the length of the link; for links l, l', let $d(l', l) \triangleq d(x(l'), r(l))$. Each link l uses a transmission power level $P(l)$. We assume the commonly used path loss models [4,10],

in which the transmission on link l is possible only if:

$$\frac{P(l)/d^\alpha(l)}{N(1+\phi)} \geq \beta, \tag{1}$$

where $\alpha > 2$ is the "path-loss exponent," $\beta > 1$ is the minimum SINR required for successful reception, N is the background noise, and $\phi > 0$ is a constant (note that α, β, ϕ and N are all constants). **Wireless Interference**. We study both graph-based interference model (based on which we propose RA-SCHED) and the physical interference model based on geometric SINR constraints (henceforth, referred to as the SINR model) (based on which we propose RA-SCHED-SINR):

(1) *Graph-based interference model.* As in [11, 27], interference is defined on a 2D binary symmetric adjacency matrix, such that the interference relationship between any two links is *binary* and *symmetric*. Each link l has its *interference sets* defined as $\mathcal{I}(l) = \{l' : l' \text{ interferes with } l\}$. This abstract interference model lacks specification of how links are connected, *e.g.*, transmission ranges; however, in order to communicate among interfering links, they need to be connected, either by infrastructure support (*e.g.*, enterprise WLANs) or by (intermediate) wireless links. For this reason, to account for the latter case we employ (but are not limited to) the widely used *k-hop interference model*: two links interfere with each other if and only if the shortest path between them on G is at most k hops. It is also possible to extend to other graph-based models. We consider *bidirectional* interference, which accounts for ACK packets that may be crucial in a distributed setting with random-access. Hence, by l' *interferes with* l we mean that the transmission of either data from $x(l)$ or ACK from $r(l)$ will fail when either $x(l')$ or $r(l')$ is transmitting, or vice versa. This gives leeway to using asynchronous ACKs during a time slot. The *interference degree* $\Delta = \max_l \Delta(l)$, where $\Delta(l)$ is the size of l's interference set. In most commonly used models of wireless interference, such as the distance-2 matching model [2], Δ is a constant. We assume that all the nodes have a common estimate of n, the size of the network, within a polynomial factor, and that each link knows the ID's of the links in its peripheral, *i.e.*, the ID's of the links in its interference set.

(2) *SINR interference model.* A subset $L' \subseteq L$ of links can be scheduled simultaneously if and only if the following condition holds for each $l \in L'$:

$$\frac{\frac{P(l)}{d^\alpha(l)}}{\sum_{l' \in L' \setminus \{l\}} \frac{P(l')}{d^\alpha(l',l)} + N} \geq \beta. \tag{2}$$

Distributed Computing Model in the SINR-based Model. Traditionally, distributed algorithms for wireless networks have been studied in the radio broadcast model [2, 23] and its variants. The SINR based computing model is relatively recent and has not been studied extensively. Therefore, we summarize the main aspects and assumptions underlying this model:

(1) For each link $l \in L$, $x(l)$ and $r(l)$ have an estimate of $d(l)$, but they do not need to know the coordinates or the direction in which the link is oriented;

(2) All nodes have a common estimate of n, the total number of nodes, within a polynomial factor;

(3) All nodes share a common estimate of d_{min} and d_{max}, the minimum and maximum possible link lengths;

(4) Each link knows the ID's of the links in its peripheral, *i.e.*, the ID's of the links in its interference set defined in Section 5.

We remark on the third above: (1) d_{min} can be set to the device dimension which is empirically at least 0.1 meter; (2) d_{max} depends on the type of the network and is usually bounded by an order of 10^5 meters. For example, the Wi-Fi transmission range is mostly below hundred meters and a long-distance Wi-Fi network [29] has an experimental transmission limit of hundred kilometers; in cellular networks, the coverage of a base station is at most tens of kilometers even in rural areas. The limit of transmission range in Bluetooth networks or 60GHz networks is much smaller. This implies that often $g(L) \leq \log 10^6$.

Traffic, Queuing and Queue-Stability. A schedule or scheduling scheme determines at each time t which links transmit. Following [11, 27], each link is associated with an exogenous arrival process $A_l(t)$, which is i.i.d. over time and independent of each other. The first moment, $\mathbb{E}\{A_l(t)\} = \lambda(l)$. We further assume that $\exists A_{max}, A_l(t) \leq A_{max} < \infty, \forall l, \forall t$.

We use Q_l to denote the queue on l. For time t, let $Q_l(t)$ denote the backlog on l at t. In the physical interference model, transmissions are related to both scheduling and power control. Define $u_l(t) \in \{0, 1\}$ as the actual amount successfully transmitted from Q_l on link l at time slot t – this is determined by a specific scheduling-power scheme $\mathcal{SP}$. We employ a stop-and-wait ARQ mechanism. After data transmission at slot t, the transmitter node $x(l)$ will wait for an ACK from the receiver $r(l)$ for the rest of the slot. If no ACK is received until the end of t, $x(l)$ will deem the transmission as a fail so that $u_l(t) = 0$. The queue backlog evolves in the following manner:

$$Q_l(t + 1) = Q_l(t) - u_l(t) + A_l(t).$$

The system is said to be *queue-stable* (or strongly stable) [7] under a scheduling-power control scheme $\mathcal{SP}$ if and only if

$$\limsup_{t \to \infty} \frac{1}{t} \sum_{\tau \leq t} \sum_{l \in L} \mathbb{E}\{Q_l(\tau)\} < \infty.$$

In this paper, we use queue-stable and stable interchangeably. The *throughput region* $\Lambda^{\mathcal{SP}}$ of a scheduling-power scheme $\mathcal{SP}$ is the closure of the set of all exogenous arrival rate vectors that can be stably supported under $\mathcal{SP}$. The *network capacity region* Λ^{OPT} is the closure of the set of all rate vectors that can be stably supported by any feasible scheduling scheme, which corresponds to optimum.

The Throughput Maximization Problem. Given an arbitrary wireless network, we make both *scheduling* and *power assignment* decisions to minimize a factor $C > 1$. so that for any traffic vector $\lambda \in \Lambda^{OPT}$, the network is stable under traffic vector $\frac{\lambda}{C}$; in other words, we maximize the throughput region. The factor C is the approximation factor, and $1/C$ is the efficiency ratio. In the paper, we study the single-hop version of the problem. Our approximation ratio is based on considering an optimum of both scheduling and power control.

4. ALGORITHM FOR GRAPH-BASED INTERFERENCE

We now describe RA-SCHED in an abstract wireless interference model defined in terms of the interference sets ($\mathcal{I}(l)$ for each link l). RA-SCHED uses queue length info to determine the random-access probabilities. RA-SCHED runs in frames, and each frame has two sub-frames: an info-exchange sub-frame and a scheduling-tx sub-frame. RA-SCHED allows the use of infrequently updated stale backlog information for scheduling, such that each link can operate on its own for a long time.

4.1 Description of Algorithm RA-SCHED

Algorithm RA-SCHED involves the following steps.

(1) **Framing**. Time is partitioned into contiguous frames of $H_0 = H_1 + H_2$ contiguous slots. A frame is further divided into two parts: an info-exchange sub-frame of H_1 slots and a scheduling-tx sub-frame of H_2 slots. In the *info-exchange sub-frame*, backlog info is exchanged among links and corresponding neighbors. Then in the *scheduling-tx sub-frame*, each link makes scheduling decisions based on the backlog info retrieved from the info-exchange sub-frame. Backlog info will not get updated during scheduling-tx.

(2) **Info-exchange**. Details of the backlog information exchange process are in Section 4.2. Let F be the current frame, and let τ be the time slot when F starts. Each link l maintains and updates its copies of values $V_{1,F}$ and $V_{2,F}$ for F, by using the backlog values at time τ:

$$V_{1,F}(l) = Q_l(\tau); \tag{3}$$

$$V_{2,F}(l) = \max_{l' \in \hat{\mathcal{I}}(l)} \sum_{l'' \in \hat{\mathcal{I}}(l')} Q_{l''}(\tau) = \max_{l' \in \hat{\mathcal{I}}(l)} \hat{Q}_{l'}(\tau). \tag{4}$$

Let ξ (addressed in Section 4.2) denote the probability that by the end of the info-exchange sub-frame, the info-exchange process is not completed, *i.e.*, at least one link has not fully updated backlog information for the current frame.

(3) **Random-access Scheduling-tx**. In each slot $t \in [\tau + H_1 + 1, \tau + H_0]$ of the scheduling-tx sub-frame, links use $V_{1,F}$ and $V_{2,F}$ values for scheduling decision. Let $q_l(t) \in \{0, 1\}$ indicate whether $Q_l(t) > 0$. Each link l attempts to access the channel with the following probability at t.

$$p_l(t) = 1 - e^{-V_{1,F}(l)q_l(t)/V_{2,F}(l)}.$$

A collision happens when multiple links from the same neighborhood are activated at the same slot. Colliding transmissions are considered unsuccessful.

Algorithm 1: RA-SCHED

```
/* On each l ∈ L, do the following */
1 while a new frame F do
       /* in info-exchange sub-frame */
2      V₁,F(l) ← Qₗ(tF(1));
3      InfoExchange;
4      V₂,F(l) ← max_{l'∈Î(l)} Σ_{l''∈Î(l')} V₁,F(l'');
5      repeat /* in scheduling-tx sub-frame */
6          pₗ(t) ← 1 − e^{−V₁,F(l)qₗ(t)/V₂,F(l)};
7          transmit with prob. pₗ(t) at a new slot t;
8      until end of frame F;
9 end
```

4.2 Info-exchange Step

During this step, backlog information is exchanged among interfering links to update the value of $V_{2,F}(l)$ on each link l before scheduling-tx sub-frame of every frame F. We have the following two procedures.

(1) Multicast(L'): for every link $l \in L'$, its sender $x(l)$ successfully sends a message to the senders of all links in $\mathcal{I}(l)$, and

(2) LocalBroadcast(V'): every node $v \in V'$ successfully sends a message to all the nodes within one-hop distance.

Therefore, info-exchange is equivalent to two rounds of Multicast(L):

(1) in the first round of Multicast(L), each link l sends its $V_{1,F}(l)$ value to all the links in $\mathcal{I}(l)$, and

(2) in the second round of Multicast(L), each link l calculates and sends its $\sum_{l' \in \hat{\mathcal{I}}(l)} V_{1,F}(l')$ value to all the links in $\mathcal{I}(l)$ after l has collected all $V_{1,F}$ values from all the links in $\mathcal{I}(l)$.

4.2.1 Implementation of Multicast

Under k-hop interference model, we implement one round of Multicast(L) by $k+2$ invocations of LocalBroadcast(V), where data is also forwarded tagged with the ID of the origin link of the data. We use $(x(l), r(l))$ pair as the ID of a link l. For the first round of Multicast(L), each link l does the following:

(1) **Phase** 0. Let $X(L)$ denote the set of the senders of all the links in L. First, each node $v \in V$ creates a packaged message $M^{(0)}(v)$. One node may serve as a sender on multiple links. If $v = x(l)$ for any link $l \in L$, then $M^{(0)}(v)$ grows and includes the $V_{1,F}(l)$ information associated with the ID (i.e., $(x(l), r(l))$ pair) of l. Then, we perform LocalBroadcast($X(L)$), such that for any link $l \in L$, the receiver $r(l)$ receives the $V_{1,F}(l)$ value, and $M^{(0)}(r(l))$ grows and includes the $V_{1,F}(l)$ information associated with the ID (i.e., $(x(l), r(l))$ pair) of l.

(2) **Phase** $i \in [1, k]$. We perform LocalBroadcast(V), such that each node $v \in V$ successfully sends message $M^{(i-1)}(v)$ to all its one-hop neighbors. At the end of phase i each node v creates a new message $M^{(i)}(v)$ by merging all the messages it receives during the phase.

(3) **Phase** $k + 1$. We repeat phase 0 in a reverse direction, such that for any link $l \in L$, the sender $x(l)$ receives the $M^{(k)}(r(l))$ message from $r(l)$. The union of message $M^{(k)}(x(l))$ and $M^{(k)}(r(l))$ corresponds to the set of $V_{1,F}$ values from all the links in $\mathcal{I}(l)$, and their sum produces $V_{2,F}(l)$.

The second round of Multicast(L) involves the same $k + 2$ phases, except that we deal with $V_{2,F}$ values and calculate a local maximum in the end. The problem is hence reduced to given a set of nodes, making successful local broadcasts from each node to their one-hop neighbors in a distributed manner.

4.2.2 Implementation of LocalBroadcast

Procedure LocalBroadcast(V').

1 for $j = 1$ to $\frac{C \log n}{p_b}$ **do** /* $\frac{C \log n}{p_b}$ slots in total */
2 | each $v \in V'$ transmits with probability p_b;
3 end

The implementation of LocalBroadcast(V) is simple: Each node $v \in V$ transmits with probability p_b in each time slot for a period of length $\frac{C \log n}{p_b}$, where $C \geq 8$ is a constant. Let $\mathcal{I}_b(v)$ of a node v denote the set of nodes, called *local broadcast interference set*, such that a local broadcast of v to all its one-hop neighbors is *successful* if and only if all the nodes in $\mathcal{I}_b(v)$ are silent. Let $BI_{max} \triangleq \max_{v \in V} |\mathcal{I}_b(v)|$ denote the maximum cardinality of the local broadcast interference sets. We discuss the value of p_b and show that LocalBroadcast(V) is completed in $\frac{C \log n}{p_b}$ time *w.h.p.* for the following two cases depending on the level of knowledge:

(1) Since each node has a common estimate of n, if we set $p_b = 1/n$, then at each time slot,

$Prob(\text{LocalBroadcast}(v) \text{ is completed})$

$= p_b \prod_{u \in \mathcal{I}_b(v)} (1 - p_b) \geq p_b (\frac{1}{4})^{\sum_{u \in \mathcal{I}_b(v)} p_b} \geq \frac{p_b}{4} = \frac{1}{4n}.$

After $Cn \log n$ slots, $Prob(\text{LocalBroadcast}(v)$ is not completed) $\leq (1 - \frac{1}{4n})^{Cn \log n} \leq 1 - \frac{1}{n^{C/4}}$. For LocalBroadcast($V$) to be completed (i.e., all the n nodes has made successful local broadcast) in $Cn \log n$ slots, the probability is at least $1/n^{C/4-1}$.

(2) If each node v knows BI_{max} (which can be computed from the conflict graph), we set $p_b = \frac{1}{BI_{max}}$. The running time becomes

$C \cdot BI_{max} \cdot \log n$. $Prob(\text{LocalBroadcast}(v) \text{ is completed}) \geq \frac{p_b}{4}$ at each time slot with the same analysis as the first case. We can also prove that LocalBroadcast(V) is completed *w.h.p.*

Since the size of a broadcast message is small, a regular slot can be divided to improve time cost. Hence, we have Lemma 4.1.

4.3 Analysis of Algorithm RA-SCHED

We now analyze the performance of RA-SCHED: the running time of each phase, and the throughput region.

Lemma 4.1 *If we set the length of the info-exchange sub-frame to $H_1 = \Theta(n \log n)$, then with prob. $\geq 1 - \xi$, where $\xi \leq 2(k + 2)/n^{C_0} \leq 1/n$ and $C_0 \geq 3$ is a constant, the $V_{1,f}$ and $V_{2,F}$ values can be updated successfully for all links by the end of info-exchange sub-frame. If each node knows BI_{max}, H_1 may be shortened to $\Theta(BI_{max} \cdot \log n)$.*

This lemma guarantees the info-exchange process is successful *w.h.p.* Rather pessimistically, failing to update $V_{1,f}$ and $V_{2,F}$ values on any link is considered to void an entire frame. That cost is fully taken into account in our analysis.

For proving stability and deriving throughput region, we use the idea of Lyapunov functions [7], which are a kind of potential functions. We define our Lyapunov function as

$$\mathbf{L}(\vec{Q}(t)) \triangleq \max_{k \in L} \sum_{k' \in \hat{\mathcal{I}}(k)} Q_{k'}(t) = \max_{k \in L} \hat{Q}_k(t),$$

where $\vec{Q}$ denotes the vector of queues.

Let F denote a frame and let $t_F(i)$ be the function that returns the time of the ith slot of F, e.g., $t_F(1)$ is the time when F begins. Therefore, $V_{1,F}(l) = Q_l(t_F(1))$, and $V_{2,F}(l) = \max_{l' \in \hat{\mathcal{I}}(l)} \hat{Q}_{l'}(t_F(1))$. We define $V_{2,F}(L) \triangleq \mathbf{L}(\vec{Q}(t_F(1)))$, as the value of our Lyapunov function at time $t_F(1)$.

Intuitively, we first lower-bound how much can be sent from an interference set probabilistically when the global and local total values of backlogs exceed a certain value; then, we argue that the expected value of the Lyapunov function decreases *w.h.p.* in a certain number of frames; and finally, we prove that the system is stable under a certain condition on the traffic load, leading to the throughput region.

Lemma 4.2 *In a frame F, given the info-exchange sub-frame is successful, RA-SCHED guarantees that for any $\epsilon > 0$ and $h > 0, h < \infty$, if $V_{2,F}(L) \geq C_1/\epsilon$, then for any link l with $\sum_{l' \in \hat{\mathcal{I}}(l)} V_{1,F}(l') \geq V_{2,F}(L) - C_1 + H_2$, where $C_1 = 2hH_0 |L| A_{max} + (h - h\epsilon - 2)H_2$, the following holds for any scheduling-tx slot $t \in [t_F(H_1 + 1), t_F(H_0)]$,*

$$\sum_{l' \in \mathcal{I}(l)} Prob\{u_{l'}(t) = 1\} \geq (1 - \epsilon)/e.$$

Lemma 4.3 *Let the current frame be F_0. If the average arrival satisfies $\forall l \in L, \sum_{l' \in \hat{\mathcal{I}}(l)} \lambda(l') \leq \frac{H_2}{H_0}(1 - 5\epsilon - \xi)/e,$*

RA-SCHED guarantees that for any constant $\epsilon' > 0$, there exists a positive finite number h that satisfies

$\epsilon'/|L| \geq e^{-\frac{(hH_2\epsilon)^2}{2e^2}} + e^{-\frac{(h\epsilon)^2}{2e^2}} + e^{-\frac{2h(H_2\epsilon)^2}{H_0|L|^2 A_{max}^2 \epsilon^2}}$, *such that when $V_{2,F_0}(L) \geq C_1/\epsilon + (h - 1)H_2$, the following holds for frame F_h, which is the hth frame after F_0:*

$$Prob\{V_{2,F_h}(L) \leq V_{2,F_0}(L) - hH_2\epsilon/e\} \geq 1 - \epsilon'. \quad (5)$$

The proofs of the above two lemmas are in Appendix. Lemma 4.3 says that there exists a finite number h, such that once the value of our Lyapunov function exceeds a finite number $C_1/\epsilon + (h-1)H_2$ at the start of a frame F, then in h frames, the expected value of the Lyapunov function decreases by at least a constant amount, *w.h.p.*

Theorem 4.4 (Sufficient condition of stability) RA-SCHED *guarantees queue-stability with arrival processes satisfying*

$$\forall l \in L, \sum_{l' \in \hat{\mathcal{I}}(l)} \lambda(l') \leq \delta(1 - 5\epsilon - \xi), \forall \delta [0, 1/e).$$

PROOF. Since H_1 is bounded by a finite number, for any positive constant $\delta < 1/e$, there exists a finite value for H_2, such that $\frac{H_2}{H_1 + H_2} = e\delta$. Then, the arrival processes satisfy

$$\forall l \in L, \sum_{l' \in \hat{\mathcal{I}}(l)} \lambda(l') \leq \frac{H_2}{H_0}(1 - 5\epsilon - \xi)/e.$$

Recall our Lyapunov function $\mathbf{L}\big(\vec{Q}(t)\big) = \max_{l \in L} \sum_{l' \in \hat{\mathcal{I}}(l)} Q_{l'}(t)$.

W.l.o.g., let t be the time when a frame starts.

Let J denote the event "$\mathbf{L}\big(\vec{Q}(t)\big) \geq C_1/\epsilon + (h-1)H_2$." From Lemma 4.3, we have: for any $\epsilon' < \frac{H_2\epsilon}{eH_0 A_{max}|L| + H_2\epsilon}$, there exists a finite number h, s.t. when conditioned by J,

$$Prob\Big\{\mathbf{L}\big(\vec{Q}(t + hH_0)\big) - \mathbf{L}\big(\vec{Q}(t)\big) \leq -hH_2\epsilon/e\Big\} \geq 1 - \epsilon'.$$

That is saying, whenever J happens,

$$\mathbb{E}\Big\{\mathbf{L}\big(\vec{Q}(t + hH_0)\big) - \mathbf{L}\big(\vec{Q}(t)\big) \mid \vec{Q}(t)\Big\}$$
$$\leq -(1 - \epsilon')hH_2\epsilon/e + hH_0 A_{max}|L|\epsilon' < 0.$$

By Foster's theorem [7], the system is queue-stable. $\square$

Theorem 4.5 (Necessary condition of stability) *For a network to be stable, any traffic load needs to satisfy:* $\sum_{l' \in \mathcal{I}(l) \cup \{l\}} \lambda(l') \leq \Delta$, $\forall l \in L$, *where Δ is the interference degree (i.e., the max. number of links that can make successful concurrent transmissions in any interference set).*

Theorem 4.5 and 4.4 together characterize a throughput region of almost $\frac{1}{e\Delta}\Lambda^{OPT}$. The efficiency ratio of RA-SCHED is therefore approaching $\frac{1}{e\Delta}$. In the proof of Theorem 4.4, through δ, it suggests the relation between the choice of the value of H_2 and the throughput region. The larger H_2 is, the closer the throughput region is to $\frac{1}{e\Delta}\Lambda^{OPT}$.

5. ALGORITHM FOR SINR INTERFERENCE

We now consider the SINR model, and redefine interference set in a form that resembles the abstract interference model of Section 4. Our approach involves considering a time-slot t to be augmented into a set of time-slots and having only links of a given length class transmit in a given time-slot. This may scale down the efficiency ratio, but allows us to keep or reuse the general structure and analysis of algorithm RA-SCHED in an SINR setting.

We start with preliminary definitions. Let $d_{min} \triangleq \min_{l \in L}\{d(l)\}$, and $d_{max} \triangleq \max_{l \in L}\{d(l)\}$. Let σ be a constant. We define *link diversity* $g(L) \triangleq \lceil \log_\sigma \frac{d_{max}}{d_{min}} \rceil$.

Partition $L = \{L_i\}, i = 1, 2, \ldots, g(L)$, where each $L_i = \{l \mid \sigma^{i-1}d_{min} \leq d(l) < \sigma^i d_{min}\}$ is a *length class*, *i.e.*, the set of links of roughly similar lengths. Let $\mathcal{G}_i$ be the graph induced by the links in L_i. Let $d_i = \sigma^i d_{min}$. Let $\gamma = \sqrt[\alpha]{3 \cdot 2^3 \beta \frac{1+\phi}{\phi} \frac{\alpha-1}{\alpha-2}} + 3$. For

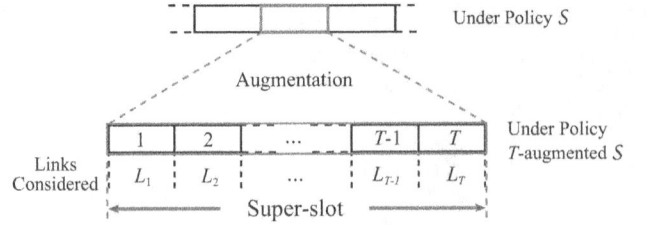

Figure 1: Structure of a T-augmented schedule: A slot under $\mathcal{S}$ extends to a super-slot, where only L_i participates at the ith slot of a super-slot.

link $l \in L_i$, define $\mathcal{I}(l) \triangleq \{l' \mid l' \in L_i, d(x(l'), x(l)) \leq \gamma d_i\}$ as its *interference set*; This notion of interference set is symmetric.

Given a scheduling policy $\mathcal{S}$ that does not discriminate link lengths, a T-*augmented* $\mathcal{S}$ is defined in the following manner (see Figure 1): under T-augmented $\mathcal{S}$, (i) each regular slot t is augmented into a super-slot that consists of T slots, and (ii) in the ith slot of a super-slot, only links in length class L_i may participate in the scheduling using the same policy as $\mathcal{S}$.

5.1 Description of Algorithm RA-Sched-SINR

RA-SCHED-SINR is a $g(L)$-augmented version of RA-SCHED over link set L for scheduling in SINR setting, with the modifications listed below this paragraph. A slot under RA-SCHED extends to a super-slot under RA-SCHED-SINR, where RA-SCHED is performed over links in L_i at the ith slot of a super-slot. As a result of the augmentation, the length of a frame is enlarged by a factor of $g(L)$. Let $H_1' = g(L)H_1$ and $H_2' = g(L)H_2$ denote respectively the lengths of the two sub-frames of a frame for RA-SCHED-SINR. Let $H_0' = H_1' + H_2'$ denote the length of a frame. In each frame each link set L_i is allocated H_1 slots for info-exchange, and H_2 slots for scheduling-tx. Below lists our modifications based on $g(L)$-augmented RA-SCHED (recall definitions in Section 4.1):

(1) **Channel-access Probability.** We reset the probability of accessing the channel for any link l to

$$p_l(t) = 1 - e^{-V_{1,F}(l)q_l(t)/\big(\Gamma V_{2,F}(l)\big)},$$

where $\Gamma > 1$ is a constant parameter whose value will be determined later (in Section 5.2.1) to achieve optimal effect.

(2) **Local Broadcast Interference Set.** We redefine the *local broadcast interference set* in SINR setting for info-exchange as: for each node $v \in X(L_i)$ (where $X(L_i)$ is the set of the senders of all the links in L_i) in the ith slot of a super-slot,

$$\mathcal{I}_b(v, i) \triangleq \{u \mid \min\{d(u, v)\} \leq 2\gamma d_i, \forall u \in X(L_i)\}.$$

Let $BI_{max} \triangleq \max_{i \in [1, g(L)]} \max_{v \in V} |\mathcal{I}_b(v, i)|$ denote the maximum cardinality of the local broadcast interference sets.

(3) **Power Assignment.** The only constraint on power assignment is that, the power values for transmission on any two links in the same set L_i are at most a constant factor away from each other. This incorporates uniform and linear power assignment schemes as special cases. For simplicity, we assume links in the same link set L_i use the same transmission power P_i. Removing this assumption may affect the results by no more than a constant factor.

(4) **Backlog Info-exchange.** Each link l boosts up the power level by a constant factor, such that the transmission range can reach the transmitter of any link in $\mathcal{I}(l)$ (a circular region). The info-exchange sub-frame consists of two equi-length rounds of local broadcast. In the first round, each link l sends the $V_{1,F}(l)$ value to $\mathcal{I}(l)$, and in the second round sends the value of $\sum_{l' \in \hat{\mathcal{I}}(l)} V_{1,F}(l')$ to $\mathcal{I}(l)$ if l has collected all $V_{1,F}(l')$ values from each $l' \in \mathcal{I}(l)$.

5.2 Correctness and Analysis of RA-Sched-SINR

Theorem 5.1 gives the conditions under which RA-SCHED-SINR is stable; it follows that our algorithm has an efficiency ratio of $\frac{1}{O(g(L))}$.

Theorem 5.1 (Sufficient condition for stability) RA-SCHED-SINR *guarantees queue-stability of the system under arrival processes that satisfy*

$$\forall l \in L, \sum_{l' \in \hat{\mathcal{I}}(l)} \lambda(l') \leq \delta(1 - 5\epsilon - \xi), \qquad (6)$$

for any positive constant $\delta < \frac{1}{g(L)} \frac{1-1/\Gamma}{e^{1/\Gamma}\Gamma}$, *where* $\Gamma = e$.

To prove Theorem 5.1, we start with the analysis of each phase, which provides insights to the algorithm.

Lemma 5.2 *For a choice of H_1' such that $H_1' = \Theta\big(g(L)n\log n\big)$, the $V_{1,F}$ and $V_{2,F}$ values are updated successfully on all links by the end of an info-exchange sub-frame, w.h.p. If each node knows BI_{max}, H_1 may be shortened to $\Theta\left(g(L) \cdot BI_{max} \cdot \log n\right)$.*

PROOF (SKETCH). The proof of the performance builds on the results from [9]. Each node needs to know when the info-exchange sub-frame ends to enter the next sub-frame. The running time of info-exchange needs to be based on some common knowledge — n or BI_{max} — resulting in different time length.

Goussevskaia, Moscibroda, and Wattenhofer [9] propose two distributed randomized local broadcasting algorithms for each node to broadcast to all nodes in its proximity with a radius of R_B. Their first algorithm requires the knowledge of the number of nodes within a distance of $R_A \geq 2R_B$. Let Δ_A denote the maximum number of nodes at most R_A away from any node. By letting each node transmits with probability $1/\Delta_A$ for $\Omega\left(\Delta_A \log n\right)$ time, they show that all the local broadcasts can finish with probability of at least $1-1/n^C$, where n is the number of nodes and C is a constant.

In our SINR model, we treat the senders of all links as nodes that participate in the local broadcasts. We set $R_A = 2\gamma d_i$ and $R_B = \gamma d_i$ so that $R_A \geq R_B$ and R_B covers all the target nodes of the local broadcast of a node. Only difference is that each node makes two rounds of broadcasts (for value $V_{1,F}$ and $V_{2,F}$ respectively). By using the result in [9], we obtain a probability guarantee of at least $1 - 2/n^C$ and $O\left(g(L) \cdot BI_{max} \cdot \log n\right)$ time for all links to finish info-exchange. When the knowledge of BI_{max} is not available, if each node transmits with probability $1/n$, in $O\left(g(L)n\log n\right)$ time all the links can finish info-exchange *w.h.p.* $\square$

Proposition 5.3 *The expected number of active (or transmitting) links in $\hat{\mathcal{I}}(l)$ of any $l \in L$ at any time t is $\leq 1/\Gamma$.*

Proposition 5.4 *We say two links k and k' are "independent" if they are not in each other's interference set. The distance between either end of k and either end of k' is at least $(\gamma - 2)d_i$.*

Lemma 5.5 *Let t be a time slot when links from L_i are scheduled. The probability for any $l \in L_i$ to make successful transmission when there is no interference from $\mathcal{I}(l)$ is*

$$Prob\Big(u_l(t) = 1 \Big| \sum_{l' \in \mathcal{I}(l)} s_{l'}(t) = 0\Big) \geq p_l(t)(1 - 1/\Gamma).$$

PROOF. We first partition the plane into concentric rings via a similar technique to that in [4, 25], and then by solving a set cover problem we develop an upper-bound on the minimum number of greater interference sets ($\hat{\mathcal{I}}$) that covers all the links that touches a ring, so that with reference to Proposition 5.3, we can bound the expected total interference from all the rings outside of $\mathcal{I}(l)$.

We partition the plane into rings all centered at $r(l)$, each of width $(\gamma - 2)d_i$. Let $R(j)$ denote the jth ring, which contains every transmitter node u of a link in L_i, that satisfy $j(\gamma - 2)d_i \leq d(u, r(l)) < (j + 1)(\gamma - 2)d_i$, where $j = 1, 2, \ldots$. When $j = 0$, $R(0)$ corresponds to a disk of radius $(\gamma - 2)d_i$, within which the nodes all belong to links in $\hat{\mathcal{I}}(l)$ due to Proposition 5.4. We say that link $k \in R(j)$, if either $x(k)$ or $r(k)$ falls in $R(j)$. Let the total number of active (or transmitting) nodes in $R(j)$ at t be $N_{R(j)}(t)$. Since each link attempts to access the channel and transmit independently, and each link can have at most one active node (either sending data from transmitter node or sending ACK from receiver node), $N_{R(j)}(t)$ may fluctuate during t; yet, we have that during t, $\mathbb{E}\{N_{R(j)}(t)\} \leq \sum_{k \in R(j)} p_k(t)$.

Let $L^{cov}(j)$ be a minimal set of links in $R(j)$, such that

(1) $\bigcup_{k \in L^{cov}(j)} \hat{\mathcal{I}}(k)$ covers all the links in $R(j)$, and

(2) all the links in L^{cov} do not interfere with each other.

By Proposition 5.3,

$$\mathbb{E}\{N_{R(j)}(t)\} \leq \sum_{k \in L^{cov}(j)} \sum_{k' \in \hat{\mathcal{I}}(k)} p_{k'}(t) \leq |L^{cov}(j)|/\Gamma. \quad (7)$$

$L^{cov}(j)$ is a set of pairwise independent links in $R(j)$. Now we pick a node set $N^{cov}(j)$ that covers all links in $L^{cov}(j)$, such that each link has one and only one node in $N^{cov}(j)$, and all the nodes in $N^{cov}(j)$ fall in $R(j)$. By Proposition 5.4, disks of radius $(\gamma - 2)d_i/2$ centered at each of the nodes in $N^{cov}(j)$ do not overlap. All these disks are contained in an extended ring $R'(j)$ of $R(j)$, with extra width of $(\gamma - 2)d_i/2$ at each side of $R(j)$.

Therefore, $|L^{cov}(j)| = |N^{cov}(j)| \leq 2^3(2j + 1)$ due to packing property. That yields from Inequality (7) that

$$\mathbb{E}\{N_{R(j)}(t)\} \leq 2^3(2j + 1)/\Gamma.$$

Since on some links it may be the transmitter nodes that are transmitting data and on others it may be the receiver node that are transmitting ACKs, the interference observed can vary in a time slot. Let $I_j^x(t)$ and $I_j^r(t)$ denote the maximum total interference from the jth ring to $x(l)$ and $r(l)$ respectively at t. By noticing that $x(l)$ is on the circle centered at $r(l)$ with radius $d(l)$, we have

$$\begin{cases} I_j^x(t) \leq N_{R(j)}(t) \cdot \frac{P_i}{(j(\gamma-2)d_i - d(l))^\alpha}; \text{ and} \\ I_j^r(t) \leq N_{R(j)}(t) \cdot \frac{P_i}{(j(\gamma-2)d_i)^\alpha}. \end{cases}$$

Then, define $I_j(t) \triangleq N_{R(j)}(t) \cdot \frac{P_i}{(j(\gamma-3)d_i)^\alpha}$. Therefore, $I_j^x(t) \leq I_j(t)$ and $I_j^r(t) \leq I_j(t)$ at the same time.

Let $I^x(t) \triangleq \sum_{j=1}^{\infty} I_j^x(t)$ and $I^r(t) \triangleq \sum_{j=1}^{\infty} I_j^r(t)$ denote the maximum total interference from outside of $\mathcal{I}(l)$ respectively. Let $I(t) = \sum_{j=1}^{\infty} I_j(t)$; then, $I(t) \geq \max\{I^x(t), I^r(t)\}$.

$$\mathbb{E}\{I(t)\} \leq \mathbb{E}\{\sum_{j=1}^{\infty} I_j(t)\} = \sum_{j=1}^{\infty} \mathbb{E}\{I_j(t)\}$$

$$= \sum_{j=1}^{\infty} \frac{2^3}{\Gamma} P_i(\gamma - 3)^{-\alpha} d_i^{-\alpha} \frac{2j + 1}{j^\alpha}$$

$$\leq \frac{3 \cdot 2^3}{\Gamma} P_i(\gamma - 3)^{-\alpha} d_i^{-\alpha} \frac{\alpha - 1}{\alpha - 2}.$$

Recall that $s_l(t) \in \{0, 1\}$ indicates whether link l chooses to access the channel at time t. Let J_1 denote the event that "$s_l(t) = 1$," J_2 the event that "$\sum_{l' \in \mathcal{I}(l)} s_{l'}(t) = 0$," and J_3 the event that "$I(t) < \Gamma \cdot \mathbb{E}\{I(t)\}$." According to Markov's Inequality:

$$Prob(J_3) = 1 - Prob\big(I(t) \geq \Gamma \cdot \mathbb{E}\{I(t)\}\big) \geq 1 - 1/\Gamma.$$

With J_1, J_2, and J_3 being true at the same time, the SINR values at $x(l)$ and $r(l)$ are at least

$$\frac{P_i d^{-\alpha}(l)}{\Gamma \mathbb{E}\{I(t)\} + N} \geq \frac{P_i d_i^{-\alpha}}{3 \cdot 2^3 \sigma^\alpha P_i (\gamma-3)^{-\alpha} d_i^{-\alpha} \frac{\alpha-1}{\alpha-2} + N} \geq \beta.$$

That means both data transmission from $x(l)$ and ACK transmission from $r(l)$ will be successful, *i.e.*, $u_l(t) = 1$.

Therefore, $Prob\big(u_l(t) = 1 | J_2\big) \geq Prob\big(J_1 \cap J_3 | J_2\big)$ leading to the statement. $\square$

We define a Lyapunov function as before in Section 4.3:

$$\mathbf{L}\left(\vec{Q}(t)\right) \triangleq \max_{k \in L} \sum_{k' \in \hat{\mathcal{I}}(k)} Q_{k'}(t) = \max_{k \in L} \hat{Q}_k(t).$$

In the same way, let F denote a frame and let $t_F(i)$ be the function that returns the time of the ith slot of F. Therefore, $V_{1,F}(l) = Q_l(t_F(1))$ and $V_{2,F}(l) = \max_{l' \in \hat{\mathcal{I}}(l)} \hat{Q}_{l'}(t_F(1))$. We use $V_{2,F}(L) = \mathbf{L}\big(\vec{Q}(t_F(1))\big)$ to denote the value of our Lyapunov function at $t_F(1)$.

Lemma 5.6 *In a frame F, given the info-exchange sub-frame is successful, RA-SCHED-SINR guarantees that for any $\epsilon > 0$ and $h > 0, h < \infty$, if $V_{2,F}(L) \geq C_1/\epsilon$, then for any i and any link $l \in L_i$ with $\sum_{l' \in \hat{\mathcal{I}}(l)} V_{1,F}(l') \geq V_{2,F}(L) - C_1 + H_2$, where $C_1 = 2hH_0'|L|A_{max} + (h - h\epsilon - 2)H_2$, the following holds in the scheduling-tx sub-frame:*

$$\sum_{l' \in \hat{\mathcal{I}}(l)} Prob\{u_{l'}(t) = 1\} \geq \frac{(1 - 1/\Gamma)}{e^{1/\Gamma}\Gamma}(1 - \epsilon), \forall i,$$

$\forall t = t_F(H_1' + i + jg(L))$, *where* $j = 0, 1, 2, \ldots, H_2 - 1$.

PROOF. (sketch) W.l.o.g., we look at a slot $t = t_F(H_1' + i + jg(L))$, when only links from L_i are being scheduled. There are in total $H_2 = H_2'/g(L)$ such slots in frame F. Let $l \in L_i$ be a link with $\sum_{l' \in \hat{\mathcal{I}}(l)} V_{1,F}(l') \geq V_{2,F}(L) - C_1 + H_2$. Let J_4 denote the event that "$\sum_{k' \in \mathcal{I}(k)} s_{k'}(t) = 0$." By applying Lemma 5.5, any link $k \in \mathcal{I}(l)$ makes a successful transmission with probability

$$Prob\{u_k(t) = 1\} \geq Prob\big(u_k(t) = 1 \cap J_4\big)$$
$$\geq Prob\big(u_k(t) = 1 \mid J_4\big) \cdot Prob(J_4)$$
$$\geq \left(e^{\frac{V_{1,F}(k)q_k(t)}{\Gamma V_{2,F}(L)}} - 1\right)\Big/e^{1/\Gamma} \geq \frac{(1 - 1/\Gamma)V_{1,F}(k)q_k(t)}{e^{1/\Gamma}\Gamma V_{2,F}(L)}.$$

Then, using the same reasoning in the proof of Lemma 4.2 will complete the proof. $\square$

Following the above lemma which is equivalent to Lemma 4.2, we have Lemma 5.7 below equivalent to Lemma 4.3.

Lemma 5.7 *Let the current frame be F_0. If the average arrival satisfies for any $l \in L$,*

$$\sum_{l' \in \mathcal{I}(l)} \lambda(l') \leq \frac{H_2'}{H_0'g(L)}(1 - 5\epsilon - \xi)\frac{(1 - 1/\Gamma)}{e^{1/\Gamma}\Gamma},$$

where $\Gamma = e$ (discussed in Section 5.2.1), RA-SCHED-SINR guarantees that for any constant $\epsilon' > 0$, there exists a finite number $h > 0$ that satisfies $\epsilon'/|L| \geq e^{-\frac{(hH_2\epsilon)^2}{2e^2}} + e^{-\frac{(h\epsilon)^2}{2e^2}} + e^{-\frac{2h(H_2\epsilon)^2}{H_0'|L|^2 A_{max}^2 e^2}}$, such that when we have $V_{2,F_0}(L) \geq C_1/\epsilon + (h-1)H_2$, the following holds for frame F_h, which is the hth frame after F_0:

$$Prob\Big\{\sum_{l' \in \hat{\mathcal{I}}(l)} V_{2,F_h}(L) \leq V_{2,F_0}(L) - hH_2\epsilon/e\Big\} \geq 1 - \epsilon'.$$

PROOF (SKETCH). Whether "$\sum_{l' \in \hat{\mathcal{I}}(l)} Q_{l'}(t_F(1)) \leq V_{2,F}(L) - hH_0'|L|A_{max} - hH_2\epsilon/e$" is true or not gives us two cases. After applying similar techniques in Lemma 4.3, the statement follows. $\square$

Theorem 5.1 can be then proved in the same way as Theorem 4.4.

5.2.1 Determining the value of Γ

We notice that $f(x) = \frac{1 - 1/x}{e^{1/x}x}$ is not a monotone function of $x > 1$. Hence, we would like to find a value of Γ to maximize $f(\Gamma)$, such that the limit of δ in Theorem 5.1 and the underlying throughput region get as large as possible. By solving "$\max_{x>1} \frac{1 - 1/x}{e^{1/x}x}$" we can find a solution x^*, where $2.5 < x^* \approx e < 2.8$. We further notice that $|f(x^*) - f(e)| \leq 0.001$. Simply taking $\Gamma = e$ yields $\frac{(1 - 1/\Gamma)}{e^{1/\Gamma}\Gamma} = \frac{1 - 1/e}{e^{1+1/e}}$.

5.2.2 Efficiency Ratio of RA-Sched-SINR

For RA-SCHED-SINR, Theorem 5.1 and Theorem 4.5 together characterize a throughput region of nearly $\frac{1}{\Delta}\frac{(1-1/e)}{g(L)e^{1+1/e}}$ of the optimum Λ^{OPT}. Δ in the SINR setting is the cardinality of a maximum independent set under optimum power assignment in $\mathcal{I}(l)$ of any link l. Due to Lemma 5.8 (whose proof is in Appendix), Δ is at most a constant. We conclude that the efficiency ratio of RA-SCHED-SINR reaches $\Omega(1/g(L))$.

Lemma 5.8 *Under any scheduling and power assignment scheme, Δ is a at most a constant in our SINR setting.*

6. CONCLUSION AND DISCUSSION

In this paper, we develop the first local distributed scheduling algorithm (RA-SCHED-SINR) that provably ensures a throughput region of $\Omega(\frac{1}{g(L)})\Lambda^{OPT}$ in the SINR model, with Λ^{OPT} being optimum considering both scheduling and power control. We also design RA-SCHED, a distributed scheduling scheme with graph-based interference, which forms the basis of RA-SCHED-SINR. Our algorithms are based on random access, and use local queue size info for scheduling. We rigorously analyze all aspects of communication complexity, including information update and exchange, whose frequency can be reduced, to limit the overall overhead.

One application of RA-SCHED and RA-SCHED-SINR is in WLANs where local info-exchange is easy. Consider the downlinks scheduling problem in [13, 32], with WLANs consisting of multiple Access Points (or APs) which are wired with access to the Internet. Downlink traffic is reported to constitute as much as 80% of the total traffic in enterprise WLANs [32]. Because in such a setting it is easier to obtain the conflict graph — *e.g.*, a micro-probing approach [1] for online conflict graph construction can obtain contention information within millisecond time scales [13, 32] — the time for info-exchange step can be improved (so that the delay performance improves) based on that. Since the length of a slot is at the level of tens to hundreds of milliseconds, our algorithms may serve as simple distributed solutions with low overhead.

7. ACKNOWLEDGMENTS

We thank all the reviewers and Dr. Saswati Sarkar for their comments. The work of the authors has been partially supported by National Science Foundation (NSF) CAREER Grant CNS-0845700, NSF NETS Grant CNS-0831633, NSF NetSE Grant CNS-1011769, Defense Threat Reduction Agency (DTRA) CNIMS Grant HDTRA1-07-C-0113, NSF HSD Grant SES-0729441 and Department of Energy (DOE) DE-SC0003957.

8. REFERENCES

[1] N. Ahmed, U. Ismail, S. Keshav, and K. Papagiannaki. Online estimation of rf interference. In *ACM CoNEXT*, pages 4:1–4:12, 2008.

[2] H. Balakrishnan, C. Barrett, V. S. A. Kumar, M. Marathe, and S. Thite. The distance 2-matching problem and its relationship to the MAC layer capacity of adhoc wireless networks. *IEEE Journal on Selected Areas in Communications*, 22(6):1069–1079, 2004.

[3] D. Chafekar, V. S. A. Kumar, M. Marathe, S. Parthasarathy, and A. Srinivasan. Approximation algorithms for computing capacity of wireless networks with sinr constraints. In *INFOCOM 2008*, pages 1166–1174, 2008.

[4] D. Chafekar, V. S. A. Kumar, M. V. Marathe, S. P. 0002, and A. Srinivasan. Cross-layer latency minimization in wireless networks with sinr constraints. In *ACM MobiHoc*, pages 110–119, 2007.

[5] D. Chafekar, D. Levin, V. S. A. Kumar, M. Marathe, S. Parthasarathy, and A. Srinivasan. Capacity of asynchronous random-access scheduling in wireless networks. In *IEEE INFOCOM*, 2008.

[6] P. Chaporkar, K. Kar, X. Luo, and S. Sarkar. Throughput and fairness guarantees through maximal scheduling in wireless networks. *IEEE Trans. on Info. Theory*, 54(2):572–594, 2008.

[7] L. Georgiadis, M. J. Neely, and L. Tassiulas. Resource allocation and cross-layer control in wireless networks. *Foundations and Trends in Netw.*, 1(1):1–149, 2006.

[8] O. Goussevskaia, M. Halldórsson, R. Wattenhofer, and E. Welzl. Capacity of arbitrary wireless networks. In *IEEE INFOCOM*, 2009.

[9] O. Goussevskaia, T. Moscibroda, and R. Wattenhofer. Local Broadcasting in the Physical Interference Model. In *ACM SIGACT-SIGOPT DialM-POMC*, 2008.

[10] O. Goussevskaia, Y. A. Oswald, and R. Wattenhofer. Complexity in geometric sinr. In *ACM MobiHoc*, 2007.

[11] A. Gupta, X. Lin, and R. Srikant. Low-complexity distributed scheduling algorithms for wireless networks. *IEEE/ACM Trans. on Netw.*, 17(6):1846–1859, 2009.

[12] P. Gupta and P. R. Kumar. The capacity of wireless networks. *IEEE Trans. on Info. Theory*, 46(2):388–404, 2000.

[13] K.-L. Hung and B. Bensaou. Distributed rate control and contention resolution in multi-cell IEEE 802.11 WLANs with hidden terminals. In *ACM MobiHoc*, pages 51–60, 2010.

[14] L. Jiang and J. Walrand. A distributed CSMA algorithm for throughput and utility maximization in wireless networks. *IEEE/ACM Trans. on Netw.*, 18:960–972, 2010.

[15] C. Joo, X. Lin, and N. Shroff. Understanding the capacity region of the greedy maximal scheduling algorithm in multihop wireless networks. *IEEE/ACM Trans. on Netw.*, 17(4):1132–1145, 2009.

[16] C. Joo and N. B. Shroff. Performance of random access scheduling schemes in multi-hop wireless networks. *IEEE/ACM Trans. on Netw.*, 17(5):1481–1493, 2009.

[17] T. Kesselheim. A constant-factor approximation for wireless capacity maximization with power control in the sinr model. In *ACM-SIAM Symposium on Discrete Algorithms (SODA)*, 2011.

[18] T. Kesselheim and B. Vöcking. Distributed contention resolution in wireless networks. In *DISC*, 2010.

[19] L. B. Le and R. R. Mazumdar. Control of wireless networks with flow level dynamics under constant time scheduling. *Wirel. Netw.*, 16:1355–1372, 2010.

[20] H.-W. Lee, E. Modiano, and L. B. Le. Distributed throughput maximization in wireless networks via random power allocation. In *WiOPT*, pages 328–336, 2009.

[21] X. Lin and S. Rasool. Constant-time distributed scheduling policies for ad hoc wireless networks. *IEEE Transactions on Automatic Control*, 54(2):231–242, 2009.

[22] X. Lin and N. B. Shroff. The impact of imperfect scheduling on cross-layer congestion control in wireless networks. *IEEE/ACM Trans. on Netw.*, 14(2):302–315, 2006.

[23] T. Moscibroda and R. Wattenhofer. Maximal independent sets in radio networks. In *ACM symposium on Principles of distributed computing (PODC)*, 2005.

[24] T. Moscibroda and R. Wattenhofer. The complexity of connectivity in wireless networks. In *IEEE INFOCOM*, 2006.

[25] T. Moscibroda, R. Wattenhofer, and A. Zollinger. Topology control meets sinr: the scheduling complexity of arbitrary topologies. In *ACM MobiHoc*, pages 310–321, 2006.

[26] B. Nardelli, J. Lee, K. Lee, Y. Yi, S. Chong, E. Knightly, and M. Chiang. Experimental evaluation of optimal CSMA. In *IEEE INFOCOM*, 2011.

[27] M. J. Neely. Delay analysis for maximal scheduling with flow control in wireless networks with bursty traffic. *IEEE/ACM Trans. on Netw.*, 17(4):1146–1159, 2009.

[28] J. Ni, B. Tan, and R. Srikant. Q-CSMA: queue-length based CSMA/CA algorithms for achieving maximum throughput and low delay in wireless networks. In *IEEE INFOCOM (Mini)*, pages 271–275, 2010.

[29] R. Patra, S. Nedevschi, S. Surana, A. Sheth, L. Subramanian, and E. Brewer. WiLDNet: Design and implementation of high performance WiFi based long distance networks. In *USENIX NSDI*, pages 87–100, 2007.

[30] G. Pei, V. S. A. Kumar, S. Parthasarathy, and A. Srinivasan. Approximation algorithms for throughput maximization in wireless networks with delay constraints. In *INFOCOM*, pages 1116–1124, 2011.

[31] A. Reddy, S. Banerjee, A. Gopalan, S. Shakkottai, and L. Ying. Wireless scheduling with heterogeneously delayed network-state information. In *Allerton Conference*, pages 1577–1584, 2010.

[32] V. Shrivastava, N. Ahmed, S. Rayanchu, S. Banerjee, S. Keshav, K. Papagiannaki, and A. Mishra. CENTAUR: realizing the full potential of centralized wlans through a hybrid data path. In *Mobicom*, pages 297–308, 2009.

[33] L. Tassiulas and A. Ephremides. Stability properties of constrained queueing systems and scheduling for maximum throughput in multihop radio networks. *IEEE Trans. on Automatic Control*, 37(12):1936–1949, 1992.

[34] Y. Xi and E. M. Yeh. Throughput optimal distributed power control of stochastic wireless networks. *IEEE/ACM Trans. on Netw.*, 18(4):1054–1066, 2010.

[35] L. Ying and S. Shakkottai. On throughput optimality with delayed network-state information. *IEEE Trans. on Info. Theory*, 57:5116–5132, 2011.

[36] C. Zhao and X. Lin. On the queue-overflow probabilities of distributed scheduling algorithms. In *IEEE CDC*, pages 4820–4825, 2009.

APPENDIX

PROOF OF LEMMA 4.2

Let k be a link in $\hat{\mathcal{I}}_l$. The probability for k to make a successful transmission at time t is:

$$Prob\left\{u_k(t)=1\right\} = p_k(t)\prod_{k'\in\mathcal{I}_k}(1-p_{k'}(t))$$

$$\geq (1-e^{-\frac{V_{1,F}(k)q_k(t)}{V_{2,F}(L)}})e^{-\frac{\sum_{k'\in\mathcal{I}_k\cup\{k\}}V_{1,F}(k')q_{k'}(t)-V_{1,F}(k)q_k(t)}{\sum_{k'\in\mathcal{I}_k\cup\{k\}}V_{1,F}(k')q_{k'}(t)}}$$

$$\geq (e^{\frac{V_{1,F}(k)q_k(t)}{V_{2,F}(L)}}-1)/e \geq \frac{V_{1,F}(k)q_k(t)}{eV_{2,F}(L)}.$$

Now let us look at link l, with $\sum_{l'\in\hat{\mathcal{I}}_l}V_{1,F}(l') \geq V_{2,F}(L) - C_1 + H_2$. Because at most H_2 packets can be transmitted from all the links in $\hat{\mathcal{I}}_l = \mathcal{I}_k \cup \{k\}$ during H_2 slots of the scheduling-tx sub-frame, we have $\sum_{l'\in\hat{\mathcal{I}}_l}V_{1,F}(l')q_{l'}(t) \geq V_{2,F}(L) - C_1$.

Therefore, $\sum_{l'\in\hat{\mathcal{I}}_l}Prob\left\{u_{l'}(t)=1\right\} \geq \sum_{l'\in\hat{\mathcal{I}}_l}\frac{V_{1,F}(l)q_l(t)}{eV_{2,F}(L)} \geq$
$\sum_{l'\in\hat{\mathcal{I}}_l}\frac{V_{2,F}(L)-C_1}{eV_{2,F}(L)} \geq (1-\epsilon)/e.$ $\square$

PROOF OF LEMMA 4.3

Let l be an arbitrary link in L. The total amount of arrival packets within any interference set is bounded as

$$\sum_{l'\in\hat{\mathcal{I}}_l}A_{l'}(t) \leq |L|A_{max}, \forall t, \forall l.$$

For any link l, there are two cases as below.

Case 1: $\sum_{l'\in\hat{\mathcal{I}}_l}Q_{l'}(t_F(1)) \leq V_{2,F}(L) - hH_0|L|A_{max} - hH_2\epsilon/e.$

Since in total at most $hH_0|L|A_{max}$ new packets arrives at all links in $\hat{\mathcal{I}}_l$, we have

$$Prob\left\{\sum_{l'\in\hat{\mathcal{I}}_l}Q_{l'}(t_{F_0}(1)+hH_0) \leq V_{2,F_0}(L) - hH_2\epsilon/e\right\} = 1.$$

Case 2: $\sum_{l'\in\hat{\mathcal{I}}_l}Q_{l'}(t_F(1)) > V_{2,F}(L) - hH_0|L|A_{max} - hH_2\epsilon/e.$

We first assume that the info-exchange process is always successful, and later we will remove this assumption. Let F_i be any one of the h frames before frame F_h starts, i.e., $i = 0, 1, 2, \ldots, h-1$. Because at most one packet can be transmitted from links in $\hat{\mathcal{I}}_l$, we have $V_{2,F_0}(L) + hH_0|L|A_{max} \geq V_{2,F_i}(L) \geq V_{2,F}(L) - (h-1)H_2 \geq C_1/\epsilon.$ Hence,

$$\sum_{l'\in\hat{\mathcal{I}}_l}V_{1,F_i}(l') = \sum_{l'\in\hat{\mathcal{I}}_l}Q_{l'}(t_{F_i}(1))$$

$$\geq \sum_{l'\in\hat{\mathcal{I}}_l}Q_{l'}(t_F(1)) - (h-1)H_2 \geq V_{2,F_i}(L) - C_1 + H_2.$$

Therefore, we have $\sum_{l'\in\hat{\mathcal{I}}_l}Prob\left\{u_{l'}(t)=1\right\} \geq (1-\epsilon)/e$, $\forall t \in [t_{F_i}(H_1+1), t_{F_i}(H_0)], \forall i \leq h-1$, according to Lemma 4.2. Then, let $u_{\hat{\mathcal{I}}_l}(t) \in \{0,1\}$ indicate whether there is a packet transmitted from any link among $\hat{\mathcal{I}}_l$; we have

$$\mathbb{E}\left\{u_{\hat{\mathcal{I}}_l}(t)\right\} \geq (1-\epsilon)/e, \forall t \in [t_{F_i}(H_1+1), t_{F_i}(H_0)], \forall i \leq h-1.$$

Now we remove the assumption that in all the h frames, all the info-exchange processes are successful. Considering the probability of a failed info-exchange process, let Z denote the total number of frames with successful info-exchange out of the h frames;

$Z = \sum_{i=0}^{h-1}z(i)$, where $z(i) \in \{0,1\}$ indicates whether the info-exchange process is successful in frame F_i. $\mathbb{E}\{z(i)\} \geq 1-\xi$, $\forall i \leq h-1$. Using the "lower tail" Chernoff's Bound yields $Prob\{Z \geq h(1-\xi-\epsilon)\} \geq 1-e^{-\frac{(h\epsilon)^2}{2e^2}}.$

W.l.o.g., we assume $h(1-\xi-\epsilon)$ to be an integer. Let X denote the total amount of transmitted packets in h frames; $X = \sum_{t=t_{F_0}(1)}^{t_{F_0}(1)+hH_0-1}u_{\hat{\mathcal{I}}_l}(t)$. Applying the "lower tail" Chernoff Bound again gives us the following conditional probability

$$Prob\{X \geq ZH_2(1-2\epsilon)/e \mid Z \geq h(1-\xi-\epsilon)\} \geq \left(1-e^{-\frac{(hH_2\epsilon)^2}{2e^2}}\right).$$

Therefore, we have

$$Prob\left\{X \geq hH_2(1-3\epsilon-\xi)/e\right\}$$
$$\geq Prob\left\{Z \geq h(1-\xi-\epsilon)\right\}\cdot$$
$$\quad Prob\left\{X \geq ZH_2(1-2\epsilon)/e \mid Z \geq h(1-\xi-\epsilon)\right\}$$
$$\geq 1-e^{-\frac{(hH_2\epsilon)^2}{2e^2}} - e^{-\frac{(h\epsilon)^2}{2e^2}}.$$

Recall that the average arrival is bounded as

$$\mathbb{E}\left\{\sum_{l'\in\hat{\mathcal{I}}_l}A_{l'}(t)\right\} \leq \frac{H_2}{H_0}(1-5\epsilon-\xi)/e.$$

Let Y denote the total amount of arrival packets in h frames; $Y = \sum_{t=t_{F_0}(1)}^{t_{F_0}(1)+hH_0-1}\sum_{l'\in\hat{\mathcal{I}}_l}A_{l'}(t)$. By using the "upper tail" Hoeffding's Inequality, we obtain

$$Prob\left\{Y \leq hH_2(1-4\epsilon-\xi)/e\right\} \geq 1-e^{-\frac{2h(H_2\epsilon)^2}{H_0|L|^2A_{max}^2e^2}},$$

implying that $Prob\left\{X-Y \geq hH_2\epsilon/e\right\} \geq 1-e^{-\frac{(hH_2\epsilon)^2}{2e^2}} - e^{-\frac{(h\epsilon)^2}{2e^2}} - e^{-\frac{2h(H_2\epsilon)^2}{H_0|L|^2A_{max}^2e^2}}$. Therefore,

$$Prob\left\{\sum_{l'\in\hat{\mathcal{I}}_l}Q_{l'}(t_{F_0}(1)+hH_0) \leq V_{2,F_0}(L) - hH_2\epsilon/e\right\} \geq 1-\frac{\epsilon'}{|L|}.$$

The two cases above together imply that $\forall l \in L$,

$$Prob\left\{\sum_{l'\in\hat{\mathcal{I}}_l}Q_{l'}(t_{F_0}(1)+hH_0) \leq V_{2,F_0}(L) - hH_2\epsilon/e\right\} \geq 1-\frac{\epsilon'}{|L|}.$$

Hence, considering $V_{2,F_h}(L) = \max_{l\in L}\sum_{l'\in\hat{\mathcal{I}}_l}Q_{l'}(t_{F_0}(1)+hH_0)$, we obtain a bound as

$$Prob\left\{V_{2,F_h}(L) \leq V_{2,F_0}(L) - hH_2\epsilon/e\right\} \geq 1-\epsilon'. \quad \square$$

PROOF OF PROPOSITION 5.3

Let $N_l(t)$ denote the number of transmitting links in $\hat{\mathcal{I}}(l)$ at t; recall that $s_l(t) \in \{0,1\}$ indicates if l chooses to transmit at time t.

$$\mathbb{E}\{N_l(t)\} = \sum_{l'\in\hat{\mathcal{I}}(l)}\mathbb{E}\{s_{l'}(t)\} \leq \frac{\sum_{l'\in\hat{\mathcal{I}}(l)}V_{1,F}(l')q_k(t)}{\Gamma\sum_{l'\in\hat{\mathcal{I}}(l)}V_{1,F}(l')q_{l'}(t)} \leq \frac{1}{\Gamma}. \quad \square$$

PROOF OF LEMMA 5.8

W.l.o.g., we look at a link $l \in L_i$. Let L' be a maximum independent set in $\mathcal{I}(l)$, where links can choose any power value to use. Let l_{minP} be a link with the smallest transmission power in L', i.e., $\forall l' \in \mathcal{I}(l), P(l_{minP}) \leq P(l')$. It is easy to see that $d(l', l_{minP}) \leq (2\gamma+1)d_i$. Using this, we can bound the SINR value at $r(l_{minP})$ when only links in L' are transmitting:

$$\beta \leq SINR(l_{minP}) \leq \frac{\sigma^\alpha(2\gamma+1)^\alpha}{(|L'|-1)+N(2\gamma+1)^\alpha d_i^\alpha/P(l_{minP})}.$$

The above inequality yields $|L'| \leq \frac{\sigma^\alpha(2\gamma+1)^\alpha}{\beta}+1$. $\square$

Stability Analyses of Longest-Queue-First Link Scheduling in MC-MR Wireless Networks

Peng-Jun Wan
Dept.of Computer Science
Illinois Institute of Technology
Chicago, IL 60616
wan@cs.iit.edu

Xiaohua Xu
Dept. of Computer Science
Illinois Institute of Technology
Chicago, IL 60616
xxu23@iit.edu

Zhu Wang
Dept. of Computer Science
Illinois Institute of Technology
Chicago, IL 60616
zwang59@iit.edu

Shaojie Tang
Dept. of Computer Science
Illinois Institute of Technology
Chicago, IL 60616
stang7@iit.edu

Zhiguo Wan
School of software
Tsinghua University
Beijing, P.R. China
wanzhiguo@tsinghua.edu.cn

ABSTRACT

Longest-queue-first (**LQF**) link scheduling is a greedy link scheduling in multihop wireless networks. Its stability performance in single-channel single-radio (SC-SR) wireless networks has been well studied recently. However, its stability performance in multi-channel multi-radio (MC-MR) wireless networks is largely under-explored. In this paper, we present a stability subregion with closed form of the **LQF** scheduling in MC-MR wireless networks, which is within a constant factor of the network stability region. We also obtain constant lower bounds on the efficiency ratio of the **LQF** scheduling in MC-MR wireless networks under the 802.11 interference model or the protocol interference model.

Categories and Subject Descriptors

C.2.1 [**Computer-Communication Networks**]: Network Architecture and Design—*wireless communication*

Keywords

Stability, multi-channel multi-radio, link scheduling

1. INTRODUCTION

With the rapid technology advances, many off-the-shelf wireless transceivers (i.e., radios) are capable of operating on multiple channels. For example, the IEEE 802.11 b/g standard and IEEE 802.11a standard provide 3 and 12 channels

respectively, and MICA2 sensor motes support more than 50 channels. The rapidly diminishing prices of the radios has also made it feasible to equip a wireless node with multiple radios. Providing each node with one or more multi-channel radios offers a promising avenue for enhancing the network capacity by simultaneously exploiting multiple non-overlapping channels through different radio interfaces and mitigating interferences through proper channel assignment. In this paper, we take a queuing-theoretic study of a well-known greedy link scheduling, called *Longest-Queue-First* (**LQF**) link scheduling, in multi-channel multi-radio (MC-MR) wireless networks under the 802.11 interference model or the protocol interference model.

We assume that time is slotted. For each $t \in \mathbb{N}$, the t-th time slot is the time interval $(t-1, t]$. Any packet arriving in a slot is assumed to arrive at the end of the slot, and may only be transmitted in the subsequent slots. In addition, the packet arrivals are assumed to be mutually independent and temporally i.i.d. processes with arrival rate vector α. In each time-slot, the **LQF** scheduling first sorts the communication links in the decreasing order of their queue lengths (ties can be broken arbitrarily) and then schedule their transmissions along this order in the following greedy manner: each link transmits as many packets as possible from its queue using the radios at its two endpoints which have not been used by any preceding links and the channels which have not been used by any preceding *conflicting* links. Let $X(t)$ (respectively, $Y(t)$) denote the vector of cumulative number of packets arriving (respectively, transmitted) in the first t time slots, and $Z(t)$ denote the vector of number of packets queued at the very end of time slot t. Then,

$$Z(t) = Z(0) + X(t) - Y(t).$$

The network is said to be *stable* if the Markov chain $(Z(t))$ is positive recurrent. The *stability region* of the **LQF** scheduling, denoted by Λ, is the set of arrival rate vectors α such that the network is stable. Let P be the maximum stability region of the network, the set of arrival rate vectors such that there exists a scheduling policy stabilizing the network. The *efficiency ratio* of the **LQF** scheduling is defined to be

$$\sup \{\sigma \in \mathbb{R}_+ : \sigma P \subseteq \Lambda\}.$$

The first main contribution of this paper is a stability subregion of the **LQF** scheduling with closed form. Such stability subregion is shown to be within a constant factor of the strict capacity subregion. In addition, it can be checked in polynomial time whether a given vector of packet arrival rates lies in this stability subregion. This computational tractability is particularly favorable for cross-layer optimization, where one needs to allocate the link rates efficiently while still ensuring the network stability under the **LQF** scheduling. The second main contribution of this paper is the discovery of constant lower bounds on the efficiency ratio of the **LQF** scheduling. Specifically, the efficiency ratio of the **LQF** scheduling is at least $1/8$ under the 802.11 interference model with uniform interference radii, at least $1/20$ under the 802.11 interference model with arbitrary interference radii, and at least

$$1/\left(2\left(\left\lceil \pi/\arcsin\frac{\varphi-1}{2\varphi}\right\rceil + 1\right)\right)$$

under the protocol interference model in which the interference radius of each node is at least φ times its communication radius for some $\varphi > 1$.

The efficiency ratio of the **LQF** scheduling in single-channel single-radio (SC-SR) wireless networks is now well understood. Joo et al. [5, 6] made a remarkable contribution to fully characterizing the throughput efficiency ratio of **LQF**. Built upon the prior works by Dimakis and Walrand [4] which presented sufficient conditions for **LQF** to achieve 100% throughput, they proved that the throughput efficiency ratio of **LQF** is exactly the *local pooling factor* (LPF) of the conflict graph of the communication links. The LPF is a pure graph-theoretic parameter. Thus, the works by Joo et al. [5, 6] built an elegant bridge between a queuing-theoretic parameter and a graph-theoretic parameter. Under the 802.11 interference model with uniform interference radii, the LPF is shown to be at least $1/6$ in [6]. Sparked by the works in [6], Leconte el al. [7] and Li el al. [8] presented some properties of LPF. Leconte el al. [7] derived tighter lower bounds on LPF in networks of size at most 28 under the 802.11 interference model with uniform interference radii. Li el al. [8] gave an alternative definition of LPF and also introduced a refined notion of LPF. Recently, Wan et al. [16] proved that the LPF is at least $1/16$ under the 802.11 interference model with arbitrary interference radii, and at least

$$1/\left(2\left(\left\lceil \pi/\arcsin\frac{\varphi-1}{2\varphi}\right\rceil - 1\right)\right)$$

under the protocol interference model in which the interference radius of the sender of each link is at least φ times the link length for some $\varphi > 1$. However, it remains computationally intractable to decide whether a given vector of packet arrival rates meets the so-called local-pooling condition. The efficiency ratios of other link scheduling algorithms were studied in [2, 10, 13, 18, 19].

In contrast, the stability of the **LQF** scheduling in MC-MR wireless networks has been under-studied. Lin and Rasool [9] derived a lower bound $1/10$ on the efficiency ratio of the **LQF** scheduling under the 802.11 interference model with uniform interference/communication radii, which is weaker than the $1/8$ lower bound derived in this paper under the same setting. The technical approach in [9] is quite different from the approach followed in this paper. In fact, the lower bound $1/10$ can be derived in a simpler manner by using the fact that the **LQF** scheduling is actually a 10-approximation algorithm for **Maximum-Weight Independent Set** under the 802.11 interference model with uniform interference/communication radii. Brzezinski et al. [1] considered the variant of the **LQF** scheduling with (temporarily) static channel assignment and the only interference assumed was the primary interference. Even in such restricted setting, no analytical bounds on the efficiency ratio were provided in [1]. The variant of the **LQF** scheduling with (temporarily) dynamic channel assignment, which is subject of this paper, was left as a subject of future research in [1].

The remainder of this paper is organized as follows. Section 2 introduces some basic results from functional analysis and probability theory. Section 3 defines the stability region of a MC-MR wireless network. Section 4 presents a stability subregion of the **LQF** scheduling. Section 5 derives the lower bounds on the efficiency ratio of the **LQF** scheduling. Finally, we conclude this paper in Section 6.

2. PRELIMINARIES

Let I be an interval in the real line $\mathbb{R}$. A function $f : I \to \mathbb{R}$ is *absolutely continuous* on I if for every $\varepsilon > 0$, there is a $\delta > 0$ such that whenever a finite sequence of pairwise disjoint sub-intervals $[s_k, t_k]$ of I satisfies

$$\sum_k |s_k - t_k| < \delta,$$

then

$$\sum_k |f(s_k) - f(t_k)| < \varepsilon.$$

If f is absolutely continuous, then f has a derivative f' almost everywhere; the points at which f is differential are called the *regular points* of f. The following property of absolutely continuous functions is implicitly used in [3].

LEMMA 2.1. *Let f be an absolutely continuous non-negative function on $\mathbb{R}_+$ and κ be a positive constant. Suppose that for every almost every regular point t, $f'(t) \leq -\kappa$ whenever $f(t) > 0$. Then, f is non-increasing, and once it reaches zero it stays zero forever. Moreover, $f(t) = 0$ for all $t \geq f(0)/\kappa$.*

A function f on $\mathbb{R}_+$ is said to be *Lipschitz continuous with Lipschitz constant C* (or simply *C-Lipschitz continuous*) for some constant $C > 0$ if for any $s, t \in \mathbb{R}_+$,

$$|f(s) - f(t)| \leq C |s - t|.$$

Lipschitz continuous functions are absolutely continuous. The following lemma may be hidden in a textbook, and we give a short proof for the sake of completeness.

LEMMA 2.2. *Suppose that $f_1, f_2, \cdots, f_k$ are k C-Lipschitz continuous functions on $\mathbb{R}_+$. Then, $\max_{1 \leq i \leq t} f_k(t)$ is also C-Lipschitz continuous.*

PROOF. For any $t \in \mathbb{R}_+$, let

$$g(t) = \max_{1 \leq i \leq t} f_i(t)$$

and

$$K(t) = \{1 \leq j \leq k : f_j(t) = g(t)\}.$$

Consider any distinct $s > t \geq 0$. If there is any $i \in K(s) \cap K(t)$, then

$$|g(s) - g(t)| = |f_i(s) - f_i(t)| \leq C(s - t).$$

Now suppose that $K(s) \cap K(t) = \emptyset$. Choose $i \in K(s)$ and $j \in K(t)$. Then,

$$f_i(s) > f_j(s), f_j(t) > f_i(t).$$

Then, there exists $r \in (t, s)$ such that $f_i(r) = f_j(r)$. So,

$$\begin{aligned}
&|g(s) - g(t)| \\
&= |f_i(s) - f_j(t)| \\
&\leq |f_i(s) - f_i(r)| + |f_j(r) - f_j(t)| \\
&\leq C(s - r) + C(r - t) \\
&= C(s - t).
\end{aligned}$$

Therefore, g is also C-Lipschitz continuous. $\square$

A function f which takes values in k-dimensional Euclidean space is said to be absolutely (respectively, Lipschitz) continuous if each of its component is absolutely (respectively, Lipschitz) continuous. For any vector x in an Euclidean Space, $\|x\|_\infty$ and $\|x\|_1$ denote the maximum norm (also called uniform norm) and the Manhattan norm of x respectively.

Let (f_n) be a sequence of functions on $\mathbb{R}_+$ and let f be a continuous function on $\mathbb{R}_+$. We say that $f_n \to f$ *uniformly on compact sets*, or simply $f_n \to f$ u.o.c., if for each $t > 0$,

$$\sup_{0 \leq s \leq t} |f_n(s) - f(s)| \to 0.$$

The following lemma was stated in Lemma 4.1 of [3].

LEMMA 2.3. *Let (f_n) be a sequence of nondecreasing real-valued functions on $\mathbb{R}_+$, and f be a continuous function on $\mathbb{R}_+$. Assume that (f_n) converges pointwise to f. Then the convergence is u.o.c.*

The following theorem on the convergence of random variables is stated in Theorem 2.2.3 of [12].

THEOREM 2.4. *Suppose that a sequence of random variables (ξ_n) converge to a random variable ξ in probability.*

1. *If ξ_n is uniformly integrable, then $\mathbf{E}[|\xi|] < \infty$ and $\lim_{n \to \infty} \mathbf{E}[\xi_n] = \mathbf{E}[\xi]$.*

2. *If $\xi_n \geq 0$, $\mathbf{E}[\xi] < \infty$, and $\lim_{n \to \infty} \mathbf{E}[\xi_n] = \mathbf{E}[\xi]$, then ξ_n is uniformly integrable.*

3. NETWORK STABILITY REGION

Consider an instance of MC-MR multihop wireless network with a set V of networking nodes and a set A of node-level communication links. Each node v has $\tau(v)$ radios, and there are λ non-overlapping channels. In the fine-grained network representation [15] of the MC-MR wireless network, each communication link is encoded by an ordered *quintuple* specifying the transmitting node, the receiver node, the radio at the transmitting node, the radio at the receiving node, and the channel. Specifically, for each node-level link (u, v) in A, we make $\lambda \cdot \tau_u \cdot \tau_v$ replications (u, v, i, j, k) for $1 \leq i \leq \tau_u$, $1 \leq j \leq \tau_v$, and $1 \leq k \leq \lambda$. A replication (u, v, i, j, k) always utilizes the i-th radio at u and the j-th radio at v over the k-th channel. For each subset B of A, we use $B^{\tau, \lambda}$ to denote the set of all replications of the links in B. In particular, $A^{\tau, \lambda}$ is the set of all replicated links of the links in A. A subset I of $A^{\tau, \lambda}$ can transmit at the same time if and only if (1) all replication links in I are radio-disjoint, in other words, no pair share a common radio, and (2) for each channel k, all the replication links in I transmitting over channel k are conflict-free. Let $\mathcal{I}^{\tau, \lambda}$ denote the collection of the subsets of $A^{\tau, \lambda}$ which can transmit successfully at the same time. For each $I \in \mathcal{I}^{\tau, \lambda}$, its *service rate* is the vector $d \in \mathbb{R}_+^A$ given by

$$d_a = \left| I_j \cap \{a\}^{\tau, \lambda} \right|$$

for each $a \in A$.

A set

$$\Pi = \left\{ (I_j, \ell_j) \in \mathcal{I}^{\tau, \lambda} \times \mathbb{R}_+ : 1 \leq j \leq m \right\}$$

is called a *(fractional) link schedule* of some $d \in \mathbb{R}_+^A$ if

$$d_a = \sum_{j=1}^m \ell_j \left| I_j \cap \{a\}^{\tau, \lambda} \right|$$

for each $a \in A$. The two values m and $\sum_{j=1}^m \ell_j$ are referred to as the *size* and *length* (or *latency*) of Π respectively. For any $d \in \mathbb{R}_+^A$, the *minimum latency* $\chi^*(d)$ of d is defined as the minimum length of all fractional link schedules of d. The stability region of the MC-MR wireless network is

$$P = \left\{ d \in \mathbb{R}_+^A : \chi^*(d) < 1 \right\}.$$

For any subset S of $A^{\tau, \lambda}$, a subset I of S is said to be a *maximal independent* of S if $I \in \mathcal{I}^{\tau, \lambda}$ and for any link $e \in S \setminus I$, $I \cup \{e\} \notin \mathcal{I}^{\tau, \lambda}$. For any $B \subseteq A$, a set $I \in \mathcal{I}^{\tau, \lambda}$ is said to be B-*maximal* if $I \cap B^{\tau, \lambda}$ is a maximal independent set of $B^{\tau, \lambda}$. We use $\mathcal{M}_B^{\tau, \lambda}$ to denote the collection of B-maximal independent sets of $B^{\tau, \lambda}$, M_B to denote the set of service rates of the sets in $\mathcal{M}_B^{\tau,\lambda}$, and Φ_B to denote the convex hull of M_B.

4. STABILITY SUBREGION

Consider an instance of MC-MR wireless network specified in Section 3. Two links in A are said to have a *conflict* if they cannot transmit at the same time over the same channel. Furthermore, a conflicting pair of distinct links in A are said to have *primary* conflict if there share one common end, and *secondary* conflict otherwise. For the sake of convenience, each link is said to have a *self-conflict* with itself. The *concise conflict graph* [20] of the MC-MR wireless network is the edge-weighted graph G on A in which there is an edge between each conflicting pair of links (a, b) whose weight denoted by $c(a, b)$, is defined as follows:

- If $b = a$ (i.e., self-conflict), then

$$c_{a,b} = 1 - \left(1 - \frac{1}{\tau_u}\right)\left(1 - \frac{1}{\tau_v}\right)\left(1 - \frac{1}{\lambda}\right)$$

  where u and v are the two endpoints of a.

- If a and b have a common endpoint u (i.e., a and b have a primary conflict), then

$$c_{a,b} = 1 - \left(1 - \frac{1}{\tau_u}\right)\left(1 - \frac{1}{\lambda}\right).$$

- If a and b have the secondary conflict, then

$$c_{a,b} = \frac{1}{\lambda}.$$

Note that $c(a, b) = c(b, a)$. Let $\mathcal{I}$ denote the collection of the independent sets in G. In other words, $\mathcal{I}$ is the collection of the subsets of A which can transmit successfully at the same time over the same channel. Note that G can be regarded as a generalization of the conventional conflict graph of the underlying SC-SR wireless network by adding a self-loop at each link and assigning each edge a weight specified by the function c. Thus, $\mathcal{I}$ is essentially the collection of the independent sets of links in the underlying SC-SR wireless network.

For any link $a \in A$, $N_G(a)$ denotes the set of neighbors of a in G. Since G has a self-loop at each vertex, a is a neighbor to itself, and hence $a \in N_G(a)$. Thus, $N_G(a)$ consists of all links in A (including itself) having conflict with a. For any link a, any subset B of links, and any $d \in \mathbb{R}_+^A$, define

$$\Gamma(B, a; d) = \sum_{b \in N_G(a) \cap B} c_{a,b} d_b.$$

By Lemma 2.3 in [20], for any $d \in M_B$ and any $a \in B$,

$$\Gamma(B, a; d) \geq 1.$$

Thus, for any $d \in \Phi_B$ and any $a \in B$,

$$\Gamma(B, a; d) \geq 1$$

as well.

Consider a link ordering $\prec$ of A. For any link $a \in A$, $N_G^{\prec}(a)$ denotes the set of neighbors of a in G preceding a in the ordering $\prec$ plus a itself. For any $d \in \mathbb{R}_+^A$, the value

$$\max_{a \in A} \Gamma\left(N_G^{\prec}(a), a; d\right)$$

is referred to as *d-weighted inductivity* of $\prec$ and is denoted by $\Delta^{\prec}(d)$. The smallest d-weighted inductivity of all possible link orderings, denoted by $\Delta^*(d)$, is called the *d-weighted inductivity* of the network. It was shown in [20] that

$$\Delta^*(d) = \max_{\emptyset \neq B \subseteq A} \min_{b \in B} \Gamma(B, b; d).$$

and $\Delta^*(d)$ is achieved by a special ordering, called *smallest-last ordering*, which is produced successively as follows: Initialize B to A. For $i = |A|$ down to 1, let a_i be a link minimizing $\Gamma(B, b; d)$ among all links b in B, and delete a_i from B. Then the ordering $\langle a_1, a_2, \cdots, a_{|A|} \rangle$ is a smallest-last ordering.

Let

$$Q^* = \left\{ d \in \mathbb{R}_+^A : \Delta^*(d) < 1 \right\}.$$

The following theorem shows that Q^* is a stability subregion of **LQF** scheduling.

THEOREM 4.1. $Q^* \subseteq \Lambda$.

We shall prove Theorem 4.1 by applying the **Malyshev-Menshikov Criterion** [11] for ergodicity of discrete-time countable-state Markov chains. For any $n \in \mathbb{N}$, we denote by $Z^{(n)}(t)$ (respectively, $X^{(n)}(t)$, $Y^{(n)}(t)$) the vector of queue length (cumulative number of arriving packets, cumulative number of transmitted packets) in a system at the end of time-slot t with its initial total queue length $\left\| Z^{(n)}(0) \right\|_1 = n$. Let

$$T = \left\lceil \frac{1 - \left(1 - \frac{1}{\|\tau\|_\infty}\right)^2 \left(1 - \frac{1}{\lambda}\right)}{1 - \Delta^*(\alpha)} \right\rceil.$$

By the **Malyshev-Menshikov Criterion** [11], Theorem 4.1 follows immediately from the theorem below.

THEOREM 4.2. *For any $\alpha \in Q^*$,*

$$\lim_{n \to \infty} \mathbf{E}\left[\left\| \frac{Z^{(n)}(nT)}{n} \right\|_1\right] = 0.$$

The proof of Theorem 4.2 utilizes Theorem 2.4. By the strong law of large numbers,

$$\left\| \frac{X^{(n)}(nT)}{n} \right\|_1 = T \left\| \frac{X^{(n)}(nT)}{nT} \right\|_1 \to T \|\alpha\|_1$$

almost surely, and

$$\mathbf{E}\left[\left\| \frac{X^{(n)}(nT)}{n} \right\|_1\right] = T \cdot \mathbf{E}\left[\left\| \frac{X^{(n)}(nT)}{nT} \right\|_1\right] = T \|\alpha\|_1.$$

By Theorem 2.4, the sequence $\left(\left\| \frac{X^{(n)}(nT)}{n} \right\|_1\right)$ is uniformly integrable. Since

$$\left\| \frac{Z^{(n)}(nT)}{n} \right\|_1 \leq \left\| \frac{X^{(n)}(nT)}{n} \right\|_1,$$

the sequence $\left(\left\| \frac{Z^{(n)}(nT)}{n} \right\|_1 \right)$ is also uniformly integrable. Again, by Theorem 2.4, Theorem 4.2 would hold if

$$\left\| \frac{Z^{(n)}(nT)}{n} \right\|_1 \to 0$$

in probability. We will actually prove a stronger result that

$$\left\| \frac{Z^{(n)}(nT)}{n} \right\|_1 \to 0$$

almost surely. Consider a sample path (i.e. realization) ω of

$$\left(Z^{(n)}(0) : n \in \mathbb{N} \right) \cup \left(X^{(n)}(t) : n, t \in \mathbb{N} \right).$$

It is is said to be *well-behaved* if

$$\lim_{t \to \infty} \frac{X^{(n)}(t, \omega)}{t} = \alpha.$$

By the strong law of large numbers, every sample path is almost surely well-behaved. We will prove that for any well-behaved sample path ω,

$$\left\| \frac{Z^{(n)}(nT, \omega)}{n} \right\|_1 \to 0,$$

from which we can conclude that

$$\left\| \frac{Z^{(n)}(nT)}{n} \right\|_1 \to 0$$

almost surely, and hence Theorem 4.2 holds.

Fix a well-behaved sample path ω. Denote $X^{(n)}(t, \omega)$ (respectively, $Y^{(n)}(t, \omega)$, $Z^{(n)}(t, \omega)$) by $x^{(n)}(t)$ (respectively, $y^{(n)}(t)$, $z^{(n)}(t)$). Then, all of them are deterministic. In order to show that

$$\left\| \frac{z^{(n)}(nT)}{n} \right\|_1 \to 0,$$

it is sufficient to show that for any infinite increasing sequence S of positive integers, there is an infinite subsequence S' of S along which

$$\left\| \frac{z^{(n)}(nT)}{n} \right\|_1 \to 0.$$

So, we further fix an infinite increasing sequence S of positive integers. For convenience, we define $x^{(n)}(0)$ and $y^{(n)}(0)$ to be the vector of zeros. We extend $x^{(n)}(t)$ (respectively, $y^{(n)}(t)$, $z^{(n)}(t)$) to all non-negative real numbers by *linear interpolation*. Then, for any $t \geq 0$ and any $n \in \mathbb{N}$,

$$\frac{z^{(n)}(nt)}{n} = \frac{z^{(n)}(0)}{n} + \frac{x^{(n)}(nt)}{n} - \frac{y^{(n)}(nt)}{n}.$$

LEMMA 4.3. *For any $t \geq 0$, $\lim_n \frac{x^{(n)}(nt)}{n} = \alpha t$.*

PROOF. The lemma holds trivially when $t = 0$. Since ω is well-behaved for any positive integer t,

$$\lim_n \frac{x^{(n)}(nt)}{n} = t \lim_n \frac{x^{(n)}(nt)}{nt} = \alpha t.$$

Now consider any non-integer $t > 0$. Since

$$\frac{x^{(n)}(\lfloor nt \rfloor)}{\lfloor nt \rfloor} \frac{\lfloor nt \rfloor}{nt} \leq \frac{x^{(n)}(nt)}{nt} \leq \frac{x^{(n)}(\lceil nt \rceil)}{\lceil nt \rceil} \frac{\lceil nt \rceil}{nt}$$

we have

$$\lim_n \frac{x^{(n)}(nt)}{nt} = \alpha.$$

Thus,

$$\lim_n \frac{x^{(n)}(nt)}{n} = \alpha t.$$

So, the lemma holds. $\square$

For each $t \in \mathbb{N}$, let $I^{(n)}(t) \in \mathcal{I}^{\tau, \lambda}$ be the set of replicated links which are scheduled to transmit in the t-th time slot, and $d^{(n)}(t)$ be the service rate of $I^{(n)}(t)$. The average service rate in a time interval $[t_1, t_2]$ is defined to be

$$\overline{y}^{(n)}(t_1, t_2) = \frac{y^{(n)}(t_2) - y^{(n)}(t_1)}{t_2 - t_1}.$$

It has the following properties.

LEMMA 4.4. *Consider any $0 \leq t_1 < t_2$.*

1. *For any $s = \varepsilon t_1 + (1 - \varepsilon) t_2$ for some $\varepsilon \in [0, 1]$,*

$$\overline{y}^{(n)}(t_1, t_2) = (1 - \varepsilon) \overline{y}^{(n)}(t_1, s) + \varepsilon \overline{y}^{(n)}(s, t_2),$$

2. *$\overline{y}^{(n)}(t_1, t_2)$ is a convex combination of*

$$\left\{ d^{(n)}(t) : \lfloor t_1 \rfloor + 1 \leq t \leq \lceil t_2 \rceil, t \in \mathbb{N} \right\}.$$

3. *$\left\| \overline{y}^{(n)}(t_1, t_2) \right\|_\infty \leq \|\tau\|_\infty.$*

PROOF. (1). The first part of the lemma holds trivially if $\varepsilon = 0$ or 1. So we assume that $\varepsilon \in (0, 1)$. Then,

$\overline{y}^{(n)}(t_1, t_2)$

$= \dfrac{y^{(n)}(t_2) - y^{(n)}(t_1)}{t_2 - t_1}$

$= \dfrac{y^{(n)}(t_2) - y^{(n)}(s)}{t_2 - t_1} + \dfrac{y^{(n)}(s) - y^{(n)}(t_1)}{t_2 - t_1}$

$= \dfrac{t_2 - s}{t_2 - t_1} \dfrac{y^{(n)}(t_2) - y^{(n)}(s)}{t_2 - s} + \dfrac{s - t_1}{t_2 - t_1} \dfrac{y^{(n)}(s) - y^{(n)}(t_1)}{s - t_1}$

$= \varepsilon \overline{y}^{(n)}(s, t_2) + (1 - \varepsilon) \overline{y}^{(n)}(t_1, s),$

and hence the first part of the lemma holds as well.

(2). If $\lfloor t_1 \rfloor + 1 = \lceil t_2 \rceil$, then

$$\overline{y}^{(n)}(t_1, t_2) = d^{(n)}(\lceil t_2 \rceil)$$

and hence the second part of the lemma holds trivially. So, we assume that Note that $\lfloor t_1 \rfloor + 1 < \lceil t_2 \rceil$. Note that for any $t \in \mathbb{R}_+$,

$$\overline{y}^{(n)}(\lceil t \rceil - 1, t) = d^{(n)}(\lceil t \rceil),$$
$$y^{(n)}(t, \lfloor t \rfloor + 1) = d^{(n)}(\lfloor t \rfloor + 1).$$

By the first part of the lemma, $\overline{y}^{(n)}(t_1, t_2)$ is a convex combination of

$y^{(n)}(t_1, \lfloor t_1 \rfloor + 1)$;

$y^{(n)}(s, s + 1), s \in \mathbb{N}$ and $\lfloor t_1 \rfloor + 1 \leq s \leq \lceil t_2 \rceil - 1$;

$y^{(n)}(\lceil t_2 \rceil - 1, t_2).$

Thus, the second part of the lemma holds.

(3). The third part of the lemma follows from the second part of the lemma and the fact that

$$\left\| d^{(n)}(t) \right\|_\infty \le \|\tau\|_\infty$$

for any $t \in \mathbb{N}$. $\quad\square$

Note that for any $0 \le t_1 < t_2$,

$$
\frac{\frac{y^{(n)}(nt_2)}{n} - \frac{y^{(n)}(nt_1)}{n}}{t_2 - t_1}
$$
$$
= \frac{y^{(n)}(nt_2) - y^{(n)}(nt_1)}{nt_2 - nt_1}
$$
$$
= \overline{y}^{(n)}(nt_1, nt_2).
$$

By the third part of Lemma 4.4, $\frac{y^{(n)}(nt)}{n}$ is $\|\tau\|_\infty$-Lipschitz continuous, and hence is equicontinuous. By the Arzela-Ascoli theorem, there is an infinite subsequence S_1 of S along which $\frac{y^{(n)}(nt)}{n}$ converges to some function $\beta(t)$. In addition, $\beta(t)$ is also $\|\tau\|_\infty$-Lipschitz continuous. Since

$$\left\| \frac{z^{(n)}(0)}{n} \right\|_1 = 1$$

for any $n \in \mathbb{N}$, there is an infinite subsequence S' of S_1 along which $\frac{z^{(n)}(0)}{n}$ converges. Therefore, along the sequence S', $\frac{z^{(n)}(0)}{n}$, $\frac{x^{(n)}(nt)}{n}$ and $\frac{y^{(n)}(nt)}{n}$ all converge. Since both $\frac{x^{(n)}(nt)}{n}$ and $\frac{y^{(n)}(nt)}{n}$ are increasing function of t for each n, they converge u.o.c along S' to αt and $\beta(t)$ respectively by Lemma 2.3. As

$$\frac{z^{(n)}(nt)}{n} = \frac{z^{(n)}(0)}{n} + \frac{x^{(n)}(nt)}{n} - \frac{y^{(n)}(nt)}{n},$$

$\frac{z^{(n)}(nt)}{n}$ also converges u.o.c. along S' to some function $\gamma(t)$. Since

$$\left\| \frac{z^{(n)}(0)}{n} \right\|_1 = 1,$$

$\|\gamma(0)\|_1 = 1$. In addition,

$$\gamma(t) = \gamma(0) + \alpha t - \beta(t).$$

Clearly, $\gamma(t)$ is also Lipschitz continuous with Lipschitz constant $\|\alpha\|_\infty + \|\tau\|_\infty$, and so is $\|\gamma(t)\|_\infty$ by Lemma 2.2.

A time $t \in \mathbb{R}_+$ is said to be a *regular point* if each component of $\gamma(t)$ and $\|\gamma(t)\|_\infty$ are differentiable at t. Since both $\gamma(t)$ and $\|\gamma(t)\|_\infty$ are Lipschitz continuous, almost every time $t \in \mathbb{R}_+$ is a regular point. Since

$$\beta(t) = \gamma(t) - \gamma(0) - \alpha t,$$

$\beta(t)$ is also differentiable at any regular point t, and

$$\gamma'(t) = \alpha - \beta'(t).$$

In the next, we derive the properties of $\|\gamma(t)\|_\infty'$ and $\beta'(t)$ at any regular point $t > 0$ with $\|\gamma(t)\|_\infty > 0$. For any regular point t, denote

$$A_0(t) = \left\{ a \in A \ : \gamma_a(t) = \max_{b \in A} \gamma_b(t) \right\},$$
$$A_1(t) = \left\{ a \in A_0(t) : \gamma_a'(t) = \max_{b \in A_0(t)} \gamma_b'(t) \right\}.$$

Then, we have the following lemma.

LEMMA 4.5. *Consider any regular point $t > 0$ with $\|\gamma(t)\|_\infty > 0$.*

1. *For any $a \in A_1(t)$, $\gamma_a'(t) = \|\gamma(t)\|_\infty'$.*

2. *$\beta'(t) \in \Phi_{A_1(t)}$.*

The proof of Lemma 4.5 is quite involved, and so is relegated to Appendix. We apply Lemma 4.5 to show that for any regular point $t > 0$ with $\|\gamma(t)\|_\infty > 0$,

$$\|\gamma(t)\|_\infty' \le \frac{\Delta^*(\alpha) - 1}{1 - \left(1 - \frac{1}{\|\tau\|_\infty}\right)^2 \left(1 - \frac{1}{\lambda}\right)}.$$

Let a be the link in $A_1(t)$ with minimum $\Gamma(A_1(t), a; \alpha)$. Then,

$$\Gamma(A_1(t), a; \alpha) \le \Delta^*(\alpha).$$

By the second part of Lemma 4.5, for each link $a \in A_1(t)$,

$$\Gamma(A_1(t), a; \beta'(t)) \ge 1$$

Thus,

$$
\Gamma\left(A_1(t), a; \gamma'(t)\right)
$$
$$
= \Gamma\left(A_1(t), a; \alpha - \beta'(t)\right)
$$
$$
= \Gamma\left(A_1(t), a; \alpha\right) - \Gamma\left(A_1(t), a; \beta'(t)\right)
$$
$$
\le \Delta^*(\alpha) - 1.
$$

On the other hand, by the first part of Lemma 4.5,

$$
\Gamma\left(A_1(t), a; \gamma'(t)\right)
$$
$$
= \sum_{b \in N_G(a) \cap A_1(t)} c_{a,b} \gamma_b'(t)
$$
$$
= \sum_{b \in N_G(a) \cap A_1(t)} c_{a,b} \|\gamma(t)\|_\infty'
$$
$$
= \|\gamma(t)\|_\infty' \sum_{b \in N_G(a) \cap A_1(t)} c_{a,b}
$$
$$
\ge \|\gamma(t)\|_\infty' \, c_{a,a}
$$
$$
\ge \|\gamma(t)\|_\infty' \left(1 - \left(1 - \frac{1}{\|\tau\|_\infty}\right)^2 \left(1 - \frac{1}{\lambda}\right)\right).
$$

Therefore,

$$\|\gamma(t)\|_\infty' \le \frac{\Delta^*(\alpha) - 1}{1 - \left(1 - \frac{1}{\|\tau\|_\infty}\right)^2 \left(1 - \frac{1}{\lambda}\right)}.$$

The property established in the previous paragraph together with Lemma 2.1 yields that $\|\gamma(t)\|_\infty = 0$ for

$$t \ge \frac{\|\gamma(0)\|_\infty}{\frac{1 - \Delta^*(\alpha)}{1 - \left(1 - \frac{1}{\|\tau\|_\infty}\right)^2 \left(1 - \frac{1}{\lambda}\right)}}.$$

Since

$$T = \left\lceil \frac{1 - \left(1 - \frac{1}{\|\tau\|_\infty}\right)^2 \left(1 - \frac{1}{\lambda}\right)}{1 - \Delta^*(\alpha)} \right\rceil$$

$$\geq \frac{1 - \left(1 - \frac{1}{\|\tau\|_\infty}\right)^2 \left(1 - \frac{1}{\lambda}\right)}{1 - \Delta^*(\alpha)}$$

$$= \frac{1}{\frac{1 - \Delta^*(\alpha)}{1 - \left(1 - \frac{1}{\|\tau\|_\infty}\right)^2 \left(1 - \frac{1}{\lambda}\right)}}$$

$$= \frac{\|\gamma(0)\|_1}{\frac{1 - \Delta^*(\alpha)}{1 - \left(1 - \frac{1}{\|\tau\|_\infty}\right)^2 \left(1 - \frac{1}{\lambda}\right)}}$$

$$\geq \frac{\|\gamma(0)\|_\infty}{\frac{1 - \Delta^*(\alpha)}{1 - \left(1 - \frac{1}{\|\tau\|_\infty}\right)^2 \left(1 - \frac{1}{\lambda}\right)}},$$

we have $\|\gamma(t)\|_\infty = 0$ whenever $t \geq T$. Consequently, $\|\gamma(t)\|_1 = 0$ whenever $t \geq T$. Therefore,

$$\lim_n \left\| \frac{z^{(n)}(nT)}{n} \right\|_1 = \|\gamma(T)\|_1 = 0.$$

This completes the proof of Theorem 4.2.

5. THE EFFICIENCY RATIO

In this section, we derive the lower bounds on the efficiency ratio of the **LQF** scheduling.

Given a link ordering $\prec$ of A, its *backward local independence number* (BLIN) is defined to be

$$\max_{a \in A} \max \left\{ |I| : I \subseteq N_G^\prec(a), I \in \mathcal{I} \right\}.$$

An *orientation* of G is a digraph obtained from G by imposing an orientation on each edge of G. Note that in each orientation D of G also has a self-loop at each vertex, and consequently, $a \in N_D^{in}(a) \cap N_D^{out}(a)$ for each $a \in A$. Given an orientation D of G, its *inward local independence number* (ILIN) is defined to be

$$\max_{a \in A} \max \left\{ |I| : I \subseteq N_D^{in}(a), I \in \mathcal{I} \right\}.$$

Since BLIN and ILIN only depend on the topology of G rather than the edge weight function c, the following properties which hold in the convectional conflict graph of the underlying SC-SR wireless network also hold in the G:

- Under the 802.11 interference model with uniform interference radii, the lexicographic ordering of A has BLIN at most 6 [6].

- Under the 802.11 interference model with arbitrary interference radii, there is an orientation D of G with ILIN at most 8 [17].

- Under the protocol interference model in which the interference radius of the sender of each link is at least φ times the link length for some $\varphi > 1$, there is an orientation D of G with ILIN $\left\lceil \pi / \arcsin \frac{\varphi - 1}{2\varphi} \right\rceil - 1$ [14].

THEOREM 5.1. *The following two statements are true:*

1. *If there is a link ordering of A with BLIN μ, then $Q^* \supseteq \frac{1}{\mu + 2} P$.*

2. *If there is an orientation of G with ILIN μ, then $Q^* \supseteq \frac{1}{2(\mu + 2)} P$.*

PROOF. (1). Consider any $d \in P$. By the first part of Corollary 2.8 in [20],

$$\Delta^*(d) \leq (\mu + 2) \chi^*(d) < \mu + 2.$$

Thus, $d \in (\mu + 2) Q^*$. Hence, $P \subseteq (\mu + 2) Q^*$, which implies that $Q^* \supseteq \frac{1}{\mu + 2} P$.

(2). Consider any $d \in P$. By the second part of Corollary 2.8 in [20],

$$\Delta^*(d) \leq 2(\mu + 2) \chi^*(d) < 2(\mu + 2).$$

Thus, $d \in 2(\mu + 2) Q^*$. Hence, $P \subseteq 2(\mu + 2) Q^*$, which implies that $Q^* \supseteq \frac{1}{2(\mu + 2)} P$. $\square$

Theorem 4.1 and Theorem 5.1 immediately implies the following lower bounds on the efficiency ratio of the **LQF** scheduling.

- 1/8 under the 802.11 interference model with uniform interference radii, the efficiency ratio of the **LQF** scheduling is at least .

- 1/20 under the 802.11 interference model with arbitrary interference radii, the efficiency ratio of the **LQF** scheduling is at least .

- $1 / \left(2 \left(\left\lceil \pi / \arcsin \frac{\varphi - 1}{2\varphi} \right\rceil + 1 \right) \right)$ under the protocol interference model in which the interference radius of the sender of each link is at least φ times the link length for some $\varphi > 1$.

6. DISCUSSIONS

Most of the recent works on the stability of **LQF** scheduling established the stability by using Theorem 4.2 of [3], which states that in the context of multiclass queuing networks the stability of fluid-limit systems imply the stability of the original system under certain conditions. One crucial condition is that the queuing service discipline is *working-conserving* (middle of pp. 65 in [3]): a server is idle only when there is no customer waiting for the service. Apparently, the **LQF** scheduling in wireless networks is not working-conserving, as a link with non-empty queue may be idle due to the interference from other nearby links. So, the direct applicability of Theorem 4.2 of [3] to wireless link scheduling is questionable. Instead, we have applied the **Malyshev-Menshikov Criterion** [11] to establish the stability of **LQF** scheduling. An attractive feature of this technical approach is that we push the deterministic (sample-path) arguments as far as possible while trying to avoid the heavy machinery of stochastic processes. The advantages of this approach are clear: the sample-path arguments are simple and intuitive; thus they provide a clear insight into the issues at hand. Sample-path analysis also helps pinpoint what and when stochastic conditions are needed to guarantee the stability. For example, the packet arrival process in this paper is only required to be mutually independent and temporally i.i.d., while the packet arrival processes in [4, 5, 6] have to meet additional conditions.

The local pooling factor (LPF) of a MC-MR wireless network can be defined as follows. For any non-empty subset B of A, let

$$\sigma_B = \min\left\{c \in \mathbb{R}^+ : \exists x, y \in \Phi_B \text{ s.t. } x \leq cy\right\}.$$

Then, the LPF is the parameter

$$\sigma^* = \min_{\emptyset \neq B \subseteq A} \sigma_B.$$

Using the same (and easier) argument as in Section 4, we can show that $\Lambda \supseteq \sigma^* P$. Thus, the LPF σ^* is also a lower bound on the efficiency ratio. We can also derive the same lower bounds on σ^*. Specifically, σ^* is at least $1/8$ under the 802.11 interference model with uniform interference radii, at least $1/20$ under the 802.11 interference model with arbitrary interference radii, and at least

$$1/\left(2\left(\left\lceil \pi/\arcsin\frac{\varphi-1}{2\varphi} \right\rceil + 1\right)\right)$$

under the protocol interference model in which the interference radius of the sender of each link is at least φ times the link length for some $\varphi > 1$.

ACKNOWLEDGEMENTS: This work is supported in part by the National Science Foundation of USA under grants CNS-0831831 and CNS-0916666, and by the National Natural Science Foundation of P. R. China under grant 61128005. The authors would like to thank Dr. Alexandre Proutiere for shepherding the revision of this paper, and the anonymous reviewers for many helpful suggestions.

7. REFERENCES

[1] A. Brzezinski, G. Zussman, and E. Modiano, Distributed throughput maximization in wireless mesh networks via pre-partitioning, *IEEE/ACM Transactions on Networking* 16(6): 1406-1419, 2008.

[2] P. Chaporkar, K. Kar, X. Luo, and S. Sarkar, Throughput and fairness guarantees through maximal scheduling in wireless networks, *IEEE Transactions on Information Theory* 54(2):572–594, 2008.

[3] J. G. Dai, On Positive Harris Recurrence of Multiclass Queueing Networks: A Unified Approach via Fluid Limit Models, *Annals of Applied Probability* 5(1):49–77, 1995.

[4] A. Dimakis and J. Walrand, Sufficient conditions for stability of longest queue first scheduling: second order properties using fluid limits, *Advances in Applied Probability* 38(2):505–521, 2006.

[5] C. Joo, X. Lin, and N. B. Shroff, Greedy Maximal Matching: Performance Limits for Arbitrary Network Graphs Under the Node-exclusive Interference Model, *IEEE Transactions on Automatic Control* 54(12): 2734-2744, 2009.

[6] C. Joo, X. Lin, and N. B. Shroff, Understanding the Capacity Region of the Greedy Maximal Scheduling Algorithm in Multi-hop Wireless Networks, *IEEE/ACM Transactions on Networking* 17(4):1132-1145, 2009.

[7] M. Leconte, J. Ni, and R. Srikant, Improved bounds on the throughput efficiency of greedy maximal scheduling in wireless networks, in *Proc. ACM MOBIHOC* 2009.

[8] B. Li, C. Boyaci, and Y. Xia, A refined performance characterization of longest-queue-first policy in wireless networks, in *Proc. ACM MOBIHOC* 2009.

[9] X. Lin and S. Rasool, A Distributed Joint Channel-Assignment, Scheduling and Routing Algorithm for Multi-Channel Ad-hoc Wireless Networks, *Proc. IEEE INFOCOM* 2009, pp. 1118-1126.

[10] X. Lin and N. B. Shroff, The impact of imperfect scheduling on cross-layer rate control in wireless networks, *IEEE/ACM Transactions on Networking* 14(2):302–315, 2006.

[11] V.A. Malyshev and M.V. Menshikov, Ergodicity, continuity and analyticity of countable Markov chains, *Trans. Moscow Math. Soc.* 39 1–48, 1981.

[12] A.V. Skorokhod, Basic principles and applications of probability theory, edited by Yu.V. Prokhorov, and translated from the 1989 Russian original by B.D. Seckler. Springer-Verlag, Berlin, 2005.

[13] L. Tassiulas and A. Ephremides, Stability properties of constrained queueing systems and scheduling policies for maximum throughput in multihop radio networks, *IEEE Transactions on Automatic Control* 37(12):1936–1948, 1992.

[14] P.-J. Wan, Multiflows in Multihop Wireless Networks, in *Proc. ACM MOBIHOC* 2009.

[15] P.-J. Wan, Y. Cheng, Z. Wang, and F. Yao, Multiflows in Multi-Channel Multi-Radio Multihop Wireless Networks, in *Proc. IEEE INFOCOM* 2011.

[16] P.-J. Wan, M. Li, L. Wang, and O. Frieder, Local Pooling Factor of Multihop Wireless Networks, in *Proc. IEEE INFOCOM* 2011.

[17] P.-J. Wan, C. Ma, Z. Wang, B. Xu, M. Li, and X. Jia, Weighted Wireless Link Scheduling without Information of Positions And Interference/Communication Radii, in *Proc. IEEE INFOCOM* 2011.

[18] X. Wu and R. Srikant, Scheduling efficiency of distributed greedy scheduling algorithms in wireless networks, in *Proc. of IEEE INFOCOM* 2006.

[19] X. Wu, R. Srikant and J. R. Perkins, Queue-Length Stability of Maximal Greedy Schedules in Wireless

Networks, *IEEE Transactions on Mobile Computing* 6(6): 595–605, 2007.

[20] P.-J. Wan et al., Concise Conflict Graph of MC-MR Wireless Networks and Its Algorithmic Applications, manuscript, available at http://www.cs.iit.edu/~wan/ccg.pdf.

Appendix

In this appendix, we prove Lemma 4.5. Fix a regular point $t > 0$ with $\|\gamma(t)\|_\infty > 0$. We make the convention that the maximum of an empty set is zero. Define $\eta(t)$ as follows: if $A_1(t) = A_0(t)$,

$$\eta(t) = \|\gamma(t)\|_\infty - \max_{a \in A \setminus A_0(t)} \gamma_a(t);$$

otherwise,

$$\eta(t) = \max_{u \in A_0(t)} \gamma_a'(t) - \max_{u \notin A_0(t) \setminus A_1(t)} \gamma_a'(t).$$

Then, $\eta(t) > 0$.

LEMMA 8.1. *There exists $\delta > 0$ such that for any $s \in [t, t + \delta]$,*

$$\min_{a \in A_1(t)} \gamma_a(s) > \max_{a \in A \setminus A_1(t)} \gamma_a(s) + \frac{s-t}{2} \eta(t).$$

PROOF. We consider two cases.

Case 1: $A_1(t) = A_0(t)$. By the continuity of γ, there exists $\delta \in (0,1)$ such that for any $s \in [t, t + \delta]$,

$$\min_{a \in A_1(t)} \gamma_a(s)$$
$$= \min_{a \in A_0(t)} \gamma_a(s)$$
$$> \max_{a \in A \setminus A_0(t)} \gamma_a(s) + \frac{1}{2}\eta(t)$$
$$= \max_{a \in A \setminus A_1(t)} \gamma_a(s) + \frac{1}{2}\eta(t)$$
$$\geq \max_{a \in A \setminus A_1(t)} \gamma_a(s) + \frac{s-t}{2}\eta(t).$$

Case 2: $A_1(t) \neq A_0(t)$. There exists $\delta > 0$ such that for any $s \in (t, t + \delta]$,

$$\min_{a \in A_0(t)} \gamma_a(s) > \max_{a \in A \setminus A_0(t)} \gamma_a(s),$$
$$\max_{a \in A} \left| \frac{\gamma_a(s) - \gamma_a(t)}{s-t} - \gamma_a'(t) \right| < \frac{\eta(t)}{4}.$$

Fix an $s \in (t, t + \delta]$. Then,

$$\max_{a \in A \setminus A_1(t)} \gamma_a(s) = \max_{a \in A_0(t) \setminus A_1(t)} \gamma_a(s).$$

For any $a \in A_1(t)$ and any $b \in A_0(t) \setminus A_1(t)$, we have

$$\frac{\gamma_a(s) - \gamma_b(s)}{s-t}$$
$$= \frac{\gamma_a(s) - \gamma_a(t)}{s-t} - \frac{\gamma_b(s) - \gamma_b(t)}{s-t}$$
$$\geq \gamma_a'(t) - \gamma_b'(t) - \left| \frac{\gamma_a(s) - \gamma_a(t)}{s-t} - \gamma_a'(t) \right|$$
$$\quad - \left| \frac{\gamma_b(s) - \gamma_b(t)}{s-t} - \gamma_b'(t) \right|$$
$$> \eta(t) - \frac{\eta(t)}{4} - \frac{\eta(t)}{4}$$
$$= \frac{\eta(t)}{2},$$

which implies

$$\gamma_a(s) > \gamma_b(s) + \frac{s-t}{2}\eta(t).$$

Hence, for any $s \in (t, t + \delta]$,

$$\min_{k \in A_1(t)} \gamma_a(s)$$
$$> \max_{a \in A_0(t) \setminus A_1(t)} \gamma_a(s) + \frac{s-t}{2}\eta(t)$$
$$= \max_{a \in A \setminus A_1(t)} \gamma_a(s) + \frac{s-t}{2}\eta(t).$$

Note the above inequality holds trivially when $s = t$. Therefore, the lemma holds. $\square$

Now, we give the proof of the first part of Lemma 4.5 using Lemma 8.1. Consider any $a \in A_1(t)$ and any $\varepsilon > 0$. By Lemma 8.1, there exists $\delta > 0$ such that for any $s \in (t, t+\delta]$,

$$\min_{a \in A_1(t)} \gamma_a(s) > \max_{a \in A \setminus A_1(t)} \gamma_a(s),$$
$$\left\| \frac{\gamma(s) - \gamma(t)}{s-t} - \gamma'(t) \right\|_\infty < \varepsilon.$$

Thus, for any $s \in (t, t + \delta]$,

$$\frac{\|\gamma(s)\|_\infty - \|\gamma(t)\|_\infty}{s-t} - \gamma_a'(t)$$
$$= \frac{\max_{b \in A_1(t)} \gamma_b(s) - \|\gamma(t)\|_\infty}{s-t} - \gamma_a'(t)$$
$$= \max_{b \in A_1(t)} \frac{\gamma_b(s) - \|\gamma(t)\|_\infty}{s-t} - \gamma_a'(t)$$
$$= \max_{b \in A_1(t)} \left(\frac{\gamma_b(s) - \|\gamma(t)\|_\infty}{s-t} - \gamma_a'(t) \right)$$
$$= \max_{b \in A_1(t)} \left(\frac{\gamma_b(s) - \gamma_b(t)}{s-t} - \gamma_b'(t) \right),$$

which implies

$$\left| \frac{\|\gamma(s)\|_\infty - \|\gamma(t)\|_\infty}{s-t} - \gamma_a'(t) \right|$$
$$\leq \max_{b \in A_1(t)} \left| \frac{\gamma_b(s) - \gamma_b(t)}{s-t} - \gamma_b'(t) \right|$$
$$< \varepsilon.$$

Therefore,

$$\|\gamma(t)\|_\infty' = \gamma_a'(t),$$

and the first part of Lemma 4.5 holds.

We move on to prove the second part of Lemma 4.5. Consider any $\varepsilon \in (0,1)$. By Lemma 8.1, there exists $\delta > 0$ such that for any $s \in (t, t + \delta]$,

$$\min_{a \in A_1(t)} \gamma_a(s) > \max_{b \in A \setminus A_1(t)} \gamma_b(s) + \frac{s - t}{2} \eta(t),$$

and

$$\left\| \frac{\beta(s) - \beta(t)}{s - t} - \beta'(t) \right\|_\infty \leq \varepsilon.$$

Let $\varepsilon_1 = \varepsilon / \|\tau\|_\infty$. By the u.o.c. convergence, there exists a sufficiently large $n \in S'$ such that

$$n \geq \max \left\{ \frac{2}{\varepsilon_1 \delta}, \frac{8}{\varepsilon_1 \cdot \eta(t)} \|\tau\|_\infty \right\}$$

and for any $s \in [t, t + \delta]$,

$$\left\| \frac{y^{(n)}(ns)}{n} - \beta(s) \right\|_\infty < \frac{\varepsilon}{2} \delta$$

$$\left\| \frac{z^{(n)}(ns)}{n} - \gamma(s) \right\|_\infty < \frac{\varepsilon_1}{16} \delta \eta(t).$$

We make the following two claims.

CLAIM 8.2. $\left\| \overline{y}^{(n)}(n(t + \varepsilon_1 \delta), n(t + \delta)) - \beta'(t) \right\|_\infty \leq 3\varepsilon.$

PROOF. By the third part of Lemma 4.4,

$$\left\| \overline{y}^{(n)}(n(t + \varepsilon_1 \delta), n(t + \delta)) - \overline{y}^{(n)}(nt, n(t + \delta)) \right\|_\infty$$

$$\leq \varepsilon_1 \|\tau\|_\infty$$

$$= \varepsilon.$$

Thus, it is sufficient to show that

$$\left\| \overline{y}^{(n)}(nt, n(t + \delta)) - \beta'(t) \right\|_\infty \leq 2\varepsilon,$$

which will be proved subsequently. Since

$$\left\| \overline{y}^{(n)}(nt, n(t + \delta)) - \frac{\beta(t + \delta) - \beta(t)}{\delta} \right\|_\infty$$

$$= \left\| \frac{y^{(n)}(n(t + \delta)) - y^{(n)}(nt)}{n\delta} - \frac{\beta(t + \delta) - \beta(t)}{\delta} \right\|_\infty$$

$$\leq \frac{1}{\delta} \left\| \frac{y^{(n)}(n(t + \delta))}{n} - \beta(t + \delta) \right\|_\infty +$$

$$\frac{1}{\delta} \left\| \frac{y^{(n)}(nt)}{n} - \beta(t) \right\|_\infty$$

$$< \frac{1}{\delta} \cdot \frac{\varepsilon}{2} \delta + \frac{1}{\delta} \cdot \frac{\varepsilon}{2} \delta$$

$$= \varepsilon.$$

we have

$$\left\| \overline{y}^{(n)}(nt, n(t + \delta)) - \beta'(t) \right\|_\infty$$

$$\leq \left\| \overline{y}^{(n)}(nt, n(t + \delta)) - \frac{\beta(t + \delta) - \beta(t)}{\delta} \right\|_\infty$$

$$+ \left\| \frac{\beta(t + \delta) - \beta(t)}{\delta} - \beta'(t) \right\|_\infty$$

$$< 2\varepsilon.$$

Therefore, our claim holds. $\square$

CLAIM 8.3. $\overline{y}^{(n)}(n(t + \varepsilon_1 \delta), n(t + \delta)) \in \Phi_{A_1(t)}.$

PROOF. By the second part of Lemma 4.4, $\overline{y}^{(n)}(n(t + \varepsilon_1 \delta), n(t + \delta))$ is a convex combination of

$$\left\{ d^{(n)}(j) : \lfloor n(t + \varepsilon_1 \delta) \rfloor + 1 \leq j \leq \lceil n(t + \delta) \rceil \right\}.$$

Thus, it is sufficient to show that $d^{(n)}(j) \in M_{A_1(t)}$ for any integer j between $\lfloor n(t + \varepsilon_1 \delta) \rfloor + 1$ and $\lceil n(t + \delta) \rceil$.

We first show that for any $s \in [t + \varepsilon_1 \delta / 2, nt + \delta]$,

$$\min_{a \in A_1(t)} z_a^{(n)}(ns) > \max_{b \in A \setminus A_1(t)} z_b^{(n)}(ns) + \|\tau\|_\infty.$$

Consider any link $a \in A_1(t)$, and any link $b \in A \setminus A_1(t)$. Then,

$$\gamma_a(s) - \gamma_b(s) > \frac{s - t}{2} \eta(t) \geq \frac{\varepsilon_1 \delta}{4} \eta(t),$$

and

$$\left| \frac{z_a^{(n)}(ns) - z_b^{(n)}(ns)}{n} - (\gamma_a(s) - \gamma_b(s)) \right|$$

$$\leq \left| \frac{z_a^{(n)}(ns)}{n} - \gamma_a(s) \right| + \left| \frac{z_b^n(ns)}{n} - \gamma_b(s) \right|$$

$$< \frac{\varepsilon_1 \delta}{8} \eta(t).$$

Therefore,

$$\frac{z_a^{(n)}(ns) - z_b^{(n)}(ns)}{n}$$

$$> \gamma_a(s) - \gamma_b(s) - \frac{\varepsilon_1 \delta}{8} \eta(t)$$

$$> \frac{\varepsilon_1 \delta}{8} \eta(t),$$

which implies

$$z_a^{(n)}(ns) > z_b^{(n)}(ns) + n \frac{\varepsilon_1 \delta}{8} \eta(t)$$

$$\geq z_b^{(n)}(ns) + \|\tau\|_\infty.$$

So, the desired inequality holds.

Consider any integer j such that $j - 1 \in [n(t + \varepsilon_1 \delta / 2), n(t + \delta)]$. Then, at the end of the $(j - 1)$-th times-slot, the queue length of each link in $A_1(t)$ is at least $\|\tau\|_\infty$ and is larger than any link not in $A_1(t)$. Thus, $I^{(n)}(j)$ is $A_1(t)$-maximal, which implies that $d^{(n)}(j) \in \Phi_{A_1(t)}$. Since

$$\lfloor n(t + \varepsilon_1 \delta) \rfloor - n(t + \varepsilon_1 \delta / 2)$$

$$> n(t + \varepsilon_1 \delta) - 1 - n(t + \varepsilon_1 \delta / 2)$$

$$= n\varepsilon_1 \delta / 2 - 1 \geq 0$$

and

$$\lceil n(t + \delta) \rceil - 1 < n(t + \delta),$$

for each integer j between $\lfloor n(t + \varepsilon_1 \delta) \rfloor + 1$ and $\lceil n(t + \delta) \rceil$, we have $d^{(n)}(j) \in \Phi_{A_1(t)}$. $\square$

Since $\Phi_{A_1(t)}$ is compact, the above two claims together with the fact that ε can be chosen arbitrarily small imply the correctness of the second part of Lemma 4.5.

Enabling Real-Time Interference Alignment: Promises and Challenges

Kyle Miller, Atresh Sanne
Department of ECE
University of Texas, Austin
Austin, TX - 78712
kmil16@gmail.com
atresh.sanne@utexas.edu

Kannan Srinivasan
Department of CSE
The Ohio State University
Columbus, OH - 43210
kannan@cse.ohio-
state.edu

Sriram Vishwanath
Department of ECE
University of Texas, Austin
Austin, TX - 78712
sriram@ece.utexas.edu

ABSTRACT

As its name suggests, "interference alignment" is a class of transmission schemes that aligns multiple sources of interference to minimize its impact, thus aiming to maximize rate in an interference network. To our knowledge, this paper presents the first real-time implementation of interference alignment. Other implementation in the literature are either done offline or assume a backchannel between participating nodes to perform alignment. On the other hand, this paper presents a blind interference alignment scheme, one that does not require channel state information at the transmitters or the knowledge of other transmitters data or the knowledge of data between receivers and functions in real-time.

Categories and Subject Descriptors

C.2 [**Computer-Communication Networks**]: Network Protocols

General Terms

Interference Alignment

Keywords

Interference Management, Wireless Networks

1. INTRODUCTION

Interference alignment (IA), a fresh and exciting transmission strategy recently introduced as a mechanism for improving the performance of interference networks, is known to have tremendous gains over traditional orthogonal coding schemes in literature [1]. Specifically, alignment is shown to achieve the optimal degrees of freedom in time varying $K > 3$ user interference channels in [2], and X channels in [3]. Here, degrees of freedom denotes the ratio of the sum-rate achieved by the channel and the logarithm of the signal

We gratefully acknowledge the support of ONR under Award # N000141010337.

to noise ratio ($\log SNR$) as $SNR \to \infty$. Mathematically, this can be summarized as:

$$DoF = \lim_{SNR \to \infty} \frac{C(SNR)}{\log SNR}$$

where $C(SNR)$ is the sum-rate achieved using any transmission scheme. Using alignment, it is shown in [2] that a DoF of $K/2$ can be achieved for the K user time-varying interference channel. Using time-division, frequency division or other orthogonal schemes, the DoF is limited to 1. Similarly, alignment can improve the DoF of X channels from 1 to 4/3 [3]. The degrees of freedom achieved by alignment in these two cases can be shown to be optimal, i.e., that no other scheme exists that can achieve a higher DoF. Overall, in general, whenever there are multiple sources of interference, IA can potentially increase the achievable sum-rate in the network over traditional orthogonal transmissions schemes.

Ever since its discovery less than four years ago, there has been a rapid growth in literature associated with alignment, including the study of interference alignment for non-time varying (constant) channel gains [4], ergodic interference alignment [5], and most recently, lattice alignment schemes have been developed for finite SNR that achieve close to the performance of asymptotic alignment schemes presented in [6]. Typically, these alignment schemes are derived under many assumptions: including perfect and instantaneous knowledge of channel state and perfect synchronism between all parties. Fortunately, there are no additional requirements such as data cooperation between the nodes (as required by cooperative MIMO). However, the assumption of perfect and instantaneous channel state information (CSI) is very stringent and is very difficult to accomplish in practice.

In view of this, recent papers have focused on alignment under relatively relaxed assumptions on CSI. One body of work focuses on alignment in the presence of limited channel state feedback [7, 8]. The second body of work studies alignment based on delayed CSI [9]. Finally, the body of work most relevant to this paper accomplishes alignment *without* any CSI [10]. This blind interference alignment (blind IA) scheme is described in greater detail in Section 2.1 of the paper.

Given the excitement created by interference alignment, there has been some work at bringing it closer to practice. In [11], the authors implement a cancellation mechanism designed for a multiple input multiple output (MIMO) local area network (LAN), thus increasing the throughput of such a system. The scheme presented in [11] is novel, however, it requires considerable cooperation over an Ethernet backbone that goes beyond the needs of an alignment algorithm as studied in literature. Furthermore, their work does not address how to carry out channel estimation needed for interference alignment. In fact, their interference requires receiver channel knowledge at the transmitters. Furthermore, they do not study the effects of transmitter synchronization mismatch. Channel estimation and synchronization are critical for IA performance. In this paper, we look at these aspects.

1.1 Related Work

An implementation of alignment with limited feedback was studied in [12]. The authors implement an iterative alignment scheme to determine alignment vectors given limited channel state information. The results in [12] present improvements in performance in terms of throughput over orthogonal schemes. However, there are two issues with the work in [12]. First, the implementation effort is done offline; second, the iterative optimization scheme presented is an approximation and may not always match the globally optimal alignment scheme.

1.2 Our Contributions

Our paper is distinct from this existing body of work in that, to the best of our knowledge, it represents the *first real-time* implementation of an interference alignment scheme. Its only requirement is symbol-level time synchronism between transmitters and receivers. It *does not* require any back-end cooperation in either data or channel state, nor does it make any idealized assumptions on the wireless medium, estimation, feedback and other associated algorithms.

The rest of this paper is organized as follows: the next section motivates the need for IA and presents the blind IA scheme implemented in this paper. Section 3 presents the main challenges in accomplishing interference alignment in practice. Section 3.3 presents the frame format that enables blind alignment, while Section 4 provides a detailed description of the experimental setup. Finally, Section 5 presents the main results of this paper while Section 6 discusses its implications for networks.

2. BACKGROUND & MOTIVATION

The number and types of wireless-featured devices has grown tremendously placing a heavy demand on limited spectral resources. Interference between these devices can significantly degrade performance and reduce the throughput per user. Therefore, the key to improving the performance of wireless systems lies in managing interference effectively. Interference alignment accomplishes this goal. Fundamentally, the concept of alignment is to extend modulated symbols to a higher dimensional space such that the interfering signals are all aligned and occupy the smallest subspace at each receiver. In this case, the remaining dimensions can be used to receive the desired signal essentially free of interference. Note that IA is not the same as spread spectrum, multiple access schemes or other interference minimization or avoidance schemes. IA makes no attempt to avoid, cancel or minimize interference. Instead, it aims to align the interference along dimensions that are different from that of the signal. As such, it has been found to perform much better than all the existing interference management schemes.

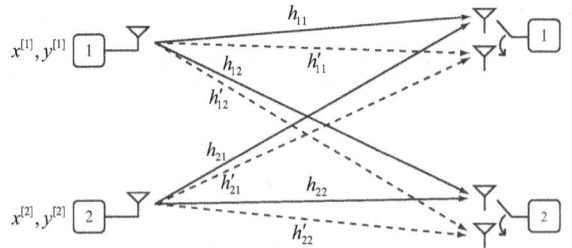

Figure 1: Blind Interference Alignment 2x2 system. Each receiver has 2 antennas to switch between.

As mentioned in the introduction, IA can achieve a DoF that increases linearly with the number of users in interference systems [1]. Note that this is quite amazing, as traditional time or frequency division multiple access schemes have no increase in DoF with users, thus each user only obtains a fraction of the total throughput which decreases with increasing number of users in the system. It is important to emphasize that this linear growth result is a theoretical one, and significant work needs to be done to bring the growing body of literature to practice.

2.1 Basic Principle

As discussed in the introduction, typical IA schemes require perfect global channel knowledge (i.e., perfect knowledge of all the channel coefficients in the system) at all the transmitters and receivers. But, this requirement is impractical in most systems. In this paper, we consider the blind (channel-state unaware) alignment scheme first presented in [10]. In this section, we provide a short description of the theory behind the blind IA technique, while challenges in practical implementation are detailed later. In our experiment setup, we consider two transmitters (TX1 and TX2) and two receivers (RX1 and RX2) as shown in Figure 1. Each transmitter $i = 1; 2$ has two messages $x^{[i]}; y^{[i]} \in \mathbb{C}$ which need to be recovered at RX1 and RX2 respectively. This is commonly referred to as the X-channel. Although the blind IA scheme is much more general, we focus our attention below to the specific 2-transmitter 2-receiver (referred to as 2x2 IA) setup for the sake of illustration. Our testbed will be more generic and allow for a higher order IA module.

In the 2x2 blind IA scheme, each receiver is equipped with a reconfigurable antenna which can switch between two different receive modes, but the receivers have only one RF chain. The receivers switch modes based on a predetermined pattern known to everyone. For $i; j \in 1; 2$, let $h_{ij} \in \mathbb{C}$ and $h'_{ij} \in \mathbb{C}$ denote the coefficient of the channel between transmitter i and receiver j with the receive mode set to 1 and 2 respectively. The channel coefficients have a generic (nondegenerate) continuous probability distribution. It is assumed that the channel stays constant across a supersymbol, which constitutes three transmission slots.

At time instant t, suppose that $u^{[1]}(t); u^{[2]}(t) \in \mathbb{C}$ are transmitted from TX1 and TX2 respectively. In this case, the received signal at user j using receive mode 1 is given by

$$z^{[j]}(t) = h_{1j}u^{[1]}(t) + h_{2j}u^{[2]}(t) + w^{[j]}(t), \qquad (1)$$

where $w^{[j]}(t) \sim \mathbb{CN}(0, \sigma^2)$ is the additive white Gaussian

Time Slot	1	2	3
Transmit	$x^{[i]} + y^{[i]}$	$x^{[i]}$	$y^{[i]}$
RX1	mode 1	mode 2	mode1
RX2	mode 1	mode 1	mode2

Table 1: Transmissions and receiver modes across 3 timeslots: mode corresponds to the antenna to which a receiver switches.

noise (AWGN). The received signal using mode 2 is obtained analogously using the channel coefficients h'_{1j} and h'_{2j} in Equation 1. The transmitters have absolutely no knowledge of the channel state. On the other hand, each receiver j is assumed to know only the local channel coefficients h_{1j}, h_{2j}, h'_{1j} and h'_{2j}. Note that for many physical layer techniques these channel estimates are available for every received packet. For example, Orthogonal Frequency Division Multiplexing (OFDM) systems use known pilot subcarriers to compute channel estimates that they later use to equalize for the channel impairments. Therefore, this requirement on the receivers is very practical.

Recall that transmitter i has two messages $x^{[i]}, y^{[i]} \in \mathbb{C}$ intended for RX1 and RX2 respectively. Assuming perfect synchronization between the transmitters and receivers, the blind IA scheme operates in 3 time slots as shown in Table 1.

In time slot 1, each transmitter sends the sum of its two messages, while in timeslots 2 and 3 the messages for RX1 and RX2 are transmitted separately. The receivers switch between the two receive modes according to the pattern shown in Table 1. Due to the symmetry, the recovery process is analogous at the two receivers, and so we focus on RX1 below. The received signal $z^{[1]}(t)$ at RX1 over the three time slots, assuming the channel remains constant, is given by

$$
\begin{aligned}
z^{[1]}(1) &= h_{11}(x^{[1]} + y^{[1]}) + h_{21}(x^{[2]} + y^{[2]}) + w^{[1]}; \\
z^{[1]}(2) &= h'_{11}x^{[1]} + h'_{21}x^{[2]} + w^{[1]}; \\
z^{[1]}(3) &= h_{11}y^{[1]} + h_{21}y^{[2]} + w_{[1]};
\end{aligned} \quad (2)
$$

Note that we have used the fact that RX1 switches to mode 2 in the second time slot. We now take advantage of the particular manner in which the interfering and desired signals are aligned in order to remove (i.e., zero-force) the interference and obtain the following system of equations:

$$
\begin{pmatrix} z^{[1]}(1) - z^{[1]}(3) \\ z^{[1]}(2) \end{pmatrix} = \begin{pmatrix} h_{11} & h_{21} \\ h'_{11} & h'_{21} \end{pmatrix} \begin{pmatrix} x^{[1]} \\ x^{[2]} \end{pmatrix} \quad (3)
$$
$$
+ \begin{pmatrix} w^{[1]}(1) - w^{[1]}(3) \\ w^{[1]}(2) \end{pmatrix}
$$

Since the channel coefficients are chosen from a continuous distribution, the channel vectors involved in Equation (3) are linearly independent with high probability. Therefore, RX1 can recover the desired messages $x^{[1]}$ and $x^{[2]}$.

A high- signal to noise ratio (SNR) performance metric of wireless networks is its degrees of freedom (DoF). DoF is defined as $lim_{SNR\to\infty} C_{sum}(SNR)/logSNR$, where $C_{sum}(SNR)$ denotes the maximum sum throughput achievable in the network. Thus, DoF denotes the asymptotic

growth rate of throughput with SNR. Using the alignment scheme described above, we find that we can achieve a DoF of $\frac{4}{3}$. It has been shown that this is, in fact, the optimal DoF for this channel [3]. Thus, alignment promises in theory to achieve rates that are significantly larger than other schemes in existence.

In this paper, we explore the implementation of this promising theoretic idea. In our implementation, we look at a generic interference alignment framework; the 2x2 system discussed above is just to showcase the benefits and powerfulness of interference alignment. Here, we wish to highlight alignment, a physical layer coding scheme. Therefore, traffic models, including burstiness, are not currently incorporated into our framework. With different loads and varying traffic patterns, we anticipate different subsets of users to be chosen across time for the alignment procedure, where users participating in the alignment algorithm have non-empty queues. In the following section, we list the challenges in this implementation effort and discuss how our implementation overcomes them.

3. BLIND IA - IMPLEMENTATION CHALLENGES

Although blind interference alignment is considerably less complex in terms of the control and coordination needed between nodes in a network, it still presents implementation challenges inherent to any interference alignment technique. In this section, we go over these challenges and then present our plans to address them.

3.1 Implementation Assumptions

The first implicit assumption made in any interference alignment system, including blind alignment, is that transmitted symbols are synchronized. It is important to point out that this does not require carrier-level synchronism or sub-microsecond precision in synchronism. Even though it is relatively coarse, synchronization is essential - the lack of synchronism can lead to undesirable combinations at each receiver, significantly impacting the ability to align interference.

The second assumption is that each of the receivers can estimate the channel between them and each of the transmitters. This receiver side-information is a standard feature of any coherent communication system.

The third assumption is that each receiver can switch between antennas in real-time, i.e., choose between different receive 'modes'. As in the case of channel estimation, this antenna-switching is a relatively standard feature of multiple antenna systems, although implementing it in a testbed in real-time can be considerably involved as described further in Section 4. Note that asynchrony in the transmitted symbols can also impact the receiver's ability to switch between modes - the receivers rely on symbol boundaries to decide which antenna to switch to.

Finally, the channel is assumed to have a larger coherence time than the rate of antenna switching to be performed. Specifically, we assume that the channel gains between a particular transmit antenna-receive antenna pair remain fixed for a duration longer than the duration needed to affect alignment in interference. For example, the channel gains in Equation (3) remain the same in time-slots 1 and 3. This is, arguably, the most non-trivial assumption

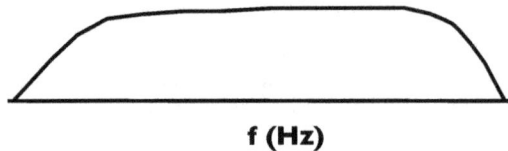

(a) A Non-OFDM System

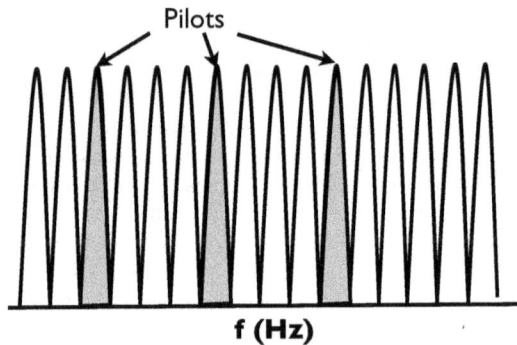

(b) An OFDM System

Figure 2: OFDM aids alignment by turning a frequency selective medium into a flat fading one per sub-carrier, easing symbol-level synchronism and channel estimation.

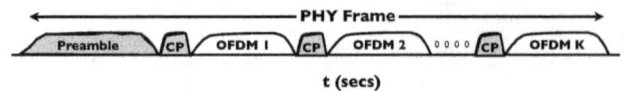

Figure 3: The physical layer frame showing multiple OFDM symbols. It starts with a preamble followed by OFDM symbols interspersed with cyclic prefixes.

necessary for blind alignment. As we find in later sections, this assumption is valid for our test-setup. In general, such an assumption would hold true when the channel coherent time is much larger than symbol duration, which is typically the case for a majority of wireless systems.

3.2 Implementation

In this paper, we work with Orthogonal Frequency Division Multiplexing (OFDM) based physical layer. This is a natural choice for both our testbed and most existing and emerging wireless standards. There are multiple advantages to using an OFDM framework for our system. First, it enables us to transform the original frequency selective wireless channel into multiple flat-fading equivalents. This allows for the blind alignment technique as described in Section 2.1 to be utilized as is within each sub-band. Moreover, the duration of an OFDM symbol is considerably longer than duration of a symbol in a non-OFDM system. This larger symbol duration in an OFDM system helps improve error tolerance on the otherwise stringent requirement of symbol-level synchronization imposed by interference alignment. Finally, OFDM systems typically assign specific sub-carriers to pilot symbols. Figure 2(b) illustrates such an assignment of pilots to subcarriers. These pilots are typically used to estimate the channel at the receiver. This channel estimation plays a critical role in the design of the receive-chain at each receiver as well as in the blind alignment algorithm used in this paper.

An OFDM transmission typically includes a cyclic prefix (CP) before every OFDM symbol. For a tutorial on details underlying the physical-layer frame structure of OFDM transmission, see [13]. Figure 3 shows a physical layer frame with multiple OFDM symbols separated by CPs. This CP proves useful in facilitating antenna switching in our blind alignment scheme. When an antenna switching process is

initiated, a transient oscillation is observed in the received signal which stabilizes on the timescale of a few microseconds. The CP offers an excellent opportunity during which the switch between modes can be performed without impacting the successful reception of the OFDM symbol that follows the CP.

The preamble in every frame is conventionally used by the receivers to carry out time synchronization and frequency offset correction. To enable IA, the preamble can also be used to synchronize all transmitters in the system. For example, to enable synchronization, a single transmitter is chosen at any one time to send a preamble while the remainder of the transmitters capture and use this preamble to synchronize transmission. Note that many other mechanisms for achieving symbol-level synchronism exist in literature [14, 13], and that the mechanism described here is fairly simple but effective in satisfying the requirements for symbol-synchronization for IA algorithms.

3.3 Frame Format

The frame format for our setup consists of a preamble, pilots, and payload data symbols. The preamble consists of 2 OFDM symbols which are used in frame detection and frequency offset correction using the Schmidl-Cox algorithm [15]. The pilots are used for channel equalization when decoding the data symbols. In our implementation of the X-channel, both transmitters transmit their respective pilots simultaneously in the same OFDM symbol, but on alternating subcarriers. Transmitter 1 sends pilots on all odd subcarriers while transmitter 2 sends pilots on all even subcarriers. The orthogonal nature of the pilot transmission scheme allows the receiver to extract the channel state of both transmitters from a single OFDM symbol. Figure 4 shows the pilot to subcarrier mapping of the transmitted and received symbols. Our choice of pilots may impose some constraints on the scalability of the algorithm. However, scaling may be accomplished by determining suitable subsets of users that align their interference. Within those subsets, the pilots need not be scaled. Furthermore, alignment as structured in our paper is designed for environments with large coherence times. In such environments, the pilots consume a smaller fraction of resources compared to the case with smaller coherence times.

3.3.1 Frame Structure

Initially, we consider a somewhat naive frame format that is overhead-intensive, but constitutes a great starting point for our implementation. This format is shown in Fig. 5. This format is useful when each receiver possesses two receive chains. With two receive chains, all channel estimates can be obtained within one OFDM timeslot. This simple frame is sufficient to perform blind IA. However, it is highly inefficient, as the overhead incurred is substantial. An al-

	TX 1 Transmitted Pilots	TX 2 Transmitted Pilots	Received Pilots
	TX1 pilot 1	Null	TX1 pilot 1
	Null	TX2 pilot 1	TX2 pilot 1
	TX1 pilot 2	Null	TX1 pilot 2
	Null	TX2 pilot 2	TX2 pilot 2
	...	...	...
	TX1 pilot N	Null	TX1 pilot N
	Null	TX2 pilot N	TX2 pilot N

Figure 4: Pilot to subcarrier mapping

ternate frame structure that minimizes overhead is highly desirable, which is presented next.

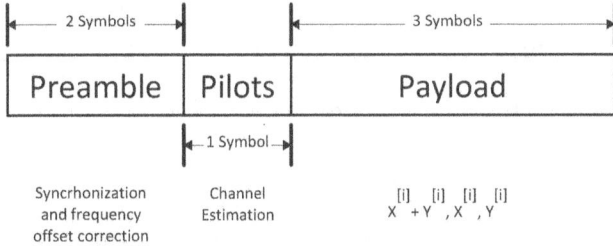

Figure 5: The first frame structure we consider to enable alignment. Here, there is a preamble and pilot for every 3 symbols, which is clearly in excess of what is needed.

3.3.2 Alternate Frame Structure

An alternate frame structure for IA includes added pilot symbols before each payload block and is shown in Figure 6. A payload block consists of a variable number of OFDM symbols of the same type. The first payload block consists of N OFDM symbols of the type $x[i] + y[i]$. The second and third payload blocks consist of N symbols of type $x[i]$ and N symbols of type $y[i]$, respectively. By grouping these symbols together in chunks, we can use the same preamble and pilots for more data, thus decreasing the frame overhead. This can increase efficiency so long as the channel coherency is maintained throughout the frame. Note that it is only necessary to have two pilot symbols per frame while our approach has three. This is because, an IA receiver only needs to get the channel estimates twice – one from each antenna. The third (optional) pilot symbol may be used to verify if the three payload pieces are still within channel coherence time.

3.4 Channel Coherence Time

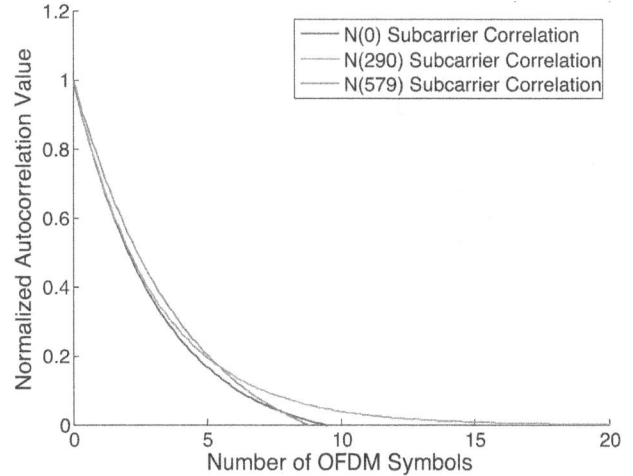

Figure 7: The channel coherence time in our environment. Showing only the phse of auto-correlation across multiple sub-carriers. The magnitude of autocorrelation (not shown here) had a coherence time of several hundreds of OFDM symbols.

As noted earlier, a key assumption made for blind IA operation is that the channel coherence time is much larger than a single OFDM symbol. To quantify coherence time, we ran an experiment in which a transmitter sends a series of OFDM symbols following a preamble. The preamble helps a receiver with synchronization and frequency offset correction. From the following OFDM symbols, a receiver computes for how long its channel to its transmitter remains correlated. We crunched this quantity for many sub-carriers over several experimentation runs. Figure 7 shows the autocorrelation across multiple sub-carriers for a transmitter-receiver pair. This result is representative of all of the other experiments. It shows only the phase of the auto-correlation and not the magnitude. The magnitude of autocorrelation showed correlation over nearly 600 OFDM symbols. The phase, however, as shown in the figure is still correlated over many multiple OFDM symbols before it becomes uncorrelated (with an autocorrelation of 0). The number of symbols in the case of phase varies from 8 to 20 depending on the sub-carriers. However, on average, the coherence time across all sub-carriers was around 12 OFDM symbols. For this reason, in the rest of the experiments, we restrict the number of OFDM symbols per payload (N in Figure 5) to 3. Thus, we have 3 pilot symbols and 9 (3x3) payload symbols equaling 12 OFDM symbols per physical layer frame. Note that IA does not require preamble to have the same channel realization as the rest of the frame.

4. EXPERIMENT DESCRIPTION

The hardware used in this experiment is the National Instruments' Software-Defined Radio (SDR) platform. This consists of a PXIe-8130 real-time controller, PXIe-7965R FlexRIO FPGA module with Xilinx Virtex 5 SX95T, NI-5781 100M samples/sec baseband transceiver, and an Ettus Research XCVR2450 daughterboard. A PXIe-1082 8-slot chassis with a high speed PXI-Express backplane bus can

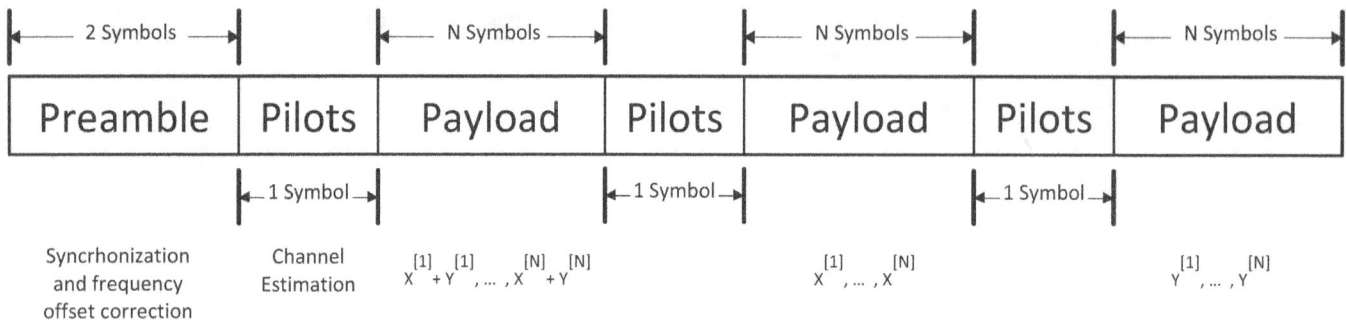

| ← 2 Symbols → | ← N Symbols → | ← N Symbols → | ← N Symbols → |

| Preamble | Pilots | Payload | Pilots | Payload | Pilots | Payload |

← 1 Symbol →   ← 1 Symbol →   ← 1 Symbol →

Synchronization and frequency offset correction   Channel Estimation $X^{[1]} + Y^{[1]}, ..., X^{[N]} + Y^{[N]}$ $X^{[1]}, ..., X^{[N]}$ $Y^{[1]}, ..., Y^{[N]}$

Figure 6: The efficient frame structure we use to enable blind IA

be mounted with up to 8 PXIe cards. Each module consisting of these components provides for a receiver or a transmitter. Figure 8 is a picture of the experimental hardware. The three white modules are the PXIe-1082 chassis, which are TX 1 & 2, RX1, and RX2 (going from the right to left). Each module is connected to an Ettus Research XCVR2450 daughterboard (two daughterboards for the transmit module) through SPI lines. All antennas are spaced such that.

The modulation scheme for this experiment is OFDM modulation with QPSK signaling. The operating frequencies are in the 2.4 GHz ISM band with a center carrier frequency of 2.437 GHz. Our OFDM modulation consists of subcarrier spacings of 15 kHz. Each symbol lasts $\frac{1}{12}$ ms in time, including a cyclic prefix length of 16.67 μs. These parameters are chosen based on the LTE standard. The 12 carriers around DC are nulled resulting in $N' = N - 12$. Furthermore, the $N = 600$ subcarriers are assigned to 580 data subcarriers (Nd) and 20 frequency pilots (Np). These frequency pilots are used to correct for a residual frequency offset.

Three PXIe-1082 chassis modules are employed in this experiment. One of these modules is fitted with two FlexRIO SDRs to control two transmitters TX1 and TX2. The remaining two modules function as receivers RX1 and RX2. Each receiver chassis has two receive modes provided by the Ettus XCVR2450 transceivers. The three chassis are connected through a local area network to development computers. LabVIEW, National Instruments graphical programming environment, is used to design and program the real-time controller. The FlexRIO FPGA module extends the LabVIEW to develop for the FPGAs in our experiment.

4.1 Transmitter

A block diagram of the transmitter is shown in Figure 9. Known psuedo-random bitstreams of length $2N'$ are extended to generate $2N$ bits. These bits are modulated with QPSK signaling to generate two streams of N samples $\boldsymbol{x}^{[i]} = (x_1^{[i]}, ..., x_N^{[i]})$ and $\boldsymbol{y}^{[i]} = (y_1^{[i]}, ..., y_N^{[i]})$. These samples are then precoded into the data symbols required for IA. The result in a payload of $3N$ samples $\boldsymbol{x}^{[i]} + \boldsymbol{y}^{[i]}$, $\boldsymbol{x}^{[i]}$, $\boldsymbol{y}^{[i]}$. Here $2N$ samples of preamble and N samples of pilots are inserted between each data sample to produce a $12N$ sample frame. We note that the second and third preambles are not neccesary to decode IA symbols. They are used for debugging purposes and are not included in the results obtained. Finally, the samples are OFDM modulated: the second and third preambles are modulated with a 3854 length and appended with a 1024 length cyclic prefix. This is done so it is not detected by the FPGA correlation algorithm. The final frame structure is 12 OFDM symbols in length as shown in Figure 6. TX1 and TX2 transmit their corresponding frames in a synchronous pattern.

4.2 Receiver

A block diagram of the reciever is shown in Figure 9. The reciever begins by aquiring two successive frames of samples. The beginning of the frame is detected by correlating over the known preamble. The preamble is also used to correct the frequency offset using the Schmidl-Cox algorithm [15]. The preambles are then removed and the rest of the frame is demodulated to result in $3N'$ payload samples and the 1 pilot symbol for each timeslot (3 total pilot symbols). Channel estimation is performed by selecting the appropriate pilot symbols depending on which reciever is operating. IA decoding then solves equation (2) to recover the $2N'$ payload data. Finally the aquired channel coefficients are used in ML detection to recover the transmitted bits.

4.3 FPGA Correlation Method and Antenna Switching

As mentioned earlier, for Blind IA to function, a receiver should be able to switch between its available antennas in

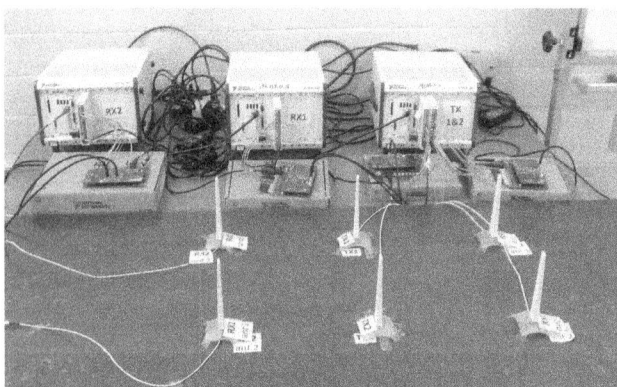

Figure 8: National Instruments SDR platform controls two transmitters (middle) and two recievers, each with two recieve modes (right and left antennas).

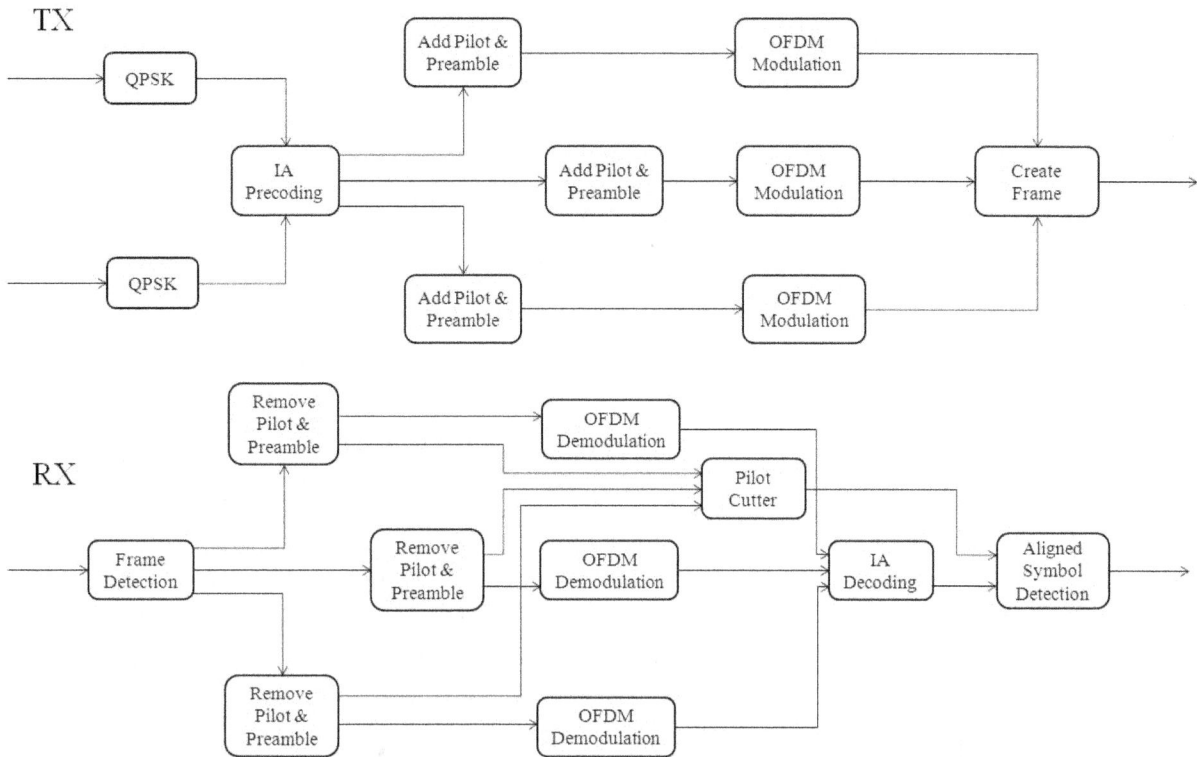

Figure 9: Block diagram of the transmitter (top) and receiver (bottom).

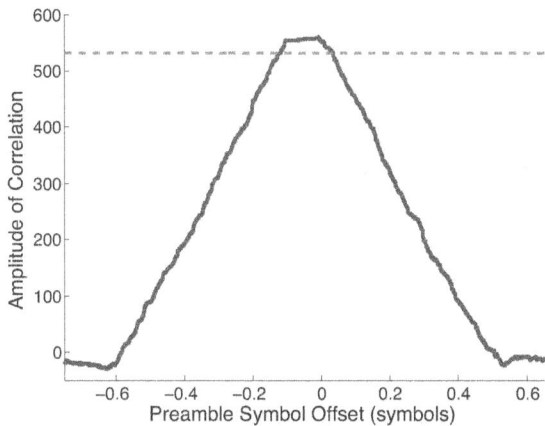

Figure 10: Symbol timing estimation performed on the FPGA. The horizontal dotted line is the user-specified threshold value for detecting peaks. The threshold value tracks the highest amplitude peak.

real-time. The accuracy of antenna switching is entirely dependent on OFDM symbol timing estimation on the FPGA. In this real-time implementation, we move the Schmidl-Cox symbol timing estimation [15] to high speed FPGAs. The challenges of moving this algorithm to FPGAs are the inherent computational restrictions imposed by the FPGA. As a result, FPGA correlation is sensitive to channel variations and SNR because we cannot normalize our estimated timing. Our method employs a variable amplitude correlation with a threshold peak detection. The largest peak over time is tracked and any amplitude that exceeds a threshold percentage of the largest peak is determined to be that frame start. A fluctuation in the channel will irreversibly spike the threshold value to a false peak, causing future preambles to evade detection. Figure 10 is a correlation plot of two identical preamble symbols with a 512 length cyclic prefix. Notice this correlation is not normalized due to previously mentioned FPGA processing constrictions.

Once the preamble is detected, the precise switching time is simply a known offset from the beginning of the frame. The FPGA issues commands to the baseband transceiver through SPI interfacing on a 40MHz clock. The antenna switching command can now be issued through a finite state machine that is configured for SPI protocol. This process automates antenna switching, making it invisible to the reciever. Recieved frames now have the proper antenna switches performed and IA decoding can run independently.

5. RESULTS

This section presents the bit error rate and throughput performance results of our blind IA system compared to a

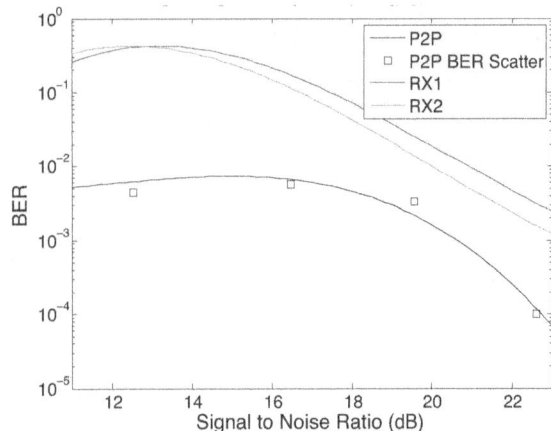

Figure 11: Signal to noise ratio (SNR) vs bit error rate (BER) performance of IA and time division point-to-point (P2P) systems. BER for IA is always worse than P2P.

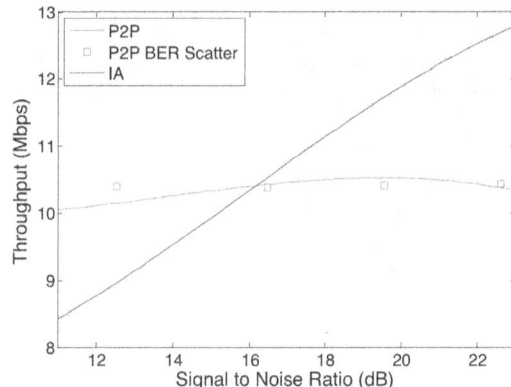

Figure 12: Signal to noise ratio (SNR) vs throughput performance of IA and point-to-point systems. Throughput of IA is worse than that of point-to-point up to an SNR of 16dB. In the very high SNR region (beyond 16dB), however, IA's throughput is better than P2P.

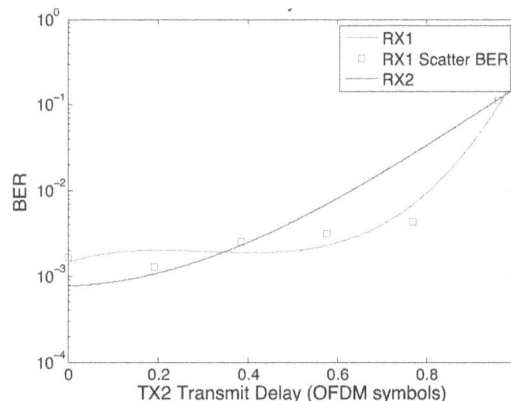

Figure 13: Signal to noise ratio (SNR) vs bit error rate (BER) performance of IA and time division systems. BER of IA is better than time division at very low and very high SNR regions. In the low and high SNR regions, pilots from two transmitters slightly improve BER performance for IA.

time division point-to-point system. It also presents the results for IA when the transmitters are not perfectly synchronized. Overall, when the two transmitters are perfectly synchronized, IA delivers superior throughput when the SNRs to the receivers are very high. As the transmitters get out-of-sync, however, IA's BER performance remains consistent up to a certain value and deteriorates sharply after that. This shows that IA is a promising strategy even in practice.

5.1 Bit Error Rate (BER) Performance

Figure 11 shows the bit error rate (BER) performance of IA and time division point-to-point (P2P) systems. BER for IA is worse than that for time division across the whole range of SNR. Note that, in our IA system, the two transmitters send the same preamble. Therefore, SNR for just the preamble is twice as high as it is for the corresponding time division P2P system. However, the rest of the frame contains symbols from both transmitters at similar power levels. Therefore, the rest of the frame would have a very low signal to interference and noise ratio (SINR). However, in the time division scenario, the SNR remains the same across the entire frame. Therefore, to illustrate the results for the same setup, our plots show the SNR for the point-to-point time division system for the same transmit power levels as in the IA case.

5.2 Throughput Performance

The measured throughput is at the link level for both IA and P2P systems. We calculated throughput as the amount of data bits recieved per frame (Figure 6), or packet, as bounded by the sampling rate $30.72 \frac{Msamples}{sec}$. The total amount of transmitted bits per second were then multiplied by the successful recovery probability, $1 - BER$, to compute the final aggregate throuput in bps. Additionally, we assume an infinite input buffer for our input traffic, and we focus on physical layer performance. Thus, we do not retransmit any packets. Therefore, there is no ARQ/HARQ in our setup. The BER and throughput values are based on the raw bits that are transmitted and received.

Figure 12 shows the aggregate (or sum) throughput results for both the time division P2P and IA systems. It

shows that the sum throughput for IA is worse than point-to-point up to an SNR of 16dB. Beyond that, however, IA starts to outperform P2P. IA achieves a throughput gain of close to 1.3 when the SNR hits 23dB. Note that this is very close to the maximum gain (its DoF corresponds to $\frac{4}{3} = 1.33$) for a 2x2 IA scheme over time-division point-to-point system. It is important to note that the 1.33x gain is under information-theoretic degrees of freedom calculations; and that in practice, the MAC and other practical coding and signaling aspects of the system impact both the non-aligned and aligned systems, reducing the performance below the information theoretic optimum.

5.3 Synchronization Mismatch

One of the aspects that is critical for IA operation is the synchronization of symbols across transmitters. Figure 13 shows the synchronization mismatch between the two transmitters. The x-axis corresponds to the extent to which transmitter 2's transmission is delayed. This delay (or mismatch) is given in terms of an OFDM symbol, which is about $\frac{1}{12}$ msec. It shows that IA's BER performance worsens as the mismatch increases. A mismatch equalling 1 OFDM symbol causes IA to completely fail. However, a small asynchronism in the order of a cyclic prefix (of 0.2 OFDM symbol) corresponds to a reasonable receiver BER. This is a mismatch of roughly 17usecs for our system. This shows that IA's performance is fairly robust to small mismatches.

5.4 Higher-Order Blind IA

This paper focuses its efforts on implementing a 2x2 blind IA system. It shows that the throughput gain promised in theory is achievable in practice using blind IA in a 2x2 system. For a higher-order (3x3 and beyond) IA system, however, the gains from IA will be higher and will increase with the order [3]. Our current RF front-end only supports switching between 2 antennas. Due to this hardware limitation, we are unable to demonstrate throughput gains for higher-order blind IA systems. However, our physical layer design including the frame format can be easily extended to a higher-order blind IA set up.

6. DISCUSSION

This section discusses the implications of employing interference alignment (IA) in existing networks and network architectures.

Implications to Access Networks: In an access network, multiple wireless clients connect to the Internet through a wireless Access Point (AP). As an AP serves multiple clients, a single AP can serve the purpose of multiple transmitters in an IA system. For example, in our 2x2 setup, an AP could perform the duties of two transmitters and its two clients could serve as its two receivers. By having two transmitters combined into a single AP, issues such as transmitter synchronization would disappear.

Higher Layer Implications: Interference alignment (IA) is a cooperative technique in which multiple transmitters send data to multiple receivers simultaneously. This is unlike MIMO, where only one transmitter is sending data to one receiver using multiple antennas. A MIMO-like point-to-point system works without any modifications to the current wireless network architecture. However, since IA requires synchronization among multiple nodes, the current network architecture needs modifications to take full advantage of the throughput gains from IA.

Channel Estimation: In IA, every receiver needs to be able to estimate the channel between itself and every IA transmitter. To achieve this, different transmitters may be assigned different subcarriers for pilots and when the pilots are transmitted no other transmitter may use that subcarrier to send anything (as discussed in Section 3.3). Thus, the number of pilots for IA increases linearly with the number of transmitters. Such a scalability requirement is inherent to any technique that needs coordination. However, this overhead can be made considerably small when channel coherence is very large (compared to an OFDM frame) or by adjusting the number of IA transmitter-receiver pairs.

MAC Layer Functionality: In a traditional wireless network, a node is responsible for doing carrier sensing before it transmits any data to avoid collisions. In an IA network, the responsibility is on all the nodes that participate in an IA session. Note that not all the nodes in a network may participate in every IA transmission. IA promises orders of magnitude throughput gains only when all of the participating receivers have excellent signal to noise ratios (SNRs) from all of the participating transmitters. Therefore, a network may need to be split into several sets. Each set still needs to do carrier sensing before it chooses to transmit a packet, to avoid collisions. So, with IA, nodes within a set share the responsibility of channel sensing. To save energy, only a subset of nodes could participate in channel sensing.

Network Layer and Above: Note that, in IA, every transmitter can potentially send packets to multiple receivers simultaneously. In a mesh network, where multiple flows crisscross, the crisscross point becomes the bottleneck for all the flows. IA may be used at the crisscross point to mitigate such bottlenecks.

Acknowledgements

We would like to thank Prof. Syed Jafar, Prasanth Anthapadmanabhan and Saul Duran for their discussions. We would also like to thank our shepherd, Supratim Deb.

7. REFERENCES

[1] S. Jafar, "Interference alignment: A new look at signal dimensions in a communication networks," *Foundations and Trends® in Communications and Information Theory*, vol. 7, no. 1, pp. 1–136.

[2] V. Cadambe and S. Jafar, "Interference alignment and degrees of freedom of the-user interference channel," *Information Theory, IEEE Transactions on*, vol. 54, no. 8, pp. 3425–3441, 2008.

[3] M. Maddah-Ali, A. Motahari, and A. Khandani, "Communication over MIMO X channels: Interference alignment, decomposition, and performance analysis," *Information Theory, IEEE Transactions on*, vol. 54, 2008.

[4] A. Motahari, S. Gharan, and A. Khandani, "Real interference alignment with real numbers," *Arxiv preprint arxiv:0908.1208*, 2009.

[5] B. Nazer, S. Jafar, M. Gastpar, and S. Vishwanath, "Ergodic interference alignment," in *Information Theory, 2009. ISIT 2009. IEEE International Symposium on*. IEEE, 2009, pp. 1769–1773.

[6] A. Jafarian, J. Jose, and S. Vishwanath, "Algebraic lattice alignment for K-user interference channels," in *Communication, Control, and Computing, 2009. Allerton 2009. 47th Annual Allerton Conference on*. IEEE, 2009, pp. 88–93.

[7] R. Krishnamachari and M. Varanasi, "Interference alignment under limited feedback for MIMO interference channels," in *Information Theory Proceedings (ISIT), 2010 IEEE International Symposium on*. IEEE, 2010, pp. 619–623.

[8] H. Bolcskei and I. Thukral, "Interference alignment with limited feedback," in *Information Theory, 2009. ISIT 2009. IEEE International Symposium on*. IEEE, 2009, pp. 1759–1763.

[9] M. Maddah-Ali and D. Tse, "On the degrees of freedom of MISO broadcast channels with delayed feedback," *EECS Department, University of California, Berkeley, Tech. Rep. UCB/EECS-2010-122, Sep*, pp. 2010–122, 2010.

[10] T. Gou, C. Wang, and S. Jafar, "Aiming perfectly in the dark-blind interference alignment through staggered antenna switching," in *GLOBECOM 2010, 2010 IEEE Global Telecommunications Conference*. IEEE, 2010, pp. 1–5.

[11] S. Gollakota, S. Perli, and D. Katabi, "Interference alignment and cancellation," in *ACM SIGCOMM Computer Communication Review*, vol. 39, no. 4. ACM, 2009, pp. 159–170.

[12] S. Peters and R. Heath, "Interference alignment via alternating minimization," in *Acoustics, Speech and Signal Processing, 2009. ICASSP 2009. IEEE International Conference on*. IEEE, 2009, pp. 2445–2448.

[13] A. B. Narasimhamurthy, C. Tepedelenlioğlu, and M. K. Banavar, *OFDM Systems for Wireless Communications*. Morgan and Claypool Publishers, 2010.

[14] D. Lee and K. Cheun, "Coarse symbol synchronization algorithms for OFDM systems in multipath channels," *Communications Letters, IEEE*, vol. 6, no. 10, pp. 446–448, 2002.

[15] T. Schmidl and D. Cox, "Robust frequency and timing synchronization for OFDM," *Communications, IEEE Transactions on*, vol. 45, no. 12, pp. 1613–1621, 1997.

Optimization Schemes for Protective Jamming

Swaminathan Sankararaman[*]
Department of Computer Science
Duke University
swami@cs.duke.edu

Karim Abu-Affash
Department of Computer Science
Ben Gurion University
abuaffas@cs.bgu.ac.il

Alon Efrat
Department of Computer Science
The University of Arizona
alon@cs.arizona.edu

Sylvester David Eriksson-Bique
CS Department
University of Helsinki
sylvester.eriksson-bique@helsinki.fi

Valentin Polishchuk
CS Department
University of Helsinki
valentin.polishchuk@helsinki.fi

Srinivasan Ramasubramanian
Department of Electrical and Computer Engineering
The University of Arizona
srini@ece.arizona.edu

Michael Segal
Department of Communication Systems Engineering
Ben Gurion University
segal@cse.bgu.ac.il

ABSTRACT

In this paper, we study strategies for allocating and managing friendly jammers, so as to create virtual barriers that would prevent hostile eavesdroppers from tapping sensitive wireless communication. Our scheme precludes the use of any encryption technique. Applications include domains such as (i) protecting the privacy of storage locations where RFID tags are used for item identification, (ii) secure reading of RFID tags embedded in credit cards, (iii) protecting data transmitted through wireless networks, sensor networks, etc. By carefully managing jammers to produce noise, we show how to reduce the $SINR$ of eavesdroppers to below a threshold for successful reception, without jeopardizing network performance.

We present algorithms targeted towards optimizing power allocation and number of jammers needed in several settings. Experimental simulations back up our results.

Categories and Subject Descriptors

C.2.0 [**Computer-Communication Networks**]: General—*Security and protection (e.g., firewalls)*; C.2.1 [**Computer-Communication Networks**]: Network Architecture and Design—*Wireless Communication*

---

[*]This work was undertaken while S.S. was at the University of Arizona.

General Terms

Algorithms, Theory

Keywords

Friendly jamming, RFID, wireless, security

1. INTRODUCTION

Wireless communicaton is especially susceptible to eavesdropping due to its broadcast nature. Ensuring private communication has typically been considered at higher layers of the network stack by using cryptographic techniques. However, in many types of communication, such as RFID communication and sensor networks, sophisticated cryptographic techniques are often impractical or impossible to implement, due to power or other hardware constraints. Therefore, it is of interest to consider physical layer-based techniques to secure the communication by exploiting the nature of the wireless channel. Such techniques rely on reducing the Signal-to-Interference-plus-Noise Ratio ($SINR$) of eavesdroppers to below a threshold required for successful reception, while taking care not to reduce the $SINR$ at legitimate receivers too much so as to prevent reception.

Consider the following scenario motivating the application of such a technique. We have a warehouse where items are stored with RFID tags embedded on them for inventory management. These items are perpetually being transported in or out and can even be moved inside the warehouse. The RFID tags on them may contain private information such as the history of transactions on the item, which must be secured form eavesdroppers. We may ensure physical security of warehouses by building a fence around the warehouse such that potential eavesdroppers may not enter the fence. However, communication security is complicated by the fact that RFID devices are limited capability precluding the implementation of cryptographic techniques.

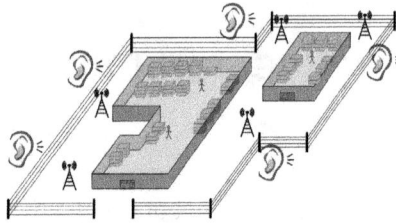

Figure 1: An example application scenario. Jammers secure communication in the warehouses gainst eavesdroppers outside the fence.

To complicate matters further, although we may be able to guess at the capabilities of eavesdroppers, we are unaware of their exact locations. Thus, to ensure the privacy of communication, friendly jammers which transmit artificial noise need to be deployed so that, (i) at *any* potential eavesdropper location, sufficient interference is caused to prevent reception, and (ii) *any* legitimate communication inside the warehouse is not disrupted; see Figure 1. Where should the jammers be placed and what should be their transmission powers such that the above requirements are satisfied?

RFID communication is an especially important application since, although the information stored may be especially sensitive, it is relatively easy to eavesdrop since the capabilities of tags are extremely limited. For example, in [7], the authors demonstrated the vulnerability of credit card RFID tags by successfully performing various attacks including eavesdropping using a device built at a cost of about $150. Although there do exist RFID tags that possess cryptographic capabilities [21], these have been shown to be weak and vulnerable to even a brute-force attack (in [17], the authors showed the weakness of the algorithms in a widely used cryptographic RFID tag).

In general, friendly jamming may be applied in any scenario where cryptographic techniques are not preferred or where we desire additional security to cryptography. Physical methods such as insulation of the environment by some means of padding or physically ensuring that eavesdroppers cannot get near may oftentimes be cost-prohibitive and therefore, friendly jamming may provide a cheaper alternative. For example, it may not be cost-effective to use such methods in hospitals, warehouses or other large areas where important communication may take place.

This paper focuses on application scenarios where communication is geographically restricted, is of short range and we may ensure some minimal physical security. One additional form of wireless communication worth mentioning is the wireless sensor network, for example, in medical applications [14] and *Ambient Assisted Living* application [27, 16]. Sensor nodes have low power requirements and frequently operate in adverse environments where packet errors may make security schemes difficult. In general, although sensor hardware may be capable of cryptography, these schemes rely on either a trusted third party or secure key management schemes (see [26, 22]). Further, the exact network topology is hard to determine due to the large size and random deployment. These properties make the application of friendly jamming suitable. Placing jammers in such a manner creates a *"virtual Faraday cage"* preventing malicious nodes outside from eavesdropping.

The Environment Model. The model of the environment is termed as a *storage/fence* model. We assume that legitimate communication takes place in the *storage* which is a geographic region physically secured by a *fence* inside which eavesdroppers may not enter. The storage is not restricted in any way apart from the requirement that it is enclosed by the fence. In particular, a wireless network when the exact topology is known or multiple warehouses inside which the communication topology is difficult or impossible to determine are both encompassed by this model. The fence may not intersect the storage, i.e., we assume some minimum gap between the storage and fence. If this requirement is removed, eavesdroppers may move arbitrarily close to legitimate transmitters which makes the problem infeasible. Friendly jammers may be located inside the fence but not in the storage, termed as *jammer space*. Further, we assume that some estimate of eavesdroppers' capabilities or some desired protection level is known.

A similar model may be developed for the case when communication outside the fence should not be eavesdropped upon inside the storage or communication from inside the storage to outside the fence should be jammed. Such a model would be applicable in scenarios such as prisons where cellphone use is not permitted inside. The algorithms in this paper may be extended to this model as well.

Contributions. We present algorithms for placing and assigning power to jammers in the jammer space satisfying two objectives, as described above: (i) at *any* potential eavesdropper location, sufficient interference is caused to prevent reception, and (ii) *any* legitimate communication inside the warehouse is not disrupted. We consider two problems. The first problem is one of assigning transmission powers to a set of fixed jammers, referred to as *power assignment* and the second is one of locating a minimum number of *fixed-power* jammers. In addition, if we are given a set of candidate jammer locations, we show how to solve both problems simultaneously, i.e., locate a number of jammers and assign transmission powers to them so that a cost function which is a weighted sum of the number of jammers and the total transmission power is minimized. In all cases, we consider the setting where jammers may be co-operative as well as when the jammer are responsible for individually preventing eavesdropping.

Power Assignment. We present a linear programming formulation for optimally assigning power to the jammers when both the possible eavesdropper locations as well as possible storage locations (communication nodes) are discrete sets of points. In the more general case, where they may be continuous regions, we present an ε-approximation algorithm which solves a linear program with $O((n^2/\varepsilon^2)(\log^2(n/\varepsilon) + \log L))$ constraints in which, given a tunable parameter $0 < \varepsilon < 1/2$, the interference at a storage location is approximated within a factor of $(1-2\varepsilon)$, while the total power assigned is approximated within factor $(1+\varepsilon)$. Here, n is the total number of vertices, edges of storage/fence plus jammers and L is the distance between the two farthest points on the fence.

Jammer Placement. We present a linear programming formulation with $O((n|\mathcal{J}|/\varepsilon^2)(\log(n/\varepsilon)\log(|\mathcal{J}|/\varepsilon)+\log L))$ when the jammer space is a discrete set of points $\mathcal{J}$ of size $|\mathcal{J}|$. The solution to the linear program yields the minimum number of jammers so that, if each jammer is assigned factor $(1+\varepsilon)$ more power, the interference in the storage is ε-

approximated, similar to above. In addition, for the case when jammers are operating individually, the storage and fence are convex polygons and the jammers' power is fixed at a specified value, we provide an almost-optimal algorithm for placing jammers anywhere in a continuous jammer space. This is interesting primarily as a theoretic contribution and serves to illustrate some of the difficulties of the problem.

We also show how to extend the algorithms to find a combined optimum solution for both power allocation and jammer location when the jammer space is discrete. In addition, when eavesdroppers use directional antennas to reduce the interference region, we show how to extend the linear programs to take this into account. Finally, we present the results of some preliminary simulations to compare individual jammers versus co-operating ones.

Prior Work. Several issues have been identified as being specific to RFID security [8, 18]. Although active jamming has been identified as a possible approach in previous works [10], to the best of our knowledge, this method has not been fully explored. This is partly because most works are interested in the security of a specific RFID tag. A similar approach to active jamming is explored in [10] where a single tag, placed in a container such as a bag, triggers a second *"blocker"* tag on the bag which sends interference to untrusted readers. This has also been extended to software approaches through *"soft blocking"* [9]. The drawback of these approaches are that they are special-purpose and require modification of RFID tags. In contrast with these approaches, we consider the region in which tags may be present for security purposes. To the best of our knowledge, such an approach is novel.

Sensor network security [19] has been mostly focused on cryptographic techniques. Asymmetric key cryptography is, in general, resource intensive and hence, the focus is on symmetric key cryptography where the primary problem is key management [22]. This still exposes vulnerabilities to eavesdropping or relay attacks during the key distribution phase.

In wireless networks, active jamming for security has been considered before, particularly in military applications. In [3], the authors formulate the problem of locating jammers with an integer program similar to the formulation in this paper. However, they do not consider the geometry of the region. In information-theoretic security, there exists a substantial number of works following the seminal work of [28], focusing on analysis of channel secrecy even when eavesdroppers have unlimited resources [12, 15, 23, 25]. Other works include game-theoretic approaches for power allocation to jammers [6] and also identifying "forbidden" regions where eavesdroppers must not be present. However, the geometry has not been fully explored and optimization schemes providing guarantees are not presented.

Outline of the paper. We begin, in Section 2, by describing the problem settings. In Section 3, we show that, under reasonable assumptions, it is sufficient to consider only the fence as possible eavesdropper locations irrespective of where eavesdroppers could lie. Section 4 describes our algorithms for power assignment and Section 5 for jammer placement. In Section 6, we show how to extend our algorithms for providing combined solutions as well as when eavesdroppers use directional antennas. Simulation results are presented in Section 7 followed by a few concluding remarks in Section 8.

2. SETTINGS

Let $S \subset \mathbb{R}^2$ be the storage region which may be a discrete set of points or a collection of polygonal regions, inside which legitimate communication takes place and let $\mathcal{F}$ be the boundary of a polygon containing S, representing the fence. Let this polygon be denoted by $P_{\mathcal{F}}$. Eavesdroppers may lie anywhere in the region $\mathbb{R}^2 \setminus P_{\mathcal{F}}$. Let $\mathcal{J}$ denote the jammer space which, typically, is the region between S and $\mathcal{F}$. We denote by n the *description complexity* of the problem. For the power assignment problem, n is the total number of vertices and edges of S and $\mathcal{F}$ plus the number of jammers and for the placement problem, n denotes the total number of vertices and edges of S and $\mathcal{F}$.

Slightly abusing notation, we refer to a node (eavesdropper, jammer or legitimate node in the storage) by its location, i.e., a jammer located at point j is referred to as j. For any two points $p_1, p_2 \in \mathbb{R}^2$, $\|p_1 - p_2\|$ indicates the Euclidean distance between them. For two sets of points (possibly infinite) $Q, Q' \subset \mathbb{R}^2$, we denote by $d(Q, Q')$, the minimum distance $\|q - q'\|$ over all points $q \in Q$ and $q' \in Q'$. Given a set of points Q and a point p, let $NN(p, Q)$ denote the point in Q closest to p. Let $d(S, \mathcal{F}) = 1$ and let L denote the distance between the two farthest points on $\mathcal{F}$. Our algorithms for power assignment run in time polynomial in n and $\log L$ and those for location depend only on n.

Communication Model. We use the *Signal to Interference plus Noise Ratio (SINR)* model (termed as physical model in [5]). Assuming all other factors are normalized and following the standard power dissipation model [20], for a transmission from p to q given a set of jammers J,

$$SINR_p(q) = \frac{P_p \|p - q\|^{-\gamma}}{\sum_{j \in J} P_j \|j - q\|^{-\gamma}}, \qquad (1)$$

where P_p is the transmission power of p, P_j is the transmission power of jammer j, and γ is the path loss exponent (typically from 2 to 4). For clarity, we assume no ambient noise throughout the paper. All our results, with the exception of that of Section 5.2, can be extended to take this into account. A receiver q is able to successfully receive a transmission from p if $SINR_p(q)$ is at least a threshold depending on the node characteristics. We refer to the *SINR* at any eavesdropper location p of transmissions from its nearest point on S as $SINR(p)$. We assume that only jammer signals cause interference, since typically, we would have some collision resolution protocol for transmissions inside the storage.

Equation (1) assumes a model in which all jammers co-operate to interfere with a node. We term this the *Fully Cooperative* interference model, denoted by Full. In addition, we define the *Nearest Jammer* interference model, denoted by NJ, where a receiver only encounters interference from the closest jammer to it. Thus, in Equation (1), the denominator would now incorporate only the interference from the nearest jammer. The NJ model may be extended to include the k closest jammers yielding the k-NJ model. In practice, we expect that the NJ model may not too far from the Full interference model, due to the path loss exponent γ in the power dissipation equation: interference from the closest jammer is most important, while interference due to farther jammers fades away fast with distance.

For the purposes of clarity, we assume that legitimate communication inside S is of short enough range so as to experience insignificant path loss, but our algorithms can be extended to the cases where we know an upper bound on the range, or if, we know the exact topology of the communicating nodes. We also assume that all transmitters in $\mathcal{S}$ have the same transmitting power (normalized to 1). This assumption may be removed if the exact topology of legitimate nodes is known in advance. Let the *SINR* threshold for successful reception by legitimate receivers be normalized to 1 and the threshold for eavesdroppers be δ. The capabilities of eavesdropper nodes may be different from those of legitimate receivers due to possibly different hardware and therefore, we use different thresholds. We note that, for an eavesdropper, it is sufficient to jam possible transmissions from its nearest point on $\mathcal{S}$.

Finally, throughout, we make the assumption that jammers may be assigned a maximum power P_{max} (due to hardware constraints, a jammer may not be assigned an arbitrarily high power) and a minimum power of $(1/\delta)$. Roughly, the minimum power assumption implies that, if eavesdroppers and legitimate receivers have similar capabilities, then jammers must transmit at a power at least that of legitimate transmitters. The greater the capabilities of eavesdroppers, the higher the jammers' minimum transmission power. We show, in Section 3, that this assumption implies that it is sufficient to consider eavesdroppers on $\mathcal{F}$, i.e., *if an eavesdropper cannot eavesdrop from any location on $\mathcal{F}$, it cannot eavesdrop from any location in $\mathbb{R}^2 \setminus \mathcal{F}$*. Although this does not look surprising, if the jammers may be assigned an arbitrarily low transmission power, it is easy to construct examples, where an eavesdropper may be able to successfully eavesdrop by moving away from $\mathcal{S}$ even though it could not eavesdrop from a closer location. We may remove the minimum power assumption if we instead assume that once an eavesdropper gets too far from any point in $\mathcal{S}$, it cannot eavesdrop (possibly due to ambient noise). In this case, our algorithms can be easily extended with running times which have an additional logarithmic dependence on this maximum distance.

Under the above communication model, assuming that eavesdroppers may lie only on $\mathcal{F}$, the following equations formalize the requirements of a set of jammers J where each jammer $j \in J$ has transmission power P_j: (i) at *any* potential eavesdropper location, sufficient interference is caused to prevent reception, and (ii) *any* legitimate communication inside the warehouse is not disrupted.

$$\frac{1}{\sum_{j \in J} P_j \|j - s\|^{-\gamma}} \geq 1, \qquad \forall s \in \mathcal{S} \qquad (2)$$

$$\frac{d(p, \mathcal{S})^{-\gamma}}{\sum_{j \in J} P_j \|j - p\|^{-\gamma}} < \delta. \qquad \forall p \in \mathcal{F} \qquad (3)$$

The above equations would be modified under the NJ model. We focus, in this paper, on the Full model and indicate the changes wherever we refer to the NJ model.

3. CONSIDERING THE BOUNDARIES IS SUFFICIENT

We show that under our communication model: (i) jamming the fence $\mathcal{F}$ is sufficient to ensure that eavesdroppers located outside the fence are also jammed successfully and (ii) ensuring that the any receiver on the boundary of $\mathcal{S}$ is not jammed is sufficient to ensure that receivers inside $\mathcal{S}$ are not jammed.

LEMMA 3.1. *Under any interference model, if $SINR(p) < \delta$ for all points p on $\mathcal{F}$, then for all points p' outside the region encapsulated by $\mathcal{F}$, $SINR(p') < \delta$.*

PROOF. We prove the lemma under the Full model. The proof for the NJ model is part of this proof. Let J be a set of jammers such that no eavesdropper on $\mathcal{F}$ is successfull and let P_j be the transmission power for any jammer $j \in J$. Let p' be a point outside $\mathcal{F}$ and let p be a point on $\mathcal{F}$ on the segment connecting p' to $\mathrm{NN}(p', \mathcal{S})$. Clearly, $\mathrm{NN}(p', \mathcal{S}) = \mathrm{NN}(p, \mathcal{S})$. Rearranging the *SINR* equation, we need to show that, to show that $(d(p, \mathcal{S}))^{-\gamma} < \delta \sum_{j \in J} P_j (\|j - p\|)^{-\gamma}$ implies that $(d(p', \mathcal{S}))^{-\gamma} < \delta \sum_{j \in J} P_j (\|j - p'\|)^{-\gamma}$.

We will show the proof by induction on the number of jammers. For any subset $X \subset J$, let a_X be a real number satisfying $a_X^{-\gamma} = \delta \sum_{j \in X} P_j (\|j - p\|)^{-\gamma}$. Consider a singleton jammer j and the corresponding a_j. Clearly, $(a_j + \|p - p'\|)^{-\gamma} < \delta P_j (\|j - p\| + \|p - p'\|)^{-\gamma}$ since $P_j \geq 1/\delta$. Thus, the base case is satisfied. This completes the proof for the NJ model.

Now, consider some subset $X \subset J$. The inductive hypothesis is that,

$$(a_X + \|p - p'\|)^{-\gamma} < \delta \sum_{j \in X} P_j (\|j - p\| + \|p - p'\|)^{-\gamma} \quad (4)$$

Now, consider than a jammer j' is added to X and let $b_{X,j'}$ be a real number satisfying

$$b_{X,j'}^{-\gamma} = a_X^{-\gamma} + \delta P_{j'} \|j' - p\|'^{-\gamma} \qquad (5)$$

Clearly, $b_{X,j'} \leq a_x$ and $b_{X,j'} \leq \|j' - p\|$ since $\delta P_{j'} \geq 1$.

$$(b_{X,j'} + \|p - p'\|)^{-\gamma} = b_{X,j'}^{-\gamma}(1 + (\|p - p'\|/b_{X,j'}))^{-\gamma}$$

$$= \frac{a_X^{-\gamma}) + \delta P_{j'} \|j' - p\|^{-\gamma}}{(1 + (\|p - p'\|/b_{X,j'}))^{\gamma}},$$

by Equation (5). Since $b_{X,j'} < a_X$ and $b_{X,j'} < \|j' - p\|$, this implies that,

$$(b_{X,j'} + \|p - p'\|)^{-\gamma} \leq (a_X + \|p - p'\|)^{-\gamma}$$
$$+ \delta P_{j'}(\|j' - p\| + \|p - p'\|)^{-\gamma}.$$

Hence, we know, for $X = J$, Equation (4) is satisfied. Now, since $a_X \leq d(p, \mathcal{S})$, the lemma is proved. $\square$

LEMMA 3.2. *Under any interference model, if for all points p on the boundary of $\mathcal{S}$, $SINR(p) \geq 1$, then for all points p' inside $\mathcal{S}$, $SINR(p') > 1$.*

PROOF. For the NJ model, select an arbitrary point p' inside $\mathcal{S}$ whose closest jammer is $j(p')$. Let p be an intersection point of the segment joining p' and $j(p')$ with $\mathcal{S}$. Since $j(p') = j(p)$, we clearly have $1 \leq SINR(p) < SINR(p')$.

For the Full model, the statement is equivalent to showing that the *SINR* attains it's minimum at the boundary of $\mathcal{S}$. This is the same as showing that the interference of the jammers attains its maximum on the boundary of $\mathcal{S}$. We do this by showing that the interference, as a function of position, is a sub-harmonic function and thus satisfies the Maximum principle known from complex analysis [1]. This is shown by differentiation:

$$\Delta_s I_s = (\partial_{s_1} + \partial_{s_2}) \sum_{j \in J} P_j |s - j|^{-\gamma} = \sum_{j \in J} \gamma^2 |s - j|^{-\gamma - 2}.$$

Clearly, the Laplacian is positive. Hence, the function is sub-harmonic and the result follows. $\square$

4. POWER ASSIGNMENT

In this section, we provide algorithms to assign powers to a set of fixed jammers J such that Equation (2) and Equation (3) are satisfied and the total power assigned is minimized. We may express the problem by means of the optimization program below, termed as JAMMING-LP.

$$\text{JAMMING-LP: } \textbf{Minimize } \sum_{j \in J} P_j$$

$$\textbf{s.t. } \forall s \in \mathcal{S}: \sum_{j \in J} P_j \|s - j\|^{-\gamma} \leq 1, \tag{I}$$

$$\forall p \in \mathcal{F}: \sum_{i \in J} P_j \|i - p\|^{-\gamma} > \frac{1}{\delta d(p, \mathcal{S})^\gamma}, \tag{II}$$

$$\forall j \in J: \quad (1/\delta) \leq P_j \leq P_{max}. \tag{III}$$

Constraints (I) and (II) are the equivalent of Equations (2) and (3). I_s and I_e are dependent on the variables P_j and are dictated by the interference models as described in Section 2. The number of constraints (I) and (II) is uncountably infinite if $\mathcal{S}$ and $\mathcal{E}$ are continuous regions in $\mathbb{R}^2$.

First note that when $\mathcal{S}$ and $\mathcal{E}$ are discrete sets of points, JAMMING-LP becomes a linear program which may be solved in polynomial time since the number of constraints depends on the cardinalities of $\mathcal{S}$ and $\mathcal{F}$.

The continuous case is a more difficult since, as mentioned before, the number of constraints is uncountably infinite. To get around this difficulty, we provide an ε-approximation algorithm based on discretizations of $\mathcal{S}$ and $\mathcal{F}$. Given a parameter ε in the range $(0, 1)$, the algorithm proceeds according to the following steps:

(1) Compute a discrete set $\mathcal{S}' \subset \mathcal{S}$ such that if Equation (2) is satisfied for $\mathcal{S}'$, then Equation (2) is satisfied for $\mathcal{S}$ with threshold $1/(1 + 2\varepsilon) \geq (1 - 2\varepsilon)$.
(2) Compute a discrete set $\mathcal{F}' \subset \mathcal{F}$ such that if Equation (3) is satisfied for $\mathcal{F}'$ for some power assignment, then, by increasing the powers of the jammers by a factor $(1+\varepsilon)$, Equation (3) is satisfied for $\mathcal{F}$.
(3) Solve the linear program JAMMING-LP with constraints corresponding to $\mathcal{S}'$ and $\mathcal{F}'$.

THEOREM 1. *Given storage region(s) $\mathcal{S}$, fence $\mathcal{F}$, a set of jammer locations J and an interference model, by solving a linear program with $O((n^2/\varepsilon^2)(\log^2(n/\varepsilon) + \log L))$ constraints, we can compute a power assignment for J such that $\sum_{j \in J} P_j \leq (1 + \varepsilon) \sum_{j \in J} P_j^*$ where P_j^* is the optimal power assignment for j and (i) for each location $p \in \mathcal{F}$, $SINR(p) < \delta$, (ii) for each location $s \in \mathcal{S}$, $SINR(s) \geq (1 - 2\varepsilon)$.*

$\mathcal{S}'$ is constructed so that the interference at the point in $\mathcal{S}$ at which interference is maximum is approximated within factor $(1 - \varepsilon)$. Similarly, for the fence $\mathcal{F}$, the point p on $\mathcal{F}$ at which $SINR$ is maximum for a transmission from $NN(p, \mathcal{S})$, does not receive more than factor $(1 + \varepsilon)$ more $SINR$ than the corresponding point in $\mathcal{F}'$. Now, if each jammer is actually assigned $(1 + \varepsilon)$ of the power assignment returned by JAMMING-LP, we can jam every point on $\mathcal{F}$ and no point on $\mathcal{S}$ will reduce its $SINR$ by more than a factor of $1/(1+\varepsilon)^2 > (1 - 2\varepsilon)$. Thus, Theorem 1 is proved. For the remainder of this section, we assume the Full interference model. However, all results may be applied to the NJ model with minimal modification. The schemes, particular the discretization of $\mathcal{S}$, use some of the ideas of Vigneron [24].

First, we briefly outline a couple of preliminary concepts which are essential for the rest of this section.

Voronoi Diagrams. The *Voronoi Diagram* (see [2] and [4, Chapter 7]) for a set of points Q, denoted by $VD(Q)$ is a decomposition of the plane into cells such that all points in a cell are closest to the same point $q \in Q$. A cell is denoted by $Vor(q)$ and edges of the Voronoi Diagram are straight-line segments (parts of bisectors between pairs of points of Q). The *generalized Voronoi Diagram* [11, 13] of a polygon P, is the generalization of the Voronoi Diagram to the vertices and edges of P. This is a decomposition of the plane into cells such that, in each cell, all points have the same closest vertex/edge. Both may be constructed in $O(|P| \log |P|)$ time where $|P|$ is the number of vertices/edges of P.

We are interested in the Voronoi Diagrams of the jammer set $VD(J)$ and the generalized Voronoi diagram $VD(\mathcal{S})$ of $\mathcal{S}$. Similar to our notation above, we denote by $Vor((u, v))$ and $Vor(u)$, the Voronoi cells of an edge (u, v) and vertex u of $\mathcal{S}$ respectively.

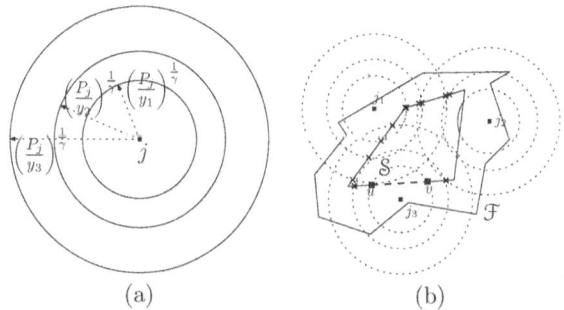

(a)   (b)

Figure 2: (a) The disks corresponding to the super-level sets for a jammer when $Y_j = \{y_1, y_2, y_3\}$. (b) The arrangement of the disks with respect to $\mathcal{S}$. Vertices are marked as ×.

Superlevel Sets and Arrangements. For a set of objects Γ and a polygon or collection of polygons P, the *arrangement* $\mathcal{A}_P(\Gamma)$ of Γ is the planar subdivision induced by Γ on the boundaries of polygons in P. Namely, its vertices are the intersection points of the boundaries of the disks and polygons in P together with original vertices of polygons in P and edges are the maximal connected portions of the boundaries of P not crossing a vertex; see Figure 2b. If the number of vertices in P is M, the objects in Γ are segments, rays or lines and the number of objects in Γ is N, the complexity, i.e., the number of vertices and intervals in $\mathcal{A}_P(\Gamma)$ is $O(MN)$.

For a jammer j, let $D[j; t]$ denote the disk of radius t centered at j. Note that at all points in $D[j; t]$, the interference due to j is atleast $P_j t^{-\gamma}$. In mathematics, $D[j; t]$ is known as a *superlevel set* of the function $f_j(x) = P_j \|j - x\|^{-\gamma}$.

Given three parameters $\rho > 0, \alpha > 0$ and $l \in \mathbb{Z}^+$, we define

$$Y(\rho, \alpha, l) = \{y_i = \rho/(1 + \alpha)^i \mid 0 \leq i \leq l\}.$$

Given a set of jammers J and $Y(\rho, \alpha, l)$ for a jammer j, let $\mathcal{D}_j = \{D[j; y_i] \mid 0 \leq i \leq l\}$; see Figure 2a for an example. Consider the arrangement $\mathcal{A}_P(\mathcal{D}_j)$ for some polygon or collection of polygons P. The intervals are all located between successive concentric disks centered at j. Clearly, the following lemma holds for $\mathcal{A}_P(\mathcal{D}_j)$

LEMMA 4.1. *Let a, b be two points lying in the same interval of $A_P(\mathcal{D}_j)$. If a, b lie outside all disks of $\mathcal{D}_j$, then $P_j \|j - a\|^{-\gamma} \leq \rho/(1+\alpha)^s$ and $P_j \|j - b\|^{-\gamma} \leq \rho/(1+\alpha)^s$. Otherwise, $P_j \|j - a\|^{-\gamma} \geq (1/(1+\alpha))P_j \|j - b\|^{-\gamma}$.*

Discretization of $\mathcal{S}$. We generate a discrete set $\mathcal{S}' \subset \mathcal{S}$ as follows. First, we set $\rho = P_j d(j, \mathcal{S})^{-\gamma}$, $\alpha = \varepsilon/9$ and $l = \lceil (1/\varepsilon) \log(n/\varepsilon) \rceil$. Next, setting $Y_j = Y(\rho, \alpha, l)$, we compute the set of disks $\mathcal{D}_\mathcal{S} = \cup_{j \in J} \mathcal{D}_{j, Y_j}$. Finally, we compute the arrangement $A_\mathcal{S} = A_\mathcal{S}(\mathcal{D}_\mathcal{S})$ and select an arbitrary point in each interval of $A_\mathcal{S}$ to add to the set $\mathcal{S}'$.

We choose $\rho = P_j d(j, \mathcal{S})^{-\gamma}$ because it is an upper bound on $P_j \|j - s\|^{-\gamma}$ for any point $s \in \mathcal{S}$, implying that there is no point of s in the smallest disk of $\mathcal{D}_{j, Y_j}$ for all jammers j. It is important to note that we do not know the values P_j to determine the value of ρ. However, we do not actually need it to compute the radii of the disks in $\mathcal{D}_{j, Y_j}$.

Correctness follows from the following two lemmas.

LEMMA 4.2. *Let s be the point selected by our algorithm in some interval of $A_\mathcal{S}$ and let s' be another point in the same interval. For any jammer $j \in J$, we have*

$$\|j - s\|^{-\gamma} \geq \begin{cases} \frac{1}{1+\alpha} \|j - s'\|^{-\gamma}, & \text{if } s \notin D[j; \frac{d(j, \mathcal{S})}{(1+\alpha)^l}], \\ \|j - s'\|^{-\gamma} - \frac{\alpha}{n} d(j, \mathcal{S})^{-\gamma}, & \text{otherwise.} \end{cases}$$

PROOF. If $s \notin D[j; d(j, \mathcal{S})/(1+\alpha)^l]$, i.e., if it lies outside the outermost disk centered at j, then by the choice of l, the lemma follows. Otherwise, there exist two consecutive concentric disks centered at j such that interval containing s and s' lies in the region between these disks. By Lemma 4.1, the proof follows. $\square$

LEMMA 4.3. *Given a power assignment for the jammers, let s^* be the point maximizing $\sum_{j \in J} P_j \|j - s\|$ over all $s \in \mathcal{S}$ and let $\hat{s}$ be the point selected by our algorithm in the same interval in $A_\mathcal{S}$ as s^*. Then,*

$$\sum_{j \in J} P_j \|j - s^*\|^{-\gamma} \leq (1 + \varepsilon/3) \sum_{j \in J} P_j \|j - \hat{s}\|^{-\gamma}.$$

PROOF. Let J_{out} be the set of jammers such that s^* and $\hat{s}$ lie outside $D[j; d(j, \mathcal{S})/(1+\alpha)^l]$ for all $j \in J_{\text{out}}$ and let J_{in} be the remaining jammers. Lemma 4.2 implies that

$$\sum_{j \in J} P_j \|j - \hat{s}\|^{-\gamma} \geq \sum_{j \in J_{\text{in}}} \frac{P_j}{1+\alpha} \|j - s^*\|^{-\gamma}$$
$$+ \sum_{j \in J_{\text{out}}} P_j \left(\|j - s^*\|^{-\gamma} - \frac{\alpha}{n} d(j, \mathcal{S})^{-\gamma} \right)$$
$$\geq \frac{1}{1+\alpha} \sum_{j \in J} P_j \|j - s^*\|^{-\gamma}$$
$$- \alpha \sum_{j \in J} P_j \|j - s^*\|^{-\gamma},$$

since s^* is the point maximizing $\sum_{j \in J} P_j \|j - s\|^{-\gamma}$ over all $s \in \mathcal{S}$. Since $\alpha = \varepsilon/9$, the lemma follows. $\square$

Lemma 4.3 implies that if the point $\hat{s}$ does not receive too much interference from the jammers, no other point in $\mathcal{S}$ would have too much interference. Since we do not actually know which point is $\hat{s}$, we take care to ensure that the entire set $\mathcal{S}'$ is not jammed. If each jammer;s power is $P_j(1+\varepsilon)$, then the approximation factor would become $(1 + 2\varepsilon)$.

There are $O((n/\varepsilon) \log(n/\varepsilon))$ level sets in our arrangement. Thus, the cardinality of $\mathcal{S}'$ which is the number of vertices of $A_\mathcal{S}$ is $O((n^2/\varepsilon^2) \log^2(n/\varepsilon))$ implying that the number of constraints (I) in JAMMING-LP would be $O((n^2/\varepsilon^2) \log^2(n/\varepsilon))$ with an equal time required to generate them.

Discretization of $\mathcal{F}$. We generate a discrete set $\mathcal{F}' \subset \mathcal{F}$ as follows. Recall that L denotes the distance between the two farthest points in $\mathcal{F}$. First, we set $\rho = 1$, $\alpha = \varepsilon/3$ and let l be the largest integer such that $1/(1+\alpha)^l \leq L^{-\gamma}$. Next, setting $Y_j = Y(\rho, \alpha, l)$, we compute the set of disks $\mathcal{D}_\mathcal{F} = \cup_{j \in J} \mathcal{D}_{j, Y_j}$. We compute the generalized Voronoi Diagram (see Section 3) VD($\mathcal{S}$) of $\mathcal{S}$ and let Γ denote the rays and lines constituting VD($\mathcal{S}$). Finally, we compute the arrangement $A_\mathcal{F} = A_\mathcal{F}(\mathcal{D}_\mathcal{F} \cup \Gamma)$ and add the vertices of $A_\mathcal{F}$ to $\mathcal{F}'$.

We note that on each interval ϕ of $A_\mathcal{F}$, $d(p, \mathcal{S})$ for all points $p \in \phi$ is a linear function since there is a corresponding segment or point on $\mathcal{S}$ on which lie all the points closest to points in ϕ. Thus, the maximum and minimum distances are at the vertices of ϕ. Contrary to the discretization of $\mathcal{S}$ where we approximate the maximum interference received by points in $\mathcal{S}$, we approximate the maximum $SINR$. The choice of l based on the diameter L is to ensure that no point on $\mathcal{F}$ lies outside the disks for any j. Also, since $P_j \geq 1/\delta$ for all $j \in J$, eavesdroppers within distance 1 from any j are always jammed, i.e., their $SINR$ is always too low. Note that we do not need to know the powers to compute the disks.

Correctness follows from the following two lemmas.

LEMMA 4.4. *Let p be a point selected by our algorithm for any interval in $A_\mathcal{F}$ and let p' be a point in the same interval. For any jammer $j \in J$, $\|j - p\|^{-\gamma} \leq (1+\alpha)\|j - p'\|^{-\gamma}$.*

PROOF. The distance from p_ϕ to j is between 1 and L. Thus, there exists two consecutive concentric disks in $\mathcal{D}_{j, Y_j}$ such that both p and p' lie between these disks. The proof follows from Lemma 4.1. $\square$

LEMMA 4.5. *Given a power assignment for the jammers, let p^* be the point on $\mathcal{F}$ at which $SINR(p)$ is maximum over all $p \in \mathcal{F}$ and let $\hat{p}$ be the vertex in $\mathcal{F}'$ in the interval of p^* such that $d(\hat{p}, \mathcal{S}) \leq d(p^*, \mathcal{S})$. Then,*

$$SINR(p^*) < (1 + \varepsilon)SINR(\hat{p}).$$

PROOF. Let $\sum_{j \in J} P_j \|j - p^*\|^{-\gamma} \leq \sum_{j \in J} P_j \|j - \hat{p}\|^{-\gamma}$ since otherwise, there is nothing to prove. By Lemma 4.4,

$$\sum_{j \in J} P_j \|j - \hat{p}\|^{-\gamma} \leq (1+\alpha) \sum_{j \in J} P_j \|j - p^*\|^{-\gamma}.$$

Since $d(p^*, \mathcal{S})^{-\gamma} \geq d(\hat{p}, \mathcal{S})^{-\gamma}$ and by our choice of $\alpha = \varepsilon/3$, the lemma follows. $\square$

Lemma 4.5 implies that the $SINR(p) < (1 + \varepsilon)\delta$ for any $p \in \mathcal{F}$. Thus, by assigning a power $(1 + \varepsilon)P_j$ for all jammers $j \in J$, we can ensure that $SINR(p) < \delta$ for all $p \in \mathcal{F}$. The number of level sets corresponding to jammers is $O((n/\varepsilon) \log L)$. The number of vertices in their arrangement on $\mathcal{F}$ is $O((n^2/\varepsilon^2) \log^2 L)$ leading to as many constraints (II) in JAMMING-LP, with an equal time required to generate $\mathcal{F}'$.

Remarks. We note that if all the jammers' powers are required to be the same, and we need to find the minimum power assignment, we may remove the dependency on the diameter L of $\mathcal{F}$. Briefly, this is due to the fact that, for the discretization of $\mathcal{F}$, we may develop an upper and lower bound on the power received at the eavesdropper with maximum $SINR$ whose ratio is independent of L.

5. PLACEMENT OF JAMMERS

In this section, we consider the problem of placing a minimum number of jammers all of which have the same transmission power $\hat{P}$.

We first give some basic definitions. Note that for every point $s \in \mathcal{S}$, according to Equation 2, if a jammer j lies in the disk $\mathrm{D}[s; \hat{P}^{1/\gamma}]$, it will prevent reception at s. We define the *forbidden region* $\varphi(\mathcal{S}) = \cup_{s \in \mathcal{S}} \mathrm{D}[s; \hat{P}^{1/\gamma}]$. This is essentially the Minkowski sum [4] of a disk with radius $\hat{P}^{1/\gamma}$ and $\mathcal{S}$; see Figure 3. Next, for a point $p \in \mathcal{F}$, according to Equation 3, a jammer must lie in the disk $\mathrm{D}[p; (\delta\hat{P})^{1/\gamma}/d(p,\mathcal{S})]$. We call this the *critical disk* and denote it by D_p and define the *visibility region* $\mathrm{Vis}(p)$ as $(P_{\mathcal{F}} \cap \mathcal{D}_p) \setminus \varphi(\mathcal{S})$. This is the region in which a jammer must lie in the jammer space to successfully jam p; see Figure 3. We call two visibility regions $\mathrm{Vis}(p_1)$ and $\mathrm{Vis}(p_2)$ *adjacent* if their intersection is exactly one point.

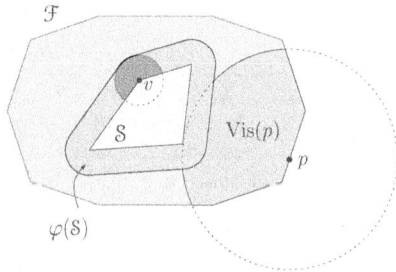

Figure 3: Forbidden region $\varphi(\mathcal{S})$ of $\mathcal{S}$ and visibility region $\mathrm{Vis}(p)$ for a point $p \in \mathcal{F}$.

Before we proceed with the algorithms, let us try and understand why this problem is challenging. Consider the simple examples in Figure 4. In both cases, we consider the NJ model. In Figure 4a, we have two disks which are concentric representing $\mathcal{S}$ and $\mathcal{F}$, while in Figure 4b, the disks are not concentric. Critical disks are also shown for various points on the fence. In both cases, an almost-optimal solution is to place the set of jammers at the points where two disks touch. In Figure 4a, since all critical disk are congruent, it is easy to characterize the optimal placement but in Figure 4b, it is not simple to characterize algebraically since the function of the distance between $\mathcal{S}$ and $\mathcal{F}$ is now more complicated. If, even in this simple example, the characterization of the problem is difficult, if we take into account all parameters such as jammer power, eavesdropper capability and possibly complicated shapes of $\mathcal{S}$ and $\mathcal{F}$, characterizing the solution seems to be particularly difficult.

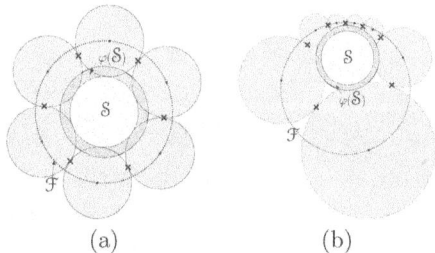

Figure 4: Two simple and similar examples where solutions differ significantly. The optimal placement of jammers is marked as $\times$.

With that in mind, we consider two basic settings: (i) when the jammer space $\mathcal{J}$ is a discrete set of points and (ii) when $\mathcal{J}$ is the entire region $P_{\mathcal{F}} \setminus \mathcal{S}$, where $P_{\mathcal{F}}$ is the polygon enclosed by $\mathcal{F}$. In the former, we give an ε-approximation algorithm and in the latter, we provide an optimal algorithm under a restricted setting.

5.1 ε-approximation given a discrete set of candidate locations

Given a discrete set of candidate locations $\mathcal{J}$ not in $\varphi(\mathcal{S})$, we discretize $\mathcal{F}$ and $\mathcal{S}$ using the scheme of Section 4. This gives us discrete sets $\mathcal{F}' \subset \mathcal{F}$ and $\mathcal{S}' \subset \mathcal{S}$. We can now design the following integer linear program JAMMING-ILP adapted from JAMMING-LP with binary variables c_i for each location $i \in \mathcal{J}$ indicating whether i is chosen or not.

$$\text{JAMMING-ILP: } \mathbf{Minimize} \sum_{i \in \mathcal{J}} c_i$$

$$\mathbf{s.t.} \ \forall s \in \mathcal{S}' : \sum_{i \in \mathcal{J}} c_i \hat{P} \|s - i\|^{-\gamma} \leq 1, \quad \text{(I)}$$

$$\forall p \in \mathcal{F}' : \sum_{i \in \mathcal{J}} c_i \hat{P} \|i - p\|^{-\gamma} > \frac{d(p,\mathcal{S})^{-\gamma}}{\delta}. \quad \text{(II)}$$

Although JAMMING-ILP is formulated for the Full model of interference, it may easily be modified for the NJ model. This gives us the following theorem:

THEOREM 2. *Given storage region(s) $\mathcal{S}$, a fence $\mathcal{F}$, an interference model, a discrete set of candidate locations $\mathcal{J}$ for the jammers and a fixed power $\hat{P}$, we can find a minimum number of jammer locations from $\mathcal{J}$ such that Equation (3) is satisfied and for every point $s \in \mathcal{S}$, $SINR(s) > (1 - 2\varepsilon)$ by solving an Integer Linear Program with $O((n|\mathcal{J}|/\varepsilon)(\log(n/\varepsilon) \log(|\mathcal{J}|/\varepsilon) + \log L))$ constraints.*

5.2 Near-optimal algorithm for a restricted setting

We consider the problem under the following restricted setting: **(i)** NJ interference model, **(ii)** $\mathcal{S}$ and $\mathcal{F}$ are convex, and **(iii)** each jammer is assigned a power $1/\delta$. Note that the assumption that each jammer has a power exactly $1/\delta$ implies that D_p for any $p \in \mathcal{F}$ will have radius exactly $d(p,\mathcal{S})$. Without this assumption, Lemma 5.1 does not hold and it is not possible to show almost-optimality for the algorithm.

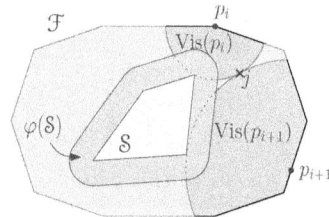

Figure 5: One step of the algorithm. p_{i+1} is selected such that $\mathrm{Vis}(p_i)$ and $\mathrm{Vis}(p_{i+1})$ are adjacent.

The algorithm proceeds as follows. We first pick an arbitrary point $p_0 \in \mathcal{F}$ as a starting point and keep finding adjacent regions by moving clockwise along the boundary until we reach a region which intersects $\mathrm{Vis}(p_0)$ again (see Figure 5). At the ith step of the algorithm, we place a jammer

j_{i+1} at the point of intersection of $\mathrm{Vis}(p_i)$ and $\mathrm{Vis}(p_{i+1})$. Let p_k be the last point. We place a jammer j_{k+1} at the point in the intersection of $\mathrm{Vis}(p_k)$ and $\mathrm{Vis}(p_0)$ which is farthest from $\mathcal{S}$. Let J denote the set of jammers obtained.

The following lemma shows correctness as well as the fact that for each i, we need atleast one jammer.

LEMMA 5.1. *Let p_{i-1}, p_i be two consecutive points selected and let j_i be the jammer placed at the point of intersection of $\mathrm{Vis}(p_{i-1})$ and $\mathrm{Vis}(p_i)$. The points jammed by j_i consists of the clockwise portion of $\mathcal{F}$ between p_{i-1} and p_i.*

PROOF. First, we note that if $\mathrm{D}_{p_{i-1}}$ and D_{p_i} are not tangential, j is placed at the their point of intersection away from $\mathcal{S}$. This is the single point of intersection of $\mathrm{Vis}(p_{i-1})$ and $\mathrm{Vis}(p_i)$. Let l denote the line passing through p_{i-1} and p_i and t denote the line passing through j perpendicular to l. l and t separate the plane into four quadrants. The portion of the fence in between p_{i-1} and p_i must clearly lie in the two quadrants which does not contain $\mathcal{S}$. Consider an eavesdropper location p between p_{i-1} and p_i on the quadrant Q_1 containing p_i; see Figure 6a. $\mathcal{S}$ must lie wholly in the sector Ψ_P formed by the lines through the points (p, p_i) and (p, j) of smaller angle. For a point $q \in \Psi_p$ which also lies on the boundary of $\mathcal{D}_{p_i}$, consider the angles $\theta = \angle p, p_i, q$ and $\phi = \angle p, p_i, j$ at p_i. It is easy to see that $\theta > \phi$ for every p. Hence, $\|p - j\| \le \|p - q\|$. The proof for the other quadrant which contains p_{i-1} is similar.

Moving on to the second part, first, let $n_{i-1} = \mathrm{NN}(p_{i-1}, \mathcal{S})$ and $n_i = \mathrm{NN}(p_i, \mathcal{S})$. Clearly, $\|p_i - n_i\| = \|j - n_i\|$ and $\|p_{i-1} - n_{i-1}\| = \|j - n_{i-1}\|$, implying that the Voronoi diagram of $\{j, n_{i-1}, n_i\}$ (see Section 4 for a description of Voronoi diagrams) must pass through p_i and p_{i-1}, implying that the portion of $\mathcal{F}$ clockwise from p_i to p_{i-1} does not lie in the Voronoi cell of j as then, $\mathcal{F}$ would not be convex. Thus, all points of $\mathcal{F}$ in the portion clockwise from p_i to p_{i-1} are closer to either n_i or n_{i-1} than j, implying that they are not jammed. □

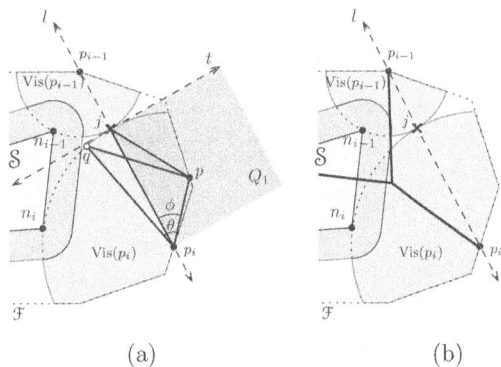

(a)   (b)

Figure 6: Illustration of the proof of Lemma 5.1

We are now ready to bound the number of jammers in J.

LEMMA 5.2. *Let OPT be the size of the optimal set of jammers. Then, $|J| \le \mathrm{OPT} + 1$.*

PROOF. For each interval along the fence $[p_i, p_{i+1}]$, by Lemma 5.1, j_{i+1} is the only location where a jammer can jam every point on $[p_i, p_{i+1}]$ and does not jam any other point, implying that we need atleast one jammer for this interval. The proof follows. □

Putting it all together, we get the following theorem.

THEOREM 3. *Given convex $\mathcal{S}$ and $\mathcal{F}$, when jammers have power $\hat{P} = 1/\delta$, we can find a set of jammer locations J in the jammer space such that Equation (2) and Equation (3) are satisfied under NJ model of interference and $|J| \le |OPT| + 1$. The time required is polynomial in $\max\{\mathrm{OPT}, n\}$ where OPT is the size of the optimal solution.*

Remarks. The solution guarantee does not hold when ambient noise is present. Further, the algorithm and its correctness proof outline the conditions under which guaranteed solutions may be obtained and serve to demonstrate the difficulty of the problem in general. Therefore, it is of primarily theoretical importance.

6. EXTENSIONS

We extend the algorithms of Sections 4 and 5 in two ways: (i) we show how to provide a solution to the combined problem of power assignment and placement of jammers while optimizing a linear combination of the total power and number of jammers, (ii) if eavesdroppers are equipped with directional antennas, we show how to extend JAMMING-LP and JAMMING-ILP to incorporate this fact while still maintaining a tractable number of constraints.

6.1 Combined Solution

We may develop a combined solution when given a discrete set of candidate locations $\mathcal{J}$ as follows. We set a weighting parameter μ, and define the cost functions $\sum_{j \in \mathcal{J}} c_j + \mu \sum_{j \in \mathcal{J}} P_j$, where $c_j \in \{0, 1\}$ and P_j is the power assigned to jammer j. If no jammer is located at j, this is simply indicated by a value of $P_j = 0$. Here, the weighting parameter μ specifies how we prefer one criteria versus the other. We substitute this in JAMMING-ILP to get the desired program.

6.2 Directional eavesdroppers

Let eavesdroppers be equipped with *directional antennas* which may be orientable. Such an antenna would enable the eavesdropper to receive more powerful signals in one direction while other directions would have reduced power. If jammers are sparse enough, eavesdroppers could avoid interference. We model the beam of a directional antenna as a cone of opening angle θ, centered at the eavesdropper. Under this (simplified) model, the eavesdropper receives a signal, from a transmitter only if it lies in this cone.

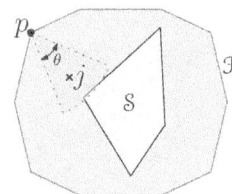

Figure 7: Jammer j needs to lie in the range of the directional antenna of eavesdropper p to affect p.

Given a discrete set of candidate locations $\mathcal{J}$, We need to find jammer locations and/or power assignment such that no such direction exists, so that for every cone orientation and location, there exists a jammer in this cone which would jam the signal from any point in $\mathcal{S}$ in the cone; see Figure 7. It is important to note that this is particularly applicable to RFID communication because, due to the low frequencies of RFID tags (13.56 Mhz), θ would be relatively large.

THEOREM 4. *Given θ, the opening angle of a directional antenna used by the eavesdropper, it is possible to find an ε-approximation to both power allocation and jammer placement problems by solving a linear program with polynomial number of constraints.*

PROOF SKETCH. First, we note that, for a point on $\mathcal{F}$, if there exists an orientation of the directional antenna where no jammer in $\mathcal{J}$ exists in the cone at this orientation, then it is not possible to jam this point. Thus, we assume that there does not exist any location on the fence with an orientation that contains no points in $\mathcal{J}$.

First, we obtain the set $\mathcal{F}'$ as in Section 4. However, to the set $\mathcal{F}'$, we add further points to obtain a new set $\mathcal{F}''$. Consider a point $p \in \mathcal{F}$. If we perform a circular sweep of a cone Ψ with p as apex, we have many "events" corresponding to some point $j \in \mathcal{J}$ or some vertex v of $\mathcal{S}$ added/deleted from Ψ. The number of such events is $O(n + |\mathcal{J}|)$. The set $\mathcal{F}''$ is constructed in such a manner that for each interval (u, v) on the fence obtained from consecutive points $u, v \in \mathcal{F}''$, two points in (u, v) have the same order of the sweep events.

For a location $p'' \in \mathcal{F}''$, we add a constraint for each "event" of the circular sweep. In between the events, the closest point in the storage is farther than at one of the events and the set of jammer candidates is the same. Since the number of events is $O(n + |\mathcal{J}|)$, we have only a polynomial number of constraints in total. □

7. SIMULATIONS

We conducted preliminary experiments to compare the NJ and Fullmodels. The setting we have chosen is the storage/fence shown in Figure 9. The fence is of dimensions 50x33 units and we placed a grid of 1x1 cells in the entire region. We simulated both JAMMING-LPand JAMMING-ILPin this setting. For the power assignment from JAMMING-LP, we investigated the difference in power and for JAMMING-ILP, we investigated the difference in number of jammers. Finally, we observed the variation in total power assigned with ε and δ and the number of jammers placed with ε, δ and $\hat{P}$. We chose the following values: (i) $\varepsilon = \{0.1, 0.2, 0.3, 0.4, 0.5\}$, (ii) $\delta = \{0.5, 0.6, \ldots, 1\}$, (iii) $\hat{P} = \{(1/\delta), (2/\delta), \ldots, (5/\delta)\}$. In both Full and NJ, we removed all grid points which were in the forbidden region.

For JAMMING-LP, we picked 10 random points from this set of grid points, repeated the simulation 50 times and calculated the mean and variance. Figure 8(a) shows the variation in total relative power with δ, which indicates how much more capable the eavesdropper is than legitimate receivers. As the eavesdropper gets more capable than storage receivers, the drop in the total relative power under NJ model is sharper than under Full model. The gap between them seems to be no more than constant-factor (approximately 2-3 times) but is definitely not negligible. However, the variance in NJ is also extremely high (ranging from aroung 60 to 100 vs 5 to 20 for the Full model). Possibly, the random selection of jammer locations leads to the large variance over different choices. The variance is likely to be much more in NJ model because each jammer contributes all the interference at a large number of nodes instead of

only being a part of the entire jammer set. This emphasizes the importance of carefully locating the jammers. We conclude that, in practical scenarios, it would be of benefit to consider the combined problem of location and power assignment rather than computing an optimal power assignment for a naive placement of jammers. Further, the graph indicates that as the eavesdropper gets more and more capable, the effectiveness of the NJ model diminishes.

For JAMMING-ILP, the candidate jammer locations were all the points on the grid. In total, there are 1121 points. Figure 8(b) and Figure 8(c) show the variation of the number of jammers located with the power assigned and with δ, respectively. In this case, we note that NJ model and Full model are not far apart thus demonstrating the benefits of NJ model in this example setting. We noted that there was no significant variation in total relative power or number of jammers with ε indicating that even choosing large values of ε would yield results better than theoretical guarantees.

8. CONCLUSION

We considered the problem of friendly jamming under the storage/fence environment model when jammers are both cooperative and non-cooperative. We presented ε-approximation algorithms for the problem of assigning transmission powers to a set of fixed jammers as well as for selecting a minimum number of jammers from a discrete candidate set. We also presented an algorithm to place a near-optimal number of jammers when the jammers may be located anywhere between the storage and fence under a restricted setting. The former algorithms were extended to provide a combined solution where we are interested in achieving a tradeoff between number of jammers and power consumption, as well as to the setting where eavesdroppers may be equipped with directional antennas. Our preliminary simulations validated the theoretical results and show that the simpler non-cooperative model may not be significantly different than the cooperative model. Further, for the power assignment problem, the simulations also show that careful location of the jammers is paramount and further emphasizes the importance of the jammer placement problem.

Finally, we mention two directions in which this work may be extended: (i) When jammers use directional antennas, although the power of the jammers is now concentrated, it is not clear whether we may obtain better results. For example, if the storage and fence were convex regions, then using directional antennas could remove the jammers' effect on the storage but in general, it is not obvious whether this would be the case. Moreover, the region of influence of jammers is reduced leading to possibly a greater number of required jammers. (ii) It would also be constructive to see if battery-power jammers may be deployed and if we may schedule their transmissions so as to successfully protect the storage while maximizing their operating time.

Acknowledgments

The authors would like to thank Paz Carmi, Esther Ezra, Matya Katz, Anantha Raman Krishnan and Zvika Lotker for enlightening discussions. The authors would also like to thank the anonymous reviewers for their useful comments.

Valentin Polishchuk is supported by Academy of Finland grant 135280.

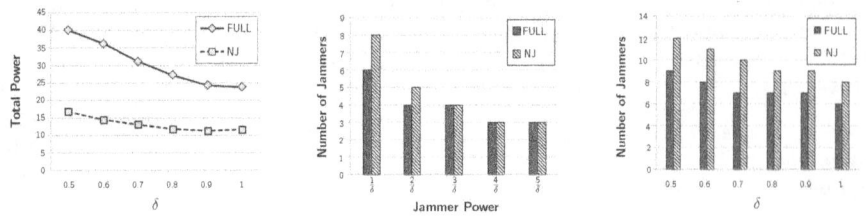

(a) JAMMING-LP: *Total Power vs eavesdroppers' capability (δ).*

(b) JAMMING-ILP: *Number vs power of jammers.*

(c) JAMMING-ILP: *Number of jammers vs eavesdroppers' capability (δ)*

Figure 8: Results of simulations with Jamming-LPand Jamming-ILP.

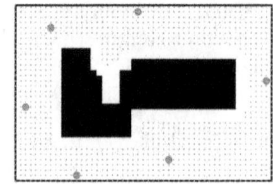

Figure 9: Storage/fence with candidate locations (small dots) and solution of Jamming-ILP(Large dots).

9. REFERENCES

[1] L. V. Ahlfors. *Complex analysis: an introduction to the theory of analytic functions of one complex variable.* International series in pure and applied mathematics. McGraw-Hill, 1979.

[2] F. Aurenhammer. Voronoi diagrams-a survey of a fundamental geometric data structure. *ACM Comput. Surv.*, 23(3):345–405, 1991.

[3] C. Commander, P. Pardalos, V. Ryabchenko, S. Uryasev, and G. Zrazhevsky. The wireless network jamming problem. *J. Comb. Optim.*, 14(4):481–498, 2007.

[4] M. de Berg, M. van Kreveld, M. Overmars, and O. Schwarzkopf. *Computational Geometry: Algorithms and Applications.* Springer-Verlag, 2000.

[5] P. Gupta and P. R. Kumar. The capacity of wireless networks. *IEEE T. Inform. Theory*, 46(2):388–404, 2000.

[6] Z. Han, N. Marina, M. Debbah, and A. Hjørungnes. Physical layer security game: Interaction between source, eavesdropper, and friendly jammer. *EURASIP J. Wirel. Comm.*, pp. 11:1–11:10, Jun. 2009.

[7] T. Heydt-Benjamin, D. Bailey, K. Fu, A. Juels, and T. O'Hare. Vulnerabilities in first-generation rfid-enabled credit cards. In *Financial Cryptography and Data Security*, vol. 4886 of *LNCS*, pp. 2–14. Springer, 2007.

[8] A. Juels. Rfid security and privacy: a research survey. *IEEE J. Sel. Area. Comm.*, 24(2):381–394, 2006.

[9] A. Juels and J. Brainard. Soft blocking: flexible blocker tags on the cheap. In *Proc. 2004 ACM Workshop on Privacy in the Electronic Society*, pp. 1–7, 2004.

[10] A. Juels, R. L. Rivest, and M. Szydlo. The blocker tag: selective blocking of rfid tags for consumer privacy. In *Proc. 8th ACM Conf. on Computer and Communications Security*, pp. 103–111, 2003.

[11] D. G. Kirkpatrick. Efficient computation of continuous skeletons. In *Proc. 20th Annual Sympos. on Foundations of Comp. Sci.*, pp. 18–27, 1979.

[12] L. Lai and H. El Gamal. The relay-eavesdropper channel: Cooperation for secrecy. *IEEE T. Inform. Theory*, 54(9):4005 –4019, 2008.

[13] D. T. Lee and R. L. Drysdale. Generalization of voronoi diagrams in the plane. *SIAM J. Comput.*, 10(1):73–87, 1981.

[14] D. Malan, T. Fulford-Jones, M. Welsh, and S. Moulton. Codeblue: An ad hoc sensor network infrastructure for emergency medical care. In *Proc. 1st Int. Workshop on Wearable and Implantable Body Sensor Networks*, pp. 55–58, 2004.

[15] R. Negi and S. Goel. Secret communication using artificial noise. In *Proc. IEEE 62nd Vehicular Technology Conf.*, pp. 1906–1910, 2005.

[16] J. Nehmer, M. Becker, A. Karshmer, and R. Lamm. Living assistance systems: an ambient intelligence approach. In *Proc. 28th Int. Conf. on Software Engineering*, pp. 43–50, 2006.

[17] K. Nohl, D. Evans, Starbug, and H. Plötz. Reverse-engineering a cryptographic rfid tag. In *Proc. USENIX Security Sympos.*, pp. 185–193, 2008.

[18] P. Peris-Lopez, J. Hernandez-Castro, J. Estevez-Tapiador, and A. Ribagorda. Rfid systems: A survey on security threats and proposed solutions. In *Personal Wireless Communications*, vol. 4217 of *LNCS*, pp. 159–170. Springer, 2006.

[19] A. Perrig, J. Stankovic, and D. Wagner. Security in wireless sensor networks. *Commun. ACM*, 47(6):53–57, 2004.

[20] T. Rappaport. *Wireless Communications: Principles and Practice.* Prentice Hall PTR, 2001.

[21] N. X. P. Semiconductors. Mifare classic 1k. http://www.nxp.com/documents/data_sheet/MF1S50YYX.pdf.

[22] M. Simplício Jr., P. Barreto, C. Margi, and T. Carvalho. A survey on key management mechanisms for distributed wireless sensor networks. *Comp. Netw.*, 54(15):2591–2612, 2010.

[23] X. Tang, R. Liu, P. Spasojevic, and H. V. Poor. Interference assisted secret communication. *IEEE T. Inform. Theory*, 57(5):3153–3167, 2011.

[24] A. Vigneron. Geometric optimization and sums of algebraic functions. In *Proc. 21st Annual ACM-SIAM Sympos. on Discrete Algorithms*, pp. 906–917, 2010.

[25] J. P. Vilela, M. Bloch, J. Barros, and S. W. McLaughlin. Wireless secrecy regions with friendly jamming. *IEEE T. Inf. Foren. Sec.*, 6(2):256–266, 2011.

[26] Y. Wang, G. Attebury, and B. Ramamurthy. A survey of security issues in wireless sensor networks. *IEEE Communications Surveys Tutorials*, 8(2):2–23, 2006.

[27] W. Weber, J. M. Rabaey, and E. H. L. Aarts, editors. *Ambient intelligence.* Springer-Verlag, 2005.

[28] A. D. Wyner. The Wire-tap Channel. *Bell Systems Technical Journal*, 54(8):1355–1387, 1975.

Taming Uncertainties in Real-Time Routing for Wireless Networked Sensing and Control

Xiaohui Liu, Hongwei Zhang, Qiao Xiang, Xin Che, Xi Ju
Department of Computer Science, Wayne State University, USA
{xiaohui,hongwei,xiangq27,chexin,xiju}@wayne.edu

ABSTRACT

Real-time routing is a basic element of closed-loop, real-time sensing and control, but it is challenging due to dynamic, uncertain link/path delays. The probabilistic nature of link/path delays makes the basic problem of computing the probabilistic distribution of path delays NP-hard, yet quantifying probabilistic path delays is a basic element of real-time routing and may well have to be executed by resource-constrained devices in a distributed manner; the highly-varying nature of link/path delays makes it necessary to adapt to in-situ delay conditions in real-time routing, but it has been observed that delay-based routing can lead to instability, estimation error, and low data delivery performance in general. To address these challenges, we propose the *Multi-Timescale Estimation (MTE)* method; by accurately estimating the mean and variance of per-packet transmission time and by adapting to fast-varying queueing in an accurate, agile manner, MTE enables accurate, agile, and efficient estimation of probabilistic path delay bounds in a distributed manner. Based on MTE, we propose the *Multi-Timescale Adaptation (MTA)* routing protocol; MTA integrates the stability of an ETX-based directed-acyclic-graph (DAG) with the agility of spatiotemporal data flow control within the DAG to ensure real-time data delivery in the presence of dynamics and uncertainties. We also address the challenges of implementing MTE and MTA in resource-constrained devices such as TelosB motes. We evaluate the performance of MTA using the NetEye and Indriya sensor network testbeds. We find that MTA significantly outperforms existing protocols, e.g., improving deadline success ratio by 89% and reducing transmission cost by a factor of 9.7.

Categories and Subject Descriptors

C.2.2 [**Computer-Communication Networks**]: Network Protocols—*Routing Protocols*

General Terms

Algorithms, Measurement, Performance

Keywords

Wireless sensing and control networks, real-time routing, delay quantile estimation, multi-timescale estimation, multi-timescale adaptation

1. INTRODUCTION

Besides deployments for open-loop sensing such as environmental monitoring, embedded wireless networks are increasingly being explored for real-time, closed-loop sensing and control. For instance, the wireless networking standard IEEE 802.15.4g has been defined for large scale process control applications such as smart grid sensing and control [3], and wireless networks are expected to serve as major communication infrastructures in neighborhood area networks and home area networks of the smart grid [18, 49, 44]. In addition, wireless networking standards such as the IEEE 802.15.4e, WirelessHART, and ISA SP100.11a have been defined for industrial monitoring and control [4, 46, 41], wireless sensor networks have been deployed for industrial automation [32, 45], and the automotive industry has also been exploring the application of wireless networks to vehicle sensing and control [19, 42]. In wireless networked sensing and control, message passing (or messaging for short) across wireless networks is a basic enabler for coordination among distributed sensors, controllers, and actuators. In supporting mission-critical tasks such as smart grid control and industrial process control, wireless messaging is required to be reliable (i.e., having high delivery ratio) and in real-time [38]. This is because packet loss and large delay usually reduce the system stability (e.g., in proportional-integral control), lengthen the settling time, and increase the maximum overshoot in control [25].

In multi-hop wireless networks, a basis for reliable, real-time messaging is real-time routing which routes data packets from their sources to destinations within specified deadlines. Nonetheless, link/path delays (i.e., the time taken to successfully deliver a packet across a link or a path) are dynamic, uncertain in wireless sensing and control (WSC) networks due to factors such as the spatiotemporal wireless link dynamics and the queueing dynamics along links/paths. The dynamics and uncertainties in link/path delays introduce fundamental *challenges* to real-time routing:

- Firstly, the dynamics and uncertainties make link/path delays probabilistic in nature. Given the delay distributions of individual links along a path, the basic problem of computing the probabilistically guaranteed path delays is NP-hard [23]; in real-time routing where nodes need to identify paths that ensure certain delay bounds, however, quantifying probabilistic path delays is a basic task, and it may well have to be executed by resource-constrained nodes in a distributed manner.

- Secondly, given that link/path delay is a highly varying metric and that it can change at a short timescale of each packet transmission, it is important to adapt to in-situ delay conditions in routing. Yet the highly-varying nature of link/path delays makes it difficult to accurately estimate path delays in a distributed and agile manner, and it has been observed that delay-adaptive routing can lead to routing instability and low data delivery performance in general [15, 53].

Despite much work in throughput- or energy-efficiency-oriented wireless routing [7, 14, 22, 51], real-time routing is much less studied. Moreover, the existing work that do consider data delivery delay in wireless routing either only try to minimize average path delay without ensuring probabilistic delay bounds [47, 21, 24, 33, 48], or they do not address the challenges that delay uncertainties pose to the task of quantifying probabilistic path delays and the task of addressing instability of delay-adaptive routing [27, 39]. Therefore, how to enable real-time routing in the presence of dynamic, uncertain link/path delays remains an important open problem for real-time wireless networked sensing and control.

Towards enabling routing with probabilistic delay bounds in WSC networks, we propose the *Multi-Timescale Adaptation (MTA)* routing protocol that addresses the aforementioned challenges of dynamic, uncertain link/path delays in real-time routing. In MTA, nodes leverage the different timescales of dynamics to accurately estimate probabilistic path delay bounds in an agile manner and to adapt spatiotemporal data flow control at the same timescales of the dynamics themselves. More specifically, we make the following contributions:

- For accurate, agile estimation of probabilistic path delay bounds, we decompose contributors to path delay uncertainties into two factors: dynamic per-packet transmission time (which we refer to as *packet-time* hereafter) and dynamic queueing along paths. Through detailed experimental analysis, we find that, given a network condition, the distribution of packet-time is quite stable despite the quick variation of instantaneous packet-time. This enables each node to accurately estimate the mean and variance of packet-time from itself to the next-hop along a path. We also observe that the packet-time for different packet transmissions, whether from the same node or from different nodes, are uncorrelated; this enables each node to compute, for a given time instant, the variance of the path delay from itself to the destination node as the sum of the variances of the packet-times for all the packets queued along the path at that instant. Based on these observations, we develop a multi-timescale approach, denoted by *multi-timescale estimation (MTE)*, to accurately estimate the highly-varying mean and variance of path delay by accurately estimating the mean and variance of packet-time and by adapting to fast-varying queueing in an accurate, agile manner. Using the mean and variance of path delay, we evaluate different methods of upper-bounding quantiles, and we identify Chebyshev Inequality as an effective basis for computing probabilistic delay bounds in constant time.

- For enabling adaptivity while addressing instability and low-performance in real-time routing, we propose the *Multi-Timescale Adaptation (MTA)* routing protocol: to facilitate the aforementioned multi-timescale estimation (MTE) and to avoid detrimental instability while ensuring data delivery performance during adaptation, a directed-acyclic-graph (DAG) is maintained at lower frequencies based on the relatively slow-varying link property ETX (i.e., expected number of transmissions taken to successfully deliver a packet),

which reflects network throughput, data delivery reliability, and the overall trend of data delivery delay [14, 16]; at higher frequencies and based on the MTE method, the data flow within the DAG is controlled on a per-packet basis to minimize ETX and to ensure packet delivery within the required probabilistic delay bound. By ensuring overall stability and performance while addressing short-term dynamics at the same time, MTA enables efficient, real-time routing in the presence of complex dynamics and uncertainties.

- We implement MTE and MTA in TinyOS, and we address the challenges of limited memory, limited CPU capability, and the lack of real-time operation support in TinyOS. Besides the running MTA protocol, these implementation strategies may well be of interest to real-time routing in general.

- We evaluate the performance of MTA and and other related work in the high-fidelity sensor network testbeds NetEye [5] and Indriya [2]. We find that MTA significantly outperforms existing protocols, e.g., improving deadline success ratio by 89% and reducing transmission cost by a factor of 9.7.

The rest of the paper is organized as follows. We briefly introduce the NetEye and the Indriya testbeds in Section 2. We present the MTE method and the MTA protocol in Sections 3 and 4 respectively, and then we present the measurement study in Section 5. We discuss related work in Section 6, and we make concluding remarks in Section 7.

2. PRELIMINARIES

Our study leverages two publicly available wireless sensor network testbeds NetEye [5] and Indriya [2]. In what follows, we briefly introduce the two testbeds and the traffic patterns we study.

NetEye testbed. NetEye [5] is deployed in a large lab space at Wayne State University. We use a subset of a 15×7 grid of TelosB motes in NetEye, where every two closest neighboring motes are separated by 2 feet. The subset of the grid forms a random network, and it is generated by removing each mote of the 15×7 grid with probability 0.2.

Each of these TelosB motes is equipped with a 3dB signal attenuator and a 2.45GHz monopole antenna. In our measurement study, we set the radio transmission power to be -25dBm (i.e., power level 3 in TinyOS) such that multihop networks can be created and the link reliability is over 90% for links up to 6 feet long. Given the high availability and high fidelity of NetEye, we mainly use NetEye in our measurement study, but we verify key observations using the Indriya testbed too.

Indriya testbed. Indriya [2] is deployed at three floors of the School of Computing at the National University of Singapore. Our measurement study uses all of its 127 TelosB motes, and we use a transmission power of -10dBm (i.e., power level 11 in TinyOS) to generate a well-connected multi-hop network where the link reliability is over 90% for links up to 20 feet long.

Traffic pattern. Using the two testbeds, we study both periodic and event traffic patterns.

For periodic traffic, we study three types of them based on the amount of queueing they introduce when the default TinyOS routing protocol CTP [22] is used: 1) light traffic: no queueing in network; 2) medium traffic: moderate queueing in network but with very rare queue overflow; 3) heavy traffic: severe queueing in network and with frequent queue overflow. To this end, we select one node as the sink and another 10 nodes as traffic sources; the sink and the sources are nearly at opposite positions in both testbeds to create as many routing hops as possible. In NetEye, more specifically, mote 15 is the sink, and motes 61, 62, 63, 64, 76, 77, 79, 91, 92, and 93 are the sources, with each source generating a packet every 1,000ms, 400ms, and 75ms for light, medium, and heavy traffic respectively. To verify observations from NetEye and given that the medium traffic is the common case in WSC networks, we also study medium traffic scenario in Indriya as follows: mote 105 at the third floor is the sink, and motes 1 . . . 10 at the first floor are the sources, with each source generating a packet every 600ms.

For event traffic, we use a publicly available event traffic trace for a field sensor network deployment [50, 1] to evaluate the performance of different protocols. The traffic trace corresponds to the packets generated in the 7×7 grid of a field mote network when a vehicle passes across the middle of the network. In NetEye, the 7×7 subgrid in the trace data corresponds to the 49 motes that are the farthest from the sink mote 15. In Indriya, the 7×7 subgrid is mapped to motes 1 . . . 49.

3. MULTI-TIMESCALE ESTIMATION OF PATH DELAYS

To enable routing with probabilistic delay guarantees, a basic task is to quantify probabilistic delays along paths. Given a path $P = v_0, v_1, \ldots, v_n, v_{n+1}$ from v_0 to v_{n+1} as shown in Figure 1 and assuming that the number of packets queued at node $v_i (i = 0 \ldots n)$ is $m_i(t)$ at time t and that the packet transmission scheduling algorithm at each node is first-come-first-serve (FCFS), the instantaneous delay along path P at time t, denoted by $d_P(t)$, is the delay that a packet arriving at v_0 at time t experiences in reaching the destination node v_{n+1} assuming that the number of queued packets at each node does not change while the packet is being delivered to v_{n+1}. Therefore,

$$d_P(t) = \sum_{i=0}^{n} \sum_{j=1}^{m_i(t)+1} d_i^j(t), \qquad (1)$$

where $d_i^j(t)$ is the time taken for v_i to deliver its j-th queued packet to v_{i+1} at time t.

Challenges of highly-varying delay distribution. Given the distributions of $d_i^j(t)(i = 0 \ldots n, j = 1 \ldots m_i(t) + 1)$, it is NP-hard to compute the distribution of $d_P(t)$ [23]. Since nodes of WSC networks tend to be resource-constrained (e.g., in CPU and memory), it is infeasible to directly compute the distribution of $d_P(t)$ in general. One alternative is to first sample $d_P(t)$ and then use non-parametric approaches (e.g., the P^2 algorithm) to estimate delay quantiles based on these delay samples [29, 35]. Nonetheless, non-parametric quantile estimation usually converges slowly, for instance, taking more than 200 samples [36]. Yet the distribution of path delays varies at a much shorter timescale than the sample size required for non-parametric quantile estimation to converge. As can be seen from (1), in particular, the distribution of path delays vary with the queueing levels (i.e., $m_i(t)$) along paths, and network queueing can vary quickly over time.[1] For the NetEye medium traffic scenario, for instance, Figure 2 show the histogram of the coherence window size for node queueing, where each coherence window of a time series is a maximal consecutive segment of the time series where all the sample values are the same. We see that the coherence window size is two orders of magnitude less

than the sample size required for non-parametric quantile estimation to converge. Therefore, the highly-varying nature of path delay distribution makes non-parametric quantile estimation unable to accurately estimate instantaneous path delay quantiles in an agile manner; that is, the distribution of the path delay changes before the estimation converges to an accurate value.

To circumvent the computational complexity and the inability of accurate, agile estimation of exact path delay quantiles, we propose to identify upper bounds on probabilistic path delays and to use the delay bounds in identifying paths for real-time data delivery. As we will discuss later in this section, upper bounds on path delays can be derived using probability inequalities such as Chebyshev Inequality [40], and the properly identified delay bounds are still orders of magnitude less than the maximum delays, thus enabling effective utilization of network real-time capacity. Most probability inequalities use the mean and/or the standard deviation of the corresponding random variable, thus we need an accurate, agile mechanism of estimating the mean and standard deviation of path delays. Nonetheless, the sample size required for accurate estimation of path delay statistics tend to be quite large, for instance, being greater than 100 in most cases and can be up to 594 [36], which are significantly greater than the coherence window size of node/path queueing levels. In addition, node queueing levels tend to have low autocorrelation, not to mention staying unchanged, for time lags greater than 100 samples, as shown in Figures 3. Thus, the large sample size requirement and the highly-varying nature of path delay distribution makes it impossible to accurately estimate instantaneous mean path delay in an agile manner; that is, the mean path delay changes before the estimation converges to an accurate value.

Multi-Timescale Estimation (MTE). Towards addressing the challenges of highly-varying distribution and statistics of path delay, we decompose contributors to path delay variations into two factors: dynamic packet-time and dynamic queueing. By leveraging the different timescales at which packet-time and queueing vary, we propose the multi-timescale-estimation (MTE) method that accurately estimates the highly-varying mean and variance of path delay by 1) accurately estimating the mean and variance of packet-time at a longer timescale and 2) by adapting to fast-varying queueing at a shorter timescale. In what follows, we elaborate on the design of the MTE method.

OBSERVATION 1. *Packet-time distribution is stable.*

Through detailed experimental analysis, we find that, given a network condition, the distribution of packet-time is quite stable despite the quick variation of instantaneous packet-time. Using the Generalized KPSS test [26], we analyze the stationarity window size of packet-time, where each stationary window of a time series is a maximal consecutive segment of the time series that is weak-sense stationary (i.e., with a constant mean and variance over time). Figure 4 shows the histogram of the stationarity window size for packet-time in the NetEye medium traffic scenario. The minimum and the maximum window size are 1,901 and 21,937 respectively, both of which are significantly greater than the sample size required to precisely estimate the mean and variance of packet-time. For instance, Figure 5 shows the empirical CDF of the sample size required for estimating the mean packet-time at 90% accuracy and 90% confidence level. The sample size requirement is less than 360 with over 99% probability, and the maximum sample size requirement is 600.

The stability of the packet-time distribution has two important implications:

---

[1] As we will discuss shortly, the distribution of packet-time (i.e., $d_i^j(t)$) is quite stable over time, and we leverage the stability of packet-time distribution in designing our multi-timescale estimation (MTE) method.

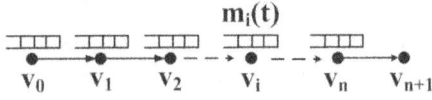

Figure 1: An example path P

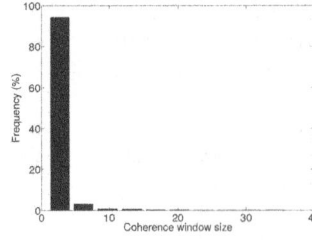

Figure 2: Histogram of the coherence window size for node queueing level

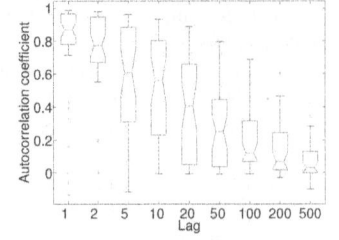

Figure 3: Autocorrelation of node queueing levels

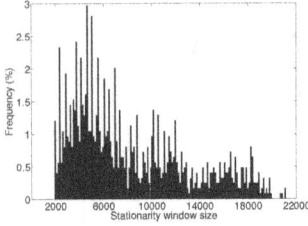

Figure 4: Histogram of stationarity window size for packet-time

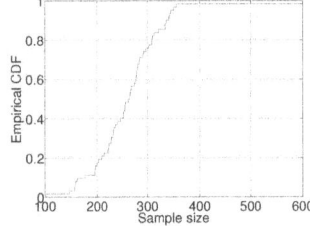

Figure 5: Empirical CDF of the sample size required for estimating the mean packet-time at 90% accuracy and 90% confidence level

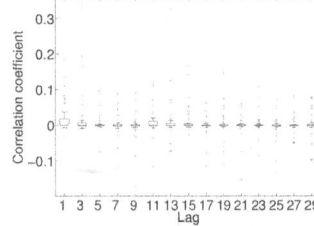

Figure 6: Autocorrelation coefficient of packet-time along the same link

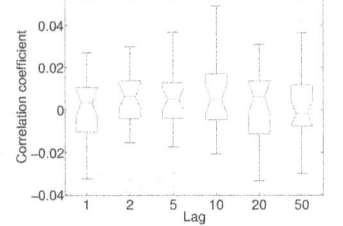

Figure 7: Correlation coefficient between the packet-time across different links along a path

- Given the real-time requirement on data delivery and the constrained memory size in embedded WSC networks, the stationarity window size tends to be greater than the maximum node queue size in general. This implies that, in Formula (1), the mean and variance for $d_i^j(t)$'s $(j = 1 \ldots m_i(t) + 1)$ are the same for a given $i(i = 0 \ldots n)$, and we denote them as $\mu(d_i(t))$ and $\sigma^2(d_i(t))$. Therefore, the mean path delay can be computed as follows:

$$\mu(d_P(t)) = \sum_{i=0}^{n} (m_i(t) + 1)\mu(d_i(t)). \tag{2}$$

- The stability of packet-time distribution enables each node to accurately estimate the mean and variance of packet-time from itself to the next-hop along a path, and, with Formula (2), the mean path delay can be accurately estimated.

OBSERVATION 2. *Packet-time is uncorrelated.*

We also observe that the packet-time for different transmissions, whether from the same node or from different nodes, tend to be uncorrelated. For instance, Figure 6 shows the small autocorrelation coefficient of packet-time along the same link, with the box-plot showing, for each lag, the distribution of autocorrelation coefficients across different links; Figure 7 shows the small correlation coefficient of packet-time along different links, where the lag is defined as the number of packets queued between two packets at different node queues as shown in Figure 1, assuming that the packets queued at time t flow through the nodes queues in a FIFO manner. We also observe that the median autocorrelation coefficient and the median cross-link correlation coefficient of packet-time is zero at 99% confidence level. An intuition for the uncorrelatedness between packet-time is that, given a network condition, the behavior

of one packet transmission does not have much impact on the behavior of another packet transmission as far as the MAC protocol is concerned.

The Bienaymé Formula [37] shows that the variance of the sum of pair-wise uncorrelated random variables is the sum of the variances of the individual random variables. Thus our observations on the uncorrelatedness of packet-time along a path enables each node to compute, for a given time instant, the standard deviation of the path delay from itself to the destination node as the square root of the sum of the variances of the packet-times for all the packets queued along the path at that instant. For path P of Figure 1, for instance, the standard deviation of $d_P(t)$ can be computed as follows:

$$\sigma(d_P(t)) = \sqrt{\sum_{i=0}^{n} (m_i(t) + 1)\sigma^2(d_i(t))}. \tag{3}$$

For a typical 5-hop path in NetEye, Figure 8 shows the histogram of the relative errors in estimating the standard deviation of path delay in the presence of different queueing levels along the path. We see that the estimation is quite accurate, with most relative errors within the range of (-0.075, 0.075). Note that, if we directly estimate the variance of the sojourn time $d_i^{so}(t) = \sum_{j=1}^{m_i(t)+1} d_i^j(t)$ at node v_i without decomposing $d_i^{so}(t)$ into its individual components $d_i^j(t)$'s $(j = 1 \ldots m_i(t) + 1)$, we cannot compute $\sigma(d_P(t))$ as $\sqrt{\sum_{i=0}^{n} \sigma^2(d_i^{so}(t))}$. This is because, for $i1 \neq i2$, $d_{i1}^{so}(t)$ and $d_{i2}^{so}(t)$ are correlated due to the correlation between queueing levels at different nodes of a path.

In the simplified scenario of Figure 1, the next-hops of all the packets in a node queue are the same. In reality, a node may well use different next-hops for different packets, for instance, depending on packet deadlines as we will discuss in Section 4. Assume that, at time t, each node $v_i(i = 0 \ldots n)$ has $N_i(t)$ number of

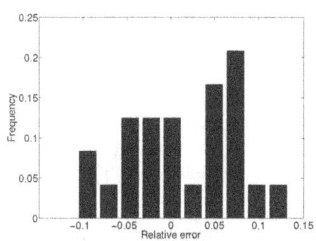

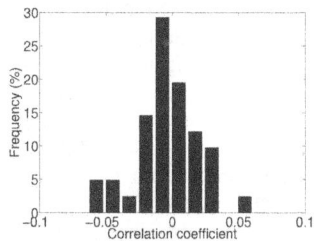

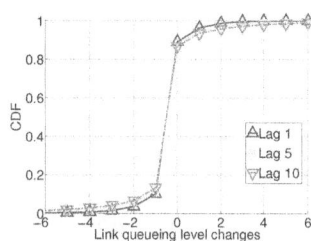

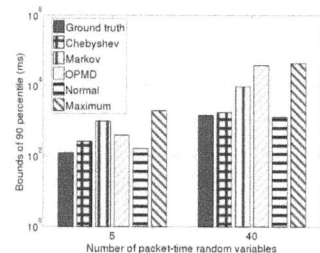

Figure 8: Histogram of relative errors in estimating the standard deviation of path delay

Figure 9: Histogram of the correlation coefficient between the packet-time across different outgoing links from the same node

Figure 10: CDF of link queueing level changes

Figure 11: Bounds on 90-percentile of path delay

next-hops, the number of packets (including the one arriving at v_0 at time t) to be forwarded to its k-th next hop is $m_i^k(t)$, and the packet-time from v_i to its k-th next-hop is $d_{i,k}(t)$. We observe that, given v_i and t, $d_{i,k}(t)$'s tend to be uncorrelated for different ks. For instance, Figure 9 shows the histogram of the correlation coefficient between the packet-time across different outgoing links from the same node. The correlation coefficient is very small and is less than 3% most of the time. We also observe that the median correlation coefficient is zero at 99% confidence level. Then, based on Formulae (2) and (3), the mean and standard deviation of the delay along a path P can be computed as follows:

$$\mu(d_P(t)) = \sum_{i=0}^{n} \sum_{k=1}^{N_i(t)} m_i^k(t)\mu(d_{i,k}(t))$$
$$\sigma(d_P(t)) = \sqrt{\sum_{i=0}^{n} \sum_{k=1}^{N_i(t)} m_i^k(t)\sigma^2(d_{i,k}(t))} \quad (4)$$

OBSERVATION 3. *Network queueing is relatively stable at short timescales.*

To leverage Formula (4) in estimating the mean and variance of the delay along path P in a distributed manner, each node v_i can compute the mean and variance of the delay from itself to the destination node v_{n+1} based on those of its next-hop along P. Denoting the mean and standard deviation of the delay from v_i to v_{n+1} by $\mu(d_P^i(t))$ and $\sigma(d_P^i(t))$ respectively ($i = 0 \ldots n$), then we have

$$\mu(d_P^i(t)) = \mu(d_P^{i+1}(t)) + \sum_{k=1}^{N_i(t)} m_i^k(t)\mu(d_{i,k}(t))$$
$$\mu(d_P^{n+1}(t)) = 0 \quad (5)$$

$$\sigma(d_P^i(t)) = \sqrt{\sigma^2(d_P^{i+1}(t)) + \sum_{k=1}^{N_i(t)} m_i^k(t)\sigma^2(d_{i,k}(t))}$$
$$\sigma(d_P^{n+1}(t)) = 0 \quad (6)$$

Formulae (7) and (8) can be implemented using a distance-vector-type routing algorithm. Due to information diffusion delay τ_i from v_{i+1} to v_i in routing, the implemented version of Formulae (7) and (8) are

$$\mu(d_P^i(t)) = \mu(d_P^{i+1}(t - \tau_i)) + \sum_{k=1}^{N_i(t)} m_i^k(t)\mu(d_{i,k}(t))$$
$$\mu(d_P^{n+1}(t)) = 0 \quad (7)$$

$$\sigma(d_P^i(t)) = \sqrt{\sigma^2(d_P^{i+1}(t - \tau_i)) + \sum_{k=1}^{N_i(t)} m_i^k(t)\sigma^2(d_{i,k}(t))}$$
$$\sigma(d_P^{n+1}(t)) = 0 \quad (8)$$

For accurate estimation of $\mu(d_P^i(t))$ and $\sigma^2(d_P^i(t))$, we need to make τ_i as small as possible. This can be achieved by having v_{i+1} piggyback $\mu(d_P^{i+1}(t))$ and $\sigma^2(d_P^{i+1}(t))$ onto its data transmissions as well as control signaling (e.g., broadcast of routing beacons) so

that v_i can overhear the values quickly. In WSC networks, sample data about physical behavior are usually generated periodically in a continuous manner, thus τ_i is at the same timescale of inter-packet arrival interval, which enables quick diffusion of path delay statistics.

We also observe that, even though network queueing varies significantly at a long timescale of hundreds of inter-packet intervals, it is much more stable at a short timescale of a few inter-packet intervals. In the NetEye medium traffic scenario and for the time lags of 1, 5, and 10 packet transmissions, for instance, Figure 10 shows the empirical cumulative distribution function (CDF) of link queueing level changes (i.e., changes in $m_i^k(.)$). We see that, at the timescale of a few inter-packet intervals, network queueing remains relatively stable, and, with more than 90% probability, the absolute changes in link queueing levels are no more than 1. To ensure enough real-time capacity for each source node, routing hops tend to be limited (e.g., less than 10) in WSC networks [6]. This fact, together with the quick information diffusion and the relative stable network queueing at short timescales, enables the MTE method to accurately estimate the mean and variance of path delays in an agile manner.

Probabilistic path delay bound. With the mean and variance of path delay estimated via the MTE method, a node can derive the probabilistic path delay bounds using probability inequalities. To this end, we have

PROPOSITION 1. *For a random variable X, if*

$$Pr\{X \geq f(x)\} \leq g(x), \quad (9)$$

then $Q_X^q = f(g^{-1}(1-q))$ is an upper bound on the q-quantile of X, where g^{-1} is the inverse function of g.

PROOF. Let $g(x) = 1 - q$, then

$$Pr\{X \geq f(x)\} \leq 1 - q.$$

Thus,

$$Pr\{X \leq f(x)\} \geq q$$

Since $x = g^{-1}(1-q)$, we have

$$Pr\{X \leq f(g^{-1}(1-q))\} \geq q$$

Thus, $f(g^{-1}(1-q))$ is an upper bound on the q-quantile of X. $\square$

For $\lambda > 0$ and a non-negative random variable X with mean μ and variance σ^2, two widely-applicable probability inequalities [43] are the Markov Inequality

$$Pr\{X \geq \lambda\} \leq \frac{\mu(X)}{\lambda}, \quad (10)$$

and the one-tailed Chebyshev Inequality

$$Pr\{X - \mu \geq \lambda\sigma\} \leq \frac{1}{1+\lambda^2}. \qquad (11)$$

Thus, we have

COROLLARY 1. *Using Markov Inequality, $Q_X^q = \frac{\mu}{1-q}$.*

COROLLARY 2. *Using one-tailed Chebyshev Inequality, $Q_X^q = \mu + \sigma\sqrt{\frac{q}{1-q}}$.*

For a typical five-hop path in NetEye, Figure 11 shows the ground truth and the estimated 90-percentile of the path delay when it is the sum of 5 and 40 packet-time random variables respectively. For comparison purpose, we present the probabilistic delay bounds estimated via the optimal-partition-minimum-delay (OPMD) method [35] or by assuming path delay is normally distributed [31]; we also present the maximum path delay. Given that Markov Inequality usually gives a looser bound than one-tailed Chebyshev Inequality, the probabilistic delay bound by Markov Inequality is greater than that by Chebyshev Inequality. Compared with the method of assuming path delay is normally distributed, the bound by Chebyshev Inequality is always greater than the actual 90-percentile path delay, whereas the former method underestimates the 90-percentile for the case of 40 packet-time random variables. Given that the OPMD method is rather conservative in estimating path delay bound, the bound by Chebyshev Inequality is also much less than the bound by the OPMD method, especially when path queueing increases. We also see that the maximum path delay is orders of magnitude greater than the bound by Chebyshev Inequality, thus using probabilistic delay bound instead of maximum path delay helps improve application-usable real-time capacity. Since the bound by Chebyshev Inequality upper-bounds and is close to the probabilistic path delay, we use the probabilistic delay bound by Chebyshev Inequality in our protocol design.

From FCFS to EDF. In real-time scheduling, the earliest-deadline-first (EDF) algorithm is a commonly used algorithm, and it can ensure a smaller deadline than what is feasible with the first-come-first-serve (FCFS) algorithm. Our MTE method of computing probabilistic path delay bounds is derived by assuming that nodes use FCFS algorithm in intra-node transmission scheduling, but we observe that the delay bounds derived via our FCFS-based MTE method can also serve as the basis of selecting real-time packet forwarding paths even if the EDF algorithm is used. Formally,

PROPOSITION 2. *If routed along a path whose probabilistic delay bound computed via the FCFS-based MTE method is less than the relative deadline of a packet, the packet will reach its destination before its deadline even if the EDF algorithm is used for intra-node transmission scheduling.*

(Interested readers can find the proof in [36].)

4. MULTI-TIMESCALE ADAPTATION FOR REAL-TIME ROUTING

Real-time routing is subject to dynamics and uncertainties at multiple timescales. At a longer timescale, link properties such as ETX (i.e., expected-number-transmissions to successfully deliver a packet) vary as a result of changing environmental conditions (e.g., temperature); at a shorter timescale, data transmission delay varies on a per-packet basis, and bursty traffic may introduce sudden changes to network conditions. Robust system design usually requires adaptation to dynamics at the same or shorter timescales of the dynamics themselves. Yet we have found that, due to the highly-varying nature of path delays, routing using delay-based metrics can introduce large estimation errors and lead to routing instability as well as low performance [53]. To ensure long-term stability and optimality while addressing short-term dynamics at the same time, we propose a multi-timescale adaptation (MTA) framework for real-time routing as follows.

At lower frequencies, a directed-acyclic-graph (DAG) is maintained for data forwarding, and any path within the DAG is a candidate path for packet delivery. Given that link/path ETX reflects network throughput, data delivery reliability, and the overall trend of data delivery delay [14, 16], and that ETX-based routing structures tend to be stable even if ETX is dynamic [52], we propose to maintain the data forwarding DAG based on link and path ETX such that the DAG reflects long-term system optimality and changes relatively slowly compared with delay variation. More specifically, there is a directed edge from node v_i to v_i' in the DAG if and only if the minimum path ETX from v_i' to the destination v_0' is less than that from v_i to v_0'. The DAG defines, for each node v_i, a set of forwarder candidates $\mathcal{R}(v_i)$ where $v_i' \in \mathcal{R}(v_i)$ if and only if link $\langle v_i, v_i' \rangle$ belongs to the DAG.

At higher frequencies, the spatiotemporal flow of packets within the data forwarding DAG is adaptively controlled to ensure reliable, real-time data delivery in the presence of short-timescale dynamics such as transient packet losses and per-packet variations of link delay. More specifically, each packet contains information about the remaining time to deadline, denoted by L, and the required real-time guarantee probability q. When the packet reaches a node v_i, v_i first finds the set of forwarder candidates within the DAG, denoted by $\mathcal{R}'(v_i, L, q)$, that can ensure the real-time requirements L and q; then v_i sets the next-hop node of the packet as the node of the smallest path ETX to v_0' among all the nodes in $\mathcal{R}'(v_i, L, q)$, and v_i puts the packet in the transmission queue. Queued packets are then scheduled for transmission using the earliest-deadline-first (EDF) algorithm.

Using the above approach, packets are always routed along the minimum-ETX path that satisfies the real-time data delivery requirement. In the presence of heavy traffic load that induces queueing, this real-time forwarding mechanism creates the water-filling effect as follows: packets are delivered to the minimum-ETX path until the path cannot ensure the required timeliness of data delivery (e.g., due to queueing), at which point the forthcoming packets are delivered to the path with the second-minimum-ETX, and so on; once the delay along paths of less ETX decreases (e.g., due to reduced queueing), more packets will be delivered to those paths to fill them up. This process repeats at the same timescale of packet arrival process, and it enables real-time data delivery while ensuring as small ETX in data delivery as possible. Unlike traditional delay-based routing that may lead to instability and low performance, this quick adaptation of spatial packet flow is enabled by 1) the MTE method of accurate, agile estimation of probabilistic path delay bound at the same timescale of the changes in path delay distribution and by 2) the overall stability of spatial packet flow along the data forwarding DAG.

Due to the limitation of space, we delegate the system architecture and the implementation strategies for the MTA framework to [36].

5. MEASUREMENT EVALUATION

5.1 Methodology

Protocols. We have implemented the MTA framework (including the MTE method) in TinyOS. To understand the design decisions of MTA, we have comparatively studied MTA with its variants; due to the limitation of space, we delegate the detailed discussions to [36]. Towards understanding the benefits of MTA, we also comparatively study MTA with the following existing protocols:

- *MCMP*: a multi-path QoS routing protocol where end-to-end QoS requirements on reliability and timeliness are uniformly divided into per-hop reliability and timeliness requirements, upon which a node chooses the minimum number of next-hops to satisfy the per-hop requirements in data delivery [27];

- *MM*: the geographic routing protocol MMSPEED [21] that routes and schedules packet transmissions based on nodes' distances to destinations, packet delivery deadlines, and mean link delays; MMSPEED also tries to improve packet delivery reliability by transmitting packets along multiple paths; (Note: we denote MMSPEED as MM for the readability of figures to be presented in the next subsection.)

- *MM-CD*: same as MMSPEED but, instead of using the mean link delay, uses a conservative estimate of link delay that equals the sum of the mean delay and three times the standard deviation of the delay;

- *SDRCS*: similar to MMSPEED but, instead of using geographic distance, uses data forwarding hop count as the measure of distance, where the hop-count is computed based on received-signal-strength (RSS) between nodes [47]; data forwarding is through receiver contention similar to that in opportunistic routing;

- *CTP* : an ETX-based non-real-time routing protocol in TinyOS [22].

(Note: for MM and MM-CD which use node locations, we configured each node with its correct location in our TinyOS programs.)

Traffic & real-time requirements. We use all periodic and event traffic discussed in Section 2. In our experiments, we use 90% as the required real-time guarantee probability by default, but we have also experimented with the real-time guarantee probability of 99% and observed similar phenomena [36]. For differentiating the performance of different protocols, the deadline for each traffic scenario is chosen so that it is neither too stringent (that no protocol can support) nor too loose (that all protocols can support). In NetEye, the deadlines for light, medium, and heavy traffic are 250ms, 2 seconds, and 7.5 seconds respectively, and the deadline for event traffic is 10 seconds. In Indriya, the deadlines for periodic and event traffic are 2 seconds and 2.5 seconds respectively.

Due to the limitation of space, here we only present results for medium periodic traffic in NetEye and Indriya. Similar phenomena have been observed for other traffic scenarios; interested readers can find them in [36]. We have also experimented with networks of lower connectivity and have observed similar phenomena [36].

Metrics. For each combination of protocol, testbed, and traffic scenario, we run it for 10 times and evaluate protocol performance in terms of the following metrics:

- *Deadline success ratio (DSR)*: ratio of packets delivered to the sink before their deadlines;

- *Packet delivery ratio (PDR)*: ratio of packets delivered to the sink;

- *Number of transmissions per packet delivered (NTX)*: total number of transmissions, including retransmissions, divided by the number of unique packets delivered to the sink.

To understand protocol behavior in more detail, we also analyze the different causes for a packet to miss its deadline:

- *Overflow*: packet discarded due to node queue overflow;

- *Transmission failure*: a packet not delivered to the next hop even after the maximum number of retransmissions at a node;

- *Rejection*: no candidate path can ensure the required real-time delivery guarantee (i.e., deadline and probability) of a packet;

- *Expiration*: deadline expired before the packet reaches the sink, whether or not the packet is delivered to the sink.

5.2 Measurement results

NetEye. For periodic medium traffic in NetEye, Figures 12, 13, 14, and 15 show the deadline success ratio, packet delivery ratio, transmission cost, and packet delivery status respectively. We see that MTA always ensures the required real-time delivery performance, and the median deadline success ratio in MTA is 76%, 89%, 86%, 11%, and 38% higher than that in MCMP, MM, MM-CD, SDRCS, CTP respectively. The median number of transmissions per packet delivered in MTA is less than that in MCMP, MM, MM-CD, SDRCS, CTP by a factor of 3.7, 9.7, 6.5, 1.2, and 1.2 respectively.

One reason why the deadline success ratio in MCMP, MM, and MM-CD is very low is because a large fraction of packets are lost due to queue overflow as can be seen in Figure 15. They all try to use multiple paths to ensure data delivery reliability: at the sources, multiple copies of a packet can be sent; at the next hop, each of these copies can be multiplied again, and so on. This multi-path routing mechanism can lead to exponentially increasing number of copies of a packet, thus causing severe queue overflow and large transmission cost as shown in Figures 15 and 14 respectively. SDRCS does not have this problem because it does not use multi-path routing, instead it uses a data forwarding mechanism similar to opportunistic routing.

Another reason why MCMP, MM, MM-CD, and SDRCS do not perform well is because, by evenly dividing end-to-end QoS requirements into per-hop requirements, they implicitly assume that network conditions are uniform across the network which is usually not the case. Among these protocols, the negative impact of this assumption is relatively less severe in SDRCS because it uses signal strength as the basis of measuring forwarding distances and signal strength is a better metric for measuring wireless link quality than geographic distance. MTA does not have this problem because MTE enables accurate, agile estimation of end-to-end delay quantiles without assuming uniform network conditions. A third reason for the low performance of MM, MM-CD, and SDRCS is because they only consider mean delays instead of the probabilistic distributions of delays.

Compared with MTA, CTP has higher packet delivery ratio, but CTP only enables a median deadline success ratio of 56% which is lower than the real-time gurantee probability of 90% and much lower than the 93% probability guarantee by MTA. This is because CTP only considers path ETX in routing, and it is delay-unaware. Even if a low-ETX path is experiencing large delay due to queueing, CTP still uses the path, thus leading to large data delivery delay and deadline miss. Through accurate, agile estimation of path delay via MTE, in contrast, MTA can switch to a less congested path whenever it detects the inability of the current low-ETX path to deliver packets before their deadlines, creating the water-filling effect as we have discussed in Section 4. For instance, Figure 16 shows the histogram of the ETX of the paths taken by all the pack-

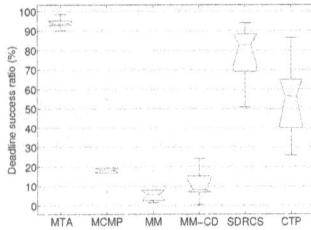

Figure 12: Deadline success ratio: MTA and existing protocols in NetEye

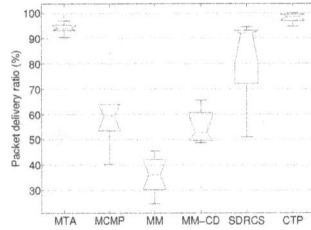

Figure 13: Packet delivery ratio: MTA and existing protocols in NetEye

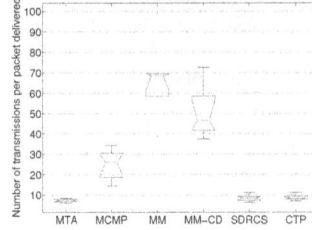

Figure 14: Number of transmissions per packet delivered: MTA and existing protocols in NetEye

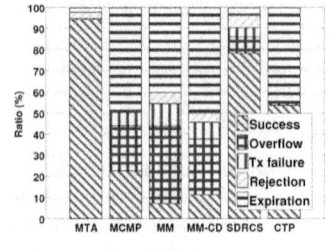

Figure 15: Packet delivery status: MTA and existing protocols in NetEye

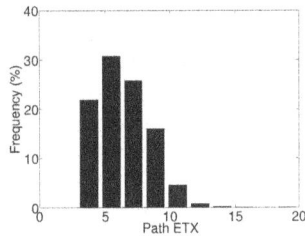

Figure 16: Histogram of the ETX of the paths taken by all the packets

ets. We see that MTA tends to use paths of lower-ETX with higher probability, even though the minimum-ETX path will not always be used with the highest probability (e.g., when the capacity of the minimum-ETX path is reduced due to the shared path segments with other paths).

Indriya. Figures 17, 18, 19, and 20 show the deadline success ratio, packet delivery ratio, transmission cost, and packet delivery status for periodic traffic in Indriya respectively. The overall relative behavior between protocols is similar to that in NetEye, but the performance of MM, MM-CD, and SDRCS become much worse compared with MTA. MTA still ensures real-time data delivery, but MM and MM-CD can hardly deliver any packet to the sink, let alone delivering packets in time; the deadline success ratio in SDRCS is also more than 72% less than that in MTA. One major cause for this is that MM, MM-CD, and SDRCS implicitly assume uniform network conditions while the degree of heterogeneity in Indriya is significant and higher than that in NetEye. As a result of the uniformity assumption, for instance, about 60% and 50% of packets are rejected in MM and MM-CD respectively as shown in Figure 20. Given a packet, more specifically, its deadline and the distance from its source to the sink determines the required forwarding speed for this packet. For the packet to reach the sink, every single hop it traverses has to provide a speed no less than the required speed. That is, if any intermediate hop cannot meet the required speed, a packet is rejected. For MM and MM-CD, it only takes about 3 or 4 hops to reach the sink from the sources in Net-Eye, while it takes about 7 or 8 hops in Indriya, making packets in Indriya more likely to be rejected. Note that there is significant queue overflow in SDRCS because SDRCS happen to use low reliability links in Indriya which reduce the network throughput and thus increase the queueing and queue overflow.

6. RELATED WORK

QoS routing has been well studied for the Internet [9, 12] and wireless networks [20, 10, 11]. But most did not consider uncertainties in link/path properties (e.g., delay). Link property uncertainties were considered in [23] and [31], and it was shown that the problem of checking probabilistically guaranteed path delays is NP-hard [23]. Focusing on Internet QoS routing, these work assumed link-state routing, and their solutions were not amenable to light-weight, distance-vector-type implementation. Since link-state routing is usually not suitable for dynamic, resource constrained WSC networks where reliable network-wide link-state update itself is a challenging issue and where nodes may only have very limited memory space (e.g., up to 4KB of RAM), the approaches of [23] and [31] are not applicable to WSC networks.

Data delivery delay was also considered in wireless and sensor network routing [21, 24, 27, 33, 48]; but they only tried to minimize average path delay without ensuring probabilistic delay bounds [47, 21, 24, 33, 48], they did not consider the probabilistic nature of link/path delays [39], they were based on geographic forwarding without addressing network non-uniformity and wireless communication irregularity [24, 21], or they uniformly partitioned multihop QoS requirements (e.g., reliability and timeliness) along the links of a path without considering network non-uniformity [47, 27]. Huang et al. [27] used Chebyshev Inequality in single-hop real-time satisfiability testing, but they did not address the challenges of accurate, agile estimation of multi-hop probabilistic delay bounds, and they did not comparatively study Chebyshev Inequality with other well-known probability inequalities. Liu et al. [35] proposed the pseudo-polynomial time algorithm optimal-partition-minimum-delay (OPMD) for upper-bounding probabilistic path delays, but, as we have shown in Section 3, the bound of the OPMD algorithm is quite loose for multi-hop paths.

Multi-timescale adaptation has been considered in Internet traffic engineering [8, 30]. Focusing on load balancing, these work did not consider QoS assurance. Liu et al. [35] also studied multi-timescale adaptation in routing, but they used the OPMD method to estimate probabilistic delay bounds which are significantly looser than the bound identified through our MTE method, thus leading to real-time capacity loss. The IETF ROLL working group [28] considered building routing trees based on directed-acyclic-graphs (DAG) for low-power wireless networks. Serving as a general reference framework, the ROLL routing proposal did not consider specific optimization methods (e.g., for real-time guarantees). For stable data delivery reliability, Lin et al. [34] proposed to route data based on long-term link properties and to address transient perturbations using power and retransmission control; they focused on data delivery reliability instead of real-time, thus they did not consider the challenges of dynamic, uncertain link/path delays in real-time routing.

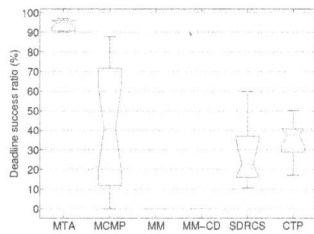

Figure 17: Deadline success ratio: MTA and existing protocols in Indriya

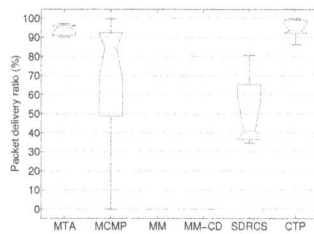

Figure 18: Packet delivery ratio: MTA and existing protocols in Indriya

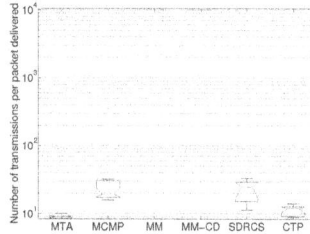

Figure 19: Number of transmissions per packet delivered: MTA and existing protocols in Indriya

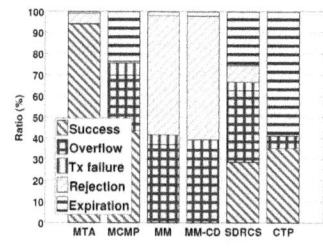

Figure 20: Packet delivery status: MTA and existing protocols in Indriya)

The WirelessHART [46] and the ISA SP100.11a [41] standards have been recently proposed for wireless networking in industrial process measurement and control. They mostly focus on high-level system frameworks instead of specific algorithms in real-time routing. In the literature of real-time wireless networking, techniques such as power control [13] as well as joint routing and scheduling [39] have been studied; energy-efficiency [17] has also been considered too. Orthogonal to these studies, our study here has focused on addressing the challenges that dynamic, uncertain link/path delays pose to two basic elements of real-time routing, i.e., determining probabilistic path delays and addressing instability in delay-adaptive routing. Integrating our results with those work will be an interesting research avenue to pursue, but detailed study of it is beyond the scope of this paper.

7. CONCLUDING REMARKS

For addressing the challenges of highly-varying path delays to distributed estimation of path delay quantiles, we have proposed the MTE method that leverages the stability of packet-time distribution and the quick diffusion of path delay statistics (i.e., mean and variance) to accurately estimate probabilistic path delay bounds in an agile manner. Based on accurate, agile characterization of path delays using MTE, our MTA routing framework enables the stability and optimality of data forwarding while adapting to fast-changing network queueing and delay. Through extensive measurement study in both the NetEye and the Indriya wireless sensor network testbeds, we have shown that MTE/MTA-based routing ensures efficient, real-time data delivery, and it significantly outperforms existing real-time routing protocols. We have mainly focused on real-time spatial flow control in this study, even though we have experimentally analyzed the benefits of using EDF instead of FCFS in intra-node scheduling; how to control temporal packet flow between neighbors and across the network and how to jointly optimize the spatial and temporal packet flow will be an important area to explore, where the MTE method and the MTA framework are expected to serve as basic systems building-blocks. The technique of leveraging different timescales of dynamics in protocol design may well be of generic interest to wireless networking in dynamic, uncertain environments too.

Acknowledgment

We thank George Yin for discussions on probability inequalities, and we also thank our MobiHoc'12 shepherd Ivan Seskar for helpful comments on our paper. The authors' work is supported in part by NSF awards CNS-1136007, CNS-1054634, GENI-1890, and GENI-1633, as well as grants from Ford Research and GM Research.

8. REFERENCES

[1] An event traffic trace for sensor networks. http://www.cs.wayne.edu/~hzhang/group/publications/Lites-trace.txt.

[2] Indriya testbed. http://indriya.comp.nus.edu.sg/.

[3] IEEE 802.15 smart utility networks task group 4g. http://www.ieee802.org/15/pub/TG4g.html.

[4] IEEE 802.15.4e Working Group. http://www.ieee802.org/15/pub/TG4e.html.

[5] NetEye testbed. http://neteye.cs.wayne.edu/neteye/home.php, 2008.

[6] S. Bapat, V. Kulathumani, and A. Arora. Analyzing the yield of exscal, a large-scale wireless sensor network experiment. In *IEEE ICNP*, pages 53–62, 2005.

[7] S. Biswas and R. Morris. ExOR: Opportunistic multi-hop routing for wireless networks. In *ACM SIGCOMM*, 2005.

[8] M. Caesar, M. Casado, T. Koponen, J. Rexford, and S. Shenker. Dynamic route computation considered harmful. *ACM SIGCOMM Computer Communication Review*, 40(2), 2010.

[9] A. Chakrabarti and G. Manimaran. Reliability constrained routing in QoS networks. *IEEE/ACM Transactions on Networking*, 13(3):662–675, 2005.

[10] S. Chakrabarti and A. Mishra. QoS issues in ad hoc wireless networks. *IEEE Communications Magazine*, pages 142–148, February 2001.

[11] L. Chen and W. B. Heinzelman. QoS-aware routing based on bandwidth estimation for mobile ad hoc networks. *IEEE JSAC*, 23(3):561–571, 2005.

[12] S. Chen, M. Song, and S. Sahni. Two techniques for fast computation of constrained shortest paths. *IEEE/ACM Transactions on Networking*, 16(1), February 2008.

[13] O. Chipara, Z. He, G. Xing, Q. Chen, X. Wang, C. Lu, J. Stankovic, and T. Abdelzaher. Real-time power-aware routing in sensor networks. In *IWQoS*, 2006.

[14] D. S. J. D. Couto, D. Aguayo, J. Bicket, and R. Morris. A high-throughput path metric for multi-hop wireless routing. In *ACM MobiCom*, 2003.

[15] R. Draves, J. Padhye, and B. Zill. Comparison of routing metrics for static multi-hop wireless networks. In *ACM SIGCOMM*, 2004.

[16] R. Draves, J. Padhye, and B. Zill. Routing in multi-radio, multi-hop wireless mesh networks. In *ACM MobiCom*, 2004.

[17] S. C. Ergen and P. Varaiya. Energy efficient routing with delay guarantee for sensor networks. *Wireless Networks*, 13(5), 2007.

[18] M. Erol-Kantarci and H. T. Mouftah. Wireless sensor networks for cost-efficient residential energy management in the smart grid. *IEEE Transactions on Smart Grid*, 2(2):314–325, 2011.

[19] Y. P. Fallah, C. Huang, R. Sengupta, and H. Krishnan. Design of cooperative vehicle safety systems based on tight coupling of communication, computing and physical vehicle dynamics. In *ACM/IEEE ICCPS*, 2010.

[20] X. Fang, D. Yang, P. Gundecha, and G. Xue. Multi-constrained anypath routing in wireless mesh networks. In *IEEE SECON*, 2010.

[21] E. Felemban, C.-G. Lee, E. Ekici, R. Boder, and S. Vural. Probabilistic QoS guarantee in reliability and timeliness domains in wireless sensor networks. In *IEEE INFOCOM*, 2005.

[22] O. Gnawali, R. Fonseca, K. Jamieson, D. Moss, and P. Levis. Collection tree protocol. In *ACM SenSys*, 2009.

[23] R. A. Guerin and A. Orda. QoS routing in networks with inaccurate information: Theory and algorithms. *IEEE/ACM Transactions on Networking*, 7(3):350–364, 1999.

[24] T. He, J. Stankovic, C. Lu, and T. Abdelzaher. SPEED: A stateless protocol for real-time communication in sensor networks. In *IEEE ICDCS*, 2003.

[25] J. Hellerstein, Y. Diao, S. Parekh, and D. M. Tilbury. *Feedback Control of Computing Systems*. Wiley-IEEE Press, 2004.

[26] B. Hobijn, P. H. Franses, and M. Ooms. Generalizations of the KPSS-test for stationarity. *Statistica Neerlandica*, 58(4):483–502, 2004.

[27] X. Huang and Y. Fang. Multiconstrained QoS multipath routing in wireless sensor networks. *Wireless Networks*, 14:465–478, 2008.

[28] IETF. Routing over low power and lossy networks (ROLL) working group. http://www.ietf.org/html.charters/roll-charter.html.

[29] R. Jain and I. Chlamtac. The P^2 algorithm for dynamic calculation of quantiles and histograms without storing observations. *Communications of the ACM*, 28(10), 1985.

[30] S. Kandula, D. Katabi, B. Davie, and A. Charny. Walking the tightrope: Responsive yet stable traffic engineering. In *ACM SIGCOMM*, 2005.

[31] T. Korkmaz and M. Krunz. Bandwidth-delay constrained path selection under inaccurate state information. *IEEE/ACM Transactions on Networking*, 11(3), 2003.

[32] A. LaJoie. Wireless sensors enjoy timely role in factory automation. *Industrial Ethernet Book*, April 2010.

[33] H. Li, Y. Cheng, and C. Zhou. Minimizing end-to-end delay: A novel routing metric for multi-radio wireless mesh networks. In *IEEE INFOCOM*, 2009.

[34] S. Lin, G. Zhou, K. Whitehouse, Y. Wu, J. A. Stankovic, and T. He. Towards stable network performance in wireless sensor networks. In *IEEE RTSS*, 2009.

[35] X. Liu, H. Zhang, and Q. Xiang. Towards predictable real-time routing for wireless networked sensing and control. In *CPS Week RealWin workshop*, 2011.

[36] X. Liu, H. Zhang, Q. Xiang, X. Che, and X. Ju. Taming uncertainties in real-time routing for wireless networked sensing and control. Technical Report DNC-TR-11-04 (https://sites.google.com/site/dnctrs/DNC-TR-11-04.pdf), Wayne State University, 2011.

[37] M. Loeve. *Probability Theory*. Springer-Verlag, 1977.

[38] J. R. Moyne and D. M. Tilbury. Control and communication challenges in networked real-time systems. *Proceedings of the IEEE*, 95(1), 2007.

[39] S. Munir, S. Lin, E. Hoque, S. M. S. Nirjon, J. A. Stankovic, and K. Whitehouse. Addressing burstiness for reliable communication and latency bound generation in wireless sensor networks. In *IEEE/ACM IPSN*, 2010.

[40] S. M. Ross. *Introduction to Probability Models*. Academic Press, 2006.

[41] ISA SP100.11a. http://www.isa.org//MSTemplate.cfm?MicrositeID=1134\&CommitteeID=6891.

[42] H.-M. Tsai, C. Saraydar, T. Talty, M. Ames, A. Macdonald, and O. K. Tonguz. Zigbee-based intra-car wireless sensor network. In *IEEE ICC*, 2007.

[43] J. V. Uspensky. *Introduction to Mathematical Probability*. McGraw-Hill Book Company Inc., 1937.

[44] W. Wang, Y. Xu, and M. Khanna. A survey on the communication architectures in smart grid. *Computer Networks*, 55:3604–3629, 2011.

[45] A. Willig. Recent and emerging topics in wireless industrial communications: A selection. *IEEE Transactions on Industrial Informatics*, 4(2), May 2008.

[46] WirelessHART. http://www.hartcomm2.org/hart_protocol/wireless_hart/wireless_hart_main.html.

[47] Y. Xue, B. Ramamurthy, and M. C. Vuran. SDRCS: A service-differentiated real-time communication scheme for event sensing in wireless sensor networks. *Computer Networks*, 55:3287–3302, 2011.

[48] S. Yin, Y. Xiong, Q. Zhang, and X. Lin. Traffic-aware routing for real-time communications in wireless multi-hop networks. *Wireless Communications and Mobile Computing*, 6:825–843, 2006.

[49] F. R. Yu, P. Zhang, W. Xiao, and P. Choudhury. Communication systems for grid integration of renewable energy resources. *IEEE Network*, pages 22–29, September/October 2011.

[50] H. Zhang, A. Arora, Y. ri Choi, and M. Gouda. Reliable bursty convergecast in wireless sensor networks. *Computer Communications (Elsevier), Special Issue on Sensor-Actuator Networks*, 30(13), 2007.

[51] H. Zhang, A. Arora, and P. Sinha. Link estimation and routing in sensor network backbones: Beacon-based or data-driven? *IEEE Transactions on Mobile Computing*, 8(5), May 2009.

[52] H. Zhang, L. Sang, and A. Arora. On biased link sampling in data-driven link estimation and routing in low-power wireless networks. *ACM Transactions on Autonomous and Adaptive Systems*, 4(3), July 2009.

[53] H. Zhang, L. Sang, and A. Arora. Comparison of data-drive link estimation methods in low-power wireless networks. *IEEE Transactions on Mobile Computing*, 9(11), November 2010.

Minimum-Energy Connected Coverage in Wireless Sensor Networks with Omni-Directional and Directional Features

Kai Han[*†]   Liu Xiang[†]   Jun Luo[†]   Yang Liu[‡]

[*]School of Computer Science
Zhongyuan University of Technology, China, 450007

[†]School of Computer Engineering
Nanyang Technological University, Singapore, 639798

[‡]School of Information Sciences and Engineering
Henan University of Technology, China, 450007

{hankai, xi0001iu, junluo}@ntu.edu.sg, enjoyang@gmail.com

ABSTRACT

Wireless Sensor Networks (WSNs) have acquired new features recently, i.e., both the sensor and the antenna of a node can be **directional**. This brings new challenges to the *Connected Coverage* (CoCo) problem, where a finite set of targets needs to be monitored by some active sensor nodes, and the connectivity of these active nodes with the sink must be retained at the same time. In this paper, we study the *Minimum-Energy Connected Coverage* (MeCoCo) problem in WSNs with Omni-directional (O) and Directional (D) features, aiming at minimizing the total energy cost of both sensing and connectivity. Considering different combinations of O and D features, we study the MeCoCo problem under four cases, namely: O-Antenna and O-Sensor (OAOS), O-Antenna and D-Sensor (OADS), D-Antenna and D-Sensor (DADS), as well as D-Antenna and O-Sensor (DAOS). We prove that the MeCoCo problem is NP-hard under all these cases, and present approximation algorithms with provable approximation ratios. In particular, we propose a constant-approximation for OAOS, and polylogarithmic approximations for all other cases. Finally, we conduct extensive simulations and the results strongly confirm the effectiveness of our approach.

Categories and Subject Descriptors

C.2.1 [**Computer-Communication Networks**]: Network Architecture and Design—*Network topology*

General Terms

Algorithm, Theory, Performance

Keywords

Wireless sensor networks, coverage, connectivity, energy efficiency, directional antenna, directional sensor

1. INTRODUCTION

Wireless Sensor Networks (WSNs) have long been considered as powerful instruments for environment monitoring, battlefield surveillance and industrial diagnostics [2]. For these applications, *sensing coverage* is a fundamental problem and has thus aroused tremendous interests from both industrial and academic communities. Roughly speaking, three types of coverage problems have been investigated: (i) *area coverage* (e.g., [3, 17, 24, 31, 32]), (ii) *target coverage* (e.g., [1, 4–7, 11, 12, 21, 30, 33]), and (iii) *barrier coverage* (e.g., [9, 18, 20, 29]). Our paper focuses on the target coverage problem. In a nutshell, after a WSN is deployed, a finite set of *targets* are identified in the network area, and we need to choose a subset of sensor nodes to monitor (or sense) these targets.

As a special case of coverage problems, *Connected Coverage* (CoCo) problem moves one step further. Since the sensing nodes need to communicate with each other, one may need to further find relaying nodes to form a connected subnet along with the sensing nodes. Depending on what performance objective a coverage problem has, the sensing and relaying sensor nodes may need to be selected jointly to yield the best performance. Although there are studies on CoCo, they are mostly concerned with maximizing network lifetime (e.g. [6, 7, 21, 33]). The problem of minimizing the total energy cost of both sensing and connectivity is, to the best of our knowledge, surprisingly unexplored. In fact, as a WSN can involve tens of hundreds nodes, its total consumed energy is large. Therefore, making WSNs energy efficient is imperative in creating a sustainable earth.

In recent years, sensors with **directional sensing ability** are increasingly adopted by WSNs (e.g., [5, 15, 30]). Moreover, nodes with **directional antennas** are also used for energy conservation (e.g., [27, 32]). However, existing proposals mostly assume omni-directional sensors and antennas. This has motivated us to consider the CoCo problem in the context of sensor nodes using both directional sensors and directional antennas.

In this paper, we study the *Minimum-energy Connected Coverage* (MeCoCo) problem in WSNs, where sensor nodes can be equipped with directional or omni-directional sensors/antennas. We aim at selecting suitable sensing/relaying nodes as well as the working directions of their sensors and antennas, such that the total energy cost for sensing and connectivity is minimized. We classify the MeCoCo problem into four sub-problems according to the different combinations of omni-directional and directional features. We show the NP-hardness of all these sub-problems and provide approximation algorithms with guaranteed approximation ratios for them. In particular, we show that, for a WSN with node set V to monitor a set D of targets:

- Omni-directional Antenna and Omni-directional Sensor (OAOS) has a $(4 + \epsilon)$-approximation.

- Omni-directional Antenna and Directional Sensor (OADS) has a randomized $\mathcal{O}(\log(\Delta|V|)\log|D|\log\Delta)$-approximation.

- Directional Antenna and Directional Sensor (DADS) has a randomized $\mathcal{O}(\log(\Delta|V|)\log|D|\log\Delta)$-approximation.

- Directional Antenna and Omni-Directional Sensor (DAOS) has a randomized $\mathcal{O}(\log(\Delta|V|)\log|D|\log\Delta)$-approximation.

We also perform extensive simulations and the simulation results demonstrate the effectiveness of our algorithms. To the best of our knowledge, we are the first to provide approximation algorithms with provable performance ratios for the MeCoCo problem in WSNs with omni-directional and directional features.

The rest of the paper is organized as follows. In Section 2, we review the related work. The preliminaries and problem definitions are provided in Section 3. We provide our approximation algorithms as well as their respective performance analysis in Section 4. In Section 5, we evaluate the performance of the proposed algorithms through simulations. Section 6 concludes the paper. In order to maintain fluency, we postpone all our proofs to the appendix.

2. RELATED WORK

As our focus is on target coverage, we only survey the related target coverage research proposals. This means that we have to leave out other fascinating topics on area coverage (e.g., [17, 19, 24, 32]) or barrier coverage (e.g., [9, 29]) due to their marginal relevance to our current focus.

2.1 Target Coverage for WSNs with Omni-Directional Sensors

Cardei et al. [7] aim at maximizing network lifetime of a randomly deployed WSN by identifying non-disjoint sensor sets in the network such that all targets can be covered by any sensor set. They prove the NP-completeness of maximizing lifetime by activating these sensor sets in turn and also propose some heuristics. In fact, Luo et al. [22] later propose a general resource allocation algorithm; it can obtain the optimal solution to this problem for relatively large scale WSNs. Zhao et.al. [33] studied the Maximum Cover Tree (MCT) problem for lifetime maximization, where a cover tree stands for a tree rooted at the sink node that can cover all the target points. Though their problem is also NP-complete, they provide an approximation algorithm with provable approximation ratio. A similar problem is also studied in [6] but only heuristics are provided. Lu et al. [21] consider the target coverage problem under the assumption that each sensor node can adjust its sensing ranges. Their

objective is to schedule the sensors' sensing ranges to cover the target points such that the network lifetime is maximized. Note that the energy cost for connectivity was neglected in [21].

All these proposals deal with lifetime maximization, which is a quite different objective from ours. Whereas Wan et al. [28] recently propose approximation algorithms for both maximum lifetime and minimum cost coverage problems, connectivity is not their concern.

2.2 Target Coverage for WSNs with Directional Sensors

Ai et.al. [1] were among the first to investigate the directional coverage problem. They aim at a Pareto optimal solution, balancing between a larger number of covered targets and a smaller number of directional sensors. They prove the NP-completeness of the problem and provide a distributed greedy heuristic. Fusco et al. [11] extend [1] by requiring that each target should be covered $k : k \geq 1$ times by the directional sensors, and propose an approximation algorithm with provable performance ratio. Neither [1] nor [11] allows the rotation of the directional sensors once their working directions are determined. Cai et al. [4, 5] study the Multiple Directional Cover Sets (MDCS) problem, which is a directional version of [7]. Lately, [12, 30] study the coverage delay problem assuming rotating directional sensor takes non-negligible time. They aim at reducing the uncovered time of the targets when directional sensors rotate their working directions. Finally, the sensor deployment problem for WSNs with directional sensors is studied in [15, 26]. However, their objective is to find optimal locations for placing the directional sensors rather than to select covering sensor nodes from a randomly deployed sensor network. Note that connectivity is never a requirement in [1, 4, 5, 11, 12, 30], and so is minimizing the total energy cost. These are exactly the problems we want to tackle in this paper.

3. MODEL AND PROBLEM

3.1 Network Models

A WSN is modeled as an undirected connected graph $G = (V, E)$, where V is the set of sensor nodes and E is the set of wireless links. There exists a sink node $\mathfrak{s} \in V$ and a set of target points $D : D \cap V = \emptyset$ to be monitored by the sensor nodes in V. Both D and V lie on a 2D Euclidean plane; their locations are assumed to be known through certain localization mechanisms [14, 23].

We assume that each sensor node $v \in V$ is equipped with a **sensor** $\curlywedge(v)$ and an **antenna** $\curlyvee(v)$. The sensor is used to monitor targets, and the antenna (along with an RF transceiver) is used to communicate with other nodes. Each sensor has a sensing radius R_s and each antenna has a transmission radius R_t (R_s can be different from R_t). We also assume that, for any target $d \in D$, there exists at least one guarding sensor node $v \in V$ such that $d \in Disk(v, R_s)$, where $Disk(v, R_s)$ is the disk with radius R_s centered at v.

A sensor $\curlywedge(v)$ can be omni-directional, such as a temperature or pressure sensor, or directional, such as a video or infrared sensor. An omni-directional sensor $\curlywedge(v)$ has $Disk(v, R_s)$ as its sensing range, while the sensing range of a directional sensor is a sector area of $Disk(v, R_s)$ with a fixed beamwidth η_s. Following a very common assumption, we assume that a directional sensor can rotate to different *working directions* θ to monitor different sector areas of its neighborhood.

Similarly, an antenna $\Upsilon(v)$ can also be omni-directional or directional. We assume that there exists an edge in E between two sensor nodes in V if their Euclidean distance is no more than R_t, regardless of what antenna is equipped. Similar to $\curlywedge(v)$, a directional antenna $\Upsilon(v)$ also has a sector area with a fixed beamwidth η_t as its transmission range, and it can rotate its working direction to cover the neighboring sensor nodes in V. An edge $(u, v) \in E$ can be used for communication iff u and v cover each other. We illustrate a WSN with directional sensors and omni-directional (or directional) antennas in Figure 1.

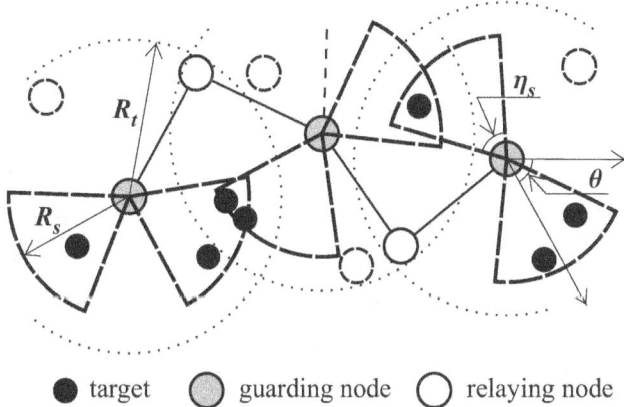

● target   ◉ guarding node   ○ relaying node

Figure 1: A connected target coverage example with a WSN using directional sensors. To avoid confusion, we refrain from drawing antenna sectors; only links enabling connectivity are shown.

Obviously, directional sensors and directional antennas share certain properties. For a working direction θ of a node v (for either $\curlywedge(v)$ or $\Upsilon(v)$), a subset of D (in terms of $\curlywedge(v)$) or of V (in terms of $\Upsilon(v)$) that lies in the sector area covered by v with working direction θ is called a *coverage set* $\kappa_\theta(v)$. Here we abuse the notation by using v to represent either $\curlywedge(v)$ or $\Upsilon(v)$. Two working directions θ_1, θ_2 of v is said to be *equivalent* if and only if $\kappa_{\theta_1}(v) = \kappa_{\theta_2}(v)$. For any working direction θ_1 of v, if there does not exist a working direction θ_2 of v such that $\theta_1 \neq \theta_2$ and $\kappa_{\theta_1}(v) \subset \kappa_{\theta_2}(v)$, then θ_1 is called an *independent* working direction of v. An *Independent Direction Set* (IDS) of v is a set of non-equivalent independent working directions of v. A *Maximum Independent Direction Set* (MIDS) of v, denoted by $\chi(v)$, is an IDS of v such that the set cardinality is maximized.

We define, for $\curlywedge(u), u \in V$, a *Direction Setting Set* (DSS) Θ^u containing all working directions used by $\curlywedge(u)$. Similarly, we also define a DSS Ξ^u for $\Upsilon(u), u \in V$. For convenience, we denote by δ_1 the maximum node degree of G, denote by δ_2 the maximum number of target points in D that lie in any $Disk(u, R_s), u \in V$, denote by δ_3 the maximum number of nodes in V that lie in any $Disk(d, R_s) : d \in D$, and let $\Delta = \max\{\delta_1, \delta_2, \delta_3\}$.

3.2 Problem Formulation and NP-hardness

Our Minimum-Energy Connected Coverage (MeCoCo) problem requires that each target in D is covered by some guarding nodes in V and that the guarding nodes must be con-

nected to the sink (probably though some relaying nodes). Therefore, there are two kinds of energy costs that we need to consider in such a scenario: the energy cost for sensing and the energy cost for connectivity. The former accounts for the cost of the guarding nodes in monitoring the targets, and the latter accounts for the cost in connecting the guarding nodes to the sink. Hence, the objective of our MeCoCo problem is to find a set of guarding and relaying nodes such that the sum of the sensing and connectivity energy costs is minimized. Because both sensor and antenna can be either omni-directional (O) or directional (D), we have the following four versions of the MeCoCo problem:

O-Antenna and O-Sensor (OAOS).

In this case, the energy cost of each sensor for monitoring its neighborhood is the same, which is denoted by ε_s. Meanwhile, the energy cost of each node for connectivity is also the same, which is denoted by ε_c. Therefore, the MeCoCo problem in the OAOS case (or MeCoCo-OAOS) is to find a set of connected nodes $S_1 \subseteq V$ that contains $\mathfrak{s}$ and a set of guarding nodes $M_1 \subseteq S_1$, so that the total energy cost $\xi_{OAOS}(S_1, M_1) = |S_1| \cdot \varepsilon_c + |M_1| \cdot \varepsilon_s$ is minimized.

O-Antenna and D-Sensor (OADS).

In this case, the directional sensor of a guarding node may need to rotate to several working directions to cover the targets. Suppose that the energy cost of a directional sensor for sensing in one working direction is $\tau_s < \varepsilon_s$, the MeCoCo problem in the OADS case (or MeCoCo-OADS) is to find a set of connected sensor nodes $S_2 \subseteq V$ that contains $\mathfrak{s}$, a set of guarding nodes $M_2 \subseteq S_2$, and a set of (sensor) DSSs $\Theta_2 = \{\Theta_2^v | v \in M_2\}$, such that $D \subseteq \bigcup_{v \in M_2} \bigcup_{\theta \in \Theta_2^v} \kappa_\theta(\curlywedge(v))$ is satisfied and the total energy cost $\xi_{OADS}(S_2, M_2, \Theta_2) = |S_2| \cdot \varepsilon_c + \sum_{v \in M_2} |\Theta_2^v| \cdot \tau_s$ is minimized.

D-Antenna and D-Sensor (DADS).

In this case, the antennas of guarding and relaying nodes may also need to rotate to several working directions to connect with other nodes. Let the connectivity energy cost of an antenna in one working direction be $\tau_c < \varepsilon_c$. The MeCoCo problem in the DADS case (or MeCoCo-DADS) is to find a set of sensor nodes $S_3 \subseteq V$ that contains $\mathfrak{s}$, a set of guarding nodes $M_3 \subseteq S_3$, a set of (antenna) DSSs $\Xi_3 = \{\Xi_3^u | u \in S_3\}$, and a set of (sensor) DSSs $\Theta_3 = \{\Theta_3^v | v \in M_3\}$, such that the following conditions are satisfied:

c1 D is covered by all the guarding nodes, i.e., $D \subseteq \bigcup_{v \in M_3} \bigcup_{\theta \in \Theta_3^v} \kappa_\theta(\curlywedge(v))$,

c2 The nodes in S_3 are connected by their directional antennas, i.e., connected by the edges in $E_3 = \{(u, v) | \exists \theta_1 \in \Xi_3^u, \exists \theta_2 \in \Xi_3^v, v \in \kappa_{\theta_1}(\Upsilon(u)) \wedge u \in \kappa_{\theta_2}(\Upsilon(v))\}$,

c3 The total energy cost $\xi_{DADS}(S_3, M_3, \Xi_3, \Theta_3) = \sum_{u \in S_3} |\Xi_3^u| \cdot \tau_c + \sum_{v \in M_3} |\Theta_3^v| \cdot \tau_s$ is minimized.

D-Antenna and O-Sensor (DAOS).

In this case, the definition of the MeCoCo-DAOS problem can be obtained by combining those of OAOS and DADS. Therefore, we omit the details but only state the objective: choose S_4, M_4, Ξ_4 in order to minimize $\xi_{DAOS}(S_4, M_4, \Xi_4) = \sum_{u \in S_4} |\Xi_4^u| \cdot \tau_c + |M_4| \cdot \varepsilon_s$.

Hardness Results.

As the *Geometric Set Cover* (GSC) problem [10] is a special case of our MeCoCo problem, we omit the proof for the following statement.

THEOREM 1. *Under all cases of OAOS, OADS, DADS, and DAOS, the MeCoCo problem is NP-hard.*

We omit the proof because, although we can prove the NP-hardness of our MeCoCo problem by a reduction from the GSC problem, we can not directly apply a GSC algorithm to solve the MeCoCo problem. This is due to the additional complexity and high dimensionality introduced by the non-trivial connectivity requirements and directional features of our problem, which are not considered in GSC or other related work. Hence, we need to design new algorithms to solve the MeCoCo problem.

4. APPROXIMATION ALGORITHMS

In this section, we present approximation algorithms for the MeCoCo problem under the OAOS, OADS, DADS, and DAOS cases.

4.1 Approximating MeCoCo-OAOS

In this case, we provide a constant-ratio approximation algorithm for the MeCoCo-OAOS problem when $0 < R_s \leq R_t/2$, as shown in **Algorithm 1**. It can be easily seen that the algorithm output is a feasible solution to MeCoCo-OAOS. The approximation ratio of **Algorithm 1** is given in THEOREM 2.

Algorithm 1: Approximating MeCoCo-OAOS

Input: The network graph $G = (V, E)$ and the target set D.
Output: The set of connected sensor nodes S_1' and the set of guarding nodes M_1'

1 Use the PTAS proposed in [28] to find a set of nodes $M_1' \subseteq V$ so that each node in D is covered by certain node in M_1' within the sensing radius R_s.
2 Use a 2-approximation Steiner Tree algorithm [25] to find a Steiner tree T_S that spans the nodes in $\{\mathfrak{s}\} \cup M_1'$. Let S_1' be the set of nodes in T_S.
3 return S_1', M_1'

THEOREM 2. **Algorithm 1** *has an approximation ratio of $4 + \epsilon$ for the MeCoCo-OAOS problem.*

As we will explain in Section 4.2, we can get another algorithm for the MeCoCo-OAOS problem with a polylogarithmic approximation ratio if $R_s > R_t/2$.

4.2 Approximating MeCoCo-OADS

We propose an approximation algorithm based on a method of graph transformation to solve the MeCoCo-OADS problem in **Algorithm 3**. However, to run it correctly, we need to identify the MIDS of any sensor/antenna. We present a polynomial-time algorithm for finding the MIDS of any sensor in **Algorithm 2**; it can be applied to antennas with minor modifications.

In **Algorithm 2**, we first sort clockwise the targets in $Disk(v, R_s)$ and get an ordered (cyclic) sequence **d** (line 1).

We then apply a greedy strategy (lines 2 to 6) to find all independent working directions of $\lambda(v)$. In each iteration, we find a working direction whose coverage set is a longest subsequence of **d** and is not contained in any other coverage set of the found working directions. If a working direction's coverage set is all the target points in $Disk(v, R_s)$, we simply return this working direction as the MIDS of $\lambda(v)$ (line 5).

Algorithm 2: Find the MIDS of a sensor

Input: A directional sensor $\lambda(v), v \in V$ and the target set D.
Output: A MIDS of $\lambda(v)$.
1 $Q \leftarrow \emptyset$; Sort the targets $\{d_k\}$ in $D \cap Disk(v, R_s)$ to an ordered sequence $\mathbf{d} : d_1, d_2, ..., d_h$, such that $0 \leq \angle d_1 v d_2 \leq \angle d_1 v d_3 \leq ... \leq \angle d_1 v d_h \leq 2\pi$.
2 **for** $k = 1$ **to** h **do**
3 $\quad$ Find the maximal $t, 1 \leq t \leq h$ such that $\angle d_{1+(k-1)\bmod(h)} v d_{1+(k+t-2)\bmod(h)} < \eta_s$.
4 $\quad$ Let θ be a working direction of $\lambda(v)$ such that $\kappa_\theta(\lambda(v)) = \{d_{1+(i)\bmod(h)} | k - 1 \leq i \leq k + t - 2\}$.
5 $\quad$ **if** $t = h$ **then** $Q \leftarrow Q \cup \{\theta\}$; **return** Q.
6 $\quad$ **else if** $\angle d_{1+(k-2)\bmod(h)} v d_{1+(k+t-2)\bmod(h)} > \eta_s$ **then** $Q \leftarrow Q \cup \{\theta\}$.
7 **return** Q

According to the definition of MIDS, we can prove the correctness of **Algorithm 2** in LEMMA 1, which immediately leads to Corollary 1.

LEMMA 1. *Algorithm 2 is correct, i.e., Q is a MIDS of $\lambda(v), v \in V$.*

COROLLARY 1. *For any node $v \in V$, we always have $|\chi(\Upsilon(v))| \leq \delta_1$ and $|\chi(\lambda(v))| \leq \delta_2$.*

After obtaining the MIDSs of all sensors and antennas using **Algorithm 2**, we use **Algorithm 3** to compute an approximate solution to MeCoCo-OADS. We first find a subset Y of V such that a node in Y may cover at least one target (line 3). Then we create a graph $\hat{G}$ where a "shadow node" is created for each node in Y (line 5). We also represent each working direction θ in the MIDS of any sensor $\lambda(u), u \in Y$ as a "sensor sector node" $\varrho_u(\theta)$ in $\hat{G}$, and we connect $\varrho_u(\theta)$ to the shadow node of u (lines 6 to 9). For any target d in the coverage set of θ, we also add the sensor sector node $\varrho_u(\theta)$ to the group $\mathcal{G}_d$ (line 8). After creating $\hat{G}$, we merge G and $\hat{G}$ into a new graph $\bar{G}$ by connecting the nodes in Y and their shadow nodes in $\hat{G}$ (lines 11 to 13).

In the new graph $\bar{G}$, only the edges connecting sensor sector nodes and shadow nodes and the edges in E are assigned with non-zero weights (τ_s and ε_c, respectively), and an approximate GST algorithm is employed to find a tree containing at least one node from every $\mathcal{G}_d : d \in D \cup \{\mathfrak{s}\}$ (line 14). Although the output of **Algorithm 3**, T_{OD}, appears to be irrelevant to MeCoCo-OADS, we can map it to a feasible solution of MeCoCo-OADS:

THEOREM 3. *The output of **Algorithm 3** can be mapped to a feasible solution $\{S_2', M_2', \Theta_2'\}$ of MeCoCo-OADS, so there exists a randomized α-approximation for MeCoCo-OADS, where $\alpha = \mathcal{O}(\log(\Delta|V|) \log |D| \log \Delta)$.*

The proof of THEOREM 3 is very similar to the proofs of THEOREM 4 and THEOREM 5, and thus it is omitted. Intu-

Algorithm 3: Approximating MeCoCo-OADS

Input: The network graph $G = (V, E)$ and the target set D.

Output: A tree T_{OD} that can be mapped to a feasible solution of MeCoCo-OADS.

1   Create a graph $\hat{G} = (\hat{V}, \hat{E})$; $\hat{V} \leftarrow \emptyset$; $\hat{E} \leftarrow \emptyset$.

2   $\mathcal{G}_\mathfrak{s} \leftarrow \{\mathfrak{s}\}$;   **foreach** $d \in D$ **do** $\mathcal{G}_d \leftarrow \emptyset$.

3   Let the area covered by $\bigcup_{d \in D} Disk(d, R_s)$ be $\mathcal{A}$. Let $Y \subseteq V$ be the set of nodes lying in $\mathcal{A}$.

4   **foreach** $u \in Y$ **do**

5   Create a "*shadow node*" $\iota(u)$ of u; $\hat{V} \leftarrow \hat{V} \cup \{\iota(u)\}$.

6   **foreach** $\theta \in \chi(\lambda(u))$ *and* **each** $d \in \kappa_\theta(\lambda(u))$ **do**

7   Create a "*sensor sector node*" $\varrho_u(\theta)$ for θ.

8       $\hat{V} \leftarrow \hat{V} \cup \{\varrho_u(\theta)\}$; $\mathcal{G}_d \leftarrow \mathcal{G}_d \cup \{\varrho_u(\theta)\}$.

9   Add an edge $(\varrho_u(\theta), \iota(u))$ with weight τ_s into $\hat{E}$.

10   Assign each edge in E a weight of ε_c.

11   Create a graph $\bar{G} = (\bar{V}, \bar{E})$; $\bar{V} \leftarrow V \cup \hat{V}$; $\bar{E} \leftarrow E \cup \hat{E}$.

12   **foreach** $u \in Y$ **do**

13   Add an edge $(u, \iota(u))$ with weight 0 into $\bar{E}$;

14   Use an approximate Group Steiner Tree (GST) algorithm to find T_{OD} in $\bar{G}$ that contains at least one node from every $\mathcal{G}_d : d \in D \cup \{\mathfrak{s}\}$.

15   **return** T_{OD}

Algorithm 4: Approximating MeCoCo-DADS

Input: The network graph $G = (V, E)$ and the target set D.

Output: A tree T_{DD} that can be mapped to a feasible solution of MeCoCo-DADS.

1   Create the node set Y, the graph $\hat{G}$ and the groups $\{\mathcal{G}_d | d \in D \cup \{\mathfrak{s}\}\}$ using lines 1–9 of **Algorithm 3**.

2   Create a graph $\check{G} = (\check{V}, \check{E})$; $\check{V} \leftarrow \{\mathfrak{s}\} \cup \hat{V}$; $\check{E} \leftarrow \hat{E}$.

3   **foreach** $u \in V$ **do**

4   **foreach** $\theta \in \chi(\Upsilon(u))$ **do**

5   Create an "*antenna sector node*" $\rho_u(\theta)$ for θ; $\check{V} \leftarrow \check{V} \cup \{\rho_u(\theta)\}$.

6   **if** $u = \mathfrak{s}$ **then**

7   Add a 0-weight edge $(u, \rho_u(\theta))$ to $\check{E}$.

8   **if** $u \in Y$ **then**

9   Add a 0-weight edge $(\rho_u(\theta), \iota(u))$ to $\check{E}$.

10   Create a 0-weight edge between each pair of antenna sector nodes of u; add these edges to $\check{E}$.

11   **foreach** $(u, v) \in E$ **do**

12   **foreach** $\theta \in \chi(\Upsilon(u))$ *and* **each** $\theta' \in \chi(\Upsilon(v))$ **do**

13   **if** $v \in \kappa_\theta(\Upsilon(u))$ and $u \in \kappa_{\theta'}(\Upsilon(v))$ **then**

14   Add an edge $(\rho_u(\theta), \rho_v(\theta'))$ with weight τ_c into $\check{E}$.

15   Use an approximate Group Steiner Tree (GST) algorithm to find T_{DD} in $\check{G}$ that contains at least one node from every $\mathcal{G}_d : d \in D \cup \{\mathfrak{s}\}$.

16   **return** T_{DD}

---

itively, the graph transformation method employed by **Algorithm 3** is intentionally designed to facilitate a quantitative mapping relationship between the solutions to MeCoCo-OADS and the GSTs in $\bar{G}$, thus we can find an approximation solution to MeCoCo-OADS based on an approximated GST in $\bar{G}$. We will illustrate this idea more in Section 4.3.

Algorithm 3 may serve as a randomized α-approximation to the MeCoCo-OAOS when $R_s > R_t/2$, because OAOS is a special case of OADS.

4.3 Approximating MeCoCo-DADS

Our algorithm for solving MeCoCo-DADS is shown in **Algorithm 4**. We again employ a graph transformation method. We create a graph $\check{G}$ that contains the graph $\hat{G}$ obtained from **Algorithm 3** (lines 1 and 2). The working directions of any antenna is represented as a "antenna sector node" in $\check{G}$ (line 5), and the antenna sector nodes are connected in a special way: each of them is first connected to its corresponding shadow node (line 9), then the antenna sector nodes of the same antenna are connected with each other to form a clique (line 10), whereas the antenna sector nodes of neighboring antennas are selectively connected with each other according to the coverage sets of the working directions represented by these antenna sector nodes (lines 11 to 14). Note that the groups $\{\mathcal{G}_d | d \in D \cup \{\mathfrak{s}\}\}$ in $\check{G}$ are inherited from **Algorithm 3**. We show an example of $\check{G}$ in Figure 2. We hereby show that the sizes of $\check{G}$ and the groups are bounded by LEMMA 2:

LEMMA 2. *There are at most $\mathcal{O}(\Delta|V|)$ nodes in $\check{V}$, and the size of the largest group in $\{\mathcal{G}_d | d \in D \cup \{\mathfrak{s}\}\}$ is no more than Δ^2.*

After $\check{G}$ is constructed, we find a GST in $\check{G}$ (line 15). This GST can be mapped to a feasible solution of MeCoCo-DADS, as proven by LEMMA 3 and LEMMA 4.

LEMMA 3. *There exists an optimal solution $S_3, M_3, \Xi_3, \Theta_3$ to MeCoCo-DADS, such that for any $u \in S_3$ and any $v \in M_3$, we have: $\Xi_3^u \subseteq \chi(\Upsilon(u))$ and $\Theta_3^v \subseteq \chi(\lambda(v))$.*

Literally, there exist optimal working directions fully contained in the MIDS of each sensor or antenna. The proof of Lemma 3 is based on the definition of MIDS. The key is that, for any working direction with non-empty coverage set in an optimal solution to MeCoCo-DADS, we can always replace it by an independent working direction without compromising the optimality of the solution. Therefore, Lemma 3 guarantees the validity of $\check{G}$, based on which we can quantify the relationships between the optimal GST in $\check{G}$ and the optimal solution of MeCoCo-DADS, as we will show in LEMMA 4 and LEMMA 5.

LEMMA 4. *Any GST T_{DD} in $\check{G}$ that contains at least one node from every $\mathcal{G}_d : d \in D \cup \{\mathfrak{s}\}$ can be mapped to a feasible solution $\grave{S}_3, \grave{M}_3, \grave{\Xi}_3, \grave{\Theta}_3$ of MeCoCo-DADS so that $\xi_{DADS}(\grave{S}_3, \grave{M}_3, \grave{\Xi}_3, \grave{\Theta}_3) \leq 2 \cdot weight(T_{DD})$.*

The proof of LEMMA 4 is constructive, and heavily depends on the special topology of $\check{G}$. Roughly speaking, the antenna sector nodes in T_{DD} can be mapped to the working directions of some sensor nodes' directional antennas, and the sensor sector nodes in T_{DD} can be mapped to the working directions of those sensor nodes' directional sensors. Thanks to the special construction method of $\check{G}$, we can prove that the result of such a mapping process is exactly a feasible solution to MeCoCo-DADS, hence proving the inequality shown in LEMMA 4. Here the value 2 appears

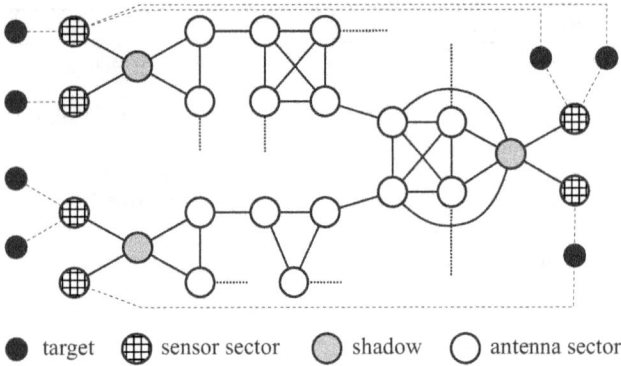

Figure 2: The $\breve{G}$ for the connected target coverage example in Figure 1, assuming directional antennas with a narrow beamwidth so that each sector only covers one neighboring node. The targets and the (thin) dashed lines do not belong to $\breve{G}$; they are used only to indicate $\{\mathcal{G}_d\}$.

because each edge with weight τ_c in T_{DD} can be mapped to two working directions of directional antennas.

Moreover, we also prove that the weight of an optimal GST in $\breve{G}$ is no more than the total energy cost of the optimal solution to the MeCoCo-DADS problem:

LEMMA 5. *Let T_{DD}^* be an optimal GST in $\breve{G}$ that contains at least one node from every $\mathcal{G}_d : d \in D \cup \{\mathfrak{s}\}$, then we have:* weight$(T_{DD}^*) \leq \xi_{DADS}(S_3, M_3, \Xi_3, \Theta_3)$.

The proof of LEMMA 5 is also constructive. Intuitively, given an optimal solution to the MeCoCo-DADS problem, we can map it to a connected sub-graph of $\breve{G}$ that contains at least one destination node from every $\mathcal{G}_d : d \in D \cup \{\mathfrak{s}\}$, and the weight of the sub-graph is no more than the weight of the optimal solution. LEMMA 5 then follows from the fact that T_{DD}^* is an optimal GST.

Summarizing all these lemmas, we now have:

THEOREM 4. *If* weight$(T_{DD}) \leq \gamma \cdot$ weight(T_{DD}^*), *i.e., the approximate Group Steiner Tree algorithm has an approximation ratio γ, then the MeCoCo-DADS problem can be approximated with ratio 2γ.*

We can use any approximate GST algorithms in our algorithms, such as those proposed in [13, 16]. In particular, Chekuri *et al.* [8] pointed out that there exists a polynomial-time randomized $\mathcal{O}(\log N \log m \log |\mathcal{V}|)$-approximation algorithm for GST on general graphs. In our case, $|\mathcal{V}| = \mathcal{O}(\Delta|V|)$, $m = |D| + 1$ and $N \leq \Delta^2$ (see Lemma 2). Using Theorem 4, Theorem 5 follows.

THEOREM 5. *There exists a polynomial-time randomized $\mathcal{O}(\log(\Delta|V|) \log |D| \log \Delta)$-approximation algorithm for MeCoCo-DADS.*

4.4 Approximating MeCoCo-DAOS

MeCoCo-DAOS can be solved by a similar method as that for MeCoCo-DADS, so we omit the details given the similarity. By similar reasoning as that for THEOREM 5, we can get COROLLARY 2:

COROLLARY 2. *MeCoCo-DAOS has a polynomial-time randomized $\mathcal{O}(\log(\Delta|V|) \log |D| \log \Delta)$-approximation.*

5. PERFORMANCE EVALUATION

We evaluate the practical performance of our algorithms in this section. As we are the first to work on the MeCoCo problem, our simulation results are not meant to compare with others. The two goals we bear in mind are: (i) demonstrating the efficacy of our algorithms, and (ii) showing the advantage (in terms of total energy cost) of exploring the directional features (for sensors or/and antennas).

We implement a simulator using C++. In the simulations, we randomly deploy $|V|$ sensor nodes in a square of $10^4 |V| \text{m}^2$ (so that the density is fixed), and the transmission radius R_t and the sensing radius R_s of each node are set to 200 meters and 100 meters, respectively.[1] We also generate targets randomly in the same square area, with the constraint that that each target is covered by at least one sensor node. The energy costs ε_c and ε_s are set to 2 Joule and 1 Joule, respectively. The beamwidths of directional sensors/antennas are set to the same value[2]. We also set the energy cost of any directional sensor to be proportional to ε_s with respect to the value of its beamwidth, i.e., $\tau_s = \varepsilon_s \cdot \eta_s / 2\pi$, and τ_c is computed in the same way.

For the approximate GST algorithm, we use the one given in [16]. Although it has a sub-linear approximation ratio $\mathcal{O}(|D|^\varepsilon)$, it is much more practical than those pointed out in [8, 13]. In fact, a sub-linear approximation algorithm virtually has no difference from a polylogarithmic approximation algorithm when coming to average cases.

In Figure 3, we study how the performance of our algorithms is affected by the the number of sensor nodes and the sensing/transmission beamwidths. The number of targets is fixed to 100, and the number of sensor nodes scales from 100 to 700 with an increment of 100. Three remarkable observations are: (i) the performance in terms of total energy cost follows a total order: DADS < DAOS < OADS < OAOS, (ii) the smaller the beamwidth, the lower the total energy cost is, and (iii) given the fact that each target is bounded to be covered by at least one sensor node, increasing the network size brings the total energy cost up.

While the first observation clearly shows the advantage of exploring the directional features, the second observation is rather straightforward to understand, given the way we set the energy costs of sensors and antennas. Although the last observation is slightly artificial given our requirement on each target to be covered by at least one node, it does confirm the advantage of *intentional deployment* over *random deployment*: if we can properly deploy nodes based on the locations of the targets, we end up with smaller networks. As smaller networks require lower energy to maintain connectivity, a lower total energy cost is achieved.

In Figure 4, we fix the network size to 500 and vary the number of targets and the beamwidth. It is natural to see that the total energy cost increases with the number of targets, given the increased number of targets. Finally, we check the effects of fine-tuning the beamwidth by further reducing it to $\pi/6$, and the results for the beamwidth rang-

---

[1]Although our polylogarithmic approximations work for $R_s > R_t/2$, the constant approximation for MeCoCo-OAOS requires $0 < R_s \leq R_t/2$. Since we need the OAOS case to serve as a benchmark for demonstrating the superiority of exploring directional features, we have to stick to $0 < R_s \leq R_t/2$.

[2]The results concerning different beamwidths for directional sensors and antennas will be studied in a future work.

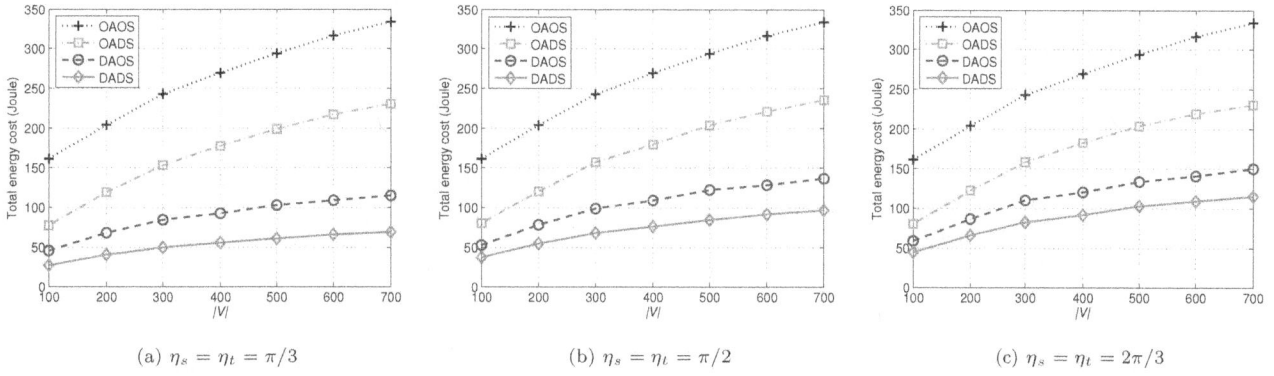

(a) $\eta_s = \eta_t = \pi/3$ (b) $\eta_s = \eta_t = \pi/2$ (c) $\eta_s = \eta_t = 2\pi/3$

Figure 3: Comparing our algorithms for different network sizes and beamwidths, with $|D| = 100$.

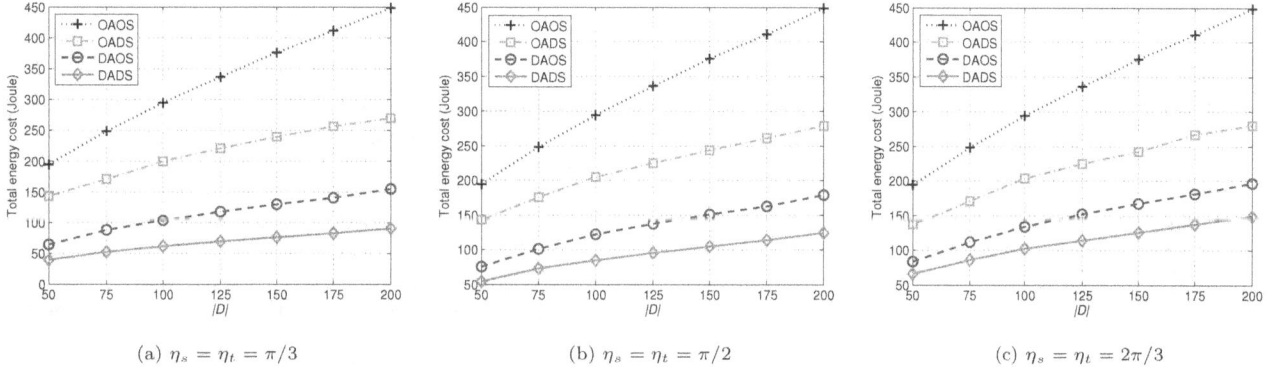

(a) $\eta_s = \eta_t = \pi/3$ (b) $\eta_s = \eta_t = \pi/2$ (c) $\eta_s = \eta_t = 2\pi/3$

Figure 4: Comparing our algorithms for different target set sizes and beamwidths, with $|V| = 500$.

ing from $\pi/6$ to $2\pi/3$ are shown in Figure 5. It clearly demonstrates that, if technology allows, the narrower the beamwidth, the more energy costs we could reduce.

6. CONCLUSION AND DISCUSSIONS

In this paper, we have studied the minimum-energy connected-coverage (MeCoCo) problem; it aims at efficiently monitoring a finite set of targets using sensor nodes with omni-directional or/and directional sensors or/and antennas. Considering different combinations of the omni-directional and directional features of sensors and antennas, we have investigated four cases of the MeCoCo problem and have provided approximation algorithms for every case. We have also proven the approximation ratios (one constant and three polylogarithmic) of our algorithms, and we have finally performed extensive simulations to demonstrate the effectiveness of our algorithms.

In this work, we assume that the sensors and antennas have the uniform covering costs and uniform connection costs in each sub-problem (i.e.,OAOS,DAOS,OADS or DADS) of the MeCoCo problem. However, our work can be easily adapted to the scenarios where these costs are heterogeneous. For example, in the MeCoCo-DADS problem, when each node u has its specific covering cost τ_s^u and connection cost τ_c^u, and only symmetric communication links are considered[3], we can revise the construction method of $\breve{G}$ by modifying τ_s to τ_s^u in line 9 of **Algorithm 3**, and modifying τ_c to

$\tau_c^u + \tau_c^v$ in line 14 of **Algorithm 4**. Under these modifications, it would not be difficult to see that our algorithm still delivers the same approximation ratio for MeCoCo-DADS. However, if the asymmetric communication links are taken into account, it may need to involve building an approximate directed group Steiner tree in our algorithm, and the resulting approximation ratio could be larger. We are on the way towards designing more efficient approximation algorithms for the case of asymmetric communication links.

At this initial investigation phase, we have neglected the energy consumption of rotating sensors and antennas. This makes sense if the sensors and antenna both use the phased array technology. However, there also exist directional sensors and antennas that are steered by motors. We will focus on MeCoCo problems under this later type of sensors and antennas in a future work, by factoring in the energy consumption of rotations.

Acknowledgments

We would like to thank the anonymous reviewers for their insightful comments and constructive suggestions. This work was supported in part by the National Natural Science Foun-

---

[3]By *symmetric* links, we mean that the transmission radii

of two nodes A and B are such that they lie within each other's transmission radius, hence they have a bi-directional link $A \leftrightarrow B$ between them and can communicate with each other. Under heterogeneous connection costs, an *asymmetric* link $A \rightarrow B$ is possible, indicating that B lies within A's transmission radius, but not vice versa.

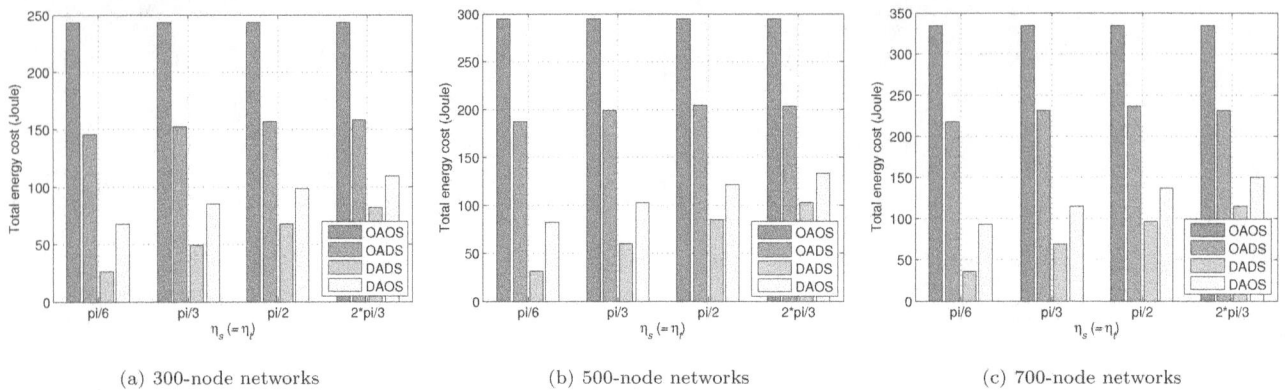

| (a) 300-node networks | (b) 500-node networks | (c) 700-node networks |

Figure 5: Fine-tuning beamwidths under different network sizes with $|D| = 100$.

dation of China (NSFC) under grant No.61103007 and grant No.61003052, the Start-up Grant of NTU, and AcRF Tier 2 Grant ARC15/11.

7. REFERENCES

[1] J. Ai and A. Abouzeid. Coverage by directional sensors in randomly deployed wireless sensor networks. *Journal of Combinatorial Optimization*, 11:21–41, 2006.

[2] I. Akyildiz, W. Su, Y. Sankarasubramaniam, and E. Cayirci. Wireless sensor networks: a survey. *Computer Networks*, 38(4):393 – 422, 2002.

[3] X. Bai, S. Kumar, D. Xuan, Z. Yun, and T. H. Lai. Deploying wireless sensors to achieve both coverage and connectivity. In *Proc. ACM MobiHoc*, pages 131–142, 2006.

[4] Y. Cai, W. Lou, M. Li, and X.-Y. Li. Target-oriented scheduling in directional sensor networks. In *Proc. IEEE INFOCOM*, pages 1550 –1558, 2007.

[5] Y. Cai, W. Lou, M. Li, and X.-Y. Li. Energy efficient target-oriented scheduling in directional sensor networks. *IEEE Trans. on Computers*, 58(9):1259 –1274, sept. 2009.

[6] I. Cardei and M. Cardei. Energy-efficient connected-coverage in wireless sensor networks. *International Journal of Sensor Networks*, 3(3):201–210, 2008.

[7] M. Cardei, M. Thai, Y. Li, and W. Wu. Energy-efficient target coverage in wireless sensor networks. In *Proc. IEEE INFOCOM*, volume 3, pages 1976 – 1984 vol. 3, 2005.

[8] C. Chekuri, G. Even, and G. Kortsarz. A greedy approximation algorithm for the group steiner problem. *Discrete Applied Mathematics*, 154(1):15 – 34, 2006.

[9] A. Chen, S. Kumar, and T. H. Lai. Designing localized algorithms for barrier coverage. In *Proc. ACM MobiCom*, pages 63–74, 2007.

[10] K. L. Clarkson and K. Varadarajan. Improved approximation algorithms for geometric set cover. *Discrete and Computational Geometry*, 37:43–58, January 2007.

[11] G. Fusco and H. Gupta. Selection and orientation of

[12] G. Fusco and H. Gupta. Placement and orientation of rotating directional sensors. In *Proc. IEEE SECON*, pages 1 –9, 2010.

[13] N. Garg, G. Konjevod, and R. Ravi. A polylogarithmic approximation algorithm for the group steiner tree problem. *Journal of Algorithms*, 37(1):66 – 84, 2000.

[14] D. Goldenberg, P. Bihler, M. Cao, J. Fang, B. Anderson, A. Morse, and Y. Yang. Localization in Sparse Networks using Sweeps. In *Proc. of the 12th ACM MobiCom*, 2006.

[15] X. Han, X. Cao, E. Lloyd, and C.-C. Shen. Deploying directional sensor networks with guaranteed connectivity and coverage. In *Proc. IEEE SECON*, pages 153 –160, june 2008.

[16] C. S. Helvig, G. Robins, and A. Zelikovsky. An improved approximation scheme for the group steiner problem. *Wiley Networks*, 37(1):8–20, 2001.

[17] G. Kasbekar, Y. Bejerano, and S. Sarkar. Lifetime and coverage guarantees through distributed coordinate-free sensor activation. In *Proc. ACM MobiCom*, pages 169–180, 2009.

[18] S. Kumar, T. H. Lai, and A. Arora. Barrier coverage with wireless sensors. In *Proc. ACM MobiCom*, pages 284–298, 2005.

[19] F. Li, J. Luo, S.-Q. Xin, W.-P. Wang, and Y. He. LAACAD: Load bAlancing k-Area Coverage through Autonomous Deployment in Wireless Sensor Networks. In *Proc. of the 32nd IEEE ICDCS*, pages 1–10, 2012.

[20] B. Liu, O. Dousse, J. Wang, and A. Saipulla. Strong barrier coverage of wireless sensor networks. In *Proc. ACM MobiHoc*, pages 411–420, 2008.

[21] M. Lu, J. Wu, M. Cardei, and M. Li. Energy-efficient connected coverage of discrete targets in wireless sensor networks. *International Journal of Ad Hoc and Ubiquitous Computing*, 4(3/4):137–147, 2009.

[22] J. Luo, A. Girard, and C. Rosenberg. Efficient algorithms to solve a class of resource allocation problems in large wireless networks. In *Proc. ICST/IEEE WiOpt*, pages 1–9, 2009.

[23] J. Luo, H. Shukla, and J.-P. Hubaux. Non-Interactive Location Surveying for Sensor Networks with

Mobility-Differentiated ToA. In *Proc. of the 25th IEEE INFOCOM*, pages 1241–1252, 2006.

[24] H. Ma, X. Zhang, and A. Ming. A coverage-enhancing method for 3d directional sensor networks. In *Proc. IEEE INFOCOM*, pages 2791 –2795, 2009.

[25] K. Mehlhorn. A faster approximation algorithm for the steiner problem in graphs. *Information Processing Letters*, 27(3):125–128, 1988.

[26] Y. Osais, M. St-Hilaire, and F. Yu. Directional sensor placement with optimal sensing range, field of view and orientation. *Mobile Networks and Applications*, 15:216–225, 2010.

[27] S. Roy, Y. Hu, D. Peroulis, and X.-Y. Li. Minimum-energy broadcast using practical directional antennas in all-wireless networks. In *Proc. IEEE INFOCOM*, pages 1–12, june 2007.

[28] P. Wan, X. Xu, and W. Zhu. Wireless coverage with disparate ranges. In *Proc. ACM MobiHoc*, pages 111–118, 2011.

[29] Y. Wang and G. Cao. Barrier coverage in camera sensor networks. In *Proc. ACM MobiHoc*, pages 119–128, 2011.

[30] Y. Wang and G. Cao. Minimizing service delay in directional sensor networks. In *Proc. IEEE INFOCOM*, pages 1790 –1798, 2011.

[31] G. Xing, X. Wang, Y. Zhang, C. Lu, R. Pless, and C. Gill. Integrated coverage and connectivity configuration for energy conservation in sensor networks. *ACM Trans. on Sensor Networks*, 1:36–72, August 2005.

[32] Z. Yu, J. Teng, X. Bai, D. Xuan, and W. Jia. Connected coverage in wireless networks with directional antennas. In *Proc. IEEE INFOCOM*, pages 2264 –2272, 2011.

[33] Q. Zhao and M. Gurusamy. Lifetime maximization for connected target coverage in wireless sensor networks. *IEEE/ACM Trans. on Networking*, 16(6):1378 –1391, dec. 2008.

APPENDIX

PROOF OF THEOREM 2. Let $C^* \subseteq V$ be the minimum set of nodes that can cover all the targets in D. Clearly $|C^*| \leq |M_1|$. Since the algorithm used in line 1 of **Algorithm 1** is a PTAS, we have:

$$|M_1'| \leq (1 + \epsilon)|C^*| \leq (1 + \epsilon)|M_1|. \qquad (1)$$

Let the optimal Steiner tree spanning the nodes in $\{\mathfrak{s}\} \cup M_1'$ be T_S^*, and let $|T_S^*|$ counts the number of nodes in T_S^*. Clearly we have:

$$|S_1'| - 1 \leq 2(|T_S^*| - 1). \qquad (2)$$

For any node $u \in M_1 \subseteq S_1$ (where M_1, S_1 are the optimal solution), we can find a node $d \in D$ such that u lies in the disk centered at d with radius R_s. Clearly, we can also find a node $v \in M_1'$ in the same disk. Therefore, the distance between u and v is no more than $2R_s \leq R_t$. This reveals that the subgraph induced by $M_1' \cup S_1$ is connected. Since $\mathfrak{s} \in S_1$ and T_S^* is the optimal Steiner tree spanning the nodes in $\{\mathfrak{s}\} \cup M_1'$, we can get:

$$|T_S^*| \leq |M_1'| + |S_1|. \qquad (3)$$

Combining (1)-(3) and $|M_1| \leq |S_1|$, we have:

$$
\begin{aligned}
|S_1'| &\leq 2(|M_1'| + |S_1|) \\
&\leq (2 + \epsilon)|M_1| + 2|S_1| \\
&\leq (4 + \epsilon)|S_1|,
\end{aligned}
$$

which in turn leads to the following:

$$
\begin{aligned}
\xi_{OAOS}(S_1', M_1') &= |S_1'| \cdot \varepsilon_c + |M_1'| \cdot \varepsilon_s \\
&\leq (4 + \epsilon)|S_1| \cdot \varepsilon_c + (1 + \epsilon)|M_1| \cdot \varepsilon_s \\
&\leq (4 + \epsilon) \cdot (|S_1| \cdot \varepsilon_c + |M_1| \cdot \varepsilon_s) \\
&= (4 + \epsilon) \cdot \xi_{OAOS}(S_1, M_1),
\end{aligned}
$$

hence the approximation ratio $4 + \epsilon$. $\square$

PROOF OF LEMMA 1. According to **Algorithm 2**, we can easily know that Q is an IDS. Here we only prove $|\chi(\lambda(v))| \leq |Q|$, so Q is an MIDS.

If there is only one working direction θ in Q, then we must have $\kappa_\theta(\lambda(v)) = \{d_1, d_2, ..., d_h\}$ (according to line 5 of **Algorithm 2**). For this case, clearly we have $\chi(\lambda(v)) = Q$, so $|\chi(\lambda(v))| \leq |Q|$.

Otherwise, let $\mathbf{d}(r, s)$ be the set $\{d_{1+(i-1) \bmod (h)} | r \leq i \leq r + s - 1, r \in \mathbb{Z}, 1 \leq s \leq h\}$. Literally, $\mathbf{d}(r, s)$ is the set of s nodes in the ordered sequence $\mathbf{d}$ starting from $d_{1+(r-1) \bmod (h)}$. For any $\theta \in \chi(\lambda(v))$, there must exist two integers r_1 and $s_1, 1 \leq s_1 \leq h$ such that $\kappa_\theta(\lambda(v)) = \mathbf{d}(r_1, s_1)$. Since θ is independent, we must have

$$\angle d_{1+(r_1-2) \bmod (h)} v d_{1+(r_1+s_1-2) \bmod (h)} > \eta_s, \text{and}$$

$$\angle d_{1+(r_1-1) \bmod (h)} v d_{1+(r_1+s_1-1) \bmod (h)} > \eta_s;$$

otherwise a contradiction to the independence of θ. According to lines 3–6 of **Algorithm 2**, a working direction θ' for which $\kappa_{\theta'}(\lambda(v)) = \mathbf{d}(r_1, s_1)$ must have been taken into Q. Hence, θ and θ' are equivalent. This implies $|\chi(\lambda(v))| \leq |Q|$. $\square$

PROOF OF THEOREM 3. The proof is similar to that of THEOREM 4 and 5 and is hence omitted. $\square$

PROOF OF LEMMA 2. There are at most $|V|$ nodes in Y, so there are at most $|V|$ shadow nodes in $\breve{V}$. According to COROLLARY 1, at most δ_2 sensor sector nodes and δ_1 antenna sector nodes are created for any node $u \in V$. Hence, the number of nodes in $\breve{V}$ is no more than

$$|V| + \delta_2|V| + \delta_1|V| + 1 = \mathcal{O}(\Delta|V|).$$

According to lines 6–8 of **Algorithm 3**, any target node in D is covered by no more than $\delta_2\delta_3 \leq \Delta^2$ sensor sector nodes in $\breve{G}$. Hence, the size of the largest group in $\{\mathcal{G}_d | d \in D \cup \{\mathfrak{s}\}\}$ is no more than Δ^2. $\square$

PROOF OF LEMMA 3. For a node $u \in V$, if there exists a working direction θ of $\Upsilon(u)$ such that $\theta \in \Xi_3^u$, and $\theta \notin \chi(\Upsilon(u))$, then θ must satisfy one of the following three conditions: (i) $\kappa_\theta(\Upsilon(u)) = \emptyset$, (ii) θ is equivalent to another working direction θ' in $\chi(\Upsilon(u))$, and (iii) θ is not independent.

If condition (i) is satisfied, then we can remove θ from Ξ_3^u such that Ξ_3 is still (part of) a feasible solution to MeCoCo-DADS, but $\xi_{DADS}(S_3, M_3, \Xi_3, \Theta_3)$ is decreased by τ_c; a contradiction to the optimality of the original Ξ_3. If condition (ii) is satisfied, then we can replace θ by θ' in Ξ_3^u. The objective value remains the same due to the equivalence between θ and θ'. If condition (iii) is satisfied, then there must exist a working direction $\theta' \in \chi(\Upsilon(u))$ such that $\kappa_\theta(\Upsilon(u)) \subset \kappa_{\theta'}(\Upsilon(u))$; otherwise $\chi(\Upsilon(u))$ is not a MIDS of $\Upsilon(u)$. In such a scenario, we can replace θ by θ' in Ξ_3^u without changing the optimal value. By the above discussions, we know that $\Xi_3^u \subseteq \chi(\Upsilon(u))$. With similar reasoning, we can also show $\Theta_3^v \subseteq \chi(\lambda(v))$. $\square$

PROOF OF LEMMA 4. According to the method used to construct $\breve{G}$ in **Algorithm 4**, the nodes in T_{DD} can be partitioned into several mutually disjoint subsets: the set $\{\mathfrak{s}\}$, the set of antenna sector nodes W, the set of shadow nodes H and the set of sensor sector nodes I. According to lines 3–10 of **Algorithm 4**, we can partition the nodes in W into mutually disjoint subsets $W_1, W_2, ..., W_q$, such that all nodes in W_i are created from the same node $u_i \in V$. Next, we will show how to construct $\grave{S}_3, \grave{M}_3, \grave{\Xi}_3, \grave{\Theta}_3$ from these sets and T_{DD}.

Let $\grave{S}_3$ be the set $\{u_i | 1 \le i \le q\}$. Note that $\mathfrak{s} \in \grave{S}_3$ due to lines 6–7. The set $\grave{\Xi}_3$ (and the implied edge set $\grave{E}_3$) can be constructed as follows. For each $u_i \in \grave{S}_3$, we initialize $\grave{\Xi}_3^{u_i}$ to an empty set, and we also initialize $\grave{E}_3$ to an empty set. Let Z be the set $\{(\breve{u}_j, \breve{v}_k) \in T_{DD} | \breve{u}_j \in W_j$ and $\breve{v}_k \in W_k, 1 \le j \ne k \le q\}$. According to lines 11–14 of **Algorithm 4**, for each $(\breve{u}_j, \breve{v}_k) \in Z$, we can find two working directions $\theta \in \chi(\Upsilon(u_j))$ and $\theta' \in \chi(\Upsilon(u_k))$ such that $u_k \in \kappa_\theta(\Upsilon(u_j))$ and $u_j \in \kappa_{\theta'}(\Upsilon(u_k))$. Then we add θ into $\grave{\Xi}_3^{u_j}$, add θ' into $\grave{\Xi}_3^{u_k}$, and add (u_j, u_k) into $\grave{E}_3$. Upon finishing this process, it is straightforward to see that the nodes in $\grave{S}_3$ are connected by the edges in $\grave{E}_3$, thus condition $c2$ (see Section 3.2) is satisfied by $\grave{S}_3$ and $\grave{\Xi}_3$. Moreover, since each node in $\grave{\Xi}_3^u : u \in \grave{S}_3'$ corresponds to a terminal node of an edge in Z, we have:

$$\sum_{u \in \grave{S}_3} |\grave{\Xi}_3^u| \cdot \tau_c \le 2|Z| \cdot \tau_c \le 2 \cdot \text{weight}(T_{DD}) \quad (4)$$

Let $\grave{M}_3$ be the nodes in V with shadow nodes in H, i.e., $\grave{M}_3 = \{u \in V | \iota(u) \in H\}$. According to line 9 of **Algorithm 4**, any shadow node $\iota(u) \in H$ is connected to the remaining part of $\breve{G}$ only through u's antenna sector nodes in W. Therefore, we have $\grave{M}_3 \subseteq \grave{S}_3$.

The set $\grave{\Theta}_3$ can be constructed as follows. For any node $u \in \grave{M}_3$, we initialize $\grave{\Theta}_3^u$ to an empty set. Then we check each target $d \in D$. According to the definition of GST and lines 8–9 of **Algorithm 3**, we can find a sensor sector node $\varrho_v(\theta)$ and a shadow node $\iota(v)$ such that $\theta \in \chi(\lambda(v))$, $d \in \kappa_\theta(\lambda(v))$ and $(\varrho_v(\theta), \iota(v))$ is an edge with weight τ_s in T_{DD}. So we add θ into $\grave{\Theta}_3^v$. When all nodes in D are checked, we have $D \subseteq \bigcup_{u \in \grave{M}_3} \bigcup_{\theta \in \grave{\Theta}_3^u} \kappa_\theta(\lambda(u))$, satisfying condition $c1$. Note that in the above process, no edges in Z are involved. Therefore, we can get:

$$\sum_{v \in \grave{M}_3} |\grave{\Theta}_3^v| \cdot \tau_s + |Z| \cdot \tau_c \le \text{weight}(T_{DD}) \quad (5)$$

With all sets $\grave{S}_3, \grave{M}_3, \grave{\Xi}_3, \grave{\Theta}_3$ ready, we can compute the total energy cost as:

$$\begin{aligned} &\xi_{DADS}(\grave{S}_3, \grave{M}_3, \grave{\Xi}_3, \grave{\Theta}_3) \\ =~& \sum_{u \in \grave{S}_3} |\grave{\Xi}_3^u| \cdot \tau_c + \sum_{v \in \grave{M}_3} |\grave{\Theta}_3^v| \cdot \tau_s \\ \le~& 2\left(|Z| \cdot \tau_c + \sum_{v \in \grave{M}_3} |\grave{\Theta}_3^v| \cdot \tau_s\right) \end{aligned}$$

hence the claimed upper bound due to (5). $\square$

PROOF OF LEMMA 5. From LEMMA 3 and line 5 of **Algorithm 4**, we know that, for any node $u \in S_3$, the working directions in Ξ_3^u are mapped to a set of antenna sector nodes of u in $\breve{G}$. This set of antenna sector nodes can be seen as a "super node" of u, or $SN(u)$, which induces a clique, according to line 10 of **Algorithm 4**. Therefore, we say that the super node $SN(u)$ is *self-connected*. In particular, we know from line 7 of **Algorithm 4** that $\mathfrak{s}$ is connected to $SN(\mathfrak{s})$.

On one hand, condition $c2$ of MeCoCo-DADS suggests that, for any edge $(u, v) \in E_3$, there exist two working directions $\theta \in \Xi_3^u$ and $\theta' \in \Xi_3^v$ such that $v \in \kappa_\theta(\Upsilon(u)) \wedge u \in \kappa_{\theta'}(\Upsilon(v))$. According to LEMMA 3, θ is also in $\chi(\Upsilon(u))$ and θ' is also in $\chi(\Upsilon(v))$. Therefore, according to lines 11–14 of **Algorithm 4**, there exists an edge $(\rho_u(\theta), \rho_v(\theta'))$ with weight τ_c between $SN(u)$ and $SN(v)$. Since S_3 is connected by E_3 (according to $c2$) and each super node is self-connected, the subgraph of $\breve{G}$ induced by $\mathfrak{s}$ and all antenna sector nodes is connected.

On the other hand, from $D \subseteq \bigcup_{v \in M_3} \bigcup_{\theta \in \Theta_3^v} \kappa_\theta(\lambda(v))$ we can know that, for any target $d \in D$, we can find a node $w \in M_3$ and a working direction $\theta \in \Theta_3^w$ such that $d \in \kappa_\theta(\lambda(w))$. According to line 1 of **Algorithm 4** and line 8–9 of **Algorithm 3**, there must exist a sensor sector node $\varrho_w(\theta) \in \breve{G}$ and a shadow node $\iota(w) \in \breve{G}$ such that $\varrho_w(\theta) \in \mathcal{G}_d$ and $\varrho_w(\theta)$ is connected to $\iota(w)$. Finally, since $M_3 \subseteq Y$ and $M_3 \subseteq S_3$, we know from line 9 of **Algorithm 4** that the shadow node $\iota(w)$ is connected to $SN(w)$.

The above discussions and the weight assignment of the edges in $\breve{G}$ suggest that we can find a connected subgraph $\breve{G}'$ of $\breve{G}$ such that $\breve{G}'$ contains at least one node from every $\{\mathcal{G}_d | d \in D \cup \{\mathfrak{s}\}\}$ and $\text{weight}(\breve{G}') \le \xi_{DADS}(S_3, M_3, \Xi_3, \Theta_3)$. Since $\text{weight}(T_{DD}^*) \le \text{weight}(\breve{G}')$, the lemma follows. $\square$

PROOF OF THEOREM 4. According to LEMMA 4, we can map the output of **Algorithm 4**, T_{DD}, to a feasible solution $\grave{S}_3, \grave{M}_3, \grave{\Xi}_3, \grave{\Theta}_3$ of MeCoCo-DADS, such that the following inequality holds

$$\xi_{DADS}(\grave{S}_3, \grave{M}_3, \grave{\Xi}_3, \grave{\Theta}_3) \le 2 \cdot \text{weight}(T_{DD}).$$

Combing this with $\text{weight}(T_{DD}) \le \gamma \cdot \text{weight}(T_{DD}^*)$ and the inequality proven in LEMMA 5, we have proven the approximation ratio 2γ. $\square$

Probabilistic Missing-tag Detection and Energy-Time Tradeoff in Large-scale RFID Systems

Wen Luo, Shigang Chen, Tao Li, Yan Qiao
Computer & Information Science & Engineering
University of Florida, Gainesville, FL, USA
{wluo, sgchen, tali, yqiao}@cise.ufl.edu

ABSTRACT

RFID (radio frequency identification) technologies are poised to revolutionize retail, warehouse and supply chain management. One of their interesting applications is to automatically detect missing tags (and the associated objects) in a large storage space. In order to timely catch any missing event such as theft, the detection operation may have to be performed frequently. Because RFID systems typically work under low-rate channels, past research has focused on reducing execution time of a detection protocol, in order to prevent excessively-long protocol execution from interfering normal inventory operations. However, when active tags are used to provide a large spatial coverage, energy efficiency becomes critical in prolonging the lifetime of these battery-powered tags. Existing literature lacks thorough study on how to conserve energy in the process of missing-tag detection and how to jointly optimize energy efficiency and time efficiency. This paper makes two important contributions: First, we propose a novel protocol design that takes both energy efficiency and time efficiency into consideration. It achieves multi-fold reduction in both energy cost and execution time when comparing with the best existing work. In some cases, the reduction is more than an order of magnitude. Second, we reveal a fundamental energy-time tradeoff in missing-tag detection. Through our analytical framework, we are able to flexibly control the tradeoff through a couple of system parameters in order to achieve desirable performance.

Categories and Subject Descriptors

C.2.1 [**Network Architecture and Design**]: Wireless communication

General Terms

Algorithms, Performance

Keywords

RFID Systems, Energy-Time Tradeoff, Missing Tag Detection

1. INTRODUCTION

RFID (radio frequency identification) technologies [1, 2, 3, 4, 5, 6, 7, 8, 9, 10, 11] are poised to revolutionize retail, warehouse and supply chain management. They are also having a profound impact on our daily lives, with important applications in automatic toll payment, access control to parking garages, object tracking, and theft prevention. Comparing with barcodes that have to be read from a very close range by a laser scanner, RFID tags have great advantages: they can be read wirelessly over a distance, and they are able to perform simple computations. Starting from August 1, 2010, Wal-Mart has begun to embed RFID tags in clothing [12].

An interesting application of RFID tags is to detect missing items in a large storage. Consider a major warehouse that keeps thousands of apparel, shoes, pallets, cases, appliances, electronics, etc. How to find out if anything is missing? We may have someone walk through the warehouse and count items. This is not only laborious but also error-prone, considering that clothes may be stacked together, goods on racks may need a ladder to access, and they may be blocked behind columns. If we attach a RFID tag to each item,[1] the whole detection process can be automated with one or multiple RFID readers communicating with tags to find out whether any tags (and their associated objects) are absent.

There are two different missing-tag detection problems: *exact detection* and *probabilistic detection*. The objective of exact detection is to identify exactly *which tags are missing*. The objective of probabilistic detection is to detect *a missing-tag event* with a certain predefined probability. An exact detection protocol [13, 14, 15] gives much stronger results, but its overhead is far greater than a probabilistic detection protocol [6, 16, 17, 18]. Hence, they both have their values. In fact, they are complementary to each other, and should be used together. For example, a probabilistic detection protocol may be scheduled to execute frequently, e.g., once every minute, in order to timely catch any loss event such as theft. Once it detects some tags are missing, it may invoke an exact detection protocol to pinpoint which tags are missing. If one execution of a probabilistic detection protocol detects a missing-tag event with 99% probability, five executions will detect the event with 99.99999999% probability. If that is not enough, we may schedule an exact detection protocol [13] every five times the probabilistic detection protocol is executed.[2]

This paper focuses on probabilistic detection. Because it is performed frequently, its performance becomes very important. Suppose a missing-tag detection protocol is scheduled to execute once

---

[1] A tag may be attached in a way that ruins the product if it is detached inappropriately, such as releasing ink onto clothing.

[2] Another exact detection protocol [14] reports which tags are missing, but does not guarantee to report all missing ones. The protocol in [13] guarantees 100% reporting.

every few minutes in a warehouse. If the execution time of the protocol is a minute, any normal operations that move goods out are likely to trigger false alarm. To reduce the chance of interfering normal operations, we want to make the protocol's execution time as small as possible. Another performance requirement is to minimize the protocol's energy cost. To cover a large area, battery-powered active tags are preferred. To prolong their lifetime, we need to make any periodically-executed protocol as energy-efficient as possible, particularly if one is scheduled to execute once every few minutes for 24 hours a day, day by day. To make the problem more challenging, energy efficiency and time efficiency are often conflicting objectives that cannot be optimized simultaneously.

Despite its importance, the problem of probabilistic missing-tag detection is relatively new and under-investigated. The basic detection method is introduced in the pioneer work [6]: A RFID reader monitors a time frame of f slots. Through a hash function, each tag pseudo-randomly selects a slot in the time frame to transmit. The reader can predict in which slot each known tag will transmit. It detects a missing-tag event if no tag transmits during a slot when there is supposed to be tag(s) transmitting. However, multiple tags may select the same slot to transmit. If a tag is missing, its slot may be kept busy by transmission from another tag. Consequently, the reader cannot guarantee the detection of a missing-tag event. The protocol in [6] only considers time efficiency, but not energy efficiency. As a follow-up work [16] points out, even for time efficiency, it is far from being optimal. Firner et al. [17] design a simple communication protocol, Uni-HB, to detect missing items for fail-safe presence assurance systems and demonstrate it can lead to longer system lifetime and higher communication reliability than several popular protocols. The protocol however does not consider time efficiency and requires all tags to participate and transmit, which will be less efficiency than a sampling-based protocol design that requires only a small fraction of the tags to participate. Similarly, the method in [18] also requires all tags to participate.

In this paper, we make two contributions. First, we propose a new, more sophisticated protocol design for missing-tag detection. It takes both energy efficiency and time efficiency into consideration. By introducing multiple hash seeds, our new design provides multiple degrees of freedom for tags to choose which slots they will transmit in. This design drastically reduces the chance of collision, and consequently achieves multiple-fold reduction in both energy cost and execution time. In some cases, the reduction is more than an order of magnitude. Second, with the new design, we reveal a fundamental energy-time tradeoff in missing-tag detection. Our analysis shows that better energy efficiency can be achieved at the expense of longer execution time, and vice versa. The performance tradeoff can be easily controlled by a couple of system parameters. Through our analytical framework for energy-time tradeoff, we are able to compute the optimal parameter settings that achieve the smallest protocol execution time or the smallest energy cost. The framework also enables us to solve the energy-constrained least-time problem and the time-constrained least-energy problem in missing-tag detection.

The rest of the paper is organized as follows: Section 2 gives the system model and problem definition, as well as the prior art. Section 3 proposes a new missing-tag detection protocol. Section 4 investigates energy-time tradeoff in protocol configuration. Section 5 evaluates the protocol through simulations. Section 6 draws the conclusion.

2. PRELIMINARIES

2.1 System Model

There are three types of RFID tags [19]. *Passive tags* — the kind used by Wal-Mart — are most widely deployed today. They are cheap, but do not have internal power sources. Passive tags rely on radio waves emitted from a RFID reader to power their circuit and transmit information back to the reader through backscattering. They have short operational ranges, typically a few meters in an indoor environment, which seriously limits their applicability. *Semi-passive tags* carry batteries to power their circuit, but still rely on backscattering to transmit information. *Active tags* use their own battery power to transmit, and consequently do not need any energy supply from the reader. Active tags operate at a much longer distance, making them particularly suitable for applications that cover a large area, where one or a few RFID readers are installed to access all tagged objects and perform management functions automatically. With richer on-board resources, active tags are likely to gain more popularity in the future, when their prices drop over time as manufactural technologies are improved and markets are expanded. They are particularly attractive for high-valued objects such as luxury bags, laptops, cell phones, TVs, etc., or when the tags are reused over and over again. In this paper, we consider active tags.

Communication between a reader and tags is time-slotted. The reader's signal synchronizes the clocks of tags. There are different types of time slots [13], among which two types are of interest in this paper. The first type is called a *tag-ID slot*, whose length is denoted as T_{tag}, during which a reader is able to broadcast a tag ID. The second type is called a *short-response slot*, whose length is denoted as T_{short}, during which a tag is able to transmit one-bit information to the reader, for instance, announcing its presence. Based on the parameters of Philips I-Code [20], T_{tag} is about 3927 μs, and T_{short} is about 321 μs. See Section 5 for details.

2.2 Problem

The problem is to design an efficient protocol for a RFID reader to detect whether some tags are missing, subject to a *detection requirement*: A single execution of the protocol should detect a missing-tag event with probability α if m or more tags are missing, where α and m are two system parameters. For example, a big shoe store may carry tens of thousands of shoes. We may set the parameters to be $\alpha = 99\%$ and $m = 10$, so that one execution of the protocol will detect any event of missing 10 or more shoes with 99% probability. If the protocol is periodically executed, the detection probability of any missing event will approach to 100%, no matter what the values of α and m are. Furthermore, as we have explained in the introduction, a low-overhead probabilistic detection protocol may be used in conjunction with a high-overhead exact detection protocol (which is scheduled much less frequently) to catch any miss.

We consider two performance metrics: execution time of the protocol and energy cost to the tags. The former is measured as the time it takes the protocol to complete one execution. We cannot find a well-accepted energy model for RFID tags. However, in our design, each tag will wake up only if it will participate in a scheduled protocol execution, and the energy expenditure for participating tags is about the same; they receive the same amount of data, stay active for a similar amount of time, and for most of them, transmit the same amount of data. Therefore, we can use the number of tags that participate in each protocol execution as an approximate measurement for energy cost.

The above two performance metrics are important due to the following reasons: First, RFID systems use low-rate communication channels. In the Philips I-Code system, the rate from a reader to a tag is about 27Kbps and the rate from a tag to a reader is about 53Kbps. Low rates, coupled with a large number of tags, often cause long execution times for RFID protocols. To apply such protocols in a busy warehouse environment, it is desirable to reduce protocol execution time as much as possible. Second, active tags carry limited battery power. Replacing tags or their batteries is a tedious, manual operation. One way to save energy is to minimize the number of tags that are needed to participate in each protocol execution. When a tag participates in protocol execution, it has to power its circuit during the execution, receive request information from the reader, and transmit back. When a tag does not participate, it goes into the sleep mode and incurs insignificant energy expenditure. The energy cost to the RFID reader is less of a concern because the readerŠs battery can be easily replaced or it may be powered by an external source.

We assume that the RFID reader has access to a database that stores the IDs of all tags. This assumption is necessary [6]. Without any prior knowledge of a tag's existence, how can we know that it is missing?

The above assumption can be easily satisfied if the tag IDs are read into a database when new objects are moved into the system, and they are removed from the database when the objects are moved out — this is what a typical inventory management procedure will do. Even if such information is lost due to a database failure, we can recover the information by executing an ID-collection protocol [21, 22, 23, 24] that reads the IDs from the tags. In this case, we will not detect missing-tag events that have already happened. However, now that we have the IDs of the remaining tags, we can detect the missing-tag events after this point of time.

2.3 Prior work

We first describe the Trusted Reader Protocol (TRP) by Tan, Sheng and Li [6]. To initiate the execution of the protocol, a RFID reader broadcasts a detection request, asking the tags to respond in a time frame of f slots. The detection request has two parameters, the frame size f and a random number r. Each tag maps itself to a slot in the frame by hashing its ID and r. It then transmits during that slot.

A slot is said to be *empty* if no tag responds (transmits) in the slot. It is called a *singleton slot* if exactly one tag responds. It is a *collision slot* if more than one tag responds. A singleton or collision slot is also called a *busy slot*.

The reader records which slots are busy and which are empty. This is binary information where each slot carries either '1' or '0'. When a tag transmits, it does not have to send any particular information. It only needs to make the channel busy. Because the reader knows the IDs of all tags, it knows which tags are mapped to which slots. More specifically, it knows which slots are expected to be busy. If an expected busy slot turns out to be empty, the tag(s) that is mapped to this slot must be missing. TRP is designed to minimize execution time by using the smallest frame size that ensures a detection probability α if m or more tags are missing. Certainly, if fewer tags are missing, the detection probability will be lower. A follow-up work [14] essentially executes TRP iteratively to identify which tags are missing.

A serious limitation of TRP is that it only considers time efficiency. It is not energy-efficient because all tags must be active and transmit during the time frame. B. Firner et al. [17] consider energy cost, but their protocol requires all tags to participate and transmit, which will be less efficient than a sampling-based solution where only a small fraction of tags participate.

The efficient missing-tag detection protocol (EMD) [16] is similar to TRP except that each tag is sampled with a probability p for participation in each protocol execution. Only a sampled tag will select a slot to transmit. Simulations show that EMD performs better than TRP. However, the paper does not give a way to determine the optimal sampling probability.

In this paper, we show TRP and EMD are special cases of a much broader protocol design space. Not only are there protocol configurations that perform much better than TRP and EMD in terms of both time and energy efficiencies, but also we reveal a fundamental energy-time tradeoff in this design space, which allows us to adapt protocol performance to suit various needs in practical systems.

3. MULTI-SEED MISSING-TAG DETECTION PROTOCOL (MSMD)

3.1 Motivation

Both TRP [6] and EMD [16] map tags to time slots using a hash function. We derive the probability θ that an arbitrary slot t will become a singleton, which happens when only one tag is sampled and mapped to slot t while all other tags are either not sampled or mapped to other slots. The probability for any given tag to be sampled and mapped to t is $\frac{p}{f}$, where f is the number of slots and p is the sampling probability, which is 100% for TRP. The probability for all other tags to be either not sampled or not mapped to slot t is $\left(1 - \frac{p}{f}\right)^{n-1}$, where n is the number of tags. Hence, we have

$$\theta = n\frac{p}{f}\left(1 - \frac{p}{f}\right)^{n-1} \approx \frac{np}{f}e^{-\frac{np}{f}} \leq \frac{1}{e} \approx 36.8\%,$$

where $\frac{np}{f}e^{-\frac{np}{f}}$ reaches its maximum value when $np = f$. This upper bound for θ is true for both TRP and EMD.

Singletons are important. If a missing tag is sampled and mapped to a singleton slot, since no other tag is mapped the same slot, this expected singleton slot will turn out to be empty, which is observed by the reader, resulting in missing-tag detection.

The problem is that the majority of all slots, 63.2% or more of them, are either empty slots or collision slots. They are mostly wasted. Obviously, empty slots do not contribute anything in missing-tag detection. If a collision slot only has missing tags, detection will be successfully made because the reader will find this expected busy slot to be actually empty. However, when the number of missing tags is small when comparing with the total number of tags, the chance for a collision slot to have only missing tags is also small.

Naturally, we want a protocol design that ensures a large value of θ, much larger than 36.8%, because more singleton slots increase detection power. However, the value of θ in TRP is in fact much smaller than 36.8% because TRP minimizes its execution time by using as fewer time slots as possible, which results in a large percentage of collision slots. More specifically, the detection probability of TRP is about $1 - (1 - \theta)^m$ because each of the m missing tags has a probability of θ to map to a singleton slot and thus be detected.[3] Now, if the requirement is to detect a missing-tag event with 99% probability when 100 tags are missing, TRP will reduce its frame size to such a level that $\theta = 4.5\%$, just enough to ensure 99% detection probability.

This leaves a great room for improvement. We show that a new protocol design, different from that of TRP and EMD, can reduce the frame size to a level that is much smaller than they can do, yet keep θ at a value much greater than 36.8%. Our idea is that if we can find a way to turn most empty/collision slots into singletons,

---

[3]To quickly get to the point without dealing with too much details, we ignore the small contribution of collision slots in detection.

we shall be able to improve time/energy efficiencies significantly. There is a compound effect of such a new design when it is coupled with sampling: Suppose $\alpha = 99\%$ and $m = 100$, same as in the previous paragraph. Under sampling, the detection probability is $1 - (1 - p \times \theta)^m$ because each of the m missing tags has a probability of $p \times \theta$ to be sampled and mapped to a singleton slot. If our protocol design can improve θ to 90%, we will be able to set $p = 5\%$. With such a sampling probability, we achieve much better energy efficiency because only 5% of all tags participate in each protocol execution. We also achieve far better time efficiency because, with much fewer tags transmitting, the chance of collision is reduced and a fewer number of time slots are needed to ensure a certain level of singletons.

Before we present our design called *Multiple-Seed Missing-tag Detection protocol* (MSMD), we first discuss how to implement a hash function efficiently for RFID tags below.

3.2 Hash Function

There exist many efficient hash functions in the literature. In order to keep the tag circuit simple, we build our hash function on top of the simple scheme in [13] using a ring of pre-stored random bits: Before a tag is deployed, an offline random number generator uses the ID of a tag as seed to produce a string of pseudo-random bits, which are stored in the tag. The bits form a logical ring. After deployment, the tag generates a hash value $H(id, s)$ by returning a certain number of bits after the sth bit in the ring, where id is the tag ID and s is a given *hash seed* that can alter the hash output. This hash output is predictable by a RFID reader that knows the tag ID and the seed s.

More sophisticated hash implementations can be designed based on a ring of pseudo-random bits. For example, we may interpret s as a concatenation of a flag x and two random numbers, r_1 and r_2. To produce a hash output, we go clockwise along the ring if $x = 0$ or counterclockwise if $x = 1$. We then output the r_1th bit on the ring, and then output one more bit after every r_2 bits on the ring. If the hash output is required to be in a range $[0, y)$, we first take a sufficient number of hash bits as described above and then perform modulo y.

This hash function is easy to implement in hardware and thus suitable for tags. But it can only produce a limited number of different hash values, depending on the size of the ring. It is not suitable for a protocol whose operations require each tag to produce a large number of different hash values, but it works well for a protocol that only requires each tag to produce a few independent hash values.

3.3 Basic Idea

We know in Section 3.1 that random mapping from tags to slots generates a limited number of singleton slots. An arbitrary slot only has a probability of up to 36.8% to be a singleton. Now, if we apply two independent random mappings from tags to slots, a slot will have a probability of up to $1 - (1 - 36.8\%)^2 \approx 60.1\%$ to be a singleton in one of the two mappings. If we apply k independent mappings from tags to slots, it has a probability of $1 - (1 - 36.8\%)^k$ to be a singleton in one of the k mappings. The value of $1 - (1 - 36.8\%)^k$ quickly approaches to 100% as we increase k.

It is easy to generate multiple mappings. In the detection request, the RFID reader can broadcast k seeds, $s_1, s_2, ..., s_k$, to tags. Each seed s_i corresponds to a different mapping where a tag is mapped to a slot indexed by $H(id, s_i)$.

A slot may be a singleton under one mapping, but a collision slot under other mappings. Different slots may be singletons under different mappings. To maximize the number of singletons, the reader — with the knowledge of all tag IDs and all seeds — selects

a mapping (i.e., a seed) for each slot, such that the slot can be a singleton. From each slot's point of view, a *specific* seed is used to map tags to it. From the whole system's point of view, multiple seeds are used to map different tags to different slots.

In our protocol, the reader determines system parameters, including the sampling probability p and the frame size f. After selecting k random seeds, the reader chooses a seed for each slot and constructs a *seed-selection vector* V (or selection vector for short), which contains f *selectors*, one for each slot in the time frame. Each selector z has a range of $[0, k]$. If $z > 0$, it means that the zth seed, i.e., s_z, should be used for its corresponding slot. If $z = 0$, it means that the slot is not a singleton under any seed. Finally, the reader broadcasts the selection vector to the tags. Based on the selectors, each tag determines which slot it should use to respond.

We will address the problems of how to choose the optimal system parameters, p and f, and how the number k of seeds will affect the protocol performance in Section 4. Before we describe the operations of the protocol, we introduce the concept of segmentation. In our design, the above idea is actually applied segment by segment.

3.4 Segmentation

The seed-selection vector has f selectors, each of which is $\lceil \log(k + 1) \rceil$ bits long. f may be too large for the whole vector to fit in a single slot. For example, if $k = 7$, each selector is 3 bits long. If we use the same slot T_{tag} for carrying a 96-bit ID to carry the selection vector, it can only accommodate 32 selectors. When f is more than that, we have to divide the selection vector into 96-bit segments, so that they can fit in T_{tag} slots. Each segment contains $\frac{96}{\lceil \log(k+1) \rceil}$ selectors. Let $l = \frac{96}{\lceil \log(k+1) \rceil}$. The total number of seed-selection segments are $\frac{f}{l}$, and the jth segment is denoted as V_j.

Since we divide the selection vector into segments, we also divide the time frame into sub-frames, each containing l slots accordingly. The jth time sub-frame is denoted as F_j. This allows our protocol to deal with one sub-frame at a time.

Our protocol consists of two phases. In Phase one, the reader identifies the set of sampled tags, and randomly assigns the sample tags to the sub-frames. The subset of sampled tags that are assigned to the jth sub-frame is denoted as N_j. For each sub-frame F_j, the reader selects a seed for each of its slots, constructs the seed-select segment V_j, and maps the tags in N_j to slots in F_j using the selected seeds.

In Phase two, the reader broadcasts the seed-selection segments one after another, each in a slot of T_{tag}. Each seed-selection segment is followed by a time sub-frame of l slots, each of which is T_{short} long. The tags in N_j will respond in these slots. Each tag only needs to be active during its sub-frame, which conserves energy.

Below we give details of the protocol design.

3.5 Phase One: Finding Sampled Tags

To implement a sampling probability p, the reader will broadcast an integer $x = \lceil p \times X \rceil$ and a prime number q, where X is a large, pre-configured constant (e.g., 2^{16}). During the ith round of protocol execution, a tag is sampled if and only if the hash result $H(id, q \times i)$, which is a pseudo-random number in the range of $[0, X)$, is smaller than x, where id is the tag's ID.

After receiving x and q, each tag can predict which rounds of protocol execution it will participate. Since the protocol is scheduled to execute periodically with pre-defined intervals, each tag knows when it should wake to participate. The reader, with the knowledge of all tag information, can predict which tags are sampled for each protocol execution.

3.6 Phase One: Assigning Sampled Tags to Sub-frames

When assigning sampled tags to time sub-frames, the reader selects an additional random seed s, which is different from $s_1, ..., s_k$. For each sampled tag, the reader produces a hash output $H(id, s)$ and assigns the tag to the $H(id, s)$th sub-frame, where id is the tag's ID and the range of $H(id, s)$ is $[0, \frac{f}{l})$. Note that each tag will know which sub-frame it is assigned to, after it receives s in the detection request broadcast by the reader at the beginning of Phase two.

3.7 Phase One: Determining Seed-selection Segments

Each seed-selection segment is determined independently. Consider the jth segment V_j, the jth time sub-frame F_j, and the set N_j of sampled tags that are assigned to F_j. All selectors in V_j are initialized to zeros. The reader begins by using the first seed s_1 to map tags in N_j to slots in F_j. For each tag in N_j, the reader produces a hash output $H(id, s_1)$ and maps the tag to the $H(id, s_1)$th slot in F_j, where id is the tag's ID and the range of $H(id, s_1)$ is $[0, l)$. After mapping, the reader finds singleton slots. Each singleton has one and only one tag mapped to it. We assign the tag to the slot so that it will transmit in the slot, free of collision, during Phase two. The reader sets the corresponding selector in V_j to be 1, meaning that the first seed s_1 should be used for this slot. The slot is now called a *used* slot, and the sole tag mapped to it will be called an *assigned* tag.

The reader repeats the above process with other seeds, one at a time, for the remaining mappings. For each mapping, we only consider the slots whose selectors have not been determined yet and only consider the tags that have not been assigned to any slots yet. In other words, the used slots and the assigned tags will not be considered. For a singleton slot that is found using seed s_i, the corresponding selector in V_j will be set to be i.

After all k mappings, if the value of a selector in V_j remains zero, it means that the corresponding slot in F_j is not a singleton under any seed. As a final attempt to utilize these unused slots, if there exist unassigned tags in N_j, the reader randomly assigns the unassigned tags to unused slots. More specifically, it chooses an additional random seed s' and produces a hash output $H(id, s')$ to assign each tag that is not assigned yet to the $H(id, s')$th unused slot, where id is the tag's ID. In case that only one tag is assigned to an unused slot, we will have an extra singleton. Since the whole tag-to-slot assignment is pseudo-random, the reader knows which unused slots will become singletons. As we will see later in Phase two, after receiving $s_1, ..., s_k$, each tag will know whether it is assigned to a slot. If not, from the received s', it will know which unused slot it is assigned to.

3.8 Phase Two

At the beginning of this phase, the reader broadcasts a detection request, which is followed by a time frame for sampled tags to respond. The detection request consists of a frame size f and a sequence of seeds, $s, s_1, ..., s_k$, and s'. The time frame is divided into sub-frames. Before each sub-frame F_j, the reader broadcasts the corresponding seed-selection segment V_j in a single tag-ID slot T_{tag}. It is followed by l short slots (T_{short}) of the sub-frame, during which the tags in N_j can respond. Recall that each selection segment is 96 bits long. If $k = 7$, a segment has $l = \frac{96}{\log_2(7+1)} = 32$ selectors, and thus each time sub-frame has 32 slots.

Consider an arbitrary tag t. It wakes up to participate in a scheduled protocol execution that it is sampled for. After t receives the

detection request from the reader, it uses $H(id, s)$ to determine which sub-frame it is assigned to. Without loss of generality, let the sub-frame be F_j. The tag sets a timer to wake up before F_j begins. After receiving the seed-selection segment V_j, tag t uses $H(id, s_1)$ to find out which time slot it is mapped by seed s_1. It then checks whether the corresponding selector in V_j is 1. If the selector is 1, according to the construction of V_j in Section 3.7, t must be the sole tag that is mapped (and assigned) to this slot under s_1. If the selector is not 1, it means that s_1 should not be used to map any tag to this slot. In the latter case, t will move on to other seeds and repeat the same process to determine if it is assigned to a slot. If so, it will transmit during that slot. Otherwise, if t is not assigned to a slot after all k seeds, it will make a final attempt by finding out all unused slots (whose corresponding selectors in V_j are zeros) and using $H(id, s')$ as index to identify an unused slot to transmit.

In summary, after Phase one, the reader knows (1) which sub-frame each sampled tag is assigned to, (2) which slot each sampled tag is expected to transmit, (3) which slots are expected to be singletons, and (4) which slots are expected to be collision slots (due to the final attempt using s'). After Phase two, if an expected singleton/collision slot turns out to be empty, the reader detects a missing-tag event. Because multiple mappings reduce the number of empty/collision slots, both energy efficiency and time efficiency are greatly improved, as we will demonstrate analytically and by simulations in the following sections.

4. ENERGY-TIME TRADEOFF IN PROTOCOL CONFIGURATION

We formally derive the detection probability and the energy-time tradeoff curve. We show how to compute system parameters and how to solve the constrained least-time (or least-energy) problem.

4.1 Detection Probability

To find the detection probability after one protocol execution, we need to first derive the probability for an arbitrary sampled tag t to be assigned to a singleton slot during Phase one. There are k mappings. Let P_i be the probability that tag t is assigned to a singleton slot after the first i mappings. Let n be the total number of tags and n' be the number of sampled tags that are mapped to the same sub-frame as t does. n' follows a binomial distribution, $Bino(n, p\frac{l}{f})$, i.e.,

$$Prob\{n' = j\} = \binom{n}{j}(p\frac{l}{f})^j(1 - p\frac{l}{f})^{n-j}. \quad (1)$$

$P_0 = 0$. We derive a recursive formula for P_i, $1 \le i \le k$. After the first $i - 1$ mappings, there are two cases. Case 1: tag t has been assigned to a slot; the probability for this to happen is P_{i-1}. Case 2: tag t has not been assigned to a slot; the probability for this case is $1 - P_{i-1}$. We focus on the second case below.

In the ith mapping, the slot that tag t is mapped to has a probability of $(1 - \frac{n'P_{i-1}}{l})$ to be unused. Each of the other $n' - 1$ tags has a probability $(1 - P_{i-1})$ to be unassigned. If it is unassigned, the tag has a probability of $\frac{1}{l}$ to be mapped to the same slot as t does. Hence, the probability p' for tag t to be the only one that is mapped to an unused slot is

$$p' = (1 - (1 - P_{i-1})\frac{1}{l})^{n'-1} \cdot (1 - \frac{n'P_{i-1}}{l}). \quad (2)$$

Recall that we are considering Case 2 here. Combining both

cases, we have

$$P_i = P_{i-1} + (1 - P_{i-1}) \cdot \sum_{j=0}^{n} Prob\{n' = j\} \cdot p'$$

$$= P_{i-1} + (1 - P_{i-1}) \cdot \sum_{j=0}^{n} \binom{n}{j} (p\frac{l}{f})^j (1 - p\frac{l}{f})^{n-j}. \quad (3)$$

$$(1 - (1 - P_{i-1})\frac{1}{l})^{j-1} \cdot (1 - \frac{jP_{i-1}}{l}),$$

where the first item on the right side is the probability for a tag to be assigned to a slot by the first $i - 1$ mappings and the second item is the probability for the tag to be assigned to a slot by the ith mapping. The probability for tag t to be assigned to a slot after all k mappings is P_k.

After the k mapping, we have a final attempt, in which an unassigned tag may be mapped to a singleton slot or a collision slot. If the tag is mapped to a collision slot, it is highly unlikely that all tags in that slot will be missing because the parameter m is typically set far smaller than n. Hence, the contribution of collision slots to missing-tag detection can be ignored. When the tag is mapped to a singleton slot, detection will be made if the tag is missing. Therefore, the final mapping has no difference from the previous mappings. The probability for tag t to transmit in a singleton slot is P_{k+1}, which can be computed recursively from (3).

Each of the m missing tags has a probability p to be sampled. When the tag is sampled, it has a probability of P_{k+1} to be assigned a singleton slot. When that happens, since a missing tag cannot transmit, the reader will observe an empty slot instead, resulting in the detection. Therefore, the detection probability of MSMD, denoted as $P_{msmd}(p, f)$, is

$$P_{msmd}(p, f) = 1 - (1 - p \times P_{k+1})^m. \quad (4)$$

The value of $P_{msmd}(p, f)$ not only depends on the choice of p and f, but also depends on n, m and k, which are not included in the notation for simplicity. The values of p and f are determined by the reader and broadcast to tags. They control the energy-time tradeoff as we will reveal shortly. The values of n, m and k are pre-known, where n is known because it is simply the number of tags that the reader expects to be in the system, m is known as a given parameter in the detection requirement, and k is determined before the tags are deployed.

EDM [16] is a special case of MSMD with $k = 1$ and without the final attempt. Hence, the detection probability of EMD, denoted as $P_{emd}(p, f)$, is

$$P_{emd}(p, f) = 1 - (1 - p \times P_1)^m. \quad (5)$$

TRP [6] is a special case of EDM with $p = 1$. Namely, sampling is turned off.

4.2 Energy-time Tradeoff Curve

The protocol's energy cost can be characterized by the sampling probability p because the expected number of tags that participate in each protocol execution is $p \times n$. A smaller value of p means a smaller number of participating tags, which in turn means a smaller energy cost. The protocol's execution time can be characterized by the frame size f. The total number of slots for tags to respond is f, and the total size of the seed-selection vector is also f. A smaller value of f generally means a shorter protocol execution time. The time for the reader to broadcast the detection request is a constant, which is negligible when comparing with the time frame and the seed-selection vector if f is large. The actual protocol execution time, measured in seconds, will be studied in the next section based on the Philips I-code specification.

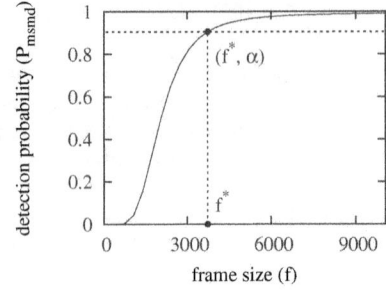

Figure 1: Detection probability $P_{msmd}(p, f)$ **with respect to the frame size** f **when** $n = 50,000$, $m = 100$, $k = 3$, **and** $p = 5\%$.

We cannot arbitrarily pick small values for p and f. They must satisfy the requirement $P_{msmd}(p, f) \geq \alpha$. Subject to this constraint, we show that the values of p and f cannot be minimized simultaneously. The choice of p and f represents an energy-time tradeoff.

If we fix the value of p, $P_{msmd}(p, f)$ becomes a function of f. The solid line in Fig. 1 shows an example of the curve $P_{msmd}(p, f)$ with respect to f when $n = 50,000$, $m = 100$, $k = 3$, and $p = 5\%$. Because $P_{msmd}(p, f)$ is an increasing function, the minimum value of f that satisfies the requirement $P_{msmd}(p, f) \geq \alpha$ can be found by solving the following equation,

$$P_{msmd}(p, f) = \alpha.$$

The solution is denoted as f^*. See Fig. 1 for illustration.

For each different sampling probability p, we can compute the smallest usable frame size f^* that satisfies $P_{msmd}(p, f) \geq \alpha$. Hence, f^* can be considered as a function of p, denoted as $f^*(p)$. A practical RFID system may consider a frame size beyond a certain upper bound U to be unacceptable due to excessively long execution time. In addition, f^* must be an integer. Considering these factors, we give a more accurate definition of f^* below.

$$f^*(p) = \min\{f | P_{msmd}(p, f) \geq \alpha \wedge f \leq U, f \in I^+\}. \quad (6)$$

The algorithm that computes $f^*(p)$ based on bi-section search is given in Algorithm 1.

Algorithm 1 Search for $f^*(p)$

INPUT: n, m, α, k, p
OUTPUT: frame size that minimizes execution time under sampling probability p

if $P_{msmd}(p, U) < \alpha$ **then** exit; \\requirement cannot be met
$f_0 = 1$, $f_1 = U$;
while $f_1 - f_0 > 1$ **do**
    $f_2 = \lceil \frac{f_0 + f_1}{2} \rceil$;
    if $P_{msmd}(p, f_2) < \alpha$ **then** $f_0 = f_2$ **else** $f_1 = f_2$;
end while
return f_1;

The left plot in Fig. 2 shows the curve of $f^*(p)$ when $n = 50,000$, $m = 75$, $k = 3$, and $\alpha = 95\%$. We call it the *energy-time tradeoff curve*. Each point, $(p, f^*(p))$, represents an operating point whose energy cost is measured as $n \times p$ participating tags and whose time frame consists of $f^*(p)$ slots. The symbols in the plot will be explained later. The energy-time tradeoff is controlled by the sampling probability p. If we decrease the value of p, we decrease the energy cost, but at the mean time the value of $f^*(p)$ may have to increase, which increases the execution time.

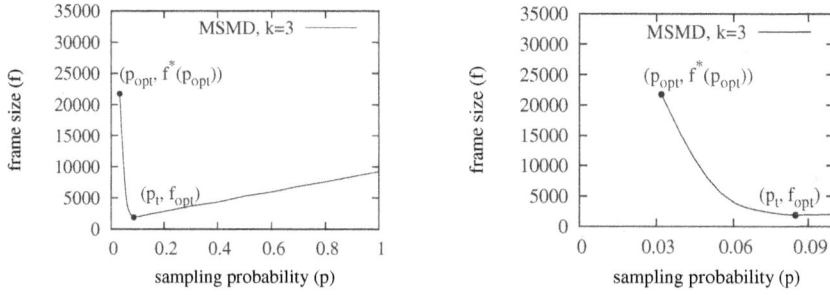

Figure 2: • *Left plot:* Energy-time tradeoff curve, i.e., frame size $f^*(p)$ with respect to sampling probability p, when $n = 50,000$, $m = 75$, $k = 3$, and $\alpha = 95\%$. • *Right plot:* Energy-time tradeoff curve in the range $p \in [p_{opt}, p_t]$.

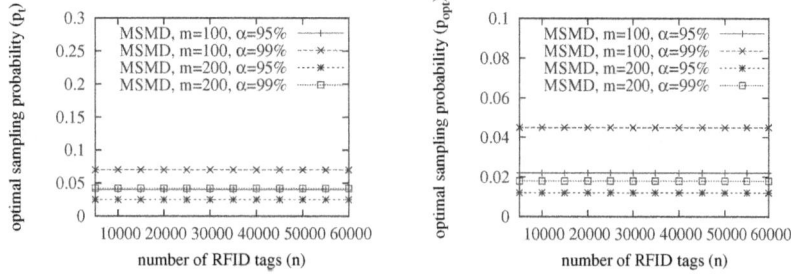

Figure 3: • *Left Plot:* the value of p_t with respect to n. • *Right Plot:* the value of p_{opt} with respect to n.

Algorithm 2 Search for p_{opt}

INPUT: n, m, α, k

OUTPUT: optimal sampling probability that minimizes energy cost

$p_0 = 0, p_1 = 1, \delta = 0.01, U = 1,000,000; \backslash\backslash 400$ seconds
while $p_1 - p_0 > \delta$ **do**
   $p_2 = \lceil \frac{p_0 + p_1}{2} \rceil$;
   if $P_{msmd}(p_2, U) < \alpha$ **then** $p_0 = p_2$ **else** $p_1 = p_2$;
end while
return p_1;

Algorithm 3 Search for f_{opt}

INPUT: n, m, α

OUTPUT: optimal frame size and sampling probability that minimize execution time

$p_0 = p_{opt}, p_1 = 1, \delta = 0.01$;
while $p_1 - p_0 > \delta$ **do**
   $p_2 = \lceil \frac{p_0 + p_1}{2} \rceil$;
   if $f^*(p_2) > f^*(p_2 + \frac{\delta}{2})$ **then** $p_0 = p_2$ **else** $p_1 = p_2$;
end while
return $f^*(p_1)$ and p_1;

4.3 Minimum Energy Cost

When the sampling probability p is too small, the detection probability $P_{msmd}(p, f)$ will be smaller than α for any value of f. Such a small sampling probability cannot be used. We design a bi-section search method in Algorithm 2 to find the smallest value of p, denoted as p_{opt}, which can satisfy $P_{msmd}(p, f) \geq \alpha$ with a frame size no greater than the upper bound U. The sampling probability returned by the algorithm is within an error of δ from the true optimal value p_{opt}, where δ is a parameter that can be set arbitrarily small. When p_{opt} and $f^*(p_{opt})$ are used, the energy cost is minimized.

4.4 Minimum Execution Time

From the energy-time tradeoff curve (the left plot in Fig. 2), we can find the smallest value of $f^*(p)$, denoted as f_{opt}, that minimizes the execution time.

$$f_{opt} = \min\{f^*(p) \mid p_{opt} \leq p \leq 1\} \qquad (7)$$

Let p_t be the corresponding sampling probability. Namely, $P_{msmd}(p_t, f_{opt}) = \alpha$. The values of f_{opt} and p_t are determined through bi-section search in Algorithm 3, where the **if** statement uses the local gradient to guide the search direction towards the minimum. When p_t and f_{opt} are used, the protocol execution time is minimized.

We amplify the segment of the energy-time tradeoff curve between point $(p_{opt}, f^*(p_{opt}))$ and point (p_t, f_{opt}) in the right plot of Fig. 2. When we increase the value of p from p_{opt} to p_t, the energy cost of the protocol is linearly increased, while the execution time of the protocol is decreased. We should not choose $p > p_t$ because both energy cost and execution time will increase when the sampling probability is greater than p_t.

4.5 Offline Computation

Because the computation of p_{opt}, $f^*(p_{opt})$, p_t, and f_{opt} relies

101

only on the values of n, m, α and k, we can calculate them offline in advance. The values of m and α are pre-configured as part of the system requirement. The value of k is determined before tag deployment. Hence, we can pre-compute p_{opt}, $f^*(p_{opt})$, p_t, and f_{opt} in a table format with respect to different values of n (for instance, from 100 to 100,000 with an increment step of 100), so that these values can be looked up during online operations.

When performing such computation, we observe that when we change n, the values of p_t and p_{opt} remain largely constants, as shown in Fig. 3. Hence, their values are actually determined by m, α and k. It means that as long as the detection requirement specified by m and α does not change, p_t and p_{opt} can be approximately viewed as constants even though the number of tags in the system changes.

Suppose the values of m and α may be changed only at the beginning of each hour. The reader picks a sampling probability p, which is p_t, p_{opt} or a value between them. It then downloads p to all tags and synchronizes their clocks. For the rest of the hour, the reader does not have to transmit the sampling probability again.

4.6 Constrained Least-time (or Least-energy) Problem

The *energy-constrained least-time problem* is to minimize the protocol's execution time, subject to a detection requirement specified by m and α and an energy constraint specified by an upper bound u on the expected number of tags that participate in each protocol execution. To minimize execution time, we need to reduce the frame size as much as possible. Our previous analysis has already given the solution to this problem, which is simply $f^*\left(\frac{u}{n}\right)$, where $\frac{u}{n}$ is the maximum sampling probability that we can use under the energy constraint.

The *time-constrained least-energy problem* is to minimize the number of tags that participate in protocol execution, subject to a detection requirement specified by m and α and an execution time constraint specified by an upper bound u' on the frame size. A solution can be designed by following a similar process as we derive $f^*(p)$ in Section 4.2: Starting from (4), if we fix $f = u'$, $P_{msmd}(p, f)$ becomes a function of p. We can use bi-section search to find p that meets $P_{msmd}(p, f) = \alpha \wedge p \leq p_t$.

4.7 Impact of k

We study how the number k of hash seeds will affect the protocol's performance. Figure 4 compares the energy-time tradeoff curves of EMD and MSMD with $k = 3, 7, 15$, respectively. Recall that EMD is a special case of MSMD with one hash seed and TRP is a special case of EMD with $p = 1$, represented by a point on the curve of EMD as shown in the figure. For MSMD, when $k = 3$, each seed selector needs 2 bits; recall that the value zero is reserved for non-singleton slots. When $k = 7$, each selector needs 3 bits. When $k = 15$, each selector needs 4 bits.[4]

In Figure 4, a lower curve indicates better performance because, for any sampling probability, its frame size is smaller, i.e., its execution time is smaller. Alternatively, it can be interpreted as, for any frame size, its sampling probability is smaller, i.e., it needs fewer tags to participate in each protocol execution. Clearly, MSMD significantly outperforms EMD and TRP. As k increases, the performance of MSMD improves. However, the amount of improvement shrinks rapidly, demonstrated by the small gap between $k = 7$ and $k = 15$. When we further increase k to 31 using 5-bit selectors, the improvement becomes negligible. Increasing the value of k does

---

[4]One may ask why we do not use $k = 8$ or other values. The reason is that each selector needs 4 bits even when $k = 8$. In that case, we should certainly choose $k = 15$ for better performance.

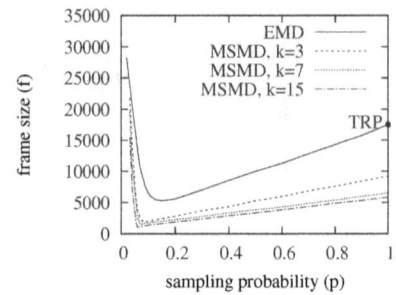

Figure 4: Energy-time tradeoff curves of EMD and MSMD under different k values, when $n = 50,000$, $m = 100$, and $p = 5\%$.

not come for free; larger selectors mean more overhead. For one, it takes more time for the reader to broadcast the seed-selection vector. Therefore, we believe $k = 7$ is a good choice in practice because the performance gain beyond that is very limited.

5. NUMERICAL RESULTS

We have performed extensive simulations to study the performance of the proposed MSMD, and compare it with EMD [16] and TRP [6]. The simulation setting is based on the Philips I-Code specification [20]. Any two consecutive transmissions (from the reader to tags or vice versa) are separated by a waiting time of 302 μs. The transmission rate from the reader to tags is 26.5 Kb/sec; it takes 37.76 μs for the reader to transmit one bit. A 96-bit slot that carriers a seed-selection segment is 3927 μs long, which includes a waiting time before the transmission. The transmission rate from a tag to the reader is 53Kb/sec; it takes 18.88 μs for a tag to transmit one bit. A single-bit slot T_{short} for a tag to respond (i.e., make the channel busy) is 321 μs, also including a waiting time.

For each set of system parameters, including m, α, and n, TRP will compute its optimal frame size. Once the frame size f is determined, the execution time is known, which is $f \times T_{short}$ plus the time for broadcasting a detection request. The energy cost of TRP is measured as n participating tags. MSMD and EMD will choose a sampling probability p, and compute the optimal frame size f^* under that sampling probability. EMD does not give a way to compute its optimal frame size. However, since EMD is a special case of MSMD, we use our analytical framework in the previous section to compute it. The energy cost of these two protocols is measured as $n \times p$ participating tags. For EMD, its execution time is $f^* \times T_{short}$, plus the time for a request. For MSMD, we need to add the time slots for the selection vector.

The design of all three protocols ensures that the detection requirement specified by m and α is always met. This is indeed what we observe in our simulations. Our comparison below is made in terms of energy efficiency and time efficiency, given a certain detection requirement.

5.1 Energy-time Tradeoff

Let $n = 50,000$, $\alpha = 95\%$, and $m = 50$. Fig. 5 shows the energy-time tradeoff curves produced by simulations. Recall that the energy cost of MSMD or EMD is proportional to p. The point at $p = 1$ on the EMD curve represents TPR. Clearly, MSMD significantly outperforms EMD. MSMD with $k = 7$ uses three-bit elements in the selection vector, while MSMD with $k = 3$ uses two-bit elements. Even though it incurs more overhead in the selection vector, MSMD with $k = 7$ slightly outperforms MSMD with $k = 3$. Further increasing k cannot bring performance gain

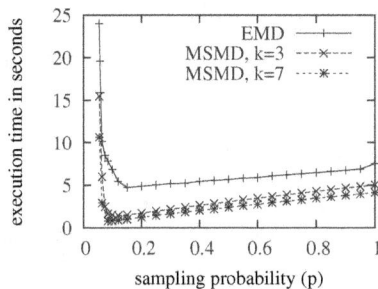

Figure 5: Protocol execution time with respect to sampling probability, when $\alpha = 95\%$, $m = 50$, and $n = 50,000$.

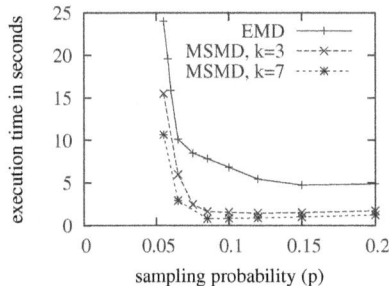

Figure 6: Zoom-in view of energy-time tradeoff in Figure 5 in the sampling probability range of [0, 0.2].

Table 1: Relative energy cost and execution time of MSMD ($k = 7$) under p_{opt} and p_t, when $\alpha = 95\%$ and $n = 50,000$

	p_t		p_{opt}	
	energy	time	energy	time
$m = 200$	2.1%	4.09%	1.4%	269.1%
$m = 100$	3.6%	5.61%	2.5%	226.2%
$m = 50$	8.1%	10.84%	5.5%	82.3%

any missing-tag event with probability 99.9%. The left plot compares the energy cost of three protocols with respect to n, and the right plot compares their execution times. MSMD has a smaller energy cost than EMD, which in turn has a much smaller energy cost than TRP. In the meanwhile, MSMD also has a much smaller execution time than EMD and TRP. Similar results can be drawn from Fig. 8 where $\alpha = 99\%$ and Fig. 9 where $\alpha = 90\%$. In the latter case, the execution time of MSMD is less than a second.

In Fig. 10-11, we keep $\alpha = 99\%$ and vary the value of m. In Fig. 10, $m = 25$. In Fig. 11, $m = 100$. The performance of MSMD remains the best among all three.

6. CONCLUSION

This paper proposes a new protocol design that integrates energy efficiency and time efficiency for missing-tag detection. It uses multiple hash seeds to provide multiple degree of freedom for tags to select time slots when they announce their presence to the RFID reader in the process of missing-tag detection. The result is a multi-fold cut in both energy cost and execution time. Such performance improvement is critical for a protocol that needs to be executed frequently. We also reveal a fundamental energy-time tradeoff in the protocol design. This tradeoff gives flexibility in performance tuning when the protocol is applied in practical environment.

7. REFERENCES

[1] T. Kriplean, E. Welbourne, N. Khoussainova, V. Rastogi, M. Balazinska, G. Borriello, T. Kohno, and D. Suciu, "Physical Access Control for Captured RFID Data," *IEEE Pervasive Computing*, 2007.

[2] Y. Liu, L. Chen, J. Pei, Q. Chen, and Y. Zhao, "Mining Frequent Trajectory Patterns for Activity Monitoring Using Radio Frequency Tag Arrays," *Proc. of IEEE PerCom*, 2007.

[3] L. M. Ni, Y. Liu, Y. C. Lau, and A. Patil, "LANDMARC: Indoor Location Sensing Using Active RFID," *ACM Wireless Networks (WINET)*, vol. 10, no. 6, November 2004.

[4] Y. Li and X. Ding, "Protecting RFID Communications in Supply Chains," *Proc. of ASIACCS*, 2007.

[5] B. Sheng, C. Tan, Q. Li, and W. Mao, "Finding Popular Categories for RFID Tags," *Proc. of ACM Mobihoc*, 2008.

[6] C. Tan, B. Sheng, and Q. Li, "How to Monitor for Missing RFID Tags," *Proc. of IEEE ICDCS*, 2008.

[7] A. Nemmaluri, M. Corner, and P. Shenoy, "Sherlock: Automatically Locating Objects for Humans," *Proc. of ACM MobiSys*, 2008.

[8] L. Ravindranath, V. Padmanabhan, and P. Agrawal, "Sixthsense: RFID-based Enterprise Intelligence," *Proc. of ACM MobiSys*, 2008.

[9] T. Li, S. Wu, S. Chen, and M. Yang, "Energy Efficient Algorithms for the RFID Estimation Problem," *Proc. of IEEE INFOCOM*, March 2010.

due to overly large overhead for the selection vector. In Fig. 6, We zoom in for a detailed look at the curve segment in the sampling probability range of [0, 0.2]. When $p = 0.08$, the execution time of MSMD with $k = 7$ is 11.2% of the time taken by EMD. When we fix the execution time at 5 seconds, the number of participating tags in MSMD with $k = 7$ is 46.7% of the number in EMD. We vary the values of n, α and m. Similar conclusions can be drawn from the simulation results, which are omitted due to space limitation.

The tradeoff curves in Fig. 6 agree with our analytical results in Fig. 4 in principle. We want to stress that our simulations do not simply reproduce the analytical results. Simulations consider system details by using a real RFID specification. Such details are not captured by analysis. In addition, simulations consider the exact impact of selection vector on execution time (measured in seconds), instead of characterizing time in an indirect way using the frame size.

In Table 1, we show the relative performance of MSMD ($k = 7$) with respect to TRP, where $n = 50,000$, $\alpha = 95\%$, and $m = 50, 100$, or 200. MSMD is operated under sampling probability p_t and p_{opt}. For example, when $m = 50$, $p_t = 0.085$ and $p_{opt} = 0.055$. The numbers in the table are ratios of MSMD's energy cost (or execution time) to TRP's energy cost (or execution time). For example, when $m = 200$, the energy cost of MSMD with p_t is 2.1% of what TRP consumes, and its execution time is 4.09% of the time TRP takes.

5.2 Performance Comparison

Next, we compare the performance of MSMD ($k = 7$), EMD, and TRP under different values of m, α and n. MSMD and EMD are operated with their optimal sampling probabilities p_t. In Fig. 7-9, we keep $m = 50$ and vary the value of α. In Fig. 7, we let $\alpha = 99.9\%$, meaning that each protocol execution should detect

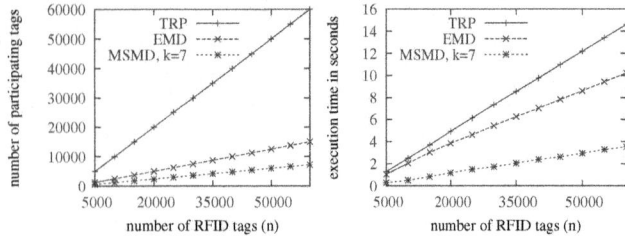

Figure 7: • *Left plot:* The number of participating tags with respect to the number of tags, when $m = 50$ and $\alpha = 99.9\%$. • *Right plot:* The protocol execution time with respect to the number of tags, when $m = 50$ and $\alpha = 99.9\%$.

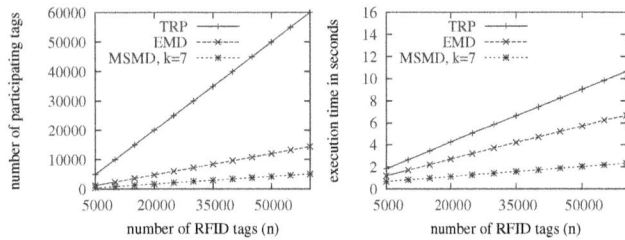

Figure 8: Same as the caption of Fig. 7 except for $\alpha = 99\%$.

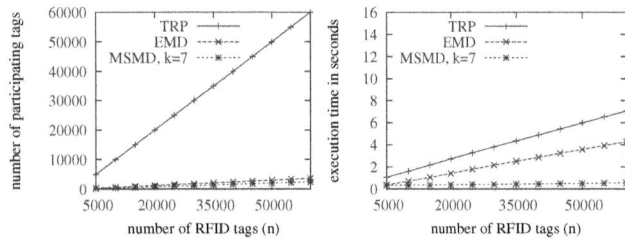

Figure 9: Same as the caption of Fig. 7 except for $\alpha = 90\%$.

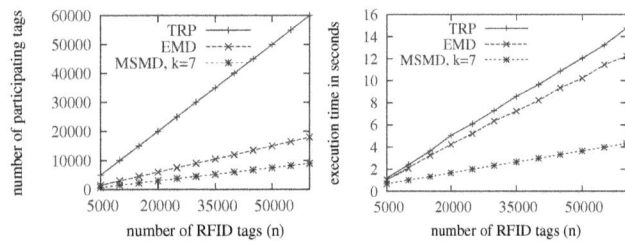

Figure 10: • *Left plot:* The number of participating tags with respect to the number of tags, when $m = 25$ and $\alpha = 99\%$. • *Right plot:* The protocol execution time with respect to the number of tags, when $m = 25$ and $\alpha = 99\%$.

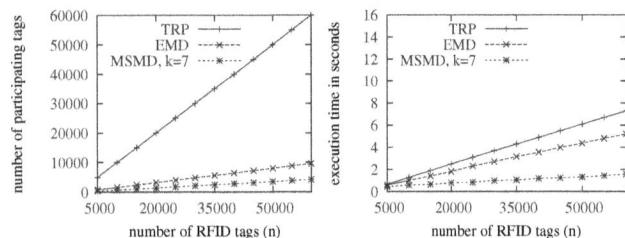

Figure 11: Same as the caption of Fig. 10 except for $m = 100$.

[10] S. Chen, M. Zhang, and B. Xiao, "Efficient Information Collection Protocols for Sensor-augmented RFID Networks," *Proc. of IEEE INFOCOM*, April 2011.

[11] Yan Qiao, Shigang Chen, Tao Li, and Shiping Chen, "Energy-efficient Polling Protocols in RFID Systems," *Proc. of ACM MobiHoc*, May 2011.

[12] J. Wolverton, "Wal-Mart to Embed RFID Tags in Clothing Beginning August 1," *http://www.thenewamerican.com/index.php/tech-mainmenu-30/computers/4157-wal-mart-to-embed-rfid-tags-in-clothing-beginning-august-1*, July 2010.

[13] T. Li, S. Chen, and Y. Ling, "Identifying the Missing Tags in a Large RFID System," *Proc. of ACM Mobihoc*, 2010.

[14] B. Sheng, Q. Li, and W. Mao, "Efficient Continuous Scanning in RFID Systems," *Proc. of IEEE INFOCOM*, 2010.

[15] R. Zhang, Y. Liu, Y. Zhang, and J. Sun, "Fast Identification of the Missing Tags in a Large RFID System," *Proc. of the 8th Annual IEEE Communications Society Conference on Sensor, Mesh and Ad Hoc Communications and Networks (SECON)*, 2011.

[16] W. Luo, S. Chen, T. Li, and S. Chen, "Efficient Missing Tag Detection in RFID Systems," *Proc. of IEEE INFOCOM, mini-conference*, 2011.

[17] B. Firner, P. Jadhav, Y. Zhang, R. Howard, W. Trappe, and E. Fenson, "Towards Continuous Asset Tracking: Low-Power Communication and Fail-Safe Presence Assurance," *Proc. of the 6th Annual IEEE Communications Society Conference on Sensor, Mesh and Ad Hoc Communications and Networks (SECON)*, 2009.

[18] P. Popovski, K. F. Nielsen, R. M. Jacobsen, and T. Larsen, "Robust Statistical Methods for Detection of Missing RFID Tags," *IEEE Communications Magazine, Wireless Communications Series*, 2010.

[19] A. Juels, "RFID Security and Privacy: A Research Survey," *IEEE Journal on Selected Areas in Communications (JASC)*, vol. 24, no. 2, pp. 381–394, February 2006.

[20] Philips Semiconductors, "Your Supplier Guide to ICODE Smart Label Solutions," *http://www.nxp.com/acrobat_download2/other/icode_supplier_list_2008_10.pdf*, October 2008.

[21] J. Myung and W. Lee, "Adaptive Splitting Protocols for RFID Tag Collision Arbitration," *Proc. of ACM MobiHoc*, May 2006.

[22] N. Bhandari, A. Sahoo, and S. Iyer, "Intelligent Query Tree (IQT) Protocol to Improve RFID Tag Read Efficiency," *Proc. of IEEE International Conference on Information Technology (ICIT)*, December 2006.

[23] V. Sarangan, M. R. Devarapalli, and S. Radhakrishnan, "A Framework for Fast RFID Tag Reading in Static and Mobile Environments," *International Journal of Computer and Telecommunications Networking*, vol. 52, no. 5, pp. 1058–1073, 2008.

[24] B. Zhen, M. Kobayashi, and M. Shimizu, "Framed ALOHA for Multiple RFID Object Identification," *IEICE Transactions on Communications*, March 2005.

Adaptive Network Coding for Scheduling Real-time Traffic with Hard Deadlines

Lei Yang
School of ECEE
Arizona State University
Tempe, AZ 85287, USA
lyang55@asu.edu

Yalin Evren Sagduyu
Intelligent Automation, Inc.
Rockville, MD 20855, USA
ysagduyu@i-a-i.com

Jason Hongjun Li
Intelligent Automation, Inc.
Rockville, MD 20855, USA
jli@i-a-i.com

ABSTRACT

We study adaptive network coding (NC) for scheduling real-time traffic over a single-hop wireless network. To meet the hard deadlines of real-time traffic, it is critical to strike a balance between maximizing the throughput and minimizing the risk that the entire block of coded packets may not be decodable by the deadline. Thus motivated, we explore adaptive NC, where the block size is adapted based on the remaining time to the deadline, by casting this sequential block size adaptation problem as a finite-horizon Markov decision process. One interesting finding is that the optimal block size and its corresponding action space monotonically decrease as the deadline approaches, and the optimal block size is bounded by the "greedy" block size. These unique structures make it possible to narrow down the search space of dynamic programming, building on which we develop a monotonicity-based backward induction algorithm (MBIA) that can solve for the optimal block size in *polynomial time*. Since channel erasure probabilities would be time-varying in a mobile network, we further develop a joint real-time scheduling and channel learning scheme with adaptive NC that can adapt to channel dynamics. We also generalize the analysis to multiple flows with hard deadlines and long-term delivery ratio constraints, devise a low-complexity online scheduling algorithm integrated with the MBIA, and then establish its asymptotic throughput-optimality. In addition to analysis and simulation results, we perform high fidelity wireless emulation tests with real radio transmissions to demonstrate the feasibility of the MBIA in finding the optimal block size in real time.

Categories and Subject Descriptors

C.2.1 [**Computer-Communication Networks**]: Network Architecture and Design—Wireless communication

General Terms

Theory, Algorithms, Design

Keywords

Network coding, real-time scheduling, wireless broadcast, deadlines, delay, throughput, resource allocation

1. INTRODUCTION

The past few years have witnessed a tremendous growth of multimedia applications in wireless systems. To support the rapidly growing demand in multimedia traffic, wireless systems must meet the stringent quality of service (QoS) requirements, including the minimum bandwidth and maximum delay constraints. However, the time-varying nature of wireless channels and the hard delay constraints give rise to great challenges in scheduling multimedia traffic flows. In this paper, we explore *network coding* (NC) to optimize the throughput of multimedia traffic over wireless channels under the hard deadline constraint.

In capacitated multihop networks, NC is known to optimize the multicast flows from a single source to the min-cut capacity [1]. NC also provides coding diversity over unreliable wireless channels and improves the throughput and delay performance of single-hop broadcast systems, compared to (re)transmissions of uncoded packets [2–8]. Nevertheless, the block NC induces "decoding delay," i.e., receivers may not decode network-coded packets until a sufficient number of innovative packets are received. Therefore, the minimization of NC delay has received much attention (e.g., [9–12]).

For multimedia traffic, meeting the deadline may be more critical than reducing the average delay. Under the *hard deadline* constraints, NC may result in significant performance loss, unless the receivers can decode the packets before the deadline. Different NC mechanisms (e.g., [13–16]) have been proposed recently to incorporate deadline constraints. An immediately-decodable network coding (IDNC) scheme has been proposed in [14] to maximize the broadcast throughput subject to deadlines. A partially observable Markov decision process (POMDP) framework has been proposed in [15] to optimize media transmissions with erroneous receiver feedback.

These works focus on optimizing network codes in each transmission; however, such an approach is typically not tractable due to the "curse of dimensionality" of dynamic programming. To reduce the complexity of optimizing network codes in each transmission, [16] has formulated a joint coding window selection and resource allocation problem to optimize the throughput in deadline-constrained flows. However, the computational complexity can be still overwhelming due to the finite-horizon dynamic programming involved in the coding window selection. To overcome this

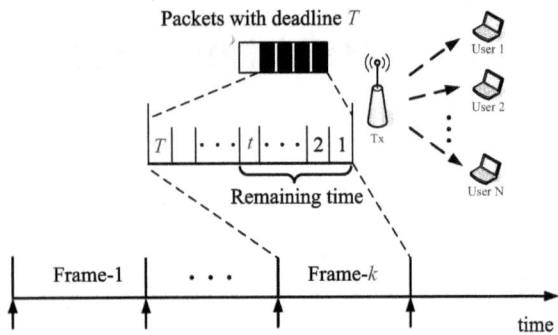

Figure 1: System model. (The arrow denotes the time instant for drops of undelivered packets and arrivals of new packets.)

limitation, [16] has proposed a heuristic scheme with fixed coding window to tradeoff between optimality and complexity.

A primary objective of this study is to (i) explore optimal adaptive NC schemes with low computational complexity, and (ii) integrate channel learning with adaptive NC over wireless broadcast erasure channels. Our main contributions are summarized as follows.

- We develop an adaptive NC scheme that sequentially adjusts the block size (coding block length) of NC to maximize the system throughput, subject to the hard deadlines (cf. [16]). We show that the optimal block size and its corresponding action space monotonically decrease as the packet deadline approaches, and the optimal block size is bounded by the "greedy" block size that maximizes the immediate throughput only. These unique structures make it possible to narrow down the search space of dynamic programming, and accordingly we develop a monotonicity-based backward induction algorithm (MBIA) that can solve for the optimal block size in *polynomial time*, compared with [15, 16]. We also develop a joint real-time scheduling and *channel learning* scheme with adaptive NC for the practical case, in which the scheduler does not have (perfect) channel information.

- We generalize the study on adaptive NC to the case with multiple flows. We develop a joint scheduling and block size adaptation approach to maximize the weighted system throughput subject to the long-term delivery ratio and the hard-deadline constraint of each flow. By integrating the MBIA in the model with multiple flows, we construct a low-complexity online scheduling algorithm. This online algorithm is shown to be throughput optimal in the asymptotic sense as the step size in iterations approaches zero.

- We implement the adaptive NC schemes in a realistic wireless emulation environment with real radio transmissions. Our high fidelity testbed results corroborate the feasibility of the MBIA in finding the optimal block size in real time. As expected, the adaptive NC scheme with the MBIA outperforms the fixed coding scheme, and the proposed scheme of joint real-time schedul-

ing and channel learning performs well under unknown and dynamic channel conditions.

The rest of the paper is organized as follows. In Section 2, we introduce the system model and present the block size adaptation problem with the hard deadlines. In Section 3, we develop the MBIA to solve for the optimal block size and building on this we devise the joint real-time scheduling and channel learning scheme with adaptive NC for the case with unknown channel information. In Section 4, we generalize the study on adaptive NC to multiple flows. In Section 5, we implement the adaptive NC schemes and test them in a realistic wireless emulation environment with hardware-in-the-loop experiments. We conclude the paper in Section 6. Due to space limitation, the details for the proofs are omitted and can be found in [26].

2. THROUGHPUT MAXIMIZATION VS. HARD DEADLINE

2.1 System Model

We consider a time-slotted downlink system with one transmitter (e.g., base station) and N receivers (users), as illustrated in Fig. 1. Time slots are synchronized across receivers and the transmission time of a packet corresponds to one time slot. The transmitter broadcasts M packets to N receivers over *i.i.d.* binary erasure channels with erasure probability ϵ.[1] We assume immediate and perfect feedback available at the transmitter. For multimedia communications, it is standard to impose deadlines for delay-sensitive data (see, e.g., [4, 14–17]). We assume that packets must be delivered to each receiver before T slots, i.e., the deadline of each packet is T slots. Any packet that cannot be delivered to all receivers by this deadline is dropped without contributing to the throughput.

Worth noting is that this model can be readily applied to finite-energy systems with NC, where the objective is to maximize the system throughput before the energy is depleted for further transmission. Therefore, the energy and delay constraints can be used interchangeably.

In Section 3, we consider the basic model with one flow and one frame of T slots. In Section 4, we generalize the model to multiple frames with multiple flows, where packets arrive at the beginning of each frame and they are dropped if they cannot be delivered to their receivers by the deadline of T slots.

2.2 Network Coding for Real-time Scheduling

As noted above, the throughput gain of NC comes at the expense of longer decoding delay (since packets are coded and decoded as a block), which may reduce the throughput of the system due to the hard deadline constraints. Let K denote the block size, i.e., the number of original packets encoded together by NC. We assume that the transmitter and each receiver know the set of coding coefficients, and the transmitter broadcasts the value of K to receivers before the NC transmissions start. The coding coefficients can also be chosen randomly from a large field size (or from a predetermined coding coefficient matrix of rank K) such that with high probability K packet transmissions deliver K innovative packets in coded form to any receiver, i.e., the entire

---

[1]The results derived in the paper can be readily applied to heterogeneous channels with different erasure probabilities.

block of packets can be decoded after K successful transmissions. As shown in [2], the probability that all receivers can decode the block of size K within T slots is given by

$$P(K,T) = \left(\sum_{\tau=K}^{T} \binom{\tau-1}{K-1} \epsilon^{\tau-K} (1-\epsilon)^K \right)^N, \quad (1)$$

where $\binom{n}{m}$ denotes the number of combinations of size m out of n elements.[2] Note that (1) strongly depends on the choice of block size K and we can show that,

Lemma 2.1. *The decoding probability (1) is monotonically decreasing with K for fixed T.*

With block NC, there is the risk that none of the packets can be decoded by the receivers before the hard deadline. By using IDNC, it may be possible to start decoding without waiting for the entire block to arrive but the complexity of finding a suitable code may be overwhelming due to the dynamic programming involved in the problem [14]. Here, we provide the throughput guarantees for the worst-case scenario, where either the whole block or none of the packets can be decoded at any slot. There is a tradeoff between the block size and the risk of decoding. In particular, we cannot greedily increase K to maximize the system throughput under the hard deadline constraints, since the risk that some receivers cannot decode the packets, i.e., $1 - P(K,T)$, also increases with K according to Lemma 2.1.

If the first block is delivered within the deadline, i.e., T slots, the size of a new block (with new packets) needs to be re-adjusted for the remaining slots. In other words, we need *real-time* scheduling of network-coded transmissions depending on how close the deadline is. For example, when there is only one slot left before the deadline, the optimal block size is 1, since for any $K > 1$, no receivers can decode the packets before the deadline. Also, the block size in a given slot statistically determines the remaining slots (before the deadline) along with the future system throughput. In Section 3, we derive the optimal block size adaptation policy to maximize the system throughput under the deadline constraints for one frame with one flow. In Section 4, we generalize the results to the case with multiple frames with multiple flows.

2.3 Problem Formulation: A Markov Decision Process View

The NC-based multimedia traffic scheduling of one frame is a sequential decision problem, which can be formulated as a Markov decision process (MDP) described as follows.

Horizon: The number of slots available before the deadline over which the transmitter (scheduler) decides the block size is the horizon. Due to the hard deadline, this MDP problem is a finite horizon problem with T slots (one frame).

State: The remaining slots $t \in \{0, 1, ..., T\}$ before the hard deadline is defined as the state,[3] where $t = 0$ denotes that there is no slot left for transmissions.

Action: Let K_t, $t \in \{1, ..., T\}$, denote the action taken at state t, which is the block size for the remaining t slots. Let M_t denote the number of packets undelivered at state t.

---

[2]We can also employ random NC with a finite field size q. This would change the decoding probability (1) to a function of q. However, the general structure of the results will remain the same.

[3]We use the terms "state" and "slot" interchangeably.

Thus, at state $t > 0$, K_t can be chosen from 1 to $\min(t, M_t)$. For $t = 0$, the transmitter stops transmitting any packet, i.e., $K_0 = 0$. In general, the action space is defined as $\mathcal{K}_t = \{0, 1, ..., \min(t, M_t)\}$.

Expected immediate reward: For the remaining t slots, the expected immediate reward is the expected number of packets successfully decoded by all receivers, which is given by

$$R_t(K_t) = K_t \, P(K_t, t), \quad (2)$$

where $P(K_t, t)$ is given by (1), denoting the probability that each receiver can decode these K_t packets within t slots.

Block size adaptation policy: A block size adaptation policy $\mathcal{P}$ is a sequence of mappings, $\mathcal{P} = \{\mathcal{P}_t\}_{t=1}^T$, from t, M_t, ϵ, and N to an action $K_t \in \{0, 1, ..., \min(t, M_t)\}$, i.e., $K_t = \mathcal{P}_t(t, M_t, \epsilon, N) = \min(\mathcal{P}_t(t, \epsilon, N), M_t)$. Without loss of generality, in Section 3, we assume that M_t is always larger than t, i.e., $K_t \in \{0, 1, ..., t\}$. This does not change the monotonicity structure of the block size with state t. We will discuss these structural properties in detail in Section 3.

Total expected reward: Given the adaptation policy $\mathcal{P}$, the total expected reward for the remaining t slots is given by

$$\begin{aligned} V_t(K_t; \mathcal{P}) &= R_t(K_t) + E[V_j(K_j; \mathcal{P})] \\ &= R_t(K_t) + \sum_{j=0}^{t} q_t^{K_t}(j) V_j(K_j; \mathcal{P}), \end{aligned} \quad (3)$$

where the probability mass function $q_t(j) = P(K_t, t-j) - P(K_t, t-j-1)$ denotes the probability that the block of size K_t is delivered over exactly j slots before the deadline.

3. NETWORK CODING WITH ADAPTIVE BLOCK SIZE

A main contribution of this paper is the development and analysis of the polynomial-time monotonicity-based backward induction algorithm (MBIA). The design of the MBIA is motivated by the structures of the optimal and the greedy policies that are formally defined as follows.

DEFINITION 3.1. *A real-time scheduling policy with adaptive network coding is optimal, if and only if it achieves the maximum value of the total expected reward given by the Bellman equation [18] in dynamic programming:*

$$\begin{aligned} V_t(K_t^*; \mathcal{P}^*) = \max_{K_t \in \{0,1,...,t\}} \{ & R_t(K_t) \\ & + \sum_{j=0}^{t-K_t} q_t(j) V_j(K_j^*; \mathcal{P}^*) \}, \end{aligned} \quad (4)$$

where K_t^* denotes the optimal block size, $\mathcal{P}^*$ denotes the optimal block size adaptation policy, and the terminal reward is given by $V_0(0; \mathcal{P}^*) = 0$.

DEFINITION 3.2. *The greedy policy maximizes only the expected immediate reward (2) without considering the future rewards and the greedy decision is given by*

$$\hat{K}_t = \arg\max_{K_t \in \{0,1,...,t\}} R_t(K_t). \quad (5)$$

3.1 Optimal Block Size Adaptation Policy

In each slot t, the optimal policy balances the immediate reward and the future reward by selecting a suitable block size K_t^*. In general, the approach of solving for the optimal block size by traditional dynamic programming [18] suffers from the "curse of dimensionality," where the complexity of computing the optimal strategy grows exponentially with t. However, the optimal block size and its corresponding action space exhibit the monotonicity structures, and the optimal block size is bounded by the greedy block size. These unique structures make it possible to narrow down the search space of dynamic programming, and accordingly we develop a monotonicity-based backward induction algorithm (MBIA) with polynomial time complexity.

The MBIA searches for the optimal block size by backward induction and provides the optimal block size for each system state. Depending on the remaining time to deadline, the scheduler transmits coded packets with the optimal block size until each user receives enough packets to decode this block. Then, the scheduler adjusts the block size based on the current state, and proceeds with the new block transmission. This continues until the packet deadline expires or all packets are delivered. We present next the structural properties of block size adaptation problem that will lead to the formal definition of the MBIA.

Lemma 3.1. *The action space $\mathcal{K}_t$ monotonically shrinks as t decreases.*

Proof outline: As the number of remaining slots t decreases, the maximum possible block size decreases as well, since $K_t \in \{0, 1, ..., t\}$; otherwise no receiver can decode the block of coded packets. $\qquad\square$

Proposition 3.1. *The expected immediate reward function $R_t(K_t)$ has the following properties:*

1. *$R_t(K_t)$ is unimodal for $K_t \in \{0, 1, ..., t\}$.[4]*

2. *$\hat{K}_t$ in (5) monotonically decreases as t decreases.*

Proof outline: To show the unimodal property, it suffices to show that $R_t(K_t)$ is log-concave, which can be shown by using induction method. The monotonicity property of $\hat{K}_t$ can be shown by invoking the contradiction argument and applying $\lim_{t \to \infty} R_t(K_t) = K_t$. $\qquad\square$

Fig. 2 shows the possible curves of $R_t(K_t)$ for different values of t, illustrating the unimodal property of $R_t(K_t)$ formally stated in Proposition 3.1. Based on Proposition 3.1, the monotonicity property of the optimal block size K_t^* is given by the following theorem.

Theorem 3.1. *The optimal block size K_t^* monotonically decreases as t decreases, i.e., $K_t^* \geq K_{t-1}^*$, for any t.*

Proof outline: Based on Proposition 3.1, we can show that if $K_t^* < K_{t-1}^*$, $V_{t-1}(K_t^*; \mathcal{P}^*) > V_{t-1}(K_{t-1}^*; \mathcal{P}^*)$, which contradicts that K_{t-1}^* is the optimal action in slot $t-1$. $\qquad\square$

As t decreases, the risk that some receivers cannot decode the given block of packets increases for a fixed block size. Therefore, the scheduler becomes more conservative in the block size adaptation and selects a smaller block size.

Remarks: 1) For $t = 1, 2$, the optimal block size is $K_t^* = 1$, which can be obtained by computing the Bellman

---

[4] $f(x)$ is a unimodal function if for some m, $f(x)$ is monotonically increasing for $x \leq m$ and monotonically decreasing for $x \geq m$. The maximum value is attained at $x = m$ and there are no other local maximum points.

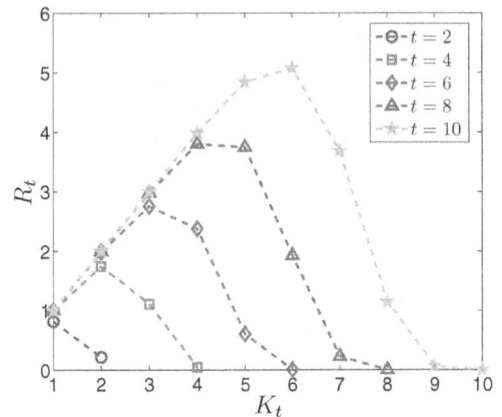

Figure 2: The unimodal property of $R_t(K_t)$.

equation (4). 2) When $N = 1$, the optimal block size is $K_t^* = 1$ for all t, since the plain retransmission policy with $K_t = 1$ is better than the block coding with $K_t > 1$ in the presence of the hard deadlines.

Theorem 3.2. *The optimal block size K_t^* is not greater than the greedy block size $\hat{K}_t$ for any t.*

Proof outline: Based on Proposition 3.1, we can show if $K_t^* > \hat{K}_t$, along any sample path, the system throughput by taking the action $\hat{K}_t$ is at least as high as that by taking the action K_t^*, which contradicts that K_t^* in this case is the optimal action in slot t. $\qquad\square$

Corollary 3.1. *At state t, if $R_t(K_t) > R_t(K_t + 1)$, then $K_j^* \leq \hat{K}_j \leq K_t$ for any $j \in \{1, ..., t\}$.*

Corollary 3.1 follows directly from Proposition 3.1 and Theorem 3.2. Based on these structural properties, we develop the MBIA, which is presented in Algorithm 1.

Algorithm 1 Monotonicity-based Backward Induction Algorithm (MBIA)

---

1) Set $t = 0$ and $V_0(0; \mathcal{P}^*) = 0$.

2) Substitute $t + 1$ for t, and compute $V_t(K_t^*; \mathcal{P}^*)$ by searching $K_t \in \mathcal{K}_t$, where $\mathcal{K}_t = \{K_{t-1}^*, K_{t-1}^* + 1, ..., \hat{K}_t\}$,

i.e., $V_t(K_t^*; \mathcal{P}^*) = \max_{K_t \in \mathcal{K}_t} \{R_t(K_t) + \sum_{j=0}^{t-K_t} q(j) V_j(K_j^*; \mathcal{P}^*)\}$,

and $K_t^* = \arg\max_{K_t \in \mathcal{K}_t} \{R_t(K_t) + \sum_{j=0}^{t-K_t} q(j) V_j(K_j^*; \mathcal{P}^*)\}$.

3) If $t = T$, stop; otherwise go to step 2.

The MBIA confines the search space at state t to the interval from K_{t-1}^* (the optimal policy at state $t - 1$) to $\hat{K}_t$ (the greedy policy at state t). Thus, the MBIA reduces the search space over time and reduces the complexity of dynamic programming as given by the following theorem.

Theorem 3.3. *The MBIA is a polynomial-time algorithm and the complexity is upper bounded by $O(T^2)$.*

Proof: Based on Proposition 3.1, $R_t(K_t)$ has the *unimodal* property and therefore $\hat{K}_t$ can be solved efficiently by the

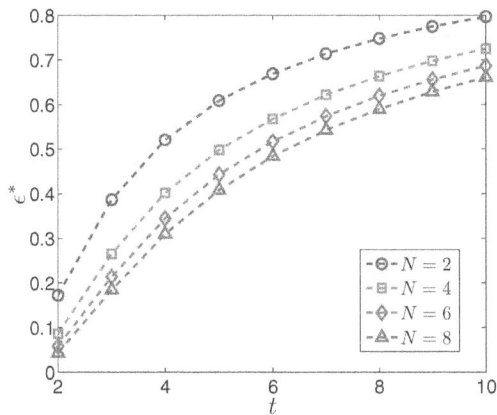

Figure 3: The monotonicity property of ϵ^*.

Fibonacci search algorithm [19], which is a sequential line search algorithm with a complexity of $O(\log(t))$ at state t. Therefore, in each iteration, it takes $O(\log(t) + \hat{K}_t - K_{t-1}^*)$ slots to find K_t^*. Based on Lemma 3.1, $\hat{K}_t - K_{t-1}^*$ is upper bounded by t. After some algebra, we show that the complexity of Algorithm 1 is bounded by $O(T^2)$ and Theorem 3.3 follows. □

Remarks: By using the MBIA, the optimal block size can be computed in polynomial time, which is a desirable property for online implementation. The optimal block size depends on the number of receivers and channel erasure probabilities. For different flows, the set of receivers may be different, which may result in different optimal block sizes, even when the number of remaining slots is the same across these flows. Therefore, without using the MBIA, offline schemes would need to compute the optimal policies for all possible receiver sets; however, this would be a computationally demanding task, as the number of receivers increases.

Based on the monotonicity properties of the greedy and optimal block sizes, the optimal policy becomes the plain retransmission, if the channel erasure probability is sufficiently large. This sufficiency condition for $K_t^* = 1$ at slot t is formally given as follows.

Theorem 3.4. *At slot t, the optimal policy switches to the plain retransmission policy, i.e., $K_t^* = 1$, when the erasure probability satisfies the threshold condition*

$$\epsilon > \epsilon^*(t, N), \tag{6}$$

where $\epsilon^*(t, N) \in (0, 1)$ is the non-trivial (unique) solution to $R_t(1) = R_t(2)$.

Proof outline: The proof follows directly by comparing $R_t(1)$ and $R_t(2)$ that are expressed as a function of ϵ. □

Note that (6) is a sufficient condition only and indicates the optimality of the greedy policy when ϵ is large enough.

Fig. 3 depicts how the threshold ϵ^* varies with t and N. The underlying monotonicity property is formally stated in Corollary 3.2.

Corollary 3.2. *The threshold $\epsilon^*(t, N)$ increases monotonically with t and decreases monotonically with N.*

Remarks: 1) When the channel is good enough (with $\epsilon < \epsilon^*$), NC with $K_t > 1$ can always improve the throughput compared to the plain retransmission policy. 2) As t increases (i.e., the deadline becomes looser), the risk of decoding network-coded packets decreases, i.e., $\epsilon^*(t, N)$ increases. 3) As N increases, it becomes more difficult to meet the deadline for each of N receivers and therefore $\epsilon^*(t, N)$ drops accordingly.

3.2 Robustness vs. Throughput

The real-time scheduling policies presented so far focus on the expected throughput without considering the variation from the average performance. Therefore, it is possible that the instantaneous throughput drops far below the expected value. To reduce this risk, we use additional variation constraints to guarantee that the throughput performance remains close to the average. In particular, for each slot t, we introduce the variation constraint to the block size adaptation problem as follows:

$$v_t(K_t) < \sigma_t^2, \ \forall K_t \in \mathcal{K}_t, \tag{7}$$

where σ_t^2 is the maximum variation allowed in slot t and the performance variation $v_t(K_t)$ under action K_t is given by

$$v_t(K_t) = \sum_{i=1}^{\infty} i^2 (P(K_t, i) - P(K_t, i-1)). \tag{8}$$

Since $v_t(K_t)$ increases with K_t, (7) can be rewritten as the maximum block size constraint for each slot t, i.e.,

$$K_t \leq K_t^{\max}, \tag{9}$$

where $K_t^{\max} = \max\{K_t | K_t = \lfloor v_t^{-1}(\sigma_t) \rfloor\}$, $v_t^{-1}(\cdot)$ is the inverse mapping of $v_t(\cdot)$, and $\lfloor x \rfloor$ denotes the largest integer smaller than x. The variation constraints do not change the monotonicity property of the optimal block size provided by Theorem 3.1. By introducing the variation constraints (9), the scheduler becomes more conservative in the block size adaptation. The additional bound $K_t^{\max}$ can be easily incorporated into the MBIA by changing the action space to $\mathcal{K}_t = \{K_{t-1}^*, K_{t-1}^* + 1, ..., \min(K_t^{\max}, \hat{K}_t)\}$ at state t.

3.3 Block Size Adaptation under Unknown Channels

So far we have discussed the real-time scheduling policies with adaptive NC, where the channel erasure probability ϵ is perfectly known to the scheduler. The throughput performance of these policies depends on ϵ; therefore, the scheduler needs to learn ϵ while adapting the block size, when it does not have (perfect) channel knowledge. Let $\hat{\epsilon}_t$ denote the estimate of the channel erasure probability in slot t. The scheduler can update $\hat{\epsilon}_t$ based on the feedback from the receivers. In slot T, if $\hat{\epsilon}_T < \epsilon$, we would expect with high probability that a block of packets with the size that is calculated with respect to $\hat{\epsilon}_T$ cannot be delivered before the deadline. Therefore, it is better to select the block size conservatively at the beginning, when the estimate $\hat{\epsilon}_t$ cannot be highly accurate yet, because of the small number of samples. As $\hat{\epsilon}_t$ improves over time, the block size can be gradually increased to improve the system throughput. Once the estimate is close enough to the actual value of ϵ after enough samples are collected, the block size should be adjusted (and reduced over time) according to the MBIA.

Clearly, there is a tradeoff between the channel learning and the block size adaptation. Here, we formulate a joint real-time scheduling and channel learning algorithm (Algorithm 2) to adapt the block size while updating the maximum likelihood estimate $\hat{\epsilon}_t$ of channel erasure probability. In slot t, based on the feedback, the scheduler can compute the packet loss ratio, $\epsilon_t = 1 - \frac{n_t}{N}$, where n_t denotes the number of receivers that successfully receive a packet in slot t. Accordingly, the estimated channel erasure probability $\hat{\epsilon}_t$ is given by the moving average

$$\hat{\epsilon}_t = \frac{(T-t)\hat{\epsilon}_{t+1} + \epsilon_t}{T-t+1}. \tag{10}$$

The scheduler decides on the block size by comparing the temporal variation $|\hat{\epsilon}_t - \hat{\epsilon}_{t+1}|$ with a threshold δ. A detailed description is given in Algorithm 2.

Algorithm 2 Joint Real-time Scheduling and Channel Learning with Adaptive Network Coding

---

Initialization: Choose threshold δ and set $K_T = 1$.
Repeat until $t = 0$.
   Update channel estimate $\hat{\epsilon}_t$ by (10).
   Compute block size K_t^* by Algorithm 1 with $\hat{\epsilon}_t$.
   If $|\hat{\epsilon}_t - \hat{\epsilon}_{t+1}| > \delta$ **then**
     If $K_t^* \geq K_{t+1} + 1$ **then**
       $K_t = K_{t+1} + 1$,
     Else
       $K_t = K_{t+1}$.
     Endif
   Else
     $K_t = K_t^*$.
   Endif

Remarks: Algorithm 2 captures the tradeoff between the channel learning and block size adaptation. There are two options for the scheduler depending on the relationship between $|\hat{\epsilon}_t - \hat{\epsilon}_{t+1}|$ and δ. If the channel estimation is not yet good enough, Algorithm 2 chooses the block size conservatively by incrementing K_t by at most 1. Otherwise, Algorithm 2 computes the block size by applying the MBI-A.

3.4 Performance Evaluation

Fig. 4 illustrates for $N = 5$ the monotonicity structure of the optimal block size (Theorem 3.1) and verifies that $K_t^* \leq \hat{K}_t$ (Theorem 3.2). Both the optimal and greedy block sizes increase when the channel conditions improve (from $\epsilon = 0.5$ to $\epsilon = 0.2$). Next, we evaluate the performance (average system throughput) of different policies. For comparison purposes, we also consider a soft delay-based *conservative* policy, where the scheduler chooses the largest block size with the expected completion time less than or equal to the number of remaining slots. The expected completion time is studied in [2], and it is given by

$$S(K) = K + \sum_{t=K}^{\infty} \left(1 - P(K,t)\right). \tag{11}$$

Fig. 5 compares the performance of the optimal, greedy, conservative and plain retransmission policies for $N = 10$ and $T = 10$. The plain retransmission policy always performs the worst, whereas the conservative policy performs worse than the greedy policy. However, as ϵ increases, all policies select smaller block sizes and their performance gap diminishes.

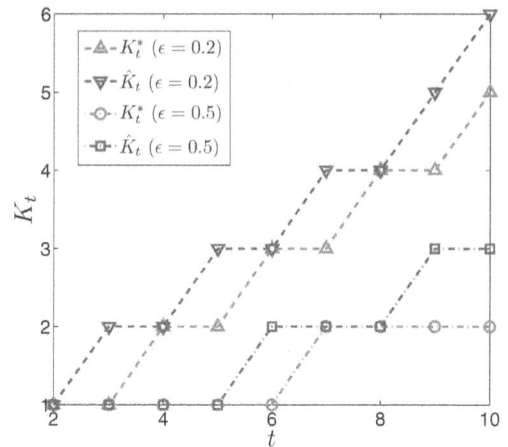

Figure 4: K_t^* is nondecreasing and $K_t^* \leq \hat{K}_t$.

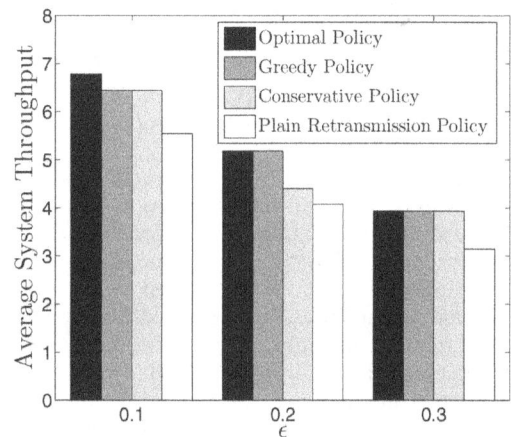

Figure 5: Performance (average system throughput) comparison of different policies.

Fig. 6 shows the tradeoff between the average system throughput and the throughput variation. When the channels are good (e.g., $\epsilon = 0.1$ in Fig. 6), the variation constraint (7) makes the scheduler choose a small block size, which reduces the average system throughput accordingly. However, there is no significant effect of (7) when channels are bad (e.g., $\epsilon = 0.5$ in Fig. 6), since the scheduler already chooses a small block size for large ϵ. Fig. 7 evaluates the performance of Algorithm 2 under channel uncertainty and show that Algorithm 2 is robust with respect to the variation of δ and achieves a reliable throughput performance close to the case with perfect channel information.

4. JOINT SCHEDULING AND BLOCK SIZE OPTIMIZATION

In this section, we generalize the study on adaptive NC to the case of multiple frames, where the scheduler serves a set $\mathcal{F}$ of flows subject to the hard deadline and the long-

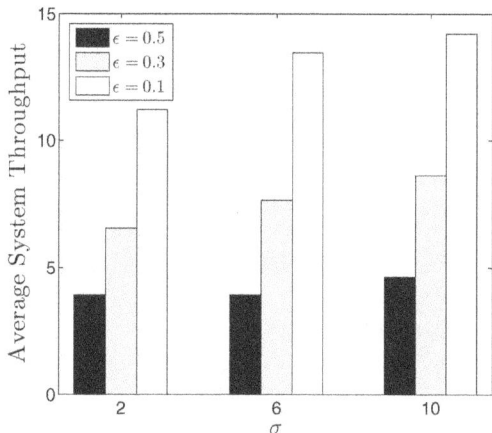

Figure 6: **Average system throughput vs. throughput variation, where $N = 10$ and $T = 20$.**

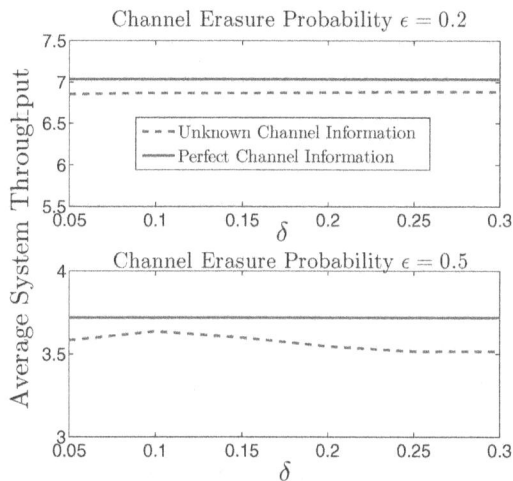

Figure 7: **Performance of Algorithm 2.**

term delivery ratio constraints. The packets of each flow f arrive at the beginning of every frame and they are dropped if they cannot be delivered to its receivers $\mathcal{N}_f$ within this frame (see Fig. 1). We impose that the loss probability for flow f due to deadline expiration must be no more than $1 - q_f$, where q_f is the delivery ratio requirement of flow f. For a given frame, the vector $a = (a_f)_{f \in \mathcal{F}}$ denotes the number of packet arrivals at each flow, where a_f is the number of packets generated by flow f. We assume that a_f is $i.i.d.$ across frames with finite mean λ_f and variance[5]. For ease of exposition, we assume perfect channel information at the scheduler and consider coding within each flow but not across different flows.

4.1 Multi-Flow Scheduling

The scheduler allocates slots for each flow and uses the

---

[5]The algorithm developed for the $i.i.d.$ case can be readily applied to non $i.i.d.$ scenarios. The analysis and performance guarantees can be obtained using the delayed Lyapunov drift techniques developed in [22,23].

optimal real-time scheduling policy with adaptive NC developed in Section 3 to transmit network-coded packets. Given the arrivals, the scheduler needs to allocate a suitable number of slots for each flow to satisfy the delivery ratio requirement. This resource allocation is defined as a feasible schedule, $s = (s_f)_{f \in \mathcal{F}}$, where s_f denotes the number of slots allocated to flow f and $\sum_{f \in \mathcal{F}} s_f \leq T$. Our goal is to maximize the weighted throughput subject to the delivery ratio and hard deadline constraints. We find the optimal schedule, i.e., the probability $Pr(s|a)$ that given the arrivals a, the schedule $s \in \mathcal{S}$ is used from the set $\mathcal{S}$ of all feasible schedules. Then, the expected service rate for flow f is upper-bounded by

$$\mu_f \leq \sum_{s,a} c_f(s_f) Pr(s|a) Pr(a), \qquad (12)$$

where $c_f(s_f)$ is the expected number of packets that can be delivered under schedule s_f, which is a constant and can be solved by the MBIA. Hence, we formulate the joint resource allocation and block size adaptation as the following optimization problem:

$$
\begin{aligned}
\text{maximize} \quad & \sum_{f \in \mathcal{F}} w_f \mu_f \\
\text{subject to} \quad & \mu_f \geq \lambda_f q_f, \ \forall f \in \mathcal{F}, \\
& \mu_f \leq \sum_{s,a} c_f(s_f) Pr(s|a) Pr(a), \ \forall f \in \mathcal{F}, \\
& Pr(s|a) \geq 0, \ \forall s \in \mathcal{S}, \ \sum_{s \in \mathcal{S}} Pr(s|a) \leq 1, \ \forall a, \\
\text{variables} \quad & \{\mu_f, Pr(s|a)\},
\end{aligned}
$$
$$(13)$$

where w_f is the weight for flow f and can be used as a fairness metric for resource allocation to each flow.[6] Note that (13) generalizes the problem studied in [21] by using adaptive NC schemes in packet transmissions.

4.2 Dual Decomposition

Since (13) is strictly convex, the duality gap is zero from the Slater's condition [20]. The dual problem is given by

$$
\begin{aligned}
\text{maximize} \quad & \sum_{f \in \mathcal{F}} (w_f \mu_f + \nu_f(\mu_f - \lambda_f q_f)) \\
\text{subject to} \quad & \mu_f \leq \sum_{s,a} c_f(s_f) Pr(s|a) Pr(a), \ \forall f \in \mathcal{F}, \\
& Pr(s|a) \geq 0, \ \forall s \in \mathcal{S}, \ \sum_{s \in \mathcal{S}} Pr(s|a) \leq 1, \ \forall a, \\
\text{variables} \quad & \{\mu_f, Pr(s|a)\},
\end{aligned}
$$
$$(14)$$

where ν_f is the Lagrangian multiplier for flow f. The objective function of (14) is linear and the upper bounds for μ_f are affine functions. Therefore, the optimization problem (14) can be rewritten as:

$$\max_{s \in \mathcal{S}} \sum_{f \in \mathcal{F}} (w_f + \nu_f) c_f(s_f). \qquad (15)$$

Thus, we have the following gradient-based iterative algorithm to find the solution to the dual problem (14),

$$
\begin{aligned}
& s^*(k) \in \arg\max_{s \in \mathcal{S}} \sum_{f \in \mathcal{F}} (w_f + \nu_f(k)) c_f(s_f), \\
& \mu_f^*(k) = c_f(s_f^*(k)), \\
& \nu_f(k+1) = \max(0, \nu_f(k) + \rho(\lambda_f q_f - \mu_f^*(k))),
\end{aligned}
$$
$$(16)$$

where k is the step index, $\rho > 0$ is a fixed step-size parameter, and $c_f(s_f^*(k))$ is the expected service rate for flow f under schedule $s_f^*(k)$. Letting $\hat{\nu}_f(k) = \frac{\nu_f(k)}{\rho}$, (16) is rewrit-

---

[6]The problem (13) can be generalized to the case with congestion control by treating the weights as virtual queues for flow rates (similar to the service deficit queues that we use later in Section 4.2).

ten as

$$s^*(k) \in \arg\max_{s \in \mathcal{S}} \sum_{f \in \mathcal{F}} (\tfrac{w_f}{\rho} + \hat{\nu}_f(k)) c_f(s_f),$$
$$\mu_f^*(k) = c_f(s_f^*(k)), \qquad\qquad (17)$$
$$\hat{\nu}_f(k+1) = \max(0, \hat{\nu}_f(k) + (\lambda_f q_f - \mu_f^*(k))).$$

Remarks: The update equation for $\hat{\nu}_f$ can be interpreted as a virtual queue for the long-term delivery ratio with the arrival rate $\lambda_f q_f$ and the service rate $\mu_f^*(k)$, which keeps track of the deficit in service for flow f to achieve a delivery ratio greater than or equal to q_f. Note that (17) provides only the static solution to (14). Next, we provide an online scheduling algorithm which takes into account the dynamic arrivals of the flows.

4.3 Online Scheduling Algorithm

The online scheduling algorithm is given by

$$s^*(k) \in \arg\max_{s \in \mathcal{S}} \sum_{f \in \mathcal{F}} (\tfrac{w_f}{\rho} + \hat{\nu}_f(k)) c_f(s_f),$$
$$\hat{\nu}_f(k+1) = \max(0, \hat{\nu}_f(k) + \hat{a}_f(k) - \hat{c}_f(s_f^*(k))), \qquad (18)$$

where $\hat{c}_f(s_f^*(k))$ denotes the actual delivered number of packets under the schedule $s_f^*(k)$ depending on the channel realizations, and $\hat{a}_f(k)$ is a binomial random variable with parameters $a_f(k)$, the number of packet arrivals of flow f in the kth frame, and q_f. This implementation for $\hat{a}_f(k)$ was proposed in [21]. At the beginning of each period, the schedule $s^*(k)$ is determined by (18). Then, the packets of each flow f are transmitted with the MBIA in the scheduled $s_f^*(k)$ slots. The virtual queue $\hat{\nu}_f$ is updated based on the number of successfully delivered packets $\hat{c}_f(s_f^*(k))$ of each flow f. With Lyapunov optimization techniques [22,23], it can be shown that (18) has the following properties.

Theorem 4.5. *Consider the Lyapunov function $L(\hat{\nu}) = \frac{1}{2} \sum_{f \in \mathcal{F}} \hat{\nu}_f^2$. If $\mu_f^* > \lambda_f q_f$ for all $f \in \mathcal{F}$, then the expected service deficit $\hat{\nu}_f$ is upper-bounded by*

$$\limsup_{k \to \infty} E[\sum_{f \in \mathcal{F}} \hat{\nu}_f(k)] \leq B_1 + \tfrac{1}{\rho} B_2,$$

for some positive constants B_1 and B_2. Furthermore, the online algorithm can achieve the long-term delivery ratio requirements, i.e., for all $f \in \mathcal{F}$ we have

$$\liminf_{K \to \infty} E[\sum_{k=1}^{K} \hat{c}_f(s_f^*(k))] \geq \lambda_f q_f.$$

Theorem 4.6. *Let $\rho > 0$ and μ_f^* be the solution to (17). If $\mu_f^* > \lambda_f q_f$ for all $f \in \mathcal{F}$, it follows for $B > 0$ that*

$$\limsup_{K \to \infty} E[\sum_{f \in \mathcal{F}} (w_f \mu_f^* - \tfrac{w_f}{K} \sum_{k=1}^{K} \hat{c}_f(s_f^*(k)))] \leq B\rho.$$

The proofs follow from the optimization framework in [22,23] and they are similar to the proofs presented in [21]. Note that the online scheduling algorithm (18) can approach within $O(\rho)$ of the optimal solution to (14) and does not require any knowledge of the packet arrival statistics.

4.4 Performance Evaluation

We consider a network with two flows, each with five receivers. The packet traffic of each flow follows Bernoulli distribution with mean λ_f packets/frame for $f = 1, 2$, and the length of each frame is 10 slots. In the simulation, we set $\lambda_f = \lambda$ for $f = 1, 2$. The channel erasure probability ϵ is 0.3, the weights w_f are 1 for all flows, the step-size ρ is 0.1, and the simulation time is 10^5 frames.

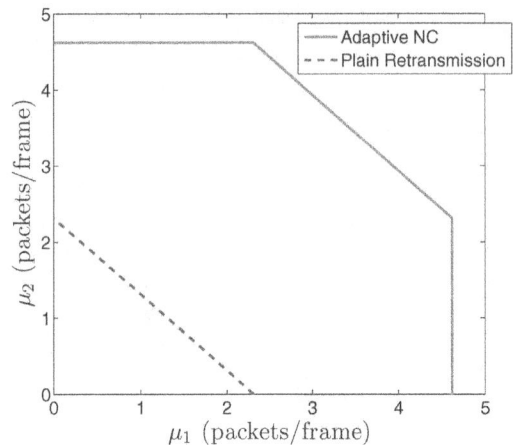

(a) $\lambda = 3$ (packets/frame)

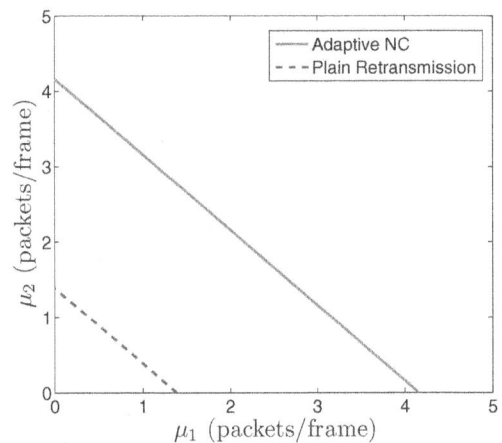

(b) $\lambda = 4.8$ (packets/frame)

Figure 8: Achievable rate regions under adaptive NC and plain retransmission policies.

We evaluate the performance of our algorithm by comparing the region of achievable rates (μ_1, μ_2) with the plain retransmission under different traffic flow rates λ, where the achievable rates denote the feasible solution to (13) for given delivery ratio requirements q_f. By varying q_f, we find the achievable rate region. As illustrated in Fig. 8, the plain retransmission only achieves a small fraction of the region with adaptive NC. By using adaptive NC, the network can support flows with heavier traffic.

Fig. 9 shows the average service deficit $\hat{\nu}$ of two flows. The delivery ratio requirement of each flow is 0.8. As λ increases, $\hat{\nu}$ grows unbounded, which means that the conditions, $\mu_f^* > \lambda_f q_f$ for all $f \in \mathcal{F}$, are not satisfied, i.e., the arrival rates are not in the "stability" region, and the online scheduling algorithm cannot meet the delivery ratio requirements.

5. HIGH FIDELITY WIRELESS TESTING WITH HARDWARE IMPLEMENTATION

We tested the adaptive NC schemes in a realistic wireless emulation environment with real radio transmissions. As illustrated in Fig. 10, our testbed platform consists of four

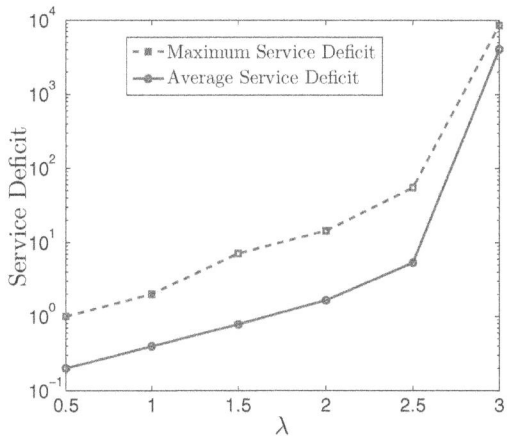

Figure 9: Service deficit vs. average arrival rate.

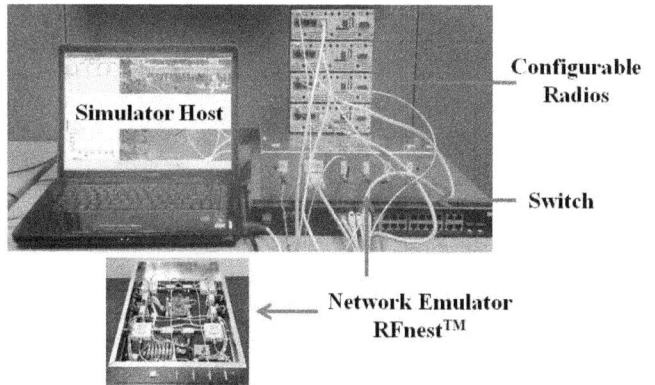

Figure 10: Programmable RFnest^{TM} testbed.

main components: radio frequency network emulator simulator tool, RFnest^{TM} [24] (developed and owned as a trademark by Intelligent Automation, Inc.), software simulator running higher-layer protocols on a PC host, configurable RF front-ends (RouterStation Pro from Ubiquiti), and digital switch. We removed the radio antennas and connected the radios with RF cables over an attenuator box. Then, real signals are sent over emulated channels, where actual physical-layer interactions occur between radios, and in the meantime the physical channel attenuation is digitally controlled according to the simulation model or recorded field test scenarios can be replayed.

In the hardware experiments, we executed wireless tests at 2.462GHz channel with 10dBm transmission power and 1Mbps rate. We used CORE (Common Open Research Emulator) [25] to manage the scenario being tested. We changed the locations of receivers through RFnest^{TM}GUI and let the signal power decay as $d^{-\alpha}$ over distance d with path loss coefficient $\alpha = 4$. By using real radio transmissions according to this model, we varied the attenuation from the transmitter to each of the receivers and generated different channel erasure probabilities. With RFnest^{TM}, we replayed the same wireless traces for each of the NC algorithms and compared them under the high fidelity network emulation with hardware-in-the-loop experiments.

Fig. 11 illustrates the performance of the optimal policy, the greedy policy and the fixed block size policy suggested by [16]. The experimental results show that the greedy policy performs close to the optimal policy in practice. Both the greedy and the optimal policies outperform the fixed block size policy, and the complexity remains low with the polynomial-time algorithm MBIA. Fig. 12 illustrates the wireless test performance for the case when the unknown channel erasure probabilities are learned over time. Algorithm 2 performs close to optimal in this case and converges quickly in several frames.

6. CONCLUSION

We considered adaptive NC for multimedia traffic with hard deadlines and formulated the sequential block size adaptation problem as a Markov decision process for a single-hop wireless network. By exploring the structural properties of

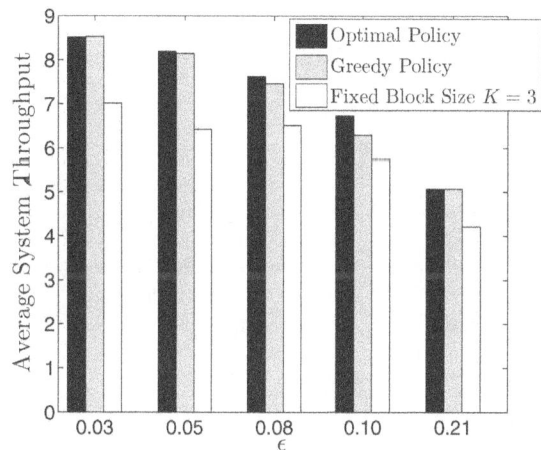

Figure 11: Performance comparison of different NC block size adaptation policies with network emulation, where $N = 10$ and $T = 10$.

the problem, we derived the polynomial time policy, MBIA, to solve the optimal NC block size adaptation problem and developed the joint real-time scheduling and channel learning scheme that can adapt to wireless channel dynamics if the perfect channel information is not available at the scheduler. Then, we generalized the study to multiple flows with hard deadlines and long-term delivery constraints, and developed a low-complexity online scheduling algorithm integrated with the MBIA. Finally, we performed high fidelity wireless emulation tests with real radios to demonstrate the feasibility of the MBIA in finding the optimal block size in real time. Future work should extend the model to integrate congestion control with adaptive NC and real-time scheduling under deadline constraints.

7. ACKNOWLEDGMENTS

We would like to thank Junshan Zhang from Arizona State University for helpful comments and discussions, and Lei Ding from Intelligent Automation, Inc. for help with hardware experiments. This material is based upon work

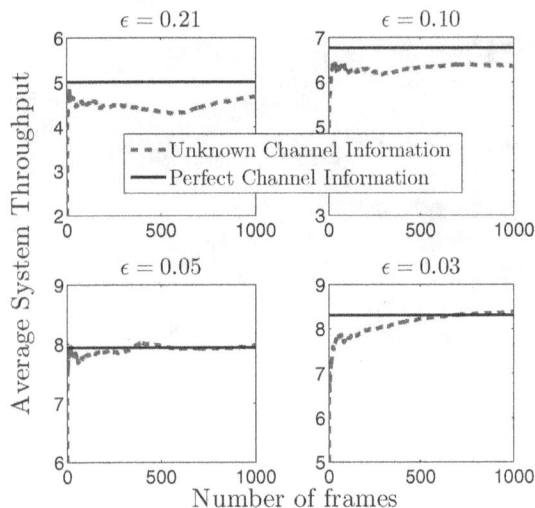

Figure 12: Average system throughput and convergence rate of Algorithm 2, where $N = 10$, $T = 10$ and the initial channel erasure probability estimation is 0.5.

supported by the Air Force Office of Scientific Research under Contracts FA9550-10-C-0026, FA9550-11-C-0006 and FA9550-12-C-0037. Any opinions, findings and conclusions or recommendations expressed in this material are those of the authors and do not necessarily reflect the views of the Air Force Office of Scientific Research.

8. REFERENCES

[1] R. Ahlswede, N. Cai, S. Li, and R. Yeung. Network information flow. *IEEE Trans. Inform. Theory*, 46(4):1204-1216, 2000.

[2] A. Eryilmaz, A. Ozdaglar, M. Medard, and E. Ahmed. On the delay and throughput gains of coding in unreliable networks. *IEEE Trans. Inform. Theory*, 54(12):5511-5524, 2008.

[3] J.-K. Sundararajan, D. Shah, and M. Medard. ARQ for network coding. In *Proc. of IEEE ISIT*, pages 1651-1655, 2008.

[4] E. Drinea, C. Fragouli, and L. Keller. Delay with network coding and feedback. In *Proc. of IEEE ISIT*, pages 844-848, 2009.

[5] Y. E. Sagduyu, and A. Ephremides. On broadcast stability of queue-based dynamic network coding over erasure channels. *IEEE Trans. Inform. Theory*, 55(12):5463-5478, 2009.

[6] D. Nguyen, T. Tran, T. Nguyen, and B. Bose. Wireless broadcast using network coding. *IEEE Trans. Veh. Technol.*, 58(2):914-925, 2009.

[7] J. Barros, R. A. Costa, D. Munaretto, and J. Widmer. Effective delay control in online network coding. In *Proc. of IEEE INFOCOM*, pages 208-216, 2009.

[8] T. Ho, R. Koetter, M. Medard, M. Effros, J. Shi, and D. Karger. A random linear network coding approach to multicast. *IEEE Trans. Inform. Theory*, 52(10):4413-4430, 2006.

[9] D. Traskov, M. Medard, P. Sadeghi, and R. Koetter. Joint scheduling and instantaneously decodable network coding. In *Proc. of IEEE Globecom Workshops*, pages 1-6, 2009.

[10] J. Heide, M. V. Pedersen, F. H. P. Fitzek, and T. Larsen. Network coding for mobile devices - systematic binary random rateless codes. In *Proc. of IEEE ICC Workshops*, pages 1-6, 2009.

[11] W. Yeow, A. Hoang, and C. Tham. Minimizing delay for multicast-streaming in wireless networks with network coding. In *Proc. of IEEE INFOCOM*, pages 190-198, 2009.

[12] A. A. Yazdi, S. Sorour, S. Valaee, and R. Y. Kim. Optimum network coding for delay sensitive applications in WiMAX unicast. In *Proc. of IEEE INFOCOM*, pages 2576-2580, 2009.

[13] R. Li, and A. Eryilmaz. Scheduling for end-to-end deadline-constrained traffic with reliability requirements in multi-hop networks. In *Proc. of IEEE INFOCOM*, pages 3065-3073, 2011.

[14] X. Li, C.-C. Wang, and X. Lin. On the capacity of immediately-decodable coding schemes for wireless stored-video broadcast with hard deadline constraints. *IEEE J. Sel. Areas Commun.*, 29(5):1094-1105, 2011.

[15] D. Nguyen and T. Nguyen. Network coding-based wireless media transmission using POMDP. In *Packet Video Workshop*, pages 1-9, 2009.

[16] H. Gangammanavar and A. Eryilmaz. Dynamic coding and rate-control for serving deadline-constrained traffic over fading channels. In *Proc. of IEEE ISIT*, pages 1788-1792, 2010.

[17] I.-H. Hou and P. R. Kumar. Scheduling heterogeneous real-time traffic over fading wireless channels. In *Proc. of IEEE INFOCOM*, pages 1-9, 2010.

[18] D. P. Bertsekas, *Dynamic Programming and Optimal Control*, Athena Scentific, Belmount, Massachusetts, 2005.

[19] J. Kiefer. Sequential minimax search for a maximum. *Proc. Amer. Math. Soc.*, 4(3):502-506, 1953.

[20] S. Boyd and L. Vandenberghe. *Convex Optimization*. Cambridge, U.K.: Cambridge Univ. Press, 2004.

[21] J. J. Jaramillo, R. Srikant, and L. Ying. Scheduling for optimal rate allocation in ad hoc networks with heterogeneous delay constraints. *IEEE J. Sel. Areas Commun.*, 29(5):979-987, 2011.

[22] L. Georgiadis, M. J. Neely, and L. Tassiulas. Resource allocation and cross-layer control in wireless networks. *Foundations and Trends in Networking*, 1(1):1-149, 2006.

[23] M. J. Neely. Stochastic network optimization with application to communication and queueing systems. *Synthesis Lectures on Communication Networks*, 3(1):1-211, 2010.

[24] J. Yackoski, B. Azimi-Sadjadi, A. Namazi, J. H. Li, Y. E. Sagduyu, and R. Levy. RF-NEST: radio frequency network emulator simulator tool. In *Proc. of IEEE MILCOM*, pages 1882-1887, 2011.

[25] J. Ahrenholz, C. Danilov, T. Henderson, and J. Kim. CORE: a real-time network emulator. In *Proc. of IEEE MILCOM*, pages 1-7, 2008.

[26] L. Yang, Y. E. Sagduyu, J. H. Li, and J. Zhang. Adaptive network coding for scheduling real-time traffic with hard deadlines. *arXiv:1203.4008v3*, 2012.

Distributed Network Coding-based Opportunistic Routing for Multicast

Abdallah Khreishah
Department of Computer &
Information Sciences
Temple University,
Philadelphia PA
akhreish@temple.edu

Issa M. Khalil
Faculty of Information
Technology
United Arab Emirates
University, UAE
ikhalil@uaeu.ac.ae

Jie Wu
Department of Computer &
Information Sciences
Temple University,
Philadelphia PA
jiewu@temple.edu

ABSTRACT

In this paper, we tackle the network coding-based opportunistic routing problem for multicast. We present the factors that affect the performance of the multicast protocols. Then, we formulate the problem as an optimization problem. Using the duality approach, we show that a distributed solution can be used to achieve the optimal solution. The distributed solution consists of two phases. In the first phase, the most reliable broadcasting tree is formed based on the ETX metric. In the second phase, a credit assignment algorithm is run at each node to determine the number of coded packets that the node has to send. The distributed algorithm adapts to the changes in the channel conditions and does not require explicit knowledge of the properties of the network. To reduce the number of feedback messages, and to resolve the problem of delayed feedback, we also perform network coding on the feedback messages. We evaluate our algorithm using simulations which show that in some realistic cases the throughput achieved by our algorithm can be double or triple that of the state-of-the-art.

Categories and Subject Descriptors

C.2.1 [**Network Architecture and Design**]: Wireless Communications

General Terms

Algorithm, Performance

Keywords

Network coding, opportunistic routing, multicast, distributed algorithms, coded-feedback

1. INTRODUCTION

Multicasting is an important operation in wireless multihop networks. Its applications range from software updates

to video/audio file downloads. Designing an efficient and reliable multicasting protocol for wireless multihop networks is not a straightforward extension from the protocols designed for wireline networks. This is due to the unique features of wireless multihop networks. These features are the lossy behavior, the diversity of the links, the broadcast nature of the links, and the correlations among the links.

Opportunistic routing [3] has been proposed as a way to exploit the unique features of wireless multihop networks. In opportunistic routing, there is no specific next-hop node. Therefore, any node that receives the packets can forward it. To avoid duplication, the receivers of a specific transmission need to coordinate to specify which one of them has to forward the packet that has been received by more than one receiver. This requires the design of a specific MAC protocol. Another shortcoming of opportunistic routing is the difficulty of the extension to the multicast case as stated in [6].

Using *intrasession network coding* [6], the shortcomings of opportunistic routing can be eliminated. In intrasession network coding, the source node divides the message it wants to send into batches, each having K packets of the form $P_1, \ldots, P_K$. The source node keeps sending coded packets of the form $\sum_{i=1}^{K} \gamma_i P_i$, where $\gamma_i, \forall i$ is a random coefficient chosen over a finite field of large enough size, typically 2^8–2^{16}. Upon receiving a coded packet, the intermediate relay node checks to see if the coded packet is linearly independent to what it has received before. If so, it keeps the coded packet, otherwise it drops the packet. When the destination receives K linearly independent packets, this means that it can decode all of the packets of the batch. Therefore, it sends a feedback to the source, using the traditional shortest path, that says: stop sending from this batch and move to the next one. The advantage of using network coding is that the destination node does not need to receive the specific K original packets, but can receive any K linearly independent ones. This resolves the problem of designing a new MAC protocol because we do not insist on receiving a specific packet. Network coding-based opportunistic routing can also be generalized to the multicast case as network coding enhances the achievable throughput for the multicast case with low complexity and in a distributed way, even for wireline networks [1, 11, 12].

Despite the attractiveness of using intrasession network coding-based opportunistic routing for the multicast case in wireless networks, most of the works in the literature focus on the unicast case [10, 15, 19, 27, 28]. The major

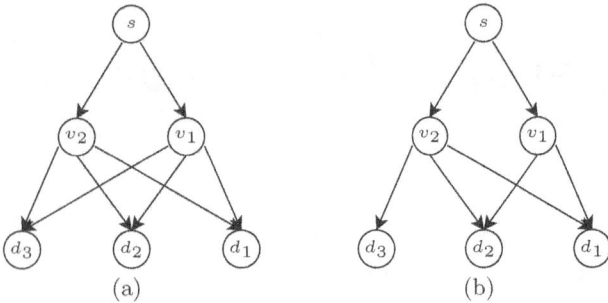

Figure 1: Examples of multicast wireless networks. The second figure is formed by removing the link from v_1 to d_3, which changes the optimal solution.

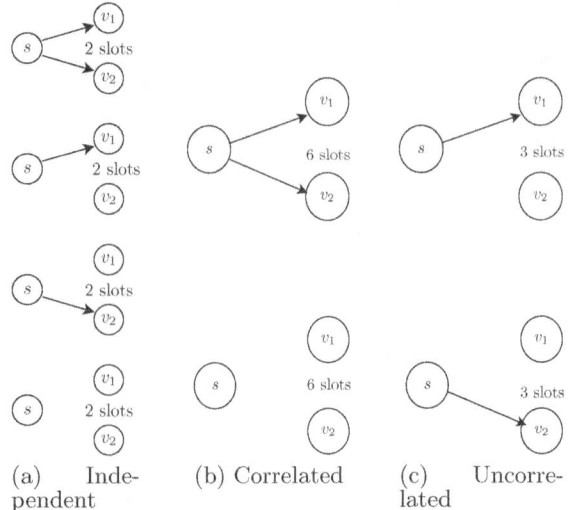

(a) Independent (b) Correlated (c) Uncorrelated

Figure 2: Illustration of the channel activation scenarios that insure that v_1 and v_2 collectively achieve full rank under different correlation conditions between the channels.

challenge that the implementation of intrasession network coding-based opportunistic routing for multicast faces is to specify the number of packets that each node has to send. The work in [6, 14] resolves this problem by using an estimation based on the link loss rates. While this approach shows improvement over the non-coded approach, the estimation performed by this approach does not take into account all of the factors that affect the optimal solution. Also, it does not adapt to the changes in the channel conditions.

Obviously, the loss rates of the links are a major factor in deciding the optimal solution. However, there are other factors that affect the optimal solution. To illustrate the other factors, take Fig. 1(a) as an example. In the figure, source s is the source that multicasts its packets to the destination nodes $\{d_1, d_2, d_3\}$. Node s should send enough packets so that the next-hop nodes have, collectively, a full rank matrix. Assume that the batch size is 6, and the loss rate of both of the output links of s is 0.5. If the two output links are correlated, i.e., they are on at the same time and off at the same time, then we need 12 transmissions to ensure that the next-hop nodes collectively achieve full rank. On the other hand, if the two links are independent, we need 8 transmissions; and we only need 6 transmissions, if the links are uncorrelated. Fig. 2 represents the three cases. While most of the work on opportunistic routing and network coding in wireless networks assume independent links, a recent study [20] has shown that the correlation among the links can be arbitrary and dynamically changing over time. Therefore, it is important to formulate and solve the problem under arbitrary correlations among the links.

In addition to the loss rate and the correlations among the links, the reachability of the nodes plays a major role in specifying the number of packets a node has to send. Fig. 1(b) represents the same network in Fig. 1(a) but after removing the link between nodes v_1 and d_3. In this case, node v_2 has to receive vectors from s that can achieve full rank as there is no other path to d_3 without going through this node. Therefore, we need 12 transmission regardless of the correlations among the links.

In this paper, we show that these three factors - the loss rate, the correlation among the links, and the reachability of the node - can be used to design optimal network coding-based opportunistic routing multicast in wireless multihop networks. Our contribution lies in the following:

- We formulate the optimal network coding-based opportunistic routing for multicast as an optimization problem. We use a wireless to wireline mapping mechanism to show that our formulation achieves the maximum possible rate with intrasession network coding.

- We develop a fully distributed algorithm for the problem such that each node only needs local information. The distributed algorithm consists of two phases: reliable multicast tree construction and distributed credit assignment. The distributed algorithm adapts to the changes in the channel conditions and converges to the optimal solution. Also, it does not need an explicit knowledge of the channel conditions or the correlation among the links.

- We integrate our algorithm with the coded feedback approach to reduce the number of feedback messages and eliminate the assumption of immediate feedback.

- Using simulations we compare our results to the state of the art opportunistic routing protocol for multicast ,MORE [6], and show the effectiveness of our protocol in maximizing the throughput.

The rest of the paper is organized as follows. In Section 2 we present our settings followed by the mapping of wireline to wireless networks in Section 3. We present the formulation of the problem in Section 4 and derive a distributed algorithm based on the formulation in Section 5. The integration of the coded-feedback approach with our algorithm is presented in Section 6. We evaluate our algorithm by using simulations in Section 7, and we conclude the paper in Section 8.

2. SETTINGS

In this paper, we consider a network that is represented by a set of nodes V. The links between the nodes are lossy and time varying. A transmission by a node can be received by any subset of next-hop nodes. We represent this by a hyperedge (u, J), where u is the node that performs the transmission and J is a subset of the set of next-hop nodes. Unlike previous work, the correlation among individual links of a given hyperedge is arbitrary, and we don't need to measure it as has been done in [20]. There are N multicast sessions in the network, each with a source s_i, a set of destination nodes D_i, a rate R_i, and a utility function $U_i(R_i)$, $\forall i \in \{1, \ldots, N\}$. Like most of the opportunistic routing protocols [6, 15, 27, 28], we are interested in the transmission of large files. Therefore, the throughput is the most important factor, and the individual packet delays are of lower importance. Also, the transmission scheme has to be reliable such that every packet sent by the source of the session has to be received by all of the destinations of that session, regardless of the network bandwidth to that receiver. This means that the same rate R_i has to be supported by all of the destinations of session i. Since we are using intrasession network coding, one important factor to determine is the rate of linearly independent packets that a node has to successfully deliver to next-hop nodes. To model this factor, we use the concept of credits similar to [6, 19]. Symbol X_{uv}^i is used to represent the rate of credits for session i that node u gives to node v, which represents the total number of linearly independent packets that node v has to forward to next-hop nodes (out of the packets it has received from node u). Therefore, the total rate of credits for session i at node u would be $\sum_{v \in V} X_{vu}^i$, and these credits will be distributed to next-hop nodes. We also use α_u^i to represent the fraction of time that node u is scheduled to send the packets of session i. Symbol R_{uJ} represents the rate of packets that are sent by node u and are received by any of the nodes in J. Since our solution is based on building multicast trees, we use $T(i)$ to represent the multicast tree for the i-th multicast session. For any node $u \in T(i)$, we use $RC(u, i)$ to represent the descendent destination nodes for node u on T(i). Therefore, if the destination node $d \in DC(u, i)$, then $\exists$ a path from u to d on $T(i)$. We also use $I(u, i)$ $(O(i, u))$ to represent the direct parents (children) of node u on tree $T(i)$. Also, for a set of nodes J, another node $d \in RC(J, i)$, if $d \in RC(v, i), \forall v \in J$.

3. WIRELESS NETWORKS AND THEIR WIRELINE COUNTERPART

For wireline networks, and for a single multicast session, it has been shown in [1] that the maximum multicast rate is the minimum of the min-cut max-flow between the source and each destination. For multiple multicast sessions in wireline networks, intrasession network coding -where coding is performed on the packets of every session separately- is used to share the bandwidth of the network [7, 26]. Therefore, the optimal algorithm in wireline networks has to achieve the minimum of the min-cut max-flow value between the source and each destination for the single source multicast problem. Also, the same optimal algorithm can use intrasession network coding to share the resources of the network in the case of multiple multicast sessions. This is due to the difficulty of using intersession network coding [9, 16], which codes packets of different flows together.

Due to the broadcast nature of wireless links and the correlations among the links, how we can find the min-cut max-flow between two nodes is not clear. In this section, we perform mapping from any wireless network with any channel conditions to its corresponding wireline network, such

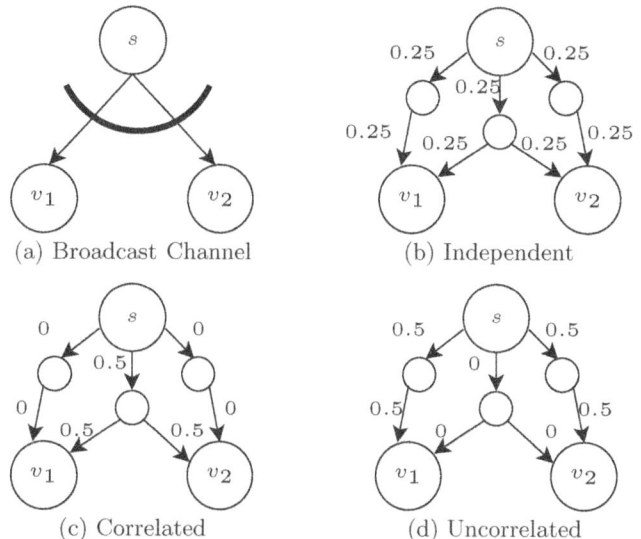

Figure 3: Wireless to wireline mapping with different correlations among the channels. The big circles represent the original nodes and the small ones represent the added auxiliary nodes.

that the capacity properties of the wireless network are preserved. A similar conversion has been done in [13, 25] for lossless wireless links. The mapping can be done on each broadcast link by introducing an auxiliary node for each set of receivers in the broadcast link and then by connecting the transmitter node of the broadcast link to each auxiliary node with a separate directed link and the auxiliary node to the set of receivers it represents with directed links. The weight assigned to each link that uses a given auxiliary node can be computed as the transmission bandwidth of the source node of the broadcast link times the probability that all of the outgoing nodes of the auxiliary node *exclusively* overhear a given transmission. Fig. 3 represents this mapping with different correlations among the broadcast links.

In Fig. 3(b), we assume that the two links of the broadcast channel in Fig. 3(a) are independent. Therefore, the probability that only v_1 (v_2) overhears a transmission would be $p_{s,v_1} \times (1 - p_{s,v_2})$ $(p_{s,v_2} \times (1 - p_{s,v_1}))$, where p_{uv} is the delivery rate from node u to v. Similarly, the probability that both of v_1 and v_2 overhear a transmission would be $p_{s,v_1} \times p_{s,v_2}$. Assuming the fraction of time where node s is scheduled equals one justifies the weights assigned to the links on Fig. 3(b). In Figs. 3(c) and 3(d), the channels are correlated and uncorrelated, respectively, which justifies the weights assigned to the channels in these Figs.

Note that this mapping allows us to make an equivalent wireline network model for any wireless network with arbitrary characteristics. For example, if we assume that the nodes in Fig. 1(a) use orthogonal channels, which means that the nodes can be scheduled for an arbitrary fraction of the time, the min-cut max-flow from s to each of the destinations will be 0.75, 0.5, and 1, respectively, if the links are independent, correlated, and uncorrelated, respectively. The mapping can also handle the case where the nodes are scheduled for limited amount of time due to interference by multiplying the weight of every link by the fraction of time that the sender node of that link is scheduled. Note also that

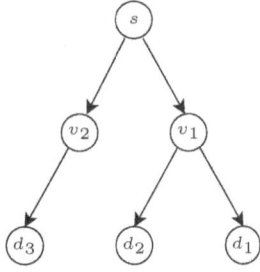

Figure 4: An example of a multicast wireless network formed by removing links from the networks in Fig.s 1(a) and 1(b).

(a) Wireless network

(b) The counterpart network

(c) Wireless network

(d) The counterpart network

Figure 5: Two examples of wireless networks and their wireline counterpart. The dotted line between two nodes means there exists a path between them. a) Both d_1 and d_2 are reacheable from v_1, v_2, and v_3. b) The wireline counterpart of part (a) with the cut C_1 that is an upperbound on the number of credits that can be forwarded to v_1, v_2, and v_3. c) Both d_1 is reacheable from v_1, v_2, and v_2 is reachable from v_3, v_3. d) The wireline counterpart network of part (c). Since not all of the d nodes can be reached by all of the v nodes, we have two different constraints on the maximum credits that can be sent to both v_1 and v_2 and to both v_2 and v_3. These constraints are represented by the two cuts C_2 and C_3 on the mapped wireline netwwork.

our algorithm does not require us to do the mapping. We use the mapping only to show that our algorithm achieves the min-cut max-flow bound of a wireline network that is equivalent to our wireless network.

4. PROBLEM FORMULATION

Our problem can be formulated as follows.
Maximize:

$$\sum_{i=1}^{N}(U(R_i))$$

Subject to:

$$\sum_{v:v\in I(u,i)} X_{vu}^i - \sum_{\substack{v:v\in O(u,i)\\ d\in RC(v,i)}} X_{uv}^i \leq \begin{cases} -R_i & u=s_i \\ 0 & else \end{cases}$$

$$\forall i, \forall u \in T(i)\backslash D_i, d\in D_i, d\in RC(u,i) \quad (1)$$

$$\sum_{\substack{v:v\in J\\ d\in RC(J,i)}} X_{uv}^i \leq \alpha_u^i R_{uJ}$$

$$\forall i, \forall (u,J), \forall d\in D_i, d\in RC(u,i) \quad (2)$$

We assume that the utility function $U_i(R_i)$ is non-decreasing and strictly concave. If the utility function is chosen properly, maximizing the objective function will achieve different kinds of fairness among the sessions [4]. Examples of $U(R_i)$ would be $w_i \log(1+R_i)$ and $w_i\frac{R_i^{1-\gamma}}{1-\gamma}$, where $0 \leq \gamma \leq 1$, and w_i is the weight assigned for session i.

Here, α_u^i depends on the underlying interference model. Typically, it corresponds to the convex hull of all of the achievable rates at all links [17]. Generally, the corresponding optimal scheduling policy is NP-hard, and approximation algorithms are used. While it is easy to extend our formulation and algorithm to include optimal scheduling, we consider that scheduling is of secondary importance, and we use IEEE 802.11 in the simulations. We do this to focus on the network coding part and to have a fair comparison with the other approaches that use IEEE 802.11.

The first set of constraints represents balance equations so that at every node, and for every destination d, the total number of received credits should be no less than the total number of credits assigned to next-hop nodes that can reach destination d. For example, in Fig. 1(a), node s can split the credits it has between nodes v_1 and v_2 because all of $\{d_1, d_2, d_3\}$ are reachable from both v_1 and v_2. On the other hand, in Fig. 1(b), node s can't split the credits it has

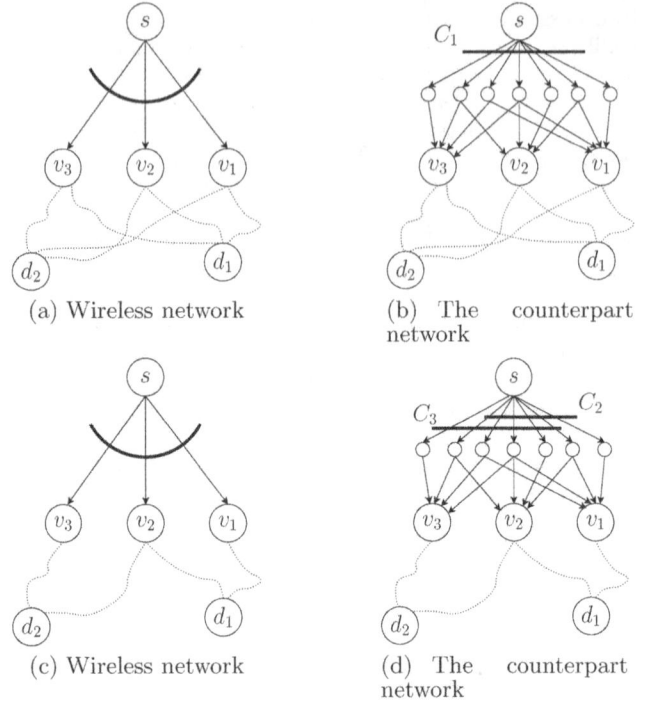

between v_1 and v_2 because d_3 is reachable by only v_2, and v_2 has to receive the same number of credits assigned to s which is confirmed by the first set of constraints.

The second set of constraints represents the fact that if a packet is received by more than one of the next-hop nodes, such that all of these nodes can reach destination d, only one of them can use this packet to increase its credits. For example, in both Figs. 1(a) and 1(b), if the links of node s are independent, and assuming that node s has sent m randomly coded packets, then both v_1 and v_2 will receive $0.5m$ linearly independent packets. Half of the received packets by v_1 are also received by v_2 due to the independence of the links. Therefore, the total credits assigned to both v_1 and v_2 can not exceed $0.75m$ because both v_1 and v_2 can reach both d_1 and d_2. However, in Fig. 3, the total credits assigned for both v_1 and v_2 can be m because there is no destination that is reachable by both v_1 and v_2, which nullifies the effect of the commonly received packets by both v_1 and v_2.

LEMMA 1. *Assuming the use of perfect scheduling for one session, the achievable rate of the formulation in ((1)-(2)*

is the minimum of the min-cut max-flow from the source to each of the destinations after converting the network to its wireline counterpart.

PROOF. It is easy to show that the first set of constraints guarantees achieving the min-cut max-flow bound from the source to the destinations for wireline networks. Therefore, what remains to show is that the second set of constraints is equivalent to mapping the wireless network to its wireline counterpart described in the previous section.

For a given broadcast channel, the rank of the matrix, represented by the packets sent by the sender of the broadcast channel and received by a subset of the receivers node of this broadcast channel, is upper bounded by the probability that any one of these nodes receives a given sent packet multiplied by the transmission rate of the sender node of that broadcast channel. This is because, if a packet is received by more than one of the receivers of the broadcast channel, the rank of the matrix that represents all of the packets at these nodes that received the transmitted packet can not be increased by more than one. If all of these nodes can reach a specific destination d_1, then the total number of linearly independent packets that these receiver nodes can push to the destination d_1 can not exceed the rank of the matrix, which limits the number of credits assigned to these nodes, i.e., for a packet that has been received by two nodes, only one credit can be assigned to both of these two nodes. However, if a group of these receiver nodes can reach destination d_1 but not d_2, and another group of these receivers nodes can reach destination d_2 but not d_1, then, if a packet is received by nodes in the two groups, the packet can increase the rank of the matrix in both of these groups because the matrices are destined to two different receivers. This means that the packet that is received by the two groups can increase the number of credits by two instead of one.

Note that the rank of the matrix at a group of receiver nodes of a broadcast channel equals the max-flow from the source of the broadcast channel to these receiver nodes on the mapped wireline counterpart of the wireless broadcast channel, as illustrated in Fig. 5. Therefore, the second constraint set is equivalent to mapping the wireless network to its wireline counterpart.

InFigure, if a destination is reacheable by all of the nodes in the set $\{v_1, v_2, v_3\}$, then the cut in Fig. 5(b) represents an upper bound on the maximum number of credits that can be assigned to the nodes in the set $\{v_1, v_2, v_3\}$. However, when there is no destination node that is reachable by all of the nodes in $\{v_1, v_2, v_3\}$, we need to consider other cuts as in Fig. 5(d), and these cuts are also represented by the second constraints. □

So far we have shown that any wireless network can be mapped to a wireline counterpart such that both of them have the same capacity characteristics. We have also shown that for one source multicast, our formulation achieves the min-cut max-flow bound on the mapped wireline counterpart network, which is the maximum achievable rate.

PROPOSITION 1. *For multiple sessions, the achievable rates of our formulation represent the optimal solution with intrasession network coding.*

PROOF. With intrasession network coding no coding is permitted between different sessions. Therefore, if the maximum rate can be achieved for every session by Lemma 1,

then by using the time sharing variables α_u^i, the optimal solution with intrasession network coding can be achieved. □

Note that the optimal multiple unicast sessions case [10, 15, 19, 27, 28] can be obtained as a special case of our formulation. This can be done by assigning every destination set to a single node. Also, the tree can be replaced by a single or multiple paths. Alternatively, we can use a back-pressure algorithm [21] to jointly assign the credits and to find the paths to the destination.

5. DISTRIBUTED ALGORITHM

5.1 Constructing Broadcast Trees

In this section, we provide a simple way for constructing the broadcasting tree and setting up $RC(u, i)$ for every node u. Each node in the network computes the ETX metric [8] to the destination nodes of every session. Also, every node u initializes $RC(u, i), \forall i$ to an empty set. For every session i, the source node s_i sets $RC(s_i, i)$ to D_i. For every destination node in its $RC(s_i, i)$, the source node selects each node u that has a lower ETX metric value to that destination than its own as a forwarder node and adds that destination to $RC(u, i)$. Every intermediate node u repeats the same comparison process, but just for the destinations in its own $RC(u, i)$ instead of $RC(s_i, i)$.

5.2 Structure of the Optimal Solution

After building the broadcast tree, the next step is to develop a distributed algorithm that assigns credits to the nodes along the tree, which we discuss in this Section.

Since the constraints are linear, we have a convex optimization problem. Therefore, there is no duality gap, and we can use the duality approach to solve the problem [2, 5].

Ignoring the scheduling constraints, we associate a Lagrange multiplier q_{ud}^i with each constraint in (1) and another one $\lambda_{(u,J)d}^i$ with each constraint in (2). This results in the following Lagrange function $L(\mathbf{R}, \mathbf{X}, \mathbf{q}, \lambda)$ that is equal to

$$\sum_{i=1}^{N} U_i(R_i) - \sum_{i,u}\Big(\sum_{\substack{d:d\in D_i \\ d\in RC(u,i)}} q_{ud}^i\Big(\sum_{\substack{v:v\in I(u,i) \\ }} X_{vu}^i - \sum_{\substack{v:v\in O(u,i) \\ d\in RC(v,i)}} X_{uv}^i\Big)\Big)$$
$$- \sum_{\substack{i,(u,j) \\ }} \sum_{\substack{d:d\in D_i \\ d\in RC(u,i)}} \lambda_{(u,j)d}^i\Big[\Big(\sum_{\substack{v:v\in J \\ d\in RC(J,i)}} X_{uv}^i\Big) - \alpha_u^i R_{u,J}\Big]$$

With simple changes of variables, the Lagrange function becomes

$$\sum_{i=1}^{N}\Big[U_i(R_i) - \sum_{d:d\in D_i} q_{s_i d}^i R_i\Big]$$
$$+ \sum_{u,i}\sum_{v}\Big(\sum_{\substack{d:d\in D_i \\ d\in RC(u,i)}}\Big(q_{ud}^i - \sum_{\substack{J:v\in J \\ d\in RC(J,i)}}(\lambda_{(u,J)d}^i)\Big)$$
$$- \sum_{\substack{d:d\in D_i \\ d\in RC(v,i)}} q_{vd}^i\Big)X_{uv}^i + \sum_{i,(u,J)}\sum_{\substack{d:d\in D_i \\ d\in RC(u,i)}} \lambda_{(u,J)d}^i R_{(u,J)}$$

The Lagrange function is separable [2], which means that the problem can be solved in a distributed way by using the gradient method as follows.

Source Algorithm: Each source s_i selects its rate at each time slot as follows:

$$R_i(t) = \arg\max_{R_i}[U(R_i) - \sum_{d:d\in D_i} q^i_{s_id}(t)R_i] \quad (3)$$

Intermediate Node Algorithm: Each intermediate node u selects the number of credits for session i to transfer to all of its next-hop nodes at each time slot as follows:

$$\{X^i_{uv}(t)\} = \arg\max_{\mathbf{X}} \sum_{v\in O(u,i)} \Big(\sum_{\substack{d:d\in D_i \\ d\in RC(u,i)}} (q^i_{ud}(t)$$

$$- \sum_{\substack{J:v\in J \\ d:d\in RC(J,i)}} (\lambda^i_{(u,J)d}(t))) - \sum_{\substack{d:d\in D_i \\ d\in RC(v,i)}} q^i_{vd}(t) \Big) X^i_{uv}$$

$$(4)$$

Dual Variables Updates: The dual variables can be updated in a distributed way as follows:

$$q^{id}_u(t+1) = [q^i_{ud}(t) + \beta^i_{ud}(\sum_{v:v\in I(u,i)} X^i_{vu}(t) - \sum_{\substack{v:v\in O(u,i) \\ d\in RC(v,i)}} X^i_{uv}(t))]^+,$$

$$(5)$$

$$\lambda^i_{(u,J)d}(t+1) = [\lambda^i_{(u,J)d}(t) \quad (6)$$

$$+ \beta^i_{(u,J)d}(\sum_{\substack{v:v\in J \\ d\in RC(J,i)}} X^i_{uv}(t) - \alpha^i_u(t)R_{uJ}(t))]^+. \quad (7)$$

Here, $[.]^+$ is a projection on the positive real numbers, and β is the step size.

THEOREM 1. *The algorithm converges to the optimal solution of the problem.*

Due to space limitations, we remove the proof.

6. INTEGRATING THE ALGORITHM WITH THE CODED FEEDBACK APPROACH

6.1 Challenges

The algorithm represented by ((3)-(7)) converges to the optimal solution, but it has the following shortcomings. Firstly, the algorithm requires a large amount of feedback messages. For example, if the batch size is 32, and the node which has l next-hop forwarders sends 32 packets from the batch, we need $(32l)$ feedback messages. Secondly, the links are lossy, which increases the number of required transmissions for the feedback messages. Thirdly, the algorithm assumes immediate hop-by-hop feedback which is not realistic due to the scheduling problem in wireless networks. Finally, it takes very long time to converge. Also, to converge, the generation size should be very large. However, for practical reasons, and in order to carry the coding coefficients, the generation size should be small, typically 32. Despite these shortcomings, the structure of the solution above inspires us to design an efficient distributed algorithm for the problem. We introduce our solutions to the above problems through the use of the coded-feedback approach.

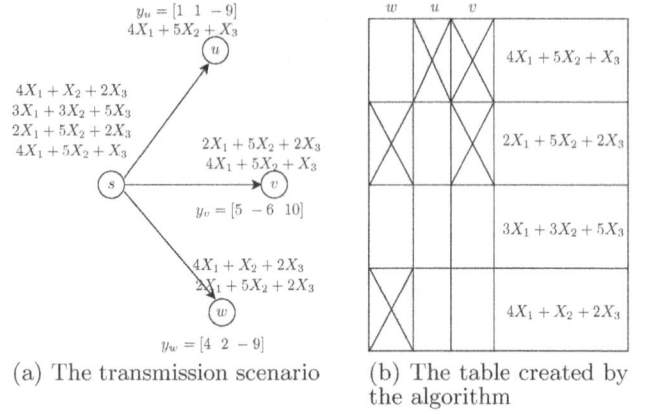

(a) The transmission scenario (b) The table created by the algorithm

Figure 6: **An example representing our coded feedback approach. An X in a cell means that the coded packets representing the row has been received by the node representing the column.**

6.2 Integrating the Coded Feedback Approach with the Algorithm

In this Section, we use the coded-feedback approach, which has been proposed recently [15, 18, 23], to resolve the previously mentioned shortcomings of our basic algorithm. The objective of the coded feedback approach is to perform network coding on the feedback messages, such that the transmitter node can learn about the linear space that each receiver node has received through the channel. The common way of performing coded feedback is through the null space. The null space of the matrix A is the linear space of vectors such that the result of multiplying anyone of these vectors by A equals zero. For example, if y belongs to the null space of A, then $y^T A = 0$, where y^T is the transpose of y.

Take Fig. 6(a) as an example, in which node s sends four coded packets. Node v receives two of them. Node v can compute the null space[1] of the space of the packets it receives, choose a vector from this space, and send it back to node s. As is illustrated in Fig. 6(b), node s can now multiply this vector with each of the packets it has sent. If the result is zero, node s can infer that the packet has been received by node v with high probability. Otherwise, node s knows that the packet has not been received by node v. By using a hash table, the work in [15] makes the false positive probability very low, about 10^{-10}. In Fig. 6(a), y_v is a randomly chosen vector from the null space of the received packets by node v. If node s multiplies the y_v vector with each of the vectors representing it's own packets, the result will be zero for the following two vectors, $4X_1 + X_2 + 2X_3$ and $4X_1 + 5X_2 + X_3$. Therefore, node s will conclude that these packets have been received by node v. On the other hand, the result of multiplying y_v with the vectors representing the other coded packets sent by s will be non-zero, and node s will conclude that these packets have not been received by node v.

In order to integrate the coded-feedback approach with

---

[1]Note that the example here is for illustrative purpose. That is why we use the negative sign for the vector in the null space. In reality, the elements of the vectors in the null space will be positive, and their values depend on the size of the finite field.

our algorithm, we first relax the Lagrange function by removing the constraints in (2) from the objective function of the dual problem, and we keep them in the constraints. The new Lagrange function becomes:

$$\sum_{i=1}^{N}[U_i(R_i) - \sum_{d:d\in D_i} q_{s_id}^i R_i] + \sum_{u,i}\sum_{v}(\sum_{\substack{d:d\in D_i \\ d\in RC(u,i)}} q_{ud}^i$$

$$- \sum_{\substack{d:d\in D_i \\ d\in RC(v,i)}} q_{vd}^i)X_{uv}^i$$

subject to (2).

Therefore, we end up with one type of queue (dual variables) q_{ud}^i. Also, the new dual problem is separable, and we have the same source algorithm with the following modified intermediate node algorithm:

Intermediate Node Algorithm: Each intermediate node u selects the packets to send and the number of credits for each session to transfer to each of its next-hop nodes at time t by solving the following optimization problem:

$$\{X_{uv}^i(t)\}_{v\in V, i\in\{1,\dots,N\}} = \arg\max_{\mathbf{X}} \sum_{i=1}^{N}\sum_{v}(\sum_{\substack{d:d\in D_i \\ d\in RC(u,i)}} q_{ud}^i(t) \tag{8}$$

$$- \sum_{\substack{d:d\in D_i \\ d\in RC(u,i)}} q_{vd}^i(t))X_{uv}^i \tag{9}$$

Subject to:

$$\sum_{\substack{v:v\in J \\ d\in RC(v,i)}} X_{uv}^i \leq \alpha_u^i R_{uJ}$$

$$\forall i, \forall(u,J), \forall d\in D_i, d\in RC(u,i) \tag{10}$$

The relay node u has to perform two decisions that lead to maximizing (8), subject to (10).

- It has to decide the session that the packet should be sent from.

- It has to also decide the number of credits to be assigned to each next-hop node.

To perform the first decision optimally, the relay node should choose session i^* that achieves the maximum value for the following among all of the sessions.

$$\{X_{uv}^i(t)\} =$$

$$\arg\max_{\mathbf{X}} \sum_{v\in O(u,i)} \left(\sum_{\substack{d:d\in D_i \\ d\in RC(u,i)}} q_{ud}^i(t) - \sum_{\substack{d:d\in D_i \\ d\in RC(v,i)}} q_{vd}^i(t)\right) X_{uv}^i \tag{11}$$

Subject to:

$$\sum_{\substack{d\in RC(v,i) \\ v:v\in J}} X_{uv}^i \leq \alpha_u^i R_{uJ} \tag{12}$$

$$\forall i, \forall(u,J), \forall d\in D_i, d\in RC(J,i)$$

To do so, for every session i, node u ranks the next-hop nodes v according to the backlog difference $(\sum_{\substack{d:d\in D_i \\ d\in RC(u,i)}} q_{ud}^i -$

$\sum_{\substack{d:d\in D_i \\ d\in RC(v,i)}} q_{vd}^i)$. Then, it gives as many virtual credits[2] to this next-hop node, subject to (12). For every sent packet, next-hop node v gets a virtual credit if node v has received the packet, and no other node w has received the packet, such that (1) w has a higher backlog difference and (2) w has a common receiver, i.e., $RC(u,i) \cap RC(w,i) \neq \phi$. This can be checked by using the coded feedback approach. Let us denote the virtual credit for session i and node v by Z_v^i; then, node u calculates $w_i = \sum_v((\sum_{\substack{d:d\in D_i \\ d\in RC(u,i)}} q_{ud}^i - \sum_{\substack{d:d\in D_i \\ d\in RC(v,i)}} q_{vd}^i)Z_v^i)$, such that all of the v nodes have positive backlog differences. Then, node u selects the session that achieves the maximum w_i. Algorithm 1 describes the above strategy.

Algorithm 1 Selecting the packet to send

1: $Z_v^i \leftarrow 0, \forall i, \forall v$
2: **for** $i \leftarrow 1$ N **do**
3:     Sort next-hop nodes according to $(\sum q_{ud}^i - \sum q_{vd}^i)$.
4:     Remove the nodes with negative backlog
5:     set T to the remaining nodes
6:     **for** Each sent packet P **do**
7:         set S to each node v in T such that $y_v^{iT} * P$ is zero.
8:         **while** S is not empty **do**
9:             Choose node v with the highest non-negative backlog difference from S
10:             Remove each node w from S such that $RC(u,i) \cap RC(w,i) \neq \phi$
11:             set $Z_v^i \leftarrow Z_v^i + 1$
12:         **end while**
13:     **end for**
14:     set $w_i \leftarrow \sum_v Z_v^i(\sum_{\substack{d:d\in D_i \\ d\in RC(u,i)}} q_{ud}^i - \sum_{\substack{d:d\in D_i \\ d\in RC(v,i)}} q_{vd}^i)^+$
15: **end for**
16: select $i^* \leftarrow \arg\max_i w_i$.
17: send a packet from session i^*

Node u can perform the second decision by assigning the credits to next-hop nodes in a batch-by-batch manner. Therefore, for each session i, node u keeps sending packets from the batch with the smallest index until it makes sure that for each receiver $d \in RC(u,i)$, the total number of linearly independent packets from that batch, received by the next-hop nodes that have paths to d is no less that the total credit it is assigned for that batch. At that time, this node assigns the credits for next-hop nodes and moves to the next batch of that session. Note that this approach might increase the delay of individual packets, but the total throughput is not affected if the size of the file is very large. This is because the source does not wait for the receiver to decode the batch in order to move to the next batch. As long as next-hop nodes are assigned credits for the batch, the node moves to the next batch. This approach for delaying the sending of packets until enough feedback has been received, is used in different works and is shown to achieve the capacity under specific conditions [22, 24].

Every time node u receives a vector in the null space from the next-hop node, it multiplies that vector with all of the

---

[2]Note that these are different from the actual credits that will be distributed as a strategy for the second decision the node has to perform. These credits are just for knowing the packet of which session should be sent.

Algorithm 2 Credits assignment algorithm for session i

1: Set $C_v^i \leftarrow 0, \forall v$
2: **for** Each sent packet P **do**
3:    $S \leftarrow$ each nex-hop node v with positive back-log difference and with $y_v^{iT} P = 0$ and $C_v^i \leq C_u^i$
4:    Sort nodes in S with respect to the backlog difference.
5:    **while** $S \neq \phi$ **do**
6:       Choose node v from S with the highest back-log difference.
7:       Set $C_v^i \leftarrow C_v^i + 1$
8:       Remove each node w from S such that $\exists d \in RC(u,i)$ s.t. $RC(v,i) \cap RC(w,i) = d$
9:    **end while**
10: **end for**

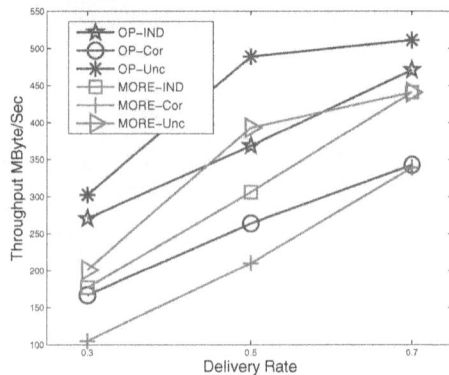

Figure 7: Simulation results for the topology in Fig. 1(a).

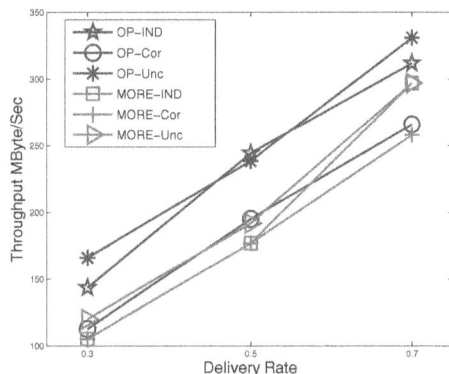

Figure 8: Simulation results for the topology in Fig. 1(b).

packets it has sent so that it can know the number of linearly independent packets that has been received by next-hop nodes. For each $d \in RC(u,i)$, once that rank for all next-hop nodes that have paths to d becomes equal to or greater than the number of credits assigned for that batch at that node, the node distributes its credits to next-hop nodes in a fashion similar to Algorithm 1. However, this time the node only focuses on one session i, and the credits that are assigned are real not virtual credits. Algorithm 2 represents the credit assignment algorithm. In the algorithm, C_u^i represents the total credits assigned to node u.

6.3 Details of the Practical Protocol

So far, we have identified the structure of the optimal solution and discussed the major challenges that face its implementation. We have then designed a back-pressure algorithm that uses the coded feedback approach to resolve these challenges. In this section, we outline the details of implementing our algorithm under practical settings.

In our protocol, every node maintains the following information: the received and sent coded packets, the available number of credits, and the batch and session number of the received and sent packets. We adopt a packet format similar to [6, 15], such that each packet has 1500 bytes of data. The packet also contains the coefficients of the coding vector along with its session and batch numbers. The packet contains the three most recent batches it has received, each with a vector from the null space of the packets in that batch. The packet contains the number of currently queued credits. The packets also contain the number of credits assigned to each next-hop node and the batch number for these credits. As can be seen, the overhead is about 1-2%, which is very small.

In every time slot, the source node computes the source rate according to (3). This adds more credits to the source node's queue. The source node moves to the next batch when the number of credits assigned to the current batch equals the size of the batch. When an intermediate node transmits a packet, it fills the null space fields with randomly chosen vectors from the null spaces of the batches it is currently receiving coded packets from. Note that the coded packets that the node sends, and the null space vectors that the node generates at any given time, could be for different batches. A node keeps sending packets from a batch until it makes sure using the coded feedback approach, that for each destination the total number of linearly independent packets received by all next-hop nodes that can reach that destina-

tion, is no less than the number of credits it has for this batch. At that time, the node transfers the credits to the next-hop nodes using Algorithm 1. An intermediate node sends the credit assignment information for the last three batches it has made assignments for. Note that there is a very small feedback or credit assignment overhead due to the integration with the data packets. Also, assigning the credits to next-hop nodes when enough packets have been sent from a batch serves two purposes. Firstly, it gives enough time for the feedback packets and the credits to reach the intended nodes due to the lossy behavior of the links. Secondly, it allows Algorithm 2 to find the optimal credit assignments.

7. EVALUATION

In this section, we provide simulation results to illustrate the effectiveness of our protocol over the state-of-the-art opportunistic routing-based multicast protocol, MORE [6]. We start by showing results for the illustrative topologies presented in the introduction section, and then we show results for a 4×4 grid network. We develop our simulations using MATLAB. We also use similar values of parameters to MORE [6] to make a fair comparison.

7.1 Results on Illustrative Topologies

We simulate both MORE and our protocol on the topology in Fig. 1(a). We choose the following values for the delivery rate of the links 0.3, 0.5, and 0.7. We also vary the correlation among the links according to the κ-factor in [20]: we set this parameter to $-1, 0$, and 1 respectively. When the value of $\kappa = -1$, this represents the uncorrelated links, and we add the symbol, Unc, to the name of the scheme in the plots. Similarly, when $\kappa = 0$ or 1, we add IND or Cor symbols, respectively, to the name of the scheme in the plots. We use a batch size of 32 packets, and we assume that transmission bandwidth of all of the nodes is 1500Mbytes/sec. We also use IEEE 802.11 to perform scheduling. We use the symbol, OP, to represent our protocol and MORE to represent MORE.

Fig. 7 represents the simulation results for the topology in Fig. 1(a). The results show that our protocol always results in more gain compared to MORE; also, the gain of our protocol is maximized when we have low delivery rate links. As illustrated in the figure, when the delivery rate is 0.3, the gain of our protocol is in the range of 50-75%, depending on the correlation among the links, while when the delivery rate is 0.7, the gain is in the range of 2-20%. Also, MORE does not exploit the benefit of having uncorrelated or independent broadcast links, as the gain of our protocol is maximized in these cases. This is alligned with the expectations in [20] that we have many coding opportunities under these cases which are not fully captured by MORE. Under the correlated case, there are not many coding opportunities, which justifies the small gain of our protocol over MORE. Fig. 8 shows the results for the topology in Fig. 1(b). Our protocol still has a gain of 5-45% depending on the delivery rate and the correlations, even with the limited coding opportunity caused by the bottleneck receiver d_1, as explained in the introduction.

7.2 Results on a 4×4 Grid Topology

We perform simulations on a 4×4 grid topology. We set up the link delivery rate to 0.5. We vary the correlation among the links to be the following three cases: independent, correlated, and uncorrelated. We place the source at one of the corners and place the destinations randomly on the two sides of the grid topology opposite to the source. We vary the number of destinations to 2, 4, and 6. We select the following utility function, $U(R_i) = log(R_i)$. We plot the results in Fig. 9.

As can be noted from the figure, our protocol results in a higher gain compared to MORE in all cases. The gain obtained by our protocol varies from 50% to 4-fold depending on the number of receivers and the correlations among the links. The gain increases as we increase the number of receivers. When we have two receivers, the gain varies from 50 to 90% while it increases to about 3 to 4-fold with six receivers. Also, the throughput decreases dramatically as we increase the number of receivers in the MORE case while decreases slowly in our case. The reason is that MORE works in a batch-by-batch manner. Therefore, its throughput is limited by the length of the path to the farthest receiver, while in our protocol, the throughput is limited to the min-cut max-flow between the source and the worst receiver, which has a smaller effect on the throughput than the length of the worst path. Our protocol takes advantage of the coding opportunities created by uncorrelated and in-

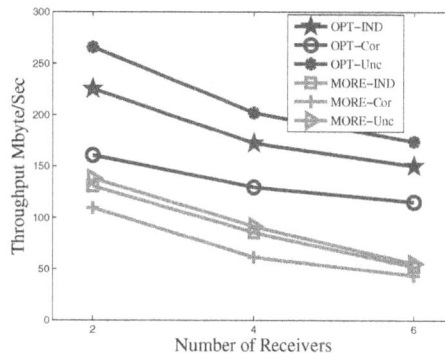

Figure 9: Simulation results for one session and different numbers of receivers on a 4×4 grid topology.

dependent links, which agrees with the conclusion in [20]. This is justified by a gain of about 70% when uncorrelated links are used compared to correlated links, and a gain of about 50% when independent links are used compared to correlated links. On the other hand, MORE throughput increases by about only 30% when uncorrelated or independent links are used compared to the correlated case.

8. CONCLUSION

In this paper we tackle the problem of optimal network coding-based opportunistic routing for multicast, which has received less attention from the community compared to the unicast problem. We identify the factors that affect the optimal solution, which are the delivery rates of the links, the correlations among the links presented recently in [20], and the reachability of the nodes to the different destinations. We formulate the problem as an optimization problem and show that it achieves the maximum possible rate by using mapping from wireless to wireline framework. We then develop a distributed solution based on the duality approach. We integrate our solution with the coded feedback approach so that it can be implemented with a delayed and lossy feedback environment. We evaluate our protocol by using simulations which show the effectiveness of our protocol.

9. ACKNOWLEDGMENT

This research was supported in part by NSF grants ECCS 1128209, CNS 1065444, CCF 1028167, CNS 0948184, and CCF 0830289.

References

[1] R. Ahlswede, N. Cai, S.-Y R. Li, and R. W. Yeung. Network information flow. *IEEE Trans. on Information Theory*, 46(4):1204–1216, 2000.

[2] D. Bertsekas and J. N. Tsitsikalis. *Parallel and Distributed Computation: Numerical Methods*. Athena Scientific, 1997.

[3] S. Biswas and R. Morris. Opportunistic routing in multi-hop wireless networks. In *Proc. ACM Special Interest Group on Data Commun. (SIGCOMM), Philadelphia, PA, USA*, Sept 2005.

[4] T. Bonald and L. Massoulie. Impact of fairness on Internet performance. In *Proc. of ACM Joint International Conference on Measurement and Modeling of Computer Systems (Sigmetrics), Cambridge, MA*, June 2001.

[5] S. Boyd and L. Vandenberghe. *Convex Optimization*. Cambridge University Press, 2004.

[6] S. Chachulski, M. Jennings, S. Katti, and D. Katabi. Trading structure for randomness in wireless opportunistic routing. In *ACM Special Interest Group on Data Commun. (SIGCOMM) Kyoto, Japan,*, Aug 2007.

[7] L. Chen, T. Ho, S. H. Low, M. Chiang, and J. C. Doyle. Optimization based rate control for multicast with network coding. In *Proc. of IEEE Conference on Computer Communications (INFOCOM), Anocharage, AK*, May 2007.

[8] D. De Couto, D. Aguayo, J. Bicket, and R. Morris. A high-throughput path metric for multi-hop wireless routing,. In *ACM MobiCom. San Diego, CA,,* Sept 2003.

[9] R. Dougherty, C. Freiling, and K. Zeger. Insufficiency of linear coding in network information flow. *IEEE Trans. on Information Theory*, 51(8):2745–2759, 2005.

[10] C. Gkantsidis, W. Hu, P. Key, B. Radunovic, P. Rodriguez, and S. Gheorghiu. Multipath code casting for wireless mesh networks. In *in Proc. of ACM CoNEXT*, 2007.

[11] T. Ho, M. Médard, R. Koetter, D. Karger, M. Effros, J. Shi, and B. Leong. A random linear network coding approach to multicast. *IEEE Trans. Inform. Theory*, 52(10):4413–4430, 2006.

[12] S. Jaggi, P. Sanders, P. A. Chou, M. Effros, S. Egner, K. Jain, and L. Tolhuizen. Polynomial time algorithms for multicast network code construction. *IEEE Trans. on Information Theory*, 51(6):1973–1982, 2005.

[13] A. Khreishah, C.-C. Wang, and N.B. Shroff. Cross-layer optimization for wireless multihop networks with pairwise intersession network coding. *IEEE Journal on Selected Areas in Communications*, 27(5):606–621, 2009.

[14] D. Koutsonikolas, Y. Hu, and C. Wang. Pacifier: High-throughput, reliable multicast without crying babies in wireless mesh networks. In *28th IEEE Conference on Computer Communications (INFOCOM). Rio de Janeiro, Brazil,,* April 2009.

[15] D. Koutsonikolas, C.-C. Wang, and Y.C. Hu. CCACK: Efficient network coding based opportunistic routing through cumulative coded acknowledgments. In *in Proceedings of the 29th Conference on Computer Communications (INFOCOM), San Diego, USA,,* March 2010.

[16] A. Lehman and E. Lehman. Complexity classification of network information flow problems. In *Proceedings of ACM-SIAM SODA*, Jan 2004.

[17] X. Lin and N. Shroff. The impact of imperfect scheduling on cross-layer congestion control in wireless networks. *IEEE/ACM Trans. Networking*, 14(2):302–315, 2006.

[18] J. Park, M. Gerla, D. Lun, Y. Yunjung, and M. Medard. Codecast: a network-coding-based ad hoc multicast protocol. *IEEE Wireless Communications*, 13(5), 2006.

[19] B. Radunovic, C. Gkantsidis, P. Key, and P. Rodriguez:. Toward practical opportunistic routing with intra-session network coding for mesh networks. *IEEE/ACM Trans. Networking*, 18(2):420–433, 2010.

[20] K. Srinivasan, M. Jain, J. Choi, T. Azim, E. Kim, P. Levis, and B. Krishnamachari. The κ-factor: Inferring protocol performance using inter-link reception correlation,. In *ACM MobiCom. Chicago, IL,,* Sept 2010.

[21] L. Tassiulas and A. Ephremides. Stability properties of constrained queueing systems and scheduling policies for maximum throughput in multihop radio networks. *IEEE Trans on Automatic Control*, 37(12), Dec 1992.

[22] C.-C. Wang. On the capacity region of 1-hop intersession network coding – a broadcast packet erasure channel analysis with general message side information,. In *in the Proceedings of IEEE ISIT*, June 2010.

[23] C.-C. Wang. Pruning network coding traffic by network coding — a new class of max-flow algorithms. *IEEE Trans. on Info. Theory*, 56(4):1909–1929, 2010.

[24] C.-C. Wang, A. Khreishah, and N.B. Shroff. On cross-layer optimizations for intersession network coding on practical 2-hop relay networks,. In *Asilomar Conference*, November 2009.

[25] Y. Wu, P. Chou, Q. Zhang, K. Jain, W. Zhu, and S.-Y. Kung. Network planning in wireless ad hoc networks: a cross-layer approach. *IEEE JSAC*, 23, 2005.

[26] Y. Wu and S.-Y. Kung. Distributed utility maximization for network coding based multicasting: a shortest path approach. *IEEE J. on Selected Areas in Communications*, 24(8):1475–1488, Apr 2006.

[27] B. Li Y. Lin and B. Liang. CodeOR: Opportunistic routing in wireless mesh networks with segmented network coding. In *in the Proceedings of the 16th IEEE International Conference on Network Protocols (ICNP 2008), Orlando, Florida*, October 2008.

[28] X. Zhang and B. Li. Optimized multipath network coding in lossy wireless networks. *IEEE Journal on Selected Areas in Communications*, 27(5):622–634, 2009.

Throughput of Rateless Codes over Broadcast Erasure Channels

Yang Yang
Department of ECE
The Ohio State University
Columbus, OH-43210, USA
yangy@ece.osu.edu

Ness B. Shroff
Department of ECE and CSE
The Ohio State University
Columbus, OH-43210, USA
shroff@ece.osu.edu

ABSTRACT

In this paper, we characterize the throughput of a broadcast network with n receivers using rateless codes with block size K. We assume that the underlying channel is a Markov modulated erasure channel that is *i.i.d.* across users, but can be correlated in time. We characterize the system throughput asymptotically in n. Specifically, we explicitly show how the throughput behaves for different values of the coding block size K as a function of n, as n approaches infinity. Under the more restrictive assumption of memoryless channels, we are able to provide a lower bound on the maximum achievable throughput for any finite values of K and n. Using simulations we show the tightness of the bound with respect to system parameters n and K, and find that its performance is significantly better than the previously known lower bound.

Categories and Subject Descriptors

C.2 [**Computer-Communication Networks**]: Miscellaneous

General Terms

Performance

Keywords

Broadcast erasure channel, Markov modulated channel, rateless erasure code, random linear network code, throughput, achievable rate

1. INTRODUCTION

In this work, we study the throughput of a wireless broadcast network with n receivers using rateless codes. In this broadcast network, channels between the transmitter and the receivers are modeled as packet erasure channels where transmitted packets may either be erased or be successfully received. This model describes a situation where packets may get lost or are not decodable at the receiver due to a variety of factors such as channel fading, interference or checksum

errors. We assume that the underlying channel is a Markov modulated packet erasure channel that is *i.i.d.* across users, but can be correlated in time. We let γ denote the steady state probability that a packet is transmitted successfully on the erasure channel.

Instead of transmitting the broadcast data packet one after another through feedback and retransmissions, we investigate a class of coding schemes called rateless codes (or fountain codes). In this coding scheme, K broadcast packets are encoded together prior to transmission. K is called the coding block size. A rateless encoder views these K packets as K input symbols and can generate an arbitrary number of output symbols (which we call coded packets) as needed until the coding block is decoded. Although some coded packets may get lost during the transmission, rateless decoder can guarantee that any $K(1 + \varepsilon)$ coded packets can recover the original K packets with high probability, where ε is a positive number that can be made arbitrarily small at the cost of coding complexity. Examples of rateless erasure codes include Raptor code [8], LT Code [4] and random linear network code [3], where the former two are used when K is very large and random linear network code is used when K is relatively small and the symbol space of packets is large. The best encoding and decoding complexity of rateless codes (e.g. Raptor codes) increase linearly as the coding block size K increases. Further, increasing the coding block size can result in large delays and large receiver buffer size. Therefore, real systems always have an upper bound on the value of K.

We consider broadcast traffic and a discrete time queueing model, where the numbers of packet arrivals over different time slots are independent and identically distributed and the packet length is a fixed value. We let λ denote the packet arrival rate and assume that the encoder waits until there are at least K packets in the queue and then encodes the first K of them as a single coding block. In this case, the largest arrival rate that can be stabilized is equal to the average number of packets that can be transmitted per slot, which we call the throughput. Therefore, we only need to characterize the throughput that can be achieved using rateless codes under parameters K and n. As described in Figure 1, the channel dynamics for the ith receiver is denoted by a stochastic process $\{X_{ij}\}_{j \in \mathbb{N}}$, where j is the index of the time slot in which one packet can be transmitted and X_{ij} is the channel state of ith receiver during the transmission of the jth packet. We capture a fairly general correlation structure by letting the current channel state be impacted by the channel states in previous l time slots, where l can be any number. As the number of receivers n approaches infinity, we show that the throughput is nonzero only if the coding block size K increases at least as fast as $\log n$. In other words, if $c \triangleq \lim_{n \to \infty} \frac{K}{\log n}$, the asymptotic[1] throughput is positive whenever $c > 0$. In Theorem 1, by utilizing large deviation techniques, we give an explicit expression of the asymptotic throughput, which is a function of K, n, γ and the channel correlation structure.

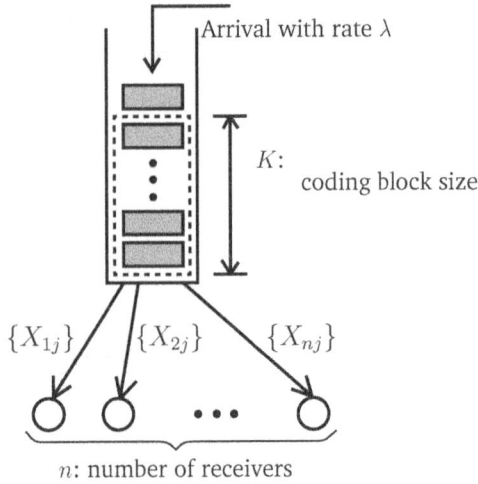

Figure 1: Broadcast with discrete time queueing model

To study the non-asymptotic behavior of the system, we make a more restrictive assumption that the channels are memoryless, which is a special case of the correlated channel model with $l = 0$. In other words, the erasure probability of every receiver channel at any time slot is $1 - \gamma$. In this case, we show that, when $\frac{K}{\log n}$ is kept to be a constant, the throughput will follow a decreasing pattern as the number of receivers n increases. By combining this result with the characterization of the asymptotic throughput, we are able to provide a lower bound on the maximum achievable throughput for any finite values of K and n. This lower bound captures the asymptotic throughput in the sense that when n approaches infinity, it coincides with the asymptotic throughput.

1.1 Related Work

Among the works that investigate the throughput over erasure channels, [1], [9] and [7] are the most relevant to this work. In [9], the authors investigate the asymptotic throughput as a function of n and K and also show that the asymptotic throughput will be non-zero only if K at least scales with $\log n$. However, they only consider the channel correlation model with $l = 1$ and use a completely different proof technique. Moreover, no explicit expression on the asymptotic throughput is provided. In [1], a lower bound is provided on the maximum achievable rate λ of this paper. However, their bound does not converge to the asymptotic throughput when n approaches infinity. Further, their bound is only valid for $K > 16$ while our result is applicable for any values of K. Moreover, our bound is shown to be better in a variety of simulation settings with finite K and n, as will be showed in Section 5. In [7], the authors consider the case when instant feedback is provided from every user after the transmission of

each decoded packets while we only assume that feedback is provided after the entire coding block has been decoded.

1.2 Key Contributions

The main contributions of this work are summarized as follows:

- We give an explicit expression of the asymptotic throughput of the system when the number of receivers n approaches infinity with different scales of K and n under the erasure channel with any levels of correlation. (Theorem 1)

- Under a more restrictive assumption that channels are memoryless, we reveal that when K grows with n in a way that the ratio $\frac{K}{\log n}$ is kept a constant, the throughput will follow a decreasing pattern as n increases, which tells us that for a quadratic increase of network size n, we need to have the coding block size K more than doubled in order to get the same throughput. (Theorem 2)

- We provide a lower bound on the maximum achievable throughput for any finite values of K and n under the memoryless channel assumption and show that its performance is significantly better than the previously known bound in [1]. (Theorem 3)

The rest of this paper is organized as follows. In Section 2 we describe our model and assumptions. In Section 3 we give the characterization of asymptotic throughput. In Section 4 we provide a lower bound on the maximum achievable rate for any finite values of K and n. In Section 5 we use simulations to verify our theoretical results. Detailed proofs on all the theorems can be found in Section 6. Finally, in Section 7, we conclude the paper.

2. SYSTEM MODEL

We consider a broadcast channel with n receivers. Time is slotted, and the numbers of broadcast packet arrivals over different time slots are i.i.d. with finite variance. We denote the expected number of packet arrivals per slot as the packet arrival rate λ. The transmission starts when there are more than K packets waiting in the incoming queue intended for all the receivers. Instead of transmitting these packets one after another using feedback and retransmissions, we view each data packet as a symbol and encode the first K of them into an arbitrary number of coded symbols as needed using rateless code (For example, Raptor Code [8] or random linear network code [3]) until the coding block is decoded. These K packets together form a single coding block with K being called block size. During the transmission, the coded symbols are transmitted one after another.

Each receiver send an ACK feedback signal after it has successfully decoded the K packets. In the following context, the term *packet* and *symbol* are used interchangeably.

We model the broadcast channel as a slotted broadcast packet erasure channel where one packet can be transmitted per slot. The channel dynamics can be represented by a stochastic process $\{X_{ij}\}_{1 \leq i \leq n, j \in \mathbb{N}}$, where X_{ij} is the state of channel between transmitter and the ith receiver during the transmission of jth packet (we also call it the jth time slot in the ith

---

[1]the asymptotic is with respect to increasing the number of receivers n

channel), which is given by

$$
X_{ij} = \begin{cases} 1 & j^{\text{th}} \text{ packet in the } i^{\text{th}} \text{ channel is} \\ & \text{successfully received} \\ 0 & \text{otherwise} \end{cases} .
$$

We assume that the dynamics of the channels for different receivers are independent and identical. More precisely, for all $1 \leq i \leq n$, $\{X_{ij}\}_{j \geq 1}$ are independent and identical processes.

Since, in practice, the channel dynamics are often temporarily correlated, we investigate the situation where the current channel state distribution depends on the channel states in the preceding l time slots. More specifically, for $\mathcal{F}_{im} = \{X_{ij}\}_{j \leq m}$ and fixed l, we define $\mathcal{H}_{im} = \{X_{im}, \ldots, X_{i(m-l+1)}\}$ for $m \geq l \geq 1$ with $\mathcal{H}_{im} = \{\varnothing, \Omega\}$ for $l = 0$, and assume that $\mathbb{P}[X_{i(m+1)} = 1|\mathcal{F}_{im}] = \mathbb{P}[X_{i(m+1)} = 1|\mathcal{H}_{im}]$ for all $m \geq l$. To put it another way, when $l \geq 1$, the state $(X_{im}, \ldots, X_{i(m-l+1)})$, $m \geq l$ forms a Markov chain. Denote by Π the transition matrix of the Markov chain $\{(X_{im}, \ldots, X_{i(m-l+1)})\}_{m \geq l}$, where

$$
\Pi = [\pi(s,u)]_{s,u \in \{0,1\}^l},
$$

with $\pi(s,u)$ being the one-step transition probability from state s to state u. Throughout this paper, we assume that Π is irreducible and aperiodic, which ensures that this Markov chain is ergodic [6]. Therefore, for any initial value $\mathcal{H}_l$, the parameter γ_i is well defined and given by

$$
\gamma_i = \lim_{m \to \infty} \mathbb{P}[X_{im} = 1],
$$

and, from the ergodic theorem [6] we know

$$
\mathbb{P}\left[\lim_{m \to \infty} \frac{\sum_{j=1}^{m} X_{ij}}{m} = \gamma_i\right] = 1.
$$

Since $\{X_{ij}\}_{j \geq 1}$ for all $1 \leq i \leq n$ are i.i.d., we denote $\gamma = \gamma_i$, for all $1 \leq i \leq n$.

Using near optimal rateless codes, such as Raptor Code [8], LT Code [4] and random linear network code [3], only slightly more than K coded symbols are needed to decode the whole coding block. For simplicity, here we assume that any combination of K coded symbols can lead to a successful decoding of the K packets.

According to the above system model, we have the following definitions:

DEFINITION 1. *The number of time slots (number of transmitted coded symbols) needed for user i to successfully decode K packets is defined as*

$$
T_i(K) = \min_{m} \left\{ \sum_{j=1}^{m} X_{ij} \geq K \right\}.
$$

DEFINITION 2. *The number of time slots (number of transmitted coded symbols) needed to complete the transmission of a single coding block to all the receivers is defined as*

$$
T(n,K) = \max\{T_i(K), i = 1, 2, \ldots, n\}.
$$

DEFINITION 3 (INITIAL STATE). *Since the current channel state depends on the channel states in the previous l time-slots, for each receiver i, by assuming that the system starts at time slot 1, we define the initial state of receiver i as*

$$
\mathcal{E}_i = \{X_{i(-l+1)}, X_{i(-l+2)}, \ldots, X_{i0}\}.
$$

The initial states for all the receivers is then denoted as $\mathcal{E} \triangleq \cup_{i=1}^{n} \mathcal{E}_i$.

DEFINITION 4 (THROUGHPUT). *Under an initial state $\mathcal{E}$, the average number of packets that can be successfully transmitted per slot is defined as*

$$
\eta(n,K,\mathcal{E}) = \frac{K}{\mathbb{E}[T(n,K)|\mathcal{E}]},
$$

which we call the throughput under initial condition $\mathcal{E}$. In the special case where the channel states are i.i.d. ($l = 0$), we know that $\mathcal{E} = \varnothing$ and we can denote the throughput as

$$
\eta(n,K) = \frac{K}{\mathbb{E}[T(n,K)]}. \tag{1}
$$

3. ASYMPTOTIC THROUGHPUT

Before presenting the main results, we need to introduce some necessary definitions. First, define a mapping f from the state space of the Markov chain $\{0,1\}^l$ to $\{0,1\}$ as

$$
f((X_{im}, \ldots, X_{i(m-l+1)})) = X_{im}.
$$

Then, given a real number θ, we define a matrix Π_θ as

$$
\Pi_\theta = \begin{cases} \left[\pi(s,u)e^{\theta f(u)}\right]_{s,u \in \{0,1\}^l} & \text{when } l \geq 1 \\ [\gamma e^\theta] & \text{when } l = 0 \end{cases} .
$$

Last, define a standard large deviation rate function $\Lambda(\beta, \Pi)$ as

$$
\Lambda(\beta, \Pi) = \sup_\theta \{\theta\beta - \log \rho(\Pi_\theta)\}, \tag{2}
$$

where $\rho(\Pi_\theta)$ denotes the Perron-Frobenious eigenvalue of Π_θ (See Theorem 3.1.1 in [2]), which is the largest eigenvalue of Π_θ.

The asymptotic throughput for any values of K as a function of n under any initial condition $\mathcal{E}$ is characterized by the theorem below:

THEOREM 1. *Assume that K is a function of n and the value of $\lim_{n \to \infty} \frac{K}{\log n}$ exists, which we denote as $c \triangleq \lim_{n \to \infty} \frac{K}{\log n}$, then for any initial state $\mathcal{E}$ we have*

$$
\lim_{n \to \infty} \eta(n,K,\mathcal{E}) = \sup\left\{\beta \,\middle|\, c \geq \frac{\beta}{\Lambda(\beta, \Pi)}, 0 \leq \beta < \gamma\right\}. \tag{3}
$$

PROOF. see Section 6.1. □

From Theorem 1, we know that, if the coding block size K is set to be a function of the network size n, then we can characterize the asymptotic throughput when n approaches infinity in an explicit form. Equation (3) shows that the asymptotic throughput is irrelevant to the initial state $\mathcal{E}$ and is a function of γ, $\lim_{n \to \infty} \frac{K}{\log n}$ and the channel correlation structure indicated by Π.

By Theorem 1, the asymptotic throughput in the special cases when $K \in o(\log n)$ and $K \in \omega(\log n)$ are given in the following corollary.

COROLLARY 1.1. *Assume K is a function of n, for any initial state $\mathcal{E}$ we have*

1. if $K \in o(\log n)$, then[2]

$$
\lim_{n \to \infty} \eta(n,K,\mathcal{E}) = 0.
$$

---

[2]We use standard notations: $f(n) = o(g(n))$ if $\lim_{n \to \infty} \frac{f(n)}{g(n)} = 0$ and $f(n) = \omega(g(n))$ if $\lim_{n \to \infty} \frac{f(n)}{g(n)}$ diverges

2. if $K \in \omega(\log n)$, then

$$\lim_{n \to \infty} \eta(n, K, \mathcal{E}) = \gamma.$$

PROOF. 1) If $K \in o(\log n)$, then $c = \lim_{n \to \infty} \frac{K}{\log n} = 0$ and we have

$$\left\{ \beta \middle| c \geq \frac{\beta}{\Lambda(\beta, \Pi)}, 0 \leq \beta < \gamma \right\} = \{0\}.$$

According to Theorem 1, we get

$$\lim_{n \to \infty} \eta(n, K, \mathcal{E}) = \sup\{0\} = 0.$$

2) If $K \in \omega(\log n)$, then $c = \lim_{n \to \infty} \frac{K}{\log n} = \infty$ and we have

$$\left\{ \beta \middle| c \geq \frac{\beta}{\Lambda(\beta, \Pi)}, 0 \leq \beta < \gamma \right\} = [0, \gamma).$$

According to Theorem 1, we get

$$\lim_{n \to \infty} \eta(n, K, \mathcal{E}) = \sup[0, \gamma) = \gamma.$$

$\square$

Corollary 1.1 says that the throughput will vanish to 0 as the increase of n, when K does not scale as fast as $\log n$. Whereas when K scales faster than $\log n$ (Or more specifically, when $K \in \omega(\log n)$), the capacity of this system γ can be approached. It should be noted that Theorem 1, together with Corollary 1.1, are a generalized version of Theorem 1 in [9], which only consider the case when $l = 1$ and does not give the explicit expression of asymptotic throughput.

For the special case when the channels are memoryless ($l = 0$), we can compute $\Lambda(\beta, \Pi)$ explicitly, as shown in the corollary below

COROLLARY 1.2. *Assume that K is function of n and the channels are memoryless ($l = 0$), we have*
if $\lim_{n \to \infty} \frac{K}{\log n} = c$, where c is a positive constant, then

$$\lim_{n \to \infty} \eta(n, K) =$$
$$\sup \left\{ \beta \middle| \log \frac{\beta}{\gamma} + \frac{1 - \beta}{\beta} \log \frac{1 - \beta}{1 - \gamma} \geq \frac{1}{c}, 0 \leq \beta < \gamma \right\}. \quad (4)$$

PROOF. When $l = 0$, $\Pi_\theta = [\gamma e^\theta]$ is a degenerate matrix with a single entry and $\rho(\Pi_\theta) = \rho(\gamma e^\theta) = \gamma e^\theta$. Therefore we have, according to Equation (2)

$$\Lambda(\beta, \Pi) = \beta \log \frac{\beta}{\gamma} + (1 - \beta) \log \frac{1 - \beta}{1 - \gamma}.$$

$\square$

4. MAXIMUM ACHIEVABLE THROUGH-PUT

For all rateless coding schemes, the encoding and decoding complexity increases linearly in K, the size of the coding block. Moreover, the value of K determines the receiver buffer size. Therefore, in reality, the value of K is often limited by the decoder buffer size or the computational power of both sender and receiver. Then we have to consider the case when K is finite and need to answer the following questions: For a given number of receivers n, channel statistics, and a maximum available coding block size K, what is maximum packet arrival rate λ that can be supported by this given system? For a specific number of receivers and channel statistics,

if we are given a target packet arrival rate λ, how can we design the value of K in the system such that the target arrival rate can be supported?

In order to answer these questions, we make a more restrictive assumption in this section that channels are memoryless, meaning that the channel states are i.i.d. across different time slots. Based on this assumption, in the theorem below, we are able to show that when we keep the value of $K/\log n$ to be a constant as we increase K or n, the throughput will follow a decreasing pattern. We leave the case when $l > 0$ for future study.

THEOREM 2. *When the channels are memoryless ($l = 0$), for any $n \in \mathbb{N}$, $K \in \mathbb{N}$ and $\alpha \in \mathbb{N}$, we have*

$$\eta(n^\alpha, \alpha K) < \eta(n, K).$$

REMARK 2.1. *While Theorem 1 tells us that in order to achieve a nonzero throughput, we can double the coding block size K for every quadratic increase of n, which is to make $K/\log n$ a fixed value, it does not tell us anything about how the throughput will converge as n approaches infinity. This theorem indicates that under the memoryless channel assumption, if we adapt the coding block size K with the increase of network size n in a way that $K/\log n$ is kept as a fixed value, then the throughput will follow a decreasing pattern before it reaches the asymptotic throughput.*

REMARK 2.2. *For the case when the channels are correlated, the expected transmission time for a single coding block varies under different initial states $\mathcal{E}$. It is technically difficult to obtain the steady state distribution of initial state $\mathcal{E}$ across different transmissions. Therefore, it is not clear whether this inequality will hold for $l > 0$.*

PROOF. see Section 6.2. $\square$

By the help of the above theorem, we can get a lower bound on the maximum stable throughput that can be achieved for any finite values of coding block size K and network size n, as shown in the theorem below.

THEOREM 3. *For a broadcast network with n receivers, coding block size K and packet arrive rate λ, when the erasure broadcast channels are memoryless ($l = 0$) with erasure probability $1 - \gamma$, the system is stable if*

$$\lambda \leq \mathcal{R} \left(\frac{\log n}{K} \right),$$

where

$$\mathcal{R}(r) = \sup \left\{ \beta \middle| \log \frac{\beta}{\gamma} + \frac{1 - \beta}{\beta} \log \frac{1 - \beta}{1 - \gamma} \geq r, 0 \leq \beta < \gamma \right\}.$$

PROOF. From Equation (10) in Lemma 1 we can see that when K and n are finite, the transmission time of a coding block $T(n, K)$ is light-tail distributed, meaning that it has finite variance. Then according to [5] we know that the queue will be stable if the traffic intensity of this queue, which is defined as the packet arrival rate λ over the service rate, is less than 1. Therefore, the queue will be stable if the arrival rate λ satisfies

$$\lambda < \sup \left\{ \mu \middle| \frac{\mu}{K/\mathbb{E}[T(n, K)]} < 1 \right\}$$
$$= \frac{K}{\mathbb{E}[T(n, K)]} = \eta(n, K). \quad (5)$$

By Theorem 2 we know that $\eta(n, K) > \eta(n^\alpha, \alpha K)$ for any integer values of α, therefore we have

$$\eta(n, K) > \lim_{\alpha \to \infty} \eta(n^\alpha, \alpha K). \qquad (6)$$

Since $\frac{\alpha K}{\log n^\alpha} = \frac{K}{\log n}$ for any value of α, then by applying Theorem 1 and Corollary 1.2, we get

$$\lim_{\alpha \to \infty} \eta(n^\alpha, \alpha K) =$$
$$\sup \left\{ \beta \left| \log \frac{\beta}{\gamma} + \frac{1-\beta}{\beta} \log \frac{1-\beta}{1-\gamma} \geq \frac{\log n}{K}, 0 \leq \beta < \gamma \right. \right\}$$
$$= \mathcal{R}\left(\frac{\log n}{K} \right),$$

which, by combining Equation (5) and Equation (6), completes the proof. $\square$

In order to compare this lower bound on the maximum achievable rate with the existing bound given in [1], we restate Theorem 2 in [1] as the following

THEOREM 4 (THEOREM 2 IN [1]). *For a broadcast network with n receivers, coding block length $K > 16$ and packet arrive rate λ, when the erasure broadcast channels are memoryless ($l = 0$) with erasure probability $1 - \gamma$, the system is stable if*

$$\lambda < \frac{(1-\gamma)K}{K + (\log n + 0.78)\sqrt{K} + 2.61}. \qquad (7)$$

For the ease of notation let us denote the bounds given in Theorem 4 as *CSE bound* using the initials of the authors' last name.

Firstly we should notice that CSE bound is only valid when $K > 16$, while there is no such restriction for our bound. Secondly, our bound converges to the asymptotic throughput in the sense that as n approaches infinity while keeping $K/\log n$ as a constant c, our bound on the maximum achievable rate will converge to the asymptotic throughput with parameter c. Or more specifically,

$$\lim_{n \to \infty} \eta(n, K) = \lim_{n \to \infty} \mathcal{R}\left(\frac{\log n}{K} \right) = \mathcal{R}\left(\frac{1}{c} \right), \qquad (8)$$

which can be seen from Theorem 1 and Theorem 3. However, in the CSE bound, when we keep the ratio $K/\log n$ to be a constant c, as n or K approaches infinity, the bound will become trial (approach 0), which can be seen from the equation below.

$$\lim_{K \to \infty} \frac{(1-\gamma)K}{K + (\log n + 0.78)\sqrt{K} + 2.61}$$
$$= \lim_{K \to \infty} \frac{(1-\gamma)}{1 + (1/c + 0.78/K)\sqrt{K} + 2.61/K}$$
$$= 0. \qquad (9)$$

Next, in Section 5, we show that our bound outperforms CSE bound under various simulation settings.

5. SIMULATION

In this Section, we conduct simulation experiments to verify our main results.

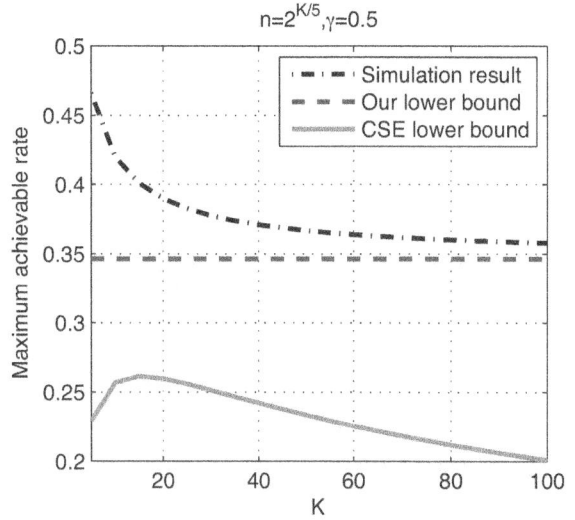

Figure 2: Illustration for example 1

5.1 Example 1

This example verifies both Theorem 1 and Theorem 2. We choose a memoryless channel with $\gamma = 0.5$. By keeping $K/\log n$ as a constant $5/\log 2$, we change K from 5 to 100 and calculate the maximum achievable rate, which is $\eta(n, K)$, through simulations for each pair of (K, n). Since the value of our bound is a function of the ratio $K/\log n$, in this case, it is a constant for all K and is equal to the asymptotic throughput with a parameter $5/\log 2$. From Figure 2 we can see that as K approaches infinity, the maximum achievable rate converges to our lower bound (which is also the asymptotic throughput in this case) in a decreasing manner, which validates both Theorem 1 and Theorem 2.

In this case, we also plotted the CSE lower bound given by Theorem 4 and we can see that CSE bound gradually approaches zero as indicated by Equation (9) while our bound is a constant value and is asymptotically tight.

5.2 Example 2

In this example, we conduct three set of experiments with different values of K as a function of n and show that our bound outperforms CSE bound in all these simulation settings.

In the first case, we set the coding block size K to be the same as the network size n and change n from 5 to 100. We plot the simulation result of the maximum achievable rate as well as our bound and CSE bound in Figure 3a, since in this case K scales faster than $\log n$, the achievable rate will approach system capacity γ as the network size n grows.

In the second case, we assume that the number of receivers is fixed to be 10 and we increase coding block size K from 5 to 100. The simulations result, together with the two bounds, are plotted in Figure 3b. In this case, the achievable rate will also approach system capacity γ as n increases.

In the final case, as shown in Figure 3c, we keep the coding block size to be a constant 20 and increase the number of receivers from 5 to 100. Since K does not increase with $\log n$ at all, the achievable rate will vanish to 0 as n grows.

From Figure 2, 3a, 3b and 3c, we can see that our bound

| (a) n=K | (b) n=10 | (c) K=20 |

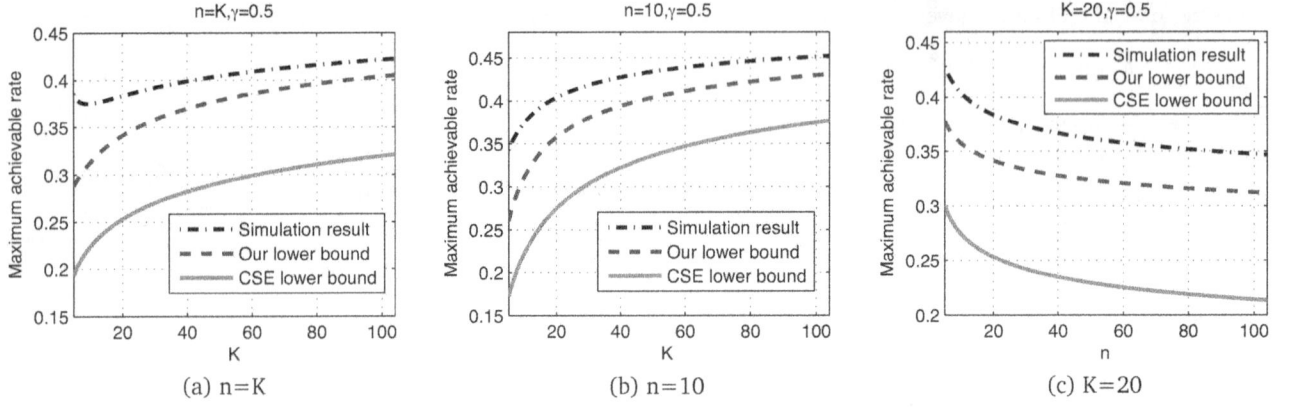

Figure 3: Illustration for example 2

obtained by Theorem 3 is significantly better than the lower bound achieved in [1] in all these four different cases.

6. PROOFS

6.1 Proof of Theorem 1

In order to prove Theorem 1, we first need the following lemmas (Lemma 1, Lemma 2 and Lemma 3).

LEMMA 1. *For any $\beta \in (0,1)$ and any values of $\mathcal{E}$, we have*

$$\mathbb{P}\left[T(n,K) > \frac{k}{\beta}\bigg|K = k, \mathcal{E}\right]$$
$$= 1 - \left(1 - e^{-\frac{k}{\beta}\Lambda(\beta,\Pi)\mathbf{1}(\beta<\gamma)+g(\beta,k,\mathcal{E})}\right)^n, \quad (10)$$

where

$$g(\beta,k,\mathcal{E}) \in \begin{cases} o(k) \text{ as } k \to \infty & \text{if } \beta < \gamma \\ o(1) \text{ as } k \to \infty & \text{if } \beta > \gamma \end{cases}.$$

PROOF OF LEMMA (1). From definition (1) and (2), we have, for any t,

$$\{T(n,K) \le t, \mathcal{E}\} = \bigcap_{i=1}^{n} \{T_i(K) \le t, \mathcal{E}_i\}.$$

Therefore, we have

$$\mathbb{P}[T(n,K) > t|K = k, \mathcal{E}]$$
$$= 1 - \mathbb{P}[T(n,K) \le t|K = k, \mathcal{E}]$$
$$= 1 - \prod_{i=1}^{n}(1 - \mathbb{P}[T_i(K) > t|K = k, \mathcal{E}_i]). \quad (11)$$

Let $t = \frac{k}{\beta}$, from definition 1 we can get, for any $1 \le i \le n$,

$$\mathbb{P}\left[T_i(K) > \frac{k}{\beta}\bigg|K = k, \mathcal{E}_i\right] = \mathbb{P}\left[\sum_{j=1}^{k/\beta} X_{ij} < k\bigg|\mathcal{E}_i\right]$$
$$= \mathbb{P}\left[\frac{\sum_{j=1}^{k/\beta} X_{ij}}{k/\beta} < \beta\bigg|\mathcal{E}_i\right].$$

and

$$\lim_{k \to \infty} \frac{\log \mathbb{P}\left[T_i(K) > \frac{k}{\beta}\bigg|K = k, \mathcal{E}_i\right]}{k/\beta} = -\Lambda(\beta,\Pi)\mathbf{1}(\beta < \gamma), \quad (12)$$

with the last equation being a direct application of Theorem 3.1.2 in [2] (Gärtner-Ellis Theorem for finite state Markov chains). Notice that the right hand side of Equation (12) is fixed for all possible values of i and $\mathcal{E}_i$ as long as the values of β and Π are fixed. Then the proof completes by combining (11) and (12). $\square$

LEMMA 2. *Assume k is a function of n and denote $k := k(n)$, and define $f(k,\beta,\mathcal{E}) := e^{\frac{k}{\beta}\Lambda(\beta,\Pi)\mathbf{1}(\beta<\gamma)-g(\beta,k,\mathcal{E})}$, then we have*

1. *For a fixed $\beta \in (0,1)$, if $\lim_{n\to\infty} \frac{n}{f(k(n),\beta,\mathcal{E})} = 0$, then*

$$\lim_{n\to\infty} \mathbb{P}\left[T(n,K) > \frac{k(n)}{\beta}\bigg|K = k(n), \mathcal{E}\right] = 0. \quad (13)$$

2. *For a fixed $\beta \in (0,1)$, if $\lim_{n\to\infty} \frac{n}{f(k(n),\beta,\mathcal{E})} = \infty$, then*

$$\lim_{n\to\infty} \mathbb{P}\left[T(n,K) > \frac{k(n)}{\beta}\bigg|K = k(n), \mathcal{E}\right] = 1. \quad (14)$$

PROOF OF LEMMA 2. According to Lemma 1 and the definition of $f(k(n),\beta,\mathcal{E})$, we have

$$\mathbb{P}\left[T(n,K) > \frac{k(n)}{\beta}\bigg|K = k(n), \mathcal{E}\right]$$
$$= 1 - \left(1 - \frac{1}{f(k(n),\beta,\mathcal{E})}\right)^n$$
$$= 1 - \left[\left(1 - \frac{1}{f(k(n),\beta,\mathcal{E})}\right)^{f(k(n),\beta,\mathcal{E})}\right]^{\frac{n}{f(k(n),\beta,\mathcal{E})}}.$$

Since the function $\left(1 - \frac{1}{x}\right)^x$ with domain $(1, +\infty)$ is a bounded and strictly increasing function with region $(0, e^{-1})$ and the fact that $f(k,\beta) > 1$, we know that if $\lim_{n\to\infty} \frac{n}{f(k(n),\beta,\mathcal{E})} =$

∞, then

$$\liminf_{n\to\infty} \mathbb{P}\left[T(n,K) > \frac{k(n)}{\beta}\Big|K = k(n), \mathcal{E}\right]$$

$$= 1 - \limsup_{n\to\infty}\left[\left(1 - \frac{1}{f(k(n),\beta,\mathcal{E})}\right)^{f(k(n),\beta,\mathcal{E})}\right]^{\frac{n}{f(k(n),\beta,\mathcal{E})}}$$

$$\geq 1 - \limsup_{n\to\infty} e^{-\frac{n}{f(k(n),\beta,\mathcal{E})}}$$

$$= 1,$$

which, together with the fact that $\mathbb{P}\left[T(n,K) > \frac{k(n)}{\beta}\big|K = k(n), \mathcal{E}\right] \leq 1$, yields Equation (14).

If $\lim_{n\to\infty}\frac{n}{f(k(n),\beta,\mathcal{E})} = 0$, then $f(k(n),\beta,\mathcal{E}) \to \infty$ as $n \to \infty$, which results in

$$\lim_{n\to\infty}\left(1 - \frac{1}{f(k(n),\beta,\mathcal{E})}\right)^{f(k(n),\beta,\mathcal{E})} = e^{-1}.$$

Then we can obtain

$$\limsup_{n\to\infty} \mathbb{P}\left[T(n,K) > \frac{k(n)}{\beta}\Big|K = k(n), \mathcal{E}\right]$$

$$-1 - \liminf_{n\to\infty}\left[\left(1 - \frac{1}{f(k(n),\beta,\mathcal{E})}\right)^{f(k(n),\beta,\mathcal{E})}\right]^{\frac{n}{f(k(n),\beta,\mathcal{E})}}$$

$$= 1 -$$

$$\liminf_{n\to\infty}\left[\lim_{n\to\infty}\left(1 - \frac{1}{f(k(n),\beta,\mathcal{E})}\right)^{f(k(n),\beta,\mathcal{E})}\right]^{\frac{n}{f(k(n),\beta,\mathcal{E})}}$$

$$= 1 - 1 = 0,$$

which leads to Equation (13). $\square$

LEMMA 3. *Let $\{h_n(x)\}$ be a set of Lebesgue measurable functions defined on $[0,\infty)$ and $h_n(x)$ converges to $\mathbf{1}(x < y)$ almost everywhere for some $y > 0$. If $h_n(x)$ is a decreasing function of x and have the range $[0,1]$ for any $n \in \mathbb{N}$, then $h_n(x)$ converges globally in measure to $\mathbf{1}(x < y)$.*

PROOF. Choose $\varepsilon > 0$. Since $h_n(x)$ converges to $\mathbf{1}(x < y)$ almost everywhere, for any $\delta > 0$, we can find $N \in \mathbb{N}$ such that for any $n > N$, we have

$$|h_n(y - \delta/2) - 1| < \varepsilon$$
$$|h_n(y + \delta/2) - 0| < \varepsilon.$$

Since $0 \leq h_n(x) \leq 1$ for any $x \in [0,\infty)$ and $h_n(x)$ is a decreasing function of x, we know that, for any $n > M$,

$$h_n(x) > 1 - \varepsilon \quad \forall x < y - \delta/2$$
$$h_n(x) < \varepsilon \quad \forall x > y + \delta/2.$$

Therefore, for any $n > N$,

$$\nu\left(\{|h_n(x) - \mathbf{1}(x < y)| > \varepsilon\}\right)$$
$$< \nu([y - \delta/2, y]) + \nu([y, y + \delta/2]) = \delta,$$

where ν is the Lebesgue measure. Since ε and δ are arbitrarily chosen, from the above inequality we know that $h_n(x)$ converges globally in measure to $\mathbf{1}(x < y)$. $\square$

With Lemma 1, 2 and 3 established, we now turn to the proof of Theorem 1.

PROOF OF THEOREM 1. Since K is assumed to be a function of n, we denote this function as $k(n)$. According to definition 4 we have, for any values of $\mathcal{E}$,

$$\lim_{n\to\infty}\left(\eta(n,K,\mathcal{E})\right)^{-1}$$

$$= \lim_{n\to\infty} \mathbb{E}\left[\frac{T(n,K)}{K}\Big|\mathcal{E}\right]$$

$$= \lim_{n\to\infty}\int_0^\infty \frac{\mathbb{P}[T(n,K) > s|K = k(n), \mathcal{E}]}{k(n)} ds$$

$$= \lim_{n\to\infty}\int_0^\infty \mathbb{P}[T(n,K) > k(n)u|K = k(n), \mathcal{E}] du. \quad (15)$$

According to the assumption that $\lim_{n\to\infty} k(n)/\log(n) = c$, we have

$$\lim_{n\to\infty} \frac{n}{e^{\frac{k(n)}{\beta}\Lambda(\beta,\Pi)\mathbf{1}(\beta<\gamma) - g(\beta,k,\mathcal{E})}}$$

$$= \begin{cases} 0 & c > \frac{\beta}{\Lambda(\beta,\Pi)\mathbf{1}(\beta<\gamma)} \\ \infty & c < \frac{\beta}{\Lambda(\beta,\Pi)\mathbf{1}(\beta<\gamma)} \end{cases}.$$

Since $\frac{\beta}{\Lambda(\beta,\Pi)\mathbf{1}(\beta<\gamma)}|_{\beta=0} = 0$, $\lim_{\beta\to\gamma^-}\frac{\beta}{\Lambda(\beta,\Pi)\mathbf{1}(\beta<\gamma)} = \infty$ and $\frac{\beta}{\Lambda(\beta,\Pi)\mathbf{1}(\beta<\gamma)}$ is a monotone increasing function on the domain $(0,\gamma)$, the equation $c = \frac{\beta}{\Lambda(\beta,\Pi)\mathbf{1}(\beta<\gamma)}$ has only one solution of β, which we denote as

$$\beta_c = \sup\left\{\beta\Big|c \geq \frac{\beta}{\Lambda(\beta,\Pi)}, 0 \leq \beta < \gamma\right\}.$$

Then by Lemma 2 we get

$$\lim_{n\to\infty} \mathbb{P}\left[T(n,K) > k(n)u\big|K = k(n), \mathcal{E}\right]$$

$$= \begin{cases} 1 & \text{if } u < \frac{1}{\beta_c} \\ 0 & \text{if } u > \frac{1}{\beta_c} \end{cases}. \quad (16)$$

Let us denote $h_n(u) = \mathbb{P}[T(n,K) > k(n)u|K = k(n), \mathcal{E}]$. Equation (16) implies that $h_n(u)$ converges to $\mathbf{1}(u < 1/\beta_c)$ pointwisely. Since $h_n(u)$ is a decreasing function of u and has the range $[0,1]$ for all n, by Lemma 3 we know that $h_n(u)$ globally converges in measure to $\mathbf{1}(u < 1/\beta_c)$. We also know that the set of function $\{h_n(u)\}$ is uniformly bounded. Then we can apply Vitali convergence theory to Equation (15) to exchange the limit and integral and obtain

$$\lim_{n\to\infty} \mathbb{E}\left[\frac{T(n,K)}{K}\Big|\mathcal{E}\right]$$

$$= \int_0^\infty \lim_{n\to\infty} \mathbb{P}\left[T(n,K) > k(n)u|K = k(n), \mathcal{E}\right] du = \frac{1}{\beta_c},$$

which, by combining Equation (15), completes the proof. $\square$

6.2 Proof of Theorem 2.

PROOF. When the channel states are i.i.d., according to definition 1 and 2, $T(n,K)$ and $T(n^\alpha, \alpha K)$ can be expressed as

$$T(n,K) = \max_{1 \leq i \leq n}\sum_{j=1}^{K} T_{ij}$$

$$T(n^\alpha, \alpha K) = \max_{1 \leq i \leq n^\alpha}\sum_{j=1}^{\alpha K} T_{ij},$$

where $\{T_{ij}\}_{i\in\mathbb{N}, j\in\mathbb{N}}$ are i.i.d. geometric random variables with parameter γ. Let us deonte

$$S_i^r = \sum_{j=1+(r-1)K}^{rK} T_{ij}.$$

Then we know that $\{S_i^r\}_{i\in\mathbb{N}, r\in\mathbb{N}}$ are also i.i.d. random variables. The above two equations can be rewritten as

$$T(n, K) = \max_{1\le i\le n} S_i^1 \tag{17}$$

$$T(n^\alpha, \alpha K) = \max_{1\le i\le n^\alpha} \sum_{r=1}^{\alpha} S_i^r. \tag{18}$$

Instead of viewing Equation (18) as a 1-dimensional maximization over n^α points, we can think of it as an α-dimensional maximization over n^α points where we can choose coordinate from 1 to n on each dimension and therefore can further rewrite Equation (18) as

$$T(n^\alpha, \alpha K) = \max_{1\le i_1\le n} \max_{1\le i_2\le n} \ldots \max_{1\le i_\alpha\le n} \sum_{r=1}^{\alpha} S_{(i_1, i_2, \ldots, i_\alpha)}^r, \tag{19}$$

where

$$S_{(i_1, i_2, \ldots, i_\alpha)}^r = S_{\sum_{u=1}^{\alpha} n^{u-1}(i_u - 1)+1}^r$$

and i_u can be viewed as the coordinate in the uth dimension.

Next, we are going to use Equation (19) to build a lower bound for the expection of $T(n^\alpha, \alpha K)$.

For fixed values of $i_2, i_3, \ldots, i_\alpha$, let us find a i_1^* such that

$$i_1^*(i_2, \ldots, i_\alpha) = \arg\max_{1\le i_1\le n} S_{(i_1, i_2, \ldots, i_\alpha)}^1, \tag{20}$$

which we denote as i_1^* for short. Then according to Equation (19), we can find a lower bound for $\mathbb{E}[T(n^\alpha, \alpha K)]$ by choosing $i_1 = i_1^*(i_2, \ldots, i_\alpha)$ for all possible values of $i_2, i_3, \ldots, i_\alpha$, which is

$$\mathbb{E}[T(n^\alpha, \alpha K)]$$
$$= \mathbb{E}\left[\max_{1\le i_1\le n} \max_{1\le i_2\le n} \ldots \max_{1\le i_\alpha\le n} \sum_{r=1}^{\alpha} S_{(i_1, i_2, \ldots, i_\alpha)}^r\right]$$
$$\overset{(a)}{\ge} \mathbb{E}\left[\max_{1\le i_2\le n} \ldots \max_{1\le i_\alpha\le n} \sum_{r=1}^{\alpha} S_{(i_1^*, i_2, \ldots, i_\alpha)}^r\right]$$
$$= \mathbb{E}\left[\max_{1\le i_2\le n} \ldots \max_{1\le i_\alpha\le n} \left(\sum_{r=2}^{\alpha} S_{(i_1^*, i_2, \ldots, i_\alpha)}^r + S_{(i_1^*, i_2, \ldots, i_\alpha)}^1\right)\right]. \tag{21}$$

Since the choice of i_1^* is only sub-optimal, the inequality (a) in Equation (21) should be strict inequality. Notice that according to Equation (20), for any values of $i_2, i_3, \ldots, i_\alpha$, we have

$$S_{(i_1^*, i_2, \ldots, i_\alpha)}^1 = \max_{1\le i_1\le n} S_{(i_1, i_2, \ldots, i_\alpha)}^1,$$

which, combining Equation (17) and the fact that $\{S_i^r\}$ are i.i.d. random variables, yields

$$\mathbb{E}\left[S_{(i_1^*, i_2, \ldots, i_\alpha)}^1\right] = \mathbb{E}\left[\max_{1\le i_1\le n} S_{(i_1, i_2, \ldots, i_\alpha)}^1\right]$$
$$= \mathbb{E}\left[\max_{1\le i\le n} S_i^1\right]$$
$$= \mathbb{E}[T(n, K)]. \tag{22}$$

As a second step, for any values of $i_3, i_4, \ldots, i_\alpha$, let us define i_2^* as

$$i_2^*(i_1^*, i_3, \ldots, i_\alpha) = \arg\max_{1\le i_2\le n} S_{(i_1^*, i_2, \ldots, i_\alpha)}^2.$$

Then similarly as Equation (21), by fixing i_2 to be i_2^*, we can obtain

$$\mathbb{E}[T(n^\alpha, \alpha K)]$$
$$> \mathbb{E}\left[\max_{1\le i_3\le n} \ldots \max_{1\le i_\alpha\le n} \left(\sum_{r=3}^{\alpha} S_{(i_1^*, i_2^*, \ldots, i_\alpha)}^r + S_{(i_1^*, i_2^*, \ldots, i_\alpha)}^1 + S_{(i_1^*, i_2^*, \ldots, i_\alpha)}^2\right)\right].$$

Also, for any values of $i_3, i_4 \ldots, i_\alpha$, we have

$$\mathbb{E}\left[S_{(i_1^*, i_2^*, \ldots, i_\alpha)}^2\right] = \mathbb{E}\left[\max_{1\le i_2\le n} S_{(i_1^*, i_2, \ldots, i_\alpha)}^1\right]$$
$$= \mathbb{E}[T(n, K)]. \tag{23}$$

By defining $i_3^*, \ldots, i_\alpha^*$ in a similar way

$$i_u^*(i_1^*, \ldots, i_{u-1}^*, i_{u+1}, \ldots, i_\alpha)$$
$$= \arg\max_{1\le i_u\le n} S_{(i_1^*, \ldots, i_{u-1}^*, i_u, \ldots, i_\alpha)}^u$$

and iterating the above step, we can get

$$\mathbb{E}[T(n^\alpha, \alpha K)]$$
$$> \mathbb{E}\left[S_{(i_1^*, i_2^*, \ldots, i_\alpha^*)}^1 + S_{(i_1^*, i_2^*, \ldots, i_\alpha^*)}^2 + \ldots + S_{(i_1^*, i_2^*, \ldots, i_\alpha^*)}^\alpha\right]$$
$$\overset{(b)}{=} \sum_{r=1}^{\alpha} \mathbb{E}\left[S_{(i_1^*, i_2^*, \ldots, i_\alpha^*)}^r\right]$$
$$\overset{(c)}{=} \alpha\mathbb{E}[T(n, K)]. \tag{24}$$

Equation (b) follows from the fact that $\{S_{(i_1^*, i_2^*, \ldots, i_\alpha^*)}^r\}_{1\le r\le\alpha}$ are independent random variables and equation (c) follows from Equation (22), (23) and iterated steps. By combining Equation (1) and Equation (24), we have

$$\eta(n^\alpha, \alpha K) = \frac{\alpha K}{\mathbb{E}[T(n^\alpha, \alpha K)]} < \frac{\alpha K}{\alpha\mathbb{E}[T(n, K)]} = \eta(n, K),$$

which completes the proof. $\square$

7. CONCLUSION

In this paper, we characterize the throughput of a broadcast network using rateless codes. The broadcast channels are modeled by Markov modulated packet erasure channels, where the packet can either be erased or successfully received and for each receiver the current channel state distribution depends on the channel states in previous l packet transmissions.

We first characterize the asymptotic throughput of the system when n approaches infinity for any values of coding block size K as a function of number of receiver n in an explicit form. We show that as long as K scales at least as fast as $\log n$, we can achieve a non-zero asymptotic throughput. Under the more restrictive assumption that the channel is memoryless ($l = 0$), we study the case when K and n are finite. We show that, by keeping the ratio $K/\log n$ to be a constant, the system throughput will converge to the asymptotic throughput in a decreasing manner as n grows. By the help of these results, we are able to give a lower bound on the maximum achievable throughput (maximum achievable rate), which is a function of K, n and erasure probability $1 - \gamma$. In contrast to the state-of-the-art, we show that our bound is asymptotically tight when $K/\log n$ is fixed as n approaches infinity. Further, through numerical evaluations, we show that our bound is significantly better than existing result.

8. ACKNOWLEDGMENTS

The authors would like to thank Dr. Yin Sun for the valuable discussion that inspired the proof of Theorem 2.

This work was supported in part by NSF grants CNS-0905408, CNS-1012700, from the Army Research Office MURI grant W911NF-08-1-0238, and an HP IRP award.

9. REFERENCES

[1] R. Cogill, B. Shrader, and A. Ephremides. Stable throughput for multicast with inter-session network coding. *Military Communications Conference, 2008*, pages 1 – 7, 2008.

[2] A. Dembo and O. Zeitouni. *Large Deviations Techniques and Applications*. Springer-Verlag, New York, second edition, 1998.

[3] T. Ho. Networking from a network coding perspective. *Ph.D. dissertation, MIT*, 2004.

[4] M. G. Luby. Lt codes, 2002.

[5] K. T. Marshall and R. V. Evans. Some inequalities in queuing. *Operations Research*, 16(3):651–668, 1968.

[6] J. R. Norris. *Markov Chain*. Cambridge University Press, Cambridge, UK, 1998.

[7] P. Sadeghi, D. Traskov, and R. Koetter. Adaptive network coding for broadcast channels. *2009 Workshop on Network Coding, Theory and Applications*, 2009.

[8] A. Shokrollahi. Raptor codes. In *IEEE Transactions on Information Theory*, pages 2551–2567, 2006.

[9] B. T. Swapna, A. Eryilmaz, and N. B. Shroff. Throughput-delay analysis of random linear network coding for wireless broadcasting. *CoRR*, abs/1101.3724, 2011.

Closing the Gap in the Multicast Capacity of Hybrid Wireless Networks

Shaojie Tang
Dept of Computer Science
Illinois Institute of Technology

Xufei Mao
TNLIST
Tsinghua University

Taeho Jung   Junze Han
Dept of Computer Science
Illinois Institute of Technology

Xiang-Yang Li
Illinois Institute of Technology
Dalian University of
Technology

Boliu Xu
Dept of Computer Science
Illinois Institute of Technology

Chao Ma
Dept of Computer Science
Illinois Institute of Technology

ABSTRACT

We study the *multicast capacity* of a random wireless network consisting of n randomly placed *ordinary wireless nodes* and m regularly placed *base stations* in a square region, known as a *hybrid network*. All ordinary wireless nodes have the uniform transmission range r and uniform interference range $R = \Theta(r)$ and they can transmit/receive at W_a-bps. Each base station can communicate with adjacent base stations directly with a data rate W_B-bps and the data transmission rate between a base station and a wireless node is assumed to be W_c-bps. Assume that there is a random set of n_s ordinary wireless nodes that will serve as the source nodes of n_s multicast flows (each has randomly selected $k - 1$ receivers). Each flow will have data rate λ_i bps. We found that the minimum per-flow multicast capacity $\min_{i=1}^{n_s} \lambda_i$ for hybrid networks has three regimes, and for each regime we derive matching asymptotic upper and lower bounds. Thus it closes the gap of previous results in the literature.

Categories and Subject Descriptors

C.2.1 [**Network Architecture and Design**]: Wireless communication, Network topology; G.2.2 [**Graph Theory**]: Network problems, Graph algorithms

General Terms

Algorithms, Design, Theory

Keywords

Hybrid wireless networks, capacity, multicast, broadcast.

1. INTRODUCTION

The main purpose of this paper is to study the *multicast capacity* of hybrid networks. A *hybrid network* consists of two types of network terminals: *base stations* and *ordinary wireless nodes*. We assume that all base stations are regularly placed as a grid in a square region with side-length a meters, and each base station is connected with adjacent (two base stations are said to be adjacent if their Voronoi region share a common boundary segment) base stations by wired lines or wireless channels. Assume that each link that connects two base stations has rate W_B-bps and a base station is neither a data source nor a data receiver; it simply serves as a relaying communication gateway. Further we assume that n ordinary wireless nodes are randomly placed in this square region and the data rate between two ordinary wireless nodes is W_a-bps. The data rate between a base station and any ordinary wireless node is W_c-bps. Given all base stations Z, the Voronoi region, denoted as $\text{Vor}(z_i, Z)$, of a base station z_i is called the service *cell* of base station z_i.

We study three different routing strategies for a hybrid network. The first one is named *Ad Hoc Routing*: given the source node and $k - 1$ receivers, a multicast tree using only ordinary wireless node is constructed and the routing is performed on this tree. This approach has the same capacity as ad hoc wireless network [11, 12]. The second one is based on service cell. For a multicast flow, for each cell that contains at least one receiver (or source node) inside, we construct a tree that spans the receivers (or source node) including the base station in this cell. Then the forest (composed of trees built for each cell) will be connected by links among base stations. This is similar to the routing in cellular networks. We call this routing strategy as *Cellular Routing*. The third routing method can use any subgraph of the original communication graph that spans the receivers and the source node for routing. We call this routing strategy as *Hybrid Routing*. Thus, hybrid networks actually present a tradeoff by combining traditional BS-oriented network with ad hoc wireless network. Compared with the previous similar work by Mao *et al.*, [19], we study more general cases rather than only studying *Cellular Routing* strategy and further close the gap between the upper bound and lower bound in multicast capacity for hybrid networks.

Our Main Contributions: In this paper we derive matching upper bounds and lower bounds on multicast capacity of a hybrid wireless network, in which base stations are distributed regularly in a grid illustrated by Figure 2. Assume that the deployment region and the transmission range r are selected such that the network is connected *w.h.p.* We always assume that $m = O(a^2/r^2)$. We show that

THEOREM 1. *The per-flow capacity $\vartheta_k(n)$ of n_s multicast ses-*

MobiHoc'12, June 11–14, 2012, Hilton Head Island, SC, USA.
Copyright 2012 ACM 978-1-4503-1281-3/12/06 ...$10.00.

sions, when Cellular Routing *strategy is used, is*

$$
\begin{cases}
\Theta(\min(\frac{W_B\sqrt{m}}{n_s\sqrt{k}}, \frac{W_c m}{n_s k}, \frac{W_a m}{n_s k})) & \text{if } k = O(m) \\
\Theta(\min(\frac{W_B}{n_s}, \frac{W_c}{n_s}, \frac{W_a}{n_s})) & \text{if } k = \Omega(m)
\end{cases} \tag{1}
$$

When the *Ad Hoc Routing* strategy is used, it was proved in [12] that the minimum per-flow multicast capacity is

$$
\lambda_k(n) = \begin{cases}
\Theta(\frac{a}{r} \cdot \frac{W_a}{n_s\sqrt{k}}) & \text{if } k = O(\frac{a^2}{r^2}) \\
\Theta(\frac{W_a}{n_s}) & \text{if } k = \Omega(\frac{a^2}{r^2})
\end{cases} \tag{2}
$$

We then proved that the *Hybrid Routing* strategy will achieve a network capacity at most the larger one between the asymptotic capacity achieved by *Cellular Routing* strategy and the asymptotic capacity achieved by the *Ad Hoc Routing* strategy. Combining the preceding results, we further prove that

THEOREM 2. *The min-per-flow capacity* $\varphi_k(n)$ *achievable by* Hybrid Routing *strategy when* $m = O(\frac{a^2}{r^2})$ *is*

$$
\begin{cases}
\Theta(\max\left[\min\left(\frac{W_B\sqrt{m}}{n_s\sqrt{k}}, \frac{W_c m}{n_s k}, \frac{W_a m}{n_s k}\right), \frac{W_a}{n_s\sqrt{k}}\frac{a}{r}\right]) & \text{if } k = O(m) \\
\Theta(\frac{a}{r} \cdot \frac{W_a}{n_s\sqrt{k}}) & \text{if } k = \Omega(m), k = O(\frac{a^2}{r^2}) \\
\Theta(\frac{W_a}{n_s}) & \text{if } k = \Omega(\frac{a^2}{r^2})
\end{cases} \tag{3}
$$

The multicast capacity of hybrid networks when using *Cellular routing* and *Ad Hoc Routing* is shown in Figure 1. The most interesting part is that the multicast capacity upper bounds of hybrid networks is the upper envelop of those two curves (*Cellular routing* and *Ad Hoc Routing*). It implies that we can either use *Cellular routing* or *Ad Hoc Routing* to "beat" any other routing strategy asymptotically.

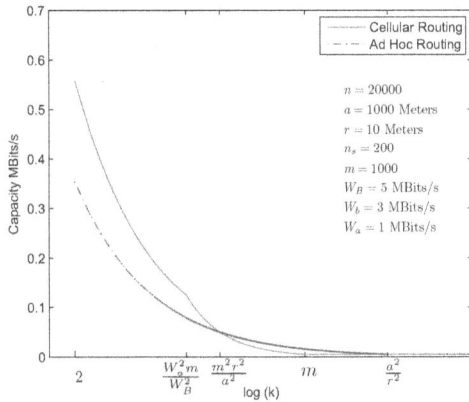

Figure 1: The capacity bounds (curves) for *Cellular Routing* and *Ad Hoc Routing*. The upper envelop of two curves is the capacity bound for *Hybrid Routing*.

Note that when the transmission range r is smaller, the achievable asymptotic capacity will be larger. However, on the other hand, the transmission range r should be at least a certain value such that the network formed by base stations and ordinary nodes will be connected *w.h.p.* It has been shown in [19] that when $\frac{a}{r} \leq \sqrt{\frac{cn\pi}{\log(c\frac{n}{m})+\beta}}$ for $\beta \to \infty$, the resulting network $G = (V \cup Z, E)$ is connected with probability at least $\frac{1}{e^{c^{-\beta}}} \to 1$, when $\beta \to \infty$ and c is constant.

The rest of the paper is organized as follows. In Section 2 we discuss in detail the network model used. In Section 3 and 4 we present the matching upper bounds and lower bounds respectively in multicast capacity when *Cellular Routing* strategy is used. In Section 5, we give the multicast capacity bound when *Hybrid Routing* strategy is used. We review the related results in Section 6 and conclude the paper in Section 8.

2. NETWORK MODEL

We assume that there is a set $V = \{v_1, v_2, \cdots, v_n\}$ of n *ordinary* wireless terminals randomly deployed in a square region with side-length a. Each wireless node has transmission range r such that nodes v_i and v_j can communicate successfully iff the Euclidean distance $|v_i - v_j| \leq r$. The data rate of every link $v_i v_j$ is W_a-bps when no interference occurs. A communication from v_i to v_j is interference-free if there is no any other node u that is transmitting and within distance R of receiving node v_j. Here $r < R = \Theta(r)$. We further assume that there are m base stations $Z = \{z_1, z_2, \cdots, z_m\}$ which are regularly placed in the region Ω. For example, the base stations are placed regularly at positions $(\frac{a}{2\sqrt{m}} + i\frac{a}{\sqrt{m}}, \frac{a}{2\sqrt{m}} + j\frac{a}{\sqrt{m}})$ with $0 \leq i \leq \sqrt{m} - 1$, and $0 \leq j \leq \sqrt{m} - 1$. We generally assume that m is a square of some integer. Figure 2(a) illustrate a simple example of hybrid networks. Clearly, these m regularly placed base stations divide the original square region into m **cells** as Voronoi diagrams with same side length $\frac{a}{\sqrt{m}}$. We use S_i to denote the cell defined by base station z_i, and for simplicity, by abusing the notation little bit, we say the cell S_i is the service region of base station z_i, *i.e.*, z_i serves as a functional gateway for all wireless nodes in cell S_i when *Cellular Routing* strategy is used. The transmission range of a base station is also assumed to be r. In other words, a base station can only directly serve nodes within distance r. The total data rate that a base station can serve all ordinary wireless nodes is at most W_c-bps with $W_c \geq W_a$. In other words, a base station can serve at most $\frac{W_c}{\lambda}$ flows if each flow requires a data rate λ.

Each base station is connected to its adjacent base stations (at most 4) by wired lines or wireless channels (using frequency different from the frequency used between ordinary wireless nodes). The links between base stations have a larger data rate W_B. We further assume that $m = o(\frac{a^2}{r^2})$ throughout the whole paper due to the following observation: when the number of base stations $m \geq \frac{a^2}{r^2}$, all these regularly distributed base stations will cover the whole square, thus a hybrid network will act as a cellular network.

The complete communication graph is a graph $G = (V \cup Z, E)$, where $V = \{v_1, v_2, \cdots, v_n\}$ is the set of ordinary wireless nodes and $Z = \{z_1, z_2, \cdots, z_m\}$ is the set of base stations, and $E = E_a \cup E_B \cup E_c$ is the set of all possible communication links

1. E_a is the set of ad hoc links uv where $u \in V$, $v \in V$, and $\|u - v\| \leq r$. Each link in E_a has data rate W_a-bps.
2. E_B is the set of backbone links $z_i z_j$ where $z_i \in Z, z_j \in Z$, and $\|z_i - z_j\| = \frac{a}{\sqrt{m}}$. The data rate of each link in E_B is W_B-bps.
3. E_c is the set of cellular links $z_i v_j$ where $z_i \in Z, v_j \in V$, and $\|z_i - v_j\| \leq r$. The data rate (both up-link and down-link) of each link in E_c is W_c-bps.

For simplicity, we use E_d to denote the set of crossing ad hoc links: $E_d = \{(v_i, v_j) \mid v_i \text{ and } v_j \text{ are from different cells}\}$.

We assume that $W_a \leq W_c \leq W_B$. Given a multicast flow with source v_i and the set of receivers U_i, the routing structure must be a subgraph of G. Three different routing strategies that will be studied can be categorized as follows

1. *Ad Hoc Routing* strategy will use only the links in E_a;

2. *Cellular Routing* strategy will *not* use links in E_d;

3. *Hybrid Routing* strategy can use any links in G.

Please see Fig 2 for illustration. In this paper, we mainly assume

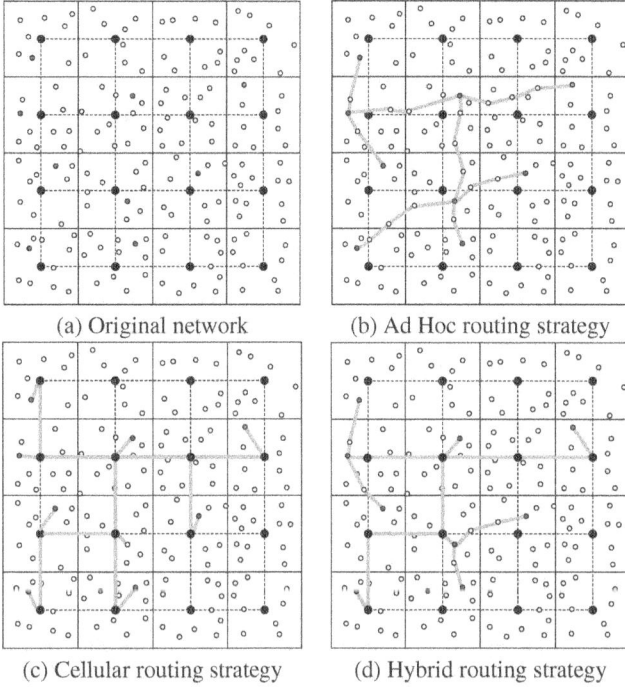

(a) Original network   (b) Ad Hoc routing strategy

(c) Cellular routing strategy   (d) Hybrid routing strategy

Figure 2: Illustration of three routing strategies. We use the red node to denote a source node and blues nodes to denote its k receivers.

that the transmission range r is fixed and thus normalized to one unit throughout the paper.

Random Multicast Flows: In this paper, we will concentrate on the *multicast capacity* of a random hybrid network, which generalizes both the unicast capacity [7] and broadcast capacity [10,14] for random networks (when $m = 0$). Assume that a subset $\mathcal{S} \subseteq V$ of $n_s = |\mathcal{S}|$ random nodes will serve as the source nodes of n_s multicast sessions. We randomly and independently choose n_s multicast sessions. To generate the i-th $(1 \leq i \leq n_s)$ multicast session, k points $p_{i,j}(1 \leq j \leq k)$ are randomly and independently chosen from the deployment region Ω. Let $v_{i,j}$ be the nearest wireless node from $p_{i,j}$ (ties are broken randomly). In the i-th multicast session, $v_{i,1}$ will be chosen as source node and multicast data to $k - 1$ nodes $U_i = \{v_{i,j} \mid 2 \leq j \leq k\}$ at an arbitrary data rate λ_i.

In this paper, we mainly focus on the protocol interference model introduced in [7]. We assume that each node v_i has a fixed interference range R which is within a small constant factor of the transmission range r, i.e., $\varrho_1 r \leq R \leq \varrho_2 r$ for some constants $1 < \varrho_1 \leq \varrho_2$. Under the protocol interference model, any node v_j will be interfered by the signal from v_k if $\|v_k - v_j\| \leq R$ where node v_k is sending signal to some node other than v_j.

Capacity Definition: The detailed definition is omitted here due to the space constraint, please refer to [12] for details. Throughout this paper, we will focus on studying the minimum per-flow multicast capacity which is defined as $\varphi_k(n) = \min_{v_i \in S} \lambda_i$.

3. UPPER BOUNDS IN MULTICAST CAPACITY BY CELLULAR ROUTING

When *Cellular Routing* is used, the capacity for a hybrid network

can be constrained due to three different congestion scenarios: (1) the backbone formed by the links E_B is congested; (2) the cellular links E_c are congested; and (3) the ad hoc links $E_a \setminus E_d$ in some cell are congested. We will derive an upper bound separately on minimum per-flow multicast capacity for each of the aforementioned three conditions.

TECHNIQUE LEMMAS: Throughout this paper, we will repeatedly use these lemmas (whose proofs are omitted here due to space limit).

LEMMA 3. *For the i-th flow, let $k_{i,j}$ be the number of terminals that will fall inside the service cell of the jth base station z_j. Then, $k_{i,j}$ is a random variable with mean $E(k_{i,j}) = \frac{k}{m}$ and variance $\text{Var}(k_{i,j}) = \frac{k}{m}(1 - \frac{1}{m})$. Note that $\boldsymbol{Pr}\left(k_{i,j} = t\right)$ is $\binom{k}{t}\left(\frac{1}{m}\right)^t\left(1 - \frac{1}{m}\right)^{k-t}$.*

LEMMA 4. *Let variable $k'_{i,j}$ denote the number of terminals of the i-th flow that fall inside the cell of the base station z_j, but not inside the communication disk of z_j (centered at z_j with radius r). Then $\boldsymbol{Pr}\left(k'_{i,j} = t\right) = \binom{k}{t}\left(\frac{1}{m} - \frac{r^2}{a^2}\right)^t\left(1 - \frac{1}{m} + \frac{r^2}{a^2}\right)^{k-t}$.*

Then its mean is $E(k'_{i,j}) = k(\frac{1}{m} - \frac{r^2}{a^2})$ and variance is $\text{Var}(k'_{i,j}) = k(\frac{1}{m} - \frac{r^2}{a^2})(1 - \frac{1}{m} + \frac{r^2}{a^2})$. Recall that in this paper, we assumed that the number of base stations $m \leq v\frac{a^2}{r^2}$ for some constant $0 < c < 1$. Thus, $\frac{1}{m} - \frac{c^2}{a^2} \geq (1-c)\frac{1}{m}$. This implies that $E(k'_{i,j}) \geq (1-c)\frac{k}{m}$ and variance $(1-c)k\frac{1}{m}(1 - (1-c))\frac{1}{m}) \leq \text{Var}(k'_{i,j}) \leq k\frac{1}{m}(1 - \frac{1}{m})$.

LEMMA 5. *Let variable $X_{i,j} \in \{0,1\}$ denote whether the j-cell (defined by base station z_j) contains some terminals from the ith flow, i.e., $X_{i,j} = 1$ if $k_{i,j} > 0$, and $X_{i,j} = 0$ if $k_{i,j} = 0$. Thus, $\boldsymbol{Pr}\left(X_{i,j} = 1\right) = 1 - (1 - \frac{1}{m})^k$. In addition, $\text{Var}(X_{i,j}) = E(X_{i,j}^2) - E(X_{i,j})^2 = (1 - (1 - \frac{1}{m})^k)(1 - \frac{1}{m})^k$.*

LEMMA 6. *Let variable f_j denote the number of flows, each of which has at least a terminal node inside the j-th cell. Then $f_j = \sum_{i=1}^{n_s} X_{i,j}$. In addition, $E(f_j) = n_s(1 - (1 - \frac{1}{m})^k)$ and variance $\text{Var}(f_j) = n_s \cdot \text{Var}(X_{i,j}) = n_s(1 - (1 - \frac{1}{m})^k)(1 - \frac{1}{m})^k$.*

LEMMA 7. *Let variable $X'_{i,j} \in \{0,1\}$ denote whether some terminals from the ith flow fall into the jth-cell, but not inside the communication disk centered at z_i, $X'_{i,j} = 1$ if $k'_{i,j} > 0$, and $X'_{i,j} = 0$ if $k'_{i,j} = 0$. Thus, $\boldsymbol{Pr}\left(X'_{i,j} = 1\right) = 1 - (1 - (\frac{1}{m} - \frac{r^2}{a^2}))^k$. In addition, $\text{Var}(X'_{i,j}) = E(X_{i,j}^2) - E(X_{i,j})^2 = (1 - (1 - (\frac{1}{m} - \frac{r^2}{a^2}))^k)(1 - (\frac{1}{m} - \frac{r^2}{a^2}))^k$.*

LEMMA 8. *Let variable f'_j denote the number of flows, each of which has at least a terminal node inside the j-th cell, but not inside the communication disk centered at z_j. Then $f'_j = \sum_{i=1}^{n_s} X'_{i,j}$. In addition, $E(f'_j) = n_s(1 - (1 - (\frac{1}{m} - \frac{r^2}{a^2}))^k)$ and variance $\text{Var}(f'_j) = n_s \cdot \text{Var}(X'_{i,j}) = n_s(1 - (1 - (\frac{1}{m} - \frac{r^2}{a^2}))^k)(1 - (\frac{1}{m} - \frac{r^2}{a^2}))^k$.*

LEMMA 9. *Let variable k_i denote the number of cells that has at least one terminal from flow i inside. Clearly, $k_i = \sum_{j=1}^{m} X_{i,j}$. Then $E(k_i) = m(1 - (1 - \frac{1}{m})^k)$ and variance $\text{Var}(k_i) = m \cdot \text{Var}(X_{i,j}) = m(1 - (1 - \frac{1}{m})^k)(1 - \frac{1}{m})^k$. Furthermore, when $k \leq m$, we have $\frac{k}{2} \leq E(k_i) \leq \min(m, 2k)$. When $k > m$, we have $\frac{m}{2} \leq E(k_i)$.*

Given a routing strategy $\mathcal{A}$, let $T_i(\mathcal{A})$ be the tree used to route the i-th flow. When $\mathcal{A}$ is clear from the context, we will simplify it as T_i by dropping $\mathcal{A}$. Hereafter, if $\mathcal{B}$ is a set, we use $|\mathcal{B}|$ to denote the cardinality of $\mathcal{B}$; if $\mathcal{B}$ is a tree, we use $|\mathcal{B}|$ to denote the total Euclidean length of tree $\mathcal{B}$. The following lemma was shown in [21].

LEMMA 10. *Given k points Q randomly placed in a square of side length a, the Euclidean minimum spanning tree $\mathrm{EMST}(Q)$ has an expected total edge length $\Theta(\sqrt{k}a)$ and its variance*

$$\mathrm{Var}(|\,\mathrm{EMST}(Q)|) \ll a^2 \cdot \log k$$

It was proved in [11, 12] that any routing tree T_i for a set Q of random k points in the square of side-length a, its total edge length is at least $\frac{1}{2}$ times the total edge length of $\mathrm{EMST}(Q)$.

3.1 Upper Bound Due to Links in E_B

The upper bound on multicast capacity due to congestion in E_B has two regimes: $k = O(m)$ and $k = \Omega(m)$.

3.1.1 When $k = O(m)$

In this case, for each flow i, we let B_i be the set of base stations whose service cell contains at least one terminal from the i-th flow. Then we need build a connected structure using only links in E_B to span B_i. Let T_B^i be the tree (covering all base stations in B_i) constructed by a given routing method. Then we know that $|T_B^i| \geq |\mathrm{EMST}(B_i)|/2$. Notice that the set B_i is a random variable and $|B_i| = k_i$, where random variable k_i is as defined before. Similar to [21], we can prove the following lemma:

LEMMA 11. *Given k_i base stations B_i randomly selected and all base stations are placed in a square region of side-length a, the Euclidean minimum spanning tree $\mathrm{EMST}(B_i)$ has an expected total edge length $c_1\sqrt{k_i}a$ for a constant $c_1 \in (0, 2\sqrt{2}]$ and its variance $\mathrm{Var}(|\mathrm{EMST}(B_i)|) \ll a^2 \cdot \log k_i$.*

THEOREM 12. *When $k \leq \theta_0 m$ for some constant θ_0, there is a constant c_3 such that, with probability at least $1 - 2e^{-n_s/8}$, the minimum data rate that can be supported using cellular routing strategy is at most $\frac{W_B\sqrt{m}}{c_3 n_s \sqrt{k}}$ for any routing strategy due to the congestion on backbone links.*

PROOF. Let $C(T_B^i)$ denote the number of cells that the routing tree T_B^i will use, *i.e.*, the number of base stations used in T_B^i. Obviously, $C(T_B^i) \geq k_i$, the number of cells that contain the receivers of the i-th flow. Notice that each base station is connected to at most 4 adjacent base stations. Then $|T_B^i|/(4\frac{a}{\sqrt{m}}) \leq C(T_B^i) \leq |T_B^i|/(\frac{a}{\sqrt{m}}) + 1$. Let variable $L = \sum_{i=1}^{n_s} C(T_B^i)$ denote the total load of all cells. Here the load of a cell resulted from a routing method is the total number of flows passing the cell. Then $L \geq \sum_{i=1}^{n_s} |T_B^i|/(4\frac{a}{\sqrt{m}}) \geq \sum_{i=1}^{n_s} |\mathrm{EMST}(B_i)|/(8\frac{a}{\sqrt{m}})$. Notice that $E(\sum_{i=1}^{n_s} |\mathrm{EMST}(B_i)|) = n_s c_1 E(\sqrt{k_i})a$. Thus $E(L) \geq c_1 n_s E(\sqrt{k_i})\sqrt{m}/8$.

We then compute the value $E(\sqrt{k_i})$. Recall that variable $X_{i,j}$ denotes whether the j-th cell contains any terminal from the i-th flow and $k_i = \sum_{j=1}^m X_{i,j}$. By definition,

$$E(\sqrt{k_i}) = E(\sqrt{\sum_{j=1}^m \mathbf{Pr}(X_{i,j} = 1)}) = \sqrt{m(1 - (1 - \frac{1}{m})^k)}$$

Then, $\sqrt{\min(m, k)/2} \leq E(\sqrt{k_i}) \leq \sqrt{\min(m, 2k)}$. When $k \leq \theta_0 m$, $E(\sqrt{k_i}) \geq c_2\sqrt{k}$ for a constant $c_2 = \sqrt{\min(\frac{1}{2\theta_0}, 1)}$.

Define random variables

$$X_q = \sum_{j=1}^q (|\,\mathrm{EMST}(B_j)| - E(|\,\mathrm{EMST}(B_j)|))$$

Then $E(X_{q+1} \mid X_1, \cdots, X_q) = X_q$, *i.e.*, variables X_i are martingale. In addition,

$$|X_q - X_{q-1}| = ||\,\mathrm{EMST}(B_q)| - E(|\,\mathrm{EMST}(B_q)|)|$$

$$\leq |\,\mathrm{EMST}(B_q)| \leq 2\sqrt{2}\sqrt{k_i}a \leq 2\sqrt{2}\sqrt{k}a$$

From Azuma's Inequality, we have

$$\mathbf{Pr}(|X_{n_s} - X_0| \geq t) \leq 2\exp(-\frac{t^2}{2\sum_{i=1}^{n_s} 8ka^2})$$

Let $t = \epsilon\sum_{i=1}^{n_s} E(|\,\mathrm{EMST}(B_i)|)$. Clearly, $\epsilon n_s c_1 c_2\sqrt{k}a \leq t \leq 2\sqrt{2}\epsilon\sqrt{k}a$. Note that $X_0 = 0$. Then,

$$\mathbf{Pr}\left(\sum_{i=1}^{n_s} |\,\mathrm{EMST}(B_i)| \leq \sum_{i=1}^{n_s} E(|\,\mathrm{EMST}(B_i)|) - t\right)$$

$$\leq \quad \mathbf{Pr}(|X_{n_s}| \geq t) \leq \exp(-\frac{t^2}{2\sum_{i=1}^{n_s} 8ka^2})$$

$$\leq \quad \exp(-\frac{(\epsilon n_s c_1 c_2\sqrt{k}a)^2}{8n_s ka^2}) = \exp(-\frac{n_s \epsilon^2 c_1^2 c_2^2}{8})$$

Consequently, for a constant $\epsilon \in (0, 1)$,

$$\mathbf{Pr}\left(\sum_{i=1}^{n_s} |\,\mathrm{EMST}(B_i)| \leq (1 - \epsilon)n_s c_1 E(\sqrt{k_i})a\right)$$

$$\leq 2e^{-\frac{n_s \epsilon^2 c_1^2 c_2^1}{8}}$$

$$\Rightarrow \quad \mathbf{Pr}\left(\sum_{i=1}^{n_s} |\,\mathrm{EMST}(B_i)| \geq \sum_{i=1}^{n_s} c_1 E(\sqrt{k_i})a/2\right)$$

$$\geq 1 - 2e^{-n_s c_1^2 c_2^2/32}$$

$$\Rightarrow \quad \mathbf{Pr}\left(L \geq n_s c_1 E(\sqrt{mk_i})/16\right) \geq 1 - 2e^{-n_s c_1^2 c_2^2/32}$$

$$\Rightarrow \quad \mathbf{Pr}\left(L \geq n_s c_1 c_2\sqrt{km}/16\right) \geq 1 - 2e^{-n_s c_1^2 c_2^2/32} \text{ if } k \leq \theta_0 m.$$

Recall that L denotes the total load of all cells. Then by Pigeonhole principle, with probability at least $1 - 2e^{-n_s c_1^2 c_2^2/32}$, there is at least one cell, that will be used by at least $\frac{n_s c_1 c_2\sqrt{km}}{m}$ flows. Thus, with probability at least $1 - 2e^{-n_s c_1^2 c_2^2/32}$, the minimum data rate that can be supported using cellular routing strategy is at most $\frac{W_B}{\frac{n_s c_1 c_2\sqrt{km}}{m}} = \frac{W_B\sqrt{m}}{c_1 c_2 n_s \sqrt{k}}$ for any routing strategy due to the congestion in backbone links. By letting $c_3 = c_1 c_2$ we finish the proof. $\square$

3.1.2 When $k = \Omega(m)$

Recall that in this case, we have shown that $E(k_i) \geq m/2$, *i.e.*, for each flow, the expected number of cells that will contain its terminals is at least $m/2$. More precisely, it is easy to show that, for any cell j, the probability, $\mathbf{Pr}(X_{i,j} = 1)$, that it will contain a terminal from flow i is at least $1 - 1/e > 1/2$. Then by using Azuma's Inequality, we can prove that, with probability at least $1 - 2e^{-n_s/8}$, the total load $L \geq n_s m/4$. Thus, by Pigeonhole principle, there is one cell such that its load (the number of flows using its base-station) is at least $n_s/4$. Consequently, we have the following theorem.

138

THEOREM 13. *When $k \geq \theta_0 m$ for some constant $\theta_0 > 1$, with probability at least $1 - 2e^{-n_s/8}$, the minimum data rate that can be supported using cellular routing strategy is at most $\frac{4W_B}{n_s}$ for any routing strategy due to the congestion in backbone.*

3.2 Upper Bound Due to Links in E_c

In this subsection, we study the minimum per-flow data rate due to the congestion when ordinary wireless nodes access the base-stations in their cells. Recall that we assume that both the uplink rate and the down-link rate between the base-station and the ordinary wireless nodes in its cell is W_c-bps. We will study upper bounds based on two subcases, whether $k = O(m)$ or not. We essentially study the number of flows f_j inside j^{th} cell that pass through a base-station z_j.

3.2.1 When $k = O(m)$

We first study the case when the number of terminals per-flow is $k = O(m)$. Notice that when $k \leq m$, $E(f_j) = n_s(1 - (1 - \frac{1}{m})^k) > \frac{k}{2m}n_s$ and $\mathrm{Var}(f_j) < \frac{2k}{m}n_s$.

LEMMA 14. *When n_s satisfies the condition (4), the variable $\max_{j=1}^{m} f_j$ is $\Theta(n_s \frac{k}{m})$ with probability at least $1 - \frac{1}{n}$.*

PROOF. We use the VC-Theorem to prove this lemma. Let the set $\mathcal{C} = \{\mathrm{Vor}(z_j, Z) \mid 1 \leq j \leq m\}$ be the class of cells defined by all base-stations. Let F_i be the i-th flow and F_i is said to "belong to" the j th cell if some of its terminals is contained inside the j-th cell $\mathrm{Vor}(z_j, Z)$, which is denoted as $F_i \in \mathrm{Vor}(z_j, Z)$. Then $f_j = \sum_{F_i} I(F_i \in \mathrm{Vor}(z_j, Z))$, where $I(F_i \in \mathrm{Vor}(z_j, Z)) = 1$ if $F_i \in \mathrm{Vor}(z_j, Z)$ and $I(F_i \in \mathrm{Vor}(z_j, Z)) = 0$ otherwise. Obviously, VC-d$(\mathcal{C}) \leq \log m$ since the cardinality of $\mathcal{C}$ is m. In addition, the probability $P(A)$ that a flow "belongs to" a cell A is $P(A) = 1 - (1 - \frac{1}{m})^k$. It is easy to show that, when $0 < k < m$, we have $\frac{k}{2m} < P(A) < \frac{2k}{m}$. Then by VC-Theorem, we know that for every $\epsilon, \delta > 0$, $\mathbf{Pr}\left(\sup_{A \in \mathcal{C}} \left| \frac{\sum_{i=1}^{n_s} I(F_i \in A)}{n_s} - P(A) \right| \leq \epsilon \right) > 1 - \delta$ whenever $n_s > \max\left\{ \frac{8 \cdot \text{VC-d}(\mathcal{C})}{\epsilon} \cdot \log \frac{13}{\epsilon}, \frac{4}{\epsilon} \log \frac{2}{\delta} \right\}$. When we choose the parameters $\epsilon = \frac{k}{4m}$, $\delta = \frac{1}{n}$, and

$$n_s > \max\left(\frac{32m \log m}{k} \log \frac{52m}{k}, \frac{16m}{k} \log(2n) \right), \qquad (4)$$

we have $\mathbf{Pr}\left(\sup_{i=1}^{m} |f_i - n_s P(A)| \leq n_s \frac{k}{4m}\right) > 1 - \frac{1}{n}$. Hence, $\mathbf{Pr}\left(\forall i \in [1, m], n_s \frac{k}{4m} \leq f_i \leq n_s \frac{9k}{4m}\right) > 1 - \frac{1}{n}$. $\square$

Based on the preceding lemma, we conclude that,

THEOREM 15. *When $k \leq m$, the rate due to the congestion of accessing the base-stations, with probability at least $1 - \frac{1}{n}$, is at most $\frac{W_c}{\max_i f_i} \leq \frac{W_c \cdot (4m)}{n_s k} = O(\frac{W_c \cdot m}{n_s k})$.*

3.2.2 When $k = \Omega(m)$

We then study an upperbound on the rate achievable due to the congestion of accessing base-stations when $k > m$. In this case, $n_s > E(f_j) = n_s(1 - (1 - \frac{1}{m})^k) > n_s(1 - \frac{1}{e})$.

LEMMA 16. *When n_s satisfies the condition (5), the variable $\max_{j=1}^{m} f_j$ is also $\Theta(n_s)$ with probability at least $1 - \frac{1}{n}$.*

PROOF. Similar as the proof for Lemma 14, except that when $k \geq m$, we have $1 - \frac{1}{e} < P(A) < 1$. Based on VC-Theorem, by choosing the parameters $\epsilon = \frac{1}{e}$, $\delta = \frac{1}{n}$, we know that when

$$n_s > \max\left(8e \log m \log(13e), 4e \log(2n) \right), \qquad (5)$$

we have $\mathbf{Pr}\left(\sup_{i=1}^{m} |f_i - n_s P(A)| \leq n_s \frac{1}{e}\right) > 1 - \frac{1}{n}$. Hence, $\mathbf{Pr}\left(\forall i \in [1, m], n_s(1 - \frac{2}{e}) \leq f_i \leq n_s\right) > 1 - \frac{1}{n}$. $\square$

Obviously, we have the following theorem.

THEOREM 17. *When $k \geq m$, with probability at least $1 - \frac{1}{n}$, the minimum per-flow rate by any Cellular Routing strategy is at most $\frac{W_c}{n_s(1 - \frac{2}{e})}$.*

3.3 Upper Bound Due to Links in $E_a \setminus E_d$

In previous subsections, we study upper bounds on the multicast capacity in hybrid networks due to the congestion at the backbone links, and due to the congestion in accessing the base-stations. We now focus on studying the capacity upper bounds due to the congestion in ad hoc links $E_a \setminus E_d$.

A trivial upper bound for total multicast capacity is $W_a \cdot n$ since there are n source nodes in total and each can send data at W_a bits/sec. However, we can make the upper bounds more tight due to the following observations. For each source node v_i, when we multicast the data from one source node v_i to all its $k - 1$ receivers in set $U_i = \{v_{i_1}, v_{i_2}, \cdots, v_{i_{k-1}}\}$, the resulting multicast tree will contain at least k nodes, and possibly more. More possibly, when a non-leaf node v in the multicast tree sends data to its children, **all** nodes that are within its transmission range will receive the data or at least they cannot transmit successfully at the same time no matter these nodes are intended receivers or not. In this case, we say all these nodes are *charged* a *copy* of the data. To study the multicast capacity, we partition the deployment square into grids of size r. Clearly, there are at most $\lceil \frac{a}{r} \rceil^2 = \Theta(\frac{a^2}{r^2})$ such grid cells. Notice that among such grid cells, some of them can be directly reached by some base-stations. Let g be the total number of grid cells that is disjoint from the union of disks $\bigcup_{j=1}^{m} D(z_j, r)$. Then obviously $g \geq \lceil \frac{a}{r} \rceil^2 - 9m = \Theta(\frac{a^2}{r^2})$ when $m \leq \frac{a^2}{10r^2}$. Here, the constant 9 comes from the fact that any base station only can over at most 9 grids of size r at the same time due to our previous assumption that the transmission range of each base station is also r. Thus, throughout this paper, we assume that $m \leq \frac{a^2}{10r^2}$.

Recall that we assume that the interference range $R > \varrho_1 r$. Then at any time instance, the distance between two active senders v_1 and v_2 is at least $R - r \geq (\varrho_1 - 1)r$. Consequently, we have

LEMMA 18. *For any grid of side-length r, there are at most a constant number (denoted as $\kappa < (1 + \frac{2}{\varrho_1 - 1})^2$) of nodes inside the grid that can send data simultaneously without causing interference to receivers.*

This lemma implies that the total data that can be sent out from any grid during any time interval t is at most $W_a \cdot \kappa t$ for a constant κ. To prove an upper bound on the capacity, we will only consider the grid cells that are disjoint from the disks defined by base-stations. In other words, for nodes located inside these grid-cells, it cannot reach the base-stations directly and its data has to be relayed by some other nodes to reach the base stations. Given a routing strategy, for the i-th flow and j-th grid cell, let $Y_{i,j}$ be the variable denoting whether the i-th flow will be routed through the j-th grid cell. Let $Y_j = \sum_{i=1}^{n_s} Y_{i,j}$ be the total number of flows that will be routed through the j-th grid cell by this routing strategy. Then from Lemma 18, we can conclude that the minimum per-flow data rate is at most

$$\frac{W_a \cdot \kappa}{\max_{j=1}^{g} Y_j} \qquad (6)$$

The rest of the subsection is devoted to give a better lower bound on $\max_{j=1}^{g} Y_j$, thus a tighter upper bound of multicast capacity of the hybrid wireless network. Notice that the bound on $\max_{j=1}^{g} Y_j$

depends on the routing strategy. We next prove that, no matter what routing strategy is used, $\max_{j=1}^{g} Y_j$ is lower bounded by certain value *w.h.p.* For simplicity, hereafter when we say k receivers, we mean that one source node pluses all its $k - 1$ receivers.

3.3.1 When $k = O(m)$

As we know, under the *Cellular Routing* strategy, all flows inside of a cell with side-length $\frac{a}{\sqrt{m}}$ will firstly go to the closest base station by one- or multi-hop. Before the traffic reach the base station, the last hop transmission is a cellular link and the second to the last hop link is an ad hoc link. Thus, the potential congestion will happen on those second to last hop ad hoc links for the following two reasons: a) Each base station only has transmission range r and can cover (touch) a relatively small area (at most 9 adjacent cells with side length r around the base station.) b) Intuitively, the cell closed to the base station will have much burden to relay ad hoc traffic to the base station. Clearly, it is equivalent to study the number of flows f'_j inside of jth cell but not inside of the communication disk centered at z_i.

LEMMA 19. *When n_s satisfies the condition (7), the variable $\max_{j=1}^{m} f'_j$ is $\Theta(n_s \frac{k}{m})$ with probability at least $1 - \frac{1}{n}$.*

PROOF. We use the similar proof used in Lemma 14 to prove this. Assume the set $\mathcal{C}' = \{\text{Vor}(z'_j, Z) \mid 1 \leq j \leq m\}$ be the class of regions (each region is the cell without the communication disk of the base station) defined by all base-stations. The ith flow F'_i is said to "belong to" the jth cell if some of its terminals is contained inside the j-th cell, but not inside the communication disk centered at z_j, $\text{Vor}(z'_j, Z)$, which is denoted as $F'_i \in \text{Vor}(z'_j, Z)$. In addition, the probability $P(A')$ that a flow "belongs to" a region A' is $P(A') = 1 - (1 - (\frac{1}{m} - \frac{r^2}{a^2}))^k$. It is easy to show that, when $0 < k < \frac{10m}{9}$ and $m \geq \frac{a^2}{10r^2}$ we have $\frac{9k}{20m} < P(A') < \frac{9k}{5m}$. By the same argument in Lemma 14 and VC-Theorem, for every $\epsilon, \delta > 0$, $\mathbf{Pr}\left(\sup_{A' \in \mathcal{C}'} \left| \frac{\sum_{i=1}^{n_s} I(F'_i \in A')}{n_s} - P(A') \right| \leq \epsilon \right) > 1 - \delta$ whenever $n_s > \max\left\{ \frac{8 \cdot \text{VC-d}(\mathcal{C})}{\epsilon} \cdot \log \frac{13}{\epsilon}, \frac{4}{\epsilon} \log \frac{2}{\delta} \right\}$. When we choose the parameters $\epsilon = \frac{k}{4m}$, $\delta = \frac{1}{n}$, and

$$n_s > \max\left(\frac{32m \log m}{k} \log \frac{52m}{k}, \frac{16m}{k} \log(2n) \right), \quad (7)$$

$\mathbf{Pr}\left(\sup_{i=1}^{m} |f'_i - n_s P(A')| \leq n_s \frac{k}{4m}\right) > 1 - \frac{1}{n}$. Hence, $\mathbf{Pr}\left(\forall i \in [1, m], n_s \frac{k}{5m} \leq f'_i \leq n_s \frac{41k}{20m}\right) > 1 - \frac{1}{n}$. $\square$

Based on the preceding lemma, we conclude that

THEOREM 20. *When $k \leq \frac{10m}{9}$ and n_s satisfies the condition (7), the rate due to the congestion of ad hoc links, with probability at least $1 - \frac{1}{n}$, is at most $O(\frac{W_a m}{n_s k})$.*

PROOF. By Lemma 19, for j^{th} cell, the number of ad hoc flows f'_j which will converge to base station z_j, $\max_{j=1}^{m} f'_j = \theta_2 \frac{n_s k}{m}$ with probability at least $1 - \frac{1}{n}$ where n_s satisfies the condition (7) and θ_2 is some positive constant. In addition, there are at most 9 cells with side-length r to relay these flows such that there exists at least one cell that has to relay at least $\frac{\theta_2}{9} \frac{n_s k}{m}$ flows by Pigeonhole principle. Therefore, by equation 6, the min-flow capacity cannot exceed $\frac{W_a \cdot \kappa}{\frac{\theta_2}{9} \frac{n_s k}{m}} = O(\frac{W_a m}{n_s k})$ where κ and θ_2 are constants. $\square$

3.3.2 When $k = \Omega(m)$ and $k = O(\frac{a^2}{r^2})$

We are going to study an upper bound on the rate achievable due to the congestion of all ad hoc flows (links) which targets to

or from the base station when $k > m$. Recall that we use f'_j to denote all the ad hoc flows which exist in j^{th} cell and as we have shown before, the expected value of $E(k'_{i,j}) \geq (1 - c)\frac{k}{m})$ for some constant c. It is not difficult to show that $n_s > E(f'_j) = n_s(1 - (1 - (\frac{1}{m} - \frac{r^2}{a^2}))^k) > n_s(1 - (\frac{1}{e})^{1-c})$. Next, we show that the maximum number of ad hoc flows inside some cell is the constant fraction of total n_s multicast flows by the following Lemma 21.

LEMMA 21. *When n_s satisfies the condition (8), the variable $\max_{j=1}^{m} f'_j$ is also $\Theta(n_s)$ with probability at least $1 - \frac{1}{n}$.*

PROOF. Similar as the proof for Lemma 19 except that when $k \geq m$ we have $1 - (\frac{1}{e})^{1-c} < P(A') < 1$. Based on VC-Theorem, by choosing parameters $\epsilon = (\frac{1}{e})^{1-c}$, $\delta = \frac{1}{n}$, we know that when

$$n_s > \max\left(8e^{1-c} \log m \log(13e^{1-c}), 4e^{1-c} \log(2n) \right), \quad (8)$$

$\mathbf{Pr}\left(\sup_{i=1}^{m} |f'_i - n_s P(A)| \leq n_s(\frac{1}{e})^{1-c}\right) > 1 - \frac{1}{n}$. Hence, $\mathbf{Pr}\left(\forall i \in [1, m], n_s(1 - \frac{2}{e^{1-c}}) \leq f'_i \leq n_s\right) > 1 - \frac{1}{n}$. $\square$

Based on *Cellular Routing* strategy, all the ad hoc flows inside of a cell will eventually approach or leave the base station. Therefore, the potential congestion will happen when the ad hoc flows converge into the communication disk of that base station. Due to Lemma 18 each cell with side length r can only have κ simultaneous transmitters. Therefore, by Lemma 21, we have

THEOREM 22. *With probability at least $1 - \frac{1}{n}$, the minimum per-flow rate for n_s multicast sessions is bounded by $O(\frac{W_a}{n_s}$ where $k = \Omega(m)$ and $k = O(\frac{a^2}{r^2})$. due to the capacity constraints of ad hoc links in $E_a \setminus E_d$.*

3.3.3 When $k = \Omega(\frac{a^2}{r^2})$

In the previous subsection, we showed an upper bound of the multicast capacity of hybrid network when $k < \theta_1 \cdot a^2/r^2$. In this subsection we will present an upper bound on multicast capacity when $k \geq \theta_1 \cdot a^2/r^2$.

In [11], Li has proved that when $k = \Omega(\frac{a^2}{r^2})$, the union of the transmission disks of these k receiver nodes in a multicast session will cover at least a constant fraction, say $0 < \rho_3 \leq 1$, of the deployment region. Thus the minimum per-flow capacity of hybrid network due to the congestion of ad hoc link will approximately be equal to the broadcast capacity, i.e., $O(\frac{W_a}{n_s})$. Combined with Theorem 13 and Theorem 17, we have the following theorem:

THEOREM 23. *When $k \geq \theta \cdot a^2/r^2$ for a constant θ, the minimum per-flow multicast capacity for the hybrid network is bounded by $O(\min(\frac{W_B}{n_s}, \frac{W_c}{n_s}, \frac{W_a}{n_s}))$ with high probability.*

4. LOWER BOUNDS IN MULTICAST CAPACITY BY CELLULAR ROUTING

In this section, we will derive asymptotically lower bound in the multicast capacity by presenting a multicast scheme.

4.1 Implement of Routing Strategies

We propose the following multicast routing strategy for *Cellular Routing* in Algorithm 1. As we have explained before, based on the *Cellular Routing* strategy, each receiver node will try to reach or be reached by the closest base station by one- or multi-hop. Assume set $U^i = \{v^i_1, v^i_2, \cdots, v^i_k\}$ is the union set of source node v^i and its randomly selected $k - 1$ receivers for the ith multicast flow, here we assume $v^i_k = v_i$ for simplicity. Assume U^i_j is the node set containing all receivers of the ith multicast flow which are falling into the jth cell. Obviously, $U^i = \bigcup_{j=1}^{m} U_j^i$. We further assume set $Z^i = \{z_1^i, z_2^i, \cdots, z_t^i\}$ contains all the base stations, each of whose cell contains at least one receiver of the ith multicast flow. Clearly, $1 \leq t \leq k$.

Algorithm 1 *Cellular Routing* strategy for i^{th} multicast flow

Input: U^i

1: Compute Z_i based on U^i, then construct a Minimum Spanning Tree (MST) which contains all nodes in Z_i (may need other base station as internal nodes) by backbone links only. Assume the root of the constructed MST is the base station (say z_s) which falls in the same cell as the source node v_i does. Then do broadcasting from z_s to the other base stations on the MST.
2: **for** each cell S_j which contains at least one receiver inside **do**
3:     if S_i contains the source node v_i, then v_i finds a shortest path connecting to z_i.
4:     if S_i contains at least one receiver from $k-1$ receivers, construct a BFS tree from the root z_i which covers all receivers inside. This may need other non-receiver nodes as internal nodes on the BFS tree. Then do multicasting from base station z_j to all wireless nodes on the constructed BFS tree.
5: **end for**

In the following section, we will analyze the lower bound multicast capacity where $k = O(\frac{a^2}{r^2})$ and $k = \Omega(\frac{a^2}{r^2})$ separately as we did in the previous sections. When the number of receivers, plus the source node, k is at most $\theta_1 \frac{a^2}{r^2}$, we will construct a multicast tree in each cell S_i which spanning k_i receivers inside and thus obtain a multicast forest which spans all k receivers. Next, we will show the lower bound capacity achievable by the Algorithm 1 under different cases.

4.2 When $k = O(m)$

When the number of receivers of each multicast session satisfies $k = O(m)$, we analyze the minimum per-flow lower bound capacity achievable by backbone links, cellular links and ad hoc links one by one. For each multicast flow, we use Algorithm 1 to do routing.

We first introduce the lower bound capacity achievable by the backbone links. We know for each multicast flow, the broadcast capacity on the MST tree constructed in Algorithm 1 is $\Theta(W_B)$ due to the result in [10]. In addition, according to the result (Theorem 31) in [11] we know that if there are n_s random multicast flows in a square region with side-length a, there is a sequence of $\delta(n) \to 0$ such that for any square cell s with side-length $\frac{a}{\sqrt{m}}$ inside of the square region,

$$\mathbf{Pr}\left(\text{\# of flows using } s \leq \frac{3\delta_3 n_s}{2} \frac{\sqrt{k} \frac{a}{\sqrt{m}}}{a} \right) = \frac{3\delta_3 n_s}{2} \frac{\sqrt{k}}{\sqrt{m}}$$

where δ_3 is some constant. Hence, *w.h.p*, the number of flows needed to be relayed by any base station is no more than $\frac{3\delta_3 n_s}{2} \frac{\sqrt{k}}{\sqrt{m}}$. Therefore, the lower bound capacity for backbone links is at least $\Omega(\frac{W_B \sqrt{m}}{n_s \sqrt{k}})$ by Algorithm 1 with a TDMA schedule.

Next, we present the lower bound capacity that is achievable by cellular links when $k = O(m)$. By Lemma 14, we know for all m cells, when n_s satisfies the condition (4), the variable $\max_{j=1}^{m} f_j$ is $\Theta(n_s \frac{k}{m})$ with probability at least $1 - \frac{1}{n}$. Here, f_j denotes the number of flows inside of jth cell that will pass through the base station z_j. Thus, for any base station, by a simple TDMA schedule, the achievable lower bound capacity for cellular links is $\Omega(\frac{W_c m}{n_s k})$.

In the remaining part of this subsection, we derive the lower bound capacity achievable by ad hoc links using *Cellular Routing* when $k = O(m)$. Recall that after applying Algorithm 1,

each multicast flow is associated with a BFS tree (down-link direction) rooted at the base station or a shortest path (up-link direction) connecting the source node to the base station in each cell if this cell contains at least one receiver of this flow. Based on the result in [10], we know that for each flow, the broadcast capacity achieved by the BFS tree constructed in Algorithm 1 is $\Theta(W_a)$ and it not difficult to show that the up-link direction shortest path which connects the source node to the base station can achieve rate $\Theta(W_a)$ as well without considering all other non-related simultaneously transmission. In addition, by Lemma 19, we know when the total number of multicast flow n_s satisfies the condition 7, the maximum number of ad hoc flows inside of any cell satisfies $\max_{j=1}^{m} f_j'$ is $\Theta(n_s \frac{k}{m})$. Assume $\max_{j=1}^{m} f_j' = c_8 n_s \frac{k}{m}$ for some constant c_8. In other words, *w.h.p*, we have at most $c_8 n_s \frac{k}{m}$ up-link flows or $c_8 n_s \frac{k}{m}$ down-link flows existing in each cell. We simply consider the down-link flows and up-link flows separately. Clearly, by a TDMA schedule, the minimum per-flow rate for both up-link flows and down-link flows can reach at least $\frac{W_a}{c_8 n_s \frac{k}{m}}$. Hence, we have

THEOREM 24. *When $k = O(m)$ and n_s satisfies the condition 7, the lower bound capacity for ad hoc links achieved by applying Algorithm 1 and TDMA schedule is $\Omega(\frac{W_a m}{n_s k})$ With probability at least $1 - \frac{1}{n}$.*

4.3 When $k = O(a^2/r^2)$ and $k = \Omega(m)$

We still use Algorithm 1 to do routing. The achievable lower bound capacity for backbone links and cellular links are easy to get (similar analysis as we did when $k = O(m)$). The only difference in this case is that the number of multicast flows which will go through some base station could be up to but no more than n_s flows due to Lemma 21. Then after applying Algorithm 1 and TDMA scheduling, the achievable lower bound capacity by backbone links and cellular links are $\Omega(\frac{W_B}{n_s})$ and $\Omega(\frac{W_c}{n_s})$ respectively.

The lower bound capacity achievable by all ad hoc links using *Cellular Routing* (Algorithm 1) when $k = \Omega(m)$ can be get by the similar proof as we did in subsection 4.2. The difference is that the possible up-link flows and down-link flows in each cell could be up to but no more than n_s flows (due to Lemma 21). By the same argument, we have the following theorem.

THEOREM 25. *With probability at least $1 - \frac{1}{n}$, the minimum per-flow rate for ad hoc links achievable by applying Algorithm 1 is $\Omega(\frac{W_a}{n_s})$ when $k = \Omega(m)$ and $k = O(\frac{a^2}{r^2})$.*

4.4 When $k = \Omega(\frac{a^2}{r^2})$

Obviously, the total multicast capacity for hybrid network is at least the lower bound of the capacity for broadcast no matter we use either *Cellular Routing* or *Ad Hoc Routing*. In [10], Keshavarz-Haddad *et al.* present a broadcast scheme to achieve capacity $\Theta(W_a)$. Thus, we have

THEOREM 26. *The minimum per-flow multicast capacity achievable by all ad hoc links is at least $c_7 \frac{W_a}{n_s}$, where $c_7 \leq \frac{1}{\Delta+1}$ is a constant.*

Obviously, the minimum per-flow multicast capacity achievable by backbone links and cellular links are $\Omega(\frac{W_B}{n_s})$ and $\Omega(\frac{W_c}{n_s})$ by the similar analysis we used in Section 4.3.

5. CAPACITY BOUND FOR HYBRID ROUTING

In this section, we will give asymptotic upper bounds for any *Hybrid Routing* strategy. The surprising implication of this results is that if we choose the one that can gain larger capacity between *Ad Hoc Routing* and *Cellular Routing* as our routing strategy, the attainable capacity is the same order of the upper bounds of any *Hybrid Routing* strategy. It implies that the upper bounds are tight and our routing strategy is asymptotically optimal.

The aforementioned result is based on the following observation. First, when *Hybrid Routing* is applied, any link in G could be used, in other words, for any multicast flow F_i, the corresponding (resultant) multicast tree T_i for *Hybrid Routing* may contain at most three types of links, the links in E_a, E_B or E_c. Assume $T_a^i = E_a \bigcap T_i$, $T_B^i = E_B \bigcap T_i$ and $T_c^i = E_c \bigcap T_i$, *i.e.*, T_a^i (T_B^i and T_c^i) contains all ad hoc links (backbone links and cellular links) used by tree T_i. Furthermore, we use sets T_a, T_B and T_c to denote the "union" of all ad hoc links, backbone links and cellular links used by all n_s multicast trees. Notice, here the reason that we quote the word union is because the link which is belong to different multicast trees will be counted multiple times in T_a, T_B and T_c, *i.e.*, if we use $|S|$ to denote the summation of the length of all links belong to link set S, then $|T_a| = \sum_{i=1}^{n_s} |T_a^i|$, $|T_B| = \sum_{i=1}^{n_s} |T_B^i|$, $|T_c| = \sum_{i=1}^{n_s} |T_c^i|$.

5.1 When $k = O(m)$

Instead of studying the upper bound for any giving routing strategy directly, we may view this problem in an alternative way: For any given routing strategy, if we can always construct a new routing tree based on it such that the upper bound of multicast capacity by using our new routing strategy is no smaller than a constant fraction of the original one, then the upper bounds for the new routing strategy must be one of the valid asymptotic upper bounds for the original routing strategy. Next, we first give an illustration of our construction method, then an upper bound for the new constructed routing tree will be derived. Finally, we use the above upper bounds as desired upper bounds. In the following contents, we will use D_i to denote a set of base stations used by the optimum hybrid routing strategy for flow i such that each base station in this set has at least one cellular link adjacent to it.

We first give an illustration of our construction approach based on the given routing strategy T_i for flow i as follows:

1. Use the minimum length tree spanning D_i to replace T_B^i, we use $T_B^{i'}$ to denote new tree.
2. Adjust the links contained in T_c^i and T_a^i such that there are no more than 19 cellular links from E_c on the resultant tree after adjustment to each base station, denoting the new trees (forests) by $T_c^{i'}$ and $T_a^{i'}$ respectively .

Next we will explain and analyze these stages in details. In the following contents, we will use λ_a^i, $\lambda_a^{i'}$, λ_B^i, $\lambda_B^{i'}$, λ_c^i and $\lambda_c^{i'}$ to denote the achievable data rate on T_a^i, $T_a^{i'}$, T_B^i, $T_B^{i'}$, T_c^i and $T_c^{i'}$ respectively. Note the data rate by optimum strategy is $\min(\lambda_a^i, \lambda_B^i, \lambda_c^i)$ and the data rate by our modified strategy from the optimum strategy is $\min(\lambda_a^{i'}, \lambda_B^{i'}, \lambda_c^{i'})$

First, we have $|T_B^{i'}| \leq |T_B^i|$, it is straight forward from the fact that $T_B^{i'}$ is a minimum length tree spanning D_i.

Second, we have $c_{11} \lambda_c^{i'} \geq \lambda_c^i$ for some constant c_{11}. This is based on the following observation: Due to the results in [22], we know that we can find at most 19 nodes as "connectors" (one hop away from the base station using cellular links) which can connect to a number of wireless nodes (say "dominators", two hop away from the base station) such that these dominators can cover all nodes which are two hop away from base station. Next, we let all receivers (exiting in the communication disk of the base station) which are not selected to the 19 base stations connect to the closest connector. On the one hand, for down-link from the base station, our construction will not decrease the cellular link rate. On the oth-

er hand, for upper link, we know that only constant number of wireless node (in our case at most 4κ nodes) can transmit at the same time such that each of 19 connectors will have addition burden at most 4κ times than before the construction. Obviously, comparing with the original scheduling period $\mathcal{T}$, after construction, $4\kappa\mathcal{T}$ time is enough for a scheduling period. Hence, we have $c_{11} \lambda_c^{i'} \geq \lambda_c^i$ for some constant $c_{11} \leq 4\kappa + 19$. See Fig. 3 for illustration.

Third, $c_{10} \lambda_a^{i'} \geq \lambda_a^i$ for some constant c_{10}. We guarantee this point by the following observation. After we get connectors during the second step (after adjust cellular links). For some nodes which are two-hop away from the base station in the routing tree before our construction, they could lose the connection to the base station when their relaying nodes to the base station are not selected as "connectors" in the second step. If so, we simply let these nodes connect to the closest "dominator". See Figure 3 for illustration. Let us take u as an example. First, for each internal node u ("dominator") close to the base station. After reconstruction, some other nodes (who lost connection to the base station) will turn to u and ask u to help to relay traffic to the base station. However, these nodes must satisfy two conditions. (1). They are in the communication range of u. (2). They can transmit simultaneously based on the original routing and scheduling strategy. Clearly, u can at most cover 4 closest square region with side length r and for each square cell with side length r, there are at most κ nodes that can transmit simultaneously as we have proved before. Hence, we can guarantee, there are at most 4κ nodes will turn to u in one time slot after construction. Therefore, node u can achieve at least $\frac{1}{4\kappa}$ rate of the original rate before construction by a TDMA scheduling, *i.e.*, $4\kappa\lambda_a^{i'} \geq \lambda_a^i$.

These three important observations guarantee that the upper bounds for the new constructed routing strategy derived from the following analysis are also valid upper bounds for the original routing strategy. See Fig. 3 for illustration.

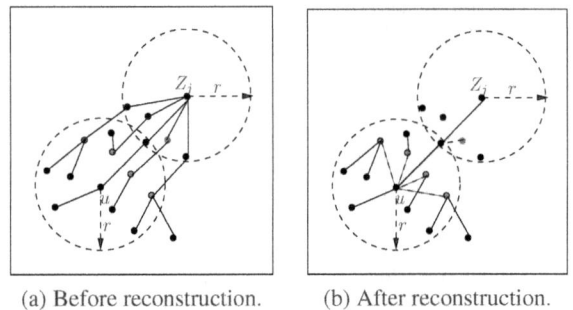

(a) Before reconstruction. (b) After reconstruction.

Figure 3: Part of the Hybrid Network which is near the base station. The dash line from red nodes to u are new added "burden" for node u. z_j is the base station for j^{th} cell. Black nodes denote internal nodes. Red nodes are nodes who will changed their routing strategy after construction. Blue node is one of receivers which is in the communication disk of the base station z_j.

From now on, we focus on studying the upper bounds for the new constructed routing tree: Due to the result in [11], we know that the total length of internal edges of n_s multicast trees spanning k receivers satisfies $\sum_{i=1}^{n_s} |T_i'| = \sum_{i=1}^{n_s} |T_a^{i'}| + \sum_{i=1}^{n_s} |T_B^{i'}| + \sum_{i=1}^{n_s} |T_c^{i'}| = |T_a'| + |T_B'| + |T_c'| \geq c_9 n_s \cdot a \cdot \sqrt{k}$ for some constant c_9. Then we discuss the following two cases respectively:

(1) If $|T_a'| \geq |T_B'| + |T_c'|$: Since the total length of T_a', T_B' and T_c' is no smaller than $c_9 n_s \cdot a \cdot \sqrt{k}$, we have $|T_a'| \geq \frac{1}{2} c_9 n_s \cdot a \cdot \sqrt{k}$.

As shown in [11], the total area covered by all of these ad hoc trees is at least $\eta_1 \cdot n_s \cdot a \cdot \sqrt{k} \cdot r$ for some constant η_1, the number of nodes covered by all ad hoc trees is at least $\eta_1 n_s \cdot a \sqrt{k} r \times \frac{n}{a^2}$ with high probability. Based on the data copy argument, it follows that: $\lambda_a^i \leq c_{10} \lambda_a^{i\,'} \leq \frac{c_{10} W_a a}{\eta_1 n_s \sqrt{k} r}$.

(2) If $|T_a'| < |T_B'| + |T_c'|$: We have $|T_B'| + |T_c'| \geq \frac{1}{2} c_9 n_s \cdot a \cdot \sqrt{k}$ According our construction approach, we know that for any routing tree T_i', each base station has at most 19 adjacent cellular links, then together with the fact that the length of each cellular link is at most r, we have the following inequation: $|T_c'| \leq \frac{|T_B'|}{a/\sqrt{m}} \times 19 \times r$. Since $m \leq \frac{a^2}{10 r^2}$, we have $|T_c'| \leq 6|T_B'|$, it follows that $|T_B'| \geq c_3 n_s a \sqrt{k}$ for some constant c_3. This implies that there is at least one base station which is used by at least $\frac{c_3 n_s a \sqrt{k}}{\sqrt{m}}$ flows. Due to the congestion on this base station, we have the following upper bound: $\lambda_B^i \leq \lambda_B^{i\,'} \leq \frac{W_B \sqrt{m}}{c_3 n_s \sqrt{k}}$. Furthermore, because $T_B^{i\,'}$ spans $|D_i|$ base stations, we have $|T_B^{i\,'}| \leq 2\sqrt{2}\sqrt{|D_i|} \cdot a$ according to the results in [11]. It follows that $c_3 n_s a \sqrt{k} \leq |T_B'| = \sum_{i=1}^{n_s} |T_B^{i\,'}| \leq \sum_{i=1}^{n_s} 2\sqrt{2}\sqrt{|D_i|} a$. Thus, $\sum_{i=1}^{n_s} \sqrt{|D_i|} \geq \eta_3 n_s \sqrt{k}$ for some constant η_3. Together with the fact that $(\sum a_i^p)^{\frac{1}{p}} (\sum b_i^q)^{\frac{1}{q}} \geq \sum a_i b_i$, we have the following:

$$\sum_{i=1}^{n_s} |D_i| > (\frac{\sum_{i=1}^{n_s} \sqrt{|D_i|}}{\sqrt{n_s}})^2 > (\frac{\eta_3 n_s \sqrt{k}}{\sqrt{n_s}})^2 = \eta_3^2 n_s k$$

We conclude that there is at least one based station which is used by at least $\frac{\eta_3^2 n_s k}{m}$ flows to connect wireless nodes directly. Due to the congestion on both ad hoc links and cellular links accessing the base station, we further have the following two upper bounds for this case:

$$\begin{cases} \lambda_a^i \leq c_{10} \lambda_a^{i\,'} \leq c_{10} \frac{W_a}{\eta_3^2 n_s k/m} = c_{10} \frac{W_a m}{\eta_3^2 n_s k} \\ \lambda_c^i \leq c_{11} \lambda_c^{i\,'} \leq c_{11} \frac{W_c}{\eta_3^2 n_s k/m} = c_{11} \frac{W_c m}{\eta_3^2 n_s k} \end{cases} \quad (9)$$

It concludes that the upper bound on capacity achieved by hybrid routing in this case is

$$O(\min\{\frac{W_B \sqrt{m}}{n_s \sqrt{k}}, \frac{W_a m}{n_s k}, \frac{W_b m}{n_s k}\})$$

The final upper bound on multicast capacity of hybrid routing is then at most the maximum one between case 1) and case 2):

LEMMA 27. *The capacity bound for any* Hybrid Routing *strategy is*

$$O(\max\left[\min\{\frac{W_B \sqrt{m}}{n_s \sqrt{k}}, \frac{W_a m}{n_s k}, \frac{W_b m}{n_s k}\}, \frac{W_a a}{n_s \sqrt{k} r})\right])$$

when $k = O(m)$ and $m = O(\frac{a^2}{r^2})$.

It implies that when $k = \Omega(m)$, it is asymptotic optimal to choose the larger one between *Ad Hoc Routing* and *Cellular Routing* as our routing strategy based on the calculated lower bound for each routing strategy.

5.2 When $k = \Omega(m)$ and $k = O(\frac{a^2}{r^2})$

Same as the proof for the previous case, we first construct a new routing tree based on any given routing strategy. Since the upper bound for the new constructed routing tree can also be considered as a valid upper bound for the original routing strategy, we will focus on studying the upper bound for the new constructed routing tree.

Similarly, we have two possible cases need to address:

(1) If $|T_a'| \geq |T_B'| + |T_c'|$: The proof is exactly same as the one shown before for the same case, we gain following upper bound: $\lambda_a^i \leq c_{10} \lambda_a^{i\,'} \leq \frac{c_{10} W_a a}{\eta_1 n_s \sqrt{k} r}$.

(2) If $|T_a'| < |T_B'| + |T_c'|$: We will prove that this case is impossible when $k = \Omega(m)$. Since $T_B^{i\,'}$ is a tree spanning at most m base stations using only backbone links, we get $|T_B'| \leq n_s(m-1) \cdot a/\sqrt{m} < n_s \sqrt{m} a$, we also know that $|T_c'| \leq 19 n_s m r$ because each base station has at most 19 adjacent cellular links. We immediately have when $k > \eta_4 m$ for some large constant η_4, it is impossible that $|T_B'| + |T_c'| > \frac{1}{2} c_9 n_s \cdot a \cdot \sqrt{k}$, in other words, $|T_a'| \not< |T_B'| + |T_c'|$.

We finally have the following lemma:

LEMMA 28. *The capacity bound for any* Hybrid Routing *strategy is at most $O(\frac{W_a a}{n_s \sqrt{k} r})$ when $k = \Omega(m)$, $k = O(\frac{a^2}{r^2})$ and $m = O(\frac{a^2}{r^2})$.*

This result implies that when $k = \Omega(m)$ and $k = O(\frac{a^2}{r^2})$, using *Ad Hoc Routing* is already asymptotic optimum.

5.3 When $k = \Omega(\frac{a^2}{r^2})$

Again, because when $k = \Omega(\frac{a^2}{r^2})$, the union of the transmission disks of these k receiver nodes in a multicast session will cover at least a constant fraction, say $0 < \rho_3 \leq 1$, of the deployment region with high probability when $k = \Omega(\frac{a^2}{r^2})$. Then based on the data copy argument stated in [12], we have the following lemma:

LEMMA 29. *The capacity bounds for any* Hybrid Routing *strategy is $O(\frac{W_a}{n_s})$ when $k = \Omega(\frac{a^2}{r^2})$, $m = O(\frac{a^2}{r^2})$.*

It is not hard to find that its asymptotic optimum to choose *Ad Hoc Routing* as our routing strategy when $k = O(\frac{a^2}{r^2})$.

5.4 Put It All Together

By concluding the previous results of this section, we get the following theorem:

THEOREM 30. *The upper bounds of the multicast capacity for any* Hybrid Routing *strategy is at most*

$$\begin{cases} O(\max\left[\min\left(\frac{W_B \sqrt{m}}{n_s \sqrt{k}}, \frac{W_b m}{n_s k}, \frac{W_a m}{n_s k}\right), \frac{W_a a}{n_s \sqrt{k} r}\right]) & \text{if } k = O(m) \\ O(\frac{a}{r} \cdot \frac{W_a}{n_s \sqrt{k}}) & \text{if } k = \Omega(m), k = O(\frac{a^2}{r^2}) \\ O(\frac{W_a}{n_s}) & \text{if } k = \Omega(\frac{a^2}{r^2}) \end{cases} \quad (10)$$

We then give a general routing strategy which can achieve the asymptotic upper bound derived before for hybrid network $N_{n,m,a}$:

- If $k = O(m)$, we choose the routing strategy that can gain larger data rate between *Ad Hoc Routing* strategy and *Cellular Routing* strategy.
- If $k = \Omega(m)$, we use *Ad Hoc Routing* strategy.

Then together with the lower bound for *Ad Hoc Routing* strategy and *Cellular Routing* strategy, we get the main result in Theorem 2.

6. LITERATURE REVIEW

Gupta and Kumar [7] studied the asymptotic unicast capacity of a multi-hop wireless networks. When each wireless node is capable of transmitting at W-bps using a constant transmission range, the throughput achievable by *each* node for a randomly chosen destination is $\Theta(\frac{W}{\sqrt{n \log n}})$ bits per second. Grossglauser and Tse recently showed that the unicast capacity can be improved by the mobility of wireless nodes regardless of delay. Gastpar and Vetterli studied the capacity of random networks using relay in [1]. Chuah *et al.* [3] studied the capacity scaling in MIMO wireless systems under correlated fading. The capacity scaling in delay tolerant networks with heterogeneous mobile devices was studied by Garetto *et al.* [4]. Keshavarz-Haddad *et al.* studied the bounds for the capacity of wireless networks imposed by topology and demand in [9]. Their techniques can be used to study unicast, multicast and broadcast capacity. Broadcast capacity of both arbitrary networks and random networks has been studied in [10, 14]. Keshavarz-Haddad *et al.* [5] studied the broadcast capacity with dynamic power adjustment for physical interference model. Multicast capacity was also studied in the literature. Jacquet and Rodolakis [6] studied the scaling properties of multicast for random wireless networks. They claimed that the maximum rate at which a node can transmit multicast data is $O(\frac{W}{\sqrt{kn \log n}})$. Recently, rigorous proofs of the multicast capacity were given in [12, 15]. Li *et al.* [12] studied asymptotic multicast capacity for a large-scale random wireless networks. They showed the total multicast capacity is $\Theta(\sqrt{\frac{n}{\log n}} \cdot \frac{W}{\sqrt{k}})$ when $k = O(\frac{n}{\log n})$ and when $k = \Omega(\frac{n}{\log n})$, the total multicast capacity is equal to the broadcast capacity, *i.e.*, $\Theta(W)$. Li *et al.* [18] studied the lower bound of multicast capacity for large scale wireless networks under Gaussian Channel model. Liu *et al.* [17] studied the unicast capacity of hybrid network (a wireless ad hoc network with infrastructure). They essentially studied the unicast capacity of hybrid wireless networks under the one-dimensional network model and two-dimensional strip model respectively. Kozat and Tassiulas [16] also studied the unicast capacity of ad hoc networks with a random flat topology under the present support of an infinite capacity infrastructure network. They showed that the per source node capacity of $\Theta(W/\log n)$. In [19], Mao *et al.*, studied the multicast capacity for hybrid networks by using *Cellular Routing* strategy. Very recently, Chen *et al.* [2] study the multicast capacity under a hybrid network model featuring both node's mobility and infrastructure support.

7. ACKNOWLEDGEMENT

The research of authors is partially supported by NSF CNS-0832120, NSF CNS-1035894, National Basic Research Program of China (973 Program) under grant No. 2010CB328101, 2010CB334707, 2011CB302705, National Natural Science Foundation of China under Grant No. 61170216, Tsinghua National Laboratory for Information Science and Technology (TNList), program for Zhejiang Provincial Key Innovative Research Team, and program for Zhejiang Provincial Overseas High-Level Talents (One-hundred Talents Program).

8. CONCLUSIONS

In this paper, we essentially studied the multicast capacity that can be achieved by hybrid networks with randomly distributed wireless nodes and regularly distributed base stations. We derived analytical matching upper bounds and lower bounds on multicast capacity of hybrid networks.

9. REFERENCES

[1] GASTPAR, M., AND VETTERLI, M. On the capacity of wireless networks: the relay case. In *IEEE INFOCOM* (2002).

[2] CHEN, X., HUANG, W., WANG, X.-B., LIN, X.-J. Multicast Capacity in Mobile Wireless Ad Hoc Network with Infrastructure Support In *IEEE INFOCOM* 2012.

[3] CHUAH, C.-N., TSE, D. N. C., KAHN, J. M., AND VALENZUELA, R. A. Capacity scaling in mimo wireless systems under correlated fading. In *IEEE TIT* (2002), vol. 48.

[4] GARETTO, M., GIACCONE, P., AND LEONARDI, E. Capacity scaling in delay tolerant networks with heterogeneous mobile nodes. In *ACM MobiHoc* (2007).

[5] KESHAVARZ-HADDAD, A., AND RIEDI, R. On the broadcast capacity of multihop wireless networks: Interplay of power, density and interference. In *IEEE (SECON)* (2007).

[6] JACQUET, P., AND RODOLAKIS, G. Multicast scaling properties in massively dense ad hoc networks. In *ICPADS* (2005).

[7] GUPTA, P., AND KUMAR, P. Capacity of wireless networks. *IEEE TIT* (1999).

[8] GROSSGLAUSER, M., AND TSE, D. Mobility increases the capacity of ad hoc wireless networks. *IEEE/ACM TON*(2002),477–486.

[9] KESHAVARZ-HADDAD, A., AND RIEDI, R. H. Bounds for the capacity of wireless multihop networks imposed by topology and demand. In *ACM MobiHoc* (2007).

[10] KESHAVARZ-HADDAD, A., RIBEIRO, V., AND RIEDI, R. Broadcast capacity in multihop wireless networks. In *ACM MobiCom* (2006).

[11] LI, X.-Y., Multicast Capacity of Wireless Ad Hoc Networks, In *IEEE/ACM ToN* (2008).

[12] LI, X.-Y., TANG, S.-J., AND OPHIR, F. Multicast capacity for large scale wireless ad hoc networks. In *ACM Mobicom* (2007).

[13] PENROSE, M. The longest edge of the random minimal spanning tree. *Annals of Applied Probability 7* (1997).

[14] TAVLI, B. Broadcast capacity of wireless networks. *IEEE Communication Letters*.

[15] SHAKKOTTAI S., LIU X. AND SRIKANT R., The multicast capacity of ad hoc networks, *Proc. ACM Mobihoc*, 2007.

[16] ULA C. KOZAT AND LEANDROS TASSIULAS, Throughput capacity of random ad hoc networks with infrastructure support, *Proc. ACM Mobihoc*, 2003.

[17] BENYUAN LIU, PATRICK THIRAN, DON TOWSLEY, Capacity of a Wireless Ad Hoc Network with Infrastructure, *ACM Mobihoc*, 2007.

[18] LI, S., LIU Y.-H., AND LI, X.-Y., Capacity of Large Scale Wireless Networks Under Gaussian Channel Model *ACM MobiCom*, 2008.

[19] MAO, X.-F., LI, X.-Y., AND TANG, S.-J., Multicast Capacity for Hybrid Wireless networks *Proc. ACM MobiHoc*, 2008.

[20] ABU-MOSTAFA, YASER S, The Vapnik-Chervonenkis Dimension: Information versus Complexity in Learning *Neural Computation*, 1989.

[21] J. M. STEELE, Growth rates of euclidean minimal spanning trees with power weighted edges. *The Annals of Probability*, 1988

[22] HUANG, S.-C., WAN, P.-J., VU, C.-T, LI, Y.-S, YAO, F.-F, Nearly Constant Approximation for Data Aggregation Scheduling in Wireless Sensor Networks. *IEEE INFOCOM* , 2007

Serendipity: Enabling Remote Computing among Intermittently Connected Mobile Devices

Cong Shi*, Vasileios Lakafosis†, Mostafa H. Ammar*, Ellen W. Zegura*
*School of Computer Science   †School of Electrical and Computer Engineering
Georgia Institute of Technology   Georgia Institute of Technology
{cshi7, ammar, ewz}@cc.gatech.edu   vasileios@gatech.edu

ABSTRACT

Mobile devices are increasingly being relied on for services that go beyond simple connectivity and require more complex processing. Fortunately, a mobile device encounters, possibly intermittently, many entities capable of lending it computational resources. At one extreme is the traditional cloud-computing context where a mobile device is connected to remote cloud resources maintained by a service provider with which it has an established relationship. In this paper we consider the other extreme, where a mobile device's contacts are only with other mobile devices, where both the computation initiator and the remote computational resources are mobile, and where intermittent connectivity among these entities is the norm. We present the design and implementation of a system, Serendipity, that enables a mobile computation initiator to use remote computational resources available in other mobile systems in its environment to speedup computing and conserve energy. We propose a simple but powerful job structure that is suitable for such a system. Serendipity relies on the collaboration among mobile devices for task allocation and task progress monitoring functions. We develop algorithms that are designed to disseminate tasks among mobile devices by accounting for the specific properties of the available connectivity. We also undertake an extensive evaluation of our system, including experience with a prototype, that demonstrates Serendipity's performance.

Categories and Subject Descriptors

C.2.4 [**Computer-Communication Networks**]: Distributed Systems —*Distributed applications*

General Terms

Algorithms, Design, Experimentation, Performance

Keywords

Remote Computing, Task Allocation, Mobile Devices, Opportunistic Networks, Energy Management

1 Introduction

Recent years have seen a significant rise in the sophistication of mobile computing applications. Mobile devices are increasingly being relied on for a number of services that go beyond simple connectivity and require more complex processing. These include pattern recognition to aid in identifying snippets of audio or recognizing images whether locally captured or remotely acquired, reality augmentation to enhance our daily lives, collaborative applications that enhance distributed decision making and planning and coordination, potentially in real-time. Additionally, there is potential for mobile devices to enable more potent "citizen science" applications that can help in a range of applications from understanding how ecosystems are responding to climate change[1] to gathering of real-time traffic information.[2]

Mobile applications have become an indispensable part of everyday life. This has been made possible by two trends. First, truly portable mobile devices, such as smartphones and tablets, are increasingly capable devices with processing and storage capabilities that make significant step improvements with every generation. While power in mobile devices will continue to be constrained relative to tethered devices, advances in battery and power management technology will enable mobile devices to manage longer-lived computations with less burden on available power [22]. A second trend that is directly relevant to our work is the availability of improved connectivity options for mobile devices. These have enabled applications that transcend an individual device's capabilities by making use of remote processing and storage.

Fortunately, a mobile device often encounters, possibly intermittently, many entities capable of lending it computational resources. This environment provides a spectrum of computational contexts for remote computation in a mobile environment. An ultimately successful system will need to have the flexibility to use a mix of the options on that spectrum. At one extreme of the spectrum is the use of standard cloud computing resources to off-load the "heavy lifting" that may be required in some mobile applications to specially designated servers or server clusters. A related technique for remote processing of mobile applications proposes the use of *cloudlets* which provide software instantiated in real-time on nearby computing resources using virtual machine technology [30]. Likewise, MAUI [12] and CloneCloud [11] automatically apportion processing between a local device and a remote cloud resource. In this paper we consider the other spectrum extreme, where a mobile device's contacts are only with other mobile devices, where both the computation initiator and the remote com-

---

[1]See http://blogs.kqed.org/climatewatch/2011/01/29/citizen-science-the-iphone-app/
[2]See http://www.crisscrossed.net/2009/08/31/citizen-scientist-how-mobile-phones-can-contribute-to-the-public-good/

putational resources are mobile, and where intermittent connectivity among these entities is the norm.

We investigate the basic scenario where an *initiator* mobile device needs to run a computational task that exceeds the mobile device's ability and where portions of the task are amenable to remote execution. We leverage the fact that a mobile device within its intrinsic motion pattern makes frequent contact with other mobile devices that are capable of providing computing resources. Contact with these devices can be intermittent, limited in duration when it occurs, and sometimes unpredictable. The goal of the mobile device is to use the available, potentially intermittently connected, computation resources in a manner that improves its computational experience, e.g., minimizing local power consumption and/or decreasing computation completion time. The challenge facing the initiator device is how to apportion the computational task into subtasks and how to allocate such tasks for remote processing by the devices it encounters.

1.1 Related Work

Our work can be viewed as enabling a truly general vision of *cyber foraging* [4, 5] which envisions mobile applications "living off the land" by exploiting nearby computational resources. At the time of the original conception of the cyber foraging idea almost a decade ago [35] it was hard to imagine the compute power of today's mobile devices (smartphones and tablets) and the vision was, therefore, necessarily limited to constant connectivity to infrastructure-based services. Today, however, it is possible to extend the flexibility of this vision to include "foraging" of the available resources in other mobile devices as we propose to do in this work.

Our work also leverages recent advances in the understanding of data transfer over intermittently-connected wireless networks (also known as disruption-tolerant networks or opportunistic networks). These networks have been studied extensively in a variety of settings, from military [26] to disasters [15] to the developing world [27]. These settings share the characteristic that fixed infrastructure is unavailable, highly unreliable, or expensive. Further, the communication links are subject to disruptions that mean network partitions are common.

Our work is also related to the efforts at developing useful applications over intermittently-connected mobile and wireless networks. Examples of this work include the work by Hanna et al. which develops mobile distributed information retrieval systems [16], and the work by Fall et al. on an architecture for disaster communications response [15] with a specific focus on situational awareness. In this latter work the authors propose an architecture that contains infrastructure-supported servers, mobile producer/consumer nodes and mobile field servers. Related, the Hastily Formed Networks (HFN) project [14] describes potential applications in disaster settings that match well with our vision requiring computation, including situational awareness, information sharing, planning and decision making.

Our work is also closely related to systems that use non-dedicated machines with cycles that are donated and may disappear at any time. In this vein, our work takes some inspiration from the Condor system architecture [31]. Our work also resembles in part those distributed computing environments that have well-connected networks but unreliable participation in the computation, such as those seen in voluntary computing efforts where users can contribute compute cycles, but may also simply turn off their machines or networks at will in the middle of a computation. Examples of these systems include BOINC [2]; other examples are SETI@home [3], and folding@home[6], all leveraging willingness on the part of individuals to dedicate resources to a large computation problem.

More recently, the Hyrax project envisions a somewhat similar capability to opportunistically use the resources of networked cellphones [25].

1.2 Paper Outline

The remainder of this paper is organized as follows: we start with the discussion of the problem context and the design challenges in Section 2; we describe the design of a job model and the Serendipity system in Section 3; the task allocation algorithms are presented in Sections 4 and 5; we describe how to enable energy-aware computing in Section 6; we undertake an extensive evaluation of our system on Emulab in Section 7; the implementation and evaluation of Serendipity on mobile devices are presented in Section 8; We conclude this paper and discuss our future work in Section 9.

2 Problem Context and Design Challenges

Network Model: We focus on a network environment that is composed of a set of mobile nodes with computation and communication capabilities. The network connectivity is intermittent, leading to a frequently-partitioned network. Every node can execute computing tasks, the number of which is constrained by its resources, such as processor capability, memory, storage size, and available energy. The period of time during which two nodes are within communication range of each other is called a *contact*. During a contact nodes can transfer data to each other. Both the duration and the transfer bandwidth of a contact are limited. There are some variants of the general network setting. For some mobile devices, a low-capacity control channel (e.g., over satellite link) is available for metadata sharing. In addition, in some special networks, such as networks with scheduled robotic vehicles or UAVs, the node mobility patterns are predictable and, thus, their future contacts are also predictable. All these variants are taken into consideration in our design.

Remote computing usually involves the execution of computationally complex jobs through the cooperation among a set of devices connected by a network. A major class of such jobs, supported by mainstream distributed computing platforms such as Condor [31], can be represented as a Directed Acyclic Graph (DAG). The vertices are programs and the directed links represent data flows between two programs. A traditional distributed computing platform maps the vertices to the devices and the links to the network so that all independent programs are executed in parallel and they transfer the output to their children. As a variant of such computing platforms, MAUI [12] and CloneCloud [11] have a simple network composed of a mobile device and the cloud.

Design Challenges: The intermittent connectivity among mobile devices poses three key challenges for remote computing. First, because the underlying connectivity is often unknown and variable, it is difficult to map computations onto nodes with an assurance that the required code and data can be delivered and the results are received in a timely fashion. This suggests a conservative approach to distributing computation so as to provide protection against future network disruptions. Second, given that the network bandwidth is intermittent, the network is more likely to be a bottleneck for the completion of the distributed computation. This suggests scheduling sequential computations on the same node so that the data available to start the next computation need not traverse the network. Third, when there is no control channel, the network cannot be relied upon to provide reachability to all nodes as needed for coordination and control. This suggests maintaining local control and developing mechanisms for loose coordination. Besides the intermittent connectivity, the limited available energy imposes

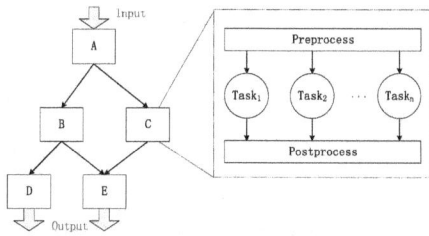

Figure 1: A job model for DTNs is a Directed Acyclic Graph (DAG), the vertices of which are PNP-blocks. Every PNP-block consists of a pre-process, a post-process and n parallel tasks.

another extra constraint on the remote computing among mobile devices.

3 Serendipity System Design

3.1 A Job Model for Serendipity

Our basic job component is called a *PNP-block*. As shown in Figure 1, a PNP-block is composed of a *pre-process* program, n parallel *task* programs and a *post-process* program. The pre-process program processes the input data (e.g., splitting the input into multiple segments) and passes them to the tasks. The workload of every task should be similar to each other to simplify the task allocation. The post-process program processes the output of all tasks; this includes collecting all the output and writing them into a single file.

The PNP-block design simplifies the data flow among tasks and, thus, reduces the impact of uncertainty on the job execution. All pre-process and post-process programs are executed on one initiator device, while parallel tasks are executed independently on other devices. The communication graph becomes a simple star graph. The data transfer delay can be minimized as the initiator device can simply choose nearby devices to execute tasks. In contrast, it is much more difficult for a complicated communication graph, such as the complete bipartite graph used in MapReduce [13], to achieve low delay among intermittently connected mobile devices because the optimization problem associated with mapping the general graph onto them is complex.

The single PNP-block job comprises an important class of distributed computing jobs often called embarrassingly parallel and useful in many applications, among which are SETI@home [3] and BOINC [2]. All jobs are graphically represented by a DAG of PNP-blocks, providing as much computational expressiveness as a regular DAG. For instance, the MapReduce model [13] can be implemented with two sequentially connected PNP-blocks, corresponding to the map phase and the reduce phase, respectively.

3.2 Serendipity System

Figure 2 shows the high-level architecture of Serendipity. A Serendipity node has a *job engine* process, a *master* process and several *worker* processes. The number of worker processes can be configured, for example, as the number of cores or processors of the node. Each node constructs its device profile and, then, shares and maintains the profiles of encountered nodes. A node's device profile includes its execution speed which is estimated by running synthetic benchmarks and its energy consumption model using techniques like PowerBooter [34]. These device profiles when combined with the jobs' execution profiles are used to estimate the jobs' execution time and energy consumption on every node, essential for task allocation. Serendipity also needs access to the contact database, if available, for better task allocation.

To submit a job, a user needs to provide a script specifying the job DAG, the programs and their execution profiles (e.g., CPU cycles) for all PNP-blocks and the input data to the job engine. Con-

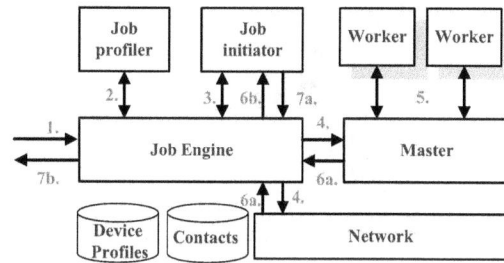

Figure 2: High-level Architecture of Serendipity. After receiving a job (1), the job engine constructs the job profile (2) and starts a job initiator, who will initiate a number of PNP-blocks and allocate their tasks (3). The job engine disseminates the tasks to either local or remote masters (4). After a worker finishes a task (5), the master sends back the results to the job initiator (6a, 6b), who may trigger new job PNP-blocks (3). After all results are collected, the job initiator returns the final results (7a, 7b) and stops.

structing accurate execution profiles of programs is a challenging problem and out of the scope of this paper. We simply follow the offline method used by both MAUI [12] and CloneCloud [11], i.e., running the programs multiple times with different input data.

The script is submitted to the *job profiler* for basic checking and constructing a complete job profile (i.e., tasks' execution time and energy consumption on every node) using its execution profiles and the device profiles. The generated job profile will be used to decide how to allocate its tasks among mobile devices.

If everything is correct, the job engine will launch a new *job initiator* responsible for the new job. It stores the job information in the local storage until the job completes. All PNP-blocks whose parents have completed will be launched by running their pre-process programs on a local worker and assigning a TTL (i.e., time-to-live), a priority and a worker to every task. The TTL specifies the time before which its results should be returned. If a task misses its TTL, it should be discarded, while a copy will be executed locally on the initiator's mobile device. The priority determines the relative importance of a job's different tasks. Section 5 will discuss how to assign the priorities.

Based on the consideration of task allocation and security, the assigned worker can be a single node, a set of candidate nodes, or a wildcard. In fact, only in the specific scenario that the future contacts are predictable while nodes have a control channel to timely coordinate the remote computing, the job initiator will use the global information to allocate tasks and assign a specific node for each task, which will be discussed in Section 4.1. Otherwise, the job initiator only specifies the set of candidate nodes it trusts and lets the job engine allocate the tasks. Finally, these tasks are sent to the job engine for dissemination.

The job engine is primarily responsible for disseminating tasks and scheduling the task execution for the local master. When two mobile nodes encounter, they will first exchange the metadata including their device profiles, their residual energy and a summary of their carried tasks. Using this information, the job engine will estimate whether it is better to disseminate a task to the encountered node than to execute it locally. Such a decision is based on the goal of reducing the job completion time (to be discussed in Section 4) or conserving the device energy (to be discussed in Section 6).

To schedule the task execution, the job engine first determines the job priority. Currently we use the first-in-first-serve policy. But it can be easily replaced by any arbitrary policy. For example, the job from a node that helps other nodes execute a lot of tasks is assigned a high priority. For the tasks of the same job, they are scheduled according to their task priorities.

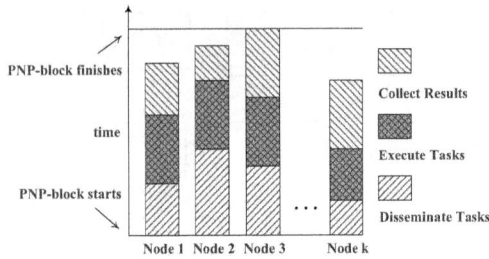

Figure 3: The PNP-block completion time is composed of a) the time to disseminate tasks, b) the time to execute tasks and c) the time to collect results, in addition to the time needed to execute pre-process and post-process programs.

The *master* is responsible for monitoring the task execution on workers. After receiving a task from the job engine, it starts a worker for it. When the task finishes, the output will be sent back to the job initiator using the underlying routing protocols like Max-Prop [7]. If the task throws an exception during the execution, the master will report it to the job initiator who will terminate the job and report to the user.

In this paper, we assume that all nodes are collaborative and trustworthy. However, there are also scenarios that some nodes are selfish (i.e., refusing to help other nodes) or even malicious (i.e., distorting the results). To motivate the selfish nodes, we can use some token-based incentive mechanism [24], making use of notional credit to pay off nodes for executing tasks. To protect the remote computing from malicious nodes, we can use reputation-based trust [8] in which nodes construct and share nodes' reputation information.

4 Task Allocation for PNP-blocks

One important goal of remote computing is to improve the performance of computationally complex jobs, especially when mobile nodes have enough energy. In this section, we will design efficient task allocation algorithms to minimize the job completion time. Specifically, since PNP-blocks are the basic blocks to allocate tasks, we will focus on the task allocation for PNP-blocks in various network settings. The problem of task scheduling for multi-processor systems [10, 18] is somewhat related to our task allocation problem. That work, however, does not deal with intermittent connectivity and cannot, therefore, be applied directly to our problem.

Figure 3 illustrates the timing and components of a PNP-block execution. Along the x-axis are the k remote nodes that will execute the parallel tasks of the block. Along the y-axis is a depiction of the time taken at each node to receive disseminated tasks from the initiator, execute those tasks, and provide the result collection back to the initiator. As illustrated, the time for each remote node to receive its disseminated tasks may vary, depending on the availability and quality of the network between the initiator and the remote node. When n tasks of a PNP-block are allocated to k nodes, each node will execute its assigned tasks sequentially, again taking a variable amount of time. After execution of all assigned tasks in the block, the node will send results back to the initiator, with time again being dependent on the network between the initiator and the remote node. Our goal for the task allocation is to reduce the completion time of the last task which equals to the PNP-block completion time.

We consider the design of task allocation algorithms in the context of three models with different contact knowledge and control channel availability assumptions.

4.1 Predictable Contacts with Control Channel

We first consider an ideal network setting where the future contacts can be accurately predicted, and a control channel is available for coordination. The performance in this type of scenarios represents the best possible performance of task allocation that is achievable among intermittently connected mobile devices. It is useful to identify the fundamental benefits and limits of Serendipity.

With future contact information a Dijkstra's routing algorithm for DTNs [20] can be used to compute the required data transfer time between any pair of nodes given its starting time. With the control channel the job initiator can obtain the time and number of tasks to be executed on the target node with which to estimate the time to execute a task on that node. Therefore, given the starting time and the target node, the task completion time can be estimated.

Using this information, we propose a greedy task allocation algorithm, *WaterFilling*, that iteratively chooses the destination node for every task with the minimum task completion time (see Algorithm 1).

Algorithm 1 Water Filling

1: **procedure** WATERFILLING(T, N)   ▷ T is task set; N is node set.
2:     current ← currentTime();
3:     rsv ← getTaskReservationInfo();
4:     inputSize ← getTaskInputSize(T);
5:     outputsize ← estimateOutputSize(T);
6:     queue ← initPriorityQueue();
7:     **for all** $n \in N$ **do**
8:         arrivalT ← dijkstra(this, n, current, inputSize);
9:         exeT ← estimateTaskExecutionTime(n,t);   ▷ $t \in T$
10:         tfinishT ← taskFinishTime(rsv[n], arrivalT, exeT);
11:         completeT ← dijkstra(n, this, tfinishT, outputSize);
12:         queue.put({n, arrivalT, exeT, completeT});
13:     **end for**
14:     **for all** $t \in T$ **do**
15:         {n, arrivalT, exeT, receiveT} ← queue.poll();
16:         updateReservation(rsv[n], t, inputSize, arrivalT, exeT);
17:         send(n, t);
18:         arrivalT ← dijkstra(this, n, current, inputSize);
19:         tfinishT ← taskFinishTime(rsv(n), arrivalT, exeT);
20:         completeT ← dijkstra(n, this, tfinishT, resultSize);
21:         queue.put({n, arrivalT, exeT, receiveT});
22:     **end for**
23:     reserveTaskTime(rsv);
24: **end procedure**

For every task, the algorithm first estimates its task dissemination time to every node. With the information of the tasks to be executed on the destination node and the estimated time to execute this task, it is able to estimate the time when this task will finish. Given that time point, the time when the output is sent back can also be computed. Among all the possible options, we choose the node that achieves the minimum task completion time to allocate the task. The allocation of the next task will take the current task into account and repeat the same process. Finally, the job initiator will reserve the task execution time on all related nodes, which will be shared with other job initiators for future task allocation.

4.2 Predictable Contacts without Control Channel

When mobile nodes have no control channels, it is impossible to reserve task execution time in advance. WaterFilling will cause contention for task execution among different jobs on popular nodes, prolonging the task execution time. To solve this problem, we propose an algorithm framework, Computing on Dissemination (CoD), to allocate tasks in an opportunistic way. The algorithm is shown in Algorithm 2.

Algorithm 2 Computing on Dissemination

```
 1: procedure ENCOUNTER(n)                    ▷ n is the encountered node.
 2:     summary ← getSummary();
 3:     send(n, summary);
 4: end procedure
 5: procedure GETSUMMARY
 6:     compute ← getNodeComputingSummary();
 7:     net ← getNetworkSummary();
 8:     tasks ← getPendingTaskSummary();
 9:     return {compute,net,tasks};
10: end procedure
11: procedure RECEIVESUMMARY(n, msg)          ▷ msg is the summary
       message of node n.
12:     updateNodes(msg.compute);
13:     updateNetwork(msg.net);
14:     toExchange ← exchangeTask(n, this.tasks, msg.tasks);
15:     isSent ← false;
16:     while n.isConnected() && !toExchange.isEmpty() do
17:         send(n, toExchange.poll());
18:         isSent ← true;
19:     end while
20:     if n.isConnected() && isSent == true then
21:         summary ← getSummary();
22:         send(n, summary);
23:     end if
24: end procedure
25: procedure RECEIVETASK(msg)                 ▷ msg contains exchanged tasks.
26:     addTasks(msg.tasks);
27: end procedure
```

The basic idea of CoD is that during the task dissemination process, every intermediate node can execute these tasks. Instead of explicitly assigning a destination node to every task, CoD opportunistically disseminates the tasks among those encountered nodes until all tasks finish. Every time two nodes encounter each other, they first exchange metadata about their status. Based on this information, they decide the set of tasks to exchange. When they move out of the communication range, they will keep the remaining tasks to execute locally or exchange with other encountered nodes in the future.

The key function of this algorithm is the *exchangeTask* function of line 14 that decides which tasks to exchange. In this subsection we assume that future contact is still predictable. Therefore, the task completion time can be estimated when the task arrives at a node as discussed in last subsection. The intuition of CoD with predictable contacts (pCoD) is to locally minimize the task completion time of every task if possible. When a node receives the summary message from the encountered node, it first estimates the execution time of its carried tasks on the other node using the job profiles and the device profiles. For each task it carries, it estimates the task completion time (i.e., the time that its result is received by the initiator) of executing locally and that of executing on the other node by using the contact information. If the local task completion time is larger than the remote one, it sends the task to the encountered node. Every node conservatively makes the decision without considering the tasks the other node will send back.

4.3 Unpredictable Contacts

Finally we consider the worst case that future contacts cannot be accurately predicted. Our task allocation algorithm, CoD with unpredictable contacts (upCoD), is still based on CoD with the constraint that future contact information is unavailable. As shown in Figure 3, minimizing the time when the last task is sent back to the job initiator will reduce the PNP-block completion time. When the data transfer time is unpredictable, we envision that reducing the execution time of the last task will also help reduce PNP-block

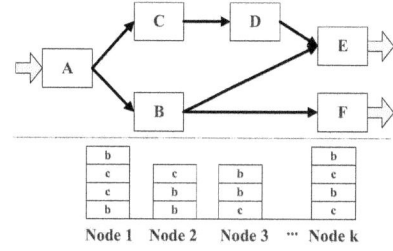

Figure 4: A job example where both PNP-block B and C are disseminated to Serendipity nodes after A completes. Their task positions in the nodes' task lists are shown blow the DAG.

completion time. This is because the locality property of CoD indicates the existence of a short time-space path between the worker node and the job initiator node. Therefore, when two nodes encounter each other, upCoD tries to reduce the execution time of every task.

In reality, historical contact information is useful to roughly estimate the future contacts [7] and, thus, should be helpful to task exchange in CoD. Its performance is probably between upCoD and pCoD. We will investigate such possibility as part of our future work.

5 PNP-block Scheduling

Our PNP block design simplifies the task allocation so that every PNP-block is treated independently. However, it is still possible to further reduce the job completion time by assigning priorities to PNP-blocks since tasks from the same job are executed according to their priority assignment.

Let's consider a simple job DAG shown in Figure 4. PNP-blocks B and C are simultaneously allocated after A completes. Their tasks arrive at the destination nodes unordered. Given a network and a task allocation algorithm, the total time required for both B and C to finish remains almost the same. However, either B or C can have a shorter PNP-block finish time if any of them is given a higher priority over the other. This will be beneficial because their children PNP-block can start earlier.

OBSERVATION 1. *It is better to assign different priorities to the PNP-blocks of a job.*

In the example shown in Figure 4, PNP-block E can only start when both B and D finish. Thus, B and D are equivalently important to E. Meanwhile, there is a time gap between the execution time of C and that of D caused by the result collection of C and task dissemination of D. During that gap, the execution of other tasks (e.g., B) will not affect the PNP-block finish time of D. Therefore, if C is assigned a higher priority than B, the total time for both B and D to finish will be shorter.

OBSERVATION 2. *All parents of a PNP-block are equivalently important to it, while parents have higher priorities than their children.*

The next question arises when B and D are in the task list of the same node, which should have higher priority. We notice that both B and D are equivalent to E, while E and F are equivalent to the job. However, if B finishes earlier, F can start earlier. This is because F only relies on B.

OBSERVATION 3. *When two PNP-blocks have the same priority, the one with more children only depending on it should be assigned a higher priority.*

If there are still PNP-blocks with the same priority, we randomly assign some different priorities to them that keep their relative priorities with other PNP-blocks. Algorithm 3 shows our priority assigning algorithm. The sort method of line 13 is based on Observation 3.

Algorithm 3 PNP-block Priority Assigning

```
 1: procedure ASSIGNPRIORITY(J)              ▷ J is the job DAG
 2:     while !J.allPNPblocksHavePriority() do
 3:         for all s ∈ J do                  ▷ s is a PNP-block
 4:             if !s.haveChild() then
 5:                 s.priority ← 0;
 6:             else if s.allChildrenHavePriority() then
 7:                 s.priority ← s.maxChildrenPriority()+1;
 8:             end if
 9:         end for
10:     end while
11:     for p = 0 → J.getMaxPriority() do
12:         PNPblocks ← J.getPNPblocksWithPriority(p);
13:         sort(PNPblocks);
14:         for i = 0 → PNPblocks.size()-1 do
15:             s ← PNPblocks.get(i);
16:             s.priority ← s.priority + i/(PNPblocks.size());
17:         end for
18:     end for
19: end procedure
```

6 Energy-Aware Computing

In the above two sections we focused on how to accelerate the job execution without any consideration of the energy consumption. Because of the limited energy available to some kinds of mobile devices (e.g., smartphones), there are also scenarios when energy conservation is at least as important as execution performance, especially when the applications can tolerate delays. In this section, we describe how to support energy-aware computing with Serendipity.

When a mobile device tries to off-load a task to another mobile device to save energy, the latter may have very limited energy, too. Meanwhile, if all nodes postpone the task execution forever, it definitely saves energy, but meaninglessly. Therefore, a reasonable objective of energy-aware computing among mobile devices makes all nodes last as long as possible while timely finishing the jobs, i.e., maximizing the lifetime of the first depleted node under the constraint that jobs complete before their deadline (i.e., TTL). Unfortunately, without information about the future jobs, it is impossible to solve this optimization problem.

An approximation to this ideal optimization is to greedily minimize a utility function when the job initiator allocates the tasks. Two factors should be considered in the utility functions, the energy consumption of all nodes involved in the remote computing of the task and the residual energy available to these nodes. A good utility function should consume less energy while avoiding nodes with small residual energy. We use a simple utility function that has been considered in energy-aware routing [9]:

$$u(T) = \sum_{i \in N_T} \frac{e_{Ti}}{R_i} \qquad (1)$$

where N_T is the set of nodes involved in the remote computing of task T, e_{Ti} is the energy consumption of node i for task T, and R_i is the residual energy of node i.

As discussed in Section 4 the task allocation algorithms, Water-Filling, pCoD and upCoD, try to optimize the job completion time. By replacing the time with the utility function $u(T)$, we can easily adapt these task allocation algorithms to be energy-aware. Specifically, the energy-aware WaterFilling algorithm iteratively chooses the destination node of every task with minimum $u(T)$ while satisfying the TTL constraint. When two nodes encounter, pCoD and upCoD will exchange a task if executing it on current node has higher utility than executing on the other node while satisfying the TTL constraint. If the future contacts are unpredictable, upCoD replaces TTL with the time that task is executed.

7 Evaluation

7.1 Experimental Setup

To evaluate Serendipity in various network settings, we have built a testbed on Emulab [33] to easily configure the experiment settings including the number of nodes, the node properties, etc. In our testbed, a Serendipity node running on an Emulab node has an emulation module to emulate the intermittent connectivity among nodes. Before an experiment starts, all nodes load the contact traces into their emulation modules. During the experiments, the emulation module will control the communication between its node and all other nodes according to the contact traces.

In the following experiments, we use two real-world contact traces, a 9-node trace collected in the Haggle project [19] and the Roller-Net trace [32]. In the RollerNet trace, we select a subset of 11 friends (identified in the metadata of the trace) among the 62 nodes so that the number of nodes is comparable to the Haggle trace. The Haggle trace represents the user contacts in a laboratory during a typical day, while RollerNet represents the contacts among a group of friends during the outdoor activity. These two traces demonstrate quite different contact properties. RollerNet has shorter contact intervals, while Haggle has longer contact durations.

We also use three mobility models to synthesize contact traces, namely the Levy Walk Model [28], the Random WayPoint Model (RWP) [29], and the Time-Variant Community Mobility Model (TVCM) [21]. We change various parameters to analyze their impact on Serendipity.

We implement a speech-to-text application based on Sphinx library [23] that translates audio to text. It will be used to evaluate the Emulab-based Serendipity. It is implemented as a single PNP-block job where the pre-process program divides a large audio file into multiple 2 Mb pieces, each of which is the task input.

To demonstrate how Serendipity can help the mobile computation initiator to speedup computing and conserve energy, we primarily compare the performance of executing applications on Serendipity with that of executing them locally on the initiator's mobile device. Previous remote-computing platforms (e.g., MAUI [12], CloneCloud [11], etc) don't work with intermittent connectivity and, thus, cannot be directly compared with Serendipity.

In all the following experiments every machine has a 600 MHz Pentium III processor and 256 MB memory, which is less powerful than mainstream PCs but closer to that of smart mobile devices. Every experiment is repeated 10 times with different seeds. The results reported correspond to the average values.

7.2 Serendipity's Performance Benefits

We initiate the experiments with the speech-to-text application using three workloads in three task allocation algorithms on both RollerNet and Haggle traces. The sizes of the audio files are 20 Mb, 200 Mb, and 600 Mb. As mentioned before, it is implemented as a single PNP-block job whose pre-process program divides the audio file into multiple 2 Mb pieces corresponding to 10, 100, and 300 tasks, respectively. The post-process program collects and combines the results. The baseline wireless bandwidth is set to 24 Mbps. We also assume that all nodes have enough energy and want to reduce the job completion time.

Figure 5 demonstrates how Serendipity improves the performance compared with executing locally. We make the following observations. First, with the increase of the workload, Serendipity achieves greater benefits in improving application performance. When the audio file is 600 Mb, Serendipity can achieve as large as 6.6 and 5.8 time speedup. Considering the number of nodes (11 for RollerNet and 9 for Haggle), the system utilization is more than 60%. More-

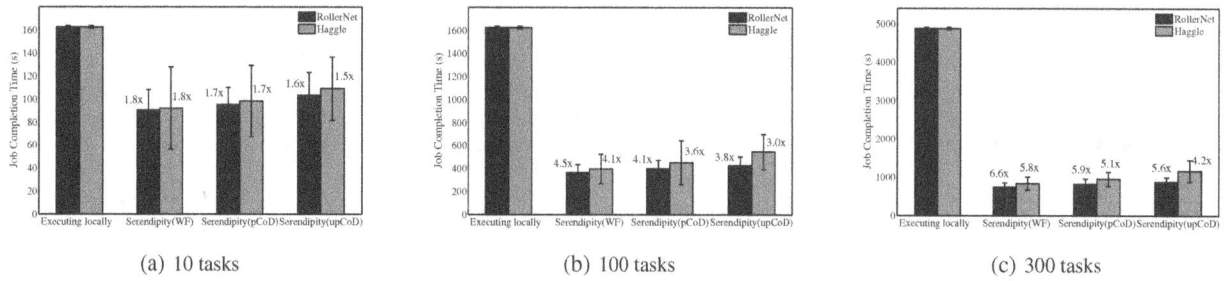

(a) 10 tasks　　　　　　　　　(b) 100 tasks　　　　　　　　　(c) 300 tasks

Figure 5: A comparison of Serendipity's performance benefits. The average job completion times with their 95% confidence intervals are plotted. We use two data traces, Haggle and RollerNet, to emulate the node contacts and three input sizes for each.

over, the ratio of the confidence intervals to the average values also decreases with the workload, indicating all nodes can obtain similar performance benefits. Second, in all the experiments WaterFilling consistently performs better than pCoD which is better than up-CoD. In the Haggle trace of Figure 5(c), WaterFilling achieves 5.8 time speedup while upCoD only achieves 4.2 time speedup. The results indicate that with more information Serendipity can perform better. Third, although Serendipity achieves similar average job completion times on both Haggle and RollerNet, their confidence intervals on Haggle are larger than those on RollerNet. This is because the Haggle trace has long contact interval and duration, resulting in the diversity of node density over the time.

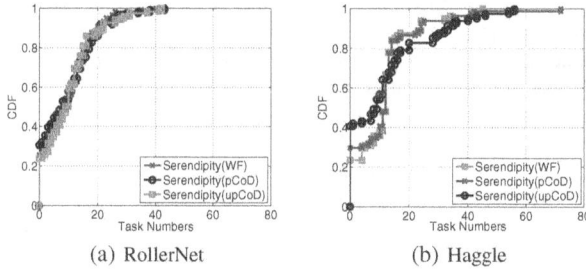

(a) RollerNet　　　　　　　　　(b) Haggle

Figure 6: The load distribution of Serendipity nodes when there are 100 tasks total, each of which takes 2 Mb input data.

To further analyze the performance diversity, we plot the workload distribution on the Serendipity nodes of Figure 5(b) in Figure 6. In the RollerNet trace, all three task allocation algorithms have similar load distribution, i.e., about 25% nodes are allocated 0 tasks while about 10% of the nodes are allocated more than 20 tasks. In the Haggle trace, WaterFilling and pCoD have similar load distribution, while upCoD's distribution is quite different from them. The long contact intervals of the Haggle trace makes the blind task dissemination of upCoD less efficient. In such an environment, the contact knowledge will be very useful to improve the Serendipity performance.

7.3 Impact of Network Environment

Next, we analyze the impact of the network environment on the performance of the three task allocation algorithms by changing the network settings from the base case.

Wireless Bandwidth: We first consider the effect of wireless bandwidth on the performance of Serendipity. The wireless bandwidth is set to be 1 Mbps, 5.5 Mbps, 11 Mbps, 24 Mbps, and 54 Mbps, which are typical values for wireless links. The audio file is 200 Mb, split into 100 tasks. We plot the job completion times of Serendipity with three task allocation algorithms in Figure 7.

We observe the following phenomena. First, in RollerNet, all three task allocation algorithms accomplish similar performance. Because these nodes have frequent contacts with each other, using the locality heuristic (upCoD) is good enough to make use of the

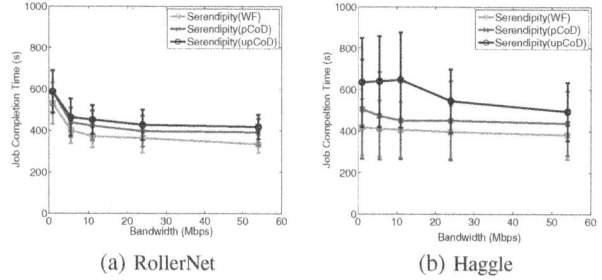

(a) RollerNet　　　　　　　　　(b) Haggle

Figure 7: The impact of wireless bandwidth on the performance of Serendipity. The average job completion times are plotted when the bandwidth is 1, 5.5, 11, 24, and 54 Mb/s, respectively.

nearby computation resource for remote computing. Second, when the bandwidth reduces from 11 Mbps to 1 Mbps, the job completion time experiences a large increase. This is because RollerNet has many short contacts which cannot be used to disseminate tasks when the bandwidth is too small. Third, in the Haggle trace, the job completion time of upCoD increases from 545.0 seconds to 647.6 seconds when the bandwidth reduces from 24 Mbps to 11 Mbps. Meanwhile WaterFilling achieves consistently good performance in all the experiments. This is because in the laboratory environment users are relatively stable and have longer contact durations. Thus, the primary factor affecting the Serendipity performance is the contact interval. On the other hand, since the contact distribution is more biased, only using locality is hard to find the global optimal task allocation.

Node Mobility: The above experiments demonstrate that contact traces impact the performance of Serendipity. To further analyze such impact, we use mobility models to generate the contact traces for 10 nodes. Specifically, we use Levy Walk Model [28], Random WayPoint Model (RWP) [29], and Time-Variant Community Mobility Model (TVCM) [21]. These models represent a wide range of mobility patterns. RWP is the simplest model and assumes unrestricted node movement. Levy Walk describes the human walk pattern verified by collected mobility traces. TVCM depicts human behavior in the presence of communities. The basic settings assume a 1 Km by 1 Km square activity area in which each node has a 100 m diameter circular communication range.

In this set of experiments we focus on the two most important aspects of node mobility, i.e., the mobility model and the node speed. The wireless bandwidth is set to 11 Mbps.

The results of this comparison are shown in Figure 8. Figure 8(a) shows that Serendipity has larger job completion time with all the mobility models than it had on Haggle and RollerNet traces. This is because their node densities are much sparser than Haggle and RollerNet traces. Thus it's harder for the job initiator to use other nodes' computation resources. We also observe that Serendipity achieves the best performance when the RWP model is used. This

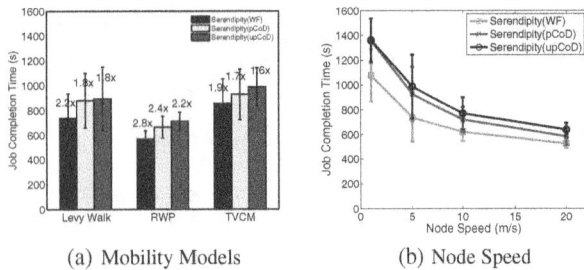

(a) Mobility Models

(b) Node Speed

Figure 8: The impact of node mobility on Serendipity. We generate the contact traces for 10 nodes in a 1 km×1 km area. In (a) we set the node speed to be 5 m/s, while in (b) we use Levy Walk as the mobility model.

is because RWP is the most diffusive [28] and, thus, results in more contact opportunities among nodes.

Node speed affects the contact frequencies and durations, which are critical to Serendipity. We vary the node speed from 1 m/s, i.e., human walking speed, to 20 m/s, i.e., vehicle speed. As shown in Figure 8(b), when the speed increases from 1 m/s to 10 m/s, the job completion times drastically decline, e.g., from 1077.1 seconds to 621.6 seconds for WaterFilling. This is because the increase of node speed significantly increases the contact opportunities and accelerates the task dissemination. When the speed further increases to 20 m/s, the job completion time is slightly reduced to 526.4 seconds for WaterFilling.

Number of Nodes: We finally examine how the quantity of available computation resources impacts Serendipity. To separate the effect of node density and resource quantity, we conduct two sets of experiments. In the first set, the active area is fixed, while in the second one, the active area changes proportionally with the number of nodes using the initial setting of 20 nodes in 1 km×1 km square area. Figure 9 shows the results where nodes follows RWP mobility model with wireless bandwidth at 2 Mbps.

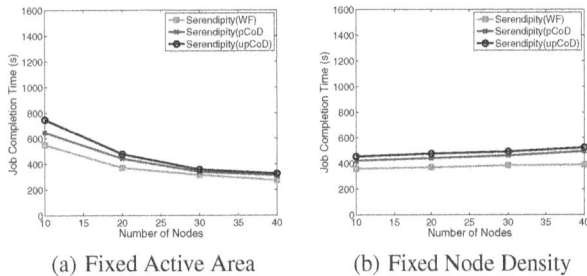

(a) Fixed Active Area

(b) Fixed Node Density

Figure 9: The impact of node numbers on the performance of Serendipity. We analyze the impact of both node number and node density by fixing the activity area and setting it proportional to the node numbers, respectively.

As shown in Figure 9(a), with the increase in the number of nodes in a fixed area, the job completion times of the three task allocation algorithms are reduced by more than 50%, from 550.0, 647.0, and 748.7 seconds to 273.0, 311.7, and 325.0 seconds for WaterFilling, pCoD and upCoD, respectively. Meanwhile, in Figure 9(b), the job completion times are almost constant despite the increase in node quantity.

7.4 The Impact of the Job Properties

Next we evaluate how the job properties affect the performance of Serendipity.

Multiple jobs: A more practical scenario involves nodes submitting multiple jobs simultaneously into Serendipity. These jobs will affect the performance of each other when their execution duration

overlaps. In this set of experiments, nodes will randomly submit 100-task jobs into Serendipity. The arrival time of these jobs follows a Poisson distribution. We change the arrival rate, λ from 0.0013 (its system utilization is less than 20%) to 0.0056 (its system utilization is larger than 90%) jobs per second. Figure 10 shows the results on the RollerNet and Haggle traces.

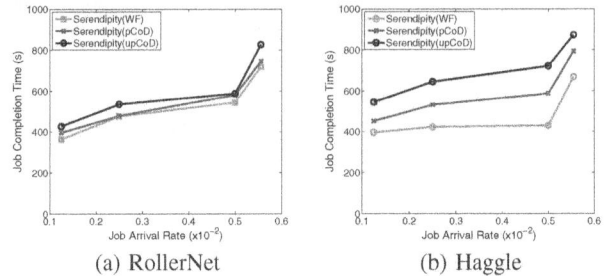

(a) RollerNet

(b) Haggle

Figure 10: Serendipity's performance with multiple jobs executed simultaneously. The job arrival time follows a Poisson distribution with varying arrival rates.

As expected, the job completion time increases with the job arrival rate. In both sets of experiments, the job completion time gradually increases with the job arrival rate until 0.005 jobs per second and, then, drastically increases when the job arrival rate increase to 0.0056 jobs per second. According to queueing theory, with the system utilization approaching 1, the queueing delay is approaching infinity. However, even when the system utilization is larger than 90% (i.e., $\lambda = 0.0056$), the job completion times of Serendipity with various task allocation algorithms are still less than 54% of executing locally, showing the advantage of distributed computation.

DAG jobs: The above experiments show that Serendipity performs well for single PNP-block jobs. Since DAG jobs are executed iteratively for all dependent PNP-blocks while parallel for all independent PNP-blocks. The above experiment results also apply to DAG jobs. In this set of experiments we will evaluate how PNP-block scheduling algorithm further improves the performance of Serendipity.

We use the job structure shown in Figure 4, where the processing of one image impacts the processing of another. We use the PNP-blocks of speech-to-text application as the basic building blocks. PNP-block A has 0 tasks; B has 200 tasks; C has 50 tasks; D has 100 tasks; E has 100 tasks; F has 0 tasks. The performance difference between our algorithm and assigning equal priority to the PNP-blocks is shown in Figure 11.

Our priority assignment algorithm achieves the job completion time of 1155.8, 1315.8 and 1383.2 seconds for WaterFilling, pCoD, and upCoD, consistently outperforming that of 1369.2, 1573.4, and 1654.4 seconds when all PNP-blocks have the same priority. These experiments demonstrate the usefulness of priority assigning. Further evaluation of our algorithm on diverse type of jobs will be part of our future work.

7.5 Energy Conservation

In this set of experiments, we demonstrate how Serendipity makes the entire system last longer by taking the energy consumption into consideration. We consider an energy critical scenario where node i has E_i% energy left, where E_i is randomly selected from [0, 20]. The energy consumption of task execution and communication is randomly selected from the measured values on mobile devices. The detailed measurement will be presented in the next section. In this set of experiments, nodes will randomly submit 100-task jobs into Serendipity. The arrival time of these jobs follows a Poisson

Table 1: A comparison of Serendipity's energy consumption. We report the number of jobs completed before at least one node depletes its battery and their average job completion time. Jobs arrive in a Poisson process with $\lambda = 0.005$ jobs per second.

	Haggle				RollerNet			
	Energy Aware		Time Optimizing		Energy Aware		Time Optimizing	
	# Completed jobs	Time (s)	# Completed jobs	Time (s)	# Completed jobs	Time(s)	# Completed jobs	Time(s)
Serendipity(WF)	17.0	2664.7	4.5	409.2	21.8	2823.9	2.5	496.2
Serendipity(pCoD)	10.0	2162.4	3.0	435.3	16.8	2173.6	4.8	539.0
Serendipity(upCoD)	9.3	2080.6	3.0	564.0	16.8	2082.6	3.5	562.7
Executing locally	N/A	N/A	1.3	1614.0	N/A	N/A	1.3	1614.0

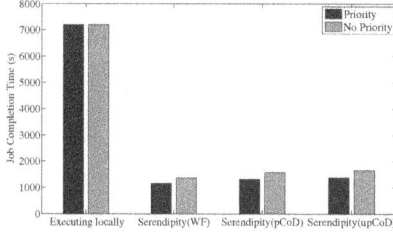

Figure 11: The importance of assigning priorities to PNP-blocks.

distribution with $\lambda = 0.005$ jobs per second. We compare energy-aware Serendipity against "time-optimizing" Serendipity and executing jobs locally. The TTL of energy-aware Serendipity is set to twice the time of executing the job locally.

Table 1 shows the number of jobs completed before at least one node depletes its energy and the average job completion time of those completed jobs. We make the following observations. First, energy-aware Serendipity completes many more jobs than executing locally and using time-optimizing Serendipity. This is because energy-aware Serendipity balances the energy consumption of all the mobile devices through adaptively allocating more tasks to devices with more residual energy. In contrast, the time-optimizing Serendipity will quickly deplete the energy of some mobile devices by allocating many tasks to them. Second, through global optimization, energy-aware Serendipity with the WaterFilling allocation algorithm completes more jobs than those with pCoD and upCoD. Third, the job completion time of energy-aware Serendipity is much larger than that of time-optimizing Serendipity. There exists a tradeoff between energy consumption and performance. Finally, compared with executing locally, time-optimizing Serendipity both completes more jobs and has smaller job completion time. This is because statistically the few devices with limited residual energy will last longer by off-loading the computation to other devices.

8 Implementation

We implemented a prototype of Serendipity on the Android OS [1]. It comprises three parts: the Serendipity worker corresponding to the worker in Fig. 2, the Serendipity controller including all other components in Fig. 2 and a user library providing the key APIs for application development.

We currently use *WifiManager*'s hidden API, *setWifiApEnabled*, to achieve the ad hoc communication between two devices, i.e., one device acts as an AP while the other device connects to it as a client.

We use the Java reflection techniques to dynamically execute the tasks. Every task has to implement the function *execute* defined in the APIs. When the Serendipity worker executes a task, it executes this function.

The separation between the Serendipity worker and the Serendipity controller is based on access control. Android's security architecture defines many kinds of permission to various resources including network, GPS, sensors, etc. The Serendipity worker is implemented as a separate application with limited access permis-

sion to these resources, acting as a sandbox for the task execution. When the Serendipity controller receives a task to execute, it will start a Senredipity worker and get the results from it.

8.1 System Evaluation

To evaluate our system, we implemented two computationally complex applications, a face detection application, and a speech-to-text application. The face detection application takes a set of pictures and uses computer vision algorithms to identify all the faces in these pictures [17]. It is implemented as a single PNP-block job where the face detection in each picture is a task. The speech-to-text application takes an audio file and translates the speech into text using the Sphinx library [23]. It is also a single PNP-block job where the pre-process program divides a large audio file into multiple pieces, each of which is input to a separate task.

We tested Serendipity on a Samsung Galaxy Tab with a 1 GHz Cortex A8 processor and a Motorola ATRIX smartphone with a dual-core Tegra 2 processor, each at 1 GHz. Both of them run the Android 2.3 OS. The face detection and speech-to-text applications are used for evaluation.

Table 2: The execution time of two applications on two devices.

	Input size (Mb)	Galaxy Tab (s)	ATRIX (s)
FaceDetection	2.2	17.9	7.2
Speech-to-text	3.0	40.3	18.8

We first executed the two applications locally on the two devices. As Motorola ATRIX smartphone has a dual-core processor, we split the input files into two parts of equal size and simultaneously executed the two tasks to fully utilize its processor. Table 2 shows their execution times. We also measured the TCP throughput between these two devices by sending 800 Mb data. We obtain 10.8 Mbps throughput on average when they are within 10 meters. In fact, they still achieve 5.9 Mbps throughput even when they are more than 30 meters away.

To assess the performance of Serendipity, we construct a simple network in which the two devices are consistently connected during the experiments. As expected, Serendipity speeds up more than 3 times than executing the applications on the Samsung Galaxy Tab.

To generate the energy consumption profiles of the two applications on these mobile devices, we repeatedly execute those applications starting with full battery until the batteries are depleted and count the number of iterations. Similarly, WiFi's energy profiles are obtained by continuously transferring data between them. Table 3 demonstrates the results.

Table 3: The energy consumption of mobile devices. The ratios of consumed energy to the total device energy capacity are reported.

	Input size (Mb)	Galaxy Tab	ATRIX
FaceDetection	2.2	4.14×10^{-4}	3.44×10^{-4}
Speech-to-text	3.0	9.32×10^{-4}	9.01×10^{-4}
WiFi	800	8.02×10^{-4}	2.04×10^{-3}

The energy required to transfer a task only accounts for 0.5 % (i.e., $\max(\frac{8.02 \times 2.2}{4.14 \times 800}, \frac{8.02 \times 3.0}{9.32 \times 800})$) and 1.6% (i.e., $\max(\frac{20.4 \times 2.2}{3.44 \times 800}$,

$\frac{20.4 \times 3.0}{9.01 \times 800}$)) of the energy required to execute the task on these devices, respectively. It indicates that Serendipity won't consume much extra energy. Instead, by delegating tasks to devices with a lot of energy, it can significantly save the job initiator's energy.

We use an extreme example to show the gains of energy-aware Serendipity. Suppose the ATRIX phone has a lot of pictures for face detection. Assume it only has 5% energy left, and the Galaxy tablet has 50% energy left. Energy-aware Serendipity can detect about 1320 pictures before the ATRIX phone depletes its battery, while time-optimizing Serendipity can only detect about 203 pictures.

9 Conclusion and Future Work

In this paper we have developed and evaluated the Serendipity system that enables a mobile device to remotely access computational resources on other mobile devices it may encounter. The main challenge we addressed is how to model computational tasks and how to perform task allocation under varying assumptions about the connectivity environment. Through an emulation of the Serendipity system we have explored how such a system has the potential to improve computation speed as well as save energy for the initiating mobile device. We have also reported on a preliminary prototype of our system on Android platforms.

In our future work we will complete our experimental evaluation of the prototype systems to include more devices and incorporate intermittent connectivity. We will also consider incentive and reputation systems that are derived from previous work in MANET and peer-to-peer systems and tailored to the Serendipity environment.

As mentioned previously we envision Serendipity as developed here to enable an extreme of a spectrum of remote computation possibilities that are available to mobile devices. Our future work will consider extending our investigation to enable hybrid remote computation where the use of cloud or cloudlet resources is augmented with the use of resources on other mobile devices.

Acknowledgments

We would like to thank our shepherd, Kannan Srinivasan, and the anonymous reviewers for their insightful feedback. This work was supported in part by the US National Science Foundation through grant CNS 0831714.

10 References

[1] Android open source project. http://source.android.com.

[2] D. P. Anderson. BOINC: A system for public-resource computing and storage. In *IEEE/ACM GRID*, 2004.

[3] D. P. Anderson, J. Cobb, E. Korpela, M. Lebofsky, and D. Werthimer. SETI@home: an experiment in public-resource computing. *Commun. ACM*, 45:56–61, November 2002.

[4] R. Balan, J. Flinn, M. Satyanarayanan, S. Sinnamohideen, and H.-I. Yang. The case for cyber foraging. In *Proceedings of the 10th workshop on ACM SIGOPS European workshop*, 2002.

[5] R. K. Balan, D. Gergle, M. Satyanarayanan, and J. Herbsleb. Simplifying cyber foraging for mobile devices. In *ACM MobiSys*, 2007.

[6] A. L. Beberg, D. L. Ensign, G. Jayachandran, S. Khaliq, and V. S. Pande. Folding@home: Lessons from eight years of volunteer distributed computing. In *IEEE IPDPS*, 2009.

[7] J. Burgess, B. Gallagher, D. Jensen, and B. N. Levine. Maxprop: Routing for vehicle-based disruption-tolerant networks. In *IEEE INFOCOM*, 2006.

[8] L. Buttyán and J.-P. Hubaux. Enforcing service availability in mobile ad-hoc wans. In *ACM MobiHoc*, 2000.

[9] J.-H. Chang and L. Tassiulas. Energy conserving routing in wireless ad-hoc networks. In *IEEE INFOCOM*, 2000.

[10] C. Chekuri, A. Goel, S. Khanna, and A. Kumar. Multi-processor scheduling to minimize flow time with ε resource augmentation. In *ACM STOC*, 2004.

[11] B.-G. Chun, S. Ihm, P. Maniatis, M. Naik, and A. Patti. Clonecloud: elastic execution between mobile device and cloud. In *ACM EuroSys*, 2011.

[12] E. Cuervo, A. Balasubramanian, D.-k. Cho, A. Wolman, S. Saroiu, R. Chandra, and P. Bahl. Maui: making smartphones last longer with code offload. In *ACM MobiSys*, 2010.

[13] J. Dean and S. Ghemawat. Mapreduce: simplified data processing on large clusters. *Commun. ACM*, 51:107–113, January 2008.

[14] P. J. Denning. Hastily formed networks. *Commun. ACM*, 2006.

[15] K. Fall, G. Iannaccone, J. Kannan, F. Silveira, and N. Taft. A disruption-tolerant architecture for secure and efficient disaster response communications. In *ISCRAM*, May 2010.

[16] K. M. Hanna, B. N. Levine, and R. Manmatha. Mobile Distributed Information Retrieval For Highly Partitioned Networks. In *IEEE ICNP*, pages 38–47, Nov 2003.

[17] E. HjelmÅěs and B. K. Lowb. Face detection: A survey. *Elsevier Computer Vision and Image Understanding*, September 2001.

[18] E. S. Hou, N. Ansari, and H. Ren. A genetic algorithm for multiprocessor scheduling. In *IEEE IPDPS*, 1994.

[19] P. Hui, J. Scott, J. Crowcroft, and C. Diot. Haggle: a networking architecture designed around mobile users. In *WONS*, 2006.

[20] S. Jain, K. Fall, and R. Patra. Routing in a delay tolerant network. In *ACM SIGCOMM*, 2004.

[21] W. jen Hsu, a. K. P. Thrasyvoulos Spyropoulos, and A. Helmy. Modeling time-variant user mobility in wireless mobile networks. In *IEEE INFOCOM*, 2007.

[22] J. G. Koomey, S. Berard, M. Sanchez, and H. Won. Assessing Trends in the Electrical Efficiency of Computation over Time. Technical report, http://www.intel.com/assets/pdf/general/computertrendsreleasecomplete-v31.pdf, 2009.

[23] K.-F. Lee, H.-W. Hon, and R. Reddy. An overview of the SPHINX speech recognition system. *IEEE Transaction on Acoustics, Speech and Signal Processing*, 1990.

[24] R. Lu, X. Lin, H. Zhu, X. Shen, and B. Preiss. Pi: A practical incentive protocol for delay tolerant networks. *IEEE Transactions on Wireless Communications*, April 2010.

[25] E. Marinelli. Hyrax: Cloud computing on mobile devices using mapreduce. Master's thesis, Computer Science Dept., CMU, September 2009.

[26] P. Marshall. DARPA progress towards affordable, dense, and content focused tactical edge networks. In *IEEE MILCOM*, 2008.

[27] A. S. Pentland, R. Fletcher, and A. Hasson. DakNet: Rethinking connectivity in developing nations. *Computer*, January 2004.

[28] I. Rhee, M. Shin, S. Hong, K. Lee, and S. Chong. On the levy-walk nature of human mobility. In *IEEE INFOCOM*, 2008.

[29] A. K. Saha and D. B. Johnson. Modeling mobility for vehicular ad-hoc networks. In *ACM VANET*, 2004.

[30] M. Satyanarayanan, P. Bahl, R. Caceres, and N. Davies. The case for VM-based cloudlets in mobile computing. *IEEE Pervasive Computing*, 2009.

[31] D. Thain, T. Tannenbaum, and M. Livny. Distributed computing in practice: the condor experience. *Concurr. Comput. : Pract. Exper.*, February 2005.

[32] P. U. Tournoux, J. Leguay, F. Benbadis, V. Conan, M. D. de Amorim, and J. Whitbeck. The accordion phenomenon: Analysis, characterization, and impact on dtn routing. In *Proc. IEEE INFOCOM*, 2009.

[33] B. White, J. Lepreau, L. Stoller, R. Ricci, S. Guruprasad, M. Newbold, M. Hibler, C. Barb, and A. Joglekar. An integrated experimental environment for distributed systems and networks. In *USENIX OSDI*, 2002.

[34] L. Zhang, B. Tiwana, Z. Qian, Z. Wang, R. P. Dick, Z. M. Mao, and L. Yang. Accurate online power estimation and automatic battery behavior based power model generation for smartphones. In *IEEE/ACM/IFIP CODES/ISSS*, 2010.

[35] J. Zhou, E. Gilman, M. Ylianttila, and J. Riekki. Pervasive service computing: Visions and challenges. In *IEEE CIT*, 2010.

Oblivious Low-Congestion Multicast Routing in Wireless Networks [*]

Antonio Carzaniga[1] Koorosh Khazaei[1] Fabian Kuhn[1,2]

antonio.carzaniga@usi.ch koorosh.khazaei@usi.ch kuhn@informatik.uni-freiburg.de

[1] Faculty of Informatics, University of Lugano, 6904 Lugano, Switzerland
[2] Department of Computer Science, University of Freiburg, 79110 Freiburg, Germany

ABSTRACT

We propose a routing scheme to implement multicast communication in wireless networks. The scheme is oblivious, compact, and completely decentralized. It is intended to support dynamic and diverse multicast requests typical of, for example, publish/subscribe and content-based communication. The scheme is built on top of a geographical routing layer. Each message is transmitted along the geometric minimum spanning tree that connects the source and all the destinations. Then, for each edge in this tree, the scheme routes a message through a random intermediate node, chosen independently of the set of multicast requests. The intermediate node is chosen in the vicinity of the corresponding edge such that congestion is reduced without stretching the routes by more than a constant factor. We first evaluate the scheme analytically, showing that it achieves a theoretically optimal level of congestion. We then evaluate the scheme in simulation, showing that its performance is also good in practice.

Categories and Subject Descriptors:
C.2.2 [Computer-Communication Networks]:
Network Protocols–*routing protocols*
F.2.2 [Analysis of Algorithms and Problem Complexity]:
Nonnumerical Algorithms and Problems—*routing and layout*

General Terms: Algorithms, Performance

Keywords: geographic routing, multicast, congestion, stretch

1. INTRODUCTION

Some modes of communication are inherently multicast, in the sense that they induce the transmission of a single message to multiple destinations. This is the case of publish/subscribe communication, where each message is transmitted from the sender (the publisher) to the set of receivers that are interested in that message (the subscribers). Furthermore, while some multicast services are based on a few and relatively stable multicast groups (e.g., video streaming over IP multicast) and therefore work well with stable routing

---

[*]This work was supported in part by the Swiss National Science Foundation under grant n. 200021-122137.

state, others are more demanding and more dynamic. For example, in *content-based* publish/subscribe, subscriptions may partially overlap, forming a large number of implicit groups—potentially, a different one for each message.

In this paper we consider a generic multicast primitive in which each message may induce a unique multicast request (m, s, T). This primitive allows a source node s to send a message m to a set of target nodes T. In particular, we consider this primitive within a wireless network. Our goal is to implement such a communication primitive through a routing scheme that is oblivious, compact, low-stretch, low-congestion, and also practical.

The scheme we propose is *oblivious* in the sense that how a request is routed does not depend on the set of requests and how the other requests are routed. We in fact prove that the scheme offers the best possible performance guarantees even in the presence of adversarial requests. The scheme is *compact* in the sense that it requires only limited state at each node, typically $O(\text{polylog } n)$ bits in a network of n nodes. The scheme is *low-stretch*, in the sense that the length of each path from a source to a target node, which roughly corresponds to the latency of each delivery, is optimal up to a small constant factor. The scheme is also *low-congestion*, in the sense that, for any given set of multicast requests, the maximum amount of traffic crossing a node is only a factor of $O(\log n)$ worse than with an ideal routing specifically optimized for that set of requests. Notice that this $O(\log n)$ factor for congestion is optimal for any oblivious scheme [5, 20], even for routing on 2-dimensional meshes. Lastly, the scheme is *practical* in the sense that the theoretical asymptotic behavior of the scheme can be realized in practice with good pre-asymptotic performance and small constants.

The scheme we propose is built on top of a geographical routing service whereby a message can be addressed to a given geographical location and therefore can be delivered, possibly through multiple hops, to the node that is closest to that location. Such geographical schemes exist and are compact and achieve low-stretch both theoretically and in practice [17]. The choice of a geographical communication primitive implies that, in its most basic form, the routing scheme we propose is *name dependent*. This means that nodes must be identified by some kind of address dictated by the communication layer (in this case, the node's geographic coordinates). However, it is also possible to extend such a basic routing scheme to be name independent, by means of a lookup service that can also be implemented efficiently [1].

In summary, we start from a compact and low-stretch geographical routing substrate, which for a request (m, s, t) can deliver a unicast message m from a source s to a target destination t, and we use it to build a low-congestion oblivious multicast scheme that can serve requests of the type (m, s, T) and deliver m from a source s to a set of target destinations T. A simple way to implement such a multicast scheme would be to implement each multicast request (m, s, T) with a series of unicast requests (m, s, t_i) for each t_i in T. However, such a scheme incurs high congestion. Intuitively, this is the case when many destinations are close to each other, even without adversarial sets of requests and instead with sources and destinations distributed uniformly over multiple requests.

A standard way to achieve low congestion with an oblivious unicast scheme is to use randomization in what is known as Valiant's trick [29]. For a unicast request (m, s, t), first route m from s to a randomly chosen intermediate destination v, and then from v to t. However, in its basic form, this trick does not work well for arbitrary worst-case sets of requests and in particular it does not work well for multicast requests. Consider for example a request (m, s, T) in which the targets $t_i \in T$ are all clustered in a small region far away from the source s. Even with Valiant's trick, a series of (unicast) copies of m going from s to a target t_i in the cluster would induce high congestion in the small perimeter around the cluster, whereas an optimal routing strategy in that case would send one copy of m from s towards the cluster, and then it would duplicate m locally to all targets within the cluster.

The scheme we propose employs a local variant of Valiant's trick, and it does that within a routing strategy that avoids congestion in the case of multicast requests. At a high level, the scheme routes a multicast request (m, s, T) along the geometric minimum spanning tree that connects the source s and all the targets in T. Then, for each edge (u, v) on that tree, the scheme uses a variant of Valiant's trick by routing m from u to an intermediate point w_{uv} chosen randomly in the vicinity of the uv segment.

In this paper we formally define this routing scheme, we then analyze its theoretical properties, and evaluate it in practice using simulation. The theoretical analysis shows that, in terms of congestion, the scheme is competitive with an ideal (non-oblivious) scheme up to a factor of $O(\log n)$, which is known to be a lower bound for congestion in oblivious schemes. The simulation study shows that the scheme is also effective in practice, with limited congestion and stretch.

2. RELATED WORK

Compared to classic wired networks, wireless ad hoc and sensor networks behave more dynamically. As a consequence, classical link-state routing protocols are often not well-suited for wireless networks and other, more reactive routing strategies are required. A standard way to do this is to combine flooding for route discovery with some caching techniques to reuse acquired routing information [7, 12, 22, 26]. While there is an abundant literature on wireless point-to-point routing, the work on wireless multicast is much less copious. In fact, Vershney claims that wireless multicast is still an important challenge [30]. Multicast protocols for wireless networks have been suggested, for example, by Royer and Perkins [27] or by Xie et al. [31].

Since the presence of wireless communication links is inherently related to the physical placement of nodes, if available, geometric information can be a powerful tool for routing. For geographic routing, it is typically assumed that all nodes are aware of their geographical position and the source node of a message knows the location of the destination. The simplest possible way to route a message that way is to proceed greedily by always forwarding a message to the neighbor closest to the destination [28]. While greedy routing is efficient in dense average-case scenarios, it might not always reach the destination. The first proposed geographic routing protocol that is guaranteed to reach the destination is face routing [15]. The delivery guarantees of the face routing protocol come at the cost of worse behavior in well-behaved settings. Therefore

greedy and face routing have been combined to obtain average-case efficient protocols with guaranteed message delivery [6, 13, 17]. All these geographic routing protocols assume that the communication network is a unit disk graph. In this paper, we extend this setting with non-uniform transmission ranges in a model similar to those proposed by others [4, 16].

To apply geographic routing, the source node of a message needs to know the location of the destination. A typical application is geocast, a variant of multicast, where all nodes in a certain geographical region have to be reached [21]. If location information of the destination is not available, geographic routing can be combined with a location service that allows to efficiently search for location information of other nodes [1, 10, 19].

All routing schemes described so far do not explicitly attempt to minimize the congestion that arises in the presence of a large number of routing requests. From an algorithmic point of view, congestion has mainly been considered in the context of oblivious routing, i.e., if each routing path is chosen independently. A seminal result by Valiant and Brebner [29] shows that in a hypercube, any permutation can be routed in $O(\log n)$ steps. The path selection is randomized and uses what is now known as Valiant's trick. Each message is first routed to a random intermediate node and from there to the destination. The technique has been applied in various other networks and in particular, it was shown by Kolman and Scheideler [14] that Valiant's trick can efficiently be used in a much more general setting. The existing work on oblivious routing culminated in a breakthrough paper by Räcke [24] that shows that there is an oblivious protocol that routes every set of routing requests with expected maximum node congestion within a logarithmic factor of the best corresponding multicommodity flow solution. In light of a lower bound that even holds for 2-dimensional meshes, this is asymptotically optimal [5, 20]. Räcke's result also applies to multicast and could also be used for our wireless network model. However, the protocol state is rather heavy-weight to set up and maintain, and the given wireless setting is amenable to specialized and much more light-weight algorithms. Most closely related to our work are two papers by Busch et al. that describe algorithms for unicast in 2-dimensional meshes [9] and for geometric networks modeling dense wireless networks [8]. For unicast, this latter algorithm [8] achieves the same asymptotic bounds as the algorithm presented here. However, we believe that our randomized scheme based on Valiant's trick is somewhat simpler and easier to use. A recent survey on oblivious routing is also due to Räcke [25]. Other papers study congestion in the context of wireless network routing, but are less related to this work [11, 18, 23, 32].

3. MODEL AND DEFINITIONS

We now formally state our assumptions about the communication network and its underlying geographic routing service.

Communication Network: We assume that n wireless network nodes are located in a bounded region in 2-dimensional Euclidean space. The nodes have unique identifiers and we denote the set of nodes by V. For simplicity, we assume that the region is a square of side length L, however, the techniques work for any "reasonable" convex region. Further, we assume that nodes are aware of their position in the plane. This can be achieved by equipping nodes with GPS devices or through some localization service. Communication in the network is characterized by two positive parameters $r_C \leq r_I$ defining communication and interference radii. Whenever two nodes u and v are at Euclidean distance at most r_C, u and v can directly communicate with each other. If two nodes u and v are within distance r_I, they can cause interference to each other. Further, we assume that there is no direct communication or even interference between two nodes at distance more than r_I. We denote the ratio between r_I and r_C by $\rho := r_I/r_C$ and typically assume ρ to be a constant (independent of n). We assume that the $L \times L$-square containing the network is reasonably densely covered by nodes. Specifically, we assume that there is a parameter r_{cov} such that for every point in the $L \times L$-square, there is a network node within distance r_{cov}. We assume r_{cov} is relatively small, such that the requirement implies that the number of nodes is at least polynomial in L/r_I.

Geographic Routing: We assume that there is a geographic routing service in place, which nodes use for communicating with each other. More formally, a node u can send a message to an arbitrary (x, y) coordinate pair within the specified geometric region that contains the wireless network nodes (i.e., the side length L square). If a message is sent to (x, y), the routing service guarantees that the node closest to (x, y) (according to Euclidean distance) receives the message. We assume that nodes populate the complete given geometric region densely enough to enable routing on almost direct paths between all pairs of nodes. We use the following definitions:

DEFINITION 3.1 (λ-PADDED PATH).
A path $P = u_1, \ldots, u_k$ connecting coordinates (x, y) and (x', y') is λ-padded if all nodes u_i of P are within Euclidean distance at most $\lambda \cdot r_I$ from the line segment connecting (x, y) and (x', y') in the plane.

DEFINITION 3.2 (σ-SPARSE PATH).
A path $P = u_1, \ldots, u_k$ is called σ-sparse if no disk of diameter r_C contains more than σ nodes u_i of P.

We assume the geographic routing service induces λ_{pad}-padded, σ-sparse paths for some positive parameters λ_{pad} and σ. Note that this in particular implies that the node distribution is dense enough so that there is a node at distance at most $\lambda_{\text{pad}} r_I$ from every point (x, y) in the geometric region covered by the network, i.e., $r_{\text{cov}} \leq \lambda_{\text{pad}} r_I$. Further note that the assumption that any two nodes within distance r_C are connected implies that nodes inside a disk of diameter r_C are fully connected and therefore, paths containing more than 2 nodes in such a disk can be shortened to contain at most 2 such nodes. Hence, if a λ_{pad}-padded path between (x, y) and (x', y') exists, then there is also a λ_{pad}-padded, 2-sparse path between the two points.

Typically, for relatively dense average-case networks, services based on greedy routing perform best. By construction, greedy routing always gives 2-sparse paths. Further, as shown in Section 7, it also gives good, $O(1)$-padded paths. For worst-case networks, geographic routing techniques [16, 17] can be used to find an $O(1)$-sparse, $O(\lambda)$-padded path, whenever a λ-padded path exists.

4. PROBLEM STATEMENT

Multicast Routing: We consider two variants of the multicast problem. A lower level geographic and a high-level name-based variant. In both cases, we are given r multicast requests $R_1, \ldots, R_r$ where request $R_i = (m_i, s_i, T_i)$ consists of a message m_i, a source node s_i and a set T_i of k_i destinations $t_{i,1}, \ldots, t_{i,k_i}$. We assume that s_i knows m_i and T_i and the objective is for s_i to send m_i to all destinations in T_i. In the case of the *geographic multicast problem*, each destination $t_{i,j}$ is given as a coordinate pair $(x_{i,j}, y_{i,j})$ and for all $i \in [r]$, message m_i has to be sent to the k_i actual network nodes closest to $(x_{i,1}, y_{i,1}), \ldots, (x_{i,k_i}, y_{i,k_i})$. In the more standard *name-based multicast problem*, each destination $t_{i,j}$ is given

as a node identifier. As usual in the context, we assume that messages m_i are large compared to the size of T_i, so that the overhead of storing all destination information in the message header is negligible [2]. The geographic multicast problem is closely related to what is generally known as geocast [21]. Unlike specifying individual destinations, typically, the destinations are given by a geographic region to which a message has to be transmitted. We note that the geographic multicast service that we present can easily be adapted to efficiently work in such a scenario. In fact, in our communication model, sending to a geographic region can be modeled by sending to a dense enough set of destinations within the area.

Congestion: As discussed in Section 3, we assume that nodes at distance at most r_I can cause interference to each other. To model congestion, we assume that whenever a node u transmits, it causes interference at all nodes within distance r_I from u. Let I_u be the set of nodes within Euclidean distance r_I from node u. Hence, whenever a node in I_u sends a message, it causes interference at node u and vice versa, whenever u transmits a message, it interferes with all nodes in I_u.

To satisfy a given multicast request $R_i = (m_i, s_i, T_i)$, message m_i has to be sent from s_i to all nodes in T_i along a subtree of the network. Given some algorithm $\mathcal{A}$, let $S_i^{\mathcal{A}}$ be the multiset of nodes that transmit message m_i in order to reach all destinations in T_i, i.e., $S_i^{\mathcal{A}}$ at least contains all the inner nodes of the tree along which m_i is sent to the destinations. Given a set of r multicast requests $R_1, \ldots, R_r$ and an algorithm $\mathcal{A}$, we define the congestion $\text{cong}_u^{\mathcal{A}}$ of a node u and the maximum node congestion $\text{cong}^{\mathcal{A}}$ of $\mathcal{A}$ as

$$\text{cong}_u^{\mathcal{A}} := \sum_{i=1}^{r} |S_i^{\mathcal{A}} \cap I_u|, \quad \text{cong}^{\mathcal{A}} := \max_{u \in V} \text{cong}_u^{\mathcal{A}}. \quad (1)$$

Our main objective will be to minimize $\text{cong}^{\mathcal{A}}$. Whenever it is clear from the context, we omit the superscript $\mathcal{A}$. In order to evaluate an algorithm, we intend to compare its behavior with the best possible maximum node congestion. Let $\text{cong}^{\star}$ be the maximum node congestion of an optimal routing solution for the given requests $R_1, \ldots, R_r$. Consider a rectangle $\mathcal{R}$ with side lengths $w(\mathcal{R})$ and $h(\mathcal{R})$. We define $\text{cut}(\mathcal{R})$ to be the set of requests R_i, $i \in [r]$ such that $\{s_i\} \cup T_i$ contains at least one node inside $\mathcal{R}$ and at least one node outside $\mathcal{R}$. To bound the optimal congestion $\text{cong}^{\star}$, we introduce the following notion:

$$\text{load}(\mathcal{R}) := \min \left\{ |\text{cut}(\mathcal{R})|, \frac{|\text{cut}(\mathcal{R})| \cdot r_I}{w(\mathcal{R}) + h(\mathcal{R})} \right\}. \quad (2)$$

The following lemma shows that asymptotically, $\text{load}(\mathcal{R})$ is a lower bound on the best possible maximum congestion $\text{cong}^{\star}$.

LEMMA 4.1. *For every set of multicast requests $R_1, \ldots, R_r$ and every rectangle $\mathcal{R}$, we have $\text{cong}^{\star} = \Omega(\text{load}(\mathcal{R}))$.*

PROOF. Consider a multicast request R_i for which $\{s_i\} \cup T_i$ contains at least one node inside $\mathcal{R}$ and at least one node outside $\mathcal{R}$. Further, let B be the geometric area defined by all points within distance r_I of the boundary of $\mathcal{R}$. In order to satisfy request R_i, a message has to be sent into or out of $\mathcal{R}$ and therefore at least one node in B has to transmit a message.

Consider a maximal independent set S of the graph defined by the nodes V_B that lie inside B and edges $\{u, v\}$ whenever u an v are at Euclidean distance at most r_I. Whenever a node in B transmits a message, it causes congestion at some node in S. Further, since nodes in S are within distance more than r_I, the number of nodes in S is at most $O(1 + (w(\mathcal{R}) + h(\mathcal{R}))/r_I)$. Hence, by the pigeonhole principle, for every solution for the given multicast problem, some node in S has congestion at least $\Omega(\text{load}(\mathcal{R}))$. □

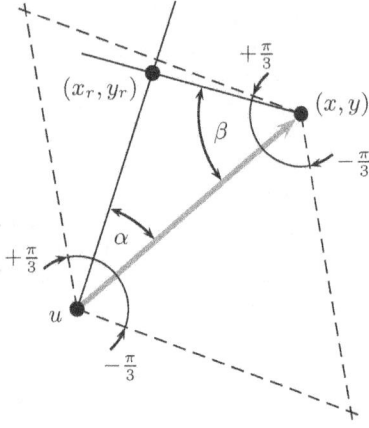

Figure 1: Choice of Intermediate Node

5. GEOMETRIC MULTICAST

Our algorithm consists of two components, which together allow to multicast a message to a set of geographical destinations based on an underlying geographic routing service as discussed in Section 3. At the core is an oblivious geographic point-to-point routing protocol with asymptotically optimal congestion properties. A multicast request is then routed on a tree by applying the point-to-point scheme.

We first describe the routing scheme to send a message from a node u to a geographical destination (x, y). The point-to-point routing algorithm is based on Valiant's classical trick of reducing overall congestion by routing messages through a randomly chosen intermediate node. To deal with worst-case collections of routing requests and to guarantee a bounded stretch factor for the routing paths, we choose the random intermediate point dependent on the source and target positions of the routing request. Specifically, a message from a node u at position (x_u, y_u) to location (x, y) is routed as follows.

1. If the Euclidean distance of (x, y) from (x_u, y_u) is at most $r_C/2$, the node closest to (x, y) is either u itself or a neighbor v of u. In that case, u directly sends the message to v.

2. Otherwise, node u chooses a random intermediate position (x_r, y_r) as follows. First, u chooses two uniform random angles $\alpha, \beta \in [0, \pi/3]$. The point (x_r, y_r) is then chosen such that the line segments from u to (x_r, y_r) and from u to the destination position (x, y) enclose an angle of α and the line segments from (x, y) to (x_r, y_r) and from (x, y) to u enclose an angle of β. There are two points (x_r, y_r) for which this is true (one to the left and one to the right of the line connecting source and destination). Node u randomly chooses one of the two points as (x_r, y_r).

   Using the underlying geographic routing protocol, the message is then routed from u to the node w closest to (x_r, y_r) and afterward from node w to the destination position (x, y).

The choice of the random point (x_r, y_r) is also illustrated in Figure 1. Note that (x_r, y_r) is chosen such that the geometric distances from (x_r, y_r) to u and (x, y) are at most as large as the distance between u and (x, y).

Based on the described scheme for point-to-point communication, we can now build the multicast routing protocol on top of it. For a given geographic multicast request $R_i = (m_i, s_i, T_i)$, let (x_i, y_i) be the position of the source node s_i and let $P_i =$ $\{(x_i, y_i)\} \cup T_i$ be the set of points of the multicast request R_i. We first construct a geometric tree spanning all the points in P_i and then use the point-to-point routing algorithm to send m_i along all the edges of the constructed spanning tree. There are different ways to choose the geometric spanning tree of the points in P_i. In terms of total routing cost, the best choice would be to choose a minimum Steiner tree w.r.t. Euclidean distances. Note that the Euclidean Steiner tree problem is NP-hard. However, there is a polynomial-time approximation scheme and thus the problem can be approximated arbitrarily well [3]. Still, since we would like our algorithm to be as simple as possible, and also since asymptotically it does not make a difference, we use the Euclidean minimum spanning tree (MST) to connect the points in P_i. Such a tree can be computed locally by the sender with an efficient algorithm.

The message m_i is sent along the edges of the Euclidean MST of P_i in a straightforward manner. The tree is directed from the source s_i at (x_i, y_i) towards the destinations T_i and slightly adapted in the following way. As long as there is a directed path u, v, w such that the Euclidean distance between u and w is at most $r_C/2$, node w is attached directly to u instead of v. This allows to reach close-by nodes by local broadcast where possible.

For each (directed) edge $((x, y), (x', y'))$, a message is sent from the node closest to (x, y) to the node closest to (x', y') by using the point-to-point routing scheme described above. Assume that a node w representing a node (x, y) in the tree needs to send messages to different neighbors (x', y') in the MST. If some of the neighbors (x', y') are at distance at most $r_C/2$ from the position of w, w sends one broadcast message to all neighbors to reach the nodes closest to these tree neighbors. For all other tree neighbors, the message is sent by using the randomized point-to-point routing scheme, i.e., the message for each edge is routed via a random intermediate point as described above.

5.1 Analysis

Recall that for the analysis, we assume that for every point in the $L \times L$-square containing the network, there is an actual node within distance r_{cov}. Further when routing from a node at point (x, y) to a node at point (x', y'), the underlying geographic routing service generates λ_{pad}-padded, σ-sparse paths. For the analysis, we require a technical lemma bounding the number of local long edges of an MST in the Euclidean plane.

LEMMA 5.1. *Let T be an Euclidean MST of a set of points $X \subseteq \mathbb{R}^2$ and consider a circle $C \subseteq \mathbb{R}^2$ of radius r. The number of edges $\{p, q\}$ of T of length at least $3r$ such that $|\{p, q\} \cap C| = 1$ (i.e., edge $\{p, q\}$ connects a point inside C with a point outside C) is at most 7.*

PROOF. Consider two edges $\{p, q\}$ and $\{p', q'\}$ of length at least $3r$ such that $p, p' \in C$, $q, q' \notin C$. Let c be the center of the circle C and let θ be the angle that is enclosed by the rays cq and cq'. Let $d_{cq}, d_{cq'}$, and $d_{qq'}$ be the Euclidean distances between c and q, c and q', as well as q and q', respectively. By the law of cosines, we have

$$\cos \theta = \frac{d_{cq}^2 + d_{cq'}^2 - d_{qq'}^2}{2 \cdot d_{cq} \cdot d_{cq'}}. \tag{3}$$

Our goal is to upper bound the above expression and therefore to get a lower bound on the angle θ. Because the edges $\{p, q\}$ and $\{p', q'\}$ have length at least $3r$ and because p and p' lie in the circle C, it follows that

$$d_{cq} \geq 2r \quad \text{and} \quad d_{cq'} \geq 2r. \tag{4}$$

158

Since both p and p' are inside C, their distance is at most $2r$. Because $\{p, q\}$ and $\{p', q'\}$ are edges of the MST T and because we assume that their length is at least $3r$, $d_{qq'}$ has to be at least as large as the length of the longer of the two edges $\{p, q\}$ and $\{p', q'\}$. W.l.o.g., assume that $d_{cq} \geq d_{cq'}$. We then get

$$d_{qq'} \geq \max\{3r, d_{cq} - r\}. \tag{5}$$

We obtain an upper bound on $\cos\theta$ by maximizing the right-hand side of (3) subject to $d_{cq} \geq d_{cq'}$ and Inequalities (4) and (5). For fixed values of d_{cq} and $d_{qq'}$, the r.h.s. of (3) is a concave function of $d_{cq'}$ and is thus maximized either for $d_{cq'} = 2r$ or for $d_{cq'} = d_{cq}$.

- $d_{cq'} = 2r$: In that case, the r.h.s. of (3) is monotonically increasing in d_{cq} and therefore maximized for $d_{cq} = d_{qq'} + r$. We then get

$$\cos\theta \leq \frac{(2r)^2 + 2rd_{qq'} + r^2}{4 \cdot (rd_{qq'} + r^2)} \leq \frac{11}{16}.$$

The second inequality follows from $d_{qq'} \geq 3r$.

- $d_{cq'} = d_{cq}$: In the second case, we get

$$\cos\theta = 1 - \frac{d_{qq'}^2}{2d_{cq}^2}.$$

The above expression gets large if $d_{qq'}$ is as small as possible and d_{cq} is as large as possible. It is maximized for $d_{qq'} = d_{cq} - r = 3r$, in which case we obtain

$$\cos\theta = 1 - \frac{(3r)^2}{2(4r)^2} = \frac{23}{32}.$$

Combining the two cases, we therefore get $\cos\theta \leq 11/16$ which implies that $\theta > 0.812 > 2\pi/8$. □

In the following, let T_i, $1 \leq i \leq r$ be the Euclidean MST corresponding to multicast request R_i and let E_i be the directed edges of T_i, where each edge is directed away from the source s_i of R_i (i.e., in the direction in which a message has to be sent). Let $E_{\mathrm{MST}} = \bigcup_{i=1}^r E_i$ be the set of all directed MST edges. For a region $A \subseteq \mathbb{R}^2$ in the plane, let $E_{\mathrm{MST}}^{\nearrow}(A)$ be the set of directed edges $(p, q) \in E_{\mathrm{MST}}$ for which $p \in A$ and let $E_{\mathrm{MST}}^{\swarrow}(A)$ be the set of directed edges $(p, q) \in E_{\mathrm{MST}}$ for which $q \in A$. For each long enough edge $(p, q) \in E_{\mathrm{MST}}$, two messages are sent by using the underlying geographic routing service, one message from p to a random intermediate destination and one message from the intermediate destination to q. Let $\mathcal{M}_{\mathrm{out}}(A)$ be the set of messages sent from p to the random intermediate destination for an edge $(p, q) \in E_{\mathrm{MST}}^{\nearrow}(A)$. Further, let $\mathcal{M}_{\mathrm{in}}(A)$ be the set of messages sent from the random intermediate node to q for an edge $(p, q) \in E_{\mathrm{MST}}^{\swarrow}(A)$.

LEMMA 5.2. *Consider a square S with side length s and let v be a node at distance at least $d \geq 3s + 2r_{\mathrm{cov}} + (\lambda_{\mathrm{pad}} + 1)r_I$ from S. The expected congestion at node v caused by messages in $\mathcal{M}_{\mathrm{in}}(S)$ and $\mathcal{M}_{\mathrm{out}}(S)$ is at most $O\left((\lambda_{\mathrm{pad}} + 1) \cdot \sigma \cdot \rho^2 \cdot \mathrm{load}(S)\right)$.*

PROOF. Let us first consider a message $m \in \mathcal{M}_{\mathrm{out}}(S)$ corresponding to some edge $(p, q) \in E_{\mathrm{MST}}^{\nearrow}(S)$. The message m is sent from the node u closest to p to a random intermediate point (x_r, y_r). Assume that the coordinates of u are (x_u, y_u). Because $p \in S$, (x_u, y_u) is at distance at most r_{cov} from S. Further, by the way the random point (x_r, y_r) is chosen, the distance from u to (x_r, y_r) is upper bounded by the distance from u to q.

Message m only causes congestion at node v if the underlying geographic routing service sends the message through a node within distance r_I from v. Because we assume that the geographic routing paths are λ_{pad}-padded, this can be the case if v is within distance $r_I(1 + \lambda_{\mathrm{pad}})$ from the line segment connecting (x_u, y_u) and (x_r, y_r). Consequently, because the distance from v to S is at least $3s + 2r_{\mathrm{cov}} + (\lambda_{\mathrm{pad}} + 1)r_I$, the distance between u and (x_r, y_r) and therefore also the distance between u and q needs to be at least $3s + r_{\mathrm{cov}}$. Because u is at distance at most r_{cov} from S, this implies that the edge (p, q) has length at least $3s$.

Further, recall that the line from u to (x_r, y_r) is at a random angle $\alpha \in [-\pi/3, \pi/3]$ from the line uq. Message m causes interference at node v only when α is such that the line connecting u and (x_r, y_r) passes within distance $(\lambda_{\mathrm{pad}} + 1)r_I$ from v. Let β be the angle between line uv and the line connecting u with (x_r, y_r). The angle β is also a uniform random angle from some interval $[\beta_0, \beta_1]$ of length $2\pi/3$. Message m can cause interference at v if $|\beta| \leq \pi/2$ and $\ell \cdot \sin\beta \leq (\lambda_{\mathrm{pad}} + 1)r_I$, where $\ell \geq 3s$ is the distance between u and v. Using $|\sin\beta| \leq |\beta|$, we get

$$|\beta| \leq \frac{(\lambda_{\mathrm{pad}} + 1)r_I}{\ell} \leq \frac{(\lambda_{\mathrm{pad}} + 1)r_I}{3s}.$$

Let $C_{m,v}$ be the event that message m causes congestion at v. The probability for this to happen is at most

$$\mathbb{P}(C_{m,v}) \leq \frac{2(\lambda_{\mathrm{pad}} + 1)r_I}{3s} \cdot \frac{1}{2\pi/3} = \frac{(\lambda_{\mathrm{pad}} + 1)r_I}{\pi \cdot s}. \tag{6}$$

We define $X_{m,v}$ to be the random variable that counts the amount of congestion caused by m at v. Hence, $X_{m,v}$ is the number of nodes in the r_I-neighborhood of v that transmit a message while sending m from u to (x_r, y_r). Clearly $X_{m,v}$ can only be positive if the event $C_{m,v}$ occurs. In this case, the value of $X_{m,v}$ is at most $O(\sigma\rho^2)$ because we assume that the paths created by the geographic routing service are σ-sparse and a disk of radius r_I can be covered with $O(\rho^2)$ disks of diameter r_C. Let $X = \sum_{m \in \mathcal{M}_{\mathrm{out}}(S)} X_{m,v}$ be the congestion at v caused by messages in $\mathcal{M}_{\mathrm{out}}(S)$. By linearity of expectation, we have

$$\mathbb{E}[X] = O\left(\frac{(\lambda_{\mathrm{pad}} + 1)r_I}{s} \cdot \sigma\rho^2 \cdot |\mathcal{M}_{\mathrm{out}}(S)|\right).$$

To bound $\mathbb{E}[X]$, it therefore remains to bound the number of messages in $\mathcal{M}_{\mathrm{out}}(S)$. We have seen that each message $m \in \mathcal{M}_{\mathrm{out}}(S)$ corresponds to some MST edge (p, q) of length at least $3s$. Consider the circle C of radius $s/\sqrt{2}$ that encloses the square S. Since $p \in S$ and q is at distance at least $3s$, we have $p \in C$ and $q \notin C$. Hence, by Lemma 5.1, for each MST, there are at most 7 such edges of length at least $3s/\sqrt{2} < 3s$. Only multicast requests that contribute to $\mathrm{load}(S)$ can have MST edges with one node inside S and one node outside S. Further, for every such multicast request there are at most 7 edges in $\mathcal{M}_{\mathrm{out}}(S)$. The expected congestion at v created by nodes in $\mathcal{M}_{\mathrm{out}}(S)$ can therefore be upper bounded as

$$\mathbb{E}[X] = O\left((\lambda_{\mathrm{pad}} + 1) \cdot \sigma \cdot \rho^2 \cdot \mathrm{load}(S)\right). \tag{7}$$

The situation for the messages in $\mathcal{M}_{\mathrm{in}}(S)$ is almost symmetric. The messages are sent from the random intermediate destination (x_r, y_r) to a position inside S. However, the actual node sending the message might be at distance r_{cov} from (x_r, y_r), therefore we must accordingly adjust the angles for which there is congestion at node v. Instead of the value obtained in (6), the probability of $C_{m,v}$ can now be upper bounded by $\mathbb{P}(C_{m,v}) \leq \frac{r_{\mathrm{cov}} + (\lambda_{\mathrm{pad}} + 1)r_I}{\pi \cdot s}$. Because $r_{\mathrm{cov}} \leq \lambda_{\mathrm{pad}}r_I$, this does not change anything asymptotically, and the congestion from messages in $\mathcal{M}_{\mathrm{in}}(S)$ can also be upper bounded by the value given in (7). The claim of the lemma therefore follows Lemma 4.1. □

We are now ready to prove the main theorem of this section, showing that the expected maximal congestion induced by our geographic multicast algorithm is within a logarithmic factor of the optimal and therefore asymptotically best possible for any oblivious algorithm [5, 20].

THEOREM 5.3. *When using the described geographic multicast algorithm to route a given set of geometric multicast requests, the expected congestion at any node v is at most*

$$O\left(\left((\lambda_{\mathrm{pad}}+1)\cdot\log n+\lambda_{\mathrm{pad}}^2\right)\cdot\sigma\rho^2\cdot\mathrm{cong}^\star\right).$$

PROOF. The multicast algorithm described at the beginning of Section 5 sends two kinds of messages. Most messages are messages sent through the underlying geographic routing layer. In addition, messages to local neighbors are sent by direct local broadcast. Node v can be affected by local broadcast messages only if they are sent by nodes within distance r_I from v. By adapting the MST structure and contracting paths of total length at most $r_C/2$, it is guaranteed that for each multicast request the number of local broadcast messages in each r_C-neighborhood is $O(1)$. Such messages must be sent by a node within range r_C. Hence, the total congestion at nodes within distance $r_I + r_C$ of v has to be within a constant factor of the congestion caused by local broadcast messages at v. Hence, for every multicast solution, there must be some node w close to v with congestion at least a constant times the congestion caused by local broadcast messages at v.

Let us therefore consider the congestion caused by messages that are sent through the underlying geographic routing layer. Note that all these messages correspond to an MST edge of length at least $r_C/2$ and they all either go from an MST node to a random intermediate destination or from a random intermediate destination to an MST node. We partition the $L \times L$-square containing the network into two parts, an area containing nodes close to v and an area with nodes far away from v. Specifically, we consider a square Q of side length $6r_{\mathrm{cov}} + 3(\lambda_{\mathrm{pad}}+1)r_I = O((\lambda_{\mathrm{pad}}+1)r_I)$ and the area $\overline{Q}$ outside Q.

The area $\overline{Q}$ can be covered with $O\left(\log L/((\lambda_{\mathrm{pad}}+1)r_I)\right) = O(\log L/r_I)$ squares S_i of side length s_i such that the distance of square S_i to v is at least $3s + 2r_{\mathrm{cov}} + (\lambda_{\mathrm{pad}}+1)r_I$ as follows. The area right around Q is covered with $O(1)$ squares of side length at most $\left(2r_{\mathrm{cov}} + (\lambda_{\mathrm{pad}}+1)r_I\right)/3$ such that Q together with these squares cover a larger square around v. The additional squares can be iteratively placed in the same way around the growing center square such that side length of the squares grows exponentially with the number of layers. By Lemma 5.2, for each of the squares S_i covering $\overline{Q}$, the expected congestion from messages in $\mathcal{M}_{\mathrm{out}}(S_i)$ and $\mathcal{M}_{\mathrm{in}}(S_i)$ is at most $O((\lambda_{\mathrm{pad}}+1)\sigma\rho^2\mathrm{cong}^\star)$. Hence, the expected congestion from messages sent from a node in $\overline{Q}$ to a random intermediate destination and from messages sent from a random intermediate destination to a node in $\overline{Q}$ is at most

$$O\left((\lambda_{\mathrm{pad}}+1)\cdot\sigma\rho^2\cdot\log n\right)\cdot\mathrm{cong}^\star. \tag{8}$$

Recall that we assume r_{cov} is small enough and thus the node density is large enough such that n is at least polynomial in L/r_I and thus $\log(L/r_I) = O(\log n)$.

To prove the lemma, it remains to bound the congestion from messages sent from a node in Q to a random intermediate destination or from a random intermediate destination to a node in Q. Let M be the set of such messages. Because we assume that the geographic routing service produces σ-sparse paths and because

the r_I-neighborhood of v can be covered by $O(\rho^2)$ disks of diameter r_C, the congestion from each message in M is at most $O(\sigma\rho^2)$. Hence, the congestion at v from messages in M is at most $O(|M|\sigma\rho^2)$.

Every message in M corresponds to an MST edge of length more than $r_C/2$ and there are at most 2 messages in M for each such MST edge. Further, for a particular multicast request, the number of MST edges of length more than $r_C/2$ with one node in Q is linear in the number of nodes in Q and at pairwise distance more than $r_C/2$. Hence, to serve all destinations in Q, in an optimal multicast protocol, nodes in Q or within distance r_C of Q need to transmit at least $\Omega(|M|)$ times. The square Q and its r_C-neighborhood can be covered with $O\left((\lambda_{\mathrm{pad}}+1)^2\right)$ disks of diameter r_I. Each message that is transmitted by a node inside this area causes congestion at all nodes in at least one of these diameter r_I disks. Hence, by the pigeonhole principle, some node in Q or its r_I-neighborhood has congestion at least $\Omega\left(|M|/(\lambda_{\mathrm{pad}}+1)^2\right)$. Thus, the congestion at v caused by messages in M can be upper bounded by

$$O\left((\lambda_{\mathrm{pad}}^2+1)\cdot\sigma\cdot\rho^2\cdot\mathrm{cong}^\star\right). \tag{9}$$

Since the congestion caused by local broadcast messages is within a constant factor of the optimal congestion, (8) and (9) together imply the claim of the theorem. □

Remarks: If the ratio $\rho = r_I/r_C$ and the parameters λ_{pad} and σ specifying the quality of the underlying geographic routing service are constants independent of n, the statement of the theorem simplifies. The theorem shows that in this case, the maximal expected node congestion of our multicast algorithm is within a factor $O(\log n)$ of the optimal maximum node congestion. Note that it is well known that this is the best achievable bound for oblivious routing. Further, since congestion contributions from different multicast requests are independent, a standard Chernoff argument shows that the bound of Theorem 5.3 does not only hold in expectation, but also with high probability. Finally, we would like to point out that within the quality guaranteed by the underlying routing layer, our multicast protocol produces routing paths and trees that are within a constant factor of the optimal.

6. NAME-BASED MULTICAST

The multicast protocol discussed in Section 5 allows to efficiently (in terms of congestion and stretch) multicast messages if the source node of a multicast request knows the positions of all the destinations. In many cases, information about the positions of destinations is not available to the node disseminating some information. In this case, a geographic routing service can be used in conjunction with a location service that allows to query the positions of nodes [1, 10, 19]. In the following, we sketch how to apply the location service LLS [1] to our context, and we show that, if for each multicast request the destination positions can be obtained with a small number of queries to the location service, then the expected maximal congestion of looking up the destination coordinates is within a constant factor of the expected maximal congestion incurred by multicasting the messages.

Let us first briefly discuss how LLS works. We describe the most basic variant of the scheme. (The authors also present a more involved scheme that takes into account update costs when nodes are moving [1].) LLS is essentially a geometric, distributed hash table. Assume that we want to store the location information for node v with identifier id_v. We assume that there is a hash function h that assigns a coordinate $h(\mathrm{id}_v) = \left(h_x(\mathrm{id}_v), h_y(\mathrm{id}_v)\right)$ in the $L \times L$-square to each node v. Using the position $h(\mathrm{id}_v)$, we define a hierarchical tiling of the plane into squares of exponentially decreasing sizes. The corners of the squares of level $\ell = 0, 1, 2, \ldots$ of the tiling are at positions $\left(h_x(\mathrm{id}_v) + i \cdot L/2^\ell, h_y(\mathrm{id}_v) + j \cdot L/2^\ell\right)$ for integers $i, j \in \mathbb{Z}$. On every level ℓ, the position information of v is stored at the four corners of the tile that contains v. Starting from the position of v in order of decreasing levels, v's information is stored in a spiral-like fashion.

To look up the coordinate information for some node v with identifier id_v, the protocol searches in the same spiral-like fashion. Assume that node u searches for v's position information. For each level ℓ, node u queries the four corners of the tile containing u in the tiling defined by $h(\mathrm{id}_v)$. The search is done in the order of decreasing ℓ, i.e., by going from small tiles to large tiles, which forms a spiral that is shown to hit a node that stores the information about v with asymptotically optimal cost [1]. The following is a list of the most important properties of the scheme for our purposes:

1. If a node u looks up the information of some node v, the distance that has to be traversed for the search is proportional to the Euclidean distance of u and v.

2. A search for node v starting at node u follows an exponentially growing spiral. The exact paths visited during the search are determined by the position $h(\mathrm{id}_v)$. Assuming that the hash function $h(\mathrm{id}_v)$ leads to a uniformly distributed position for the origin of the coordinate system defining the tiling, it can be shown that a search from node u causes interference at a node at distance d with probability proportional to $(\lambda_{\mathrm{pad}} + 1)r_I/d$. Here, we assume that the search messages are sent through the geographic routing layer described in Section 3.

3. Assuming that the distribution of nodes is sufficiently dense, the scheme is compact. Each node only needs to store the position information of a logarithmic number of other nodes.

The next theorem shows that if at most κ look-ups are necessary for each multicast request, the expected look-up cost is asymptotically upper bounded by the expected cost for multicasting all message using our algorithm using the geometric protocol of Section 5. For the theorem, we assume that the hash function h leads to uniformly distributed positions $h(\mathrm{id}_v)$ that are independent of the given multicast requests. Due to lack of space, we only give a very rough sketch of the proof of the theorem.

THEOREM 6.1. *If each multicast request requires to look up at most κ positions, at every node v, the expected congestion caused by all look-ups is at most*

$$O\left(\kappa \cdot \left((\lambda_{\mathrm{pad}} + 1) \cdot \log n + \lambda_{\mathrm{pad}}^2\right) \cdot \sigma \rho^2 \cdot \mathrm{cong}^\star\right).$$

PROOF SKETCH. The proof follows a similar reasoning to the one in Lemma 5.2 and Theorem 5.3, where the congestion of the geometric multicast algorithm is analyzed. According to the first property of LLS listed above, a search from a node u for a node v stays within distance $O(d(u, v))$ of u, where $d(u, v)$ is the Euclidean distance between u and v. Let us therefore assume that all the κ searches of the source s_i of some multicast request R_i stay within distance $c \cdot d(s_i, t_i)$, where t_i is the destination of request R_i that is farthest away from s_i.

Let us first consider the congestion at v caused by multicast requests with a source node that is relatively far away from v. Consider a square Q of side length d that is at distance at least $2c \cdot d + (\lambda_{\mathrm{pad}} + 1)r_I$ from v. Assume that the source node s_i of multicast request R_i is inside Q. For a search of s_i to contribute to the congestion at node v, the farthest destination of R_i needs to be at least at distance $2d$ from s_i. Hence, R_i is a multicast request that

has the source node in Q and at least one destination node outside Q and R_i therefore contributes to $\mathrm{load}(Q)$ of Q. By the second property of LLS described above, the probability that a search of s_i causes congestion at v is at most $O((\lambda_{\mathrm{pad}} + 1)r_I/d)$ and therefore by a similar argument as in the proof of Lemma 5.2, the expected total congestion at v from searches of source nodes in Q can be upper bounded by

$$O\left(\kappa \cdot (\lambda_{\mathrm{pad}} + 1) \cdot \sigma \cdot \rho^2 \cdot \mathrm{load}(Q)\right).$$

By Lemma 4.1, this is within a factor $O(\kappa(\lambda_{\mathrm{pad}} + 1)\sigma\rho^2)$ of the optimal maximal node congestion. As in the proof of Theorem 5.3, the congestion caused by source nodes at distance at least $3(\lambda_{\mathrm{pad}} + 1)r_I$ from v can be bounded by $O(\log n)$ times the above value because that part of the network can be covered with $O(\log n)$ squares to which the above argument can be applied. Also for the congestion from searches of sources within distance $3(\lambda_{\mathrm{pad}} + 1)r_I$ from v, a similar argument to the one in the proof of Theorem 5.3 can be applied. Together, the bounds imply the statement of the theorem. $\qquad\square$

7. SIMULATION ANALYSIS

We now evaluate our routing scheme through simulation. This experimental analysis is intended to assess the performance of the scheme in practice, and also to characterize the effects of specific variants and parameters of the scheme itself as well as of the underlying geographical routing service. We consider three high-level research questions: (1) How does the scheme perform with various underlying routing algorithms? (2) How does the scheme perform with various selections of the random intermediate point? (3) How does the scheme perform in general under various workloads?

We first describe the implementation of the scheme and the underlying routing, and then present the simulation analysis.

7.1 Variants of the Routing Algorithms

We implemented two variants of the selection of the random intermediate point. The first variant corresponds exactly to the algorithm we describe and analyze formally in Section 5 and that is illustrated in Figure 1. This variant is parametrized by the range from which the source chooses the two random angles α and β that determine the intermediate point (x_r, y_r). In particular, we analyze the scheme when α and β are chosen uniformly in the ranges $[0, \pi/3]$, $[0, \pi/4]$, and $[0, \pi/6]$. Intuitively, wider angles would disperse traffic and therefore reduce congestion, at the expense of slightly longer paths and therefore worse total traffic.

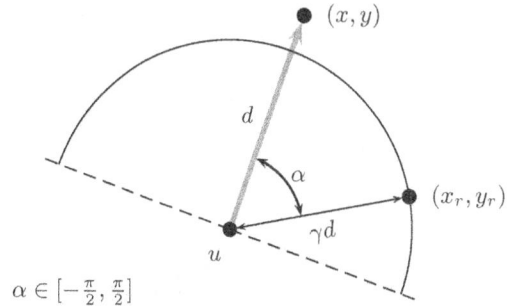

Figure 2: Alternative Selection of Intermediate Node

The second variant, illustrated in Figure 2, is a bit different: u selects an intermediate point (x_r, y_r) uniformly on a circular arc with center in u and radius $\gamma \cdot d(u, (x, y))$, where γ is a parameter of this method, and is between 0 and 2.

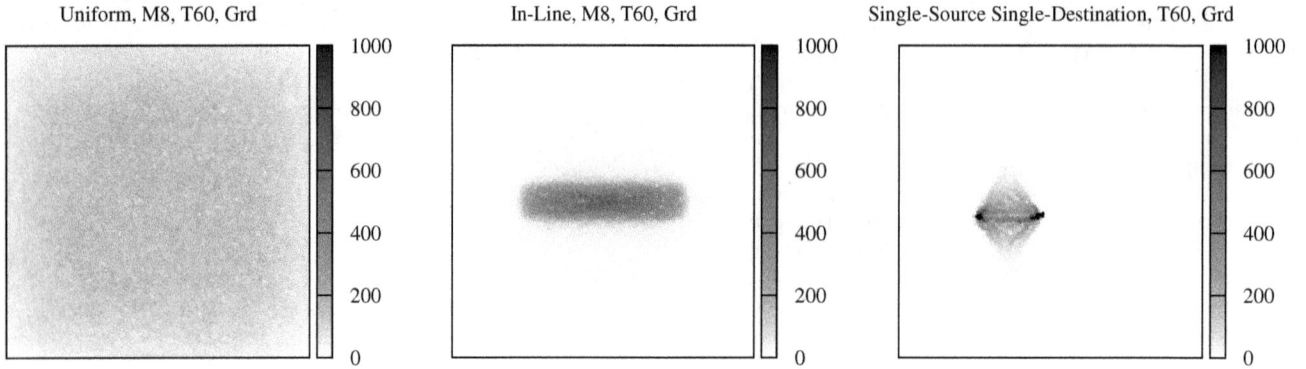

Figure 3: Examples of the Three Classes of Workloads

We also tested our scheme with various underlying routing algorithms. Recall that the geographical routing layer sends a message from a source node u to the node closest to the destination (x, y). The algorithms we considered are:

Grd Greedy routing. Each node v forwards the message to the next-hop neighbor w that is the closest to the destination (x, y).

GSP Geometric shortest path. The path between u and (x, y) is minimal in terms of geometric length.

DSP Hop-count (or "Dijkstra") shortest path. The path between u and (x, y) is minimal in terms of number of hops.

GrdRnd1 A randomized variant of greedy routing. In this case a node v forwards a message to a next-hop neighbor w chosen uniformly among the ones that advance towards the destination by at least half of the communication radius r_C.

GrdRnd2 Another randomized variant of greedy routing. A node v forwards a message to a next-hop neighbor w chosen uniformly among the ones that are within half of the communication radius r_C from neighbor $\overline{w}$, which is the closest to the destination.

7.2 Experimental Setup and Parameters

Network: We simulate a network of 80000 nodes spread uniformly over a square area of 100×100 units of length. (We also experimented with lower densities, obtaining consistent results that we do not report here for lack of space.) We set the communication radius to be equal to the interference radius ($r_C = r_I$) and we run simulations with $r_C = 1$ and $r_C = 2$ units of length. These settings correspond to a network that is dense enough to guarantee connectivity and to satisfy the more specific requirements of the underlying geographic routing, namely that it guarantees λ_{pad}-padded paths for a small constant λ_{pad}.

Table 1: λ_{pad} in practice

r_C	GSP	DSP	Grd	GrdRnd1	GrdRnd2
1	3.677	10.395	6.169	12.589	8.213
2	0.537	3.859	2.913	4.889	3.920

Table 1 shows the actual values for λ_{pad} for all five geographical routing algorithms. These values were computed over 10000 randomly selected paths. Notice that these are *maximum* values (as

per the definition of λ_{pad}-padded path) but at the chosen density the *average* distance between a routing path and a straight line between source and destination is much smaller. For lack of space, in the rest of the paper we discuss only the simulation with $r_C = 1$.

Workloads: We consider two classes of scenarios for multicast requests. One, which we denote as **uniform** in which requests involve sources and destinations chosen uniformly over the whole network, and one, which we denote as **in-line**, in which sources and destinations are chosen on a line, or more specifically on a narrow band in the middle of the network. The first class is intended to represent a generic traffic load. The second class is intended to represent a worst-case scenario for congestion. We also experimented with absolute worst-case workloads in which all requests are between the same source and the same destination. We initially show some results for all three cases for illustrative purposes, but then we focus on the **uniform** and **in-line** only because the third class is not very informative, since it incurs unavoidable congestion around the source and destination nodes.

Figure 3 shows three "heat-map" graphs representing one simulation run for each of the three classes of workloads, respectively. The graphs represent the square region covering the simulated network. Each point in the graph represents a node in the network whose color represents the total traffic (number of wireless transmissions) affecting that node, which corresponds to *load* or *congestion* of that node.

Analysis: In our analysis, we refer to a fixed set of all independent simulation parameters as a *scenario*. Thus, in a scenario we simulate all nodes running the same configuration of the geographic routing and the same configuration of our multicast routing scheme. We then simulate 1000 multicast requests, each with a fixed number of destinations chosen according to one of the scenario classes (**uniform** or **in-line**).

For each scenario we run 50 simulations to account for the variability that is due to the randomized nature of our scheme and possibly of the underlying routing. Then, for each node we compute the average load over the 50 runs, obtaining an approximation of the expected load of that node for that particular scenario. We then compute the *network congestion* as the maximum over all nodes of the per-node expected load. This is the primary metric of interest in this simulation analysis.

In summary, to answer our evaluation questions, we explore scenarios covering all combinations of the following parameters:

Intermediate point selection: type of algorithm and parameters

162

used to select the intermediate point. We use the angle-based selection with bounds $\pi/3$, $\pi/4$, and $\pi/6$ denoted with **T60**, **T45**, and **T30**, respectively. We then use the circular-arc selection with distance multiplier $\gamma = 0, 0.5, 1, 1.5, 2$, which we denote as **C0**, **C0.5**, **C1**, **C1.5**, and **C2**. Notice that **C0** corresponds to using a deterministic straight-line routing scheme. This degenerate case is useful for comparison.

Geographical routing: type of algorithm used in the underlying routing layer. We use the algorithms described in Section 7.1, denoted as **GSP**, **DSP**, **Grd**, **GrdRnd1**, **GrdRnd2**.

Multicast size: size of multicast requests (incl. source node). We use 2 (unicast), as well as 4, 8, and 16 (true multicast requests), which we denote as **M2**, **M4**, **M8**, and **M16**, respectively.

Workload class: location of sources and targets in multicast requests, chosen according to the **uniform** and **in-line** model.

7.3 Results

We now report the most important results of the simulation analysis. We first focus on the performance of the underlying geographic routing layer. We found that in all our experiments, the greedy algorithm yields the best results in terms of congestion. As

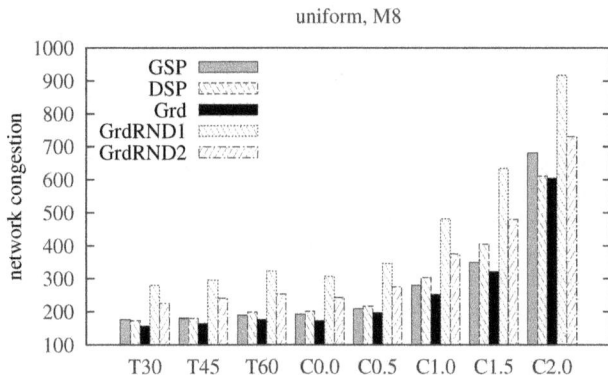

Figure 4: Comparison of Geographic Routing Algorithms

an example, Figure 4 shows the network congestion incurred by the various geographic routing primitives under a workload of uniform multicast requests of size 8, in combination with every variant of our scheme. In these scenarios, the greedy algorithm (**Grd**) is always the one that causes the lowest congestion, and as it turns out, all other scenarios show similar results. This result is particularly interesting and positive because **Grd** is also the simplest geographic algorithm available. Therefore, we dismiss all other underlying routing algorithms for the rest of our analysis.

The next question we consider is how the network congestion is affected by the selection of the intermediate point. The histogram of Figure 4 already indicates that the angle-based selection methods **T30**, **T45**, and **T60** work better than the method based on the circular arc for distance factor $\gamma > 1$.

Figure 5 confirms this result. The two graphs show the network congestion incurred by the various selection methods as a function of the size of the multicast requests, for **uniform** and **in-line** workloads, respectively. The conclusion we can draw from these experiments is that the angle-based schemes achieve the best results, only slightly better than the circular-arc method with radius less than the distance ($\gamma < 1$), and that the circular-arc method shows definitely

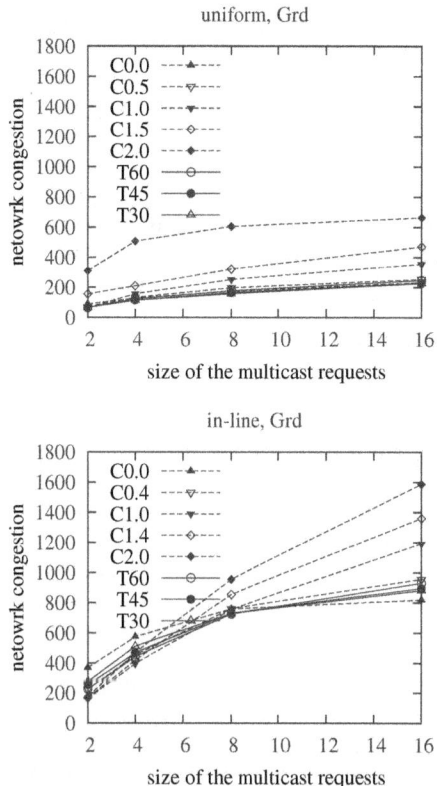

Figure 5: Comparison of Intermediate-Point Selection Methods

worse performance with higher radii ($\gamma > 1$) with proportionally worse outcomes in the case of uniform workloads. Also note that for the **in-line** model, the deterministic **C0** algorithm is the worst one for small multicast requests and it becomes the best algorithm for large multicast requests. For small requests, when routing deterministically, in the **in-line** scenario many routing paths overlap and by routing around, the congestion can be reduced. For large requests with all destinations on a line, the message has to be sent along the whole line anyway, so that sending it directly along the line becomes cheaper.

The graphs of Figure 5 also demonstrate that our multicast routing scheme performs well in an absolute sense and in particular they seem to indicate that the scheme scales gracefully with a sublinear relation between the size of the multicast requests and congestion. Recall that all workloads consist of 1000 requests, so, for example, in the case of requests of size 16, that means that each of the 1000 messages must be delivered to 15 destinations. Consider this scenario in the extreme case of requests in which all destinations lay on a line (or a narrow band) in the network, which corresponds to the case of the **in-line** workloads. It is interesting to notice that in this case, the scheme is capable of routing all requests in such a way that the maximally-loaded node sees the equivalent of a worst-case set of *unicast* requests.

8. REFERENCES

[1] I. Abraham, D. Dolev, and D. Malkhi. LLS: A locality aware location service for mobile ad hoc networks. In *Proc. 2nd Workshop on Foundations of Mobile Comp. (DIALM-POMC)*, pages 75–84, 2004.

[2] I. Abraham, C. Gavoille, and D. Ratajczak. Compact multicast routing. In *Proc. of 23rd Symp. on Distributed Computing (DISC)*, pages 364–378, 2009.

[3] S. Arora. Polynomial time approximation scheme for Euclidean TSP and other geometric problems. In *Proc. 37th Symp. on Found. of Comp. Sc. (FOCS)*, pages 2–11, 1996.

[4] L. Barrière, P. Fraigniaud, L. Narayanan, and J. Opatrny. Robust position-based routing in wireless ad hoc networks with irregular transmission ranges. *Wireless Communication and Mobile Computing*, 3:141–153, 2003.

[5] Y. Bartal and S. Leonardi. On-line routing in all-optical networks. *Theor. Comp. Sc.*, 221(1-2):19–39, 1999.

[6] P. Bose, P. Morin, I. Stojmenovic, and J. Urrutia. Routing with guaranteed delivery in ad hoc wireless networks. In *Proc. Discrete Algorithms and Methods for Mobility (DIALM)*, pages 48–55, 1999.

[7] J. Broch, D. Maltz, D. Johnson, Y.-C. Hu, and J.Jetcheva. A performance comparison of multi-hop wireless ad hoc network routing protocols. In *Mobile Computing and Networking*, pages 85–97, 1998.

[8] C. Busch, M. Magdon-Ismail, and J. Xi. Oblivious routing on geometric networks. In *Proc. 17th Symp. on Parallelism in Algorithms and Architectures (SPAA)*, pages 316–324, 2005.

[9] C. Busch, M. Magdon-Ismail, and J. Xi. Optimal oblivious path selection on the mesh. In *Proc. 19th Int. Parallel and Distributed Processing Symp. (IPDPS)*, 2005.

[10] R. Flury and R. Wattenhofer. MLS: An efficient location service for mobile ad hoc networks. In *Proc. 7th Symp. on Mobile Ad Hoc Networking and Computing (MOBIHOC)*, pages 226–237, 2006.

[11] J. Gao and L. Zhang. Trade-offs between stretch factor and load-balancing ratio in routing on growth-restricted graphs. *IEEE Trans. on Parallel and Distributed Systems*, 20(2):171–179, 2009.

[12] D. B. Johnson and D. A. Maltz. Dynamic source routing in ad hoc wireless networks. In *Mobile Computing*, volume 353, chapter 5. 1996.

[13] B. Karp and H. Kung. GPSR: greedy perimeter stateless routing for wireless networks. In *Proc. 6th Int. Conf. on Mobile Computing and Networking (MOBICOM)*, pages 243–254, 2000.

[14] P. Kolman and C. Scheideler. Improved bounds for the unsplittable flow problem. *J. of Algorithms*, 61(1):20–44, 2006.

[15] E. Kranakis, H. Singh, and J. Urrutia. Compass routing on geometric networks. In *Proc. 11th Canadian Conference on Computational Geometry*, pages 51–54, 1999.

[16] F. Kuhn, R. Wattenhofer, and A. Zollinger. Ad-hoc networks beyond unit disk graphs. *Wireless Networks*, 14(5):715–729, 2008.

[17] F. Kuhn, R. Wattenhofer, and A. Zollinger. An algorithmic approach to geographic routing in ad hoc and sensor networks. *IEEE/ACM Transactions on Networking*, 16:51–62, 2008.

[18] F. Li and Y. Wang. Circular sailing routing for wireless networks. In *Proc. 27th Int. Conf. on Computer Communications (INFOCOM)*, pages 1346–1354, 2008.

[19] J. Li, J. Jannotti, D. De Couto, D. Karger, and R. Morris. A scalable location service for geographic ad-hoc routing. In *Proc. 6th Int. Conf. on Mobile Comp. and Networking (MOBICOM)*, pages 120–130, 2000.

[20] B. Maggs, F. Meyer auf der Heide, B. Vöcking, and M. Westermann. Exploiting locality for networks of limited bandwidth. In *Proc. 38th Symp. on Foundations of Comp. Science (FOCS)*, pages 284–293, 1997.

[21] C. Maihofer. A survey of geocast routing protocols. *Communications Surveys & Tutorials*, 6(2):32–42, 2004.

[22] C. E. Perkins and E. M. Royer. Ad hoc on-demand distance vector routing. In *Proc. of 2nd IEEE Workshop on Mobile Computing Systems and Applications*, pages 90–100, 1999.

[23] L. Popa, A. Rostamizadeh, R. M. Karp, C. H. Papadimitriou, and I. Stoica. Balancing traffic load in wireless networks with curveball routing. In *Proc. 8th Symp. on Mobile Ad Hoc Networking and Computing (MOBIHOC)*, pages 170–179, 2007.

[24] H. Räcke. Optimal hierarchical decompositions for congestion minimization in networks. In *Proc. 40th Symp. on Theory of Computing (STOC)*, pages 255–263, 2008.

[25] H. Räcke. Survey on oblivious routing strategies. In *Proc. 5th Conf. on Computability in Europe (CiE)*, pages 419–429, 2009.

[26] E. Royer and C. Toh. A review of current routing protocols for ad-hoc mobile wireless networks. In *IEEE Personal Communications*, volume 6, April 1999.

[27] E. M. Royer and C. E. Perkins. Multicast operation of the ad hoc on-demand distance vector routing protocol. In *Proc. 5th Int. Conf. on Mobile Comp. and Networking (MOBICOM)*, pages 207–218, 1999.

[28] H. Takagi and L. Kleinrock. Optimal transmission ranges for randomly distributed packet radio terminals. *IEEE Transactions on Communications*, 32(3):246–257, 1984.

[29] L. G. Valiant and G. J. Brebner. Universal schemes for parallel communication. In *Proc. of 13th Symp. on Theory of computing (STOC)*, pages 263–277, 1981.

[30] U. Varshney. Multicast over wireless networks. *Communications of the ACM*, 45(12):31–37, 2002.

[31] J. Xie, R. R. Talpade, A. Mcauley, and M. Liu. AMRoute: ad hoc multicast routing protocol. *Mobile Networks and Applications*, 7(6):429–439, 2002.

[32] X. Yu, X. Ban, W. Zeng, R. Sarkar, X. Gu, and J. Gao. Spherical representation and polyhedron routing for load balancing in wireless sensor networks. In *Proc. 30th Int. Conf. on Computer Communications (INFOCOM)*, pages 621–625, 2011.

Dissemination in Opportunistic Social Networks: The Role of Temporal Communities

Anna-Kaisa Pietiläinen and Christophe Diot
Technicolor
1 rue Jeanne d'Arc
92443, Issy-les-Moulineaux, France
{annakaisa.pietilainen,christophe.diot}@technicolor.com

ABSTRACT

Epidemic content dissemination in opportunistic social networks (OSN) has been analyzed in depth, theoretically and empirically. Most related works have studied the pairwise contact history among nodes in conference or campus environments. We claim that given the nature of these networks, this approach leads to a biased understanding of the content dissemination process. We design a methodology to break OSN traces down into "temporal communities", *i.e.*, groups of people who meet periodically during an experiment. We show that these communities correlate with people's social communities. As in previous works, we observe that efficient content dissemination is mostly due to high contact rate nodes. However, we show that high contact rate nodes that are more frequently involved in temporal communities contribute less to the dissemination process, leading us to conjecture that social communities tend to limit dissemination in OSNs.

Categories and Subject Descriptors

C.2.1 [**Computer-Communication Networks**]: Network Architecture and Design—*Store and forward networks*

Keywords

Opportunistic Network, Social Network, Temporal Network, Community Detection

1. INTRODUCTION

Opportunistic networks exploit human mobility and consequent device-to-device contacts to disseminate content. Previous research has shown that the performance of such networks and applications depends on the users' social behavior [13, 17, 18] as opportunistic networks and human social networks share many key characteristics.

Human social networks are known to be extremely clustered [20]; we are more likely to meet and spend time with

our immediate social circles such as family, friends or co-workers while we meet infrequently with more casual acquaintances and accidentally with strangers. However, despite this high clustering, any pair of nodes is typically separated by a surprisingly small number of hops, known as the "small world" phenomenon [20]. These properties have been shown to appear in opportunistic networks [3, 18, 25].

The majority of data-driven research on opportunistic networks has focused on the analysis of individual node encounters (*i.e.*, pairwise contacts). We claim that this approach leads to partial or even misleading understanding of the structure of opportunistic networks, and as a result, of the dissemination process. Intuitively, pairwise contacts do not take place in isolation in real life. Instead, people tend to gather in small groups. At the same time, they are usually in proximity of some number of "strangers". As a result, the contact graph at any point in time consists of larger (possibly sparsely) connected components.

Nevertheless, the temporal community structure of opportunistic networks has received little attention in the literature. In this paper, we provide such analysis on four mobile user contact traces. We analyze the network dynamics using a temporal contact graph model [3, 24] and make the following contributions.

All four traces are structured around a large connected component that is present throughout the measurement period. Both the nodes and the edges in this connected component are highly variable. This suggests that many contacts in the connected component are due to proximity and are not related to the social behavior of its nodes.

We combine techniques for community detection in dynamic graphs [9, 26] to identify clusters of nodes that meet more frequently and for longer periods of time, that we name *temporal communities*. In particular, we identify temporal communities that appear multiple times during an experiment.

We use social information about the participants in our traces to correlate the detected temporal communities to *social communities*. We find that our temporal communities exhibit a high level of correlation with experimentalists' social characteristics such as friendship, shared affiliation, home city or country of origin.

We analyze last the impact of the above observations on the content dissemination process. As in previous works [5, 8, 11, 18], we observe that efficient content dissemination is mostly due to high contact rate nodes. However, we find that high contact rate nodes that are more frequently involved in temporal communities contribute less to the dissemination process. Instead, the truly mobile high contact rate nodes are responsible of maintaining fast and efficient content dissemination paths in a clustered opportunistic network.

The remainder of this paper is organized as follows. We review related work in Section 2. We introduce our data sets in Section 3, including a new experiment performed at SIGCOMM 2009, for which we have detailed social information of the participants. In Section 4, we propose a methodology to break the temporal contact graph into temporal communities. We discuss the properties and role of temporal communities in content dissemination in Section 5.

2. RELATED WORK

The structural similarities between social networks and opportunistic contact networks have been studied theoretically and empirically. Node degree [17], and contact and inter-contact times [2] in opportunistic networks have been found to exhibit heavy-tailed statistics. The "small world" phenomenon, *i.e.*, the existence of short paths among the vertexes of a locally clustered graph, has also been observed in opportunistic networks [3, 25]. The aggregated contact statistics show also high clustering: the majority of contacts take place among smaller groups of nodes [5,7,11,17]. These clusters are often linked by so called "weak ties" as discussed in [12,17].

We depart from the pairwise contact analysis and leverage a temporal graph model [15] to analyze the contact graph structure and evolution over time. Kossinets et. al [16] use such a graph to study the pathways followed by epidemic dissemination over a social network. Recently, similar approaches have been used in to quantify the efficiency of information dissemination in opportunistic networks [24, 25]. In contrast, our work focuses on the study of temporal communities and provides further insight into the efficiency of content dissemination.

The existence of a large connected component in opportunistic networks over time has been initially observed in [10]. We study in detail this temporal clustering phenomenon and show how it relates to human social interaction.

Our approach to correlate social and contact graphs is similar to [18], where a strong relationship is established between the centrality of a user in the social network and his role in content dissemination. More recently, Zyba and al. [27] propose three methods to divide the social graph in two populations, namely *socials* who are devices exhibiting some kind of periodic behavior in a given area, and *vagabonds* who are users visiting the area very rarely. The definition of social users in [27] implies that such users necessarily have higher contact rates than vagabonds; nevertheless, the authors observed that, in most cases, vagabonds significantly outnumber socials, and therefore have a greater collective effect in message dissemination. The main result of this paper is that the social status of mobile users does not really matter; what matters is the contact frequency. Our results complement the findings in [27]. With a definition of social users based on the notion of belonging to a *temporal community* (in contrast to being present periodically in a given area in [27]), we observe that the contact rate of the users does not depend on whether they are social or not. Specifically, we show that high contact rate users who spend more time in temporal communities are less efficient content forwarders. This is a new result as the previous works have considered all high contact rate users equal and have not looked in detail to the nature of the contacts.

Finally, there is a large body of work on community detection in graphs reviewed for example in [9]. Dynamic communities and clustering have been studied in the context of evolving social networks (see, *e.g.*, [26], and references within). We leverage these works to design our method for temporal community detection.

3. DATA SETS

We use four real-life human mobility experiments summarized in Table 1. In each experiment, participants carry a Bluetooth device that logs opportunistic contacts among the experimental devices. The data sets are chosen to cover various environments from sparse (campus) to dense (conference). In addition to physical contacts, all data sets contain social information about the participants. The details of the data sets are discussed below.

3.1 Experimental Conditions

INFOCOM. This is a trace of Bluetooth device encounters among a group of users carrying small sensor devices at INFOCOM 2006 conference, in Barcelona, Spain[1]. The measurement device is Intel mote (iMote), a small battery operated Bluetooth v1.1 radio with approximately 30 meter range. The devices perform a device discovery every 120 ± 5 seconds. Each participant is asked to fill a short questionnaire about their basic social information such as affiliation(s), home city, country and nationality. In this work, we study the subset of 63 users who answered the full questionnaire. We limit the observation period to the first two days of the trace as participants gradually leave the experiment during the last days.

SIGCOMM. We collect the second data set in a similar conference setting[2]. One hundred smartphones running Mobiclique [23], an opportunistic mobile social networking application, were distributed to SIGCOMM 2009 participants (Barcelona, August 2009). Each device is initialized with the participants Facebook profile including the following information: home city, country, affiliation, and the list of friends in Facebook. The application uses Bluetooth for opportunistic contact discovery and data communications. The device discovery is performed every 120 ± 10 seconds. The experimental devices have a class 2 Bluetooth v2.1 radio with a range of 10 to 20 meters. In this work we use the 76 devices that sustain activity throughout the observation period, the first two days of the trace.

Reality. The MIT's Reality Mining data set[3] follows 100 subjects at MIT over the course of an academic year. The data is collected using smartphones and includes call logs, Bluetooth devices in proximity, cell tower IDs, application usage, and phone status (such as charging and idle). The Bluetooth device discovery interval is 5 minutes. The data set includes the affiliation of each participant. In addition, we infer social relationships from the aggregated communications graph which is a graph where two nodes are connected if they have called or sent an SMS to each other at

---

[1] http://crawdad.cs.dartmouth.edu/meta.php?name=cambridge/haggle

[2] The data set will be made available at Crawdad.

[3] http://crawdad.cs.dartmouth.edu/meta.php?name=mit/reality

	INFOCOM	SIGCOMM	Reality	Strathclyde
Year	2006	2009	2004-2005	2010-2011
Full trace	4 days	4 days	9 months	5 months
Obs. period	2 days	2 days	3 weeks	3 weeks
Setting	conference	conference	campus	campus
Devices	63	76	97	27
Device	iMote	phone	phone	phone
Radio range (m)	30	10-20	10-20	10-20
Inquiry length (s)	5	10.24	-	-
Inquiry interval (s)	120±5	120±10.24	300	60

Table 1: The data sets.

least once. The edges are weighted by the total number of calls and messages (in either direction). Both the contact trace and social information are collected from a 3-week period during the first semester where we observe the highest contact activity.

Strathclyde. The most recent data set we analyze comes from the University of Strathclyde[4]. The data set includes logs of phone calls, text messages, Bluetooth proximity detection, WiFi access point, and cell tower ID. We use the Bluetooth proximity trace that records nearby devices every 60s. Similarly to Reality, we infer the participants social graph using the call and SMS logs. The contact trace and social information we study come from a 3-week period with high contact activity.

3.2 Social Networks

We group the participants of each experiment in social communities based on their affiliation, home city or nationality. In addition, we use the initial Facebook friendship graph (given by the participants) for SIGCOMM, and the aggregated communications graph for Reality and Strathclyde, to form clusters of participants using a well-known graph partitioning algorithm [4]. Table 2 summarizes the total number of unique communities of each social category in the data sets and their sizes.

INFOCOM		
	Total	Size Distribution
Affiliation	37	12, 5, 4, 3, 3, 2(x5), 1(x26)
City	33	10, 10, 5, 4, 2(x6), 1(x22)
Nationality	25	10, 7, 7, 5, 5, 4, 2(x7), 1(x11)
SIGCOMM		
	Total	Size Distribution
Affiliation	46	7, 5, 4, 4, 4, 3, 2(x9), 1(x32)
City	39	12, 6, 5, 4, 4, 4, 3, 3, 2(x4), 1(x27)
Friend clusters	15	20, 20, 13, 6, 6, 2, 1(x9)
Reality		
	Total	Size Distribution
Affiliation	10	30, 27, 20, 7, 6, 3, 1(x4)
Calls&SMS	61	13, 8, 7, 5, 3, 2(x5), 1(x51)
Strathclyde		
	Total	Size Distribution
Calls&SMS	13	8, 4, 3, 2, 2, 1(x8)

Table 2: Social communities.

Note that the social network data is collected independently of the contact trace in contrast to some related works where the social network formed during the experiment is used [18], or where authors rely uniquely on the contact trace properties to infer social relationship [5, 11, 17].

---

[4]http://crawdad.cs.dartmouth.edu/meta.php?name=strath/nodobo

3.3 Contact Traces

The contact traces record pairwise contacts among experimental devices. We post-process SIGCOMM and INFOCOM traces to overcome some of the limitations of Bluetooth device discovery that can fail to discover nearby devices due to simultaneous discoveries and on-going data communications. We define a *contact* between two devices A and B as the time between the first response from B to A's Bluetooth device inquiry until the time there are no more responses from B during **two** consecutive inquiry periods. In addition, we consider that if A can contact B, B should be able to contact A, *i.e.*, contacts are symmetric. The Strathclyde trace is processed similarly with a limit of **four** consecutive inquiry periods due to shorter inquiry interval. The Reality data set does not include raw device discovery logs, and we only verify that contact events are symmetric.

The two conference traces are dense containing close to 30 000 and 16 000 contacts at INFOCOM and SIGCOMM, respectively. A majority of the participants, 91% at INFOCOM and 71% at SIGCOMM, meet at least once during the experiment. In contrast, the campus traces are sparser. We observe only around 10 000 contacts at Reality and Strathclyde experiments over a much longer observation period. In the Strathclyde data set 79% of the nodes meet during the experiment. The Reality trace differs from all the others in that only a small subset, 36% of all the pairs, meet at least once.

We plot the empirical distributions of pairwise contact and inter-contact times in Figure 1. The contact and inter-contact times follow both an approximate power law in each data set which has been shown and analyzed by many prior works (*e.g.*, [2]). The larger number and longer duration of contacts at INFOCOM compared to SIGCOMM can be explained by the fact that (i) the iMotes' battery lasts for the duration of the experiment while the phones at SIGCOMM required daily recharge, (ii) the lack of data communications interfering with the discovery, and (iii) the longer radio range of the experimental devices. The contacts of the campus traces are, in general, longer with a median of 8 minutes than in the conference environment where the median contact time is 2 to 3 minutes.

4. TEMPORAL COMMUNITIES

We analyze the contact graph structure and evolution using a temporal graph model. We show that each trace is characterized by the existence of a single large connected component during the daytime hours. This component is sparse and composed of smaller and denser clusters of nodes. We propose a methodology to extract persistent clusters of nodes that we name *temporal communities*.

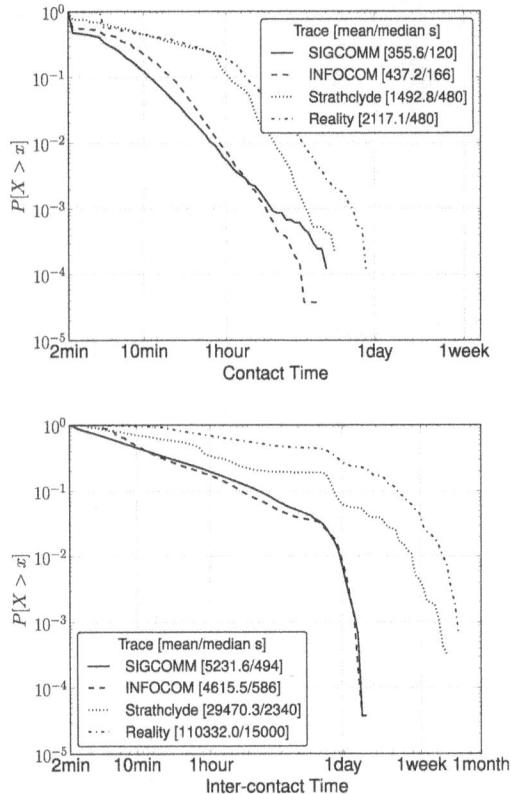

Figure 1: The distribution (complementary CDF) of contact (top) and inter-contact (bottom) durations.

4.1 Temporal Contact Graph

We define the *temporal contact graph* as a series of *snapshots* of the contact trace. Each snapshot is a static graph, $G_t(V, E_t)$, where nodes V are the experimental devices and edges represent a contact between two nodes during a time window $[t, t+1]$. We take a snapshot every 60 seconds in the conference and Strathclyde traces, and every 120 seconds for Reality (due to much longer device discovery interval). The snapshot interval is chosen to be equal or shorter than the device discovery interval in order to obtain a detailed view of the graph evolution and to account for the asynchronous device discovery periods.

The temporal contact graph is characterized in all traces by the existence of a *single large connected component* during the daytime hours. A connected component is a sub-graph where each node is reachable from every other node either directly or through multiple hops [6]. We plot the size of the largest and the second largest connected components in the temporal contact graphs for SIGCOMM and Reality in Figure 2(a) as examples of the two types of environment (complete results are available in [22]).

At SIGCOMM, the size of the largest component varies between 10 to 45% during the conference and the evening hours (due to social events). The size of the second largest component varies from 5 to 15% of the nodes while rest of components contain only one to two nodes most of the time. Similarly, at INFOCOM 35 to 80% of the nodes belong to the largest connected component during the conference

hours. The second largest component (and others) are much smaller, typically in the order of one or two nodes. The Strathclyde trace is similar to the conference traces in many aspects: a large number of participants meet everyday at the university and the largest connected component contains 38 to 67% of all the nodes. The second largest component has at most five nodes. The Reality trace is less clustered than the other data sets because of the nature of the experiment, but we observe a large connected component of ten to twenty participants daily. Other components are smaller.

Interestingly, we find that the large connected component is not stable over time. We study the variation of edges and nodes within the largest connected component over time in Figure 2(b) using SIGCOMM (other traces are similar). We plot the similarity between the set of nodes (edges) of the largest component from one time step to another using the Jaccard Index, $Jac(A, B) = \frac{|A \cap B|}{|A \cup B|}$ [14]. Value 1.0 means the two sets are identical and 0.0 that they are completely disjoint.

The internal connectivity of the largest component, captured by its set of edges, varies constantly during the daytime. The similarity between edges from a time step to another is constantly below 0.8. The node sets are more stable, *i.e.* the similarity is around 0.8 to 1.0. Bigger variations are observed during the beginning and end of conference hours due to node mobility.

We study the local structure of the connected components and find that they are sparsely connected, but show high local clustering (detailed results are available in a technical report [22]). The largest connected component has, in general, a low edge density, around 30% or less of all the possible node pairs are connected. The edge density becomes more variable as the size of the component decreases. The structure of the smallest components varies from fully connected meshes (100% edge density) to components with single link per node. The mean clustering coefficient of the connected components is approximately between 0.3 and 0.5 for the majority of the components. This value is high compared to what would be expected from a similar random network [20]. The results are somewhat different for the Strathclyde data where a majority of the components are small and dense. This can be explained by the small and homogeneous group of participants.

These observations on the structure and evolution of the connected components suggest that the nodes in most of the connected components do not belong to a single, stable, cohesive community (*i.e.*, group of people that meet). Instead, the network is build of small inter-connected clusters of nodes. We validate these conjectures in the next section.

4.2 Temporal Community Detection

We propose a methodology to break the temporal contact graph into clusters of nodes that meet more frequently and for longer periods of time during the experiment. We call these clusters *temporal communities*. The temporal community detection method proceeds in two steps. First, we partition each snapshot graph into smaller and denser *clusters* of nodes. Second, we apply a hierarchical clustering algorithm to aggregate the snapshot clusters into relevant *communities*.

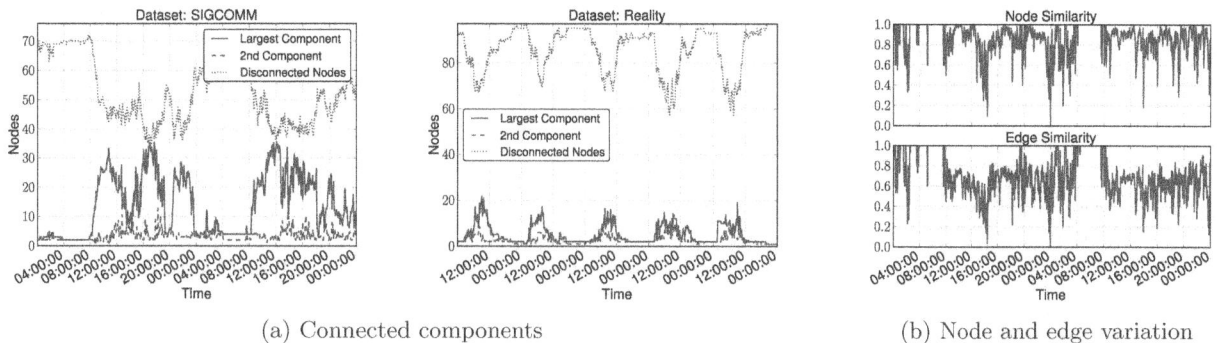

(a) Connected components   (b) Node and edge variation

Figure 2: Temporal contact graph evolution.

4.2.1 Step 1: Cluster Detection

The problem of partitioning a static graph into clusters has been studied widely [9]. We test several popular algorithms including fast modularity optimization algorithm by Clauset et al. [4], an alternative to modularity optimization called Louvain [1] and k-clique clustering [21]. All these algorithms have been shown to perform well on many different types of static graphs, including biological and social networks. Another advantage of each of these methods compared to other well-known clustering algorithms such as k-means or spectral clustering [9] is that they do not require an advance knowledge of the number of communities in a graph. Instead, they determine this value based on the graph structure.

These methods do not result in identical partitions and the selection of the "best" partition is hard in the absence of a ground truth. We choose the Louvain algorithm due to its simple and fast implementation that could be adapted for distributed computation. Moreover, it does not use any preconfigured parameters such as the clique size in k-clique. Finally, while the modularity optimization schemes are known to suffer from a resolution limit [9], the hierarchical nature of the Louvain algorithm allows to discover partitions at multiple resolution levels [1]. However, we do not investigate this further in this paper. The complete results for all of the above mentioned methods are available in a companion technical report [22].

A cluster in a graph is commonly defined as a cluster of nodes that have more links among each other than with any other nodes of the graph. This definition is directly formalized in a metric called *modularity* [19]. Modularity measures the fraction of edges in a graph that connect vertexes within clusters minus the expected value of the same quantity in a graph with the same cluster partition, but random connections between the vertexes. Modularity is calculated as follows:

$$Q = \frac{1}{2m} \sum_{ij} \left[A_{ij} - \frac{k_i k_j}{2m} \right] \delta(c_i, c_j)$$

where m is the number of edges of the graph, A is the adjacency matrix ($A_{ij} = 1$ if there is an edge (i, j), else 0), k is the node degree and $\delta(c_i, c_j)$ is 1 if $c_i == c_j$, i.e., i and j belong to the same cluster, else 0. The values of the modularity score are always below 1. If the fraction of within-cluster edges is no different from random, the mod-

ularity score is zero. Cluster detection based on modularity finds a division of the graph that maximizes the global modularity.

Cluster detection based on modularity finds a division of the graph that maximizes the global modularity. Finding the optimal solution is a NP-hard problem. Louvain [1] is a heuristic solution that approximates the optimal solution. Initially, each node forms its own cluster. In phase one, the algorithm loops over each node i. Then for each neighbor j of i, it calculates the local gain in modularity if the node i would be placed into the cluster of j. Finally, the node i is placed into the cluster with the maximum gain if the gain is *positive*, else it remains in its current cluster. This process is repeated until no further increase of gain is possible. In the second phase, the algorithm builds a new graph where the nodes are the clusters found in the first phase. The edges between the new nodes are weighted by the sum of the weights of the edges linking the two clusters in the previous graph. The algorithm re-iterates using this new graph as input. The algorithm stops once no new clusters can be formed in the new graph.

4.2.2 Step 2: Cluster Aggregation

We define a temporal community as a set of snapshot clusters that are *persistent*, i.e., they appear at several time steps (not necessarily consecutive) within a contact trace. To account for minor variations due to Bluetooth inaccuracies (some contacts are missed) or due to volatile contacts with nodes in the proximity of a cluster, we assume that two clusters belong to the same temporal community if they are sufficiently "similar".

We measure the similarity of two clusters using the above presented Jaccard Index. We define Jaccard distance between two sets of nodes A and B as $1 - Jac(A, B)$. For a given δ, a *temporal community* C is then a set of clusters whose Jaccard distance does not exceed δ. We use hierarchical clustering to aggregate similar clusters [9]. The algorithm proceeds as follows:

1. **Initialization.** Place each cluster in its own temporal community.

2. **Distance Calculation.** Compute the distance between all temporal communities. For communities that consist of only one cluster, this is the Jaccard distance defined above. For communities with more than one aggregated cluster, the distance is the average of all

pairwise Jaccard distances of clusters in these two communities. In all cases, if two communities contain clusters with the same time stamp, their distance is set to infinity.

3. **Community Merge.** Pick the two closest temporal communities. If the closest distance is larger than δ, the algorithm terminates; otherwise, the two closest communities are merged, and the process is repeated again from step 2.

The temporal community detection algorithm terminates when the shortest distance between any two temporal communities becomes larger than a threshold value δ (see step two above). The parameter δ controls the amount of variation we tolerate in the composition of a temporal community. With $\delta = 0.0$, the algorithm aggregates fully identical clusters, while setting $\delta = 1.0$ would merge any two clusters that have distinct timestamps, irrespectively of how many members they have in common.

4.3 Sessions, Cover and Core

Temporal communities are sequences of similar clusters that do not necessarily occur at consecutive snapshot time steps. We call a sequence of consecutive snapshots within a temporal community a *session*. Each temporal community may comprise of several sessions of variable duration.

The aggregation method results sometimes in sessions that are separated only by a single snapshot interval. We combine these sessions into one single session as these short gaps are most likely due to the Bluetooth inaccuracies or asynchronous device discoveries. Moreover, we exclude all sessions that are shorter than two times the device discovery interval, *i.e.*, four minutes for the conference traces and Strathclyde, and ten minutes for Reality. The device discovery interval defines the shortest possible contact observable in the data sets. Setting the minimum session length to twice the minimum contact duration filters out sets of similar clusters that are only recorded during one device discovery (by several nodes). These communities may not correspond to any meaningful social structure but appear randomly due to variations in the temporal contact graph.

The δ parameter has an impact on the total number of discovered temporal communities. Initially, increasing δ lets us discover more temporal communities as the aggregation rule is less strict. However, after a certain distance, the aggregation algorithm starts combining more and more existing communities together and the total number of temporal communities decreases. We illustrate this in Figure 3. The curves tip at $\delta = 0.2$ for the campus traces and at $\delta = 0.4$ for the conference traces.

The members of a temporal community can be defined either by the union of the aggregated clusters (*i.e.*, members include all the nodes present in the aggregated clusters), or by the intersection of the clusters, (*i.e.*, members that appear in every cluster). We refer to the union of all clusters in a temporal community as the *cover* of the temporal community. Similarly, we refer to the intersection of all clusters as the *core* of the temporal community.

The δ parameter has also an impact on the size of the cover and the core. As expected, when the δ increases, the cover becomes larger and has a higher variation as more and more distant clusters of nodes are being aggregated together. We plot the distribution of the core size as a function of δ

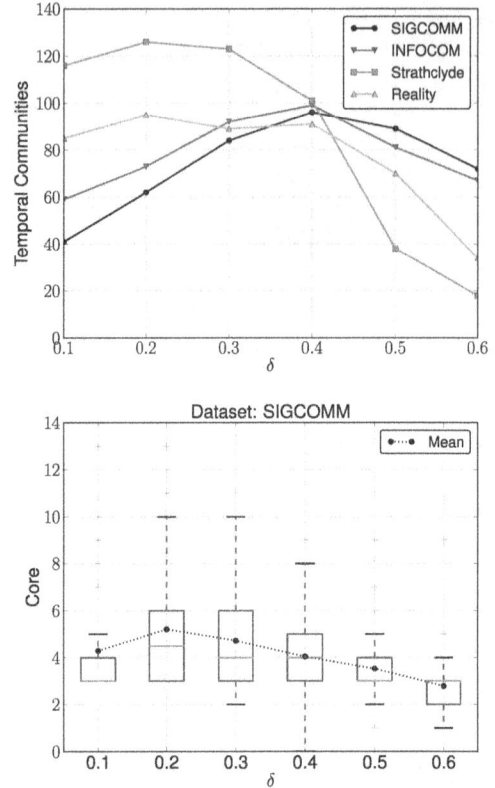

Figure 3: **Impact of δ on the number of temporal communities and the core size.**

in Figure 3 taking SIGCOMM as an example. The median number of core members remains constantly low, 3 to 7 in the conference traces and 2 to 5 in the campus traces. The median is initially higher (or increases by one) until $\delta <= 0.4$ ($\delta <= 0.3$ in Strathclyde), and decreases then rapidly as a result of more and more distant clusters being aggregated together.

The choice of δ presents various trade-offs as shown above and there is no clear "best" value. In general, the points where the core size and the number of temporal communities peak seem reasonable choices (*i.e.*, $\delta \in [0.2, 0.3, 0.4]$). In the following analysis, we will use $\delta = 0.3$ for all the data sets. We also use the most conservative strategy for the temporal community membership and consider only the core members in the rest of this paper. Moreover, we exclude any temporal communities with less than three core members (approximately 25% of the final temporal communities for $\delta = 0.3$) as our focus is on understanding group dynamics.

5. RESULTS

In this section we discuss the properties of the temporal communities. We correlate them with participants' social characteristics. We identify the presence of high contact rate (HCR) nodes inside and outside of temporal communities. We show that nodes that spend more time in temporal communities do not impact content dissemination as much as nodes outside temporal communities.

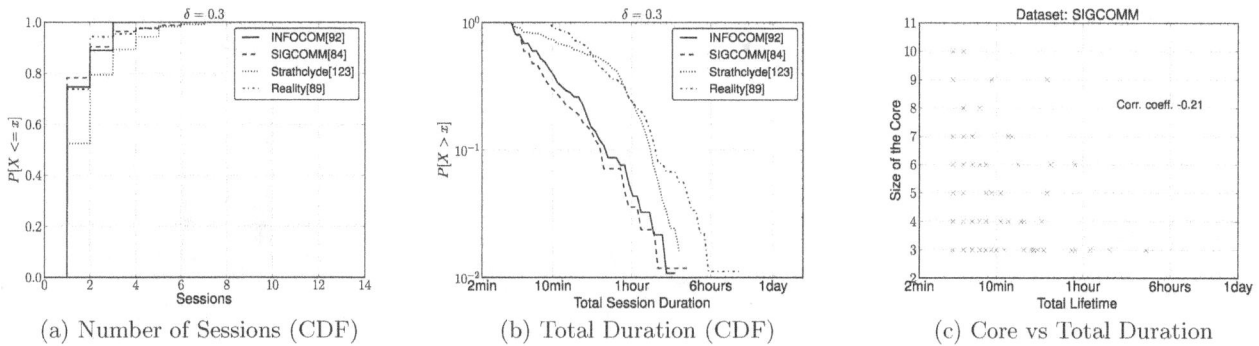

(a) Number of Sessions (CDF) (b) Total Duration (CDF) (c) Core vs Total Duration

Figure 4: Characteristics of temporal community sessions.

5.1 Session Characteristics

We find 84 to 123 temporal communities that contain three or more participants and meet once or several times for the above described minimum time. We define the total community duration as the sum of the sessions durations for each temporal community. The distributions of the total number of unique sessions and the total duration per temporal community are shown in Figure 4(a) and Figure 4(b).

Approximately 25-50% of the temporal communities meet more than once in all data sets. The majority of the communities in the conferences are short-lived with the median total community duration being 6 (SIGCOMM) and 8 (IN-FOCOM) minutes. However, approximately 30-40% of the temporal communities meet for 10 minutes or more. In contrast, in the campus traces the temporal communities are long lived, the median being 28 (Reality) and 32 (Strathclyde) minutes, and with 25% meeting for more than 1 hour. The total community duration is not correlated with the number of sessions; the long-lasting temporal communities may meet only once or few times and vice versa.

The long-lasting temporal communities are typically small as shown in Figure 4(c) where we plot the correlation between the temporal community size and total meeting time. Temporal communities that exist for longer time, 10 minutes or more, have most of the time less than 10 members.

We also look into the distribution of the sessions over time. The temporal communities appear throughout the day in all traces. However, we observe that the longer sessions occur typically during the conference breaks (lunch, coffee) or dinner time in the conference traces, and around the working hours in the campus traces.

In summary, we observe a significant number of temporal communities in all data sets. 10% to 50% of the temporal communities last 30 minutes or more, some smaller ones lasting up to several hours. 25% to 50% of the temporal communities are also observed multiple times which makes it unlikely for these communities to have a random nature. In the next section, we study the correlation the temporal communities with the participants' social information.

5.2 Social Correlation

In order to correlate social characteristics and mobility, we measure the similarity between a temporal community and a social community as the fraction of temporal community nodes that belong to a given social community. Each temporal community is compared to all social communities

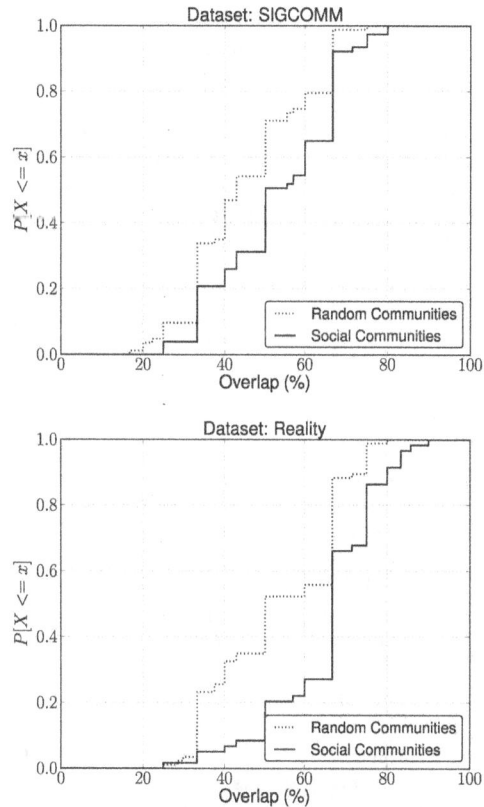

Figure 5: Correlation of the temporal and social communities.

with more than one member and the social community that is most similar to the temporal community is selected.

We do not expect a perfect match between the temporal and the social communities as (1) we do not have full knowledge of the participants' social characteristics, (2) social interactions are more complex than meeting a single community, and (3) the experimental conditions do not capture all contacts as explained earlier. Yet, some correlation should be observable. To get an idea on how "good" the similarity scores are, we create randomized communities that match the size and number of the social communities but consist of

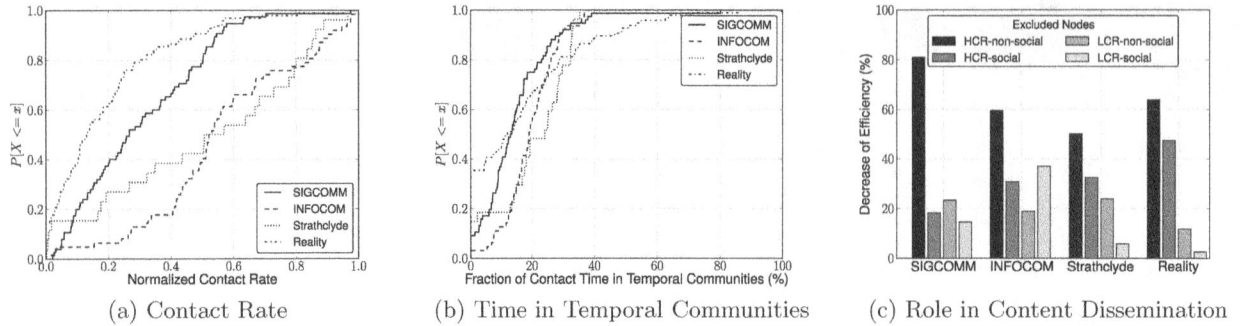

(a) Contact Rate   (b) Time in Temporal Communities   (c) Role in Content Dissemination

Figure 6: Node characterization.

nodes selected uniformly at random among the participants and calculate the similarity score for them for comparison.

We present the distribution of similarity scores for SIG-COMM and Reality in Figure 5. 50% of the temporal communities share 50% or more of their members with some social community at SIGCOMM. In the Reality data set, the overlap is even higher: 80% of the temporal communities share 50% or more of their members with a social community. Strathclyde and INFOCOM show similar results to SIGCOMM. The correlation with random communities is lower in all traces, and in particular for the campus data sets. Given the limitations introduced, we consider the correlation for SIGCOMM and INFOCOM representative of a social structure of the temporal communities.

5.3 Content Dissemination

Having shown that temporal communities are built on social relationships, we want to understand the role of these "social" communities in epidemic content dissemination. Previous work has established that the nodes that are the most "central" contribute most to the content dissemination in an opportunist network [5, 8, 11, 18]. One of the most popular ways to measure centrality is through the contact rate that is defined as the number of contacts per time unit [8]. We plot the distributions of normalized contact rates for each data set in Figure 6(a). The distribution is close to uniform for each trace. We define a *high contact rate node (HCR)* as a node with a higher than median contact rate (similar definition has been used in previous works [8]). The remaining nodes are called *low contact rate nodes (LCR)*.

In addition, we define a *social node* as a node that spends a significant portion of its contact time in temporal communities. We measure this as a fraction of the total contact time and the distributions are shown in Figure 6(b). Considering typical human activities at conferences and on campuses, we define that the significance is more than 20% of the time for the conference traces and 30% for the campus traces. In our analysis, a day is 16 hours. Therefore, by this definition social nodes meet other people for more than 3.0 hours per day (social events) at the conferences, and for more than 5 hours per day on the campuses (social events plus courses)[5]. The contact rate of the nodes does not correlate with the time spent in temporal communities. Hence, we end up with four categories of nodes depending on whether a node is high or

low contact rate, and spends more or less time in temporal communities. Our objective is to understand the role of each of these groups on epidemic dissemination.

We analyze the opportunistic content dissemination paths using vectorclocks as defined in [16]. Vectorclocks can be used to measure the *age of information* among all participating nodes upon each contact event and can be used as an approximation of the network efficiency[6]. In order to account for missing paths, we measure the network efficiency as the sum of inverted delays [24]:

$$E = \frac{1}{T} \sum_{ij} \sum_{k} \frac{1}{d_{ij}^{k}}$$

where T is the duration of the contact trace and d_{ij}^{k} is the length (in seconds) of the kth shortest path from node i to node j.

We calculate the efficiency for each contact trace including all nodes. Then we repeat the same experiment removing each of the four node categories identified above one at a time. Figure 6(c) summarizes the results for all the traces. The y-axis represents the decrease in efficiency compared to the base case, *i.e.*, higher bar means lower efficiency.

The first and expected observation is that HCR nodes are responsible for most of the epidemic dissemination in opportunistic social networks. Both HCR social and non socials have the most important impact on network efficiency when removed. However, this is the first time that a large difference of impact between social and non-social HCR nodes is observed. Removing non-social HCR nodes decreases the efficiency of the network by 50% to 80% at the conference traces, while removing social HCR nodes has a significantly more limited impact in all traces. This indicates that non-social HCR nodes are crucial in maintaining the network connectivity and creating fast paths in the temporal network. In contrast, HCR social nodes being more "static", they are not as effective at disseminating content to other nodes. The results for LCR nodes are more difficult to interpret as these nodes are impacted by the experimental conditions (they include nodes that have no – or almost no – contacts, and hence, the selected definition for social is less meaningful).

---

[5]Conference session attendance is based on technical interest, not on social relationship.

[6]This method ignores things like number and size of messages and message creation times that are not important for this study.

6. CONCLUSION

We study four experimental data sets on human mobility that combine a contact graph among a group of people (or nodes) and the social network of the nodes collected before the experiment. We make the following contributions. We observe that all data sets are well-connected, and that a large connected component appears during day time. We design a method to break down the connected components into smaller and denser sub-graphs of nodes, that we call temporal communities. We show that the temporal communities correlate with the social communities of the participants. Finally, we show that high contact rate nodes can be divided into nodes that spend a large fraction of their time in temporal communities and to truly mobile nodes rarely seen in temporal communities. We find that these high contact rate, non-social nodes are mostly responsible of efficient content dissemination. This result demonstrates the well-known social networks principle of the importance of contacts *outside* of social communities for efficient content dissemination in opportunistic social networks.

7. ACKNOWLEDGEMENTS

We are grateful to our shepherd, Bryan Parno, for his advice in improving the presentation of this work, and to the anonymous reviewers for their comments. We would also like to thank Stratis Ioannidis for his useful remarks on earlier drafts of the paper. This work was partially funded by the European Commission under the FIRE SCAMPI (FP7-IST-258414) project and by the French National Research Agency (ANR) under the PROSE project (ANR-09-VERS-007).

8. REFERENCES

[1] V. Blondel, J.-L. Guillaume, R. Lambiotte, and E. Lefebvre. Fast unfolding of communities in large networks. *J. Stat. Mech.*, 10, 2008.

[2] A. Chaintreau, P. Hui, J. Scott, R. Gass, J. Crowcroft, and C. Diot. Impact of human mobility on opportunistic forwarding algorithms. *IEEE Transactions on Mobile Computing*, 6(6), June 2007.

[3] A. Chaintreau, A. Mtibaa, L. Massoulié, and C. Diot. Diameter of opportunistic mobile networks. In *CoNext'07: Proceedings of ACM Sigcomm CoNext*, 2007.

[4] A. Clauset, M. Newman, and C. Moore. Finding community structure in very large networks. *Physical Review E.*, 70:066111, 2004.

[5] E. M. Daly and M. Haahr. Social network analysis for routing in disconnected delay-tolerant manets. In *MobiHoc'07: Proceedings of the 8th ACM international symposium on Mobile ad hoc networking and computing*, 2007.

[6] R. Diestel. *Graph Theory*. Springer, 2006.

[7] N. Eagle and A. Pentland. Reality mining: sensing complex social systems. *Personal Ubiquitous Comput.*, 10(4):255–268, 2006.

[8] V. Erramilli, A. Chaintreau, M. Crovella, and C. Diot. Diversity of forwarding paths in pocket switched networks. In *IMC'07: Proceedings of the 7th ACM SIGCOMM Conference on Internet Measurement*, 2007.

[9] S. Fortunato. Community detection in graphs. *Physics Reports*, 486:75–174, 2010.

[10] S. Heimlicher and K. Salamatian. Globs in the primordial soup: the emergence of connected crowds in mobile wireless networks. In *MobiHoc'10: Proceedings of the eleventh ACM international symposium on Mobile ad hoc networking and computing*, 2010.

[11] P. Hui, J. Crowcroft, and E. Yoneki. BUBBLE rap: Social based forwarding in delay tolerant networks. In *MobiHoc'08: Proceedings of the 9th ACM international symposium on Mobile ad hoc networking and computing*, 2008.

[12] S. Ioannidis and A. Chaintreau. On the strength of weak ties in mobile social networks. In *SNS'09: Proceedings of the Second ACM EuroSys Workshop on Social Network Systems*, 2009.

[13] S. Ioannidis, A. Chaintreau, and L. Massoulié. Optimal and scalable distribution of content updates over a mobile social network. In *Proceedings of IEEE INFOCOM*, 2009.

[14] P. Jaccard. Distribution de la flore alpine dans le bassin des dranses et dans quelques régions voisines. *Le Bulletin de la Société Vaudoise des Sciences Naturelles*, 37:241–272, 1901.

[15] D. Kempe, J. Kleinberg, and A. Kumar. Connectivity and inference problems for temporal networks. *J. Comput. Syst. Sci.*, 64, June 2002.

[16] G. Kossinets, J. Kleinberg, and D. Watts. The structure of information pathways in a social communication network. In *Proceeding of the 14th ACM SIGKDD international conference on Knowledge discovery and data mining*, 2008.

[17] A. Miklas, K. Gollu, K. Chan, S. Saroiu, K. Gummadi, and E. de Lara. Exploiting social interactions in mobile systems. In *Proceedings of UbiComp*, 2007.

[18] A. Mtibaa, A. Chaintreau, J. LeBrun, E. Oliver, A.-K. Pietiläinen, and C. Diot. Are you moved by your social networks application? In *WOSN'08: Proceedings of the 1st ACM SIGCOMM Workshop on Online Social Network*, 2008.

[19] M. Newman and M. Girvan. Finding and evaluating community structure in networks. *Physical Review E.*, 69:026113, 2004.

[20] M. E. J. Newman. The structure and function of complex networks. *SIAM Review*, 45:167–256, 2003.

[21] G. Palla, I. Derenyi, I. Farkas, and T. Vicsek. Uncovering the overlapping community structure of complex networks in nature and society. *Nature*, 435:814–818, June 2005.

[22] A.-K. Pietiläinen and C. Diot. Dissemination in opportunistic social networks: the role of temporal communities. Technical Report TR-PRL-2012-03-0001, Technicolor, March 2012. Available online: `http://www.thlab.net/~apietila/pubs/TR-PRL-2012-03-0001.pdf`.

[23] A.-K. Pietiläinen, E. Oliver, J. LeBrun, G. Varghese, and C. Diot. Mobiclique: Middleware for mobile social networking. In *WOSN'09: Proceedings of the 2nd ACM SIGCOMM Workshop on Online Social Networks*, August 2009.

[24] J. Tang, M. Musolesi, C. Mascolo, and V. Latora. Characterizing temporal distance and reachability in

mobile and online social networks. *SIGCOMM Comput. Commun. Rev.*, 40:118–124, January 2010.

[25] J. Tang, S. Scellato, M. Musolesi, C. Mascolo, and V. Latora. Small-world behavior in time-varying graphs. *Physical Review E.*, 81(5), 2010.

[26] C. Tantipathananandht, T. Berger-Wolf, and D. Kempe. A framework for community identification in dynamic social networks. In *KDD'07: Proceedings of the 13th ACM SIGKDD international conference on Knowledge discovery and data mining*, 2007.

[27] G. Zyba, S. Ioannidis, C. Diot, and G. M. Voelker. Dissemination in opportunistic mobile ad-hoc networks: the power of the crowd. In *Proceedings of IEEE INFOCOM*, 2011.

Optimal Energy-Aware Epidemic Routing in DTNs

MHR. Khouzani
Dreese Labs
The Ohio State University
khouzani@ece.osu.edu

Soheil Eshghi
ESE Department
University of Pennsylvania
eshghi@seas.upenn.edu

Saswati Sarkar
ESE Department
University of Pennsylvania
swati@seas.upenn.edu

Ness B. Shroff
ECE and CSE Departments
The Ohio State University
shroff@ece.osu.edu

Santosh S. Venkatesh
ESE Department
University of Pennsylvania
venkatesh@seas.upenn.edu

ABSTRACT

In this work, we investigate the use of epidemic routing in energy constrained Delay Tolerant Networks (DTNs). In DTNs, connected paths between source and destination rarely materialize due to the mobility and sparse density of nodes. Epidemic routing is well-suited for these environments due to its simplicity and fully distributed implementation. In epidemic routing, messages are relayed by intermediate nodes at contact opportunities, i.e., when pairs of nodes come within transmission range. Each node needs to decide whether to forward its message upon contact with a new node based on its residual energy level and the age of that message.

We mathematically characterize the fundamental trade-off between energy conservation and forwarding efficacy as a heterogeneous dynamic energy-dependent optimal control problem. We prove, somewhat surprisingly given the complex nature of the problem, that in the mean field regime, the optimal dynamic forwarding decisions follow simple threshold-based structures in which the forwarding threshold for each node depends on its current remaining energy. We analytically establish this result under generalized classes of utility functions for DTNs. We then characterize the dependence of these thresholds on current energy reserves in each node.

Categories and Subject Descriptors

C.2.2 [**Computer Communication Networks**]: Network Protocols—*Routing Protocols*

Keywords

DTN, Limited Battery, Energy-Based Epidemic Routing, Stratified Optimal Control, Threshold-Based Forwarding.

1. INTRODUCTION

Motivation

Delay Tolerant Networks (DTNs) are comprised of spatially distributed mobile nodes whose communication range is much smaller than their roaming area; hence, end-to-end connectivity is rare. In such networks, messages are typically relayed by intermediate nodes through random contacts, which are instances of spatial proximity of pairs of nodes. Specifically, a time-stamped message from a source node can flood the network for a chance to contact its destination node within the time frame of relevance of the message [18].

Examples of DTNs include disaster response and military/tactical networks where communication devices are carried by disaster relief personnel and soldiers, and environmental surveillance, where sensors can be mounted on roaming animals. Opportunistic networking based on DTNs is envisioned to assist ad hoc or infrastructure based communication in next generation networks whenever end-to-end connectivity is hard to achieve. In most of these cases, the intermediate mobile nodes are constrained in their battery reserves. Simple epidemic forwarding schemes that rely on flooding as a method of message propagation may have a detrimental impact on the energy reserves of the intermediate nodes, adversely affecting the performance of the network. A depleted node, or one with critically low battery reserve, will not be able to relay new messages in the future. This reduction in the number of relay nodes will in turn undermine the long term throughput of the network, specially for time sensitive messages. On the other hand, an overly conservative strategy would compromise the delivery of the message to the destination in a timely manner. Hence, there is an inherent trade-off between delay-sentive message throughput and energy conservation. In such networks, replacing/recharging the batteries of drained nodes is usually infeasible and/or not cost-efficient. Also, in practice, nodes have distinct levels of remaining energy, hence a forwarding decision that is optimal for a node may be harmful for another. Furthermore, each node can readily measure its own remaining energy at any given time. The trade-off between energy conservation in each node and message throughput, along with the heterogeneity of the remaining battery reserves of the nodes motivates heterogeneous energy-dependent control, which is the subject of this paper. The state of the art either ignores energy constraints [4,11,18], or does not directly utilize current energy-state information in making forwarding decisions [1,5,13,16].

Contributions

We formulate the trade-off between energy conservation and forwarding efficacy as a dynamic energy-dependent optimal control problem: at any given time, each node decides on its forwarding probability based on its current remaining energy. Since residual energy reserves decrease with transmissions and receptions, the forwarding decisions vary with time. They must, therefore, be determined so as to control the evolution of network states that capture the fraction of nodes holding a copy of the message and the remaining battery reserves of the nodes. We consider two generalized classes of objective functions. These functions characterize metrics for end-to-end quality of service and general penalties for the final distribution of energies. These functions are defined in the context of an epidemiological model based on mean-field approximation of Markov processes to represent the evolution of the network states (§§2.1, 2.2).

Our main contribution is to prove that dynamic optimal strategies follow simple threshold-based rules (§3, Theorem 1). That is, a node in possession of a copy of the message forwards the message to nodes it encounters that have not yet received it until a certain threshold time that depends on its current remaining energy. As the node forwards the message, it loses energy, and its forwarding time threshold changes accordingly. If the age of the message is past a threshold time corresponding to a nodes current level of energy, it stops forwarding the message to others and will only transmit a copy to the destination of the message. The simplicity of the analytically-derived strategies is somewhat surprising given that the system dynamics involve non-linear transitions and a vector of controls, and our objective functions are non-linear and time-dependent (the system is therefore *non-autonomous*). The proofs for optimality of threshold-type policies in such cases do not follow from existing optimal control results.

Our third contribution is to characterize the nature of the dependence of the thresholds on the energy levels. Intuitively, the less energy a node has, the more reluctant it should be to transmit the message, as the transmission will drive it closer to critically low battery levels (which in turn will impair timely delivery of future messages). However, surprisingly, our investigations reveal that this intuition can only be confirmed when the penalty associated with low final remaining energies is convex (§3, Theorem 2), and does not hold in general otherwise. That is, in the former case, higher remaining energy levels lead to longer forwarding durations, but the monotonicity of the thresholds in energy levels is not necessarily preserved otherwise.

Finally, our optimal dynamic policy provides a missing *benchmark* for forwarding policies in large networks in which no information about the mobility pattern of the individual nodes is available. This benchmark allows us to identify some even simpler heuristic policies that perform close to the optimal, and also those that substantially compromise performance for simplicity (§4).

Related Literature

Heuristic based routing policies for DTNs involving mobile nodes are proposed in [3, 4, 11, 12, 14, 17]. In the routing protocol PROPHET [11], each node maintains a vector of probabilities of delivery and the message is forwarded from lower delivery probability nodes to nodes with higher probability of delivery. Heuristics that take energy consumption into account include NECTAR [5], which tries to find a desirable path based on the contact history of the nodes, Bannerjee *et al.* [3], which proposes the introduction of fixed nodes ("throwboxes") for energy efficient routing, and Spray and Wait [17], which proposes spreading a specific number of copies of the message initially and then waiting for delivery. Encounter-based Routing [14] builds upon the above

by having nodes make the decision to spread the limited number of copies of a message contingent on the contact history of an encountered node. Finally, Lu and Hui [12] limit transmissions to times when a node has a minimum number of neighbours, limiting its use when contacts are sparse. Such protocols may lead to large calculations and messaging overhead in the network, while also providing no analytical guarantee for QoS or energy usage.

In [10], Krifa *et al.* consider another limitation of DTNs: the storage buffers of nodes. Here, if the storage capacities are not constrained, then the best policy is to keep a copy of *all* of the messages (to be delivered to their destinations upon future contact) before they are dropped at expiration of their time-to-live value (TTL). However, if the storage buffer is full when a new message arrives, a decision needs to be made about which message (from the set of the existing messages in the buffer plus the newly arrived one) should be dropped. The decision rule is referred to as a *buffer management* policy. The paper derives policies that approximately optimize for average delay and average throughput.

The problem of finding optimal dynamic forwarding policies in DTNs considering the resource overhead of replications has been investigated in [1, 2, 13, 16, 19] among others. These papers either impose an indirect constraint, e.g., restricting the total number of copies of the message in the network to control the energy overhead, or directly consider a cost for overall energy usage. However, the forwarding rules in these papers do not utilize the current energy levels of nodes, and are identical for all nodes irrespective of their remaining energy. Specifically, the parts of their objective functions that consider the energy overhead only represent the total units of energy used during the process of forwarding copies of a message before its TTL expires, and the solution is indifferent to the distribution of the residual energy reserves. In state of the art models, if a node has started with low energy, or has lost a large portion of its battery reserves during multiple transmissions, it still has to abide by the general rule that is identical for all nodes. However, it is of importance to the network how the aggregate remaining energy is distributed among nodes, as nodes with critically low remaining energy will compromise the long-run performance of the network. Motivated by these two observations, we propose a new framework that yields optimal forwarding policies attaining custom trade-offs between QoS and the desirability of the distribution of residual energy reserves among nodes.

2. SYSTEM MODEL

In §2.1, we develop our system dynamics model based on mean-field deterministic ODEs. Subsequently, in §2.2 we consider two general classes of utility functions that cogently combine a measure of QoS with a penalty for the impact of the policy on the residual energy of the nodes. We present the model for a *single-delivery* setting. Particularly, the message is destined for a single destination and it is sufficient for only one copy of the message to be delivered to its destination.[1]

---

[1]The single-delivery setting can also capture cases where there are multiple destinations that are, unlike the source and the intermediate nodes, inter-connected through a backbone network. An example of such a setting is where the final destination is (a group of) base stations in a cellular network. Note that in the latter case there is no additional benefit in delivering more than one copy of the message to one of the destinations, since once one destination receives

2.1 System Dynamics

We begin with some definitions: a node that has received a copy of the message and is not its destination is referred to as an *infective*; a (non-destination) node that has not yet received a copy of the message is called a *susceptible*. The maximum energy capacity of a node is B units for each node. Message transmission between a pair of nodes consumes τ units of energy in the transmitter and r units in the receiver. Naturally, $r \le \tau$. When an infective node contacts a susceptible at time t, the message is transmitted with a certain forwarding probability if the infective (transmitter) and susceptible (receiver) have at least τ and r units of reserve energy, respectively.

Two nodes contact each other at rate $\hat{\beta}$. We assume that inter-contact times are exponentially distributed and uniform among nodes, an assumption common to many mobility models (e.g., Random Walker, Random Waypoint, Random Direction, etc. [7]). Moreover, it is shown in [7] that

$$\hat{\beta} \propto \frac{\text{average rel. speed of nodes} \times \text{communication ranges}}{\text{the roaming area}}. \tag{1}$$

We define $S_i(t)$ (resp., $I_i(t)$) to be the *fraction* of susceptible (resp., infective) nodes that have i energy units at time t. Hence, for $t \in [0, T]$: $\sum_{i=0}^{B} (S_i(t) + I_i(t)) = 1$.

At any given time, each node can directly observe its own level of available energy, and its forwarding decision should, in general, utilize such information. Hence, upon an instance of contact between a susceptible node with i units of energy and an infective node with j units of energy at time t, as long as $i \ge r$ **and** $j \ge \tau$, the message is passed with probability $u_i(t)$ ($0 \le u_i \le 1$). Consequently, the susceptible node transforms to an infective node with $i - r$ energy units, and the infective node to an infective node with $j - \tau$ energy units. We assume that upon contact between an infective and another node, the infective can identify (through a low-load exchange of control messages) whether the other node has a copy of the message (i.e., is infective), or does not (i.e., is susceptible), and also whether the contacted node is a destination. We assume that each instance of such exchanges consumes an insignificant amount of energy.

Let N be the total number of nodes and define $\beta := N\hat{\beta}$. Following (1), $\hat{\beta}$ is inversely proportional to the roaming area, which scales with N. Hence, if we can define a density of nodes, β has a nontrivial value. The system dynamics in the mean-field regime (i.e., for large N) over any finite interval can be approximated as follows ([6, Theorem 1]):

$$\dot{S}_i = -\beta S_i \sum_{j=\tau}^{B} u_j I_j \qquad r \le i \le B \tag{2a}$$

$$\dot{I}_i = -\beta u_i I_i \sum_{j=r}^{B} S_j \qquad B - r < i \le B \tag{2b}$$

$$\dot{I}_i = \beta S_{i+r} \sum_{j=\tau}^{B} u_j I_j - \beta u_i I_i \sum_{j=r}^{B} S_j \quad B - \tau < i \le B - r \tag{2c}$$

$$\dot{I}_i = \beta S_{i+r} \sum_{j=\tau}^{B} u_j I_j + \beta u_{i+\tau} I_{i+\tau} \sum_{j=r}^{B} S_j - \beta u_i I_i \sum_{j=r}^{B} S_j$$
$$\tau \le i \le B - \tau \tag{2d}$$

$$\dot{S}_i = 0 \qquad i < r, i = 0 \tag{2e}$$

$$\dot{I}_i = \beta S_{i+r} \sum_{j=\tau}^{B} u_j I_j + \beta u_{j+\tau} I_{i+\tau} \sum_{j=r}^{B} S_j \qquad i < \tau \tag{2f}$$

Note that in the above differential equations and in the rest of the paper, whenever not ambiguous, the dependence on t is made implicit. Each differential equation of the above set is explained in the following:[2]

(2a): The rate of decrease in the fraction of susceptible nodes with energy level i is proportional to the rate of contacts between those nodes and infective nodes with energy level equal to or higher than τ.

(2b): The rate of decrease in the fraction of infective nodes with energy level i such that $i > B - r$ is proportional to their rate of contact with any susceptible with more than r units of energy. No susceptible or infective can be transformed to an infective with such a high level of energy.

(2c): Similarly, the rate of change in the fraction of infectives with energy level i such that $B - \tau < i \le B - r$ is due to the transformation of susceptibles with energy level $i + r$ upon contact with infectives that have at least τ units of energy, along with the mechanism in (2b). No infective can be transformed to an infective of such a high energy level.

(2d): This is the non-marginal equation for the evolution of the infectives. Here, three mechanisms are in place: (2b),(2c) and one more: infectives of energy level $i + \tau$ convert to infectives with energy level i upon contact with susceptibles that have sufficient energy for message exchange.

(2e): Susceptibles with less than r units of energy cannot convert to infectives.

(2f): Infectives with less than τ units of energy cannot convert to any other type.

The initial conditions are

$$\mathbf{S}(0) = \mathbf{S}_0 := (S_{01}, \ldots, S_{0B}) \ \& \ \mathbf{I}(0) = \mathbf{I}_0 := (I_{01}, \ldots, I_{0B}) \tag{3}$$

and the state constraints are

$$\mathbf{S} \succeq 0, \quad \mathbf{I} \succeq 0, \quad \sum_{i=0}^{B} (S_i(t) + I_i(t)) = 1, \quad \forall t \in [0, T]. \tag{4}$$

2.2 Objective functions

The objective function of the network represents both a measure of the efficacy of the policy in ensuring the timely delivery of the message, and the effect of the policy on the residual energy reserves of the nodes. In what follows, we first develop each of the components of the objective function separately, and then we combine them to yield the overall objective function of the network.

---

the message, it can instantly inform the other destinations of its reception using the high-speed backbone network. So we can assume all destination nodes receive the message virtually simultaneously. The only modification in our modeling would be a re-scaling of the rate of contact between a mobile node and the destination by the number of these inter-connected sinks (Base Stations).

[2]The system dynamics for the single-delivery case can ignore the single instance of the delivery of the message to the destination. This is because in the mean-field regime, i.e. for large N, and when the state is represented as the fraction of nodes of each type, the change in the energy distributions as a result of a single transmission of the message is negligible. Note that in the single-delivery scenario, once the destination receives the message, subsequent contacts between infectives and the destination will not result in a transmission of the message.

Measure of timely delivery. One plausible measure of QoS in the context of DTNs is to maximize the probability of delivery of a message to the destination before a deadline time T. At time T, which is the time after which the message is irrelevant, all infectives will drop the message. We will go further than just representing the probability of delivery in this fixed time window: we assign more reward for earlier message delivery. In an alternative scenario, the time-frame of message delivery is flexible, but instead, a minimum probability of delivery is mandated on the message. In this case, the goal is to meet this requirement as soon as possible. In what follows, we formally present these two cases.

Let $t = 0$ mark the moment of message generation. The network achieves $g(t)$ units of reward if a copy of the message is delivered to the destination at time t, where $0 \le t \le T$. There is no reward for deliveries later than T, hence $g(t)$ can be taken to be zero for $t > T$. $g(t)$ is a non-increasing function of t over $[0, T]$, since we assign more reward for earlier delivery of the message, i.e., the sooner the message is delivered, the better. We also assume $g(t)$ to be differentiable; hence $g'(t) \le 0$ for $0 < t < T$.

Mathematically, let the random variable σ represent the time at which a copy of the message is delivered to the destination. The reward associated with the delivery of the message can be represented by the random variable $g(t)\mathbf{1}_{\sigma=t}$, where $\mathbf{1}$ is the indicator function. Let $\hat{\beta}_0$ be the rate of contact of a node with the destination, potentially different from $\hat{\beta}$, and define $\beta_0 := N\hat{\beta}_0$. Following from the exponential distribution of the inter-contact times, the *expected reward*, R, to be maximized is given by:

$$R := \mathbb{E}\{g(t)\mathbf{1}_{\sigma=t}\} = \int_0^T g(t) \, P(\sigma = t) \, dt =$$

$$\int_0^T g(t) \exp\left(-\hat{\beta}_0 \int_0^t \sum_{i=\tau}^B N I_i(\xi) \, d\xi\right) \cdot \hat{\beta}_0 \sum_{i=\tau}^B N I_i(t) \, dt.$$

Note that similar to (1), $\hat{\beta}_0$ is inversely proportional to the roaming area, which itself scales with N. Theferore, as long as we can define a meaningful density (number of nodes divided by the total roaming area of nodes), this probability is nontrivial. Another point to notice is that the summation inside the integral starts from index τ, since infective nodes with less than τ units of energy cannot forward their message to the destination upon potential contact. In order to change the form of the integration to something more conducive to analysis, we use integration by parts:

$$R = g(0) - g(T)e^{-\beta_0 \int_0^T \sum_{i=\tau}^B I_i(t) \, dt}$$
$$+ \int_0^T g'(t)e^{-\beta_0 \int_0^t \sum_{i=\tau}^B I_i(\xi) \, d\xi} \, dt.$$

The alternative measure of timely delivery of the message, instead of maximizing the probability of the delivery within a given time interval, involves the stricter notion of enforcing a minimum probability of delivery but in a flexible window of time. Subject to this probability, the goal is to minimize a penalty associated with the time it takes to satisfy such a requirement (along with the adverse effects on the residual energy of the nodes, which we will discuss next). This can be interpreted as an *optimal stopping time* problem. In what follows, we mathematically represent this alternative framework. The constraint $\mathbb{P}(\text{delivery}) \ge p$ in our case is:

$1 - \exp\left(-\int_0^T \beta_0 \sum_{i=\tau}^B I_i(t) \, dt\right) \ge p$, which is equivalent to:

$$\int_0^T \sum_{i=\tau}^B I_i(t) \, dt \ge -\ln(1-p)/\beta_0. \tag{5}$$

Let us represent the cost associated with the time T it takes to satisfy the above *throughput constraint* in general to be $f(T)$. The only necessary property for $f(T)$ to be a meaningful penalty function for *delay* is that it should be *nondecreasing* in T. We further assume that $f(t)$ is differentiable w.r.t T (hence, $f'(T) \ge 0$).

Energy cost of the policy. In the simplest representation of the trade-off with the energy overhead, one can think of maximizing the aggregate remaining energy in the network at T, irrespective of how it is distributed. But, as we mentioned in the introduction, it is desirable for the network to avoid creating nodes with critically low energy reserves. Specifically, nodes with lower residual energy can contribute in relaying and/or generating future messages for shorter durations. In the extreme case, a sizable fraction of depleted nodes can gravely jeopardizes the functionality of the network. We capture the impact of a forwarding policy on the residual energy reserves of the nodes by penalizing the nodes that have lower energy levels. Specifically, the overall penalty associated with the distribution of the residual energies of nodes is captured by: $\sum_{i=0}^B a_i (S_i(T) + I_i(T))$, in which, a_i is a *decreasing* sequence in i, i.e., a higher penalty is associated with lower residual energies at T.

The trade-offs should now be clear: by using a more aggressive forwarding policy (i.e., higher $u_i(t)$s and for longer durations), the message propagates at a faster rate and hence there is a greater chance of delivering the message to the destination earlier. However, this will lead to lesser overall remaining energy in the nodes upon delivery of the message, and it will potentially push the energy reserves of some nodes toward critically low levels, degrading the future performance of the network.

Overall Objective and Problem Statements. The overall utility of the system is a (weighted) summation of the above two components (a measure of quality of service along with the effect of the policy on residual energies). We now concisely state the two optimization problems.

Problem 1: Fixed Terminal Time The system seeks to maximize the following overall utility function:

$$R = g(0) - g(T)e^{-\beta_0 \int_0^T \sum_{i=\tau}^B I_i(t) \, dt} \tag{6}$$

$$+ \int_0^T g'(t)e^{-\beta_0 \int_0^t \sum_{i=\tau}^B I_i(\xi) \, d\xi} \, dt - \sum_{i=0}^B a_i (S_i(T) + I_i(T))$$

by dynamically selecting the vector $(u_\tau(t), \ldots, u_B(t))$ subject to the state dynamics (2), initial state conditions (3), state constraints (4) and control constraints $0 \le u_i \le 1$ for all $\tau \le i \le B$ and all $0 \le t \le T$.

Problem 2: Optimal Stopping Time Similarly, the system's objective is to maximize the overall utility function:

$$R = -f(T) - \sum_{i=0}^B a_i (S_i(T) + I_i(T)) \tag{7}$$

by dynamically regulating $(u_\tau(t), \ldots, u_B(t))$ subject to satisfying the QoS requirement (5), state equations (2), state constraints (4), initial state conditions (3) and control constraints $0 \leq u_i \leq 1$ for all $\tau \leq i \leq B$ and all $0 \leq t \leq T$.

3. OPTIMAL FORWARDING POLICIES

In what follows, for both of the problems developed in the previous section, we establish that the optimal dynamic forwarding decisions follow a simple structure. Specifically, we show that nodes should opportunistically forward the message to any node that they encounter until a threshold time that depends on their current remaining energy. Once the threshold is passed, they should completely cease forwarding until the time-to-live of the message is reached.[3] In the language of control theory, we show that optimal controls for each energy level are bang-bang with at most one jump from one to zero.[4]

THEOREM 1. *For all i, an optimal control u_i is in the class of $\mathcal{U}(t - t_i)$ where $\mathcal{U}(t)$ is the reverse step function[5] and $0 \leq t_i < T$.[6, 7]*

In what follows we provide the proof of the theorem using tools from classical optimal control theory, specifically Pontryagin's Maximum Principle. We provide the full proof for the fixed terminal time scenario (6) in §3.1, and specify the modifications for the optimal stopping time problem in §3.2.

3.1 Fixed Terminal Time Scenario

This theorem is proved in the following two steps:
1- Using optimal control theory we are able to show that each optimal control assumes the maximum value (1) when a *switching function* is positive, and the minimum value (0) when the switching function is negative. Standard optimal control results, however, do not specify the nature of the optimal control when the corresponding switching function is at 0. It is also not a priori clear whether these switching functions even have a finite number of zero-crossing points.
2- The main contribution in this part is to establish, using the specifics of the problem, that each switching function is 0 only (at most) at one point, is positive before its (potential) zero-crossing epoch, and is negative subsequently. This is achieved by showing that the time derivative of the switching function will be strictly negative at all (potential)

---

[3]Implicit is the assumption that any node with sufficient energy that contacts the destination node at any time delivers the message if the destination has not yet received it.

[4]Note that as infective nodes transmit, their energy level sinks; the threshold of each infective should therefore be measured with regards to the *current* level of energy (and not, for example, the starting level). The optimal control is indeed expected to be non-increasing in time: if the control is increasing over a segment, just flipping that part of the control in time would result in earlier propagation of the message and a higher throughput with the same final state energies. The following result, however, goes beyond that intuition in that it establishes that the optimal controls are at their maximum value for a period and then drop abruptly to zero, none of which is a priori clear.

[5]A function that is 1 from $[0, t)$ and 0 from $[t, T]$

[6]Note that the theorem does not exclude the possibility of $t_i = 0$ but excludes $t_i = T$. That is, nodes of some energy levels might never forward the message, but nodes of all energy levels cease forwarding it before the time-to-live.

[7]At times when $I_i(t) = 0$, the optimal control $u_i(t)$ can take any arbitrary value, so our $u_i(t)$ can still have the above structure. As $I_i(t) = 0$ makes $u_i(t)$ trivial, henceforth we concern ourselves with $u_i(t)$ at times when $I_i(t) \neq 0$.

zero-crossing points. Hence, each optimal control will have a *bang-bang* structure with one drop from 1 to 0 in $[0, T]$.

PROOF. Consider the system in (2) and the objective function in (6). To make the formulation better suited to Pontryagin's Maximum Principle, we introduce the following new state variable:

$$\dot{E} = \sum_{i=\tau}^{B} I_i, \qquad E(0) = 0.$$

Step 1: Using the above defined state, the Hamiltonian is

$$
\mathcal{H} := g'(t)e^{-\beta_0 E} - \sum_{i=r}^{B}[\beta\lambda_i S_i \sum_{j=\tau}^{B} u_j I_j]
$$
$$
+ \sum_{i=r}^{B}[\beta\rho_{i-r} S_i \sum_{j=\tau}^{B} u_j I_j] + \sum_{i=\tau}^{B}[\beta u_i \rho_{i-\tau} I_i \sum_{j=r}^{B} S_j] \quad (8)
$$
$$
- \sum_{i=\tau}^{B}[\beta u_i \rho_i I_i \sum_{j=r}^{B} S_j] + \lambda_E \sum_{i=\tau}^{B} I_i
$$

where the co-state functions λ_i, ρ_i and λ_E satisfy

$$
\dot{\lambda}_i = -\frac{\partial \mathcal{H}}{\partial S_i} = \beta\lambda_i \sum_{j=\tau}^{B} u_j I_j - \beta\rho_{i-r} \sum_{j=\tau}^{B} u_j I_j
$$
$$
- \beta \sum_{j=\tau}^{B} u_j \rho_{j-\tau} I_j + \beta \sum_{j=\tau}^{B} u_j \rho_j I_j \qquad r \leq i \leq B
$$
$$
\dot{\lambda}_i = -\frac{\partial \mathcal{H}}{\partial S_i} = 0 \qquad\qquad\qquad i < r
$$
$$
\dot{\rho}_i = -\frac{\partial \mathcal{H}}{\partial I_i} = -\lambda_E + \beta u_i \sum_{j=r}^{B} \lambda_j S_j - \beta u_i \sum_{j=r}^{B} \rho_{j-r} S_j
$$
$$
- \beta u_i \rho_{i-\tau} \sum_{j=r}^{B} S_j + \beta u_i \rho_i \sum_{j=r}^{B} S_j \qquad \tau \leq i \leq B
$$
$$
\dot{\rho}_i = -\frac{\partial \mathcal{H}}{\partial I_i} = 0 \qquad\qquad\qquad i < \tau
$$
$$
\dot{\lambda}_E = -\frac{\partial \mathcal{H}}{\partial E} = g'(t)\beta_0 e^{-\beta_0 E} \qquad\qquad (9)
$$

with the final constraints:
$$
\lambda_i(T) = -a_i, \quad \rho_i(T) = -a_i, \quad \forall i = 0, \ldots, B
$$
$$
\lambda_E(T) = \beta_0 g(T)e^{-\beta_0 E(T)}. \qquad\qquad (10)
$$

Maximization of the Hamiltonian yields:

$$u_i = 1 \text{ for } \varphi_i > 0 \quad \& \quad u_i = 0 \text{ for } \varphi_i < 0 \quad (11)$$

where the φ_i's, called *switching functions*, are defined as:

$$
\varphi_i := \frac{\partial \mathcal{H}}{\partial u_i} = \beta I_i \Big[-\sum_{j=r}^{B} \lambda_j S_j + \sum_{j=r}^{B} \rho_{j-r} S_j + \rho_{i-\tau} \sum_{j=r}^{B} S_j - \rho_i \sum_{j=r}^{B} S_j \Big]
$$

for $\tau \leq i \leq B$, or more simply:

$$
\varphi_i = \beta I_i \Big(\sum_{j=r}^{B} (-\lambda_j + \rho_{j-r} + \rho_{i-\tau} - \rho_i) S_j \Big) \quad \tau \leq i \leq B.
$$
$$(12)$$

This reveals an accessible intuition about the logic behind the decision process: at any given time, by activating u_i, infectives with energy level i contact susceptibles of any energy level greater than r and turn into infectives with $i - \tau$

energy units, with susceptibles of energy level j turning into infectives of energy level $j - r$. The optimal control provides the answer to whether such an action is *beneficial*, taking into account the advantages (positive terms) and disadvantages (negative terms).

At $t = T$, for $\tau \leq i \leq B$, we have:

$$\varphi_i(T) = \beta I_i(T) \sum_{j=r}^{B} (a_j - a_{j-r} - a_{i-\tau} + a_i) S_j(T). \quad (13)$$

Recall that a_i is a decreasing sequence in i. Hence, for all i, $\varphi_i(T) < 0$.[8] This shows that $u_i(t) = 0$ in a subinterval that extends till $t = T$.

Step 2: This step is accomplished in 2 parts: First, the derivative of the switching function at a (potential) point where its value is equal to zero is computed, and after simplification, it is upper-bounded using the definition of the switching function. Then, the upper-bound is shown to be negative, thus forcing the time derivative of the switching function (at such a point) to be negative. The last part follows from the key insight that it is not possible to convert all of the susceptibles to infectives in a finite time interval, and hence the total fraction of infectives (with sufficient energy reserves) at terminal time is strictly less than the fraction of susceptibles and infectives with energy reserves greater than r, τ at any time before T.

The theorem is deduced from the following lemma and (11).

LEMMA 1. *For all i, φ_i is never zero on an interval of non-zero length; it crosses zero (at most) at one point and ends at a strictly negative value.*

PROOF. All φ_i are continuous and piecewise differentiable functions of time, with potential points of non-differentiability at their potential zero-crossing points.

In what follows, we show that the time derivative of φ_i at a potential zero point is strictly negative. Note that this, together with $\varphi_i(T) < 0$, is sufficient to yield the statement of the lemma. We have:

$$\dot{\varphi}_i = \dot{I}_i \frac{\varphi_i}{I_i} - \varphi_i \beta \sum_{j=\tau}^{B} u_j I_j + \beta I_i \sum_{j=r}^{B} (-\dot{\lambda}_j + \dot{\rho}_{j-r} + \dot{\rho}_{i-\tau} - \dot{\rho}_i) S_j.$$

Therefore, at a time at which $\varphi_i = 0$, we have:

$$\dot{\varphi}_i|_{\varphi_i=0} = \beta I_i \sum_{j=r}^{B} (-\dot{\lambda}_j + \dot{\rho}_{j-r} + \dot{\rho}_{i-\tau} - \dot{\rho}_i) S_j. \quad (14)$$

From the expressions for the time derivative of the co-states in (9) combined with the expression for the switching functions in (12), we can write:

$$\dot{\varphi}_i|_{\varphi_i=0} = \beta I_i \sum_{j=r}^{B} \left(-\dot{\lambda}_j - \lambda_E - \frac{\varphi_{j-r} u_{j-r}}{I_{j-r}} - \lambda_E \right.$$
$$\left. - \frac{\varphi_{i-\tau} u_{i-\tau}}{I_{i-\tau}} + \lambda_E + \frac{\varphi_i u_i}{I_i} \right) S_j$$
$$= \beta I_i \sum_{j=r}^{B} \left(-\dot{\lambda}_j - \lambda_E - \frac{\varphi_{j-r} u_{j-r}}{I_{j-r}} - \frac{\varphi_{i-\tau} u_{i-\tau}}{I_{i-\tau}} \right) S_j$$
$$\leq \beta I_i \sum_{j=r}^{B} (-\dot{\lambda}_j - \lambda_E) S_j$$

---

[8] To see this, note that each term is negative as $a_{j-r} \geq a_j$ and $a_{i-\tau} \geq a_i$.

$$= \beta I_i(t) \left(\mathcal{H}(t) - g'(t) e^{-\beta_0 E} - \lambda_E \sum_{j=\tau}^{B} I_i - \lambda_E \sum_{j=r}^{B} S_j \right).$$

The inequality follows because terms $\varphi_{j-r} u_{j-r}/I_{j-r}$ and $\varphi_{i-\tau} u_{i-\tau}/I_{i-\tau}$ are non-negative, as imposed by the optimizations in (11)– to see this, note that $\underline{u}_i = 0$ is a feasible solution of the optimization. The equality follows from expanding the summation and regrouping, noting that (from (8)),

$$\sum_{j=r}^{B} -\dot{\lambda}_j(t) S_j(t) = \mathcal{H}(t) - g'(t) e^{-\beta_0 E(t)} - \lambda_E(t) \sum_{j=\tau}^{B} I_j(t).$$

Therefore, all we need to show in order to finish the proof of this lemma, is the following lemma, which we will prove next.[9]

LEMMA 2. *For all $t \in [0, T)$, we have:*

$$\mathcal{H}(t) - g'(t) e^{-\beta_0 E} - \lambda_E \sum_{j=\tau}^{B} I_i - \lambda_E \sum_{j=r}^{B} S_j < 0. \quad (15)$$

PROOF. Note that for a general $g(t)$, the system is **not** *autonomous*[10] as the Hamiltonian has explicit dependence on t. Nevertheless, in general we have [15, p.86] that the Hamiltonian is continuous in time and $\frac{d\mathcal{H}}{dt} = \frac{\partial \mathcal{H}}{\partial t}$, i.e., the time derivative of $\mathcal{H}$ is equal to the partial derivative of $\mathcal{H}$ with respect to t (when only explicit dependence on t is considered). Therefore, in our case: $\frac{d\mathcal{H}}{dt}(t) = g''(t) e^{-\beta_0 E(t)}$. This yields:

$$\mathcal{H}(t) = \mathcal{H}(T) - \int_t^T g''(\nu) e^{-\beta_0 E(\nu)} \, d\nu$$
$$= \mathcal{H}(T) + g'(t) e^{-\beta_0 E(t)} - g'(T) e^{-\beta_0 E(T)}$$
$$+ \int_t^T g'(\nu) \beta_0 \dot{E}(\nu) e^{-\beta_0 E(\nu)} \, d\nu \quad (16)$$

where the second equality follows from integration by parts. Following the discussion after (13), $u_i(T) = 0$ for all i, and therefore from (8), $H(T)$ simplifies to

$$\mathcal{H}(T) = g'(T) e^{-\beta_0 E(T)} + \lambda_E(T) \sum_{i=\tau}^{B} I_i(T). \quad (17)$$

Also, for $\lambda_E(t)$ we have:

$$\lambda_E(t) = \lambda_E(T) - \int_t^T g'(\nu) \beta_0 e^{-\beta_0 E(\nu)} \, d\nu. \quad (18)$$

From (16), (17) and (18), the expression in (15) becomes

$$\lambda_E(T) \left(\sum_{i=\tau}^{B} I_i(T) - \sum_{i=r}^{B} S_i(t) - \sum_{i=\tau}^{B} I_i(t) \right) \quad (19a)$$

$$+ \int_t^T g'(\nu) \beta_0 \sum_{i=\tau}^{B} I_i(\nu) e^{-\beta_0 E(\nu)} \, d\nu \quad (19b)$$

$$+ \left(\sum_{j=r}^{B} S_j + \sum_{i=\tau}^{B} I_i \right) \int_t^T g'(\nu) \beta_0 e^{-\beta_0 E(\nu)} \, d\nu. \quad (19c)$$

---

[9] Note that for $I_i(t) = 0$, the value of $u_i(t)$ is irrelevant.

[10] An autonomous optimal control is one whose dynamic differential equations and its objective function do not have parameters which explicitly vary with time t.

The lemma now follows from the observations below:
(A) $\sum_{i=\tau}^{B} I_i(T) - \sum_{i=r}^{B} S_i(t) - \sum_{i=\tau}^{B} I_i(t) < 0$. This the result whose intuition was given in the preliminary outline of the proof. To see why this holds mathematically, observe that we have

$$\sum_{i=\tau}^{B} I_i(T) \le \sum_{i=\tau}^{B} I_i(t) + \sum_{i=r}^{B} S_i(t) - \sum_{i=r}^{B} S_i(T), \quad \text{and}$$

$$\frac{d}{dt} \sum_{i=r}^{B} S_i(t) = -\beta \sum_{i=r}^{B} S_i \sum_{j=\tau}^{B} u_j I_j \le -\beta \sum_{i=r}^{B} S_i$$

$$\Rightarrow \sum_{i=r}^{B} S_i(T) \ge \left(\sum_{i=r}^{B} S_i(t) \right) e^{-\beta(T-t)} > 0,$$

where the first inequality results from the decreasing nature of energy in time. This demonstrates the negativity of (19a).
(B) $g'(t) \le 0$, which shows that both (19b) and (19c) are non-positive. $\square$

This concludes the proof of the theorem.

We now investigate the relationship between the threshold-times for optimal controls corresponding to different energy-levels. Since lower levels of residual energies are penalized more and the energy consumed in each transmission and reception is the same irrespective of the energy levels of the nodes, it may appear that threshold-times will be monotonically increasing functions of the energy levels. We now prove that this is indeed the case if the terminal-time penalty sequence is strictly convex (i.e., the difference between the penalties associated with consecutive energy levels increases with a decrease in energy levels). Interestingly enough, in §4 we construct counter-examples, using partly convex and partly concave and also fully concave sequences, such that this intuition is negated and the monotonically increasing order of the threshold times is violated. This demonstrates that naive intuition is misleading and the order predicted by the theorem does not extend in general when the strict convexity condition is absent.

THEOREM 2. *Assuming that sequence $\{a_i\}$ in (6) is non-negative, decreasing and strictly convex, then the sequence of t_i in Theorem 1 is increasing in i.*[11]

PROOF. It suffices to show that if $\varphi_i(t) = 0$, we have $\varphi_k(t) \le 0$ for any $k \le i$. It then follows from the proof of the previous theorem that the threshold time for optimal control $u_k(\cdot)$, if any, precedes that of $u_i(\cdot)$.

To do this, we show that if $\varphi_i(t) = 0$ for $t = \sigma_i$, then $\varphi_k(\sigma_i) := \beta I_k \left(\sum_{j=r}^{B} (-\rho_{i-\tau} + \rho_i + \rho_{k-\tau} - \rho_k) S_j \right)$. We subsequently show that for each j, the coefficient of S_j is non-positive at each time later than or equal to σ_i. We show that the above is non-positive at $t = T$ (utilizing the convexity of the penalty sequence $\{a_i\}$), and subsequently prove that this holds for all times after, and including, $t = \sigma_i$. This argument does not follow from standard optimal control theory, and is therefore one of our theoretical contributions.

We now proceed with the proof: From (12) we have:

$$\varphi_i(\sigma_i) = \beta I_i \left(\sum_{j=r}^{B} (-\lambda_j + \rho_{j-r} + \rho_{i-\tau} - \rho_i) S_j \right) \bigg|_{t=\sigma_i} = 0.$$

Therefore, at $t = \sigma_i$ we can write[12]

$$\sum_{j=r}^{B} (-\lambda_j + \rho_{j-r}) S_j = -\sum_{j=r}^{B} (\rho_{i-\tau} - \rho_i) S_j.$$

Using the above replacement, we obtain:

$$\varphi_k(\sigma_i) = \beta I_k \left(\sum_{j=r}^{B} (-\lambda_j + \rho_{j-r} + \rho_{k-\tau} - \rho_k) S_j \right) \bigg|_{t=\sigma_i}$$

$$= \beta I_k \left(\sum_{j=r}^{B} (-\rho_{i-\tau} + \rho_i + \rho_{k-\tau} - \rho_k) S_j \right) \bigg|_{t=\sigma_i}.$$

For $\tau \le k \le i$, define:

$$\psi_{i,k}(\sigma_i) := -\rho_{i-\tau} + \rho_i + \rho_{k-\tau} - \rho_k \qquad \tau \le k$$

The theorem now follows from the following lemma.

LEMMA 3. *For any $\tau \le k \le i$, we have $\psi_{i,k}(\sigma_i) \le 0$.*

PROOF. We present the proof for $k \ge 2\tau$. The case of $\tau \le k \le 2\tau$ follows similarly. At $t = T$ following (10), we have:

$$\psi_{i,k}(T) = -\rho_{i-\tau}(T) + \rho_i(T) + \rho_{k-\tau}(T) - \rho_k(T)$$

$$= a_{i-\tau} - a_i - (a_{k-\tau} - a_k)$$

which following the properties assumed for a_i (a_i being decreasing and strictly convex in i), yields $\psi_{i,k}(T) < 0$. This also holds on a sub-interval of nonzero length that extends to $t = T$, owing to the time-continuity of $\psi_{i,k}$. We prove the lemma by contradiction:Going backward in time from $t = T$ towards $t = \sigma_i$, suppose the lemma is violated first at time $\bar{\sigma}$, that is, for at least one $k < i$ we have:

$$(-\rho_{i-\tau} + \rho_i + \rho_{k-\tau} - \rho_k) < 0 \text{ for } \sigma_i < \bar{\sigma} < t \le T; \text{ and}$$

$$(-\rho_{i-\tau} + \rho_i + \rho_{k-\tau} - \rho_k) = 0 \text{ at } t = \bar{\sigma},$$

with the non-strict inequality holding for the rest of the levels at $t = \bar{\sigma}$. In what follows we show that the time derivative of $\psi_{i,k}$ is non-negative over the interval $[\bar{\sigma}, T]$. Note that this leads to a contradiction with the existence of $\bar{\sigma}$ and hence proves the lemma, since:

$$\psi_{i,k}(\bar{\sigma}) = \psi_{i,k}(T) - \int_{t=\bar{\sigma}}^{T} \dot{\psi}_{i,k}(\nu) \, d\nu \Rightarrow \psi_{i,k}(\bar{\sigma}) \le \psi_{i,k}(T) < 0.$$

We now investigate $\dot{\psi}_{i,k}$ over $[\bar{\sigma}, T]$:

$$\dot{\psi}_{i,k} = -\dot{\rho}_{i-\tau} + \dot{\rho}_i + \dot{\rho}_{k-\tau} - \dot{\rho}_k$$

$$= (\lambda_E + \frac{\varphi_{i-\tau} u_{i-\tau}}{I_{i-\tau}}) + (-\lambda_E - \frac{\varphi_i u_i}{I_i})$$

$$+ (-\lambda_E - \frac{\varphi_{k-\tau} u_{k-\tau}}{I_{k-\tau}}) + (\lambda_E + \frac{\varphi_k u_k}{I_k})$$

$$= \frac{\varphi_{i-\tau} u_{i-\tau}}{I_{i-\tau}} - \frac{\varphi_i u_i}{I_i} - \frac{\varphi_{k-\tau} u_{k-\tau}}{I_{k-\tau}} + \frac{\varphi_k u_k}{I_k}$$

$$\ge -\frac{\varphi_i u_i}{I_i} - \frac{\varphi_{k-\tau} u_{k-\tau}}{I_{k-\tau}}.$$

The last inequality follows from (11). For the remaining terms, note that following from the definition of σ_i and as we showed in the proof of Theorem 1, we have $\varphi_i(t) \le 0$

---

[11]Ignoring energy levels i for which $I_i \equiv 0$.

[12]unless $I_i = 0$, for which the control, u_i, is irrelevant.

over the interval of $[\sigma_i, T]$. Now we show that $\varphi_{k-\tau}(t) \leq 0$ over the interval of $[\bar{\sigma}, T]$. From (12), we have:

$$\begin{cases} \varphi_i & = \beta I_i \left(\sum_{j=r}^{B} (-\lambda_j + \rho_{j-r} + \rho_{i-\tau} - \rho_i) S_j \right) \\ \varphi_{k-\tau} & = \beta I_{k-\tau} \left(\sum_{j=r}^{B} (-\lambda_j + \rho_{j-r} + \rho_{k-2\tau} - \rho_{k-\tau}) S_j \right) \end{cases}$$

From the definition of $\bar{\sigma}$, we have $\rho_{k-2\tau} - \rho_{k-\tau} \leq \rho_{i-\tau} - \rho_i$ over the interval of $[\bar{\sigma}, T]$. Hence:

$$\sum_{j=r}^{B} (-\lambda_j + \rho_{j-r} + \rho_{i-\tau} - \rho_i) S_j \leq 0$$

$$\Rightarrow \sum_{j=r}^{B} (-\lambda_j + \rho_{j-r} + \rho_{k-2\tau} - \rho_{k-\tau}) S_j \leq 0,$$

and therefore, $\varphi_i \leq 0 \Rightarrow \varphi_{k-\tau} \leq 0$. This concludes the lemma, and hence the theorem. $\square$

3.2 Optimal Stopping Time Scenario

We will use the variable final time formulation of Pontryagin's Maximum Principle for the optimal stopping time version of the problem. First, we transform the *path* constraint in (5) to a final state constraint, which is better suited to PMP. This can be done simply by introducing a new variable, E, such that:

$$\dot{E} = \sum_{i=\tau}^{B} I_i(t), \quad E(0) = 0, \quad E(T) \geq -\ln(1-p)/\beta_0.$$

Now consider a new Hamiltonian identical to (8) with the exception that the first term is removed. According to PMP, there exist absolutely continuous co-state functions $\vec{\lambda}, \vec{\rho}, \lambda_E$ and a constant $\lambda_0 \geq 0$ such that at any given time, an optimal u_i is a maximizer of the new Hamiltonian. The co-state functions satisfy the same differential equations as in (9), with the exception that here we have $\dot{\lambda}_E = 0$.

The final conditions for the co-states also change to the following:

$$\lambda_i(T) = -\lambda_0 a_i, \quad \rho_i(T) = -\lambda_0 a_i, \quad \forall i = 0, \dots, B \quad (20a)$$
$$\lambda_E(T) \geq 0, \quad \lambda_E(T)(E(T) + \ln(1-p)/\beta_0) = 0 \quad (20b)$$

together with $H(T) = \lambda_0 f'(T)$ and for every $t \in [0, T]$:

$$(\lambda_0, \vec{\lambda}(t), \vec{\rho}(t), \lambda_E(t)) \neq \vec{0}. \quad (21)$$

The last condition along with $\lambda_0 \geq 0$ leads to $\lambda_0 > 0$. Maximization of the Hamiltonian yields:

$$u_i = 1 \text{ for } \varphi_i > 0 \quad \& \quad u_i = 0 \text{ for } \varphi_i < 0 \quad (22)$$

where φ_is, the switching functions, are defined as before, and with a similar argument it can be shown that $\varphi_i(T) < 0$. Therefore, we have $H(T) = \lambda_E(T) \sum_{i=\tau}^{B} I_i(T)$. Hence,

$$\lambda_0 f'(T) = H(T) = \lambda_E(T) \sum_{i=\tau}^{B} I_i(T). \quad (23)$$

Now, we claim that both Theorems 1 and 2 apply in this case as well. The proofs are almost identical. Here we only list the alterations.

▷ **Changes in the proof of Theorem 1:**
The proof is identical up to the following point, which we develop onward:

$$\dot{\varphi}_i|_{\varphi_i=0} \leq \beta I_i \sum_{j=r}^{B} (-\dot{\lambda}_j - \lambda_E) S_j$$

$$= \beta I_i (\mathcal{H} - \lambda_E \sum_{i=\tau}^{B} I_i - \lambda_E \sum_{j=r}^{B} S_j)$$

$$= \beta I_i (\mathcal{H}(T) - \lambda_E \sum_{i=\tau}^{B} I_i - \lambda_E \sum_{j=r}^{B} S_j)$$

$$= \beta I_i (\lambda_E(T) \sum_{i=\tau}^{B} I_i(T) - \lambda_E \sum_{i=\tau}^{B} I_i - \lambda_E \sum_{j=r}^{B} S_j),$$

where the equality in line 3 comes from the fact that $H(t)$ is constant in an autonomous system [15]. Note that $\dot{\lambda}_E = 0$ and hence λ_E is a constant, i.e., $\lambda_E = \lambda_E(T)$. Furthermore, it is a strictly positive constant because first $\lambda_E(T) \geq 0$ from (20b); and second, λ_E cannot be zero, because if it is, following (23), it implies $\lambda_E = \lambda_0 \equiv 0$, which together with the ODE of the co-state functions and (20a), leads to $(\lambda_0, \vec{\lambda}(T), \vec{\rho}(T), \lambda_E) = \vec{0}$. This would contradict (21). Hence $\dot{\varphi}_i|_{\varphi_i=0} < 0$, and the rest of the proof is similar to the one presented in the previous section.

▷ **Changes in the proof of Theorem 2:**
Everything remains the same except that the final value of $\psi_{i,k}(T)$ is multiplied by λ_0, that is

$$\psi_{i,k}(T) = \lambda_0 (a_{i-\tau} - a_i - (a_{k-\tau} - a_k)).$$

As we argued in the previous item, we have $\lambda_0 > 0$ and henceforth, the rest of the arguments follow identically.

Practical Issues and Implementation

A corollary of Theorem 1 is that the forwarding policy can be completely represented as a vector of threshold times corresponding to different energy levels. This vector is of size $B - \tau$ and can be calculated once at the source node of the message and added to it as a small overhead. Each node that receives the message simply retrieves the threshold levels and forwards the message if its age is less than the threshold entry corresponding to its current energy level. The one-time calculation of the threshold levels for each message at the origin can be done by estimating the current distribution of the energy levels in the network. Note that the required information is the fractions of nodes with each level of energy and not their identity. This estimation can be done if the energy distribution at the time at which the network starts operation is known (e.g., all nodes start with full batteries) and origin nodes keep a history of the past messages. The robustness of our policy with respect to inaccuracy in the estimation of the initial energy profile of the network is an interesting direction for future research. The search for optimum thresholds is now an optimization with only $B - \tau$ variables. Moreover, following Theorem 2, the search can be limited to a small subset of the space of $[0, T]^{B-\tau}$. Finally, as we show in our numerical section, heuristically, a common threshold for all energy levels can be first optimized and its solution used as a good initial solution for the multi-variable optimization. Qualitatively, our results show that each node should have different *modes of action* depending on its residual energy, and that these modes of actions should themselves vary for each newly generated message. These observations should give provide pointers for improving state-of-the-art DTN routing policies.

In developing our theoretical model, as for any analytical model, we made a series of technical assumptions for tractability and/or avoidance of unnecessary clutter. (a) We ignored the energy dissipated in scanning the media in search of new nodes. Our model can be generalized to incorporate the media scanning; however it would have unduly complicated the model and it is left to our future research. (b) We assumed *homogeneous mixing*, i.e., the inter-contact times are similarly distributed for all pairs, which may not hold in practice. This assumption may be partly relaxed by using the ideas in [9], which suggest that our results likely generalize to spatially inhomogeneous cases. (c) We assumed that during the interval $[0, T]$, only one message is routed in the network. This assumption is valid if the load in the network is low and the routing time intervals of different messages do not overlap (i.e., the interval between the generation of new messages is longer than T). Generalization to the routing of multiple messages with overlapping routing intervals can be an interesting future direction of research.

4. NUMERICAL INVESTIGATIONS

In this section, we investigate the structure of the optimal control in systems with the above dynamics. We investigated the single-delivery fixed terminal time problem with $g(t) = 1$ (i.e., no time discrimination in $[0, T]$) in the simulations, and unless otherwise stated, our test system used parameters: $B = 5$ (i.e., five energy levels), $\tau = 2$, $r = 1$, and $T = 10$. Note that $\tau > r$, as demanded by our system model. For β (and β_0), our benchmark is 0.223 (motivated by [8]). It is instructive to note that βT denotes the average number of contacts of each node in the system in the time interval $[0, T]$, and therefore our choice of β is limited by the time-to-live (T) of the message. In our case, each node contacts more than two other nodes on average within the TTL of the message, a not-unreasonable assumption.

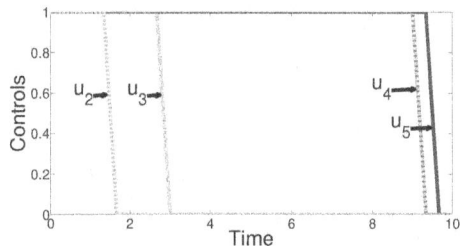

Figure 1: An illustrative example for Theorems 1 and 2. The controls are plotted for a system with parameters: $B = 5$, $r = 1$, $t = 2$. The initial distribution is $\mathbf{I_0} = (0, 0, 0.25, 0.02, 0.02, 0.01)$ and $\mathbf{S_0} = (0, 0, 0, 0.2, 0.3, 0.2)$, and the battery penalties are $a_i = \kappa(B - i)^2$, with $\kappa = 0.005$.

First, we illustrate Theorems 1 and 2 by examining a system with 5 energy levels and convex final-state cost coefficients (fig. 1). As can be seen, the optimal control for each energy level demonstrates bang-bang behavior with one drop-off point, and furthermore, because of the convexity of the final state costs and in accordance with Theorem 2, the drop-off times of the different energy level controls follow the ordering of the energy levels.

Then, we investigate the case where the final state costs are not convex. One sample configuration is when we have a sharp drop-off between two terminal-time coefficients, with coefficients on either side being close to each other. The

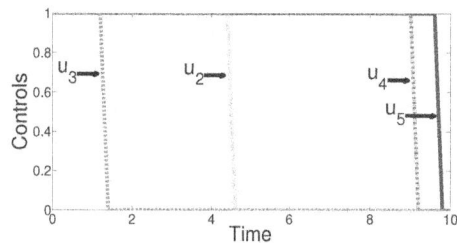

Figure 2: In this example, the parameters were exactly the same as those used in fig. 1, with the difference that the final state coefficients were $a_0 = 4.4$, $a_1 = 4.2$, $a_2 = 4$, $a_3 = 1.2$, $a_4 = 1.1$, $a_5 = 1$. As can be seen, the drop-off in final state costs between energy levels 3 and 2 motivates nodes in level 3 to be more conservative in propagating the message.

	Drop-off Times of Controls	
	Energy Level 3	Energy Level 2
$\kappa = 0.5$	9.6	9.8
$\kappa = 1.5$	9.6	9
$\kappa = 2$	9.6	8.4

Table 1: An example for non-ordered drop-off points of the optimal controls for concave final state coefficients in the settings of Theorem 2. Here we construct the final-state cost coefficients in the form $a_i = \kappa(B - i)^\alpha$ with $T = 10$, $B = 3$, $r = 0$, $t = 1$ and $\kappa = 0.01$, and vary α over the values $\{0.5, 1.5, 2\}$. The initial distribution is $\mathbf{I_0} = (0.1, 0.1, 0.1, 0.2)$ and $\mathbf{S_0} = (0, 0, 0.2, 0.3)$. As can be seen, for $\alpha = 0.5$, where the final cost coefficients become concave, the ordering of the drop-off in energy levels 2 and 3 is reversed. For $\alpha = \{1.5, 2\}$, where the final costs are strictly convex, the ordering is preserved, as predicted by Theorem 2.

motivation for such a setting could be the case where we only care about having a certain fixed amount of energy at terminal time, and any variation above or below that value is of very little importance to us. It can be seen in fig. 2 that Theorem 2 does not necessarily hold for such a setting, as nodes on either side of the drop-off would be incentivised to propagate the message (because of the low loss incurred for propagation in terms of final states), but those nodes in states on the cusp of the drop-off in final state coefficients would be extremely conservative, as there is a large penalty associated with any further propagation of the message.

Subsequently, it is shown that Theorem 2 does not hold for even wholly concave terminal-time state coefficients. To this end, a concave cost function is constructed that has an optimal control whose drop-off times are not ordered (Table 1). Therefore, the convexity of the final cost coefficients is integral to the result of Theorem 2.

To better illustrate the efficacy of the bang-bang controls, the performance of the system is compared with that of 3 heuristic algorithms:
1. The control is constant throughout $[0, T]$ and uniform for all energy levels (Static in Time and Across Energy Levels).
2. The control can vary within the space of uniform one-jump bang-bang controls for all energy levels (Static Across Energy Levels).
3. The control can vary between energy levels, but the controls of each are constant in time (Static in Time)
In fig. 3, the system utility of the optimal control and the

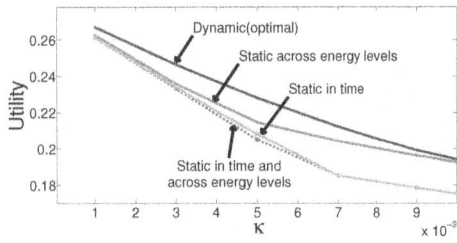

Figure 3: The performance of the heuristics is compared with the dynamic (bang-bang) optimum for different relative weights κ of the utility function for a system with parameters $B = 5$, $r = 1$, $t = 2$ and with the same initial distribution and utility function as the system in fig. 1. Here, we took β to be 0.78, so that each user has, on average, around 8 contacts in the time period. As κ goes to both extremes (0 and infinity), the utility function becomes trivial and the performance of the heuristics matches the optimal.

heuristics are plotted as a function of the relative weighting of the two parts of the utility function for a sample system. It turns out that the difference in utility becomes large whenever the drop-off times of the optimal controls are spread out across T, forcing the heuristics into choosing overly conservative policies. It is instructive to note that here, the heuristic that is static across energy levels outperforms the other heuristics, which emphasizes the importance of considering the dynamics of the system. It can be seen that here, this policy can be chosen as a *simple approximation* to the optimal without a significant loss in performance.

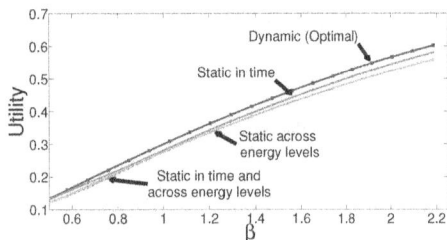

Figure 4: In this figure, β (the mean rate of contact) is varied and the performance of the optimal control and the heuristics are studied. The same parameters as those outlined in fig. 1 are used for the utility function and the initial states. For low rates of contact, the optimality of the control is less important as contacts are too infrequent to affect the energy distribution significantly. As β increases, any constraint on the policy translates to sub-optimality in the controls. The sub-optimality of even the best heuristic increases with an increase in β.

Finally, to better understand the structure of the optimal control, the relative performance of the optimal policy and the heuristics are illustrated in fig. 4 for a case where β is varied. It can be seen that the sub-optimality of all the heuristics increases with β, which illustrates the added importance of optimal decision-making at each instance in situations where there are more contacts.

5. CONCLUSION

We formulated the problem of optimal energy-dependent message forwarding in energy-constrained DTNs as a multi-variable optimal control problem using a deterministic stratified epidemic model. We analytically established that optimal forwarding decisions for two generalized notions of QoS are composed of simple threshold-based policies, where the thresholds depend on the current remaining energy in each node. Finally, we analytically characterized the nature of this dependence.

6. ACKNOWLEDGEMENTS

The work is partially supported by the Army Research Office MURI Awards W911NF-08-1-0238 and W911NF-07-1-0376, and NSF grants CNS-0831919, CNS-0721434, CNS-1115547, CNS-0915697, CNS-0915203 and CNS-0914955.

7. REFERENCES

[1] E. Altman, A. Azad, T. Basar, and F. De Pellegrini. Optimal activation and transmission control in delay tolerant networks. In *Proc. of IEEE INFOCOM*, pp. 1–5, 2010.

[2] A. Balasubramanian, B. Levine, and A. Venkataramani. DTN routing as a resource allocation problem. In *ACM SIGCOMM Computer Communication Review*, vol. 37, pp. 373–384, 2007.

[3] N. Banerjee, M. D. Corner, and B. N. Levine. Design and field experimentation of an energy-efficient architecture for dtn throwboxes. *IEEE/ACM Tr. on Net.*, 18(2):554–567, 2010.

[4] E. de Oliveira and C. de Albuquerque. NECTAR: a DTN routing protocol based on neighborhood contact history. In *Proc. of the ACM symposium on Applied Computing*, pp. 40–46, 2009.

[5] F. De Pellegrini, E. Altman, and T. Bascar. Optimal monotone forwarding policies in delay tolerant mobile ad hoc networks with multiple classes of nodes. In *Proc. of IEEE WiOpt*, pp. 497–504, 2010.

[6] N. Gast and B. Gaujal. A mean field approach for optimization in discrete time. *Discrete Event Dynamic Systems*, 21(1):63–101, 2011.

[7] R. Groenevelt, P. Nain, and G. Koole. Message delay in manet. In *Proc. of the ACM SIGMETRICS*, pp. 412–413, 2005.

[8] P. Hui, A. Chaintreau, J. Scott, R. Gass, J. Crowcroft, and C. Diot. Pocket switched networks and human mobility in conference environments. In *Proc. of the ACM SIGCOMM workshop on Delay-Tolerant Networking*, pp. 244–251, 2005.

[9] M. Khouzani, S. Eshghi, S. Sarkar, and S. S. Venkatesh. Optimal patching in clustered malware epidemics. In *Proc. of ITA*, 2012.

[10] A. Krifa, C. Baraka, and T. Spyropoulos. Optimal buffer management policies for delay tolerant networks. In *Proc. of IEEE SECON*, pp. 260–268, 2008.

[11] A. Lindgren, A. Doria, and O. Schelén. Probabilistic routing in intermittently connected networks. *ACM SIGMOBILE Mobile Computing and Communications Review*, 7(3):19–20, 2003.

[12] X. Lu and P. Hui. An energy-efficient n-epidemic routing protocol for delay tolerant networks. In *IEEE Fifth International Conference on Networking, Architecture and Storage (NAS)*, pp. 341–347, 2010.

[13] G. Neglia and X. Zhang. Optimal delay-power tradeoff in sparse delay tolerant networks: a preliminary study. In *Proc. of the ACM SIGCOMM workshop on Challenged networks*, pp. 237–244, 2006.

[14] S. Nelson, M. Bakht, and R. Kravets. Encounter-based routing in dtns. In *Proc. of IEEE INFOCOM*, pp. 846–854, 2009.

[15] A. Seierstad and K. Sydsaeter. *Optimal control theory with economic applications*, vol. 20. North-Holland Amsterdam, 1987.

[16] C. Singh, A. Kumar, and R. Sundaresan. Delay and energy optimal two-hop relaying in delay tolerant networks. In *Proc. of IEEE WiOpt*, pp. 256–265, 2010.

[17] T. Spyropoulos, K. Psounis, and C. S. Raghavendra. Spray and wait: an efficient routing scheme for intermittently connected mobile networks. In *Proc. of the ACM SIGCOMM workshop on Delay-tolerant networking*, WDTN, pp. 252–259, 2005.

[18] X. Zhang, G. Neglia, J. Kurose, and D. Towsley. Performance modeling of epidemic routing. *Computer Networks*, 51(10):2867–2891, 2007.

[19] Y. Wang and H. Wu. DFT-MSN: The delay fault tolerant mobile sensor network for pervasive information gathering. In *Proc. of IEEE INFOCOM*, pp. 1021–1034, 2006.

Transmission Delay in Large Scale Ad Hoc Cognitive Radio Networks

Zhuotao Liu
Dept. of Elec. Engin.
Shanghai JiaoTong Univ.
Shanghai, China
zhuotaoliu@sjtu.edu.cn

Xinbing Wang
Dept. of Elec. Engin.
Shanghai JiaoTong Univ.
Shanghai, China
xwang8@sjtu.edu.cn

Wentao Luan
Dept. of Elec. Engin.
Shanghai JiaoTong Univ.
Shanghai, China
lwt1104@sjtu.edu.cn

Songwu Lu
Dept. of Comp. Sci.
Univ. of California, LA
Los Angeles, USA
slu@cs.ucla.edu

ABSTRACT

There has been recent interest within the networking research area to understand the transmission delay in Cognitive Radio (CR) Networks with overlapping primary network and secondary network. In this paper, we investigate the scaling behavior of transmission delay in large scale ad hoc CR networks. We take different scenarios of CR networks into consideration and thus obtain a wind range of results. We first neglect propagation delay and study the ratio of transmission delay to distance, denoted by $\gamma(\lambda_S, \mathcal{A}_P)$, using the Poisson Boolean Model and Poisson Random Connection Model of continuum percolation theory. We show that $\gamma(\lambda_S, \mathcal{A}_P)$ is a constant and figure out its exact value in *supercritical* secondary network. In case of *subcritical* secondary network, we introduce a *multi-cluster hop transmission process* to get the lower bound of $\gamma(\lambda_S, \mathcal{A}_P)$. Then we take propagation delay into consideration to obtain further results. Finally, we use simulation results to verify our theoretical analysis. The results present the scaling behavior of transmission delay in CR networks and provide the design guidelines for large scale wireless networks.

Categories and Subject Descriptors

C.2.1 [**Computer-Communication Networks**]: Network Architecture and Design—*Wireless Communications*

General Terms

Theory and Performance

Keywords

Connectivity, Delay, Cognitive Radio

1. INTRODUCTION

Most of the frequency spectra suitable for wireless communication have already been licensed by the Federal Communications Commission (FCC) in the USA. Today, as wireless communication witnesses a dramatic growth, efficient usage of spectrum becomes necessary. However, recent measurements [4] have shown that over 90% of the time, a large percentage of the licensed bands remain unused. This under utilization of statically allocated spectra motivates the study of a new type of wireless communication network, the Cognitive Radio (CR) networks, composed of the overlapping primary network and secondary network. By carefully sensing and learning the primary users' presence through their cognitive devices, secondary users can identify and utilize spectrum opportunities while avoiding unacceptable interference to primary users. Therefore, cognitive radio has dramatically improved the spectrum efficiency.

Since the Gupta and Kumar's landmark work [5], the study of wireless networks has received great attention. Besides capacity, coverage [7] and connectivity [8] of wireless networks are also well studied. These works mainly concern the homogenous networks, where nodes are distributed according to *Poisson Point Process*. However, the heterogeneity of CR networks, caused by the heterogeneity of spectrum opportunities over time and space, makes the study of CR networks different from that of homogenous networks. Specifically, secondary users within the transmission range of any primary users that are broadcasting or receiving information can not have spectrum opportunities. Therefore, the secondary users will not be able to send or receive massages even if they are topologically connected with the outside secondary network. From the perspective of secondary network, it has "vacuum" space. Hence, the initial homogenous *Poisson Point Process* becomes heterogenous due to the impact of primary network. Therefore, the previous results can not be efficiently applied into CR networks

As CR networks receive more and more attention, some recent works have put them on priority. In [9], L. Ding *et al.* proposed a routing and spectrum allocation algorithm for CR networks to maximize the throughput. By using percolation theory, especially continuum percolation [1], W. Peng *et al.* [10] and W. Ren *et al.* [11] have studied the connectivity in CR networks in terms of density pairs of primary users and secondary users $(\lambda_S, \lambda_{PT})$. They both show that the overall connectivity exits if and only if $\lambda_S > \lambda_S^*$ and $\lambda_{PT} < \lambda_{PT}^*$, where λ_S^* and λ_{PT}^* are two critical density values.

When we consider network design and architecture, one of the most significant issues is transmission delay. In [12], S. Bodas *et al.* have considered the problem of designing scheduling algorithms for small delay in multi-channel wireless networks. In [13], Z. Kong *et al.* have answered what fraction of the network eventually receives the message when the network is *subcritical* and how long the delay is. In [14], S. Zhao *et al.* studied the relationship between node density and transmission delay. Again we noticed that these works mainly considered the homogenous networks. This motivates us to answer the following question in this paper:

- How does transmission delay scale with the distance from source to destination in large scale ad hoc Cognitive Radio networks?

To make our analysis meaningful, we first introduce the connectivity of networks, the prerequisite for communication. Full connectivity [6] can ensure the successful communication between any randomly chosen node pairs. However, the huge power consumption of maintaining such strong connectivity makes it impractical in some situations where the total throughput of energy for each node is limited. Thus, it is necessary to introduce a slightly weaker connectivity paradigm, i.e., an infinite connected component containing a high proportion of the networks' nodes exists.

Using the continuum percolation theory, it is possible to achieve this weaker connectivity in large scale networks. The two most general models in [1] are Poisson Boolean Model (BM) and Poisson Random Connection Model (RCM). The two models describe the behaviors of connected component in a random geometric graph in which nodes are distributed according to *Poisson Point Process* with node density λ, and two nodes share a link according to a connection function. A fundamental result of RCM or BM points out a *phase transition* effect in percolation of networks. For $\lambda > \lambda_c$ (supercritical), there exists a unique connected component containing an infinite number of nodes, i.e., the networks percolate. For $\lambda < \lambda_c$ (subcritical), the unique connected component is broken into infinite number of mutually disconnected finite components almost surely. Thus, the network cannot achieve overall connectivity and messages may not be able to reach the destination if it resides in the different component with the source. Therefore, we need to introduce mobility of nodes into the network as mobile nodes which have been informed the information may meet the source even if they are within different components initially. Consequently, the messages can see a probability to disseminate among the network when nodes are mobile.

We will study the transmission delay under the weaker version of connectivity. *Waiting delay* and *propagation delay* [1] are two major contributions to transmission delay. In supercritical networks, secondary users within the unique infinite component have instantaneous connectivity. Then waiting delay consists only of the waiting time for spectrum opportunities. Whereas in subcritical networks, the waiting delay is composed by the waiting time for both instantaneous connectivity and spectrum opportunities. Usually, such a waiting delay may be in the order of seconds, minutes or even larger time units. As for the propagation delay, it is in the order of milliseconds when the network is not heavily loaded. Therefore, compared to the waiting delay, it is negligibly small. Given spectrum opportunities, messages can disseminate among one connected component instantaneously if ignoring the propagation delay.

In our approach to the scaling behavior of transmission delay, we consider different scenarios of CR networks. First, we neglect the propagation delay and investigate the impact of secondary user's

density and primary network on the transmission delay in supercritical secondary network. Then we study the influence of mobility to obtain further results in subcritical secondary network. Finally, we take the propagation delay into consideration to reach more general conclusions. The major contributions of this paper are as follows:

- We investigate the scaling behavior of transmission delay in large scale ad hoc CR networks in both Boolean Model and Random Connection Model of continuum percolation theory.

- We present either exact value or the lower bound of the transmission delay to distance ratio $\gamma(\lambda_S, \mathcal{A}_p)$ as the distance goes to ∞. To the best of our knowledge, this is the first work that ever gives a precise description of delay's scaling behavior in CR networks.

- We do considerable amount of simulations to verify our theoretical analysis.

The rest of the paper is organized as follows. In section 2, we introduce our system model. In section 3, we present our main results on $\gamma(\lambda_S, \mathcal{A}_p)$ while ignoring the propagation delay. In sections 4 and 5, we give detailed proof of our results. In section 6, we take the propagation delay into consideration to obtain further results on $\gamma(\lambda_S, \mathcal{A}_p)$. Simulation results will be presented in section 7 to verify our theoretically analysis. Finally, we conclude in section 8.

2. SYSTEM MODEL

We model the large scale CR ad hoc networks as a random geometric graph. Both primary users and secondary users are distributed according to the *Poisson Point Process* with density λ_{PT} and λ_S, respectively, in an infinite two-dimensional space $\mathbb{R}^2$. Primary nodes, in other words spectrum owners, have higher priority over secondary nodes for using communication spectrum. Secondary users buy spectra and can use them while they are not used by primary users. It is clear that the waiting delay of primary network does not conclude the waiting time for spectrum opportunities. Then, the analysis of delay in primary network is a special case of that in secondary network. Therefore, we will focus on investigating the transmission delay of secondary network in the following parts. [2]

2.1 The Primary Network Model

We model the primary transmitters according to the *Poisson Point Process* $\mathbb{P}_{PT}(t)$ with density λ_{PT}. For one primary transmitter, its corresponding receivers are uniformly distributed within its transmission range R_p. Based on the displacement theorem in [3], we have that the primary receivers also form a two-dimensional *Poisson Point Process* $\mathbb{P}_{PR}(t)$. Note that the two Poisson processes are correlated with each other.

Assume that time is slotted with slot length T_S. Without lose of generality, we set $T_S = 1$. Each primary transmitter is associated with an independent identically distributed (i.i.d.) switching renewal process, denoted by $S_P(t)$, which changes between two states: the *active* state, when it is broadcasting messages, and the *sleep* state, when it is inactive. A primary receiver is active or inactive identically to its corresponding transmitter. By the Markovian property of behaviors of primary network, the process $S_p(t)$ can be viewed as a discrete time Markov on-off *(active-sleep)* process with i.i.d. *active* periods $Q_p^1(t)$, and i.i.d. *sleep* periods $Q_p^0(t)$.

---

[1] As is usual with percolation-theory based analysis for delay, the queueing delay in the transmission of messages is not accounted.

[2] In the following sections, we will use transmission delay of CR networks and transmission delay of secondary network interchangeably.

Let us denote the probability of going from *active* state to *sleep* state and probability of going from *sleep* state to *active* state by p_{10} and p_{01}, respectively. And the probability of sustaining *active* and *sleep* state in next slot are denoted by p_{11} and p_{00}, respectively. Then $p_{01} = 1/E[Q_p^0(t)]$ and $p_{10} = 1/E[Q_p^1(t)]$. The stationary distribution of $S_p(t)$ is:

$$\beta_0 \triangleq \lim_{t \to \infty} Pr(S_P(t) = 0) = \frac{E[Q_p^0(t)]}{E[Q_p^1(t)] + E[Q_p^1(t)]},^3$$

$$\beta_1 \triangleq \lim_{t \to \infty} Pr(S_P(t) = 1) = \frac{E[Q_p^0(t)]}{E[Q_p^0(t)] + E[Q_p^0(t)]}.$$

The transmission range of primary users is R_p. Spectrum opportunities are not available for secondary network within a circle centered at one *active* primary user with radii R_p. Then the primary network at time slot t can be represented by $\mathcal{A}_P(\lambda_{PT}; S_P(t); R_p)$.

2.2 The Secondary Network Model

As we have discussed above, in [1] two models, i.e., Boolean Model (BM) and Random Connection Model (RCM), are used to describe the connectivity of a random geometric graph. If we model the connectivity of secondary network by BM, two secondary nodes share a link whenever they are within each other's transmission range. However, if connectivity of secondary users are modeled by RCM, we say two that secondary nodes share a link, with a positive probability, if and only if their distance is smaller than r_s. The positive probability is not one since the severe enemy attack, natural hazards, or energy depletion may cause link failure. In other words, such a positive probability indicates that the link is unreliable, which is the key difference between BM and RCM. Since BM and RCM are both non-trivial model, this paper will contain the results for both of them.

Two secondary users can exchange information only when they have at least one open path π. The following two conditions are necessary for the openness of π:

1. π is established through multiple links;

2. messages, broadcasted by the source node, can be forwarded by other nodes in π to the destination node.

In RCM model, we will define a probability function $f(r)$ indicating the connection probability of two secondary nodes with distance r. In reality, the farther two nodes are apart, the higher risk for a successful communication connectivity. Moreover, when $r > r_s$, we assume there exists no link and thus r_s can be regarded as secondary user's transmission range. Therefore, it is reasonable to assume that $f(r)$ is a monotone decreasing function with respect to r and $f(r) = 0$ whenever $r > r_s$. Formally, the constraint of $h(r)$ can be written as:

$$0 < f(r_s) < f(r) < f(0) < 1,\ 0 < r < r_s.^4$$

In BM, two secondary nodes are connected only when they are within each other transmission range r_s. The probability function $h(r)$ is:

$$h(r) = 1,\ 0 < r \le r_s.$$

The problem of transmission delay will be discussed using both the BM and RCM in this article. We denote the BM and RCM modeled secondary network by $\mathcal{S}_t(\lambda_S; h(r))$ and $\mathcal{S}_t(\lambda_S; f(r))$, respectively, in each slot t. The subscript t indicates that the network is

dynamic. Similarly, $\mathcal{G}_t(\lambda_S; h(r); \mathcal{A}_P)$ and $\mathcal{G}_t(\lambda_S; f(r); \mathcal{A}_P)$ denotes the CR networks modeled by BM and RCM, respectively. [5]

2.3 Percolation of CR Networks

Percolation theory, especially continuum percolation, is our major theoretical tool for investigating the transmission delay. Since we focus on delay problem in secondary network, we will use percolation of secondary network and percolation of CR networks interchangeably in the rest of the paper. Suppose at the beginning, time slot 0, secondary nodes are distributed uniformly according to $\mathbb{P}_S(t)$ with density λ_S in an infinite 2-dimensional space with initial position $\mathfrak{X}^0 = \{\mathcal{X}_0^0, \mathcal{X}_1^0, \dots, \mathcal{X}_n^0, \dots\}$. We denote the component containing the origin $\mathcal{O}$ by $\mathcal{C}_\mathcal{O}^0$ and the critical percolation probability by $\lambda_c(h(r))$. Note that if $\lambda_S > \lambda_c(h(r))$, the BM network $\mathcal{G}_t(\lambda_S; h(r); \mathcal{A}_P)$ is percolated for all time slot t and $\mathcal{C}_\mathcal{O}^0$ contains infinite nodes. Whereas if $\lambda_S < \lambda_c(h(r))$, $\mathcal{G}_t(\lambda_S; h(r); \mathcal{A}_P)$ is not percolated for all t. Formally, we give the following definition of the critical density $\lambda_c(h(r))$:

DEFINITION 1. *For $\mathcal{G}_t(\lambda_S; h(r); \mathcal{A}_P)$, the percolation probability p_λ is the probability that the connected component containing the origin $\mathcal{C}_\mathcal{O}^0$ has infinite secondary users. The critical density of secondary users is defined as $\lambda_c(h(r)) = \inf\{\lambda_S > 0 : p_\lambda > 0\}$.*

Similarly, the RCM network $\mathcal{G}_t(\lambda_S; f(r); \mathcal{A}_P)$ is percolated if and only if $\lambda_S > \lambda_c(f(r))$. The critical density $\lambda_c(f(r))$ can be defined similarly to Definition 1. Note that probability function $h(r) > f(r)$ for $\forall r \in [0, r_s]$, thus it is clear that the following lemma holds.

LEMMA 1. *For the two critical percolation probabilities $\lambda_c(f(r))$ and $\lambda_c(h(r))$, they satisfy $\lambda_c(f(r)) > \lambda_c(h(r))$.*

We will further illustrate this lemma through simulation results in section 7.

2.4 Useful Notations

To make it more readable, we list some useful notations and symbols used in this paper as follows:

- $\mathcal{A}_P$: primary network. $\mathcal{G}_t(\lambda_S; h(r); \mathcal{A}_P), \mathcal{G}_t(\lambda_S; f(r); \mathcal{A}_P)$: CR network modeled by BM and RCM, respectively.

- $\mathcal{D}(u, R)$: a two-dimensional circular region centered at u with radii R. $d(u, v)$: the Eustachian distance between node u and v;

- $S_p(t)$: i.i.d. switching renewal process of each primary user. $p_{00}, p_{01}, p_{10}, p_{11}$: four state switching probabilities of process $S_p(t)$;

- $H_m(t)$: a discrete time Markov on-off *(active-sleep)* process with i.i.d. *active* periods $Z_p^1(t)$, and i.i.d. *sleep* periods $Z_p^0(t)$;

- $\pi(u, v)$: a path between node u and v. $\mathcal{T}_p(l_k)$: passing time of one link $l_k \in \pi(u, v)$. $\mathcal{I}(l_k)$: the interfered region of l_k. $\mathcal{T}(\pi)$: transmission delay of $\pi(u, v)$ in supercritical secondary network.

- $\mathcal{C}_{t,u,h(r)}$: one connected cluster containing node u at slot t in subcritical secondary network. $\mathcal{M}\{\mathfrak{X}^0, d\}$: the Constrained Circular I.I.D. Mobility Model for secondary network. $\mathcal{M}_{h(r)}(\lambda_S, \mathcal{A}_P)$: a random variable characterized the physical size of one cluster.

---

³In this paper, we use $Pr(\cdot)$ to denote the probability of one certain event and $E[\cdot]$ denotes the expectation of one random variable.
⁴$h(r)$ also satisfies $\int_{R^2} f(r)dr < \infty$ according to [1].

---

⁵In following sections, we will use $\mathcal{A}_P$ to denote the primary network $\mathcal{A}_P(\lambda_{PT}; S_P(t); R_p)$ for simplicity.

- Υ: multi-cluster hop transmission process in a subcritical secondary network. $\mathcal{T}_P(\Upsilon)$: cluster to cluster transmission delay of Υ. $\mathcal{T}_S(\Upsilon)$: waiting delay for spectrum opportunities in Υ.

3. MAIN RESULTS

After introduction of the system model, we are ready to present our main results. Before that, we first give some basic properties of $\gamma(\lambda_S, \mathcal{A}_P)$ to make the subsequent analysis more smooth.

3.1 Basic Properties of $\gamma(\lambda_S, \mathcal{A}_P)$

$\gamma(\lambda_S, \mathcal{A}_P)$ is the ratio of transmission delay to distance as the distance from source to destination goes to ∞. We use the symbol $\gamma(\lambda_S, \mathcal{A}_P)$ to indicate that γ is impacted by λ_S and $\mathcal{A}_P$. First, λ_S determines that whether the network contains a unique infinite connected component or only finite connected clusters. Second, $\mathcal{A}_P$ imposes interference on communication in secondary network because information can be forwarded by a secondary node only when it has spectrum opportunities.

W. Peng *et al.* [10] and W. Ren *et al.* [11] proved that percolation of supercritical secondary network still exists if the density of primary users λ_{PT} is less than a critical number λ_{PT}^*. However, if $\lambda_{PT} > \lambda_{PT}^*$, although percolation phenomenon still exists topologically, the infinite connected component has broken into mutually disconnected finite clusters from the perspective of spectrum opportunities. In other words, although two disjoint clusters are connected topologically, they are not able to exchange messages due to lack of available communication channels. Therefore, to avoid confusion, it is necessary to redefine the supercritical and subcritical property of CR networks.

DEFINITION 2. *For $\mathcal{G}_t(\lambda_S; h(r); \mathcal{A}_P)$,*

1. *If $\lambda_S < \lambda_c(h(r))$, the secondary network is subcritical.*

2. *If $\lambda_S > \lambda_c(h(r))$, the secondary network is supercritical. And if $\lambda_{PT} < \lambda_{PT}^*$, it is com-supercritical ("com" is short for "communication"). Whereas the secondary network is top-supercritical ("top" is short for "topology") when $\lambda_{PT} > \lambda_{PT}^*$.*

For $\mathcal{G}_t(\lambda_S; f(r); \mathcal{A}_P)$, we have the similar definition. Note that if the secondary network is supercritical, the unique infinite connected component $\mathcal{C}_{\mathcal{O}}^0$ always exists from the perspective of topology structure and all nodes with within it share instantaneous connectivity. However, even two topologically connected nodes may not be able to communicate with each other if they lack spectrum opportunities. The difference between the two kinds of supercritical network is that all nodes within $\mathcal{C}_{\mathcal{O}}^0$ have spectrum opportunities in com-supercritical network, whereas a high fraction of nodes in $\mathcal{C}_{\mathcal{O}}^0$ do not have spectrum opportunities in top-supercritical network. Then we are ready to present the basic properties of $\gamma(\lambda_S, \mathcal{A}_P)$ in different categories of secondary network.

THEOREM 1. *$\gamma(\lambda_S, \mathcal{A}_P)$ is characterized by the following basic properties:*

1. *$\gamma(\lambda_S, \mathcal{A}_P)$ increases with respect to λ_{PT}.*

2. *$\gamma(\lambda_S, \mathcal{A}_P) = 0$ for any com-supercritical secondary network.*

3. *$\gamma(\lambda_S, \mathcal{A}_P)$ is a constant for a given top-supercritical secondary network.*

4. *$\gamma(\lambda_S, \mathcal{A}_P)$ has both lower and upper bound in subcritical secondary network.*

We will present the detailed proof of these four items in the following sections.

3.2 Main Results on $\gamma(\lambda_S, \mathcal{A}_P)$

THEOREM 2. *Given a BM Network $\mathcal{G}_t(\lambda_S; h(r); \mathcal{A}_P)$ with $\lambda_S > \lambda_c(h(r))$, consider two secondary users u (source) and v (destination) connected by the path $\pi(u, v)$ in $\mathcal{C}_{\mathcal{O}}^0$, the corresponding $\gamma(\lambda_S, \mathcal{A}_P) \triangleq \lim_{d(u,v) \to \infty} \frac{\mathcal{T}(\pi)}{d(u,v)}$ satisfies:*
(i) $\gamma(\lambda_S, \mathcal{A}_P) = 0$ if $\lambda_{PT} < \lambda_{PT}^*(\lambda_S)$; [6]
(ii) if $\lambda_{PT} > \lambda_{PT}^*(\lambda_S)$,

$$\gamma(\lambda_S, \mathcal{A}_P) = \kappa \cdot \frac{1 - \beta_0^{\overline{\psi}}}{\beta_0^{\overline{\psi}}},$$

where κ is a constant independent on λ_S. $\overline{\psi}$ is a constant satisfying $\lceil \pi R_p^2 \lambda_{PT} \rceil \leq \overline{\psi} \leq \lfloor [2\pi R_p^2 - \mathcal{F}(r_s, R_p)]\lambda_{PT} \rfloor$, [7] *where $\mathcal{F}(r_s, R_p) = 2R_p^2 \arccos \frac{r_s}{2R_p} - \frac{r_s \sqrt{4R_p^2 - r_s^2}}{2}$.*

THEOREM 3. *For a RCM Network $\mathcal{G}_t(\lambda_S; f(r); \mathcal{A}_P)$ with $\lambda_S > \lambda_c(f(r))$, the conclusion on $\gamma(\lambda_S, \mathcal{A}_P)$ in Theorem 2 still holds.*

THEOREM 4. *For a BM Network $\mathcal{G}_t(\lambda_S; h(r); \mathcal{A}_P)$ with density $\frac{\lambda_c(h(r))}{\mathcal{R}^2} < \lambda_S < \lambda_c(h(r))$, under the Constrained Circular I.I.D. Mobility Model $\mathcal{M}\{\mathfrak{X}^0, d\}$, the corresponding $\gamma(\lambda_S, \mathcal{A}_P) \triangleq \lim_{d(u,v) \to \infty} \frac{\mathcal{T}_P(\Upsilon) + \mathcal{T}_S(\Upsilon)}{d(u,v)}$ satisfies* [8]:

$$\gamma(\lambda_S, \mathcal{A}_P) \geq \frac{1}{\beta_0^{\eta} \cdot E[\mathcal{M}_{h(r)}(\lambda_S, \mathcal{A}_P) + r_s]},$$

where $\mathcal{R} = \frac{4d + r_s}{r_s}$ and η is a constant satisfying $\lfloor \pi R_p^2 \lambda_{PT} \rfloor < \eta \leq \lfloor \pi(\frac{\mathcal{M}_{h(r)}(\lambda_S, \mathcal{A}_P)}{2} + R_p)^2 \lambda_{PT} \rfloor$.

4. $\gamma(\lambda_S, \mathcal{A}_P)$ IN SUPERCRITICAL NETWORKS

Given supercritical network, $\mathcal{C}_{\mathcal{O}}^0$ contains infinite secondary users for all t. Two randomly chosen secondary users within $\mathcal{C}_{\mathcal{O}}^0$ share instantaneous connectivity and the transmission delay consists of only the waiting time for spectrum opportunities if we ignore propagation delay.

4.1 $\gamma(\lambda_S, \mathcal{A}_P)$ for Supercritical $\mathcal{G}_t(\lambda_S; h(r); \mathcal{A}_P)$

If $\lambda_{PT} < \lambda_{PT}^*(\lambda_S)$, the network is com-supercritical. Then (i) of Theorem 2 is clear since $\mathcal{T}(\pi) = 0$.

If $\lambda_{PT} > \lambda_{PT}^*(\lambda_S)$, the network is top-supercritical. Now, given two secondary users u (source) and v (destination), an open path $\pi(u, v)$ in $\mathcal{C}_{\mathcal{O}}^0$ may be demolished when its node fails to forward the information to the destination due to the impact of primary network. Thus, v can not receive the messages immediately and $\mathcal{T}(\pi)$ is no longer zero. Next, we will study $\gamma(\lambda_S, \mathcal{A}_P)$ according to the following computation flow.

---

[6] According to [11], we use the notation $\lambda_{PT}^*(\lambda_S)$ to indicate that the critical density λ_{PT}^* varies with λ_S.

[7] $\lfloor A \rfloor$ denotes the largest integer less than or equal to A.

[8] Here, the expression of $\gamma(\lambda_S, \mathcal{A}_P)$ is different from that in Theorem 2 since u and v can not see any path connecting them when the network is subcritical. The messages broadcasted by u is forwarded in *multi-cluster hop transmission process*, which will be introduced in section 5.

	Computation Flow: Spectrum-to-Hop Process (SHP)
1:	$\forall x, y \in \mathbb{P}_S(t)$, they share a linkage $l(x,y)$ whenever $d(x,y) \leq r_s$.
2:	Messages can be forwarded through $l(x,y)$ when it has spectrum opportunities.
3:	The expected delay before $l(x,y)$ owning spectrum opportunities is $E[l(x,y)]$.
4:	The minimum number of hops (linkages) in an open path $\pi(u,v)$ is $\mathcal{Q}(u,v)$.
5:	The scaling behavior of hops number $\mathcal{Q}(u,v)$ is $\lim_{d(u,v)\to\infty}\frac{\mathcal{Q}(u,v)}{d(u,v)} = \kappa$.
6:	Obtain $\gamma(\lambda_S, \mathcal{A}_P) = \kappa \cdot E[l(x,y)]$.

It is clear that after the step 1 in the above flow, the topological structure of $\mathcal{G}_t(\lambda_S; h(r); \mathcal{A}_P)$ is established. When a BM modeled network is top-supercritical, we can find a topologically connected path $\pi(u,v)$ in $\mathcal{C}_\mathcal{O}^0$ after step 1. For one link l_k in $\pi(u,v)$, with two endpoints u_k, u_{k+1} and $d(u_k, u_{k+1}) = r_k$, its passing time $\mathcal{T}_p(l_k)$ is zero when none of the *active* primary users have imposed interference on it. However, $\mathcal{T}_p(l_k) > 0$ if l_k is interfered by at least one *active* primary user. Therefore $\mathcal{T}_p(l_k)$ is delay before the messages can be forwarded through l_k.

For l_k, we can define its interfered region $\mathcal{I}(l_k)$. The transmission range of primary transmitters within this interfered region covers at least one of the endpoints of l_k. Hence, if any of these primary transmitters are *active*, no spectrum opportunities will be available for u_k or u_{k+1} and messages can not be forwarded through l_k successfully. Formally, we give the definition of $\mathcal{I}(l_k)$ as follows.

DEFINITION 3. *For any point w in the same 2-dimensional space as the secondary network, if (i): $d(w, u_k) \leq R_p$ or (ii): $d(w, u_{k+1}) \leq R_p$ or both (i) and (ii) are satisfied, then we have $w \in \mathcal{I}(l_k)$.*

Assume that l_k is interfered by ψ primary transmitters. By the *Poisson Point Process* $\mathbb{P}_{PT}(t)$ formed by primary transmitters with density λ_{PT}, we have the following lemma.

LEMMA 2. *(Step 2 in Flow SHP) Given the* Poisson Point Process $\mathbb{P}_{PT}(t)$ *formed by primary users with density λ_{PT}, for a link l_k, with length $r_k \leq r_s$ in $\pi(u,v)$, the expected number of primary transmitters ψ within $\mathcal{I}(l_k)$ satisfies*

1. *if $R_p \geq \frac{r_k}{2}$,*

$$\psi = \lfloor [2\pi R_p^2 - \mathcal{F}(r_k, R_p)]\lambda_{PT} \rfloor, \quad (1)$$

where $\mathcal{F}(r_k, R_p) = 2R_p^2 \arccos\frac{r_k}{2R_p} - \frac{r_k\sqrt{4R_p^2 - r_k^2}}{2}$.

2. *if $R_p < \frac{r_k}{2}$*

$$\psi = \lfloor 2\pi R_p^2 \lambda_{PT} \rfloor.$$

PROOF. If $R_p \geq \frac{r_k}{2}$, illustrated in Figure 1 a, two circular region $\mathcal{D}(u_k, R_p)$ and $\mathcal{D}(u_{k+1}, R_p)$ are intersected with each other. According to Definition 3,

$$\mathcal{I}(l_k) = \mathcal{D}(u_k, R_p) \bigcup \mathcal{D}(u_{k+1}, R_p).$$

Considering $\mathbb{P}_{PT}(t)$ with density λ_{PT}, the expected number of primary transmitters ψ within $\mathcal{I}(l_k)$ is

$$\psi = \lfloor \lambda_{PT} \cdot |\mathcal{I}(l_k)| \rfloor, \quad (2)$$

where $|\mathcal{I}(l_k)|$ denotes the Lebesgue measure (or area) of $\mathcal{I}(l_k)$. By tools in plane geometry, we have

$$|\mathcal{I}(l_k)| = 2\pi R_p^2 - \mathcal{F}(r_k, R_p), \quad (3)$$

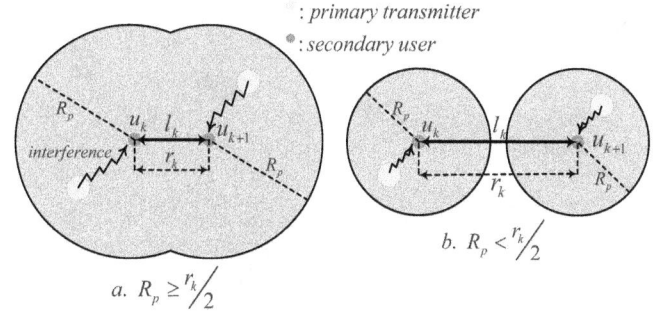

Figure 1: Illustration of Lemma 2.

where $\mathcal{F}(r_k, R_p) = 2R_p^2 \arccos\frac{r_k}{2R_p} - \frac{r_k\sqrt{4R_p^2 - r_k^2}}{2}$. Combining Equations (2) and (3), item 1 is proved.

If $R_p < \frac{r_k}{2}$, $\mathcal{D}(u_k, R_p)$ and $\mathcal{D}(u_{k+1}, R_p)$ are two independent areas and the expected number of primary transmitters in the two areas are also independent. Then item 2 can be proved by simply by substituting $|\mathcal{I}(l_k)| = 2\pi R_p^2$ into Equation (2). $\square$

Note that in practical networks, R_p is larger than r_s since the transmission power of primary transmitter may be several times larger than that of secondary user. To avoid the triviality in our analysis, we assume that $R_p > r_s$. Then the first item of Lemma 2 is always satisfied since $r_s \geq r_k$. According to Equation (1), it is clear that ψ depends on r_k. However, due to homogeneity of the *Poisson Point Process* and Boolean Model, we can use the expectation of $E[\psi]$ to approximate ψ since ψ converges to $E[\psi]$ when the subscript k goes to ∞. By $0 < r_k \leq r_s$, we have

$$\lfloor \pi R_p^2 \lambda_{PT} \rfloor \leq E[\psi] \leq \lfloor [2\pi R_p^2 - \mathcal{F}(r_s, R_p)]\lambda_{PT} \rfloor. \quad (4)$$

Thus $E[\psi]$ is a constant, independent on k, between the lower and upper bound in Equation (4). For simplicity, we will use $\overline{\psi}$ to denote $E[\psi]$ in the rest of paper.

We associate one random variable $H_m(t)$ for each link $l_m \in \pi(u,v)$. For l_k, $H_k(t) = 0$ only when all the primary transmitters in $\mathcal{I}(l_k)$ are *sleep* and $H_k(t) = 1$ if at least one of these primary transmitters is *active*. By the Markovian property behaviors of each primary user, the process $H_m(t)$ can be viewed as a discrete time Markov on-off *(active-sleep)* process with i.i.d. *active* periods $Z_p^1(t)$, and i.i.d. *sleep* periods $Z_p^0(t)$. Thus, $\mathcal{T}_p(l_k)$ is the delay before $H_k(t) = 0$ for the first time, i.e.,

$$\mathcal{T}_p(l_k) = \inf_{t \geq 0}\{t : H_k(t) = 0\}.$$

The expected delay for l_k (*step 3 in Flow SHP*) is:

$$E[\mathcal{T}_p(l_k)] = \sum_{n=0}^{\infty} n(1 - \beta_0^{\overline{\psi}})^n \cdot \beta_0^{\overline{\psi}} = \frac{1 - \beta_0^{\overline{\psi}}}{\beta_0^{\overline{\psi}}}. \quad (5)$$

Now we have obtained the expected delay for each hop in $\pi(u,v)$. Next we will investigate the scaling behavior of the number of hops in the path.

LEMMA 3. *(Steps 4 and 5 in Flow SHP) Given the top-supercritical Boolean Model Network $\mathcal{G}_t(\lambda_S; h(r); \mathcal{A}_P)$, the minimum number of hops $\mathcal{Q}(u,v)$ in one path $\pi(u,v)$ scales linearly with the distance $d(u,v)$, i.e.,*

$$\lim_{d(u,v)\to\infty} \frac{\mathcal{Q}(u,v)}{d(u,v)} = \kappa.$$

PROOF. The proof of lemma 3 uses Liggett's subadditive ergodic theorem [2]. This technique has already been explored in [13] and [14]. Here we will omit the detailed proof because of the space limitation. □

Combining Equation (5) and Lemma 3, we will finish the Computation Flow SHP after the step 6.

$$\gamma(\lambda_S, \mathcal{A}_P) = \lim_{d(u,v)\to\infty} \frac{\mathcal{T}(\pi)}{d(u,v)} = \lim_{d(u,v),k\to\infty} \frac{\sum \mathcal{T}_p(l_k)}{d(u,v)}$$

$$= \lim_{d(u,v)\to\infty} \frac{\mathcal{Q}(u,v) \cdot E[\mathcal{T}_p(l_k)]}{d(u,v)}$$

(the i.i.d. random variable sequence

$$\sum \mathcal{T}_p(l_k) \text{ converges to its expectation})$$

$$= \kappa \cdot \frac{1 - \beta_0^{\overline{\psi}}}{\beta_0^{\overline{\psi}}}.$$

Then we have proved the Theorem 2. □

4.2 $\gamma(\lambda_S, \mathcal{A}_P)$ for Supercritical $\mathcal{G}_t(\lambda_S; f(r); \mathcal{A}_P)$

To show the non-triviality of Theorem 3, we need to introduce a algorithm named Removing and Extracting Process (REP) to reveal the relation between $\mathcal{G}_t(\lambda_S; f(r); \mathcal{A}_P)$ and $\mathcal{G}_t(\lambda_S; h(r); \mathcal{A}_P)$. According to our system model, it is clear that the primary network of the two CR networks are the same. So we can simply study the correlation between $\mathcal{S}_t(\lambda_S; h(r))$ and $\mathcal{S}_t(\lambda_S; f(r))$.

Algorithm: Removing and Extracting Process
Input: $\mathcal{S}_t(\lambda_0; h(r))$ with $\lambda_0 > \lambda_c(h(r))$
Output: $\mathcal{G}'$, $\mathcal{S}_t(\lambda^\dagger, f(r))$ and $\mathcal{G}^\dagger$
1:    **For** a direct linkage $l(x, y) \in \mathcal{S}_t(\lambda_S; h(r))$
2:      $l(x, y)$ is open according to function $f(r)$.
3:      **If** $l(x, y)$ is close
4:        Remove the linkage from $\mathcal{S}_t(\lambda_S; h(r))$;
5:        Put the pair of nodes (x, y) into the set $\mathcal{Z}$.
6:      **End**
7:    **End**
8:    Return a temporal graph $\mathcal{G}'$.
9:    **For** a node pair $(x_0, y_0) \in \mathcal{Z}$
10:     **If** x_0 is connected with y_0 in $\mathcal{G}'$ through some hops
11:       Go back to step 9.
12:     **Else**
13:       **While** (x_0 is disconnected with y_0)
14:        Put another node in $\mathcal{G}'$;
15:        Connect the new node with the other nodes in $\mathcal{G}'$ according to RCM.
16:       **End**
17:     **End**
18:    **End**
19:    Return the graph $\mathcal{S}_t(\lambda^\dagger, f(r))$.
20:    **For** a direct linkage $l(x_0, y_0) \in \mathcal{S}_t(\lambda_0; h(r))$
21:     **if** $l(x_0, y_0) \in \mathcal{S}_t(\lambda^\dagger; f(r))$
22:       Put $l(x_0, y_0)$ in $\mathcal{G}^\dagger$.
23:     **Else**
24:       Put indirect linkage between x_0 and y_0 in $\mathcal{G}^\dagger$.
25:     **END**
26:    **END**
27:    Return the graph $\mathcal{G}^\dagger$.

Steps 1 to 8 are the removing process and we get a temporal graph $\mathcal{G}'$. It is clear that $\mathcal{G}' = \mathcal{S}_t(\lambda_0; f(r))$ because we have reconstructed $\mathcal{S}_t(\lambda_0; h(r))$ according to the probability function $f(r)$. However, $\mathcal{G}'$ may not be top-supercritical since $\lambda_c(h(r)) < \lambda_c(f(r))$ according to Lemma 1. The purpose of steps 9 to 19 is to construct a top-supercritical network $\mathcal{S}_t(\lambda^\dagger; f(r))$. Steps 10 and

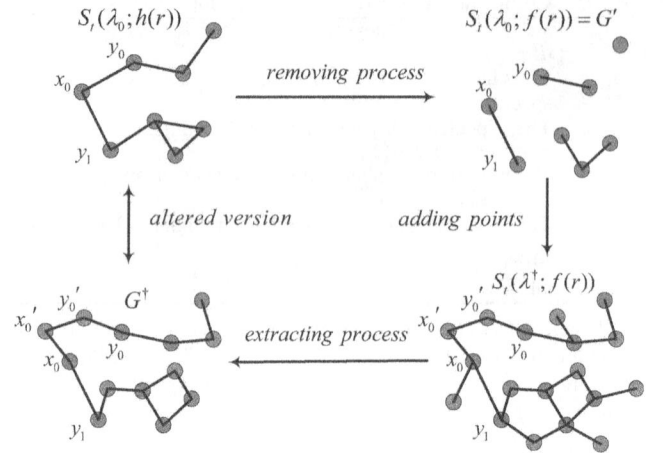

Figure 2: Illustration of Algorithm REP.

11 show a special case that even the direct linkage between x_0 and y_0 is removed, they are still connected through some hops. When x_0 and y_0 are disconnected, we need to put some new nodes into $\mathcal{G}'$ through steps 12 to 17. These new nodes are connected with each other and with the previous nodes in $\mathcal{G}'$ according to RCM. Thus, when the node density is sufficiently large, i.e., the network is supercritical, we can find a newly established path between x_0 and y_0 with probability one. In steps 20 to 26, we execute the extracting process to get a new graph $\mathcal{G}^\dagger$. The process is illustrated in Figure 2.

The purpose of Algorithm REP is to show that the new graph $\mathcal{G}^\dagger$ is an altered version of $\mathcal{S}_t(\lambda_0; h(r))$. Formally, we can write this in the following Lemma:

LEMMA 4. $\mathcal{G}^\dagger$ is an altered version of $\mathcal{S}_t(\lambda_0; h(r))$. Specifically, $\forall l(x, y) \in \mathcal{S}_t(\lambda_0; h(r))$, x and y are connected either through a direct linkage $l(x, y)$ or through an indirect linkage [9] in $\mathcal{G}^\dagger$.

PROOF. In Figure 2, $l(x_0, y_1)$ is kept in both $\mathcal{S}_t(\lambda_0; h(r))$ and $\mathcal{S}_t(\lambda^\dagger; f(r))$. But $l(x_0, y_0)$ does not exist in $\mathcal{S}_t(\lambda^\dagger; f(r))$ and x_0 is connected to y_0 through an indirect linkage between x_0 and y_0. Therefore, if we extract all these nodes and linkages (direct and indirect ones) in $\mathcal{S}_t(\lambda^\dagger; f(r))$, we will get a new graph $\mathcal{G}^\dagger$. The only difference between $\mathcal{G}^\dagger$ and $\mathcal{S}_t(\lambda_0; h(r))$ is that a direct linkage $l(x, y) \in \mathcal{S}_t(\lambda_0; h(r))$ may be an indirect linkage in $\mathcal{G}^\dagger$. Thus, we can define $\mathcal{G}^\dagger$ as an altered version of $\mathcal{S}_t(\lambda_0; h(r))$. □

Therefore, if we can find a path $\pi(u, v)$ from source u to destination v in $\mathcal{S}_t(\lambda_0; h(r))$ or $\mathcal{G}_t(\lambda_S; h(r); \mathcal{A}_P)$, we can also find a path $\pi'(u, v)$ in $\mathcal{G}^\dagger$. And the difference between $\pi'(u, v)$ and $\pi(u, v)$ is that $\pi'(u, v)$ might have more hops than $\pi(u, v)$. Fortunately, this does not make any difference when we consider their scaling behavior as only finite number of hops have been added in $\pi'(u, v)$. Thus, we can conclude that Lemma 3 still holds for $\mathcal{G}^\dagger$. Since $\mathcal{G}^\dagger$ is a subgraph of $\mathcal{S}_t(\lambda^\dagger; f(r))$, Lemma 3 also holds for $\mathcal{S}_t(\lambda^\dagger; f(r))$. Formally, we can write this conclusion as follows:

LEMMA 5. *Given the top-supercritical Random Connection Model Network* $\mathcal{G}_t(\lambda_S; f(r); \mathcal{A}_P)$, *Lemma 3 still holds.*

---

[9] An indirect linkage between x_0 and y_0 includes x_0, y_0 and the other nodes that are used to connected x_0 to y_0. For instance, the indirect linkage between x_0 and y_0 in $\mathcal{G}^\dagger$, illustrated in Figure 2, consists of x_0, y_0, x_0', y_0' and their linkages.

Our simulation results also verify this lemma. Then the proof of Theorem 3 is finished. □

5. $\gamma(\lambda_S, \mathcal{A}_P)$ IN SUBCRITICAL NETWORKS

When it is subcritical, secondary network is broken into mutually disjoint finite clusters. Then the messages can not reach destination node if it is in different clusters with source node and the network's connectivity is fixed over time.

However, when the secondary network is mobile, there is a probability that the messages can be forwarded to destination in subcritical network. For instance, a source node broadcasts a message in a subcritical network at time 0. Ignoring propagation delay, all nodes in the same cluster with the source node will receive the message instantaneously. The nodes in this cluster are only a fraction of all the nodes since the network is not percolated. As time goes on, however, nodes move, and there is a probability that the message can pass from one of these message-carrying nodes in the cluster to a new node if they are within each other's communication range. Then all other nodes in the same cluster with the newly informed node will also receive the information. As this process goes on, a large proportion of nodes will be informed of the message after some delay. In other words, the messages can spread and reach the destination in subcritical network if it is mobile. Thus, we need to introduce a mobility model for secondary network.

5.1 Mobility Model

We model the mobility of each secondary user by the constrained circular i.i.d. mobility model, which has been introduced in [13]. Formally, we give the definition of mobility model as follows:

DEFINITION 4. *The Constrained Circular I.I.D. Mobility Model* $\mathcal{M}\{\mathfrak{X}^0, d\}$: *Given the initial position of secondary nodes* $\mathfrak{X}^0 = \{\mathcal{X}_0^0, \mathcal{X}_1^0, \ldots, \mathcal{X}_n^0, \ldots\}$, *at each time slot* $k = 1, 2, 3, \ldots$, $\mathcal{X}_u^k$ *is uniformly distributed randomly within* $\mathcal{D}(\mathcal{X}_u^{k-1}, d)$: *a circular region centered at* $\mathcal{X}_u^{k-1}$ *with radii d. The positions* $\mathcal{X}_u^k$ *are mutually independent among all secondary users and independent of all previous positions* $\mathcal{X}_u^{t'}$ *where* $t' = \{0, 1, \ldots, k-1\}$. [10]

Note Theorems 2 and 3 still hold if we apply the mobility model in supercritical secondary network since mobility does not impact the percolation phenomenon in the network. However, in subcritical secondary network, mobility increase the probability of percolation. Specifically, two nodes with initial distance less than $r_s + 4d$ has a strictly positive probability that they can share link after finite delay. This indicates that the largest possible transmission range of each node is $r_s + 4d$ under the mobility model. Thus, the percolation property of the network under mobility model is the same as that of the static network in which secondary user's transmission range is $r_s + 4d$. According to the scaling property of BM in [1], we have the following lemma.

LEMMA 6. *Given network* $\mathcal{G}_t(\lambda_S; h(r); \mathcal{A}_p)$ *under the mobility model* $\mathcal{M}\{\mathfrak{X}^0, d\}$, *if we scale up* [11] *the network by ratio* $\mathcal{R} = \frac{r_s + 4d}{r_s}$, *the critical density for percolation after scaling is* $\frac{\lambda_c(h(r))}{\mathcal{R}^2}$.

For a subcritical network with density $\lambda < \lambda_c(h(r))$, according to Lemma 6, if $\lambda > \frac{\lambda_c(h(r))}{\mathcal{R}^2}$, the network still percolates. In other words, the weaker version of overall connectivity may be still established in a subcritical network.

---

[10]Note in our mobility model, the time it takes for a node to move to new position is not accounted. Thus, delay is independent on the mobility speed.

[11]Here, scaling up the network means increasing secondary user's transmission range while keeping their positions the same.

Up to now, we have showed that under Constrained Circular I.I.D. Mobility Model, information can spread among a subcritical network. However, the dissemination process is quite different from that in supercritical network. It is done through the process that we have briefly explained at the beginning of this section. We define this process as *multi-cluster hop transmission process*. To study the transmission delay, we first need to investigate this *multi-cluster hop transmission process* in the following subsection.

5.2 Multi-Cluster Hop Transmission Process

Given a cluster containing the secondary node u_0, denoted by $\mathcal{C}_{t, u_0, h(r)}$, all secondary nodes in it share instantaneous connectivity with u_0. We assume that the source node u is in the same cluster with u_0. At the beginning, time slot $t_0 = 0$, all secondary nodes within $\mathcal{C}_{t_0, u_0, h(r)}$ are informed of the message, broadcasted by u, instantaneously since we have ignored the propagation delay. However, the transmission process stops within $\mathcal{C}_{t_0, u_0, h(r)}$ because no other clusters are connected with it. This transmission process will restart at time slot $t_1 > t_0$, when at least one node u_1 is able to receive the information from $\mathcal{C}_{t_0, u_0, h(r)}$ and then all nodes in cluster $\mathcal{C}_{t_1, u_1, h(r)}$ will instantaneously informed of the messages. This process goes on, until at time slot t_M, node u_M receives the information and the destination node v is in the same cluster with u_M. Therefore, v is informed of the information instantaneously at this time slot. Then messages have been transmitted from source node u to destination node v after this process. We name this process as *multi-cluster hop transmission process*, denoted by $\Upsilon\{\mathcal{C}_{t_0, u_0, h(r)}, \mathcal{C}_{t_1, u_1, h(r)}, \ldots, \mathcal{C}_{t_M, u_M, h(r)}\}$.

We associate each cluster $\mathcal{C}_{t_k, u_k, h(r)}$ with one random variable $\mathcal{M}_{t_k, u_k, h(r)}(\lambda_S, \mathcal{A}_P)$, which is defined as:

DEFINITION 5. *For one cluster* $\mathcal{C}_{t_k, u_k, h(r)}$,

$$\mathcal{M}_{t_k, u_k, h(r)}(\lambda_S, \mathcal{A}_P) = \sup_{u_k', v_k' \in \mathcal{C}_{t, u, h(r)}} \{d(u_k', v_k')\}.$$

From the definition, it is clear that $\mathcal{M}_{t_k, u_k, h(r)}(\lambda_S, \mathcal{A}_P)$ is the description of physical size of the cluster $\mathcal{C}_{t_k, u_k, h(r)}$. Thus, it is true that $\mathcal{M}_{h(r)}(\lambda_S, \mathcal{A}_P)$ increases with respect to λ_S and decreases as the interference from primary network is heavier. Besides, $\mathcal{M}_{t_k, u_k, h(r)}(\lambda_S, \mathcal{A}_P)$ is independent on u_k and t_k. So we will rewrite it as $\mathcal{M}_{h(r)}(\lambda_S, \mathcal{A}_P)$ for simplicity.

5.3 $\gamma(\lambda_S, \mathcal{A}_P)$ for Subcritical Networks

When $\mathcal{G}_t(\lambda_S; h(r); \mathcal{A}_P)$ is subcritical, the messages must be forwarded in *multi-cluster hop transmission process*. Now given the source u and destination v, consider the process Υ stating with $\mathcal{C}_{t_0, u_0, h(r)}$ and ending with $\mathcal{C}_{t_M, u_M, h(r)}$, as illustrated in Figure 3. Once we can find one cluster containing the v receives the messages as Υ goes on, v will be successfully informed of the information broadcasted by u. For one cluster $\mathcal{C}_{t_m, u_m, h(r)}$ $(1 \leq m < M)$ in Υ, it will be able to forward the messages to the next cluster $\mathcal{C}_{t_{m+1}, u_{m+1}, h(r)}$ only when both of the following two conditions are satisfied.

(i) For $v_m \in \mathcal{C}_{t_m, u_m, h(r)}, v_{m+1} \in \mathcal{C}_{t_{m+1}, u_{m+1}, h(r)}$

$$d_{min} = \inf\{d(v_m, v_{m+1})\} < r_s$$

(ii) For the link $l_{v_m, v_{m+1}}$ in condition (i), no *active* primary users are residing within the interfered region $\mathcal{I}(l_{v_m, v_{m+1}})$ given in Definition 3.

Therefore, the total transmission delay $\mathcal{T}_C^m(\Upsilon)$ in the m'th hop $\mathcal{C}_{t_m, u_m, h(r)}$ of Υ are composed by cluster to cluster transmission delay (condition i) and the waiting delay for spectrum opportunities (condition ii), denoted by $\mathcal{T}_P^m(\Upsilon)$ and $\mathcal{T}_S^m(\Upsilon)$, respectively. We

The starting cluster   The second cluster   The Mth cluster

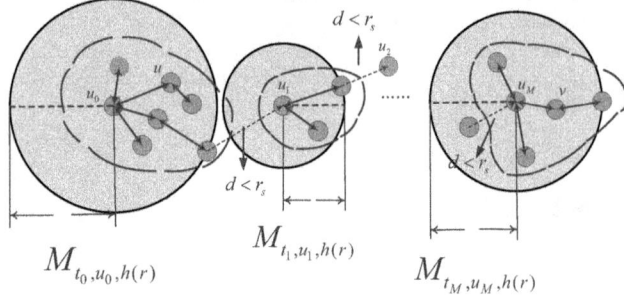

Figure 3: Illustration of the *multi-cluster hop transmission process* Υ.

assume that the expected number of primary transmitters impacting $\mathcal{C}_{t_m, u_m, h(r)}$ is η. Using the same technique in Lemma 2, we can see η is a constant satisfying:

$$\lfloor \pi R_p^2 \lambda_{PT} \rfloor < \eta \leq \lfloor \pi (\frac{\mathcal{M}_{h(r)}(\lambda_S, \mathcal{A}_P)}{2} + R_p)^2 \lambda_{PT} \rfloor.$$

Similar as the previous section, we use process $H_m(t)$ to denote the state transition process of these η primary users. Then we have

$$\mathcal{T}_C^m(\Upsilon) = \mathcal{T}_P^m(\Upsilon) + \mathcal{T}_S^m(\Upsilon)$$
$$= \mathcal{T}_P^m(\Upsilon) + \inf_{t \geq \mathcal{T}_P^m(\Upsilon)} \{H_m(t) = 0\}.$$

For $\mathcal{T}_P(\Upsilon)$, we have

$$\mathcal{T}_P(\Upsilon) = \sum_{m=0}^{M-1} \mathcal{T}_P^m(\Upsilon) = \sum_{m=0}^{M-1} (t_{m+1} - t_m)$$
$$= t_M - t_0 \geq M.$$

Note that for $\forall m = 0, 1, 2, \ldots, M-1$, u_{m+1} is connected with one node w_m in $\mathcal{C}_{t_m, u_m, h(r)}$ at slot t_{m+1}. Then their Euclidean distance satisfies the following inequality according to the triangle inequity:

$$\|u_{m+1} - u_m\| \leq \|u_m - w_m\| + \|u_{m+1} - w_m\|$$
$$\leq \mathcal{M}_{h(r)}(\lambda_S, \mathcal{A}_P) + r_s.$$

By iteration of the inequalities as Υ goes on, the distance between u and v $d(u,v)$ satisfies:

$$d(u,v) \leq \sum_{m=0}^{M-1} \|u_{m+1} - u_m\| + \|u_M - v\|$$
$$\leq \sum_{m=0}^{M-1} (\mathcal{M}_{h(r)}(\lambda_S, \mathcal{A}_P) + r_s) + \mathcal{M}_{h(r)}(\lambda_S, \mathcal{A}_P)$$
(since u_M and v are within one cluster)
$$= \sum_{m=0}^{M} (\mathcal{M}_{h(r)}(\lambda_S, \mathcal{A}_P) + r_s).$$

Moreover, according to the spatial independence of *Poisson Point Process*, $\mathcal{M}_{h(r)}(\lambda_S, \mathcal{A}_P)$ is an i.i.d. random variable. As $d(u,v)$ goes to ∞, the number of clusters M in Υ also goes to ∞. Then

we have

$$\frac{\mathcal{T}_P(\Upsilon)}{d(u,v)} \geq \lim_{M \to \infty} \frac{M}{\sum_{m=0}^{M} (\mathcal{M}_{h(r)}(\lambda_S, \mathcal{A}_P) + r_s)}$$
(since $d(u,v) \leq \sum_{m=0}^{M} (\mathcal{M}_{h(r)}(\lambda_S, \mathcal{A}_P) + r_s)$
and $\mathcal{T}_P(\Upsilon) \geq M$)
$$= \frac{1}{E[\mathcal{M}_{h(r)}(\lambda_S, \mathcal{A}_P) + r_s]}$$
(the average of an i.i.d. random variable sequence converges to its expectation).

$$\frac{\mathcal{T}_S(\Upsilon)}{d(u,v)} \geq \lim_{M \to \infty} \sum_{m=0}^{M} \frac{\mathcal{T}_S^m(\Upsilon)}{(\mathcal{M}_{h(r)}\lambda_S, \mathcal{A}_P) + r_s}$$
(since $d(u,v) \leq \sum_{k=0}^{M} (\mathcal{M}_{h(r)}(\lambda_S, \mathcal{A}_P) + r_s)$)
$$= \lim_{M \to \infty} \sum_{m=0}^{M} \frac{\mathcal{T}_S^m(\Upsilon)/M}{(\mathcal{M}_{h(r)}(\lambda_S, \mathcal{A}_P) + r_s)/M}$$
$$= \frac{E[\mathcal{T}_S^m(\Upsilon)]}{E[\mathcal{M}_{h(r)}(\lambda_S, \mathcal{A}_P) + r_s]}$$
$$= \frac{1 - \beta_0^\eta}{\beta_0^\eta \cdot E[\mathcal{M}_{h(r)}(\lambda_S, \mathcal{A}_P) + r_s]}.$$

Summing up $\frac{\mathcal{T}_P(\Upsilon)}{d(u,v)}$ and $\frac{\mathcal{T}_S(\Upsilon)}{d(u,v)}$, we can obtain

$$\gamma(\lambda_S, \mathcal{A}_P) = \frac{\mathcal{T}_P(\Upsilon) + \mathcal{T}_S(\Upsilon)}{d(u,v)}$$
$$\geq \frac{1}{E[\mathcal{M}_{h(r)}(\lambda_S, \mathcal{A}_P) + r_s]}$$
$$+ \frac{1 - \beta_0^\eta}{\beta_0^\eta \cdot E[\mathcal{M}_{h(r)}(\lambda_S, \mathcal{A}_P) + r_s]}$$
$$= \frac{1}{\beta_0^\eta \cdot E[\mathcal{M}_{h(r)}(\lambda_S, \mathcal{A}_P) + r_s]}.$$

Then the proof of Theorem 4 is finished. However, we fail to give a tight upper bound of $\gamma(\lambda_S, \mathcal{A}_P)$ in subcritical network. But intuitively $\gamma(\lambda_S, \mathcal{A}_P)$ is not ∞. $\square$

6. IMPACT OF PROPAGATION DELAY ON $\gamma(\lambda_S, \mathcal{A}_P)$

In previous sections, while studying the $\gamma(\lambda_S, \mathcal{A}_P)$, we ignored the propagation delay. However, propagation delay may become dominant factor in some cases especially when the load of networks is heavy. In this section, we will consider the impact of propagation delay on $\gamma(\lambda_S, \mathcal{A}_P)$. For ease of analysis, we assume that the propagation delays τ is the same for different links. Moreover, we assume that $\tau < 1$ since propagation delay is relatively small compared to the length of one time slot. Then we present our results on $\gamma(\lambda_S, \mathcal{A}_P)$ as follows:

THEOREM 5. *Given a BM Network $\mathcal{G}_t(\lambda_S; h(r); \mathcal{A}_P)$ with node density $\lambda_S > \lambda_c(h(r))$, its $\gamma(\lambda_S, \mathcal{A}_P)$ satisfies:*
(i) $\gamma(\lambda_S, \mathcal{A}_P) = \kappa \tau$ if $\lambda_{PT} < \lambda_{PT}^*(\lambda_S)$;
(ii) if $\lambda_{PT} > \lambda_{PT}^*(\lambda_S)$,

$$\gamma(\lambda_S, \mathcal{A}_P) = \kappa \cdot (\frac{1 - \beta_0^{\overline{\psi}}}{\beta_0^{\overline{\psi}}} + \tau)$$

THEOREM 6. *For a RCM Network $\mathcal{G}_t(\lambda_S; f(r); \mathcal{A}_P)$ with $\lambda_S > \lambda_c(f(r))$, the conclusion on $\gamma(\lambda_S, \mathcal{A}_P)$ in Theorem 5 still holds.*

THEOREM 7. *For a BM Network $\mathcal{G}_t(\lambda_S; h(r); \mathcal{A}_P)$, with density $\frac{\lambda_c(h(r))}{\mathcal{R}^2} < \lambda_S < \lambda_c(h(r))$, we have:*

$$\gamma(\lambda_S, \mathcal{A}_P) \geq \frac{1}{\beta_0^\eta \cdot E[\min\{\mathcal{M}_{h(r)}(\lambda_S, \mathcal{A}_P), \frac{r_s}{\tau}\} + r_s]},$$

where η is a constant satisfying

$$\lfloor \pi R_p^2 \lambda_{PT} \rfloor < \eta \leq \lfloor \pi (\frac{\min\{\mathcal{M}_{h(r)}(\lambda_S, \mathcal{A}_P), \frac{r_s}{\tau}\}}{2} + R_p)^2 \lambda_{PT} \rfloor.$$

PROOF. It is clear than $E[\mathcal{T}_p(l_k)] = \tau$, then item (i) of Theorem 5 can be proved using Lemma 3. Item (ii) is also straightforward. We can obtain it by simply adding a constant τ in the results of Theorem 2.

The proof of Theorem 7 is similar to that of Theorem 4. The only difference is the constraint on the physical size of $\mathcal{M}_{h(r)}(\lambda_S, \mathcal{A}_P)$. The largest range of one connected cluster is less than $\frac{r_s}{\tau}$ since the largest distance that the message can be forwarded is $\frac{r_s}{\tau}$ in one slot,. Therefore, the cluster size is $\min\{\mathcal{M}_{h(r)}(\lambda_S, \mathcal{A}_P), \frac{r_s}{\tau}\}$. Then we have:

$$\|u_{m+1} - u_m\| \leq \|u_m - w_m\| + \|u_{m+1} - w_m\|$$
$$\leq \min\{\mathcal{M}_{h(r)}(\lambda_S, \mathcal{A}_P), \frac{r_s}{\tau}\} + r_s.$$

By iteration of this inequalities, we obtain

$$d(u, v) \leq \sum_{m=0}^{M-1} \|u_{m+1} - u_m\| + \|u_M - v\|$$
$$\leq \sum_{m=0}^{M} (\min\{\mathcal{M}_{h(r)}(\lambda_S, \mathcal{A}_P), \frac{r_s}{\tau}\} + r_s).$$

Using the same method in previous section, we immediately get

$$\gamma(\lambda_S, \mathcal{A}_P) = \frac{\mathcal{T}_P(\Upsilon) + \mathcal{T}_S(\Upsilon)}{d(u, v)}$$
$$\geq \frac{1}{\beta_0^\eta \cdot E[\min\{\mathcal{M}_{h(r)}(\lambda_S, \mathcal{A}_P), \frac{r_s}{\tau}\} + r_s]}.$$

Then the proof of Theorem 7 is finished. $\square$

7. SIMULATIONS

In this section, we will present simulation results to verify our theoretical analysis of $\gamma(\lambda_S, \mathcal{A}_P)$. To begin with, we study the three critical densities for percolation, $\lambda_c(h(r))$, $\lambda_c(f(r))$, and $\lambda_{PT}^*(\lambda_S)$. Then we investigate the two constants $\overline{\psi}$ and κ, which have been defined in Equation (4) and Lemma 3, respectively. Finally, we present the simulation results of $\gamma(\lambda_S, \mathcal{A}_P)$ in Theorems 2 and 3. However, due to the space limitation, we neglect the simulation for Theorem 4 and the impact of propagation delay.

According to continuum percolation theory, there exists a *phase transition* in their connectivity property: $\mathcal{C}_\mathcal{O}^0$ has infinite nodes when $\lambda_S > \lambda_c(h(r))$. Our simulation results verified the *phase transition* of secondary network in both BM and RCM model. In Figure 4, we can conclude that $\lambda_c(h(r)) \approx 1.03$ and $\lambda_c(f(r)) \approx 1.63$. The presence of a tail in the result is due to the finiteness of simulation process.[12] Lemma 1 is also verified by the simulation result.

[12]Our simulation parameters are $r_s = 1.40$, $R_p = 2.87$. Since primary users tend to use continuous slots to finish transmission, we set $p_{11} = 0.75$ and $p_{00} = 0.9$. $\mathbb{P}_{PT}(t)$, $\mathbb{P}_{PR}(t)$ and $\mathbb{P}_S(t)$ are distributed in a square $[-20,20] \times [-20,20]$ according to homogeneous *Poisson Point Process*.

When $\lambda_{PT} < \lambda_{PT}^*(\lambda_S)$, the secondary network will be com-supercritical, which implies that $\mathcal{C}_\mathcal{O}^0$ still contains infinite secondary nodes, i.e. the unique infinite connected component $\mathcal{C}_\mathcal{O}$ still exists, in spite of the activities of primary network. However, $\mathcal{C}_\mathcal{O}$ is broken into mutually disconnected finite clusters when $\lambda_{PT} > \lambda_{PT}^*(\lambda_S)$. In Figure 5, simulation result shows that the critical primary density $\lambda_{PT}^*(2.4) \approx 0.03$.

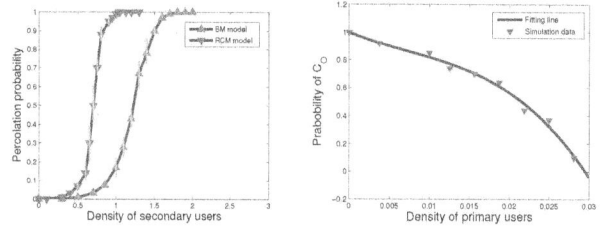

Figure 4: Critical density of secondary user for percolation in BM and RCM model: $\lambda_c(h(r))$, $\lambda_c(f(r))$. We can see that $\lambda_c(h(r)) \approx 1.03$ and $\lambda_c(f(r)) \approx 1.63$. The tail in the figure is due to the finiteness of simulation.

Figure 5: Critical density of primary user for percolation: $\lambda_{PT}^*(\lambda_S)$. It is clear that when $\lambda_{PT} > 0.03$, the infinite connected component $\mathcal{C}_\mathcal{O}$ is broken into mutually disconnected finite clusters almost surly.

According to Lemma 2, $\overline{\psi}$ is a constant that features the number of primary users within the interfered region $\mathcal{I}(l_k)$. In Figure 6, we study $\overline{\psi}$ of two hundred links in one path and the simulation result shows that $\overline{\psi} \approx 3$, well between the lower and upper bound given in Equation (4).

In Lemma 3, we show that κ, describing the scaling behavior of hops number in one path, is a constant independent on λ_S. From Figure 7, we can see $\kappa \approx 0.79$ for both BM and RCM model, which also verifies our Lemma 5. Note that when λ_S is small, κ is relatively large. This is due to that Lemma 3 holds only when the network is supercritical. For the subcritical network (λ_S is less than critical density), κ may not be a constant.

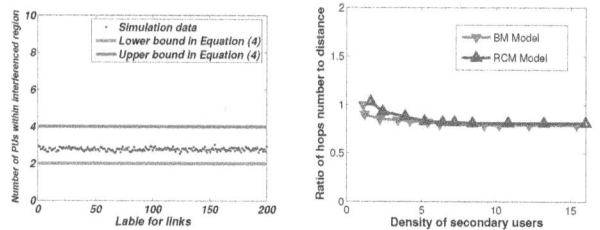

Figure 6: Average number of primary transmitters in interfered region $\mathcal{I}(l_k)$: $\overline{\psi}$. Given $\lambda_{PT} = 0.09375$, it is clear that $\overline{\psi}$ is a constant and $\overline{\psi} \approx 3$.

Figure 7: Ratio of hops number to distance in one path: κ. It is clear that κ is a constant independent on λ_S in supercritical network.

Now we are ready to present the simulation results for $\gamma(\lambda_S, \mathcal{A}_P)$. In Figure 8, it is clear that $\gamma(\lambda_S, \mathcal{A}_P)$ converges to a constant when distance is sufficiently large. The simulation parameters are $\lambda_S = 2.4$, $\lambda_{PT} = 0.035$ and the corresponding $\overline{\psi} \approx 1$. According to Theorem 2, we have $\gamma(\lambda_S, \mathcal{A}_P) \approx 0.316$, which is roughly the same as our simulation result. Figure 9 illustrates the $\gamma(\lambda_S, \mathcal{A}_P)$ for different λ_{PT}. Simulation results also fit our analysis well.

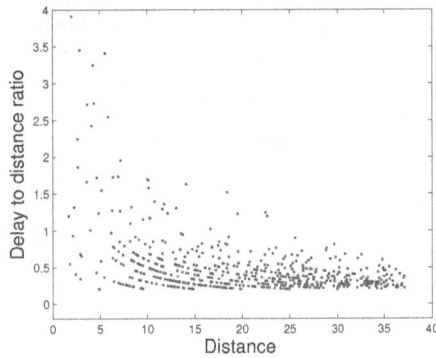

Figure 8: Illustration for Theorem 2. It is true that $\gamma(\lambda_S, \mathcal{A}_P)$ converges to a constant as the distance increases. Given $\lambda_{PT} = 0.035$, $\gamma(\lambda_S, \mathcal{A}_P) \approx 0.3$.

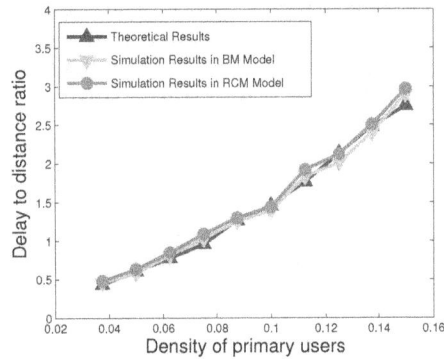

Figure 9: Illustration of $\gamma(\lambda_S, \mathcal{A}_P)$ for different λ_{PT}. It is clear that $\gamma(\lambda_S, \mathcal{A}_P)$ increases with respect to λ_{PT}.

8. CONCLUSION AND DISCUSSION

In this paper, we have studied the transmission delay in large scale ad hoc CR networks by analyzing the ratio of delay to distance $\gamma(\lambda_S, \mathcal{A}_P)$ as the distance goes to infinite. $\gamma(\lambda_S, \mathcal{A}_P)$ in three criteria of CR networks have been investigated, among which we present exact value of $\gamma(\lambda_S, \mathcal{A}_P)$ in both supercritical BM and RCM networks and a lower bound of $\gamma(\lambda_S, \mathcal{A}_P)$ in subcritical B-M network. We proved that for com-supercritical CR networks, $\gamma(\lambda_S, \mathcal{A}_P) = 0$, and $\gamma(\lambda_S, \mathcal{A}_P) = \kappa \cdot \frac{1-\beta_0^{\overline{\psi}}}{\beta_0^{\overline{\psi}}}$ for both the top-supercritical $\mathcal{G}_t(\lambda_S; h(r); \mathcal{A}_P)$ and $\mathcal{G}_t(\lambda_S; f(r); \mathcal{A}_P)$. Moreover, we showed $\gamma(\lambda_S, \mathcal{A}_P) > \frac{1}{\beta_0^{\eta} \cdot E[\mathcal{M}_{h(r)}(\lambda_S, \mathcal{A}_P) + r_s]}$ in the subcritical network $\mathcal{G}_t(\lambda_S; h(r); \mathcal{A}_P)$. Finally, we use simulation results to verify our theoretical analysis.

As to the future work, one of the main challenges is analyzing $\gamma(\lambda_S, \mathcal{A}_P)$ in subcritical RCM network. Because we have to prove that the message can also have a probability to be forwarded in a subcritical RCM network under mobility models, which needs more intensive and in-depth study of continuum percolation theory. Furthermore, we hope to find a more precise description of $\gamma(\lambda_S, \mathcal{A}_P)$ in subcritical BM network in our future works. Finally, network capacity or throughput has not been discussed in this pa-per. Our future works hope to find the capacity and delay tradeoff in CR networks.

Acknowledgment

This paper is supported by National Fundamental Research Grant (No. 2011CB302701); NSF China (No. 60832005); China Ministry of Education New Century Excellent Talent (No. NCET-10-0580); China Ministry of Education Fok Ying Tung Fund (No. 122002); Qualcomm Research Grant; Shanghai Basic Research Key Project (No. 11JC1405100).

9. REFERENCES

[1] R. Meester and R. Roy, "Continuum Percolation," *NewYork: Cambridge University Press*, 1996.

[2] T. Liggett, "An Improved Subadditive Ergodic Theorem," *Annals of Prob.*, vol. 13, pp. 1279-1285, 1985.

[3] J. F. C. Kingman, "Poisson Processes," *Clarendon Press, Oxford*, 1993.

[4] Federal Communications Commission Spectrum Policy Task Force, "Report of the spectrum efficiency working group," Nov. 2002.

[5] P. Gupta and P. R. Kumar, "The capacity of wireless networks," *IEEE Trans. on Information Theory*, vol. 46, no. 2, pp. 388-404, 2000.

[6] P. Gupta and P. R. Kumar, "Critical Power for Asymptotic Connectivity in Wireless Networks," *in Stochastic Analysis, Control, Optimization and Applications:* A Volume in Honor of W.H. Fleming, W. M. McEneany, G. Yin, Q. Zhang, Eds. Boston, MA: Birkhauser, pp. 547-566, 1998.

[7] C. F. Huang, and Y. C. Tseng, "The coverage problem in a wireless sensor network," in *Mobile Networks and Applications*, vol. 10, no. 4, pp. 519-528, 2005.

[8] C. Bettstetter, "On the minimum node degree and connectivity of a wireless multihop network," in *Proceedings of the 3rd ACM international symposium on Mobile ad hoc networking and computing*, pp. 80-91, 2002.

[9] L. Ding, T. Melodia, S. Batalama, J. Matyjas, M. Medley, "Cross-layer Routing and Dynamic Spectrum Allocation in Cognitive Radio Ad Hoc Networks," in *IEEE Transactions on Vehicular Technology*, vol. 59, no. 4, pp. 1969-1979, May 2010

[10] P. Wang, and I. F. Akyildiz, A. M. and Al-Dhelaan, "Dynamic connectivity of cognitive radio ad-hoc networks with time-varying spectral activity," in *Proceedings of Globecom*, 2010

[11] W. Ren, Q. Zhao, and A. Swami, "Connectivity of heterogeneous wireless networks," in *Information Theory, IEEE Transactions on*, vol. 57, no. 7, pp. 4315-4332, 2011.

[12] S. Bodas, S. Shakkottai, L. Ying and R. Srikant, "Scheduling for Small Delay in Multi-rate Multi-channel Wireless Networks," in *Proceedings of IEEE Infocom*, Shanghai, China, 2011

[13] Z. Kong and E. M. Yeh, "On the Latency for Information Dissemination in Mobile Wireless Networks," in *Proc. ACM MobiHoc'08*, Hong Kong SAR, China, May. 2008.

[14] S. Zhao, L. Fu, X. Wang, Q. Zhang, "Fundamental Relationship between Node Density and Delay in Wireless Ad Hoc Networks with Unreliable Links," in *ACM MobiCom 2011*, Sept. 2011.

Enforcing Dynamic Spectrum Access with Spectrum Permits

Lei Yang*, Zengbin Zhang, Ben Y. Zhao, Christopher Kruegel and Haitao Zheng
Department of Computer Science, University of California, Santa Barbara, CA, USA
*Intel Labs, Hillsboro, OR, USA
lei.t.yang@intel.com, {zengbin, ravenben, chris, htzheng}@cs.ucsb.edu

ABSTRACT

Dynamic spectrum access is a maturing technology that allows next generation wireless devices to make highly efficient use of wireless spectrum. Spectrum can be allocated on an on-demand basis for a given geographic location, time duration and frequency range. However, a major obstacle to adoption remains. There are no effective solutions to protect licensed users from *spectrum misuse*, where users transmit without properly licensing spectrum, and in doing so, interfere and disrupt legitimate flows to whom the spectrum is assigned. Given the flexibility of today's cognitive radios, an application can easily transmit on frequencies outside of its allocated range, either accidentally due to misconfiguration, or intentionally to avoid spectrum licensing costs. In this paper, we propose a system to secure dynamic spectrum transmissions, where authorized users embed secure *spectrum permits* into data transmissions, thus enabling patrolling trusted devices to detect devices transmitting without authorization. We focus our attention on the development of spectrum permits, and describe *Gelato*, a spectrum misuse detection system that minimizes both hardware costs and performance overhead on legitimate data transmissions.

Categories and Subject Descriptors

C.2 [**Computer-Communication Networks**]: Network Architecture and Design

General Terms

Design, Experimentation, Security

Keywords

Dynamic spectrum access, spectrum permits, spectrum misuse detection, cognitive radios

1. INTRODUCTION

Dynamic spectrum access, where cognitive radios are used to access unused spectrum ranges on demand, is the clear and widely-accepted solution to the spectrum shortage problem. A significant volume of recent research has built the core algorithms and techniques necessary for the deployment of dynamic spectrum networks in various frequency regions. Particular emphasis has been given to algorithms that maximize spectrum utilization through highly efficient, short-term, local spectrum allocations. While the FCC uses multi-day auctions that define nation-wide usage of large spectrum bands for many years, the vision of the ideal system is the opposite [4]: "local" spectrum owners allocate spectrum segments for a geographic location for short time periods, generally through automated, short-term auctions [36]. By letting realistic short-term demands dictate the size and duration of spectrum allocations, this approach would significantly increase utilization.

A major obstacle remains on the way to adopting current proposals of dynamic spectrum networks. Thus far, policy makers and researchers have not been able to find an effective solution to the problem of *spectrum misuse*. Specifically, we must allow users who have spectrum licenses to transmit, while preventing unauthorized users from transmitting and interfering with authorized transmissions. Without effective protection, users have no assurances their transmissions would operate without interference, and would have no incentive to pay for this type of spectrum access.

There are two general approaches to address this problem. A system can either seek to completely prevent unauthorized access, or it can detect and locate the misbehaving device during the offense. Here, prevention implies building tamper-proof mechanisms into each device to prevent it from operating without a valid spectrum license [2,9,33]. Given the ever-increasing flexibility of software defined radios, it is difficult to envision a completely tamper-proof prevention mechanism. Such changes would likely be costly, and inflexible to varying local conditions or spectrum policies.

We can draw an effective analogy between our problem and the problem of enforcing vehicle speed limits on roads and highways. Building a speed control into each vehicle would be difficult and costly, but a selective detection and punishment scheme in the form of highway patrols can be a very effective deterrent against speeding. Another similar problem is deterring illegal car parking, where authorization to park is dependent on the specific time and geographic location. Instead of a costly and complex per-vehicle solution, parking patrols (*e.g.* meter-maids) provide a much lower-cost and more practical deterrent.

Similarly, we believe a probabilistic system that detected and punished unauthorized transmitters is the approach most likely to succeed in practice. The solution should avoid prohibitively high hardware costs, such as those from densely deployed spectrum sensors [21, 33], and per-device identifiers or signatures, which can be duplicated with sophisticated hardware [33]. Like highway patrols or meter maids, our solution involves a number of trusted mobile devices that patrol transmission areas to detect unauthorized users. Authorized users display time-varying one-time keys that are easily verified but cannot be duplicated. Once an unauthorized transmission is detected, trusted devices can use secure localization techniques [20, 29] to locate the misbehaving devices and stop the unauthorized transmissions.

In this paper, we propose a system for securing dynamic spectrum transmissions through detection of spectrum misuse. When a user purchases a license to transmit on a given spectrum frequency, at a specific time and location, it receives from the spectrum owner a *spectrum permit*, a secure sequence of keys that prove its authorization to transmit in its operating spectrum. Users transmit their spectrum permits on a low-bit rate control channel embedded inside their data transmissions, while trusted *police devices* patrol transmission areas to detect misbehaving devices whose transmissions lack the necessary spectrum permits. We believe a successful spectrum permit system will help pave the way for wide-spread adoption of dynamic spectrum networks.

Spectrum permits have three key requirements: a) they must be flexible enough to specify a license for a given location, time and spectrum range; b) they must be intrinsically linked with the data transmission; and c) they must be readable by other devices without having to decode the data.

A potential solution is to insert permit into packets either as packet header or watermark within the data. But this is insufficient. First, observers must decode the packet to extract the permit, which creates serious privacy and performance issues. But more importantly, this method does not intrinsically link the permit to spectrum usage, and is thus vulnerable to spectrum misuse. An adversary can allow the permit to transmit successfully, and then override the data segment with its own transmission payload.

To meet these requirements, we propose to build a new control channel *physically embedded* inside the data transmission to carry spectrum permits. We leverage "cyclostationary signatures," a PHY layer feature where by intentionally repeating values of a range of frequency subcarriers, we can construct artificial signal peaks in the Spectral Correlation Function (SCF), which is easily detectable by external receivers. Where prior work used this to send a single constant bit between devices [27], we develop techniques to get fine grain control over positions of these signal peaks, effectively converting any arbitrary permit bit streams into features. Thus we construct a signaling channel that is embedded in the data transmission, but can be read by external receivers without decoding the data stream. We show that authorized spectrum users can repeatedly broadcast a secure certificate on this in-band signaling channel, proving their authorized status to any nearby police devices.

Building a robust spectrum permit system out of cyclostationary features faces several challenges. *First*, different hardware transmitters use different feature encoding schemes, and a receiver must understand the transmitter's scheme to decode its signals. *Second*, since each feature is transmitted in a single packet, we must discretize the signal stream into packets identifiable by observers. *Finally*, reading spectrum permits must be robust against frequency offset artifacts at each transmitter, as well as channel impairments and external malicious attacks.

The Gelato Spectrum Permit System. We develop *Gelato*, a robust spectrum permit system that embeds spectrum permits into the data channel in a way that is universally and reliably decodable by observer devices. Specifically, we make the following key contributions:

1. *Permit transmission.* A novel method to encode spectrum permits as features in the data channel, and a bootstrapping feature preamble that allows any observer to decode the permit.

2. *Permit decoding.* Time and frequency-domain tracking mechanisms to accurately detect features, despite loose synchronization and frequency offsets.

3. *Attack detection.* Addressing and detecting potential attacks, by estimating signal strength from features.

4. *Prototyping.* USRP2 GNU radio implementation, with both narrowband and wideband experiments.

2. THE SPECTRUM MISUSE PROBLEM

Despite efforts to improve spectrum allocation techniques, a dynamic spectrum network cannot function correctly without a way to enforce spectrum allocations and detect spectrum misuse. An application can easily transmit on frequencies outside of its allocated range, either accidentally due to bugs or misconfiguration, or intentionally to avoid paying spectrum license costs. Unchecked, interference from these "misuse" events will disrupt legitimate transmissions, ultimately destabilizing the system and preventing adoption.

In this section, we examine approaches to address spectrum misuse, describe our assumptions and goals, and define the threat model it must protect against. A spectrum user is "licensed" if she is authorized to transmit on a given spectrum range, at a particular location and time. We do not specify a transmit power limit in our definition, but assume it is specified by the operating spectrum range, and thus hard-coded into radio hardware. We also assume that any spectrum allocation system maintains spectrum exclusivity, *i.e.* no two users can be authorized to transmit on the same frequency, time and location. More specifically, we focus on spectrum misuse detection at the Access Point or Base Station. For potentially misbehaving client devices inside these networks, we rely on existing client registration mechanisms to identify such behavior [16].

2.1 Candidate Solutions

One general approach is *per-device prevention* using secure hardware or firmware. By deploying a spectrum enforcement module in each radio device, this approach aims to directly prevent each radio from accessing unauthorized spectrum. The enforcement module can be built into the radio hardware [33], or placed in the kernel and user space of the radio software [9]. Given the power and flexibility of software defined radios, however, studies assert that a per-device prevention mechanism would be costly and difficult to perfect [12]. This is particularly true when the allocation of spectrum varies over time, *e.g.* when spectrum is allocated in small time segments. In addition, attackers can modify software and firmware to bypass any enforcement modules. Subsequent advances on both sides can lead to an arms race between designers and attackers.

A second approach is to *detect spectrum misuse* in real time, after which police nodes can use secure localization mechanisms to locate and terminate unauthorized transmissions. Prior work has proposed solutions that rely on dense deployments of spectrum sensors, which would record local RF signal measurements, along with a device identifier for each transmission [21, 33]. The unique per-device identifiers can be used to distinguish licensed users from unauthorized users. These approaches have two significant limitations. First, they require a dense deployment of costly spectrum sensors for any geographic area using this system. This is because radiometric signatures can change over time and space [5], and it is very difficult to maintain and distribute per-device identifiers without a dense sensor deployment. Second, per-device unique identifiers are insecure, as hardware and MAC addresses can be forged, and even intrinsic hardware signatures can be replicated given the right equipment [7].

2.2 The Need for Spectrum Permits

Our goal is to build a system that protects authorized spectrum users by detecting spectrum misuse. In this context, we introduce the concept of *spectrum permits*, secure, verifiable keys demonstrating authorization to use a spectrum. An authorized user receives a secret from the spectrum owner or authority, uses it to generate a sequence of one-time cryptographic keys, and announces them sequentially over time to any nearby observers. Spectrum permits have several advantages over prior solutions. First, permits are simple to read and verify, thereby simplifying and reducing the cost of the detection infrastructure. Second, permits are implemented as one-time, cryptographic keys. As a result, they are tamperproof, and not vulnerable to attacks leveraging sophisticated hardware.

Our work makes several assumptions:

- *Spectrum allocation granularity.* Spectrum assignments are made on three dimensions: frequency, geographic area, and time. Geographic areas are no less than the transmission range of a device operating at max power.

- *Secure communication between users and spectrum owners.* Spectrum owners can securely disseminate secrets to users via a secure communication channel without fear of compromise.

- *Loosely synchronized clocks.* All transmitters should have clocks loosely synchronized with clock servers, *e.g.* NTP.

Based on our assumptions and constraints of prior approaches, we define three key goals for spectrum permits:

- *Universally decodable.* Spectrum permits must support devices operating on a variety of spectrum ranges. To avoid costly hardware requirements for police nodes, spectrum permits should be decodable by any wide-band receiver without decoding data packets.

- *In-band permit transmission.* Spectrum permits can be transmitted on a dedicated control channel. But securely associating a permit with a data transmission is difficult, especially if police nodes cannot decode the data transmission. Instead, our goal is to send spectrum permits *embedded* inside the data channel.

- *Reliability.* Spectrum permits must be transmitted reliably even in the presence of lossy data channels.

2.3 Threat Model

Our goals specify intended properties in the absence of adversaries. We now consider the types of adversaries and attacks that a wireless spectrum permit system is designed to detect. We define "attackers" to include both users who transmit without license either by accident or misconfiguration, and users who do so intentionally to avoid the costs of spectrum licenses, possibly modifying their software defined radios in the process. In either case, we assume attackers' data traffic resemble legitimate transmissions, but can be altered to avoid detection.

To detect attackers, spectrum owners (*e.g.* the FCC) deploys trusted, highly mobile devices (*police nodes*) to monitor a transmission area for spectrum misuse.

Attackers in our model have these properties. First, each attacker has full control of its software defined radio, and can use it to eavesdrop on legitimate transmissions and transmit arbitrary data. Second, they can tune parameters such as transmission power and operating frequency, but are limited by device hardware constraints, *e.g.* finite transmission power. Third, attackers have reasonable resource limitations that prevent them from computationally revealing the secret keys, *i.e.* they cannot break strong cryptography via brute force. Finally, police nodes are mobile devices, do not transmit data, and cannot be found or compromised by attackers.

3. SPECTRUM PERMITS VIA GELATO

We propose Gelato, a *spectrum permit* system for dynamic spectrum networks. The idea is that an authorized user of a spectrum range receives a secure key that allows it to generate valid permits for a fixed time period and a specific location. In Gelato, each user broadcasts its valid spectrum permit once during each time window. Mobile "spectrum police" nodes can scan different spectrum ranges, passively listen to each transmitter's permit, and verify its validity in real time with the help of an online spectrum allocation server.

The Gelato system consists of two key components, a *permit authentication* mechanism that generates and authenticates spectrum permits at the application layer, and a *permit attachment* mechanism at the physical layer that allows each user to broadcast its valid spectrum permit in its physical transmissions, and each police device to reliably detect and decode permits without decoding actual data packets. In the following, we present the permit authentication design and leave the detailed description of permit attachment to Section 4.

3.1 Spectrum Permit Authentication

The spectrum owner runs an online spectrum allocation database on a trusted server. It allocates spectrum in small time blocks of fixed-size T_{int}. Given a geographic location, time and frequency range, if the spectrum is allocated, then the spectrum database returns in real time a secret K_n that represents the tail of a secure, one-way hash chain [19].

Our license verification scheme uses a secure one-way hash chain scheme, similar to authentication mechanisms used for broadcast authentication [24]. When a user U is allocated a spectrum range for n time blocks from t_0 to t_{n-1}, it is given a secret K_0. The user then computes a chain of hash codes by applying a secure one-way hash (*e.g.* SHA-1) recursively n times, producing:

$$K_0 \xrightarrow{SHA-1} K_1 \xrightarrow{SHA-1} K_2 \xrightarrow{SHA-1} \cdots K_{n-1} \xrightarrow{SHA-1} K_n$$

Starting at time t_0, the user U transmits key K_x on the embedded control channel, where x is a counter starting from n-1 that decrements once per time block. That is, the keys are transmitted sequentially in time in reverse order of the one-way hash chain, $K_n, K_{n-1}, \cdots, K_1, K_0$. Since the one-way hash function SHA-1 cannot be reversed, a node can only generate K_i from K_{i-1}. This means that attackers cannot generate valid keys for successive time windows using past key observations.

To verify the authenticity of a transmitter, a police node uses its location, time and spectrum range of the observed transmission to obtain from the database a hash chain tail K_n and a start time t_0. It computes the number of time blocks elapsed since t_0 to get the current index x of the hash chain. Assuming the key sent on the Gelato channel is K_x, the police node applies the SHA-1 hash recursively n-x times to generate the rest of the chain. If the final result matches K_n, it proves the transmitter knows K_0, and is therefore authorized to transmit on this spectrum, location and time.

An authorized user U transmits its key K_x once per time block. Since the key can be copied and retransmitted by any nearby device, an observing police node will only consider the first transmission of K_x as valid. Even if U is not transmitting, an attacker cannot replay a previously used key K_r, because K_r does not match the correct key in the hashchain corresponding to the current time block. A police node can detect a replayed key K_r, because the number of hashes between K_r and K_n does not match the number of time blocks between the current time and t_{n-1}.

Choosing T_{int}. The choice of time block size is a tradeoff between permit efficiency and effectiveness. Because a legal user transmits one key per time block, the overhead of the permit scales inversely with T_{int}. On the other hand, since the time to transmit each permit is less than T_{int}, an attacker can transmit in between permits to evade detection. To make permit verification more reliable within a fixed time, Gelato uses small sized time blocks, e.g. 1 minute. We will also discuss in Section 5.2 ways to detect these attackers.

Using small time blocks can produce a longer hash chain in the verification process. For example, in the worst case, a police node verifying a one day spectrum permit has to perform 1440 hash operations. There are several ways to address this. First, the spectrum hash chain can be refreshed on a small fixed interval, a new hash chain secret sent to the user, and the hash chain tail sent to the allocation database. Alternatively, a police node can cache an already announced key K_{ann}, and verify a new key by terminating the hash once it reaches K_{ann}. We can also speed up verification by either embedding additional information with the chain tail [10], or trading space for verification speed by using hash trees [23].

4. ATTACHING SPECTRUM PERMITS

We now examine the issue of attaching permits to transmissions. One straightforward solution is to transmit permits on an out-of-band control channel. This solution, however, suffers two disadvantages. First, it usually requires an extra radio to transmit on the out-of-band control channel, leading to higher hardware cost and complexity. Second, the out-of-band transmission makes it highly difficult to associate spectrum permits with data transmissions. Upon detecting a permit is being transmitted by a legitimate user, an attacker can transmit comfortably on the radio frequency covered by the permit. In this way, it hides behind the legal transmissions and evades detection. Thus, for a permit to be effective, it must be intrinsically linked to the current data transmission.

In Gelato, our solution is to build on a technique in the wireless physical layer called *Cyclostationary Features*. At a high level, cyclostationary features are created when signals across some sequence of wireless frequency segments are repeated, thus generating an easy to detect energy peak in the signal's spectral correlation function (SCF) map. A transmitter embeds license stream into the data transmission by controlling where it inserts energy peaks into the SCF map. The result is visible to any police device that can sense signals on the transmitter's frequency, without decoding data content on the frequency. And more importantly, the spectrum permit is intrinsically linked to the data transmission, reflecting the actual spectrum usage.

Next, we present the detailed design of Gelato's permit attachment. We first briefly introduce cyclostationary features, and then describe how Gelato devices embed permits to their data transmissions, and how Gelato police decodes the permits.

4.1 Background on Cyclostationary Features

A cyclostationary signal $x(t)$ is a digital signal whose autocorrelation function is periodical in t for any time lag [27]. This property manifests into unique features in the frequency domain – a signal peak at a specific location in $x(t)$'s spectral coherence function (SCF). External devices can detect each feature by capturing the RF signal on the transmitter's frequency and applying a N-point FFT to compute a normalized, discretized version of the SCF, as $S_x(\alpha, k)$. Here α defines the *cyclic frequency* and k defines the *spectral frequency*, both in the unit of frequency subcarriers.

Our design leverages a fact in wireless communications: OFDM is the prevailing scheme for data communication. It is widely adopted

in current and upcoming wideband wireless technologies, such as 802.11a/g/n, LTE, DAB and Bluetooth 4.0. Using OFDM, we can intentionally introduce a cyclostationary feature into a digital signal by organizing its symbols [27]. Each OFDM symbol consists of N frequency subcarriers. We select w contiguous subcarriers indexed from p to $p + w - 1$, and repeat their signals at subcarriers indexed $p + D$ to $p + D + w - 1$. This new arrangement generates a group of w contiguous peaks in the SCF map at locations (α^*, k^*):

$$\alpha^* = D, \quad k^* = p + D/2 + i, \quad i = 1, 2, ..., w \quad (1)$$

Thus using a set of subcarrier repetition parameters (w, D, p), we can produce a distinct cyclostationary feature as a vertical strip of width w, centered at position $(\alpha = D, k = p + W/2 + D/2)$. Figure 1(a) illustrates a sample feature generated using $(12, 64, 64)$. In this paper, we assume all Gelato transmitters use the same w/N. The peak strength s of the vertical strip depends on the received signal to noise ratio (SNR) of the data packet:

$$s = \frac{SNR}{1 + SNR} \quad (2)$$

Finally, each feature needs to be transmitted continuously for a period of time (by a group of OFDM symbols). This is to ensure that the receiver can build a stable characterization of the SCF map, and suppress the impact of frequency-selective multipath fading [26]. Therefore, in Gelato, each data packet carries a single feature to maximize its robustness.

Cyclostationary features can be decoded using standard signal processing techniques without demodulating and decoding data packets. To detect cyclostationary features, each receiver computes the discrete SCF map from raw OFDM symbols and locates feature peaks. The correlation-based feature detection method has been shown to be optimal [11]. It computes the correlation between the SCF map and an ideal peak pattern (a vertical strip of width w), producing a new SCF map. This step eliminates noise in the system, as well as random occurrences of cyclostationary property in the packet data itself. Using the new SCF map, we can easily detect the feature location (α^*, k^*) by detecting peak on the projected cyclic and spectral frequency domain.

While injecting cyclostationary features requires modifying OFDM subcarriers, the decoding process can be made completely transparent to normal data transmissions. Each receiver can first detect and extract the feature, and proceed with data decoding by ignoring all subcarriers that have been identified to carry redundant data as part of the feature. Permit transmissions will not interfere with data packet delivery, because permit decoding is more robust than packet decoding (we confirm this via testbed experiments in Section 7). Of course there is a cost to transmit these control signals – a certain number (w) of subcarriers are no longer able to carry data. Our testbed experiments show that we can achieve reliable feature delivery with per-packet overhead as small as 5% for packets carrying cyclostationary features. As we discussed in the previous section, an authorized user only transmits its spectrum permit once per time block (*e.g. 1 minute*), thus the overall throughput overhead is $\ll 5\%$.

4.2 Displaying Spectrum Permits

The goal of Gelato is allowing each transmitter to display a stream of its spectrum permit bits as cyclostationary features. Thus the permit is intrinsically linked to its data transmission and readable by police devices without decoding data. To do so, Gelato faces two key challenges. First, we need an effective method to convert any arbitrary permit bit streams into features. Where prior works create signal peak to send a single constant bit between devices, we must

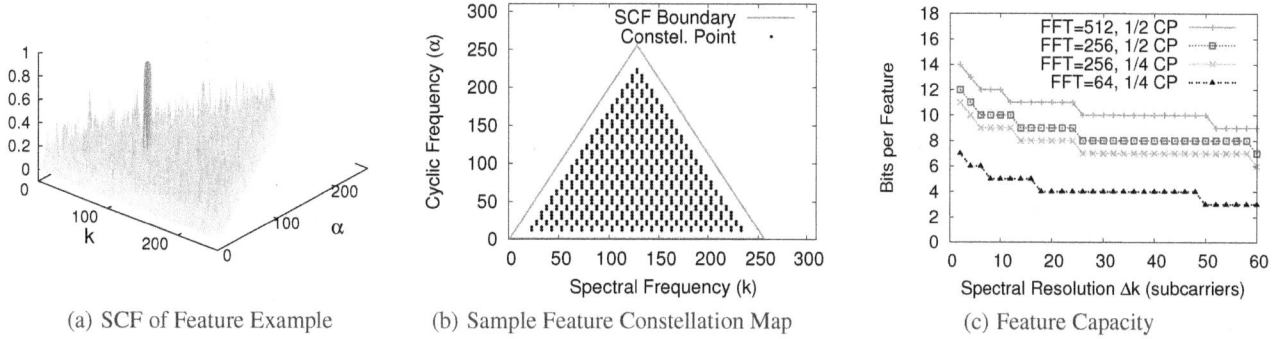

| (a) SCF of Feature Example | (b) Sample Feature Constellation Map | (c) Feature Capacity |

Figure 1: Building and encoding cyclostationary features. (a) A cyclostationary feature at ($\alpha = 64$, $k = 102$). (b) A sample feature constellation map for transmitters with FFT size of 256, CP length of $1/4$ and $\Delta k = 10$, mapping to 9 bits per feature. (c) Feature encoding capacity under different OFDM configurations (FFT size and CP length) and Δk.

develop systematic techniques to get fine grain control over positions of these peaks. Second, the content carried by each feature depends on the underlying OFDM configuration, which can differ across devices. Police devices must first obtain the configuration in order to decode the permit.

Gelato addresses these challenges via two novel solutions. First, it builds a *feature constellation map* to associate each bit pattern with a feature peak location. Second, it introduces a *feature preamble* to "broadcast" the OFDM configuration, therefore bootstrapping the permit decoding process.

A Feature Constellation Map. We encode bit patterns by associating a given bit pattern with a feature at a specific SCF map location. Since this approach is reminiscent of "constellation maps" used in digital signal modulation, *e.g.* QPSK or QAM, we refer to the collection of feature locations as the *feature constellation map*, and each location as the *feature constellation point*. Figure 1(b) illustrates a sample feature constellation map. To decode the feature, the receiving device first locates the feature peak from the SCF map, then computes encoded data as the bit pattern associated with the constellation point closest to the detected location.

The number of bits a feature can carry depends on the total number of distinct constellation points that can be reliably distinguished on an SCF map. This depends on the resolutions in the spectral frequency (k) and cyclic frequency (α) domains, *i.e.* the minimum spacing between adjacent constellation points to make them uniquely separable at the feature detector. We use Δk and $\Delta\alpha$ to represent the two.

Choosing Δk. For two reasons, the spectral frequency k is more sensitive to noise and channel artifacts compared to the cyclic frequency α. First, each cyclostationary feature maps to a single peak at a cyclic frequency α^*, but w consecutive peaks in the frequency domain (see Eq. 1). To decode k, we must accurately identify the center of the peak, which is sensitive to noise and channel artifacts. Second, frequency offsets between transmit and receive devices introduce more variability in the value of k. Therefore, Δk should be large enough to compensate noise and frequency offsets.

Choosing $\Delta\alpha$. The cyclic resolution $\Delta\alpha$ is determined by the transmitter's CP configuration. CPs are used to prefix each symbol with a repetition of its end, eliminating cross-symbol interference and mitigating multipath fading [13]. The use of CPs, however, changes the resolution of the SCF map, thus the detectable α positions. For transmissions with CP length of $1/M$, we can only

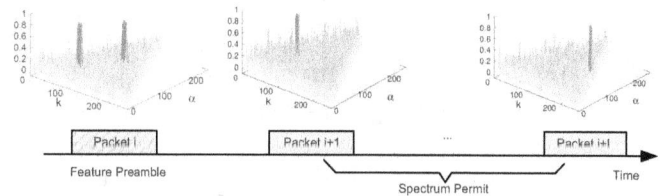

Figure 2: Each Gelato permit consists of its feature preamble and a group of features carrying the permit bit stream. The preamble carries two features, one on each half of the k-axis, carrying information on the FFT size and CP length required to decode the subsequent feature packets.

detect features at $\alpha = i \cdot M$, $i = 1, 2, 3, ...$, mapping to a cyclic resolution of $\Delta\alpha = M$. Typical values of M are 1, 2, 4 and 8 for existing OFDM systems, *e.g.* $M = 4$ in 802.11a/g [17].

For a given combination of Δk and $\Delta\alpha$, we create the optimal feature constellation map that places the maximum number of constellation points on the SCF map while satisfying the minimum spacing defined by (Δk, $\Delta\alpha$). For example, for OFDM configurations with FFT size of 256, cyclic prefix (CP) length of $1/4$, and $\Delta k = 10$, we can encode 616 distinctive features on the SCF map (shown in Figure 1(b)). This means that each feature will carry $n = \lfloor \log_2(616) \rfloor = 9$ bits. Figure 1(c) shows the maximum number of bits each feature can carry under different configurations of FFT size, CP-length, and spectral resolution Δk.

A Feature-bootstrapping Preamble. Embedding bit patterns into data packets is not enough to produce a signaling channel to embed spectrum licenses. We face an additional challenge. Different transmitters can encode their data using very different values of FFT size and CP length, both of which must be known to define a feature constellation map.

Our solution is to introduce a feature preamble carrying the transmitter's FFT size and CP length to bootstrap the receiver. Figure 2 shows an example where each spectrum license message of size M is split across a group of $L + 1$ ($L = M/n$) data packets. We embed inside the first packet of the sequence a feature preamble that "broadcasts" the FFT size and CP length. Each of the next L packets carries a n-bit cyclostationary feature.

The preamble must be decodable by all devices regardless of their OFDM configuration, and easily distinguished from normal spectrum license signal features. We encode the preamble as a set of two features, as shown in the SCF map of Figure 2. First, to make them even more easily distinguishable from normal spectrum license features, we make the width of preamble features twice the normal size. Second, we observe that common OFDM systems use a very limited number of FFT and CP length configurations, which can easily be represented by 4-5 bits. The two features in the preamble represent values for the FFT size and CP length, and the position of each feature is associated with a particular value for that parameter. For example, the feature on the left can represent one of four possible FFT sizes (64, 128, 256, 512), by dividing the left half of the SCF map into four quadrants and assigning a value to each quadrant. Knowing the values associated with each quadrant, the receiver can determine the FFT size by looking at the relative position of the left feature on the SCF map.

The FFT used in this step to build the SCF map is independent of the encoding FFT size, because the feature decoding only depends on its relative position on the SCF map. We can apply the same technique to encode one of four possible CP length values (1, 1/2, 1/4 and 1/8). Decoding the preamble allows the receiver to decode subsequent features at a higher rate with a fine-resolution constellation map.

4.3 Decoding Spectrum Permits

Intuitively, the basic feature decoding method (described in Section 4.1) should be sufficient. However, preliminary efforts to evaluate our system on a GNU radio prototype revealed several additional challenges. Next, we explain these challenges and our mechanisms to address them.

Tracking Permit in Time & Frequency. Our first challenge comes from the lack of time synchronization between the transmitter and the police. Because Gelato's feature embedding works on a per-packet basis, lack of time synchronization between devices means police receivers cannot accurately detect the beginning and end of discrete packets, leading to significant decoding errors.

To address this issue, Gelato police devices detect packet boundaries in time using an edge-detection based technique. This is done by identifying the sudden rise and drop of received signal strength that correspond to the beginning and end of each packet transmission. Specifically, a Gelato police monitors the raw received energy e on a given frequency band and computes its first-order directional derivative $\partial_t(e)$ in time. It detects a rising edge if $\partial_t(e) > \beta$, and starts to compute the SCF map. It detects a dropping edge if $\partial_t(e) < -\beta$ which marks the end of a feature transmission. If the time lag between the rise and the drop is greater than a threshold, it moves to detect and decode the feature using the captured time-averaged SCF map.

Our second challenge comes from the fact that per-device hardware artifacts produce *frequency offsets*, differences between devices' carrier frequencies that introduce large errors in the decoded spectral frequency k. Unaddressed, this would force us to use very large values for Δk, resulting in much lower bit-rates for embedding license permits. We address this challenge by applying an edge-detection technique to detect changes in the frequency domain, effectively removing the majority of frequency offsets [34]. Our testbed experiments show that the proposed tracking method effectively reduces the frequency offset to $< \pm 1$ subcarrier.

Coping with Frequency Selective Fading. Wideband transmissions often experience frequency-selective fading [15] where frequency subcarriers are attenuated differently. Thus we adjust Gelato's correlation-based feature detection to explicitly consider channel fading: we compute a new pattern for each feature location using the correlation of the channel fading pattern at the corresponding subcarriers.

Extracting Interleaved Features. Gelato is designed for dynamic spectrum networks where only one transmitter displays the spectrum license in a specific location and frequency. In rare cases, however, the police node may hear multiple permit transmissions at network boundaries. Since each spectrum license permit spreads over multiple features (thus over multiple packets), to correctly decode spectrum license permits, Gelato police nodes need to differentiate features (and packets) from different transmitters. To address this challenge, our solution separates transmitters by continuously comparing their radiometric features, including frequency offset, signal amplitude and radio transient shape developed by prior works [3, 6]. Note that we only use radiometric features as temporary radio identifiers. This differs from per-device identifier solutions that require dense sensor deployments to record radiometric features for every authenticated device [21, 33].

5. DEFENDING POTENTIAL ATTACKS

In this section, we examine in detail adversarial attacks against the Gelato spectrum permit system, and describe Gelato mechanisms to address and detect each type of attack.

Since the primary goal of our spectrum permit system is to detect spectrum misuse, we *explicitly do not seek to prevent or defend against denial of service attacks*, where an attacker sends unauthorized signals to intentionally disrupt an ongoing legitimate transmission. Such attacks are easy to detect and localize. In addition, while physically locating and punishing attackers are essential steps following attack detection, this paper focuses on spectrum permits and leave those topics as subjects for ongoing work.

5.1 The Copycat Attack

To use spectrum without a permit, attackers can eavesdrop on a legitimate transmission, extract its spectrum permit, and then attempt to use the permit for its own data transmissions. This attack is relatively easy to detect, since each legitimate user only transmits her permit once during each time block. The police node can easily detect an attacker if the same permit is transmitted twice.

Within the allocated geographic area for a given permit, there might be regions where the legitimate transmission signal is weak, and the copycat transmission will go undetected. However, since each spectrum allocation request is for a given usage area, such regions are likely small compared to areas where both transmissions overlap, and the attacker can be detected as police nodes move around the area.

5.2 The Free-rider Attack

This attacker hides behind legitimate users, *i.e.* by sending data packets in parallel without embedding spectrum permits. If the interference from the attacker is moderate, a casual observer would only observe a legitimate permit and a single transmission formed by the union of the legitimate transmission and the free-riding transmission.

Gelato police nodes can detect this attack by comparing the signal strength of the embedded control features to the raw received signal strength to detect the contribution of hidden free-riders. If the raw signal strength is significantly higher than the signal strength observed on the control features, then one or more hidden transmitters are close by. To detect this, Gelato offers a tool that estimates the received signal strength of a transmitter from the peak strength values of its features. Specifically, Gelato estimates the signal strength S^* of a transmitter from its feature strength s,

$$S^* = \left(\frac{1}{\rho/s - 1} \right) \cdot N_0 \qquad (3)$$

where N_0 is the thermal noise power, and $\rho \leq 1$ is a device-dependent parameter, *e.g.* 0.9 for the USRP2 radios that we use to prototype Gelato. If S^* is less than the raw signal strength beyond a threshold, we claim a free-rider is present.

Addressing Frequency Selective Fading. Frequency-selective fading creates an additional challenge in extracting signal strengths from features. Since each feature is carried by a subset of subcarriers, the feature strength s only depends on the received signal strength on these subcarriers, rather than the overall received signal strength S^*.

Gelato addresses this by utilizing the fading profile observed by the police node. Since feature strength s is affected only by the subcarriers used to generate the feature, S^* estimated by (3) only reflects the average signal strength at the corresponding subcarriers. Hence, we can compensate the overall signal strength estimate by a factor η,

$$\eta = \frac{\sum_{i=1}^{N} \psi_i^2 / N}{\sum_{i=1}^{w} \psi_{p+i} \cdot \psi_{p+D+i} / w} \qquad (4)$$

where N is the total number of data subcarriers and ψ_i is the channel response at subcarrier i observed by the police node. We then use the compensated signal strength ηS^* to detect free-riders.

Some wireless technologies, such as LTE and WiMAX, apply transmit power adaptation on a per-subcarrier basis. As a result, transmit power can differ across subcarriers which leads to non-uniform receive signal strength. Similarly, Gelato police nodes compensate by measuring p_i, the receive power level at each subcarrier i, and applying a compensation factor η' according to (4), except by replacing ψ_i with p_i.

Free-riders in Transmission Gaps. If the transmission of a legitimate user U has a gap, *i.e.* it does not use the entire time block, an attacker could transmit for the remainder of the block, since U has given the current key. While this does not interfere with U's transmission, we can still detect the unauthorized transmission by observing changes in transmission properties such as signal strength and thermal noise power. For example, since frequency-selective fading is unique for each transmitter location, Gelato police can detect the attack by monitoring channel fading profiles.

5.3 The Bad-mouth Attack

Another type of intelligent attackers can seek to "bad-mouth" a legitimate user, *i.e.* frame an innocent user to look like she is transmitting illegally. The attack can be performed by "replacing" the victim's features with false ones. Specifically, the attacker occasionally transmits one or more false features at high power in parallel to the legitimate transmissions, which overpower and override the legitimate features. The police node would only observe replacement bits, thus corrupting the legitimate permit.

Gelato police can detect the presence of bad-mouth attacks by comparing the observed raw signal strength and the one estimated from the detected feature. In order to overpower the legitimate feature, the attacker receive power must be no less than that of the legitimate user. Thus if the observed permit is false, and the raw signal strength is occasionally more than twice the average feature-estimated strength, then a bad-mouth attacker is likely to be present.

We note that a legitimate permit can also be corrupted due to channel fading or unexpected interference. These impairments in general prevent a feature from being detected, rather than producing a false feature. Thus police nodes can examine the length of a received permit and separate these scenarios from the above bad-mouth attack.

6. IMPLEMENTATION

We implemented a Gelato prototype on USRP2 GNU Radios. It includes Gelato transmitter and receiver pairs for normal data communication, and police nodes for verifying spectrum permits and detecting attackers. While we chose GNU Radios for their availability, our design can be ported to other platforms [14, 28, 30] for improved frequency bandwidth and processing speed.

Gelato Transmitter & Receiver. Each Gelato transmitter consists of two processing paths: the normal data path and the permit displaying path. To display a permit, we modify the OFDM subcarrier mapping module in the data path to create subcarrier repetition. We implemented pilot tones following the same pilot/data ratio of WiFi. These pilot subcarriers do not follow Gelato's repetition rule, and can degrade the feature strength.

Gelato receivers are like normal data receivers, except that we add a permit detection and removal path. This is because bits from subcarriers carrying repetitive information to display spectrum permits should be removed from the data packet. Therefore, we modify each receiver to add a feature detection module. After locating the feature, the receiver's subcarrier demapping module simply removes the w duplicated subcarriers.

To determine the proper feature width w, we used different w to understand the tradeoff between feature robustness and packet overhead. Overall, we set the feature width $w = 12$ (5% overhead), and $\Delta k = 6$ such that each feature carries 9 bits of permit information. Each Gelato receiver then applies a detection threshold to distinguish real feature peaks from those caused by random factors including noise, interference, and more importantly, *inherent cyclostationary features displayed by data packets themselves*. To support a wide range of SNR values (0–20dB in our experiments), we choose a threshold of 0.4, which can reliably detect real features, but is still well above the "noise floor."

Gelato Police. We implement each Gelato police node as a standalone spectrum permit detector. The police reads OFDM signals from its USRP radio's A/D converter, and applies our proposed mechanisms to track packet boundaries and compensate for frequency offsets. We implement the proposed cyclostationary feature detection module to identify feature peaks and extract bits. The decoded permit bit stream is then validated using the proposed permit authentication process.

Gelato police nodes are much less complex compared to typical OFDM receivers. In addition to not performing packet demodulation/decoding, they require no synchronization in time and frequency. Both are among the most complex blocks in typical OFDM receivers [13]. We show in Section 7 that Gelato police can decode features reliably without any FFT symbol level synchronization.

Frequency Configuration. Given the hardware limitation of USRP2 radios, our implementation currently supports a maximum FFT size of 2048 and a frequency bandwidth of 2MHz. To overcome this narrow-band limitation in our evaluation, we performed extensive wideband channel measurements in both indoor and outdoor environments using the tool provided by [15], and fed the measurement traces to our USRP2 transmitters to emulate wideband transmissions (20MHz) with frequency-selective fading.

7. EXPERIMENTAL EVALUATION

We evaluate Gelato using the aforementioned prototype implemented using USRP2 GNU radios. Since there are no existing

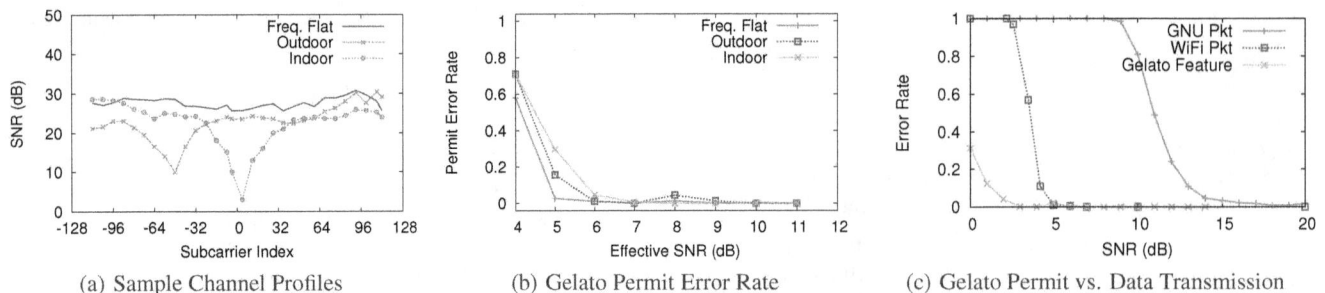

(a) Sample Channel Profiles (b) Gelato Permit Error Rate (c) Gelato Permit vs. Data Transmission

Figure 3: Reliability of Gelato spectrum permits. (a) Sample channel profiles for narrowband (frequency flat), outdoor wideband outdoor (frequency selective) and indoor wideband transmissions (frequency selective). (b) Gelato permits achieve a less than 5% permit error when the effective SNR is greater than 6dB for all three channel environments. (c) Gelato's feature detection is much more reliable than packet decoding.

comparable systems to Gelato, we instead focus on verifying its performance under various network configurations.

Experiment Setup. For each experiment, we build a set of 1600 permits, each 160-bit long. We embed each permit into a set of 18 randomly generated data packets. Each packet contains 32 OFDM symbols and carries a single cyclostationary feature. We also inject random gaps between packets. We focus on two representative indoor/outdoor scenarios in our experiments: complex indoor environments with furniture and walls, and outdoor environments with surrounding buildings, where both experiments are performed on our university campus. To examine the impact of channel fading, we also experiment with static/mobile scenarios: a static scenario where devices were placed statically, and a low-mobility scenario where we walked around the room with the feature receiver at a normal pedestrian speed. For both scenarios, there were random human movements throughout the experiments. Finally, while our prototype supports various transmission configurations on transmit power, FFT size and CP length, we observe in our experiments that these configurations lead to similar conclusions. Thus in the following we only show the results for 256 FFT and 1/4 CP length.

Our evaluation seeks to answer three questions:

1. *Can Gelato permits serve as a reliable method to authenticate spectrum usage in the presence of channel impairments and interference?*

2. *Will Gelato's feature transmissions be more reliable than data packet reception, so that they stay transparent to data transmissions?*

3. *Can Gelato police detect the presence of attackers, using the proposed feature-based signal strength estimation?*

7.1 Reliability of Gelato Spectrum Permits

Permit Error Rate. We first examine Gelato's permit error rate under different wireless transmission profiles. Specifically, we consider narrowband (1MHz frequency band with frequency-flat fading) and wideband channels (20MHz frequency band with frequency-selective fading). Figure 3(a) shows three illustrative examples of these channel profiles: one frequency-flat and two frequency-selective fading channels measured in indoor and outdoor. For the latter two, we observe large deviations across subcarrier SNRs. Figure 3(b) shows the error rate of Gelato permit reception. Since each permit is delivered by multiple features, it can only be successfully retrieved if *all* the features are received correctly. Overall, we see that the error rate reduces to <5% when

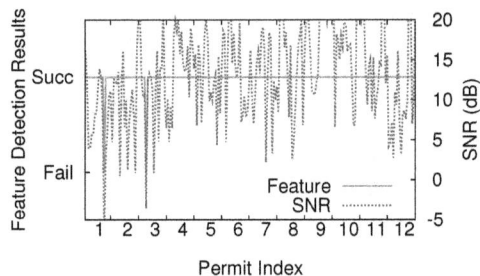

Figure 4: Impact of mobile police nodes. When walking around a large 12m×7m room with a Gelato receiver, we observe very few feature decoding errors caused by deep channel fades.

the effective SNR (ESNR)* exceeds beyond 6dB. For outdoor WiFi access points, this requirement typically maps to 200-300 meters of detectable range from the police node to the transmitting access point [25]. This result implies that Gelato police might need to move around a legitimate user to get a "clearer" view of its permit.

Impact on Data Transmission. A key requirement (and advantage) of Gelato is to guarantee that data transmission will not be affected by the permit display except the expected throughput loss due to subcarrier repetition. To do so, the intended receiver of each data packet needs to detect the cyclostationary feature embedded in the data packet, and uses the corresponding subcarrier repetition pattern to correct the subcarrier demapping, *i.e.* removing the repeated subcarriers. This requires that the feature decoding is at least as robust as the packet decoding at each intended data receiver.

To verify this requirement, in Figure 3(c) we plot the feature decoding error compared to the packet decoding error for the packets containing no features, both implemented using USRP2 radios. For a fair comparison, we ignore feature errors caused by inaccurate packet locking, because it also prevents packet reception. Thus the corresponding feature error rate is better than that in Figure 3(b). Overall, we see that Gelato's feature detection is much more robust than packet decoding. Considering the fact that our implementation of data transmission may not be as sophisticated as that of commodity wireless transceivers, this comparison might not be representative. As a reference, we also compare our feature detection performance with the empirical result obtained from a recent

---

*With frequency-selective fading, the average SNR does not accurately reflect channel quality. Thus we quantify channel condition using the Effective SNR metric [15], which is biased towards weaker subcarrier SNRs that contribute to most of the bit errors.

WiFi study [15]. Again the feature detection outperforms the WiFi packet decoding. These confirm that Gelato permit is transparent to data transmission.

Mobile Police Nodes. To capture the impact of police mobility, we carried the police node and walked around to generate a low-mobility scenario. We used the same configuration as the above static experiments and repeated it 10 times. We found that mobility has very little impact on Gelato. For example, after sending 12 permits (216 features), only two 2 features that suffer very low SNRs were not decoded, leading to 2 corrupted permits (shown in Figure 4). We believe that this can be compensated by adding a low-level of error-correction coding [8] redundancy into each spectrum permit.

7.2 Attack Detection

Next, we examine Gelato's ability to detect adversarial attacks. Since reliable permit transmissions and verification already enable the detection of copycat attacks, we focus on examining free-riders and badmouth attacks.

Accuracy of Feature-based Signal Strength Estimation. Since Gelato detects attacks by comparing the observed signal strength with the feature-estimated signal strength, we first verify the proposed signal strength estimation. To explore the impact of channel noise and interference (from other transmitters or attackers), we activate another transmitter to inject interference to the police node in the presence of the legal transmitter's transmission, and record the SINR observed at the police node. Figure 5 compares the estimated signal strength with the true value. We see that the estimation is quite accurate when the SINR is less than 8dB, but the accuracy drops at larger SINR values. This is due to the non-linear mapping between the SINR and peak strength. At high SINRs, a small deviation in peak strength computation manifests into larger errors in the estimated signal strength. Furthermore, we observe that frequency selective fading has negligible effect on the estimation accuracy *after* using our compensation method.

Attack Configuration. We implement both attacks and vary the attacker power to emulate different physical distance or power profile. Our experiments consists of an attacker, a legitimate transmitter (victim) and a police node. For both attacks, we use the *Relative Attacker Power* $(S_A(dB) - S_V(dB))$ to capture the difference between the received power of the attacker $S_A(dB)$ and that of the victim observed at the police node $S_V(dB)$. Because the legitimate receiver can be at any location within the legitimate transmitter's coverage area, as the police node moves around the network, the relative attacker power it observes also reflects the one observed at the legitimate receivers. The higher the relative attacker power observed at the victim receiver, the higher the performance degradation to the legitimate transmissions.

Detecting Free-riders. Figure 6(a) shows that in indoor settings Gelato can reliably detect almost all (95+%) of free-riders whose signal strength is no more than 6dB weaker than the legitimate user. This means that the attacker needs to transmit at a very low power level to evade the detection, thus producing much less harmful interference to the legitimate user. Detecting weaker attackers is less reliable due to increased errors in feature based signal strength estimation at high SINRs (see Figure 5). The presence of a weak attacker only leads to a small drop in feature peak strength, which could also be caused by random noise and interference. This ambiguity increases false negatives (or miss detections). We also repeat the above experiments in outdoor scenarios and observe a slightly degraded accuracy (80% detection rate). This is because outdoor transmissions suffer higher temporal variations from dynamic surroundings such as vehicles passing by. These variations introduce additional noise to feature peaks, degrading the detection accuracy.

For both scenarios the rate of false positives remains insensitive to attacker power settings. This is because false positives are mainly caused by the use of pilot tones which degrades feature peak strength and leads to false alarms. The impact depends on pilot locations rather than attacker power, and thus remains constant throughout the experiments.

Detecting Bad-Mouthers. To overwrite the victim's feature, a bad-mouther must transmit false features at a sufficiently high power. Figure 6(b) shows the performance of detecting bad-mouth attacker as a function of the attacker's relative power level for indoor scenarios. We see that Gelato's attack detection is highly effective – it forces the attacker to transmit at a significantly higher power (6+dB over the victim) in order to evade detection. These high-power attacks, however, are more visible and can be easily detected by checking signal strength and data transmission consistency over space and time, such as those proposed by [32]. Finally, we observe similar trends on false negatives and false positives like those of the free-rider attacks.

8. RELATED WORK

Spectrum Authentication and Misuse Detection. Existing work can be divided into two categories: *per-device prevention* and *external monitoring & detection*. Proposals in the first category apply on-device enforcement to prevent devices from operating without a valid spectrum license [2,9,33]. The second category includes diverse solutions designed for different network contexts. In the context of opportunistic spectrum access that contains primary and secondary users, prior works can authenticate each primary user using its unique link transmission characteristics created via a "helper" node [22], detect extra (illegal) transmitters by examining received signal strength [21], or apply extensive signal measurements to locate each transmitter and comparing their locations with those of legitimate users to identify violators [5]. These solutions require dense and costly deployments of monitoring sensors and helpers, and often assume ideal propagation models. More importantly, they place the burden of misuse detection completely on the detection infrastructure, making it costly and highly complex to perfect. Gelato takes a different direction - by forcing legitimate users to display their spectrum permits, Gelato shifts the responsibility to the users, significantly reducing the complexity and cost of the detection infrastructure.

Gelato also targets a different context, where wireless devices receive spectrum allocations on a short-term basis. In this context, the most relevant work is [1], where authenticated spectrum users are assigned unique slotted transmission patterns that serve as their authentication identities. This approach however, requires devices to share spectrum in the time domain, and also requires precise time synchronization.

Signal Embedding. Research efforts in this area have developed strategies to embed "side" information either directly into raw data bits (*i.e.* digital watermarking), or into physical-layer signals [18, 31,35]. These solutions all require demodulation/decoding of the original data transmission, which is infeasible in our scenario.

Gelato is motivated by prior work on cyclostationary features [27], but applies the concept in the context of displaying spectrum permits within transmissions. Unlike prior work, Gelato proposes a novel feature constellation map that allows features to carry arbitrary control information, and a robust detection framework to decode features in the presence of transmission artifacts and attacks.

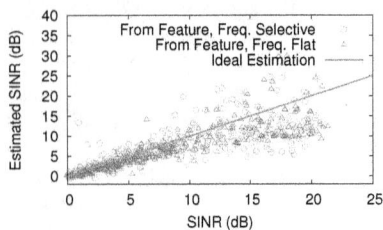

Figure 5: Estimating the signal strength of a legitimate user from its feature strength, for both indoor and outdoor environments.

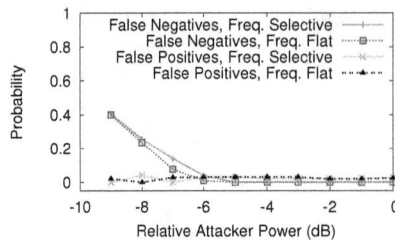

(a) Detecting Free-riders

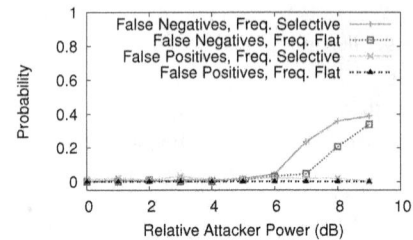

(b) Detecting Badmouth Attackers

Figure 6: Performance of Gelato's attacker detection in indoor environments.

9. CONCLUSION

We present Gelato, an initial step towards a robust spectrum permit system for authenticating spectrum usage and detecting misuse. Gelato devices transmit spectrum permits as cyclostationary features embedded inside their data transmissions, while trusted police devices patrol transmission areas to detect misbehaving devices. Gelato permits are reliable and "universally" decodable without requiring packet decoding. Detailed testbed experiments show that Gelato is a feasible, practical and cost-effective method for enforcing spectrum allocation.

Acknowledgments

We thank our shepherd Srikanth Kandula and the anonymous reviewers for their helpful suggestions. This work is supported in part by NSF Grants CNS-0916307 and CNS-0905667. Any opinions, findings, and conclusions or recommendations expressed in this material are those of the authors and do not necessarily reflect the views of the National Science Foundation.

10. REFERENCES

[1] ATIA, G., SAHAI, A., AND SALIGRAMA, V. Spectrum enforcement and liability assignment in cognitive radio systems. In *Proc. of IEEE DySPAN* (2008).

[2] BRIK, V., ET AL. Towards an architecture for efficient spectrum slicing. In *Proc. of HotMobile* (2007).

[3] BRIK, V., ET AL. Wireless device identification with radiometric signatures. In *Proc. of MobiCom* (2008).

[4] FCC. Connecting America: The National Broadband Plan, March 16, 2010. http://www.broadband.gov/plan/.

[5] CHEN, R., PARK, J.-M., AND REED, J. Defense against primary user emulation attacks in cognitive radio networks. *IEEE JSAC* (Jan. 2008), 25–37.

[6] DANEV, B., AND CAPKUN, S. Transient-based identification of wireless sensor nodes. In *Proc. of IPSN* (2009).

[7] DANEV, B., ET AL. Attacks on physical-layer identification. In *Proc. of WiSec* (2010).

[8] DAVEY, M., AND MACKAY, D. Reliable communication over channels with insertions, deletions, and substitutions. *IEEE Trans. on Information Theory 47*, 2 (Feb. 2001), 687–698.

[9] DENKER, G., ET AL. A policy engine for spectrum sharing. In *Proc. of IEEE DySPAN* (2007).

[10] FISCHLIN, M. Fast verification of hash chains. In *Proc. of CT-RSA* (2004).

[11] GARDNER, W. A. Signal interception: a unifying theoretical framework for feature detection. *IEEE Trans. on Commun. 36*, 8 (1988), 897–906.

[12] GIACOMONI, J., AND SICKER, D. Difficulties in providing certification and assurance for software defined radios. In *Proc. of IEEE DySPAN* (2005).

[13] GOLDSMITH, A. *Wireless Communications*. Cambridge University Press, New York, NY, USA, 2005.

[14] GUMMADI, R., NG, M., FLEMING, K., AND BALAKRISHNAN, H. Airblue: A system for cross-layer wireless protocol development and experimentation. In *MIT Tech. Report* (2008).

[15] HALPERIN, D., ET AL. Predictable 802.11 packet delivery from wireless channel measurements. In *Proc. of SIGCOMM* (2010).

[16] HAMID, M., ISLAM, M., AND HONG, C. S. Misbehavior detection in wireless mesh networks. In *Proc. of ICACT* (2008).

[17] IEEE 802.11. http://www.ieee802.org/11/.

[18] KLEIDER, J., ET AL. Radio frequency watermarking for OFDM wireless networks. In *Proc. of IEEE ICASSP* (2004).

[19] LAMPORT, L. Password authentication with insecure communication. *Comm. of the ACM 24*, 11 (Nov 1981).

[20] LAZOS, L., AND POOVENDRAN, R. Serloc: secure range-independent localization for wireless sensor networks. In *Proc. of WiSe* (2004).

[21] LIU, S., CHEN, Y., TRAPPE, W., AND GREENSTEIN, L. ALDO: An anomaly detection framework for dynamic spectrum access networks. In *Proc. of INFOCOM* (2009).

[22] LIU, Y., NING, P., AND DAI, H. Authenticating primary users' signals in cognitive radio networks via integrated cryptographic and wireless link signatures. In *Proc. of IEEE S&P* (2010).

[23] MERKLE, R. A digital signature based on a conventional encryption function. In *Proc. of Crypto* (1987).

[24] PERRIG, A., ET AL. Efficient and secure source authentication for multicast. In *Proc. of NDSS* (2001).

[25] ROBINSON, J., ET AL. Assessment of urban-scale wireless networks with a small number of measurements. In *Proc. of MobiCom* (2008).

[26] SUTTON, P., LOTZE, J., NOLAN, K., AND DOYLE, L. Cyclostationary signature detection in multipath rayleigh fading environments. In *Proc. of Crowncom* (2007).

[27] SUTTON, P., NOLAN, K., AND DOYLE, L. Cyclostationary signatures in practical cognitive radio applications. *IEEE JSAC* (Jan. 2008), 13–24.

[28] TAN, K., ET AL. SORA: High performance software radio using general purpose multicore processors. In *Proc. of NSDI* (2009).

[29] ČAPKUN, S., ET AL. Secure location verification with hidden and mobile base stations. *IEEE Trans. on Mobile Computing* (April 2008), 470–483.

[30] Wireless open-access research platform, http://warp.rice.edu.

[31] WU, K., ET AL. Free Side Channel: Bits over Interference. In *Proc. of MobiCom* (2010).

[32] XU, W., ET AL. The feasibility of launching and detecting jamming attacks in wireless networks. In *Proc. of MobiHoc* (2005).

[33] XU, W., KAMAT, P., AND TRAPPE, W. TRIESTE: A trusted radio infrastructure for enforcing spectrum etiquettes. In *Proc. of SDR workshop* (2006).

[34] YANG, L., HOU, W., CAO, L., ZHAO, B. Y., AND ZHENG, H. Supporting demanding wireless applications with frequency-agile radios. In *Proc. of NSDI* (2010).

[35] YU, P., BARAS, J., AND SADLER, B. Physical-layer authentication. *IEEE Trans. on Info. Forensics and Security 3*, 1 (2008), 38–51.

[36] ZHOU, X., GANDHI, S., SURI, S., AND ZHENG, H. eBay in the sky: strategy-proof wireless spectrum auctions. In *Proc. of MobiCom* (2008).

Spatial Spectrum Access Game: Nash Equilibria and Distributed Learning

Xu Chen
Department of Information Engineering
The Chinese University of Hong Kong
Shatin, Hong Kong
cx008@ie.cuhk.edu.hk

Jianwei Huang
Department of Information Engineering
The Chinese University of Hong Kong
Shatin, Hong Kong
jwhuang@ie.cuhk.edu.hk

ABSTRACT

A key feature of wireless communications is the spatial reuse. However, the spatial aspect is not yet well understood for the purpose of designing efficient spectrum sharing mechanisms. In this paper, we propose a framework of spatial spectrum access games on directed interference graphs, which can model quite general interference relationship with spatial reuse in wireless networks. We show that a pure strategy equilibrium exists for the two classes of games: (1) any spatial spectrum access games on directed acyclic graphs, and (2) any games satisfying the congestion property on directed trees and directed forests. Under mild technical conditions, the spatial spectrum access games with random backoff and Aloha channel contention mechanisms on undirected graphs also have a pure Nash equilibrium. We then propose a distributed learning algorithm, which only utilizes users' local observations to adaptively adjust the spectrum access strategies. We show that the distributed learning algorithm can converge to an approximate mixed-strategy Nash equilibrium for any spatial spectrum access games. Numerical results demonstrate that the distributed learning algorithm achieves up to 100% performance improvement over a random access algorithm.

Categories and Subject Descriptors

C.2.1 [**Network Architecture and Design**]: Wireless communication

Keywords

Cognitive Radio, Distributed Spectrum Sharing, Nash Equilibrium, Distributed Learning

1. INTRODUCTION

Cognitive radio is envisioned as a promising technique to alleviate the problem of spectrum under-utilization [1]. It enables unlicensed wireless users (secondary users) to opportunistically access the licensed channels owned by legacy spectrum holders (primary users), and thus can significantly improve the spectrum efficiency [1].

A key challenge of the cognitive radio technology is how to resolve the resource competition by selfish secondary users in a decentralized fashion. If multiple secondary users transmit over the same channel simultaneously, severe interferences or collisions might occur and the data rates of all users may get reduced. Therefore, it is necessary to design efficient spectrum sharing mechanism for cognitive radio networks.

The competitions among secondary users for common spectrum have often been studied as a noncooperative game theory (e.g., [6, 9, 18, 19, 23]). Nie and Comniciu in [18] designed a self-enforcing distributed spectrum access mechanism based on potential games. Niyato and Hossain in [19] studied a price-based spectrum access mechanism for competitive secondary users. Chen and Huang in [6] investigated stable spectrum sharing mechanism design based on evolutionary game theory. Fĺelegyhĺczi et al. in [9] proposed a two-tier game framework for medium access control (MAC) mechanism design.

When not knowing spectrum information such as channel availability, secondary users need to learn the network environment and adapt the spectrum access decisions accordingly. Han et al. in [11] used no-regret learning to solve this problem, assuming that the users' channel selections are common information. When users' channel selections are not observable, authors in [2, 13] designed multi-agent multi-armed bandit learning algorithms to minimize the expected performance loss of distributed spectrum access.

A common assumption of the above results is that secondary users are close-by and interfere with each other when they transmit on the same channel simultaneously. However, a unique feature of wireless communication is spatial reuse. If users who transmit simultaneously are located sufficiently far away, then simultaneous transmissions over the same channel may not cause any performance degradation to any user (see Figure 1 for an illustration). Such spatial effect on spectrum sharing is less understood than many other aspects in existing literature [24].

Recently, Tekin et al. in [22] and Southwell et al. in [21] proposed a novel spatial congestion game framework to take spatial relationship into account. The key idea is to extend the classical congestion game upon an *undirected* graph, by assuming that the interferences among the players are symmetric and a player's throughput depends on the number of players in its neighborhood that choose the same resource. As illustrated in Figure 1, however, the interference relationship among the secondary users can be asymmetric due to

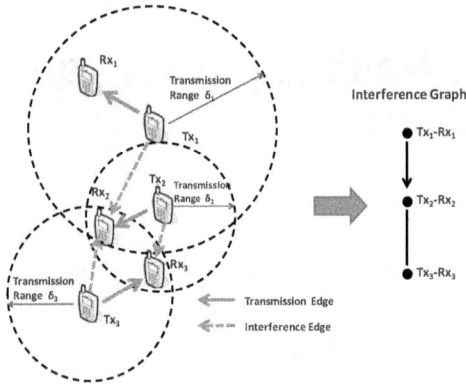

Figure 1: Illustration of distributed spectrum access with spatial reuse under the protocol interference model. Each user n is represented by a transmitter Tx_n and receiver Rx_n pair. Users 2 and 3 can not generate interference to user 1, since user 1's receiver Rx_1 is far from user 2 and 3's transmitters. On the other hand, user 1 can generate interference to user 2, since user 2's receiver Rx_2 is within the transmission range of user 1's transmitter Tx_1. Similarly, user 2 and user 3 can generate interferences to each other.

the heterogeneous transmission powers and locations of the users. We hence propose a more general framework of spatial spectrum access game on *directed* interference graphs, which take users' heterogeneous resource competition capabilities and asymmetric interference relationship into account. The congestion game on directed graphs has also been studied in [4], with the assumption that players have linear and homogeneous payoff functions. The game model in this paper is more generic and allows linear/nonlinear player-specific payoff functions. Moreover, we design a distributed algorithm for achieving the equilibria of the game. The main results and contributions of this paper are as follows:

- *General game formulation*: We formulate the distributed spectrum access problem as a spatial spectrum access game on directed interference graphs, with user-specific channel data rates and channel contention capabilities.

- *Existence of Nash equilibria*: We show by counterexamples that a general spatial spectrum access game may not have a pure Nash equilibrium. We then show that a pure strategy equilibrium exists for the following two classes of games: (1) any spatial spectrum access games on directed acyclic graphs, and (2) any games satisfying the congestion property on directed trees and directed forests. We also show that under mild conditions the spatial spectrum access games with random backoff and Aloha channel contention mechanisms on undirected graphs are potential games and have pure Nash equilibria.

- *Distributed learning for achieving approximate Nash equilibrium*: We propose a distributed learning algorithm that can converge to an approximate mixed Nash equilibrium for any spatial spectrum access games by utilizing users' local observations only. Numerical re-

sults demonstrate that the distributed learning algorithm achieves up-to 100% performance improvement over the random access algorithm.

The rest of the paper is organized as follows. We introduce the system model and the spatial spectrum access game in Sections 2 and 3, respectively. We investigate the existence of Nash equilibria in Section 4. Then we present the distributed learning algorithm in Section 5. We illustrate the performance of the proposed algorithm through numerical results in Section 6, and finally conclude in Section 7.

2. SYSTEM MODEL

We consider a cognitive radio network with a set $\mathcal{M} = \{1, 2, ..., M\}$ of independent and stochastically heterogeneous primary channels. A set $\mathcal{N} = \{1, 2, ..., N\}$ of secondary users try to access these channels distributively when the channels are not occupied by primary (licensed) transmissions. Here we assume that each secondary user is a dedicated transmitter-receiver pair.

To take users' spatial relationship into account, we denote $\boldsymbol{d}_n = (d_{Tx_n}, d_{Rx_n})$ as the **location vector** of secondary user n, where d_{Tx_n} and d_{Rx_n} denote the location of the transmitter and the receiver, respectively. Each secondary user n has a **transmission range** δ_n. Then given the location vectors of all secondary users, we can obtain the **interference graph** $G = \{\mathcal{N}, \mathcal{E}\}$ to describe the interference relationship among the users (see Figure 1 for an example). Here the vertex set $\mathcal{N}$ is the same as the secondary user set. The edge set is defined as $\mathcal{E} = \{(i, j) : ||d_{Tx_i}, d_{Rx_j}|| \leq \delta_i, \forall i, j \neq i \in \mathcal{N}\}$, where $||d_{Tx_i}, d_{Rx_j}||$ is the distance between the transmitter of user i and the receiver of user j. As illustrated in Figure 1, an interference edge can be directed or undirected. If an interference edge is directed from secondary user i to user j, then user j's data transmission will be affected by user i's transmission on the same channel, but user i will not be affected by user j. If the interference edge is undirected[1] between user i and user j, then the two users can affect each other. Note that a generic directed interference graph can consist of a mixture of directed and undirected edges. In the sequel, we call an interference graph undirected, if and only if all the edges of the graph are undirected. We also denote the set of users that can cause interference to user n as $\mathcal{N}_n = \{i : (i, n) \in \mathcal{E}, i \in \mathcal{N}\}$.

Based on the interference model above, we describe the cognitive radio network with a slotted transmission structure as follows:

- *Channel State*: the channel state for a channel m during time slot t is

$$S_m(t) = \begin{cases} 0, & \text{if channel } m \text{ is occupied} \\ & \text{by primary transmissions,} \\ 1, & \text{if channel } m \text{ is idle.} \end{cases}$$

- *Channel State Transition*: for a channel m, the channel state $S_m(t)$ is a random variable with a probability density function as ψ_m. In the following, we denote the channel idle probability θ_m as the mean of $S_m(t)$, i.e., $\theta_m = E_{\psi_m}[S_m(t)]$.

---

[1] Here the edge is actually bi-directed. We follow the conventions in [22] and [21] and ignore the directions on the edge.

- *User Specific Channel Throughput*: for each secondary user n, its realized data rate $b_m^n(t)$ on an idle channel m in each time slot evolves according to a random process with a mean B_m^n, due to users' heterogeneous transmission technologies and the local environmental effects such as fading.

- *Time Slot Structure*: each secondary user n executes the following stages synchronously during each time slot:

  - *Channel Sensing*: sense one of the channels based on the channel selection decision made at the end of previous time slot.

  - *Channel Contention*: Let a_n be the channel selected by user n, and $\boldsymbol{a} = (a_1, ..., a_N)$ be the channel selection profile of all users. The probability that user n can grab the chosen idle channel a_n during a time slot is $g_n(\mathcal{N}_n^{a_n}(\boldsymbol{a})) \in (0, 1)$, which depends on the subset of user n's interfering users that choose the same channel $\mathcal{N}_n^{a_n}(\boldsymbol{a}) \triangleq \{i \in \mathcal{N}_n : a_i = a_n\}$. Here are two examples:
    1) *Random backoff mechanism*: the contention stage of a time slot is divided into $\lambda_{\max}$ mini-slots (see Figure 2). Each contending user n first counts down according to a randomly and uniformly generated integer backoff time counter (number of mini-slots) λ_n between 1 and $\lambda_{\max}$. If there is no active transmissions till the countdown timer expires, the user monitors the channel and transmits RTS/CTS messages on that channel. If multiple users choose the same backoff counter, a collision will occur and no users can grab the channel successfully. Once successfully gets the channel, the user starts to transmit its data packet. In this case, we have

$$g_n(\mathcal{N}_n^{a_n}(\boldsymbol{a})) = Pr\{\lambda_n < \min_{i \in \mathcal{N}_n : a_i = a_n} \{\lambda_i\}\}$$

$$= \sum_{\lambda=1}^{\lambda_{\max}} Pr\{\lambda_n = \lambda\}$$
$$\times Pr\{\lambda_n < \min_{i \in \mathcal{N}_n : a_i = a_n} \{\lambda_i\} | \lambda_n = \lambda\}$$
$$= \sum_{\lambda=1}^{\lambda_{\max}} \frac{1}{\lambda_{\max}} \left(\frac{\lambda_{\max} - \lambda}{\lambda_{\max}} \right)^{K_{a_n}^n(\boldsymbol{a})}, \quad (1)$$

where $K_{a_n}^n(\boldsymbol{a}) = |\mathcal{N}_n^{a_n}(\boldsymbol{a})| = \sum_{i \in \mathcal{N}_n} I_{\{a_i = a_n\}}$ denotes the number of user n's interfering users choosing the same channel as user n.
    2) *Aloha mechanism*: user n contends for an idle channel with a probability $p_n \in (0, 1)$ in a time slot. If multiple interfering users contend for the same channel, a collision occurs and no user can grab the channel for data transmission. In this case, we have

$$g_n(\mathcal{N}_n^{a_n}(\boldsymbol{a})) = p_n \prod_{i \in \mathcal{N}_n^{a_n}(\boldsymbol{a})} (1 - p_i). \quad (2)$$

Note that for the random backoff mechanism, the channel grabbing probability $g_n(\mathcal{N}_n^{a_n}(\boldsymbol{a}))$ is *user homogeneous* since it only depends on the number of contending users $K_{a_n}^n(\boldsymbol{a})$. For the Aloha mechanism, the channel grabbing probability $g_n(\mathcal{N}_n^{a_n}(\boldsymbol{a}))$

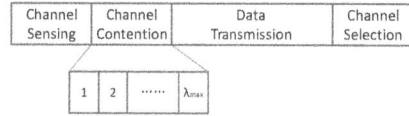

Figure 2: Time slot structure with random backoff mechanism

is *user heterogeneous* since it depends on who (instead of how many users) contend the channel.

  - *Data Transmission*: transmit data packets if the user successfully grabs the channel.

  - *Channel Selection*: choose a channel to access during next time slot according to the distributed learning algorithm in Section 5.

Under a fixed channel selection profile $\boldsymbol{a}$, the long-run expected throughput of a secondary user n choosing channel a_n can be computed as

$$U_n(\boldsymbol{a}) = \theta_{a_n} B_{a_n}^n g_n(\mathcal{N}_n^{a_n}(\boldsymbol{a})). \quad (3)$$

Since our analysis is from the secondary users' perspective, we will use the terms "secondary user" and "user" interchangeably. Due to page limit, the detailed proofs are given in our technical report [5].

3. SPATIAL SPECTRUM ACCESS GAME

We now consider the problem that each user tries to maximize its own throughput by choosing a proper channel distributively. Let $a_{-n} = \{a_1, ..., a_{n-1}, a_{n+1}, ..., a_N\}$ be the channels chosen by all other users except user n. Given other users' channel selections a_{-n}, the problem faced by a user n is

$$\max_{a_n \in \mathcal{M}} U(a_n, a_{-n}), \forall n \in \mathcal{N}. \quad (4)$$

The distributed nature of the channel selection problem naturally leads to a formulation based on the game theory, such that users can self organize into a mutually acceptable channel selection (**pure Nash equilibrium**) $\boldsymbol{a}^* = (a_1^*, a_2^*, ..., a_N^*)$ with

$$a_n^* = \arg \max_{a_n \in \mathcal{M}} U(a_n, a_{-n}^*), \forall n \in \mathcal{N}. \quad (5)$$

We thus formulate the distributed channel selection problem on an interference graph G as a **spatial spectrum access game** $\Gamma = (\mathcal{N}, \mathcal{M}, G, \{U_n\}_{n \in \mathcal{N}})$, where $\mathcal{N}$ is the set of players, $\mathcal{M}$ is the set of strategies, G describes the interference relationship among the players, and U_n is the payoff function of player n.

It is known that not every finite strategic game possesses a pure Nash equilibrium [17]. We then introduce a more general concept of mixed Nash equilibrium. Let $\boldsymbol{\sigma}_n \triangleq (\sigma_1^n, ..., \sigma_M^n)$ denote the mixed strategy of user n, where $0 \le \sigma_m^n \le 1$ is the probability of user n choosing channel m, and $\sum_{m=1}^M \sigma_m^n = 1$. For simplicity, we use the same payoff notation $U_n(\boldsymbol{\sigma}_1, ..., \boldsymbol{\sigma}_N)$ to denote the expected throughput of user n under the mixed strategy profile $(\boldsymbol{\sigma}_1, ..., \boldsymbol{\sigma}_N)$, and it can be computed as

$$U_n(\boldsymbol{\sigma}_1, ..., \boldsymbol{\sigma}_N) = \sum_{a_1=1}^M \sigma_{a_1}^1 ... \sum_{a_N=1}^M \sigma_{a_N}^N U_n(a_1, ..., a_N). \quad (6)$$

Similarly to the pure Nash equilibrium, the mixed Nash equilibrium is defined as:

DEFINITION 1 (**Mixed Nash Equilibrium** [17]). *The mixed strategy profile* $\boldsymbol{\sigma}^* = (\boldsymbol{\sigma}_1^*, ..., \boldsymbol{\sigma}_N^*)$ *is a mixed Nash equilibrium, if for every user* $n \in \mathcal{N}$, *we have*

$$U_n(\boldsymbol{\sigma}_n^*, \boldsymbol{\sigma}_{-n}^*) \geq U_n(\boldsymbol{\sigma}_n, \boldsymbol{\sigma}_{-n}^*), \forall \boldsymbol{\sigma}_n \neq \boldsymbol{\sigma}_n^*,$$

where $\boldsymbol{\sigma}_{-n}^*$ *denote the mixed strategy choices of all other users except user* n.

Note that the pure Nash equilibrium is a special case of the mixed Nash equilibrium, wherein every user chooses a single channel with probability one. One critical issue in game theory is the existence of both mixed and pure Nash equilibria, which motivates the study in the following Section 4.

4. EXISTENCE OF NASH EQUILIBRIA

In this part, we study the existence of Nash equilibria in a spatial spectrum access game. Since a spatial spectrum access game is a finite strategic game (i.e., with finite number of players and finite number of channels), we know that it always admits a mixed Nash equilibrium according to [17].

On the other hand, not every finite strategic game possesses a pure Nash equilibrium [17]. A pure Nash equilibrium is much preferable than a general mixed strategy Nash equilibrium, as in a pure strategy equilibrium users can achieve mutually acceptable channel selections without randomly picking and switching channels all the time. This motivates us to further investigate the existence of pure Nash Equilibria of the spatial spectrum access games.

4.1 Existence of Pure Nash Equilibria on Directed Interference Graphs

We first study the existence of pure Nash Equilibria on directed interference graphs.

First of all, we can construct a game which does not have a pure Nash equilibrium.

THEOREM 1. *There exists a spatial spectrum access game on a directed interference graph not admitting any pure Nash equilibrium.*

Figure 3 shows such an example. It is easy to verify that for all 8 possible channel selection profiles, there always exists one user (out of these three users) having an incentive to change its channel selection unilaterally to improve its throughput.

We then focus on identifying the conditions under which the game admits a pure Nash equilibrium. To proceed, we first introduce the following lemma.

LEMMA 1. *Consider any spatial spectrum access game on a given directed interference graph G that has a pure Nash equilibrium. Then we can construct a new spatial spectrum access game by adding a new player, who can not generate interference to any player in the original game and may receive interference from one or multiple players in the original game. The new game also has a pure Nash equilibrium.*

We know that any directed acyclic graph (i.e., a directed graph contains no directed cycles) can be given a topological sort (i.e., an ordering of the nodes), such that if node $i < j$ then there are no edges directed from the node j to node i in the ordering [3]. From Lemma 1, we know that

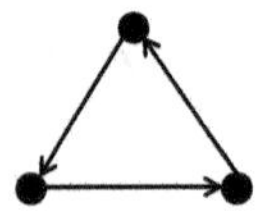

Figure 3: An example of spatial spectrum access game without pure Nash equilibria. There are two channels available and the throughput of a user n is given as $U_n(\boldsymbol{a}) = p \prod_{i \in \mathcal{N}_n^{a_n}(\boldsymbol{a})}(1 - p)$. If all three players (nodes) choose channel 1, then each player has the incentive of choosing channel 2 to improve its throughput assuming that the other two players do not change their channel choices. We can show that such derivation will happen for all 8 possible strategy profiles $\boldsymbol{a} = (a_1, a_2, a_3)$, where $a_i \in \{1, 2\}$ for $i \in \{1, 2, 3\}$.

COROLLARY 1. *Any spatial spectrum access game on a directed acyclic graph has a pure Nash equilibrium.*

To obtain more insightful results, we next impose the following property on the spatial spectrum access games:

DEFINITION 2 (**Congestion Property**). *User n's channel grabbing probability $g_n(\mathcal{N}_n^{a_n}(\boldsymbol{a}))$ satisfies the congestion property if for any $\tilde{\mathcal{N}}_n^{a_n}(\boldsymbol{a}) \subseteq \mathcal{N}_n^{a_n}(\boldsymbol{a})$, we have*

$$g_n(\tilde{\mathcal{N}}_n^{a_n}(\boldsymbol{a})) \geq g_n(\mathcal{N}_n^{a_n}(\boldsymbol{a})). \tag{7}$$

Furthermore, a spatial spectrum access game satisfies the congestion property if (7) holds for all users $n \in \mathcal{N}$.

The congestion property means that the more contending users exist, the less chance a user can grab the channel. Such a property is natural for practical wireless systems such as the random backoff and Aloha systems. We can show that

LEMMA 2. *Consider any spatial spectrum access game satisfying the congestion property on a given directed interference graph G that has a pure Nash equilibrium. Then we can construct a new spatial spectrum access game by adding a new player, whose channel grabbing probability satisfies the congestion property and who can have interference relationship with at most one player $n \in \mathcal{N}$ in the original game. The new game also has a pure Nash equilibrium.*

DEFINITION 3 (**Directed Tree** [3]). *A directed graph is called a directed tree if the corresponding undirected graph obtained by ignoring the directions on the edges of the original directed graph is a tree.*

Note that a (undirected) tree is a special case of directed trees. Since any spatial spectrum access game over a single node always has a pure Nash equilibrium, we can then construct the directed tree recursively by introducing a new node and adding an (directed or undirected) edge between this node and one existing node. From Lemma 2, we obtain that

COROLLARY 2. *Any spatial spectrum access game satisfying the congestion property on a directed tree has a pure Nash equilibrium.*

DEFINITION 4 (**Directed Forest** [3]). *A directed graph is called a directed forest if it consists of a disjoint union of directed trees.*

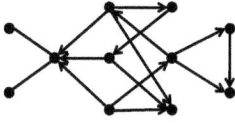

Figure 4: An interference graph that consists of directed acyclic graphs and directed trees

Similarly, we can obtain from Lemma 2 that

COROLLARY 3. *Any spatial spectrum access game satisfying the congestion property on a directed forest has a pure Nash equilibrium.*

As illustrated in Figure 4, according to Lemmas 1 and 2, we can construct more complicated directed interference graphs over which a spatial spectrum access game satisfying the congestion property has a pure Nash equilibrium.

4.2 Existence of Pure Nash Equilibria on Undirected Interference Graphs

We now study the case that the interference graph is undirected. This is a good approximation of reality if the transmitter of each user is close to its receiver, and all users' transmit powers are roughly the same.

When an undirected interference graph is a tree, according to Corollary 2, any spatial spectrum access game satisfying the congestion property has a pure Nash equilibrium. However, for those non-tree undirected graphs without a topological sort, the existence of pure Nash equilibrium can not be proved following the results in previous Section 4.1. This motivates us to further study the existence of pure Nash equilibria on generic undirected interference graphs.

First of all, [15] showed that a 3-players and 3-resources congestion game with user-specific congestion weights may not have a pure Nash equilibrium. Such a congestion game can be considered as a spatial spectrum access game on a complete undirected interference graph (by regarding the resources as channels). When all users have homogeneous channel contention capabilities and all channels have the same mean data rates, [22] showed that the spatial spectrum access game on any undirected interference graphs has a pure Nash equilibrium. Clearly, the applicability of such a channel-homogeneous model is quite limited, since the channel throughputs in practical wireless networks are often heterogeneous. We hence next focus on exploring the random backoff and Aloha systems with user-specific data rates, which provide useful insights for the user-homogeneous and user-heterogeneous channel contention mechanisms, respectively.

Here we resort to a useful tool of potential game[2], which is defined as

DEFINITION 5 (**Potential Game** [16]). *A game is called a potential game if it admits a potential function $\Phi(\boldsymbol{a})$ such that for every $n \in \mathcal{N}$ and $a_{-n} \in \mathcal{M}^{N-1}$,*

$$\text{sgn}\left(\Phi(a_n', a_{-n}) - \Phi(a_n, a_{-n})\right)$$

---

[2]Note that it is much more difficult to find a proper potential function to take into account users' asymmetric relationships (i.e., directions of edges on graph) when the interference graph is directed. Hence in this study we only apply the tool of potential game in the undirected case.

$$= \text{sgn}\left(U_n(a_n', a_{-n}) - U_n(a_n, a_{-n})\right),$$

where $\text{sgn}(\cdot)$ is the sign function.

DEFINITION 6 (**Better Response Update** [16]). *The event where a player n changes to an action a_n' from the action a_n is a better response update if and only if $U_n(a_n', a_{-n}) > U_n(a_n, a_{-n})$.*

An appealing property of the potential game is that it always admits a pure Nash equilibrium and the finite improvement property, which is defined as

DEFINITION 7 (**Finite Improvement Property** [16]). *A game has the finite improvement property if any asynchronous better response update process (i.e., no more than one player updates the strategy at any given time) terminates at a pure Nash equilibrium within a finite number of updates.*

Based on the potential game theory, we first study the random backoff mechanism. We show in Theorem 2 that when the undirected interference graph is complete, there exists indeed a pure Nash equilibrium.

THEOREM 2. *Any spatial spectrum access game on a complete undirected interference graph with the random backoff mechanism is a potential game with the potential function*

$$\Phi(\boldsymbol{a}) = \prod_{n=1}^{N} \theta_{a_n} B_{a_n}^n \prod_{m=1}^{M} \prod_{c=1}^{K_m(\boldsymbol{a})} g_n(c), \tag{8}$$

where $K_m(\boldsymbol{a})$ is the number of users choosing channel m under the strategy profile $\boldsymbol{a}$, and hence has a pure Nash equilibrium.

We then consider the random backoff mechanism in the asymptotic case that $\lambda_{\max}$ goes to infinity. This can be a good approximation of reality when the number of backoff mini-slots is much greater than the number of interfering users, and collision rarely occurs. In this case, we have

$$
\begin{aligned}
g_n(\mathcal{N}_n^{a_n}(\boldsymbol{a})) &= \lim_{\lambda_{\max} \to \infty} \sum_{\lambda=1}^{\lambda_{\max}} \frac{1}{\lambda_{\max}} \left(\frac{\lambda_{\max} - \lambda}{\lambda_{\max}}\right)^{K_{a_n}^n(\boldsymbol{a})} \\
&= \int_0^1 x^{K_{a_n}^n(\boldsymbol{a})} dx = \frac{1}{1 + K_{a_n}^n(\boldsymbol{a})},
\end{aligned} \tag{9}
$$

here $K_{a_n}^n(\boldsymbol{a})$ denotes the number of users that choose channel a_n and can interfere with user n. Equation (9) implies that the channel opportunity is equally shared among $1 + K_{a_n}^n(\boldsymbol{a})$ contending users (including user n). We consider the user specific throughput as

$$U_n(a_n, a_{-n}) = h_n \theta_{a_n} B_{a_n} \frac{1}{1 + K_{a_n}^n(\boldsymbol{a})}, \tag{10}$$

where h_n is regarded as a user-specific transmission gain. For example, a user n can possess a higher transmission gain if the distance between its transmitter and receiver is shorter than other users given that all the users transmit with the same power level. We show that

THEOREM 3. *Any spatial spectrum access game on any undirected interference graph with user-specific transmission* gains and the random backoff mechanism in the asymptotic case is a potential game with the potential function

$$\Phi(\boldsymbol{a}) = -\sum_{n=1}^{N} \left(\frac{1 + \frac{1}{2} K_{a_n}^n(\boldsymbol{a})}{\theta_{a_n} B_{a_n}} \right), \qquad (11)$$

and hence has a pure Nash equilibrium.

We now consider the Aloha mechanism. According to (2), we have the throughput function as

$$U_n(\boldsymbol{a}) = \theta_{a_n} B_{a_n}^n p_n \prod_{i \in \mathcal{N}_n^{a_n}(\boldsymbol{a})} (1 - p_i). \qquad (12)$$

We can show that

THEOREM 4. *Any spatial spectrum access game on any undirected interference graph with the Aloha mechanism is a potential game with the potential function*

$$\Phi(\boldsymbol{a}) = \sum_{i=1}^{N} -\log(1 - p_i)$$
$$\times \left(\frac{1}{2} \sum_{j \in \mathcal{N}_i^{a_i}(\boldsymbol{a})} \log(1 - p_j) + \log\left(\theta_{a_i} B_{a_i}^i p_i\right) \right), \quad (13)$$

and hence has a pure Nash equilibrium.

When a spatial spectrum access game is a potential game, we can design a distributed algorithm such that each user asynchronously updates the channel selection myopically to increase its throughput. According to the finite improvement property of the potential game, such an algorithm can achieve a pure Nash equilibrium within finite number of iterations. However, asynchronous better response updates require each user to have the complete information of other users' channel selections. This can only be achieved with extensive information exchange among the users, which may not be always feasible. It will be very nice to design a distributed algorithm that achieves the equilibrium without information exchange.

5. DISTRIBUTED LEARNING FOR SPATIAL SPECTRUM ACCESS

In this part, we discuss how to achieve an equilibrium for the spatial spectrum access games. As shown in Section 4, a generic spatial spectrum access game does not necessarily have a pure Nash equilibrium, and thus it is impossible to design a mechanism achieving pure Nash equilibria in general. We hence target on approaching the mixed Nash equilibria. Govindan and Wilson in [10] proposed a global Newton method to compute the mixed Nash equilibria for any finite strategic games. This method hence can be applied to find the mixed Nash equilibria for the spatial spectrum access games. However, such an approach is a centralized optimization, which requires that each user has the complete information of other users and compute the solution accordingly. This is often infeasible in a cognitive radio network, since acquiring complete information requires heavy information exchange among the users, and setting up and maintaining a common control channel for message broadcasting demands high system overheads [1]. Moreover, this approach is not incentive compatible since some users may not be willing to share their local information due to

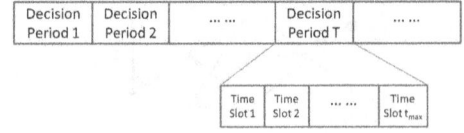

Figure 5: Time structure of a decision period

the energy consumption of information broadcasting. We thus propose a distributed learning algorithm for any spatial spectrum access games, and the algorithm does not require any information exchange among users. Each user only estimates its expected throughput locally, and learns to adjust its channel selection strategy adaptively. We show that the distributed learning algorithm can converge to a mixed Nash equilibrium approximately.

5.1 Expected Throughput Estimation

We first introduce the estimation of user's expected throughput based on local observations. To achieve an accurate estimation, a user needs to gather a large number of local observation samples. This motivates us to divide the spectrum access time into a sequence of *decision periods* indexed by $T(= 1, 2, ...)$, where each decision period consists of $t_{\max}$ time slots (see Figure 5 as an illustration). During a single decision period, a user accesses the *same* channel in all $t_{\max}$ time slots. Thus the total number of users accessing each channel does not change within a decision period, which allows users to better learn the environment.

Suppose user n chooses channel m to access at decision period T. According to (3), a user's expected throughput during period T depends on the probability of grabbing the channel $g_n(\mathcal{N}_n^{a_n}(\boldsymbol{a}(T)))$ on that period, the channel idle probability θ_m, and the mean data rate B_m^n. Similarly to our work in [7] on the expected throughput estimation for imitation-based spectrum sharing mechanism design, we will apply the maximum likelihood estimation (MLE) to get accurate estimations of there parameters for the distributed learning mechanism design, due to MLE's efficiency and ease of implementation.

5.1.1 Maximum Likelihood Estimation

At the beginning of each time slot $t(= 1, ..., t_{\max})$ of a decision period T, a user n will sense the same channel $a_n(T) = m$. If the channel is idle, the user will compete to grab the channel according to a specified channel contention mechanism. At the end of each time slot t, a user observes $S_m^n(T, t)$, $I_m^n(T, t)$, and $b_m^n(T, t)$. Here $S_m^n(T, t)$ denotes the state of the chosen channel m (i.e., whether occupied by the primary traffic), $I_m^n(T, t)$ indicates whether the user has successfully grabs the channel, i.e.,

$$I_m^n(T, t) = \begin{cases} 1, & \text{if user } n \text{ successfully} \\ & \text{grabs the channel } m, \\ 0, & \text{otherwise,} \end{cases}$$

and $b_m^n(T, t)$ is the received data rate on the chosen channel m by user n at time slot t. At the end of each decision period T, each user n can collect a set of local observations $\Omega_n(T) = \{S_m^n(T, t), I_m^n(T, t), b_m^n(T, t)\}_{t=1}^{t_{\max}}$. Note that if $S_m^n(T, t) = 0$ (i.e., the channel is occupied by the primary traffic), we also set $I_m^n(T, t)$ and $b_m^n(T, t)$ to be 0.

When the channel m is idle (i.e., no primary traffic), user

n will grab the channel with probability $g_n(\mathcal{N}_n^{a_n}(\boldsymbol{a}(T)))$. Since there are a total of $\sum_{t=1}^{t_{\max}} S_m^n(T,t)$ rounds of channel contentions in the period T and each round is independent and identically distributed (i.i.d.), the total number of successful channel captures $\sum_{t=1}^{t_{\max}} I_m^n(T,t)$ follows the Binomial distribution. A user n can then compute the likelihood of $g_n(\mathcal{N}_n^{a_n}(\boldsymbol{a}(T)))$, i.e., the probability of the realized observations $\Omega_n(T)$ given the parameter $g_n(\mathcal{N}_n^{a_n}(\boldsymbol{a}(T)))$ as

$$\mathcal{L}[\Omega_n(T)|g_n(\mathcal{N}_n^{a_n}(\boldsymbol{a}(T)))]$$
$$= \binom{\sum_{t=1}^{t_{\max}} S_m^n(T,t)}{\sum_{t=1}^{t_{\max}} I_m^n(T,t)} g_n(\mathcal{N}_n^{a_n}(\boldsymbol{a}(T)))^{\sum_{t=1}^{t_{\max}} I_m^n(T,t)}$$
$$\times (1 - g_n(\mathcal{N}_n^{a_n}(\boldsymbol{a}(T))))^{\sum_{t=1}^{t_{\max}} S_m^n(T,t) - \sum_{t=1}^{t_{\max}} I_m^n(T,t)}.$$

Then MLE of $g_n(\mathcal{N}_n^{a_n}(\boldsymbol{a}(T)))$ can be computed by maximizing the log-likelihood function $\ln \mathcal{L}[\Omega_n(T)|g_n(\mathcal{N}_n^{a_n}(\boldsymbol{a}(T)))]$, i.e., $\max_{g(k(T))} \ln \mathcal{L}[\Omega_n(T)|g_n(\mathcal{N}_n^{a_n}(\boldsymbol{a}(T)))]$. By the first order condition, we obtain the optimal solution as $\tilde{g}_n(\mathcal{N}_n^{a_n}(\boldsymbol{a}(T))) = \frac{\sum_{t=1}^{t_{\max}} I_m^n(T,t)}{\sum_{t=1}^{t_{\max}} S_m^n(T,t)}$, which is the sample averaging estimation. When length of decision period $t_{\max}$ is large, by the central limit theorem, we know that

$$\tilde{g}_n(\mathcal{N}_n^{a_n}(\boldsymbol{a}(T)))$$
$$\sim \mathcal{N}\left(g_n(\mathcal{N}_n^{a_n}(\boldsymbol{a}(T))), \frac{g_n(\mathcal{N}_n^{a_n}(\boldsymbol{a}(T)))(1 - y_n(\mathcal{N}_n^{a_n}(\boldsymbol{u}(T))))}{\sum_{t=1}^{t_{\max}} S_m^n(T,t)} \right),$$

where $\mathcal{N}(\cdot)$ denotes the normal distribution.

Similarly, we can apply the MLE to estimate the channel idle probability θ_m and the mean channel data rate B_m^n. More specifically, when the channel state $S_m(t)$ and the realized data rate $b_m^n(t)$ are i.i.d. random variables, we can easily obtain the closed-form estimations as $\hat{\theta}_m = \frac{\sum_{t=1}^{t_{\max}} S_m^n(T,t)}{t_{\max}}$ and $\hat{B}_m^n = \frac{\sum_{t=1}^{t_{\max}} b_m^n(T,t)}{\sum_{t=1}^{t_{\max}} I_m^n(T,t)}$, respectively[3]. By the MLE, we can obtain the estimation of $g_n(\mathcal{N}_n^{a_n}(\boldsymbol{a}(T)))$, θ_m and B_m^n as $\tilde{g}_n(\mathcal{N}_n^{a_n}(\boldsymbol{a}(T)))$, $\tilde{\theta}_m$ and $\tilde{B}_m^n$, respectively, and then estimate the true expected throughput $U_n(\boldsymbol{a}(T))$ as $\tilde{U}_n(T) = \tilde{\theta}_m \tilde{B}_m^n \tilde{g}_n(\mathcal{N}_n^{a_n}(\boldsymbol{a}(T)))$. Since according to the central limit theorem $\tilde{g}_n$, $\tilde{\theta}_m$, and $\tilde{B}_m^n$ follow independent normal distributions with the mean $g_n(\mathcal{N}_n^{a_n}(\boldsymbol{a}(T)))$, θ_m, and B_m^n, respectively, we thus have

$$E[\tilde{U}_n(\boldsymbol{a}(T))] = E[\tilde{\theta}_m \tilde{B}_m \tilde{g}_n(\mathcal{N}_n^{a_n}(\boldsymbol{a}(T)))] = U_n(\boldsymbol{a}(T)),$$

i.e., the estimation of expected throughput $U_n(\boldsymbol{a}(T))$ is unbiased.

5.2 Distributed Learning Algorithm

Based on the expected throughput estimation, we now propose the distributed learning algorithm for spatial spectrum access games. The idea is to extend the principle of single-agent reinforcement learning to a multi-agent setting. Such multi-agent reinforcement learning algorithm has also been applied to the classical congestion games on complete graphs [14, 20]. Here we apply the learning algorithm to the generalized spatial congestion games on any generic graphs and derive the convergence conditions accordingly. The algorithm works as follows.

---

[3]When $S_m(t)$ and $b_m^n(t)$ are non-i.i.d. random variables, the MLE can also be derived based on the specific probability distribution functions by following the similar procedure as introduced in Section 5.1.1.

At the beginning of each period T, a user $n \in \mathcal{N}$ chooses a channel $a_n(T) \in \mathcal{M}$ to access according to its mixed strategy $\boldsymbol{\sigma}_n(T) = (\sigma_m^n(T), \forall m \in \mathcal{M})$, where $\sigma_m^n(T)$ is the probability of choosing channel m. The mixed strategy is generated according to $\boldsymbol{P}_n(T) = (P_m^n(T), \forall m \in \mathcal{M})$, which represents its *perceptions* of the payoff performance of choosing different channels based on local estimations. Perceptions are based on local observations in the past and may not accurately reflect the expected payoff. For example, if a user n has not accessed a channel m for many decision intervals, then perception $P_m^n(T)$ can be out of date. The key challenge for the learning algorithm is to update the perceptions with proper parameters such that perceptions equal to expected payoffs at the equilibrium.

Similarly to the single-agent learning, we choose the Boltzmann distribution as the mapping from perceptions to mixed strategies, i.e.,

$$\sigma_m^n(T) = \frac{e^{\gamma P_m^n(T)}}{\sum_{i=1}^{M} e^{\gamma P_i^n(T)}}, \forall m \in \mathcal{M}, \qquad (14)$$

where γ is the temperature that controls the randomness of channel selections. When $\gamma \to 0$, each user will choose to access channels uniformly at random. When $\gamma \to \infty$, user n always chooses the channel with the largest perception value $P_m^n(T)$ among all channel $m \in \mathcal{M}$. We will show later on that the choice of γ trades off convergence and performance of the learning algorithm.

At the end of a decision period T, a user n computes its estimated expected payoff $\tilde{U}_n(\boldsymbol{a}(T))$ as in Section 5.1 (i.e., by using the MLE method based on the set of local observations $\Omega_n(T)$ during the period), and adjusts its perceptions as

$$P_m^n(T+1) = \begin{cases} (1 - \mu_T)P_m^n(T) + \mu_T \tilde{U}_n(\boldsymbol{a}(T)), & \text{if } a_n(T) = m, \\ P_m^n(T), & \text{otherwise,} \end{cases} \qquad (15)$$

where $(\mu_T \in (0,1), \forall T)$ are the smoothing factors. A user only changes the perception of the channel just accessed in the current decision period, and keeps the perceptions of other channels unchanged.

Algorithm 1 summarizes the distributed learning algorithm. Next we study the convergence of the learning algorithm based on the theory of stochastic approximation [12].

5.3 Convergence of Distributed Learning Algorithm

We now study the convergence of the proposed distributed learning algorithm.

First, the perception value update in (15) can be written in the following equivalent form,

$$P_m^n(T+1) - P_m^n(T) = \mu_T[Z_m^n(T) - P_m^n(T)], \forall n \in \mathcal{N}, m \in \mathcal{M}, \qquad (16)$$

where $Z_m^n(T)$ is the update value defined as

$$Z_m^n(T) = \begin{cases} \tilde{U}_n(\boldsymbol{a}(T)), & \text{if } a_n(T) = m, \\ P_m^n(T), & \text{otherwise.} \end{cases} \qquad (17)$$

For the sake of brevity, we denote the perception values, update values, and mixed strategies of all the users as $\boldsymbol{P}(T) \triangleq (P_m^n(T), \forall m \in \mathcal{M}, n \in \mathcal{N})$, $\boldsymbol{Z}(T) \triangleq (Z_m^n(T), \forall m \in \mathcal{M}, n \in \mathcal{N})$, and $\boldsymbol{\sigma}(T) \triangleq (\sigma_m^n(T), \forall m \in \mathcal{M}, n \in \mathcal{N})$, respectively.

Algorithm 1 Distributed Learning Algorithm For Spatial Spectrum Access Game

1: **initialization:**
2:     **set** the temperature γ.
3:     **set** the initial perception values $P_m^n(0) = \frac{1}{M}$ for each user $n \in \mathcal{N}$.
4:     **set** the period index $T = 0$.
5: **end initialization**

6: **loop** for each decision period T and each user $n \in \mathcal{N}$ in parallel:
7:     **select** a channel $m \in \mathcal{M}$ according to (14).
8:     **for** each time slot t in the period T **do**
9:         **sense and contend** to access the channel m.
10:         **record** the observations $S_m^n(T,t)$, $I_m^n(T,t)$ and $b_m^n(T,t)$.
11:     **end for**
12:     **estimate** $g_n(\mathcal{N}_n^{a_n}(\boldsymbol{a}(T)))$, θ_m, and B_m^n by the maximum likelihood estimation.
13:     **compute** the estimated expected payoff $\tilde{U}_n(\boldsymbol{a}(T))$.
14:     **update** the perceptions value $\boldsymbol{P}_n(T)$ according to (15).
15:     **set** the period index $T = T + 1$.
16: **end loop**

Let $Pr\{\mathcal{N}_n^m(\boldsymbol{a}(T))|\boldsymbol{P}(T), a_n(T) = m\}$ denote the conditional probability that, given that the users' perceptions are $\boldsymbol{P}(T)$ and user n chooses channel m, the set of users that choose the same channel m in user n's neighborhood $\mathcal{N}_n$ is $\mathcal{N}_n^m(\boldsymbol{a}(T)) \subseteq \mathcal{N}_n$. Since each user independently chooses a channel according to its mixed strategy $\boldsymbol{\sigma}_n(T)$, then the random set $\mathcal{N}_n^m(\boldsymbol{a}(T))$ follows the Binomial distribution of $|\mathcal{N}_n|$ independent non-homogeneous Bernoulli tries with the probability mass function as

$$Pr\{\mathcal{N}_n^m(\boldsymbol{a}(T))|\boldsymbol{P}(T), a_n(T) = m\}$$
$$= \prod_{i \in \mathcal{N}_n^m(\boldsymbol{a}(T))} (\sigma_m^i(T)) \prod_{i \in \mathcal{N}_n \setminus \mathcal{N}_n^m(\boldsymbol{a}(T))} (1 - \sigma_m^i(T))$$
$$= \prod_{i \in \mathcal{N}_n} (\sigma_m^i(T))^{I_{\{a_i(T)=m\}}} (1 - \sigma_m^i(T))^{1 - I_{\{a_i(T)=m\}}}, \quad (18)$$

where $I_{\{a_i(T)=m\}} = 1$ if user i chooses channel m, and $I_{\{a_i(T)=m\}} = 0$ otherwise.

Since the update value $Z_m^n(T)$ depends on user n's estimated payoff $\tilde{U}_n(\boldsymbol{a}(T))$ (which in turn dependents on $\mathcal{N}_n^m(\boldsymbol{a}(T))$), thus $Z_m^n(T)$ is also a random variable. The equations in (16) are hence stochastic difference equations, which are difficult to analyze directly. We thus focus on the analysis of its *mean dynamics* [12]. To proceed, we define the mapping from the perceptions $\boldsymbol{P}(T)$ to the expected payoff of user n choosing channel m as $Q_m^n(\boldsymbol{P}(T)) \triangleq E[U_n(\boldsymbol{a}(T))|\boldsymbol{P}(T), a_n(T) = m]$. Here the expectation $E[\cdot]$ is taken with respective to the mixed strategies $\boldsymbol{\sigma}(T)$ of all users (i.e., the perceptions $\boldsymbol{P}(T)$ of all users due to (14)). We show that

LEMMA 3. *For the distributed learning algorithm, if the temperature satisfies*

$$\gamma < \frac{1}{2 \max_{m \in \mathcal{M}, n \in \mathcal{N}} \{\theta_m B_m^n\} \max_{n \in \mathcal{N}} \{|\mathcal{N}_n|\}}, \quad (19)$$

the mapping from the perceptions to the expected payoff $Q(\boldsymbol{P}(T)) \triangleq$

$(Q_m^n(\boldsymbol{P}(T)), m \in \mathcal{M}, n \in \mathcal{N})$ *forms a maximum-norm contraction.*

Note that the condition (19) is a sufficient condition to form a contraction mapping, which is in turn is a sufficient condition for convergence. Simulation results show that a slightly larger γ may also lead to the convergence of the mapping. Based on the property of contraction mapping, there exists a fixed point $\boldsymbol{P}^*$ such that $Q(\boldsymbol{P}^*) = \boldsymbol{P}^*$. By the theory of stochastic approximations [12], we show that the distributed learning algorithm also converges to the same limit point $\boldsymbol{P}^*$.

THEOREM 5. *For the distributed learning algorithm, if the temperature γ satisfies (19), $\sum_T \mu_T = \infty$ and $\sum_T \mu_T^2 < \infty$, then the sequence $\{\boldsymbol{P}(T), \forall T \geq 0\}$ converges to the unique limit point $\boldsymbol{P}^* \triangleq (P_m^{n*}, \forall m \in \mathcal{M}, n \in \mathcal{N})$ of the differential equations ($\forall m \in \mathcal{M}, n \in \mathcal{N}$)*

$$\frac{dP_m^n(T)}{dT} = \sigma_m^n(T) (Q_m^n(\boldsymbol{P}(T)) - P_m^n(T)), \quad (20)$$

with probability one. Further, the limit point $\boldsymbol{P}^*$ satisfies

$$Q_m^n(\boldsymbol{P}^*) = P_m^{n*}, \forall m \in \mathcal{M}, n \in \mathcal{N}. \quad (21)$$

We next explore the property of the equilibrium $\boldsymbol{P}^*$ of the distributed learning algorithm. From Theorem 5, we see that

$$Q_m^n(\boldsymbol{P}^*) = E[U_n(\boldsymbol{a}(T))|\boldsymbol{P}^*, a_n(T) = m] = P_m^{n*}. \quad (22)$$

It means that the perception value P_m^{n*} is an accurate estimation of the expected payoff in the equilibrium. Moreover, we show that the mixed strategy $\boldsymbol{\sigma}^*$ is an approximate Nash equilibrium.

DEFINITION 8 (**Approximate Nash Equilibrium** [8]). *A mixed strategy profile $\bar{\boldsymbol{\sigma}} = (\bar{\boldsymbol{\sigma}}_1, ..., \bar{\boldsymbol{\sigma}}_N)$ is a ξ- approximate Nash equilibrium if*

$$U_n(\bar{\boldsymbol{\sigma}}_n, \bar{\boldsymbol{\sigma}}_{-n}) \geq \max_{\boldsymbol{\sigma}_n} U_n(\boldsymbol{\sigma}_n, \bar{\boldsymbol{\sigma}}_{-n}) - \xi, \forall n \in \mathcal{N},$$

where $U_n(\bar{\boldsymbol{\sigma}}_n, \bar{\boldsymbol{\sigma}}_{-n})$ denotes the expected payoff of player n under mixed strategy $\bar{\boldsymbol{\sigma}}$, and $\bar{\boldsymbol{\sigma}}_{-n}$ denotes the mixed strategy profile of other users except player n.

Here $\xi \geq 0$ is the gap from a (precise) mixed Nash equilibrium. For the distributed learning algorithm, we show that

THEOREM 6. *For the distributed learning algorithm, the mixed strategy $\boldsymbol{\sigma}^*$ in the equilibrium $\boldsymbol{P}^*$ is a ξ-approximate Nash equilibrium, with $\xi = \max_{n \in \mathcal{N}} \{-\frac{1}{\gamma} \sum_{m=1}^M \sigma_m^{n*} \ln \sigma_m^{n*}\}$.*

The gap ξ can be interpreted as the *weighted entropy*, which describes the randomness of the learning exploration. A larger ξ means worse learning performance. When each user adopts the uniformly random access, the gap ξ reaches the maximum value and results in the worst learning performance. Theorems 5 and 6 together illustrate the trade-off between the convergence and performance through the choice of γ. A small enough γ is required to explore the environment (so that users are not getting stuck in channels with the *current* best payoffs) and guarantee the convergence of distributed learning. If γ is too small, however, then the performance gap ξ is large due to over-exploration.

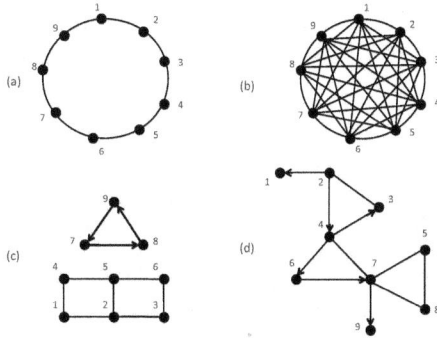

Figure 6: Interference Graphs

Figure 7: The system performance of the distributed learning algorithm with different temperature γ

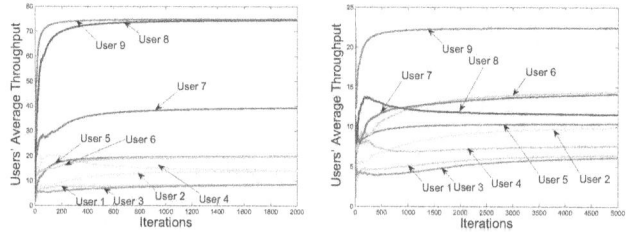

6. NUMERICAL RESULTS

We now evaluate the proposed distributed learning algorithm by simulations. We consider a Rayleigh fading channel environment. The data rate of user n on an idle channel m is given according to the Shannon capacity, i.e., $b_m^n(t) = V_m \log_2 \left(1 + \frac{\zeta_n g_m^n(t)}{N_0} \right)$, where V_m is the bandwidth of channel m, ζ_n is the power adopted by user n, N_0 is the noise power, and $g_m^n(t)$ is the channel gain, which is a random variable that follows the exponential distribution with the mean $\bar{g}_m^n$. In the following simulations, we set $V_m = 10$ MHz, $N_0 = -100$ dBm, and $\zeta_n = 100$ mW. By choosing different mean channel gains $\bar{g}_m^n$, we have different mean data rates $B_m^n = E[b_m^n(t)]$ for different channels and users. For simplicity, we set channel state $S_m(t)$ as an i.i.d. Bernoulli random variable with the idle probability $\theta_m = 0.5$.

We consider a network of $M = 5$ channels and $N = 9$ users with four different interference graphs (see Figure 6). Graphs (a) and (b) are undirected, and Graphs (c) and (d) are directed. Let $\vec{B}_n = \{B_1^n, ..., B_M^n\}$ be the mean data rate vector of user n. We set $\vec{B}_1 = \vec{B}_2 = \vec{B}_3 = \{2, 6, 16, 20, 30\}$ Mbps, $\vec{B}_4 = \vec{B}_5 = \vec{B}_6 = \{4, 12, 32, 40, 60\}$ Mbps, and $\vec{B}_7 = \vec{B}_8 = \vec{B}_9 = \{10, 30, 80, 100, 150\}$ Mbps. We implement both the random backoff and Aloha mechanisms for channel contention. For the random backoff mechanism, we set the number of backoff mini-slots in a time slot $\lambda_{\max} = 10$. For the Aloha mechanism, the channel contention probabilities of the users are $\{0.7, 0.7, 0.7, 0.5, 0.5, 0.5, 0.3, 0.3, 0.3\}$, respectively. Notice that in this study we focus on channel choices instead of the adjustment of contention probabilities.

For the distributed learning algorithm initialization, we set the length of each decision period $t_{\max} = 200$, which can achieve a good estimation of the mean data rate. We set the smooth factor $\mu_T = \frac{1}{200+T}$, which satisfies the condition $\sum_T \mu_T = \infty$ and $\sum_T \mu_T^2 < \infty$.

We first evaluate the distributed learning algorithm with different choices of temperature γ on the interference graph (d) in Figure 6. We run the learning algorithm sufficiently long until the time average system throughput does not change. The result in Figure 7 verifies the trade-off between the convergence and performance, and demonstrates that a proper temperature γ can offer the best performance. When is γ small, the gap ξ in Theorem 6 can be large. When γ is very large, the algorithm may get stuck in local optimum and the performance is again negatively affected. We set $\gamma = 5.0$ in the following simulations since it achieves

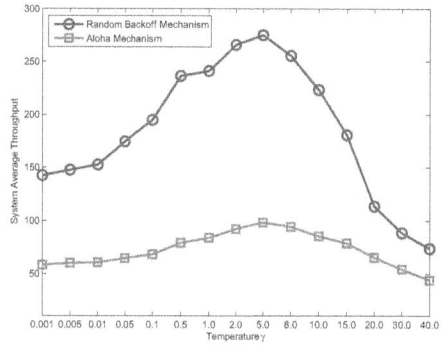

Figure 8: Users' average throughput on interference graph (d) with random backoff mechanism

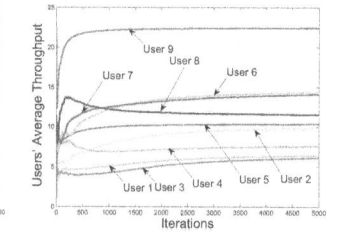

Figure 9: Users' average throughput on interference graph (d) with Aloha mechanism

good system performance in both random backoff and Aloha mechanisms as in Figure 7.

We then look at the learning dynamics. Figures 8 and 9 show the dynamics of users' time average throughputs with random backoff and Aloha mechanisms, respectively. These results demonstrate the convergence of the distributed learning algorithm in both mechanisms.

To benchmark the performance of the distributed learning algorithm, we compare it with the solutions obtained by the following two algorithms:

- Random Access: each user chooses a channel to access purely randomly.

- Centralized Optimization: the solution obtained by solving the centralized global optimization of $\max_a \sum_{n \in \mathcal{N}} U_n(a)$.

We implement these algorithms together with the distributed learning algorithm on the four types of interference graphs in Figure 6. The results are shown in Figures 10 and 11. For the random backoff (Aloha, respectively) mechanism, we see that the distributed learning algorithm achieves up-to 100% (65%, respectively) performance improvement over the random access algorithm. Compared with the centralized optimal solution, the performance loss of the distributed learning in the full-interference graph (b) is 28% (34%, respectively). Such performance loss is not due to the algorithm design; instead it is due to the selfish nature of the users. In the partial-interference graphs (a), (c), and (d), the performance loss can be further reduced to less than 10% (17%, respectively). This shows that the negative

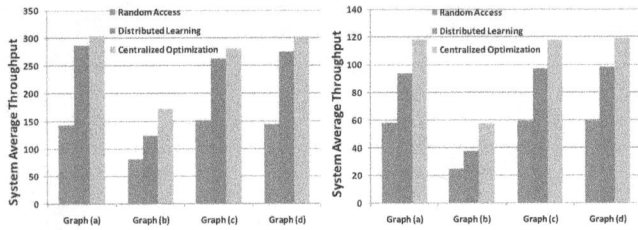

Figure 10: Comparison of distributed learning, random access, and centralized optimization with the random backoff mechanism

Figure 11: Comparison of distributed learning, random access, and centralized optimization with the Aloha mechanism

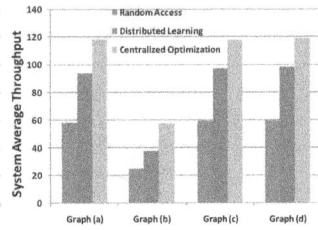

impact of users' selfish behavior is smaller when users can share the spectrum more efficiently through spatial reuse.

7. CONCLUSION

In this paper, we explore the spatial aspect of distributed spectrum sharing, and propose a framework of spatial spectrum access game on directed interference graphs. We investigate the critical issue of the existence of pure Nash equilibria, and develop a distributed learning algorithm converging to an approximate mixed Nash equilibrium for any spatial spectrum access games. Numerical results show that the algorithm is efficient and has significant performance gain over a random access algorithm that does not take the spatial effect into consideration.

For the future work, we are going to design distributed algorithms that maximize the system-wide throughput.

ACKNOWLEDGMENTS

This work is supported by the General Research Funds (Project Number 412509, 412710, and 412511) established under the University Grant Committee of the Hong Kong Special Administrative Region, China.

8. REFERENCES

[1] I. Akyildiz, W. Lee, M. Vuran, and S. Mohanty. Next generation/dynamic spectrum access/cognitive radio wireless networks: a survey. *Computer Networks*, 50(13):2127–2159, 2006.

[2] A. Anandkumar, N. Michael, and A. Tang. Opportunistic spectrum access with multiple users: learning under competition. In *Proc. of IEEE INFOCOM*, San Diego, USA, March 2010.

[3] N. Biggs, E. Lloyd, and R. Wilson. *Graph Theory*. Oxford University Press, 1986.

[4] V. Bilò, A. Fanelli, M. Flammini, and L. Moscardelli. Graphical congestion games with linear latencies. In *Proc. of ACM SPAA*, Munich, Germany, June 2008.

[5] X. Chen and J. Huang. Spatial spectrum access game: Nash equilibria and distributed learning. Technical report, The Chinese University of Hong Kong, 2012. Available at http://jianwei.ie.cuhk.edu.hk/publication /MobihocTech.pdf.

[6] X. Chen and J. Huang. Evolutionarily stable open spectrum access in a many-users regime. In *Proc. of IEEE GLOBECOM*, Houston, USA, Dec. 2011.

[7] X. Chen and J. Huang. Imitative spectrum access. In *Proc. of IEEE WiOpt*, Paderborn, Germany, May 2012.

[8] C. Daskalakisa, A. Mehtab, and C. Papadimitriou. A note on approximate Nash equilibria. *Theoretical Computer Science*, 410(17):1581–1588, 2009.

[9] M. Félegyhĺczi, M. Cagalj, and J.-P. Hubaux. Efficient mac in cognitive radio systems: A game-theoretic approach. *IEEE Transactions on wireless Communications*, 8:1984–1995, 2009.

[10] S. Govindan and R. Wilson. A global newton method to compute nash equilibria. *Journal of Economic Theory*, 110:65–86, 2003.

[11] Z. Han, C. Pandana, and K. J. R. Liu. Distributive opportunistic spectrum access for cognitive radio using correlated equilibrium and no-regret learning. In *Proc. of IEEE WCNC*, Hong Kong, March 2007.

[12] H. Kushner and G. Yin. *Stochastic Approximation and Recursive: Algorithms and Applications*. Springer-Verlag, New York, 2003.

[13] K. Liu and Q. Zhao. Decentralized multi-armed bandit with multiple distributed players. In *Proc. of IEEE IAT*, San Diego, USA, Jan. 2010.

[14] E. Melo. Congestion pricing and learning in traffic network games. *Journal of Public Economic Theory*, 13(3):351–367, 2011.

[15] I. Milchtaich. Congestion games with player-specific payoff functions. *Games and Economic Behavior*, 13:111–124, 1996.

[16] D. Monderer and L. S. Shapley. Potential games. *Games and Economic Behavior*, 14:124–143, 1996.

[17] J. Nash. Equilibrium points in n-person games. *Proceedings of the National Academy of Sciences*, 36:48–49, 1950.

[18] N. Nie and C. Comniciu. Adaptive channel allocation spectrum etiquette for cognitive radio networks. In *Proc. of IEEE DySPAN*, Baltimore, USA, Nov. 2005.

[19] D. Niyato and E. Hossain. Competitive spectrum sharing in cognitive radio networks: a dynamic game approach. *IEEE Transactions on Wireless Communications*, 7:2651–2660, 2008.

[20] D. Shah and J. Shin. Dynamics in congestion games. *ACM SIGMETRICS Performance Evaluation Review*, 38(1):107–118, 2010.

[21] R. Southwell and J. Huang. Convergence dynamics of resource-homogeneous congestion game. In *Proc. of ICST GameNets*, Shanghai, China, April 2011.

[22] C. Tekin, M. Liu, R. Southwell, J. Huang, and S. Ahmad. Atomic congestion games on graphs and their applications in networking. *to appear in IEEE/ACM Transactions on Networking*, 2012.

[23] B. Wang, Y. Wua, and K. R. Liu. Game theory for cognitive radio networks: An overview. *Computer Networks*, 54:2537–2561, 2010.

[24] M. Weiss, M. Al-Tamaimi, and L. Cui. Dynamic geospatial spectrum modelling: taxonomy, options and consequences. In *Proc. of TPRC Conference*, Fairfax, USA, Sep. 2010.

Design and Implementation of an Integrated Beamformer and Uplink Scheduler for OFDMA Femtocells

Mustafa Y. Arslan [*]
University of California,
Riverside
marslan@cs.ucr.edu

Karthikeyan Sundaresan
NEC Labs America, Inc.
Princeton, NJ, USA
karthiks@nec-labs.com

Srikanth V.
Krishnamurthy
University of California,
Riverside
krish@cs.ucr.edu

Sampath Rangarajan
NEC Labs America, Inc.
Princeton, NJ, USA
sampath@nec-labs.com

ABSTRACT

Beamforming is a signal processing technique with numerous benefits. Unlike with omni-directional communications, it focuses the energy of the transmitted and/or the received signal in a particular direction. Although beamforming has been extensively studied on conventional systems such as WiFi, little is known about its *practical* impact on OFDMA femtocell deployments. Since OFDMA schedules multiple clients (users) in the same frame (in contrast to WiFi), designing intelligent scheduling mechanisms and at the same time leveraging beamforming, is a challenging task.

Unlike downlink, we show that the integration of beamforming with uplink scheduling projects an interesting trade-off between beamforming gain on the one hand, and the power pooling gain resulting from joint multi-user scheduling on the other hand. This, in turn, makes the uplink scheduling problem even hard to approximate. To address this, we propose algorithms that are simple to implement, yet provably efficient with a worst case guarantee of half. We implement our solutions on a real WiMAX femtocell platform integrated with an eight-element phased array beamforming antenna. Evaluations from both prototype implementation and trace-driven simulations show that our solution delivers throughput gains of over 40% compared to an omni-directional scheme.

Categories and Subject Descriptors

C.2.1 [**Network Architecture and Design**]: Wireless communication

Keywords

OFDMA, Femtocells, WiMAX, Beamforming

---

[*]Mustafa Arslan interned in the Mobile Communications and Networking department at NEC Laboratories America Inc., Princeton during this work.

1. INTRODUCTION

To meet the demands for increased capacity driven by the exponential growth in mobile data traffic [1], broadband network deployments are moving towards smaller cells – called femtocells – that use Orthogonal Frequency Domain Multiple Access (OFDMA). Femtocells inherit OFDMA and their synchronous access nature from macrocells, which allows for easier deployment and seamless operations with mobile clients.

By focusing the energy in specific directions using antenna arrays, beamforming can serve as a valuable tool in enhancing the performance of OFDMA femtocells. While beamforming is extensively studied for WiFi systems (e.g., [2, 3]), the fundamental differences between WiFi and OFDMA – scheduling of multiple users in the same frame and the synchronous access in OFDMA – render such studies inapplicable to femtocells. The future releases of OFDMA systems (e.g., WiMAX 802.16m [4]) support beamforming as part of the standard. They allow for multiple beams at the base-band level (one beam per user) to be used on the data payload within the same frame (*user-level beamforming*). However, potential beamforming gains are limited as the standards typically support a limited number of antennas at femtocells (e.g., four in WiMAX and LTE-Rel.8). In addition, the specific beamforming implementations vary from one standard to another, making a globally compatible beamforming solution impractical.

A more flexible alternative to user-level beamforming is to employ *frame-level beamforming*, which applies a single beam (common to all users) to the entire frame (not only the data payload but also the control part) at the RF level. Since user-specific beams are not leveraged, such an approach may incur sub-optimal performance for the data payload. This is due to the fact that a femtocell has to find a beam pattern that "fits" all of its clients. However, we identify that frame-level beamforming still has desirable properties: **(a)** it is realizable as a plug-and-play external RF beamformer and hence, is agnostic of the access technology (WiMAX, LTE, etc.); and **(b)** beamforming gain is not limited by the number of antennas supported by the standards. Even more interestingly, we show that frame-level beamforming also offers a unique, complementary benefit for interference mitigation, in a multicell context (discussed in detail later in §3). However in realizing the true potential of frame-level beamforming, an essential first step is to integrate it with the scheduler at each femtocell.

The integration of frame-level beamforming with downlink scheduling can be accomplished by extending existing solutions [5, 6] - by running the scheduler for each beam and picking the one yielding the best objective. However, the corresponding uplink problem encounters two aspects unique to the uplink that make the problem challenging:

Power pooling vs. beamforming: When multiple users are jointly scheduled in a frame (i.e., the frame resources are shared among clients), each user pools his transmit power on a smaller set of sub-channels (instead of using all sub-channels). This results in a higher user rate per sub-channel and hence higher frame throughput called *power pooling gain*. While beamforming improves a user's rate, using a single beam (common to all users) for a frame may not yield the optimal performance for every user scheduled in the frame. Hence, the higher the number of users scheduled in a frame, the higher is the power pooling gain but lower is the beamforming gain, and vice versa.

Batch scheduling: For a femto base station (BS) to compute the uplink allocation, the scheduling requests have to come from the clients (as is the case in macrocells). Clients contend for making such requests, which they send in response to changes in their buffer occupancies. This incurs both contention delay and overhead, where multiple requests are made for a transmission buffer worth of data. A better option is to allow the BS to poll clients for their buffer occupancies at periodic intervals (e.g., called the polling service in WiMAX [7]). Polling avoids contention and reduces overhead. However, scheduling is now done for a group of frames jointly (batch scheduling), based on client buffer occupancies. More generically, even without polling, batch scheduling allows a user's data to be spread across multiple frames, thereby accentuating power pooling gains and hence, aggregate throughput (details in §4.1).

In this study, we consider the batch scheduling problem on the uplink for non-real-time traffic. While the problem is optimally solvable for omni-directional communications, we show that it is hard to even approximate when integrated with frame-level beamforming, where it also requires the determination of a beam pattern for each frame. In addressing this problem, we make the following contributions:

- We propose simple but efficient algorithms for both continuous and discrete rate functions with a worst case guarantee of $\frac{1}{2}$.

- We implement our solutions on a real WiMAX femtocell platform that is integrated with an eight-element phased array antenna used for beamforming.

- We conduct comprehensive over-the-air evaluations with both a prototype implementation and trace-driven simulations; our results indicate an average gain of 40% over an omni-directional scheme, in practical settings.

The rest of the paper is organized as follows. §2 presents a brief WiMAX overview and related work. In §3 and 4, we describe the motivation and the design of our system. §5 describes our algorithms. §6 contains implementation details and §7 shows evaluation results. We conclude in §8.

2. BACKGROUND

WiMAX: While our solutions apply to OFDMA femtocells in general, our implementation is on WiMAX. Hence, we provide brief background on relevant WiMAX components (details in [7]). OFDMA divides the spectrum into multiple frequencies (sub-carriers) and several sub-carriers are grouped to form a sub-channel. The sub-carriers can be grouped in a contiguous, partially contiguous or

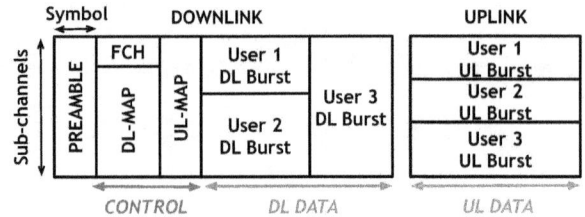

Figure 1: WiMAX Frame Structure.

fully distributed manner. Partially contiguous grouping, being the mandatory model, is alone implemented on most WiMAX devices and is hence, considered in our work. Here, sub-carriers forming a sub-channel are permuted to allow for a common transmission rate for a user on all sub-channels. This requires only a single channel feedback per user. Permuting also helps average out inter-cell interference.

A WiMAX frame is a two-dimensional template that is scheduled at the MAC with data to/from multiple mobile stations (MSs), across both time (symbols) and frequency (sub-channels). Data to/from users are allocated as rectangular bursts of resources in a frame (see Fig. 1). For e.g., the uplink allocation in Fig. 1 can be visualized to be at the sub-channel granularity. The frame consists of the preamble, control and data payload. While the preamble is used by the MS to lock on to the BS, the control consists of FCH (frame control header) and MAP. The BS, using MAP, indicates the schedule of transmissions both on the downlink and the uplink. The DL-MAP specifies the location of each burst, which MS it is intended for, and what modulation decodes it. Similarly the UL-MAP tells the MS where to place its data in the uplink and how to modulate it.

Beamforming: The ability to transmit (receive) signal energy in (from) specific directions is called beamforming. This is achieved by weighing the signals transmitted (received) from an antenna array in both magnitude and phase. Each applied weight vector generates a specific beam pattern. Such weight vectors can be determined and stored a priori (switched beamforming), or can be computed on-the-fly based on instantaneous channel feedback from the clients (adaptive beamforming).

To achieve low complexity, standards use code-book based beamforming, whereby weight vectors (pre-determined by the code-book) are employed and the one yielding the best SNR at the client is chosen. Since beams are enabled only for the data part, a beam vector suited for each client scheduled in a frame can be chosen, to encode data in the digital signal space (user-level beamforming). In contrast, applying a single beam physically at the RF level (frame-level beamforming), will provide beamforming for the entire frame, including the control part (advantageous in a multicell context as explained in §3). While its beamforming gain for the data part may be sub-optimal compared to user-level beamforming, frame-level beamforming is agnostic of the standards and thus, is not limited by them. Indeed, this has allowed us to integrate an eight-element phased array antenna with a WiMAX BS, although the current femtocells do not support beamforming and the next generation ones (e.g., 802.16m) support only four antennas. Given our experience, we believe that frame-level beamforming has potential in enabling universal beamforming solutions compatible across different standards. We consider frame-level beamforming for uplink, where the beam patterns emphasize reception in desired directions and could either correspond to physical directions covering the 360° azimuth (e.g., [3]), or be based on Gaussian code-books (e.g., [8]).

Related Work: Several works have studied the design of throughput [5, 6, 9, 10] and QoS [11, 12] schedulers for OFDMA. How-

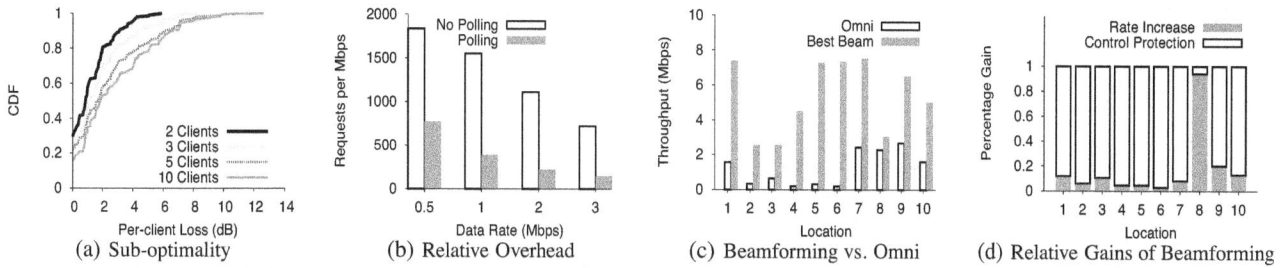

(a) Sub-optimality	(b) Relative Overhead	(c) Beamforming vs. Omni	(d) Relative Gains of Beamforming

Figure 2: Sub-optimality of frame-level beamforming (a) and relative overhead with batch scheduling (b).

ever, their focus has been on per-frame scheduling. While batch scheduling is not very different from per-frame scheduling on the downlink (in terms of optimization), the ability to leverage power pooling on the uplink makes the problem challenging and has not been addressed. While some studies (e.g., [13]) have looked at the design of polling intervals specifically for the uplink of WiMAX, they have not addressed the batch scheduling problem.

User-level beamforming has been jointly optimized with per-frame scheduling for OFDMA systems [14, 15]. However, frame-level beamforming and its various benefits have not been explored; its integration with uplink batch scheduling and the challenges it encounters have also not been considered. We contribute by designing and implementing an efficient uplink scheduler that integrates frame-level beamforming with batch scheduling for OFDMA.

3. MOTIVATION

We motivate the importance of frame-level beamforming and batch scheduling for the uplink using measurements from an experimental testbed. We defer however, a detailed description of the testbed to §6.

Small Loss Compared to User-level Beamforming: One drawback of frame-level beamforming is the constraint of using a single beam for all the users scheduled in a frame, thereby potentially limiting the per-user beamforming gain. In strong line-of-sight environments, where very few beams work well for each client, the sub-optimality of such a common beam could be significant. To understand this sub-optimality indoors, we placed a client at thirty locations in the building and recorded the uplink SNR (as measured at the BS) with various beams for each of these locations. Then, different subsets of client locations were grouped together to emulate different sets of clients being scheduled in the same frame. For each of these subsets, we chose a single beam (i.e., frame-level beamforming) with the following criteria: the chosen beam minimizes the aggregate loss (in terms of SNR) compared to the case where each client operates with its best beam (i.e., user-level beamforming). Here, the best beam of a client refers to the beam pattern yielding the highest received SNR from that particular client.

We show the CDF of the loss per client (in dB) in Fig. 2(a). It is seen that even when five clients (reasonably high for femtocells) are multiplexed in the same frame, the loss is less than 2 dB for over 60% of the scenarios, indicating that the sub-optimality of using a common beam for a frame is not significant. This can be attributed to the multipath nature of the indoor environment, which allows for multiple beams to perform reasonably well for a given location, thereby increasing the possibility of finding a mutually *good* beam for multiple clients. However, as more and more clients are multiplexed together, the sub-optimality will tend to increase.

Reduced Overhead with Batch Scheduling: Client requests for resources on the uplink typically experience contention, delay and overhead, with redundant requests being sent out prior to transmission. To overcome these issues, typical OFDMA systems

allow the BS to poll clients for data periodically (the period adjusted based on traffic variations). We profile the relative overhead between client-initiated and BS-polled requests, by measuring the number of requests generated for a fixed amount of data for a single client, in Fig. 2(b). We see that BS polling reduces the overhead by as much as five folds. Further, the overhead does not grow with input rate unlike with client-initiated requests. With polling, buffer occupancy information of clients is available only once every polling interval. During this interval, frames can be scheduled one at a time or jointly as a batch, with the latter yielding better performance owing to joint optimization.

Improved Decoding of Control Part: One not-so-obvious benefit of frame-level beamforming is its ability to improve the decoding of the control part, through improved beamforming gain. As described earlier in §2, the control part carries vital information that helps clients decode and transmit their data payload. If the control part in a frame cannot be decoded by the clients (e.g., due to interference from other cells), the frame fails to deliver even a single byte of data. To understand the impact of interference on the control part, we experiment with two BSs that have one client associated with each. To make sure that the interference impact is only on the control part, we configure the two BSs to use orthogonal sets of sub-channels for their data payloads (i.e., data payload is protected from interference). While one BS is omni-directional, the other is equipped with a beamforming antenna. The client associated with the beamforming BS is placed at various locations to generate varying interference scenarios. In each of the locations, we measure the downlink throughput observed by the client using each beam pattern.

The throughputs delivered by the best beam and the omni-directional beam are presented in Fig. 2(c). We see that there is a lot of room for improvement with beamforming ($\approx 7X$ on average) even when the data parts are immune to interference. There are two factors behind the observed throughput: (i) the rate (modulation) increase due to beamforming gain on the data part, and (ii) improved resilience of the control part and hence, reduced frame losses. Fig. 2(d) presents the relative contributions of these two components. We see that the improved decoding of the control part owing to frame-level beamforming is the dominant component and provides the bulk of the gain in most scenarios. This clearly indicates that the performance of existing resource isolation solutions such as [16] can be enhanced even further, if the control part decoding is improved with the help of frame-level beamforming[1]. To realize the benefits of frame-level beamforming[2] and batch scheduling, we next present our design of an integrated beamformer and uplink scheduler – *iBUS*.

---

[1]While a scheme that combines an external common beam on the control part with user-level beamforming on the data part may be ideal, the varying size of the control part prevents its realization due to lack of standards support.

[2]Hereafter referred to as simply beamforming.

4. DESIGN OF IBUS

4.1 Overview

In batch scheduling, given a polling interval (F frames) and the buffer occupancies of the clients, the problem is to effectively pack (schedule) the out-standing client data jointly over F frames, with potentially different schedules across frames. When batch scheduling is integrated with beamforming, in addition to packing, one also needs to determine a beam for each frame in the batch. Picking a beam for a frame and a subset of clients (for data packing) are inter-twined since the beam choice for a frame will affect the rate supported by clients in that frame and hence, the amount of data that can be packed. Similarly picking a subset of clients will influence the beam choice since the sub-optimality of beamforming will vary depending on the subset of clients.

Two aspects make the integrated scheduling problem challenging: **(a) Power Pooling:** In OFDMA, when multiple clients are multiplexed in the same uplink frame, they pool their powers on a smaller set of sub-channels, thereby potentially supporting higher rates per sub-channel. Batch scheduling across multiple frames enhances power pooling further - spreading a client's data across multiple frames will require fewer sub-channels per frame, resulting in higher power pooling gain per frame. **(b) Beamforming:** Spreading data across multiple frames for each client will maximize the power pooling gain. However, this results in more clients being scheduled in the same frame, making it harder to find a *good* common beam for each frame; this increases the sub-optimality of beamforming. The scheduler's objective is to strike the right trade-off between these two components.

Remarks: We note that unlike uplink, in the downlink, there is no notion of power pooling (due to fixed BS total transmit power) and no signaling overhead that may necessitate batch scheduling. Hence, downlink is limited to per-frame scheduling, solutions for which exist even when a user's rate varies across sub-channels [5], which is typically considered a hard problem. Further, the downlink can be integrated with beamforming by simply running existing schedulers for various beams and picking that schedule and beam, which maximizes an objective function. However for the uplink, the combination of batch scheduling and power pooling with beamforming, makes the problem challenging even when a user's rate does not vary across sub-channels.

4.2 Formulation and Hardness

The integrated uplink scheduling problem can be formally stated as the following non-linear integer program.

$$\text{ISP:} \quad \text{Maximize} \sum_i U_i(R_i)$$

$$R_i = \sum_{j,k} x_{i,j,k} \left\{ \sum_\ell y_{j,\ell} \cdot k \cdot r_{i,k,\ell} \right\} \leq B_i, \ \forall i \in \mathcal{K}$$

$$\sum_\ell y_{j,\ell} \leq 1, \ \forall j \in \mathcal{F}$$

$$\sum_{i,k} x_{i,j,k} \cdot k \leq N, \ \forall j$$

$$\sum_k x_{i,j,k} \leq 1, \ \forall i,j$$

where $x_{i,j,k}$ and $y_{j,\ell}$ are binary indicator (output) variables indicating the assignment of k contiguous sub-channels in frame j to user i and the assignment of beam ℓ to frame j respectively. $\mathcal{K}, \mathcal{F}, \mathcal{L}$ indicate the set of users, frames in the batch and the available beams,

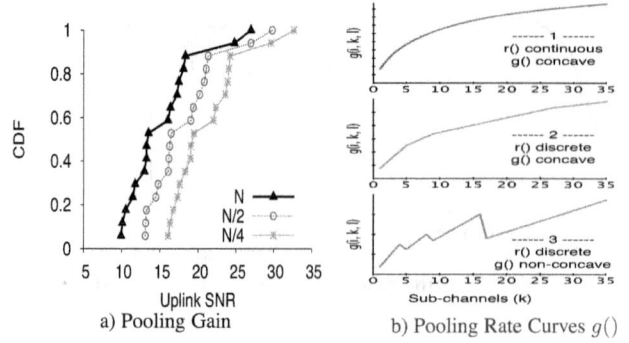

Figure 3: Power Pooling.

respectively. The objective is to maximize the aggregate client utility, where the utility function can be any concave function (e.g., logarithmic function captures proportional fairness [9]) of the data scheduled and can vary from one client to another. The first constraint limits the net allocation to a user to be limited to its buffer occupancy B_i (in bits). Further, it indicates the inter-dependence of power pooling (# sub-channels assigned, k) and beam chosen (ℓ) on the user's rate per sub-channel ($r_{i,k,\ell}$ in bits). The remaining constraints are all conservation constraints indicating the (i) assignment of a single beam per frame, (ii) net allocation of sub-channels being limited to N in a frame, and (iii) assignment of one contiguous set of sub-channels to a user in a frame, respectively.

We now show that it is hard to even obtain a PTAS (polynomial time approximation scheme) for the ISP problem. Specifically, we have the following result.

THEOREM 1. *For some $\epsilon > 0$, there is no $(1 - \epsilon)-$ approximation algorithm for ISP unless P=NP.*

PROOF. Scheduling of multiple sub-channels in a frame across users with finite buffers and varying rates across sub-channels (SCF) has been studied in [5] and shown to not admit a PTAS, using a reduction from 3-bounded 3-matching problem. We next show that SCF is a special case of ISP.

Consider a simpler version of ISP, where the beam choices for the frames do not have to be determined but are given a priori (ℓ_j for frame j). Further, consider only one sub-channel per frame ($N = 1$) available for allocation in each of the F frames. Since a user's (i) rate will vary from one beam to another and across frames ($r_{i,1,\ell_j}, \forall j$), the resulting ISP problem is now equivalent to an SCF problem with F sub-channels (mapping from frames in ISP) in a frame and users with finite buffers and varying rates across the sub-channels, where the rate of user i on sub-channel c is $r_{i,c,\ell_c}, c \in [1, F]$. Hence, the desired result. □

4.3 Components of iBUS

To address the above problem, we design and implement iBUS, which consists of the following key components.

Measurement of Beam SNRs: If there are $|\mathcal{L}|$ beam patterns, $|\mathcal{L}|$ frames are used for measurement (one for each pattern). On the uplink part of each frame, the BS schedules all the active users such that their data spans all the sub-channels (number of time symbols allocated varies across users). Such a schedule can be visualized as user bursts being arranged vertically rather than horizontally as in Fig. 1 (see uplink). The resulting SNR (ρ) for each user with each beam is measured directly at the BS and corresponds to that when power is split on all sub-channels ($\rho(i, N, \ell), \forall i, \ell$).

Rate Estimation using Power Pooling: Rate tables are determined separately to identify the best modulation and coding rate

(MCS) to employ for a given SNR that yields a specific loss rate (e.g., < 0.1). However, since the rate would vary with the number of sub-channels allocated (due to power pooling), a direct SNR measurement for each possible set of sub-channels for each beam ($\rho(i, k, \ell)$) would constitute significant overhead. Hence, based on the SNR measurement from all sub-channels ($\rho(i, N, \ell)$), we extrapolate the SNR for other subsets of sub-channels, taking power pooling into account. For e.g., if half of the sub-channels are allocated to a user, then its resulting SNR $\rho(i, \frac{N}{2}, \ell) = \rho(i, N, \ell) + 3$ (in dB). Since the power gets distributed over $\frac{1}{2}$ of sub-channels, the per sub-channel power is doubled, corresponding to $+3$ in dB scale. In general, $\rho(i, \frac{N}{\alpha}, \ell) = \rho(i, N, \ell) + 10 \log_{10}(\alpha)$, $\alpha \in [1, \frac{N}{N-1}, \ldots, \frac{N}{1}]$. Fig. 3(a) validates the modeling of power pooling by measuring the SNR in practice for a client, when data is transmitted on different sets of sub-channels. As expected, as we reduce the number of sub-channels by a factor of half, the SNR at each client location increases by ≈ 3 dB.

From $\rho(i, k, \ell)$, one can obtain $r_{i,k,\ell}$ using the rate table. Essentially, every user (i) has a power-pooling rate curve ($g(i, k, \ell)$) as a function of number of sub-channels (k) and the beam chosen (ℓ). Depending on the nature of SNR-rate mapping, these curves may or may not be concave. If continuous Shannon rates are employed, then we have $g(i, k, \ell) = k \cdot \log_2(1 + \frac{N}{k}\rho(i, N, \ell))$, which is a concave function (see Fig. 3(b)). If discrete rate tables are used then $g(i, k, \ell) = k \cdot r_{i,k,\ell}$ results in a piece-wise linear (between SNR thresholds) function, where the mapping $r_{i,k,\ell} \leftarrow \frac{N}{k}\rho(i, N, \ell)$ is determined by the rate table. Depending on the rate-SNR thresholds, $g(i, k, \ell)$ may still be concave or not (Fig.3(b) shows $g()$ for two such examples). As we shall see in §5, depending on the nature of $g()$ (concave or arbitrary), different algorithms will be required.

Buffer Estimation using Polling: The BS polls all its clients towards the end of the current polling interval to estimate their buffer occupancies for scheduling during the next polling interval. Note that if the total buffer occupancy of all clients is less than the total frame resources in the polling interval, then there will be under-utilization. Further, the schedules computed for the polling interval will be relevant only if the SNRs are relatively static during that interval. Since we focus on indoor femtocell deployments with stationary clients, the coherence time of SNRs are reasonably high (several frames) as we have experimented on our prototype (details in §6).

Schedule Determination: Once the BS has all the information needed to determine the batch schedule, it executes its algorithms. The batch schedule consists of two parts. The first part is a schedule for $F - |\mathcal{L}|$ frames ($|\mathcal{L}| << F$), where the ISP problem is solved over $F - |\mathcal{L}|$ frames. The latter part is a schedule for the remaining $|\mathcal{L}|$ frames, and serves as the measurement interval to obtain the SNR information for the next polling interval. Hence, in the latter part, the beam choices for the $|\mathcal{L}|$ frames are fixed (one beam each) during measurement, and allocation is done only across time symbols by employing the same algorithm used for solving ISP, albeit with some fixed variables. While the main purpose of the second part is measurement, one can also optimize the packing of client data that remains after the first $F - |\mathcal{L}|$ frames, given the beam choices. We next describe our solution to the ISP problem, where we determine the data packing along with the beam choices over a given set of frames.

5. ALGORITHMS IN IBUS

Given the hardness of the scheduling problem ISP, we focus on algorithms that have a provable worst case guarantee and are simple

---

Algorithm 1 Integrated Scheduler: iBUS1

1: INPUT: Buffer occupancy B_i, $\forall i \in \mathcal{K}$; rates $g(i, k, \ell)$, $\forall i, k \in [1, N], \ell \in \mathcal{L}$
2: OUTPUT: Beam choices ℓ_j, $\forall j \in \mathcal{F}$; per-frame user allocation A_{ij}, $\forall i, j$
3: **for** $f \in [1 : |\mathcal{F}|]$ **do**
4:    $O_{max} = 0$
5:    **for** $\ell_f \in [1 : |\mathcal{L}|]$ **do**
6:       $D_i \leftarrow B_i, \forall i$; $a_{ij} = A_{ij}, \forall i, j \in [1, f-1]$ and $a_{ij} = 0$ $\forall i, j = f$
7:       **for** $k = 1 : N$ **do**
8:          $i^* = \arg\max_{i \in \mathcal{K}} \Big\{ U_i\Big(\sum_{j=1}^{f} g(i, a_{ij}, \ell_j) + \min\big(g(i, a_{if} + 1, \ell_f)$ $-g(i, a_{if}, \ell_f), D_i\big)\Big) - U_i(\sum_{j=1}^{f} g(i, a_{ij}, \ell_j)) \Big\}$
9:          **if** $i^* \neq \emptyset$ **then**
10:             $D_{i^*} \leftarrow D_{i^*} - \min\big(g(i^*, a_{i^*f} + 1, \ell_f)$ $-g(i^*, a_{i^*f}, \ell_f), D_{i^*}\big)$
11:             $a_{i^*f} \leftarrow a_{i^*f} + 1$
12:          **else** break **endif**
13:       **end for**
14:       $O_f = \sum_{i \in \mathcal{K}} U_i\Big(\sum_{j=1}^{f} g(i, a_{ij}, \ell_j)\Big)$ $-U_i\Big(\sum_{j=1}^{f-1} g(i, a_{ij}, \ell_j)\Big)$
15:       **if** $O_f > O_{max}$ **then**
16:          $O_{max} = O_f; \ell_{max} = \ell_f; A_{if} = a_{if}, \forall i$
17:       **endif**
18:    **end for**
19:    $\ell_f = \ell_{max};;$ $B_i \leftarrow B_i - \min(g(i, A_{if}, \ell_f), B_i), \forall i$
20: **end for**

---

to implement. Our algorithms can be classified based on the nature of the power pooling rate curve.

5.1 Concave Rate Curves

We present two greedy algorithms called iBUS1 and iBUS2, outlined in Algs. 1 and 2, respectively. In iBUS1, resource allocation and beam selection are performed one frame at a time in the batch, sequentially. For a given frame (iteration), based on the remaining data available for the users, resource allocation is performed for every choice of the beam, and the beam yielding the best aggregate marginal utility for the frame is chosen (steps 14-16). For resource allocation within each frame given a beam (steps 6-13), iBUS1 assigns each sub-channel to the user who yields the highest marginal utility taking the users' buffer status into account (step 8). For concave utility functions, such an allocation is indeed optimal. Further, allocation based on the marginal utility also helps multiplex multiple users in the same frame thereby maximizing the benefits of power pooling. Once the best beam and its corresponding resource allocation are determined (steps 14-17) for the current frame, the user buffers are updated (step 19) and the procedure is repeated for the remaining frames in the batch, sequentially. iBUS1 has a time complexity of $O(|\mathcal{F}||\mathcal{L}||\mathcal{K}|N)$.

The key difference between iBUS1 and iBUS2 is that while iBUS1 performs resource allocation within each frame in isolation, iBUS2 performs joint resource allocation across all frames considered in an iteration. Hence, in addition to enabling power pooling within each frame, it allows a user's data to be spread across multiple frames thereby pooling the user's power across frames as well. This, in turn, results in a more efficient data packing and hence, higher aggregate throughput. Specifically, when considering a beam for a given frame (f) during a considered iteration, resource allocation is performed for all frames till the current frame jointly (steps 6-13) to determine the resulting utility for the beam. The beam yielding the highest utility is then chosen for that frame (steps 15-19). Hence, each iteration of resource allocation now involves de-

---
Algorithm 2 Integrated Scheduler: iBUS2

1: INPUT: Buffer occupancy B_i, $\forall i \in \mathcal{K}$; rates $g(i,k,\ell)$, $\forall i, k \in$ $[1, N], \ell \in \mathcal{L}$
2: OUTPUT: Beam choices ℓ_j, $\forall j \in \mathcal{F}$; per-frame user allocation A_{ij}, $\forall i, j$
3: **for** $f \in [1 : |\mathcal{F}|]$ **do**
4:     $O_{max} = 0$
5:     **for** $\ell_f \in [1 : |\mathcal{L}|]$ **do**
6:         $D_i \leftarrow B_i, \forall i$; $A_{ij} = 0 \, \forall i, j = [1, f]$; $k_j = N, \forall j \in [1, f]$
7:         **while** $k_j \neq 0, \exists j \in [1, f]$ **do**
8:             $(i^*, j^*) = \arg\max_{i \in \mathcal{K}, j \in [1,f]} \Big\{ U_i \Big(\sum_{m=1}^{f} g(i, A_{im}, \ell_m)$
                $+ \min\{g(i, A_{ij} + 1, \ell_j) - g(i, A_{ij}, \ell_j), D_i\} \Big)$
                $- U_i(\sum_{m=1}^{f} g(i, A_{im}, \ell_m)) \Big\}$
9:             **if** $(i^*, j^*) \neq \emptyset$ **then**
10:                $D_{i^*} \leftarrow D_{i^*} - \min \big(g(i^*, A_{i^*j^*} + 1, \ell_{j^*})$
                $- g(i^*, A_{i^*j^*}, \ell_{j^*}), D_{i^*} \big)$
11:                $A_{i^*j^*} \leftarrow A_{i^*j^*} + 1$; $k_{j^*} \leftarrow k_{j^*} - 1$
12:             **else** break **endif**
13:         **end while**
14:     **end for**
15:     $O_f = \sum_{i \in \mathcal{K}} U_i \Big(\sum_{j=1}^{f} g(i, A_{ij}, \ell_j) \Big)$
16:     **if** $O_f > O_{max}$ **then**
17:         $O_{max} = O_f$; $\ell_{max} = \ell_f$
18:     **endif**
19:     $\ell_f = \ell_{max}$
20: **end for**
---

termining not only the user to whom a sub-channel must be allocated but also in which frame ($\in [1, f]$, step 8). Note that once the beam is chosen for a given frame, user buffers do not have to be updated for the next frame since resource allocation will then be re-performed. Thus, for every newly considered frame, user buffers are re-initialized (step 6). Essentially, resource allocation for all frames till the one under consideration is done mainly for the purpose of determining the beam yielding the highest utility for the considered frame. The actual resource allocation is the one that is computed during the final frame of iteration, where the allocation for all the frames in the batch is jointly determined along with the beam for that frame. While iBUS2 enables power pooling both within and across frames, the joint resource allocation across frames results in an additional time complexity of $O(|\mathcal{F}|)$ and hence a net time complexity of $O(|\mathcal{F}|^2 |\mathcal{L}| |\mathcal{K}| N)$.

While iBUS2 results in a better performance than iBUS1, we now show that *even* iBUS1's worst case performance can be bounded. We provide some definitions on matroid and sub-modularity that are relevant for the proof.

Partition Matroid: Consider a ground set Ψ and let S be a set of subsets of Ψ. S is a matroid if, (i) $\emptyset \in S$, (ii) If $P \in S$ and $Q \subseteq P$, then $Q \in S$, and (iii) If $P, Q \in S$ and $|P| > |Q|$, there exists an element $x \in P \backslash Q$, such that $Q \cup \{x\} \in S$. A partition matroid is a special case of a matroid, wherein there exists a partition of Ψ into components, $\phi_1, \phi_2, \ldots$ such that $P \in S$ if and only if $|P \cap \phi_i| \leq 1$, $\forall i$.

Sub-modular function: A function $f(\cdot)$ on S is said to be sub-modular and non-decreasing if $\forall x, P, Q$ such that $P \cup \{x\} \in S$ and $Q \subseteq P$ then,

$$f(P \cup \{x\}) - f(P) \leq f(Q \cup \{x\}) - f(Q)$$
$$f(P \cup \{x\}) - f(P) \geq 0, \quad \text{and } f(\emptyset) = 0$$

THEOREM 2. *iBUS1's worst case performance is within $\frac{1}{2}$ of the optimum.*

PROOF. The sub-optimality of maximizing a sub-modular function over a partition matroid using a greedy algorithm of the form

$x = \arg\max_{x \in \phi_i} f(P \cup \{x\}) - f(P)$ in every iteration was shown to be bounded by $\frac{1}{2}$ in [17]. We will now show that iBUS1 is such an algorithm, with our scheduling objective corresponding to a submodular function to obtain the desired result.

Let the ground set be composed of the following triplets.

$$\Psi = \{(j, \ell_j, \mathbf{a}_j) : \quad j \in [1 : |\mathcal{F}|], \ell_j \in [1 : |\mathcal{L}|],$$
$$\mathbf{a}_j = \cup_{i \in \mathcal{K}} a_{ij} \text{ s.t. } \textstyle\sum_i a_{ij} \leq N \}$$

Now Ψ can be partitioned into $\phi_j = \{(j, \ell_j, \mathbf{a}_j) : \ell_j \in [1 : |\mathcal{L}|], \mathbf{a}_j = \cup_{i \in \mathcal{K}} a_{ij} \text{ s.t. } \sum_i a_{ij} \leq N\}$, $\forall j$. Let S be defined on Ψ as a set of subsets of Ψ such that for all subsets $P \in S$, we have (i) if $Q \subseteq P$, then $Q \in S$; (ii) if element $x \in P \backslash Q$, then $Q \cup \{x\} \in S$; and (iii) $|P \cap \phi_j| \leq 1$, $\forall j$. This means that S is a partition matroid. Further, any $P \in S$ will provide a feasible schedule with at most one feasible allocation and beam choice for each frame. Note that, every feasible schedule is contained in S by definition since it allows for the possibility of all feasible channel allocations and beam choices to be selected for each frame in the schedule. This, in turn, allows the partition matroid to capture our scheduling problem. Our scheduling objective is given as,

$$f(P) \;=\; \sum_{i \in \mathcal{K}} \mu_i(P)$$

$$\text{where, } \mu_i(P) \;=\; U(\min\{ \sum_{j:(j,\ell_j,\mathbf{a}_j) \in P} g(i, a_{ij}, \ell_j), B_i\})$$

It can be seen that if $Q \subseteq P$, then $\mu_i(Q) \leq \mu_i(P)$. Hence, for an element $(j, \ell_j, \mathbf{a}_j)$ such that $P \cup \{(j, \ell_j, \mathbf{a}_j)\}$ forms a valid schedule, it follows that $f(P \cup \{(j, \ell_j, \mathbf{a}_j)\}) - f(P) \leq f(Q \cup \{(j, \ell_j, \mathbf{a}_j)\}) - f(Q)$. This results both from the potential buffer limitation in subsequent frames as well as the concave nature of the utility function. This establishes that the function $f(P)$ is indeed sub-modular. Further, our scheduling problem aims to maximize this non-decreasing sub-modular function over a partition matroid. Hence, if the optimal allocation corresponding to every beam were given by some oracle for each frame, then picking the beam yielding the highest marginal utility for a frame in iBUS1 (steps 14-17) would correspond to determining

$$(j, \ell_j^*, \mathbf{a}_j^*) = \arg \max_{(j, \ell_j, \mathbf{a}_j) \in \phi_j} \{f(P \cup \{(j, \ell_j, \mathbf{a}_j))\}) - f(P)\}$$

Thus, the sub-optimality of $\frac{1}{2}$ would then follow from the result in [18].

Note that, there are exponential allocations possible that satisfy $\sum_i a_{ij} \leq N$ for every beam choice. Hence, bypassing the oracle assumption would require us to find the optimal allocation for a given beam, which is a problem in itself. However, since the utility functions are concave, this would correspond to a concave optimization problem, which iBUS1 solves optimally by employing a steepest gradient-like approach based on maximum marginal utility (steps 7-13). Hence, iBUS1 is able to bound its sub-optimality by $\frac{1}{2}$. □

5.2 Arbitrary Rate Curves

For arbitrary power pooling rate curves, we present the following algorithm iBUS3, which is a modified version of iBUS1. Specifically, iBUS3 is similar to iBUS1 in picking the beam yielding the highest marginal utility for each frame. However, its frame allocation for a given beam choice is significantly different. This is because if the power pooling rate curves are not concave, then allocations based on marginal utilities will not work. Hence, allocations to users in each frame must be made as a subset of sub-channels instead of individual sub-channels. Thus, we model the allocation problem for a given frame and beam choice (f, ℓ_f) with arbi-

Algorithm 3 Integrated Scheduler: iBUS3
___
1: INPUT: Buffer occupancy B_i, $\forall i \in \mathcal{K}$; rates $g(i,k,\ell)$, $\forall i,k \in [1,N], \ell \in \mathcal{L}$
2: OUTPUT: Beam choices ℓ_j, $\forall j \in \mathcal{F}$; per-frame user allocation A_{ij}, $\forall i,j$
3: **for** $f \in [1:|\mathcal{F}|]$ **do**
4:      $O_{max} = 0$
5:      **for** $\ell_f \in [1:|\mathcal{L}|]$ **do**
6:          $D_i \leftarrow B_i, \forall i$; $a_{ij} = A_{ij}, \forall i,j \in [1,f-1]$ and $a_{ij} = 0$ $\forall i,j=f$
7:          Run an FPTAS for multiple-choice knapsack problem $FA(f,\ell_f)$; Obtain $a_{if}, \forall i$
8:          $O_f = \sum_{i \in \mathcal{K}} U_i \left(\sum_{j=1}^{f} g(i, a_{ij}, \ell_j) \right)$
                $- U_i \left(\sum_{j=1}^{f-1} g(i, a_{ij}, \ell_j) \right)$
9:          **if** $O_f > O_{max}$ **then**
10:             $O_{max} = O_f$; $\ell_{max} = \ell_f$; $A_{if} = a_{if}, \forall i$
11:          **endif**
12:      **end for**
13:      $\ell_f = \ell_{max}$;; $B_i \leftarrow B_i - \min(g(i, A_{if}, \ell_f), B_i), \forall i$
14: **end for**
___

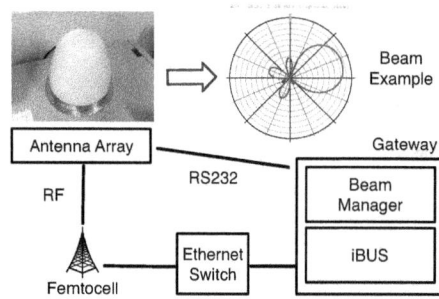

Figure 4: iBUS system architecture.

trary rate curves as the following multiple choice knapsack problem (MCKP), which is a NP-hard problem in itself.

$$FA(f,\ell_f): \quad \text{Maximize} \quad \sum_i \sum_{k \in \mathcal{N}_i} u_{ik}(f,\ell_f) x_{ik}(f)$$

$$\text{s.t.} \quad \sum_i \sum_{k \in \mathcal{N}_i} k \cdot x_{ik}(f) \leq N$$

$$\sum_{k \in \mathcal{N}_i} x_{ik}(f) = 1; \; \forall i$$

where $x_{ik}(f) \in \{0,1\}$ and $\mathcal{N}_i \in \{0,1,\ldots,N\}$.

Essentially, there are $N+1$ allocations ($\mathcal{N}_i$) possible for each user and the MCKP problem is to pick exactly one allocation for each user such that the total frame allocation does not exceed N. The user's profit or utility for each allocation is computed based on the current and previous frames' allocations as follows:

$$u_{ik}(f,\ell_f) = \sum_{i \in \mathcal{K}} U_i \left(\sum_{j=1}^{f-1} g(i, a_{ij}, \ell_j) + \min\{g(i, x_{ik}(f), \ell_f), B_i\} \right)$$

where $a_{ij} = k :$ s.t. $x_{ikj} = 1$, $\forall i,j \in [1,f-1]$. Note that the user's buffer occupancy is automatically accommodated in its profit. Now a FTPAS based on dynamic programming from [19] can be employed to efficiently solve $FA(f,\ell_f)$ to within $(1-\epsilon)$ of the optimal at a time complexity of $O(\frac{|\mathcal{K}|N}{\epsilon})$. This results in a net time complexity of $O(\frac{|\mathcal{F}||\mathcal{L}||\mathcal{K}|N}{\epsilon})$ for iBUS3.

In establishing a bound on the worst case performance of iBUS3, we employ the following lemma proved in [20].

LEMMA 1. *If the sub-modular function is only α- approximable, then the approximation guarantee of greedy maximization changes to $\frac{\alpha}{p+\alpha}$, where the maximization is subject to a $p-$ independence system.*

THEOREM 3. *iBUS3 provides an approximation guarantee of $\frac{1}{2} - \epsilon$.*

PROOF. Note that matroids are 1-independence systems (see [21] for an exposition). Given that the FPTAS yields a $(1-\epsilon)$ approximation in computing the frame allocation and hence the sub-modular function, we have $\alpha = 1 - \epsilon$ and hence the resulting performance guarantee of iBUS3 reduces to $\frac{1}{2} - \epsilon$. $\quad \square$

Remarks: While we have shown that our algorithms have an approximation guarantee of half for both concave and arbitrary power pooling rate curves, one might wonder how much further can this guarantee be improved without increasing the complexity significantly. Unfortunately, the answer is not optimistic. We had shown that ISP does not admit a PTAS and hence a $(1-\epsilon)$ approximation. Further, recall from §4.2 that single frame scheduling with varying sub-channel rates and finite user buffers (SCF) is a special case of our ISP problem. In [5], it was shown that one can improve the guarantee for SCF to 0.63 albeit through a complex LP relaxation and rounding procedure, where an exponential number of subsets is involved in the LP formulation. In light of this, we believe that our greedy algorithms strike a good balance between both performance and complexity. Further, §7 reveals their close-to-optimal performance in practice, on average, compared to their worst case guarantees of half.

6. IMPLEMENTATION

Fig. 4 depicts our system design. The testbed consists of a Wi-MAX femtocell platform and commercial clients (USB dongles attached to laptops). Our experiments are conducted over-the-air (10 MHz bandwidth) with an experimental license from FCC. We have an eight-element phased array antenna designed by Fidelity Comtech, attached via a RF cable to the BS. The array generates 16 beam patterns of $45°$ each, spaced $22.5°$ apart to cover the entire azimuth of $360°$. We also have a Linux PC for the gateway functionality required for WiMAX. In WiMAX, the gateway manages the service flows needed to transmit/receive data to/from the clients.

The BS and the gateway communicate using sockets via an Ethernet switch. When the BS decodes user data on the uplink, it passes it to the application at the gateway (*iperf*) via the switch. In addition, the switch is used to pass the client buffer information and the measured beam SNRs to the gateway. The gateway both implements our algorithms in Java, and at the same time controls the array by a serial port (RS232) application that we have developed in C.

When the gateway receives the buffer and SNR information, it executes the iBUS algorithms and sends the computed batch schedule to the BS, one frame at a time. A beam corresponding to a frame in the batch is applied to the array just before the corresponding schedule is sent to the BS. However, note that there is inevitably a delay before a particular beam is applied by the antenna following the gateway command. We measured this delay to be $\approx$ 6ms on average (8ms max.). Since each WiMAX frame is 5ms, applying one beam for a single frame is difficult. To circumvent this, each schedule corresponding to a frame in the batch is *extended* over 20 frames at the BS, by taking into account the resources from the 20 frames

Modulation	Bits per Carrier	SNR
$QPSK^{1/2}$	1	10 dB
$QPSK^{3/4}$	1.5	12 dB
$16QAM^{1/2}$	2	15 dB
$16QAM^{3/4}$	3	18 dB

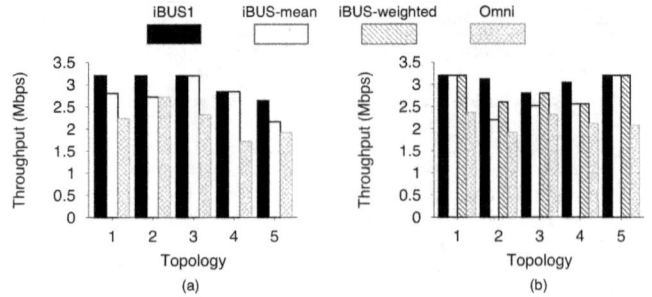

(a) MCS Thresholds (b) $g()\,Transformation$ (c) Equal rates (left), different rates (right)

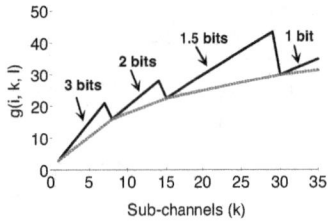

Figure 5: Parameters and performance results with iBUS in real experiments.

jointly. In addition, each beam l_i is applied 10ms (two frames) prior to schedule i. With this a priori beam application, when the BS starts executing schedule i, the hardware is ready to transmit with beam l_i. One can alternatively implement the algorithms and the beam management component at the BS, for tighter beam and schedule synchronization. However, this would add complexity and overhead to the BS, making the deployment less practical. Our design realizes a light-weight, standards-compatible OFDMA beamforming system that does not result in significant processing overhead at the BS. In addition, since the gateway is agnostic of the underlying technology, we believe that it can flexibly be extended to other OFDMA systems such as LTE, which use a similar notion of sub-channels.

In our implementation, the BS polls the clients for buffer occupancies every 100 frames, thereby forming the batch scheduling interval. Since there are 16 beam patterns, 16 frames are used to measure the beam SNRs of each client. Selecting a large polling interval will not sufficiently capture the channel diversity (e.g., the SNRs can change in the interim), making the computed schedule sub-optimal. Conversely, a short polling interval will not sufficiently reduce the contention overhead for buffer requests. We have verified in our experiments that the SNRs with each beam are relatively stable during a 100 frame (500ms) interval, and that the client buffers are large enough to hold pending data.

Practicality of our Implementation: We wish to point out here that our implementation is compatible with off-the-shelf clients since it does not require any modification on the standard client functions. Polling is supported by the standard and each client, by default, responds to the BS with the amount of bits in their buffer. Further, the clients are oblivious to the beam selection by the BS. Note that the clients are omni-directional and it is the BS that applies different beam patterns to the signals it receives from these clients and measures the SNRs. We also keep the modifications required at the BS to a minimum. Indeed, the only change we made to the BS was to disable the stock scheduler and make the BS accept and apply schedules provided externally by the gateway (14 lines of code where the entire BS code is around 10K lines). We believe that this small modification can easily be realized as a firmware update. Naturally, the "brains" of our solution resides on the gateway, which can be configured with minimal deployment effort (e.g., the scheduler and the beam manager can be downloaded from a repository).

7. PERFORMANCE EVALUATION

In this section, we evaluate iBUS using both our prototype implementation and trace-driven simulations. To isolate the gains achieved with power pooling (i.e., packing) and beam selection,

we propose two simpler versions of iBUS1 that do not require beam adaptation across frames: **(a) iBUS-mean** schedules multiple clients in the same frame like iBUS1 but uses a fixed beam for every frame. The fixed beam is chosen to minimize the mean loss from the best beams of each client. Each client i has SNR ρ^i_{max} with some beam yielding the max. SNR from its perspective. If client i has SNR ρ^i_l for beam l, then the loss is $\rho^i_{max} - \rho^i_l$. iBUS-mean picks beam $l \in \mathcal{L}$ that minimizes the mean loss over all clients, **(b) iBUS-weighted** uses a fixed beam as in **iBUS-mean** but the loss is weighted based on the buffer occupancy of each client, to account for asymmetric application rates. In this case, the loss for a client is calculated as $(\rho^i_{max} - \rho^i_l) * B_i$. Again, the beam $l \in \mathcal{L}$ that minimizes this weighted mean loss over all clients is fixed for all frames in the batch. We compare our algorithms to an omni-directional scheme (labelled **omni**) that leverages power pooling by employing the same packing procedure as iBUS1 (but each frame is transmitted omni-directionally).

7.1 Prototype Evaluation

In WiMAX, the modulation to be used by the clients for uplink communications is determined by the BS. Since the BS directly measures the received SNR from clients, it uses a threshold-based table to determine the modulation for each client. Fig. 5(a) enlists the thresholds used in our platform.

With such discrete rate tables, all SNRs in between two consecutive thresholds map to the MCS with the lower threshold (similar to WiFi). For example in Fig. 5(a), the BS instructs the clients to use $QPSK^{\frac{3}{4}}$ for all SNRs between 12dB and 15dB. Depending on the thresholds (determination of which is not specified by the standard and is left to the vendor), this may result in non-concave client power-pooling curves ($g()$). Indeed with our femtocells, we observed that such non-concavity occurs in practice. While iBUS3 can address non-concavity, we choose to implement iBUS1, iBUS-mean and iBUS-weighted on our prototype because of their low complexity and ease of implementation.

To make iBUS1 compatible with non-concave $g()$, we transform a given $g()$ into a corresponding concave function. Fig. 5(b) depicts an example transformation where $g()$ is computed as the product of number of sub-channels and the number of bits that a given MCS encodes. To do the transformation, we first determine the number of sub-channels where each MCS transition occurs. This can easily be done by computing the SNR for a given number of sub-channels using the power pooling formula. Then, the transition points are connected by lines with an appropriate slope. Note that the resulting concave function assumes that the error rate at a given MCS level increases in conjunction with the SNR drop as the power is distributed over more sub-channels. Even if the table maps

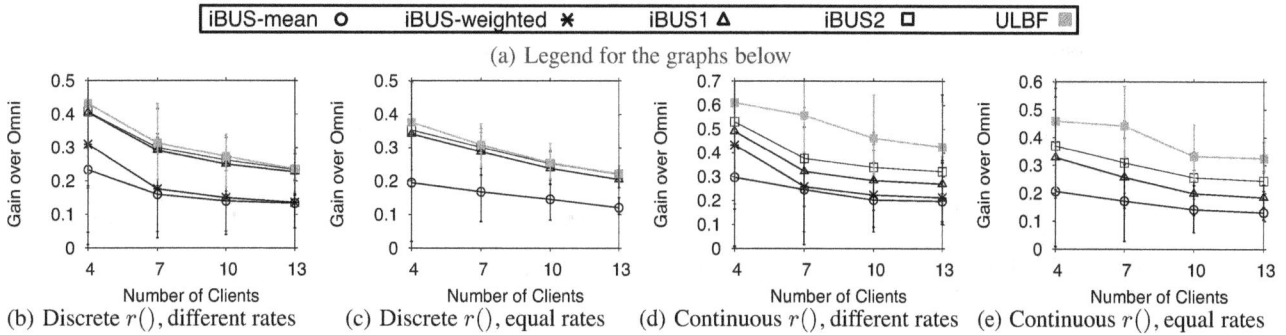

(a) Legend for the graphs below

(b) Discrete $r()$, different rates   (c) Discrete $r()$, equal rates   (d) Continuous $r()$, different rates   (e) Continuous $r()$, equal rates

Figure 6: Performance of iBUS obtained with trace-driven simulations.

two SNRs to the same MCS, the lower SNR will likely result in a higher bit (or symbol) error rate, reducing the throughput in practice. This observation is also leveraged by prior studies (e.g., [22]). In our experiments, we observed that the throughput predicted by the concave transformation matched the observed client throughput in practice reasonably well. On average, the predicted throughput was 94% of the actual throughput.

We run each scheme on several topologies, created by placing three clients in different locations. The utility function of a user is equal to the total data scheduled for that user in the batch (i.e., $U_i(R_i) = R_i$). The clients initiate UDP flows to the BS, generated by iperf. Fig. 5(c)(a) shows the aggregate throughput for five sample topologies of clients with equal application rates (iBUS-weighted yields the same fixed beam as iBUS-mean and hence, is not considered). We observe that iBUS1 outperforms omni by 42%, on average. In addition, the beam adaptation component in iBUS1 helps outperform iBUS-mean by 15% - 25%. However, cases exist where iBUS-mean performs as well as iBUS1, indicating that the proper choice of a *fixed* beam may provide significant gains, depending on the topology. Fig. 5(c)(b) shows the aggregate throughput for clients with different application rates. We observe that iBUS1 outperforms omni by 45% on average. For iBUS-weighted, we see that it can improve performance over iBUS-mean in some cases (e.g., topology 2 and 3). In addition, there are again cases where iBUS-mean and iBUS-weighted perform as well as iBUS1. To summarize, while iBUS1 consistently outperforms the omni-directional scheme by ≈45%, an appropriate choice of a fixed beam can also yield significant gains in practice.

7.2 Trace-driven Simulations

To evaluate with a large client population, we resort to SNR traces measured at 30 different client locations in our testbed. We consider 100 topologies, generated by picking random subsets of the client locations from our traces. To evaluate iBUS1 and iBUS2 with MCS tables in practice (i.e., discrete $r()$), we employ the same $g()$ transformation as before. We also consider a continuous $r()$ function (Shannon). We introduce an alternative scheme labelled **ULBF**, that mimics user-level beamforming and serves as a loose upper bound. In ULBF, packing is executed by spreading a client's data over multiple frames as in iBUS2. However, rather than using a common beam for a frame, each client can operate using its best beam (i.e., beam with the max. SNR.) even within a frame.

Fig. 6 shows the the mean gains achieved over omni for a varying number of clients. With discrete rate tables (Fig. 6(b) and 6(c)), we observe that iBUS1 delivers around 40% gain over the omni scheme for four clients (inline with our prototype evaluation). However, iBUS2 and ULBF do not significantly improve the per-

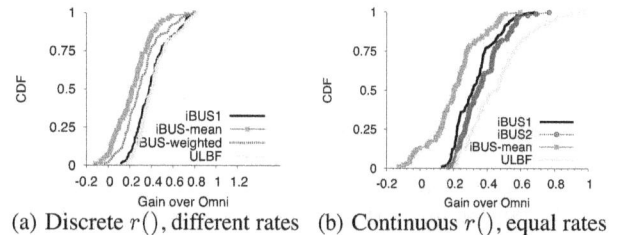

(a) Discrete $r()$, different rates   (b) Continuous $r()$, equal rates

Figure 7: CDF of the gain distribution with four clients. iBUS2 is not shown in (a) for clarity (it achieves similar performance as iBUS1).

formance further. Even though there is increased power pooling and hence higher SNR, the effective rate is not substantially higher than iBUS1 due to the nature of discrete MCS tables. In addition for all the schemes, the gains tend to drop with increasing numbers of clients. With a larger client population, finding a common beam comes at the cost of significant SNR loss from their best beam for some clients. On the other hand, the omni benefits from client diversity (increased chance of a client having high SNR with the omni beam) and is able to deliver increased throughput. However, since the femtocell client populations are typically small (<8), appreciable gains can still be attained. Thus, we conclude that iBUS1 is a low-complexity alternative to user-level beamforming in practical deployments. Since the increased power pooling and beamforming gains of user-level beamforming cannot fully be harnessed with discrete MCS tables, iBUS1 serves as a flexible alternative (i.e., number of antennas is not limited) with similar performance. Fig. 6(d) and 6(e) show the mean gains over the omni scheme for the continuous rate (Shannon) function. This time, we observe that iBUS2 outperforms iBUS1 by 10% - 20% due to its better power pooling capability across frames. Further, the gap with respect to ULBF is around 30% - 40%. This behavior is expected since the continuous rate function will benefit more from better power pooling and proper beam selection, compared to a discrete rate table.

In the above experiments, we have demonstrated the mean gains over the omni scheme. To better visualize the distribution, we plot the CDF of the gain for four clients. Fig. 7(a) shows that the maximum gain with iBUS1 can be as high as 60% - 80% for 20% of the scenarios. Further, an interesting observation that is revealed is that in 10% - 15% of the scenarios, iBUS-mean performs worse than omni. Recall that iBUS-mean picks a common beam that minimizes the mean loss from the best beam of each client. In such scenarios, some clients may have to operate with a beam, yielding a SNR that is several dB less than their optimal beam and even worse than the omni-directional beam. However, adapting the beam in iBUS1 addresses this issue to deliver improved performance. Similar observations hold for both types of rate functions (see Fig. 7(b)).

8. CONCLUSIONS

In this paper, we investigate the joint problem of uplink batch scheduling and frame-level beamforming (shown to not admit a PTAS) in OFDMA femtocells. We propose a set of algorithms with a worst case approximation guarantee of $\frac{1}{2}$. Our algorithms address the unique tradeoff between scheduling multiple users in a frame (benefiting from power pooling) and harnessing the gains from directional beams (by applying a common beam for multiple users). We implement the algorithms on a real WiMAX femtocell platform. Our prototype evaluation and trace-driven simulations demonstrate that our algorithms provide benefits of over 40% over an omni-directional scheme. To our best knowledge, this is the first study that integrates and evaluates frame-level beamforming on an actual OFDMA femtocell platform.

For future work, we would like to investigate the uplink scheduling problem, relaxing the assumption that the best user beams remain stable during a batch interval (i.e., the users are not necessarily static and can be mobile). In this paper, we have demonstrated the motivation for using beamforming for interference mitigation in a multi-cell context considering downlink traffic. We would also like to further investigate this direction, while considering beam selections for downlink scheduling in each femtocell.

Acknowledgements

We thank our shepherd, Dr. I-Hong Hou, for his guidance through the camera-ready submission process. The authors would also like to thank Meilong Jiang and Rajesh Mahindra from NEC Laboratories America Inc. for their invaluable support and suggestions during the preparation of this work. This work was partially supported by the Multi University Research Initiative (MURI) grant W911NF-07-1-0318.

9. REFERENCES

[1] Cisco, "Cisco Visual Networking Index: Global Mobile Data Traffic Forecast Update, 2010-2015," Feb 2011.

[2] V. Navda, A. P. Subramanian, K. Dhansekaran, A. Timm-Giel, and S. Das, "Mobisteer: Using Steerable Beam Directional Antenna for Vehicular Network Access," in *ACM MobiSys*, Jun 2007.

[3] M. Blanco et. al., "On the effectiveness of switched beam antennas in indoor environments," in *Passive and Active Measurements (PAM)*, Apr 2008.

[4] IEEE 802.16m 2011 Part 16, "Air Interface for Broadband Wireless Access Systems - Advanced Air Interface," *IEEE 802.16m standard*.

[5] M. Andrews and L. Zhang, "Scheduling Algorithms for Multi-carrier Wireless Data Systems," in *ACM MobiCom*, Sept 2007.

[6] S. Lee, I. Pefkianakis, A. Meyerson, S. Xu, and S. Lu, "Proportional Fair Frequency-Domain Packet Scheduling for 3GPP LTE Uplink," in *IEEE INFOCOM Mini-conference*, Apr 2009.

[7] IEEE 802.16e 2005 Part 16, "Air Interface for Fixed and Mobile Broadband Wireless Access Systems ," *IEEE 802.16e standard*.

[8] D.J. love, R.W. Heath, and T. Strohmer, "Grassmannian Beamforming for Multiple-Input Multiple-Output Wireless Systems," *IEEE Trans. on Information Theory*, vol. 49, no. 10, pp. 2735–2747, Oct 2003.

[9] G. Song and Y. Li, "Cross-Layer Optimization for OFDM Wireless Networks - Part I: Theoretical Framework," *IEEE Transactions on Wireless Communications*, vol. 4, no. 2, Mar 2005.

[10] A. Pokhariyal, G. Monghal, K. I. Pedersen, P. E. Mogensen, I. Z. Kovacs, C. Rosa, and T. E. Kolding, "Frequency Domain Packet Scheduling Under Fractional Load for the UTRAN LTE Downlink," in *IEEE VTC*, 2007.

[11] S. Deb, S. Jaiswal, and K. Nagaraj, "Real-Time Video Multicast in WiMAX Networks," in *IEEE INFOCOM*, 2008.

[12] B. Bai, W. Chen, Z. Cao, and K. B. Letaief, "Uplink Cross-Layer Scheduling with Differential QoS Requirements in OFDMA Systems," *EURASIP Journal on Wireless Communications and Networking*, vol. 2010, no. 168357, 2010.

[13] C. Nie, M. Venkatachalam, and X. Yang, "Adaptive polling service for next-generation ieee 802.16 wimax networks," in *IEEE GLOBECOM*, 2007.

[14] P. Svedman, S. K. Wilson, Jr. L. J. Cimini, and B. Ottersten, "Opportunistic Beamforming and Scheduling for OFDMA Systems," *IEEE Trans. on Communications*, vol. 55, no. 5, May 2007.

[15] N. Wei, A. Pokhariyal, T. B. Sorensen, T. E. Kolding, and P. E. Mogensen, "Performance of MIMO with Frequency Domain Packet Scheduling in UTRAN LTE Downlink," in *IEEE VTC*, 2007.

[16] M. Y. Arslan, J. Yoon, K. Sundaresan, S. V. Krishnamurthy, and S. Banerjee, "FERMI: A Femtocell Resource Management System for Interference Mitigation in OFDMA Networks," in *ACM MobiCom*, Sept 2011.

[17] L. Fleischer, M. X. Goemans, V. S. Mirrokni, and M. Sviridenko, "Tight Approximation Algorithms for Maximum General Assignment Problems," in *ACM SODA*, 2006, pp. 611–620.

[18] M. Fisher, G. Nemhauser, and G. Wolsey, "An analysis of approximations for maximizing submodular set functions-II," in *Mathematical Programming Study*, 1978.

[19] M. Bansal and V. Venkaiah, "Improved Fully Polynomial time Approximation Scheme for the 0-1 Multiple-choice Knapsack Problem," in *International Institute of Information Technology Tech Report*, 2004, `http://people.csail.mit.edu/mukul/01MCKP.pdf`.

[20] G. Calinescu, C. Chekuri, M. PǍl, and J. VondrǍk, "Maximizing a Submodular Set Function subject to a Matroid Constraint (Extended Abstract)," in *Proc. of 12th Conf. on Integer Programming and Combinatorial Optimization*, 2007.

[21] P.R. Goundan and A.S. Schulz, "Revisiting the Greedy Approach to Submodular Set Function Maximization," in *Optimization Online Pre-print*, 2007, `http://www.optimization-online.org/DB_FILE/2007/08/1740.pdf`.

[22] M. Vutukuru, H. Balakrishnan, and K. Jamieson, "Cross-Layer Wireless Bit Rate Adaptation," in *ACM SIGCOMM*, 2009.

A Case for Adaptive Sub-carrier Level Power Allocation in OFDMA Networks

Shailendra Singh
University of California, Riverside
singhs@cs.ucr.edu

Moloud Shahbazi
University of California, Riverside
mshah008@cs.ucr.edu

Konstantinos Pelechrinis
University of Pittsburgh
kpele@pitt.edu

Karthikeyan Sundaresan
NEC Labs America Inc., NJ
karthiks@nec-labs.com

Srikanth V. Krishnamurthy
University of California, Riverside
krish@cs.ucr.edu

Sateesh Addepalli
Cisco Systems, Inc.
sateeshk@cisco.com

ABSTRACT

In today's OFDMA networks, the transmission power is typically fixed and the same for all the sub-carriers that compose a channel. The sub-carriers though, experience different degrees of fading and thus, the received power is different for different sub-carriers; while some frequencies experience deep fades, others are relatively unaffected. In this paper, we make a case of redistributing the power across the sub-carriers (subject to a fixed power budget constraint) to better cope with this frequency selectivity. Specifically, we design a joint power and rate adaptation scheme (called JPRA for short) wherein power redistribution is combined with sub-carrier level rate adaptation to yield significant throughput benefits. We further consider two variants of JPRA: (a) JPRA-CR where, the power is redistributed across sub-carriers so as to support a maximum common rate (CR) across sub-carriers and (b) JPRA-MT where, the goal is to redistribute power such that the transmission time of a packet is minimized. While the first variant decreases transceiver complexity and is simpler, the second is geared towards achieving the maximum throughput possible. We implement both variants of JPRA on our WARP radio testbed. Our extensive experiments demonstrate that our scheme provides a 35% improvement in total network throughput in testbed experiments compared to FARA, a scheme where only sub-carrier level rate adaptation is used. We also perform simulations to demonstrate the efficacy of JPRA in larger scale networks.

Categories and Subject Descriptors

C.2.1 [**Network Architecture and Design**]: Wireless communication

General Terms

Algorithms, Experimentation, Measurement, Performance

Keywords

Rate Adaptation, Power Allocation, Orthorgonal Frequency Division Multiplexing (OFDM), Frequency Selective Fading

1. INTRODUCTION

Due to its inherent ability to cope with fading, OFDMA is employed today in many commercial wireless systems. The current implementations as specified by the IEEE standards [2] for such systems do not include schemes for adaptively assigning powers to subcarriers based on CSI (Channel State Information); instead, these systems evenly distribute the total transmission power budget across all subcarriers (waterfilling [21] with equal power without CSI knowledge).

However, at a receiver, since each sub-carrier typically experiences a different fade (as seen both by our work and in prior related efforts such as [16]) the transmission rates that can be supported by the different sub-carriers can differ. Spatial and temporal diversity can further complicate the communications between a single sender and multiple receivers at different times. Each receiver will likely have a different multipath fading profile given the differences in receiver locations. Even for the same receiver, the fading effects experienced by the different sub-carriers are expected to differ in time.

Traditional rate-adaptation algorithms try to cope with fading by varying the transmission rate in response to either packet losses (e.g., [23]) or signal-to-noise ratio (SNR) variations (e.g., [1]). However, these are not done at a per sub-carrier level. Thus, poor received quality on a few sub-carriers can affect the supported transmission rate. To cope with this, Rahul *et al.* [16], propose FARA, a scheme that supports per sub-carrier rate adaptation. While FARA provides gains over traditional rate-adaptation schemes, it does not address the problem of frequency-selectivity directly and does not utilize the available power budget efficiently.

In this paper, we propose redistributing the power across the sub-carriers to better cope with frequency selectivity. Specifically, we argue for an adaptive, scenario specific, uneven distribution of transmission powers across sub-carriers. By appropriately assigning the transmission powers to the different sub-carriers (while adhering to the total power budget for the transmission), and by combining this with the choice of appropriate per sub-carrier transmission rates, we envision achieving significant gains in throughput. As our main contribution, we design a joint power and rate adaptation (JPRA) scheme towards realizing these envisioned benefits.

Calibration phase in JPRA: In order to determine the right power levels for each sub-carrier, a calibration phase is necessary for JPRA. The goal of calibration is twofold; (i) to identify the *practical* decoding thresholds for each transmission rate, and (ii) to correlate the required change in the received signal of a sub-carrier (in order to support a rate) with an appropriate change in the corresponding transmit power at the sender. We perform several measurements to understand if such a calibration phase is viable. We use the error vector magnitude (EVM) [13] to determine the minimum power level needed for sustaining a specific transmission rate. We find that these minimum powers are consistent over a large number of scenarios and node locations. This suggests that infrequent benchmark measurements are sufficient to calibrate JPRA and utilize it for adaptive sub-carrier level power allocation during data-transfers.

Variants of JPRA: We propose two variants of JPRA. The first variant, named JPRA-CR (where CR stands for common rate), selects the power-assignments such that a single *best* rate can be used on an appropriately chosen set of sub-carriers. This best rate corresponds to the highest rate for which the EVM threshold is satisfied for the set of sub-carriers at the receiver. Since we operate on a fixed power budget, the scheme entails the transfer of power from relatively good sub-carriers (unaffected by fading) to those that are in deep fades. The advantage of this approach is that it is simple and reduces the transceiver complexity (since they only need to decode at a single rate). However, this may not yield the maximum achievable throughput with power re-allocation. As shown later however, it does offer an additional benefit in ensuring that transmission rate transitions are less likely than in traditional approaches (stabilizes the rate in use). In addition, it can potentially provide significant power savings for a target PDR (packet delivery ratio), compared to traditional systems.

Towards maximizing throughput via sub-carrier level rate allocation, we propose a second variant JPRA-MT. Our goal with this scheme is to minimize the total transmission time of a packet by appropriate power re-distribution. Essentially, with JPRA-MT sub-carrier powers are assigned to sub-carriers such that the total number of bits (as mapped to symbols on a constellation) transmitted on the sub-carriers in each symbol-duration is maximized. To illustrate with an example, consider a case where we have just two sub-carriers. The power assignments can be such that either both sub-carriers can be modulated using QPSK (2 bits per symbol) or, one sub-carrier can be modulated with BPSK (1 bit per symbol) while the other can be modulated with 16-QAM (4 bits per symbol). In this case, the packet airtime is minimized with the latter choice since more bits are transmitted per symbol duration (5 bits instead of 4 bits) and is consequently chosen by JPRA-MT. We wish to point out here that in order to support a higher modulation, a higher received power is required. Thus, in the latter case the power on the second sub-carrier (supporting 16-QAM) is increased compared to the power used in the former case (when it supported QPSK). Similarly, the power on the first sub-carrier has to be decreased in the latter case compared to the former one (and thus, it can support only BPSK as opposed to QPSK).

We implement both the versions of JPRA on our six node WARP radio testbed. We also implement FARA [16] for comparison. We perform extensive experiments which show that JPRA-MT outperforms FARA by as much as 35 % in terms of throughput. The gains are much more significant compared to traditional OFDMA systems (75 %). We also show the efficacy of JPRA in larger scale settings via simulations.

The main properties of JPRA are summarized below:

- *Throughput Efficiency*: JPRA-MT achieves 75% more network throughput than standard SNR based rate adaptation. It also outperforms the state of the art OFDMA rate adaptation scheme, FARA[16] by 35% in terms of total network throughput.

- *Potential Power savings*: JPRA-CR is more power efficient compared to systems with no power re-distribution. Speci-

fically, for a fixed BER (bit error rate), we observe upto 4.5 dBm in power savings.

- *Stability of Transmission Rates* : JPRA-CR reduces the number of rate changes by 27% in comparison to standard SNR based rate adaptation.

- *Impact on Carrier Sensing*: JPRA does not affect the RSS (received signal strength) of packets since the total power budget remains fixed.

We acknowledge that our system is currently applicable only on (quasi) static topologies. The calibration phase required for the deployment of our scheme cannot be as effective in scenarios that include mobility. However, we seek to examine mobility scenarios as part of our future work.

Organization: The paper is organized as follows. In Section 2, we provide relevant background and overview related work. The calibration phase of JPRA is described in Section 3. Details of the two JPRA variants for power/rate allocation are in Section 4. Section 5 describes both our experimental and simulation results. Section 6 concludes the paper.

2. BACKGROUND AND RELATED WORK

In this section, we first describe studies related to our work and then we briefly provide background on the EVM calculation required by our system.

Frequency Selective Fading: There are studies that employ rate adaptation to cope with frequency selective fading. For example, in [16] the authors propose Frequency Aware Rate Adaptation (FARA) to improve system performance. They assign subsets of sub-bands to each sender-receiver pair and based on the SNR reported by the receiver on these sub-bands, the sender performs rate adaptation. However, the authors do not propose a solution to improve the performance of sub-carriers experiencing frequency selective fading or low SNR. Barthia *et al* [5] propose a smart mapping of symbols to sub-carriers. This supports partial recovery of symbols if they are lost due to frequency selective fading. They also propose an extra layer of FEC codes on top of Physical layer FEC. One of the main limitations of this work is that the proposed solution is only compatible with block FEC schemes and its not clear how it will work with convolutional or turbo code FECs [2]. A large amount of feedback information is also required for the partial symbol recovery. In contrast our scheme is not limited by the choice of FECs and the amount of feedback information is low (we only need to send sub-carrier power values and rates).

Rate Adaptation : There is a large volume of studies on rate adaptation (e.g., [6, 1]). SampleRate [6], proposed by Bicket *et al.*, probes the performance at a random rate every 10 frames, and selects the rate that minimizes the expected transmission time (including retransmissions). Wong *et al.* [23] develop Robust Rate Adaptation Algorithm (RRAA), which uses short term loss ratios to opportunistically change rate. It further incorporates an adaptive RTS filter to prevent collision losses from lowering data rates. All these (and many similar) existing schemes adapt rate according to frame loss rates. RBAR [8] uses the RTS/CTS exchange to estimate the SNR at a receiver, and picks the transmission bit rate accordingly. OAR [17] further builds on RBAR, by opportunistically transmitting back-to-back frames when the channel quality is good. CHARM [10] leverages the reciprocity of the wireless channel to estimate the average SNR at the receiver using packets overheard from the latter. The overhead of RTS/CTS (present with

RBAR and OAR) is thus avoided and implementation on commodity cards is enabled. Sen *et al* [18] propose the use of EVM (Error Vector Magnitude) to perform rate adaptation. Since the above rate adaptation schemes use information such as loss ratio, SNR and EVM averaged over a packet, they fail to capture the effects of frequency selectivity. On the contrary, we try to directly address issues related to frequency selective fading by using per subcarrier EVM measurements.

Multi-User OFDM : There exist a few studies on subcarrier, power and rate assignments in multi-user scenario. In this case a single channel is shared among multiple users. In [12] Javidi *et al* propose a scheme to outperform "water filling based multi-user subcarrier assignment" by introducing a subcarrier allocation scheme which takes the queue lengths of different users into account. Adaptive power allocation for a multiuser OFDM environment has been proposed in [22, 9]. An OFDM channel is divided into multiple subbands and these subbands are assigned to different users. To alleviate frequency selective fading experienced by different users, redistribution of power and modulation on these subbands is done according to the SNR experienced by the users. However, the majority of these studies are evaluated only via simulations by making assumptions about channel conditions. Furthermore, power adaptation is done for each user and not on a per OFDM sub-carrier basis.

Error Vector Magnitude(EVM): In this work we use the EVM per subcarrier as the CSI (carrier state information) feedback from the receiver, for the sender to perform power redistribution and rate selection. EVM is a vector measurement taken in terms of peak (or rms) percentages between the ideal symbol position and the actual measured position in the constellation space for a particular modulation. The error vector is a vector in the I-Q plane between the ideal constellation point and the data interpretation by the receiver[13]. In other words, it is the difference between actual received symbols and ideal symbols. The average power of the error vector, normalized to the signal power, is the EVM. It can be expressed as a percentage:

$$EVM(\%) = \sqrt{\frac{P_{error}}{P_{reference}}} * 100$$

where, P_{error} is the RMS power of the error vector and $P_{reference}$ is defined as the reference constellation average power. In contrast to SNR, higher EVM values correspond to bad channel conditions, while lower EVM values represent good channel conditions.

3. CALIBRATING PHASE OF JPRA

JPRA is a measurement driven system and requires a set of calibrating measurements. These measurements are:

1. *Per sub-carrier EVM for each received packet.* This provides the system, knowledge with respect to both the sub-carriers that are experiencing deep fade and those that are relatively unaffected.

2. *EVM threshold for supporting each available rate.* This identifies the modulation that a sub-carrier can support.

3. *A mapping between a power adjustment and the corresponding EVM change.* This is critical in determining the effect of a power change on the rate that can be supported by a sub-carrier. In other words, it seeks to answer: By increasing (or decreasing) the transmission power on a sub-carrier by a certain magnitude, what is the higher (or lower) rate that that can be achieved on that sub-carrier ?

Note that the the per sub-carrier EVM should be obtained for every received packet (both during the calibration phase and when JPRA is being employed). The latter two measurements are to be obtained prior to applying power redistribution in JPRA, for the purposes of calibrating the system.

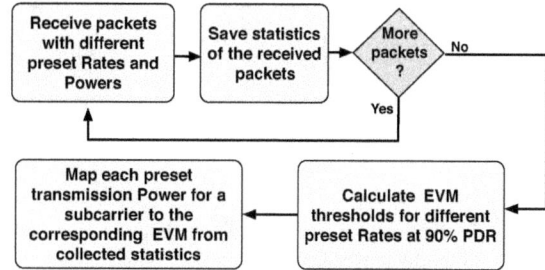

Figure 1: Calibration phase of JPRA

In what follows we provide more details on the way we obtain these measurements. Figure 1 outlines the main steps involved in calibrating JPRA. We defer a detailed description of our testbed to later; simply put, we have a six node WARP testbed that we use for all of these calibrations.

Calculating per sub-carrier EVM: EVM is an indicator of modulator or demodulator performance in the presence of impairments. For a receiver to calculate the EVM, both received and transmitted packets are required. The WARP boards provide the per sub-carrier EVM for every correctly received packet. If the packet passes the CRC check, the receiver can reconstruct the exact signal that was transmitted. The bits are re-mapped on to a constellation space to obtain ideal symbol positions based on the modulation used for the transmission. With the ideal symbol positions, it is easy to calculate the EVM as described in Section 2. With the above process it is not possible to calculate the EVM for packets that fail the CRC check; the receiver cannot deduce what was transmitted and thus, cannot obtain the ideal symbol positions. Thus, the EVM estimates are updated only based on packets that are correctly received.

Mapping rate to the EVM threshold: In order for a symbol on a sub-carrier to be demodulated with high probability, its EVM at the receiver needs to be below a specific threshold. This threshold is different for different modulation schemes. Similar thresholds has been determined in terms of SNR [16], but cannot be directly used with JPRA. The mapping between the EVM threshold and rate are calculated empirically via measurements, by varying the transmission power and the locations of the sender-receiver pair. The maximum EVM value that yields a target PDR (90% in our case) provides us with the required information. We observe that this mapping does not change from one setting to another. However, we recognize that it may vary for different hardware platforms. Thus, although we use one time pre-deployment measurements for calibration, in practice, they may have to be periodically repeated. The mappings that we obtain are presented in Table 1 and used to select the transmission bit rates with JPRA.

Estimating power adjustments for specific EVM changes: To perform joint rate and power adaptation we need to estimate the magnitude of increase/decrease in the EVMs of the received subcarriers, for given changes in the transmit power. This mapping will facilitate the power redistribution across the available sub-carriers. In the current implementation of OFDM on WARP Boards, the total transmission power (called the power budget) is distributed equally among the available sub-carriers. The radio board applies power to the analog baseband/RF waveforms. In other words, the time domain signal after the IFFT stage is processed for power alloca-

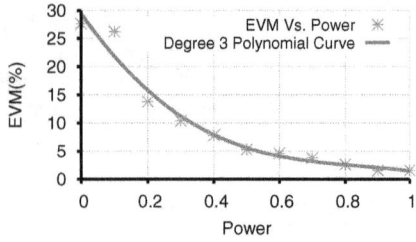

Figure 2: Average change in EVM(%) with change in sub-carrier Power from 0 to 1 for QPSK.

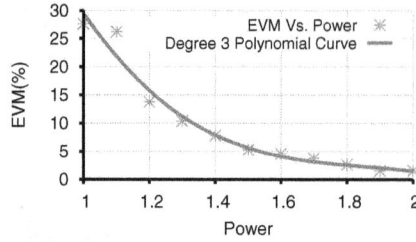

Figure 3: Average change in EVM(%) with change in sub-carrier Power from 1 to 2 for QPSK.

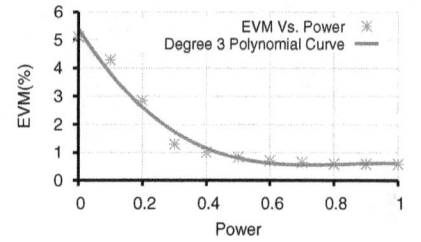

Figure 4: Average change in EVM(%) with change in sub-carrier Power from 0 to 1 for 16QAM.

tion. Since, in the time domain the individual sub-carriers are not explicitly visible, the power is uniformly distributed across all frequencies used. To achieve a non-uniform power allocation across the sub-carriers, we process each sub-carrier prior to applying the IFFT. The coefficient used to establish the power level of a sub-carrier does not correspond to actual transmission value but is a scaling factor. It takes real values between 0 and 2, with 1 representing the default power (uniform distribution of the power budget across sub-carriers), 0 mapping to zero transmission power and 2 to the maximum transmit power allowed (double the default transmission power for that sub-carrier).

In order to determine the scaling factor described above, we perform another set of calibration measurements. We use a pair of WARP board transceivers and we vary the transmission power of a sub-carrier by scaling its transmission power as explained above, with a step size of 0.1. At each step the transmitter sends back to back packets for 3 minutes; we perform 10 trials for each step. Figures 2 and 3, show the average EVM(%) change when QPSK is used when decreasing and increasing the transmission power of a sub-carrier, respectively. Analyzing the obtained data we find that a third-degree polynomial can fit the data fairly accurately (the corresponding R^2 value[1] is 96%). Consequently, this polynomial can be used to predict the change in EVM(%) with respect to sub-carrier power. We also looked at different modulations and we observe that qualitatively the results do not change with the change in modulation. To illustrate, Figure 4, depicts the results obtained when 16-QAM is used. A polynomial of degree 3 can again be used to fit the data (R^2 value is 95%). As one might expect though, the coefficients of the polynomial are different as compared to the QPSK case since the EVM(%) range is different. Finally, we repeated the above experiments using different sub-carriers obtaining similar results. In particular, the fitting polynomial coefficients for different sub-carriers for the same modulation are very similar. The above measurements are obtained for static links and the results do not significantly change for small variations in the transceivers' locations. However, for large deviations from the initial measurements positions the results will not hold; in particular, the polynomial coefficients will be different. Therefore, we need to perform these calibrating measurements when a new link appears or when the topology changes significantly. This limits the applicability of the current version of JPRA to static or slowly changing topologies. Extending JPRA to more dynamic settings is deferred to the future.

4. POWER/RATE ALLOCATION WITH JPRA

In this section we describe the two versions of JPRA in detail. In a nutshell, for both schemes, each receiver upon packet reception calculates the per sub-carrier EVM. Using this information JPRA

Maximum EVM(%)	Modulation	Coding
18.0	BPSK	1/2
10.2	BPSK	3/4
6.6	QPSK	1/2
4.0	QPSK	3/4
1.67	16-QAM	1/2
1.26	16-QAM	3/4
1.1	64-QAM	1/2

Table 1: EVM to bit-rate(modulation + coding) mapping

redistributes the power among the sub-carriers depending on the objective (i.e., sub-carriers use the same maximum possible common rate with JPRA-CR *or* the the packet air time is minimized while allowing sub-carriers to be modulated at different rates with JPRA-MT). Figure 5 depicts the high level functionalities of JPRA.

Upon executing the appropriate JPRA version, the receiver obtains a tuple for each sub-carrier specifying (i) the magnitude of increase or decrease in transmission power and (ii) the corresponding bit rate to use; this is represented as $\langle carrier\#, TXpower, Rate \rangle$. This is transmitted to the sender in the ACK (or NACK). The sender can then adjust its sub-carrier's transmission powers and bit-rates for the next packet. Note that in the case of JPRA-CR, all tuples will have the same rate information. If the channel conditions remain constant for a train of packets, the receiver may omit sending this information with every packet and the sender will use the latest sub-carrier settings for the next packet.

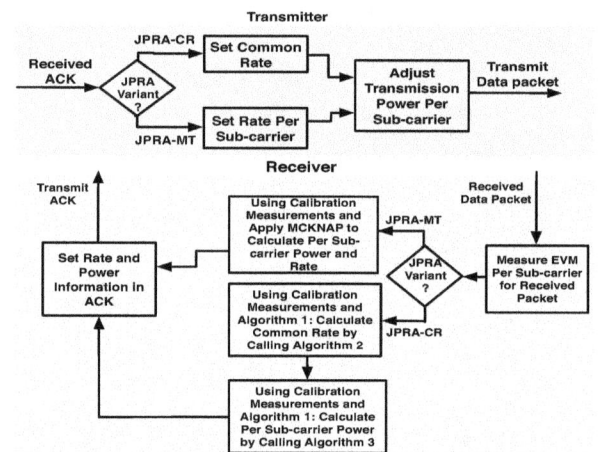

Figure 5: Power and Rate allocation with JPRA

---

[1] R^2 value[19] shows the goodness of fit of a model.

4.1 Achieving a common rate with JPRA-CR

Due to frequency selectivity, for a specified bit-rate, some of the sub-carrier's EVMs will be higher than the maximum tolerable EVM for that rate (class 1 sub-carriers); on the other hand, some others will have EVMs lower than what can be tolerated (class 2 sub-carriers). In conventional OFDMA systems, the modulation used is conservative. In particular, all sub-carriers are modulated with the minimum supported rate from among them. One can make a case for redistributing the *extra* power from class 2 sub-carriers to class 1 sub-carriers. This in turn, will reduce the EVM of the latter and can potentially allow the use a higher bit-rate modulation on all the sub-carriers. However, selecting a common rate as above for all the sub-carriers may not be optimal since there may exist carriers (in deep fades) that may be unable to carry data at acceptable rates. Thus, with JPRA-CR we propose to find a common rate on a *subset* of sub-carriers such that the aggregate rate is maximized. This constitutes removing sub-carriers which require high powers per bit delivered, and redistributing their powers to other sub-carriers; at the same time, it is possible that powers from some of the best sub-carriers are also carried over to other sub-carriers towards achieving the above objective. Note that there is an inherent trade-off here; while eliminating poor sub-carriers can increase the common rate that can be supported on the other sub-carriers, it also decreases the number of available sub-carriers and thus, contributes to a lowering of the capacity. JPRA-CR tries to find the best point of operation while accounting for this trade-off.

Let us assume, for simplicity, that all sub-carriers are considered for setting the common rate. Upon receiving a packet, the receiver iterates over the list of available rates that are greater than the currently supported lowest rate, to identify the maximum bit-rate that all the sub-carriers can satisfy after power redistribution. In each iteration, it has to solve the problem of per sub-carrier power redistribution for the specific rate considered. One can map this problem to a form of the well-known Knapsack problem (which we describe below) and based on this, we design JPRA-CR.

The solution to the considered problem is as follows. In every iteration, the receiver first subtracts the EVM of the sub-carriers under consideration from the EVM threshold of the rate under consideration (obtained from the calibration phase). For some sub-carriers this value will be negative, while for others it will be positive. If there are no negative values (the EVM of every carrier is lower than the rate's EVM threshold), all sub-carriers can be modulated using this rate. If not, it maps the EVM differences to the corresponding sub-carrier transmission power increases required in order to meet the threshold (negative weights), or the excess transmission powers present in the sub-carriers whose EVM adheres to the required threshold (positive weights). Again, this mapping is done based on the measurements from the calibration phase. The positive weights are summed to get the total extra energy available (in Knapsack problem terminology, the total weight that can be carried by the knapsack). Each of the absolute values of the negative weights is considered as the weight of the corresponding sub-carrier (item). Our objective is to select as many items as possible with the given total weight. Ideally, we want all items to be included in our Knapsack. If this can be achieved, all sub-carriers can be modulated with the rate considered and we further examine the next available higher rate. If this cannot happen, we terminate our iterations, since it is guaranteed that no other higher rate can be supported (the proof is trivial and we omit this for space reasons).

The problem of finding the maximum number of sub-carriers that can satisfy the given EVM threshold requirement (for the considered rate j) at each iteration, can be formalized by the following optimization problem. Let us assume that n sub-carriers are above

---

Input: Total number of sub-carriers n, Per sub-carrier EVM_i, Rate $j \in S \equiv \{6, 12...\}$, Rate threshold $EVMTH_j$, EVM to Power Mapping, Initial Per sub-carrier Power p_i'

Output: Rate $R(m)$, Per sub-carrier Power $p_k''(m)$

$R(n) =$ **Algorithm-2**(EVM_i, Rate $j \in S$, $EVMTH_j$, EVM to Power Mapping) ;

Aggregate Rate $AG_n = R(n) \times n$;

Remove Sub-carrier which requires highest amount of power per bit delivered and add it's power to total excess power W ;

$m \leftarrow n - 1$;
while $m > 1$ **do**
    $R(m) =$ **Algorithm-2**(EVM_i, Rate $j \in S$, $EVMTH_j$, EVM to Power Mapping);
    Calculate $AG_m = R(m) \times m$;
    if $AG_m < AG_{m+1}$ **then**
        $p_k''(m+1) =$ **Algorithm-3**(p_i', List $L_{j,1}$, List $L_{j,2}$);
        return $R(m+1)$ and $p_k''(m+1)$;
        break;
    end
    else
        Update $S = S - R(m)$;
        Remove Sub-carrier which requires highest amount of power per bit delivered add it's power to total excess power W ;
        Update $m \leftarrow m - 1$;
    end
end
$p_k''(m) = \texttt{Algorithm-3}(p_i', \texttt{List } L_{j,1}, \texttt{List } L_{j,2})$;
return $R(m)$ and $p_k''(m)$

Algorithm 1: Common rate selection and Power redistribution using JPRA-CR

---

the EVM threshold for rate j and have negative weights w_i^j (these are the items to choose from). We introduce a slack variable for each such item i, x_i^j, which is 1 if the sub-carrier is chosen for placement in the knapsack (assignment of power for rate j), and 0 if not. Then we need to solve the following problem:

$$maximize \sum_{i=1}^{n} x_i^j \qquad (1)$$

$$subject \ to \quad \sum_{i=1}^{n} x_i^j \cdot w_i^j <= W^j \qquad (2)$$

where $x_i^j \in \{0, 1\}$ for each $i \in \{1, 2, ..., n\}$ and, W^j is the total excess power available.

However, with JPRA-CR we are not interested in finding the maximum number of sub-carriers that can be assigned additional powers to meet the rate requirement (items that can be put in the knapsack); our goal is to only check if all such sub-carrier requirements can be accommodated. Specifically, we need to check if the value of the objective function is equal to n. If so, rate j can be supported; if not, rates greater than or equal to j cannot be supported. Thus, the receiver only needs to check if upon setting $x_i = 1, \forall i \in \{1, 2, ..., n\}$, the constraint 2 is satisfied. If so, it moves to the next higher rate; if not, it chooses the immediately lower rate and is done with rate selection. Algorithm 2 summarizes the steps of our common rate selection algorithm.

Algorithm-1 is the main algorithm in JPRA-CR. It iteratively invokes Algorithm 2, to identify a *set* of sub-carriers that support the maximum number of bits to be transmitted per symbol, using a common rate. To begin with it considers all sub-carriers (as described above), and calls Algorithm 2, to find the maximum common rate that can be supported in this case (say $R(n)$, where n is the number of sub-carriers). The total number of transmitted bits per symbol is then $n \times R(n)$ (after performing power redistribution). It then removes the sub-carrier that requires the highest power per bit delivered (*poorest* sub-carrier) and adds its power to the excess power budget (as discussed); the common aggregate bit-rate with the remaining $(n-1)$ sub-carriers is now calculated by invoking Algorithm 2 to be $(n-1) \times R(n-1)$, where $R(n-1)$ is the maximum common rate that is now supported. If $(n-1) \times R(n-1) < n \times R(n)$, it is easy to see that the algorithm has converged, since no further optimization is possible. Note that $R(n)$ and $R(n-1)$ are calculated after performing power redistribution using Algorithm 2; thus these are only available as this algorithm is executed iteratively. If not, from among the remaining sub-carriers, the poorest sub-carrier is removed and the process is repeated. The iterations continues until a point is reached, say with the number of sub-carriers m, where $m \times R(m) < (m+1) \times R(m+1)$. This implies that the optimal set of sub-carriers are the $(m+1)$ sub-carriers in the prior to last iteration and, $R(m+1)$ is the maximum common rate to be used.

Input: Sub-carrier EVM_i, Rate $j \in S$, Rate threshold $EVMTH_j$, EVM to Power Mapping
Output: Rate j, List $L_{j,1}$, List $L_{j,2}$

foreach *Rate j* **do**
    foreach *Sub-carrier i* **do**
        $diff_i = EVMTH_j - EVM_i$;
        Map $diff_i$ to power p_i;
        if $p_i > 0$ **then**
            $W =+ p_i$;
            Put p_i in List $L_{j,1}$;
        end
        else
            $w =+ mod(p_i)$;
            Put $mod(p_i)$ in List $L_{j,2}$;
        end
    end
    if $w \leq W$ **then**
        return Rate j , $L_{j,1}, L_{j,2}$;
    end
    else
        return Rate $j-1$, $L_{j-1,1}, L_{j-1,2}$;
    end
end

Algorithm 2: Rate selection using JPRA-CR

Once the maximum common rate and the sub-carrier set to be used have been determined as above, the receiver performs the power redistribution. It transfers power from the *fading-immune* sub-carriers, to the affected ones. Algorithm 3 presents the steps taken. The considered sub-carriers are divided into two classes (lists). The first list contains the sub-carriers with the excess power, while second list contains sub-carriers that require more power for meeting the EVM threshold for the chosen rate. Note that these lists were already returned by Algorithm 2 previously and are now used by Algorithm 3 to perform power redistribution. Upon executing this algorithm, the receiver first subtracts the excess power from the sub-carriers in List-1. It then adds the extra power as required to each sub-carrier in List-2 (in decreasing order of the required power). If total extra power is more than the required power then left over power is distributed evenly among all the sub-carriers.

Computational Complexity: The run time for JPRA-CR is $O(n^2 \cdot l)$, where n is the number of sub-carriers and l is the number of available data rates. In brief, JPRA-CR (i.e., Algorithm 1) iterates

over the n sub-carriers and executes algorithm 2 in every iteration. The latter includes a nested loop which executes over all the sub-carriers and over all the l transmission rates, thus, requiring $n \cdot l$ running time. Since, algorithm 3 (whose complexity is $O(n)$), is executed outside the loop in Algorithm 1, the time complexity of JPRA-CR is $O(n^2 \cdot l)$.

Input: Initial sub-carrier Power p_i', List $L_{j,1}$, List $L_{j,2}$
Output: Adjusted sub-carrier Power p_i''

foreach *sub-carrier in List $L_{j,1}$* **do**
    $x =+ p_i$;
    $p_i'' = p_i' - p_i$;
end
Sort List $L_{j,2}$ in decreasing order ;
foreach *Sub-carrier in List $L_{j,2}$* **do**
    $p_i'' = p_i' + p_i$;
    $y =+ p_i$;
end
if $x > y$ **then**
    $l = (x - y)/ n$;
    where n is total number of sub-carriers ;
    foreach *sub-carrier in List $L_{j,1}$ and $L_{j,2}$* **do**
        Update $p_i'' =+ l$;
        return p_i'' ;
    end
end

Algorithm 3: Power Redistribution after selecting rate through JPRA-CR

4.2 Minimizing Transmission Air Time with JPRA-MT

The transmission time of a packet depends on the modulation and coding rate used, Higher modulation schemes lead to shorter packet times. In turn, shorter packet transmission times lead to throughput improvements. Ideally, in order to minimize the packet air time one would select the highest modulation on all the sub-carriers. However, due to frequency selectivity in fading all sub-carriers might be unable to support this high rate.

Adaptive modulation and coding schemes have been proposed to cope with frequency selectivity (e.g., [16]). With such schemes, the bit-rate for each sub-carrier is selected based on its CSI (Channel State Information). However, power redistribution is not performed. Power redistribution helps in maximizing the total number of bits transmitted by the sub-carriers in a symbol duration. Let us consider a toy example with two sub-carriers A and B, initially with default power allocations. A can only support BPSK while B can support QPSK. In total A and B can carry 3 bits since BPSK and QPSK can modulate 1 and 2 bits, respectively. If one completely shuts down A and transfers its power to B, B may be able to use 16QAM with the extra power. Thus, the total number of bits transmitted in a symbol duration is now 4 instead of 3 (a 16QAM symbol maps on to 4 bits). Thus, we gain 1 extra bit per symbol by doing power redistribution. With a large number of sub-carriers, where each sub-carrier has a different EVM and requires a different level of extra energy to support a higher modulation scheme, the gains can be (and actually are) more significant.

Formalizing the problem: The problem of appropriately allocating powers (and thus, transmission bit-rates) to sub-carriers in order to minimize the packet air-time can be formally defined as follows: Lets assume that there are N sub-carriers $x_1, x_2 \ldots x_N$, and R+1 (one for each rate level and a null rate level for no assignment) bit-rate levels for each sub-carrier. On each sub-carrier only one specific bit-rate out of the R+1 available bit-rates can be used; for each bit-rate $j \in x_i$ there is a corresponding *profit* $r_{i,j}$, which is the bit-rate itself and a weight that corresponds to the power p_{ij}, needed to achieve this bit-rate. We estimate this power using our measurements from the calibration phase. Maximizing the profit, essentially translates to using the highest bit rates on the sub-carriers; as one can easily see, this maximizes the number of bits transmitted per symbol and as a consequence, minimizes the transmission air time. However, the maximization of the profit as above is constrained by the total power budget available (P).

The above problem can be mapped to the well known Multiple Choice Knapsack Problem (MCKP). In MCKP, we are given m classes $N_1, N_2, \ldots, N_m$ of items that are to be packed in a knapsack of capacity C. Each item $j \in N_i$ has a profit a_{ij} and a weight w_{ij}, and the problem is to choose at most one item from each class such that the profit sum is maximized without the weight sum exceeding C.

With our problem, a sub-carrier x_i corresponds to a class, that consists of items that correspond to the $R + 1$ bit rates. The profit is the bit-rate value r_{ij} and weight is the required power p_{ij}. The total capacity of the knapsack is the total transmission power P, available to the transmitter.

Using this mapping, our problem can be formulated as the following MCKP:

$$\text{Maximize} \quad \sum_{i=1}^{N} \sum_{j \in x_i} r_{ij} k_{ij} \qquad (3)$$

$$\text{Subject to} \quad \sum_{i=1}^{N} \sum_{j \in x_i} p_{ij} k_{ij} \leq P \qquad (4)$$

$$\sum_{j \in x_i} k_{ij} = 1, \quad i = 1, .., N$$

$$k_{ij} \in \{0, 1\}, \quad i = 1, ..., N, \quad j \in x_i$$

Here k_{ij} is a slack variable for each r_{ij}; if a particular r_{ij} is selected to be included in the knapsack it is 1 and otherwise 0. The total number of items considered is $n = \sum_{i=1}^{N} R + 1 = N \cdot (R + 1)$.

Solving the problem: The MCKP problem is NP-hard [11] as it contains the knapsack problem as a special case [11]. However, it has been shown that it can be solved in pseudo-polynomial time [7]. Pisinger [15] proposes an algorithm (MCKNAP) which has been shown to outperform other algorithms for solving the MCKP problem. The algorithm first extends the partitioning algorithm of Balas and Zemel [4] for the original knapsack problem, to the case of MCKP. The partitioning algorithm identifies a set of classes (called the *core*) that include the optimal solution with high probability; this can be thought of as a reduction phase. A dynamic programming algorithm is then applied on the core in order to identify the optimal solution in pseudo linear time. This can be thought as an expanding phase, since it might add classes that do not belong to the core. The resulting solution defines the classes and the corresponding elements that are included in the knapsack (i.e., the elements with $k_{ij} = 1$). Based on the assigned rates, the total transmit power is allocated among the carriers.

Computational Complexity: The computational complexity of MCKNAP is $O(n + c \sum_{x_i \in c} r_i)$ for a minimal core c; thus, the algorithm executes in linear time for small cores and pseudo polynomial time for large cores. The proofs are available in Section 7 of [15].

Carrier Frequency	2427MHz
RF Bandwidth	10MHz and 20MHz
Number Of Sub-carriers	64
Modulation Schemes	BPSK, QPSK, 16QAM
Payload Length	1470 bytes
Tx Power	19dBm
MAC Protocol	CSMA/CA

Table 2: MAC/PHY parameters for JPRA implementation

5. PERFORMANCE EVALUATION

In this section, we examine the performance of JPRA via extensive real testbed as well as simulation experiments.

5.1 Testbed Setup and Implementation

Our testbed utilizes the Wireless Open-Access Research Platform (WARP) developed at Rice University. The WARP platform consists of three main components; (a) A Xilinx Virtex-II Pro FPGA, (b) A 2.4/5GHz Radio Board, which supports wideband applications such as OFDM, and (c) A 10/100 Ethernet port, which serves as the interface between the board and the wired Internet. The Xilinx module implements the MAC and PHY layer protocols. MAC protocols can be implemented in C, while the PHY layer protocols are implemented within the FPGA fabric using MATLAB Simulink. The current physical layer design uses an OFDM implementation that is loosely based on the PHY layer of the 802.11a standard. The WARPMAC/WARPPHY modules provide basic building blocks towards implementing more advanced MAC/PHY protocols. The basic configuration of WARP's OFDM design is given in Table 2.

We conduct our experiments on a six node WARP indoor testbed. We randomly selects three sender/receiver pairs, and send saturated traffic with packets of 1472 bytes. We consider cases where the connections interfere with each other and cases where they don't. Each experiment lasts for 6 minutes and the reported results are the averages of 5 runs unless otherwise stated. Each node is connected to a laptop through an Ethernet switch, which acts as a controller. The controller is also used for analyzing and collecting traces during the experiment.

Implementation: We implement JPRA on top of WARP's standard OFDM design, which runs on the board's FPGA in real time. We run JPRA at the MAC layer to calculate the rates to be used and the power redistribution. The receiver communicates this information to the transmitter through ACKs or NACKs (as mentioned in Section 4). We modify the transmitter to perform power redistribution and rate adaptation per sub-carrier, upon obtaining this information from the receiver. We also implement (for comparison), a standard EVM based rate adaptation scheme, which selects a single rate for all the sub-carriers. This scheme referred to as *Standard rate*, selects the appropriate rate for each packet by comparing the average packet EVM with the threshold values in Table 1. Standard rate adaptation does not perform any kind of power adaptation. Finally, for comparison purposes as well, we also implement FARA as described in [16]. In brief, FARA selects a rate based on the EVM or SNR threshold for each sub-carrier. In our implementation, FARA examines the individual sub-carrier EVM and selects the rate which can be supported. If a carrier cannot satisfy even the basic (minimum) rate it is left unused (no symbols are mapped on to that sub-carrier).

5.2 Experimental Results

Effects of power redistribution on RSSI value: We first want to ensure that a power redistribution with JPRA does not affect key

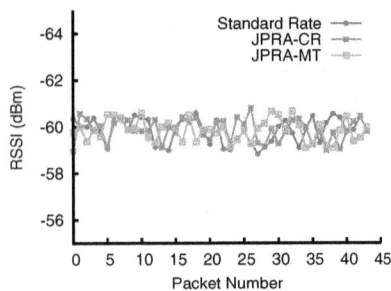

Figure 6: RSSI values per packet

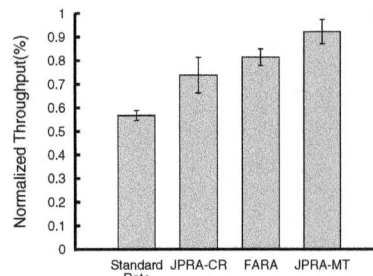

Figure 7: Normalized network throughput with 10MHz channel width.

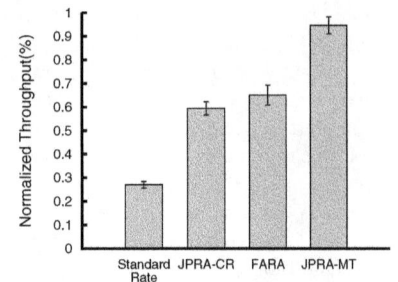

Figure 8: Normalized network throughput with 20MHz channel width.

network functionalities like carrier sensing. Typically the RSSI value is used in most of the wireless systems to determine the amount of energy on the channel. It is used as an indicator of whether the channel is idle or not. In WiFi, packet detection is based on RSSI values as well. Thus, we measure the RSSI values per packet with and without JPRA. Figure 6 depicts a part of a representative trace that we collected with Standard Rate adaptation, JPRA-CR and JPRA-MT. As one can see the RSSI values are similar in all scenarios, thus leading us to believe that JPRA will leave key higher functions such as carrier sensing or packet detection unaffected. The reason behind this observation can be explained as follows. The RSSI value is essentially the average energy present on the antenna during the reception of the preamble. Since JPRA is not applied on the preamble, we do not observe any significant RSSI variations.

Network throughput: Next, we examine the total network throughput. Figure 7 compares the throughput with JPRA with that using Standard Rate Adaptation and FARA. The throughputs are normalized by the highest throughput value among all observations (highest throughput = 1).This provides an immediate assessment of the relative performance with the schemes that we compare. The channel bandwidth in this experiment was 10MHz. We observe that JPRA-CR outperforms standard rate adaptation by up to 28%. As one may expect, FARA outperforms JPRA-CR, since it opportunistically employs higher rates on a subset of sub-carriers. However, the gains are marginal ($<$ 8%); further, JPRA-CR only requires feedback corresponding to the common rate. JPRA-MT outperforms both standard rate and FARA by, above 42% and 20% respectively. The main source of gain for JPRA-MT in comparison to FARA and Standard rate adaptation, is the per sub-carrier power redistribution. The power redistribution allows the transmission of a higher number of bits per symbol as discussed earlier. Note here that we show the 95% confidence intervals for all results.

Wider channels exhibit a higher degree or frequency selectivity. We also perform the same set of experiments with a 20 MHz channel; this channel width is typical in WiFi systems. With the 20 MHz channel the observed normalized throughput with JPRA-CR is 43% higher as compared with that of the Standard Rate adaptation (Figure 8). However, FARA still outperforms JPRA-CR by about 7%. With JPRA-MT we observed higher gains with the 20MHz channel width. In particular, our results (Figure 8) indicate that JPRA-MT delivers upto 75% more network throughput than Standard Rate Adaptation. It also outperforms FARA by 35%. Thus, *per sub carrier power adaptation enhances network performance in comparison to schemes without power adaptation.* With the increase in channel width, gains also increase. This becomes especially noteworthy, given the increasing number of technologies that use wide

channels (e.g., 802.11n with 40 MHz channels and white space networking with channels of at least 100MHz).

Stability of rate: One of the main reasons for the design of JPRA-CR is the stability of rate that it can offer. In addition JPRA-CR requires a much simpler transceiver design since it chooses a single rate for a given transmission which can be supported on all or a chosen subset of sub-carriers; therefore a single rate or modulation is applied to all the sub-carriers like standard OFDMA. In contrast FARA and JPRA-MT in the worst case use l (total number of rates) different modulations on the different sub-carriers for a given transmission, which leads to a more complex transceiver design. In particular the transceiver has to implement additional bookkeeping to track the rates on the different sub-carriers with these other schemes.

To reiterate, a single rate is chosen, which all sub-carriers can support. Since some carriers might experience fading while others may not, on average with power redistribution, the maximum common rate that can be supported is not expected to change significantly from one transmission to the next (even though the sub-carriers that experience fading might be different). In order to examine the performance of JPRA-CR with respect to rate stability, we created 10 diverse links in our lab by changing the positions of the sender and the receiver. The sender transmits 1472 byte size packets for 3 minutes and we monitor the throughput on the receive side. We also log the number of rate changes applied by the transmitter. The results are shown in Figure 9. We plot the average number of rate changes over 3 trials with JPRA-CR, Standard Rate, JPRA-MT and FARA. As one can see JPRA-CR resulted in up to 27% fewer rate changes in comparison to Standard Rate Adaptation, it also outperforms the other schemes. JPRA-MT performs upto 8% and 14% better than FARA and Standard Rate Adaptation. With FARA and JPRA-MT, we consider a rate change to have occurred if the transmission bit-rate on any sub-carrier is changed between packet transmissions. We believe this is reasonable since the receiver has to now decode packets with the new rate on the specific sub-carrier. Note here that FARA also performs better than standard rate because it shuts down sub-carriers which are in deep fades, these are the carriers where the fluctuations are most likely. *To summarize, power redistribution helps in stabilizing rate changes since it improves performance on sub-carriers experiencing selective fading (which results in increased rate changes).*

Power savings: In our next experiments, we examine the possible power saving gains with JPRA-CR. In particular, we seek to answer the following question "Can we use lower total transmission power in order to achieve a target PDR?". We set a target PDR of 90% and we vary the total transmission power of the sender. On an average over 5 trials we observe that with JPRA-CR we need 4.5dBm less power compared to Standard Rate adaptation in order

232

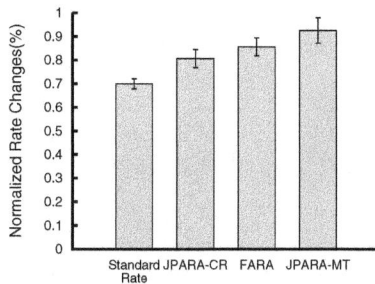

Figure 9: Normalized number of rate changes with 20 Mhz.

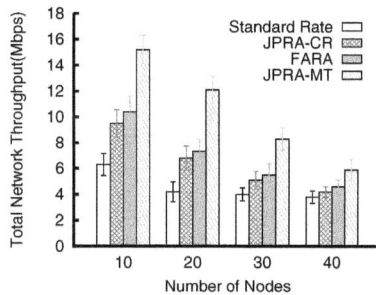

Figure 10: Network throughput with packet size 512 bytes.

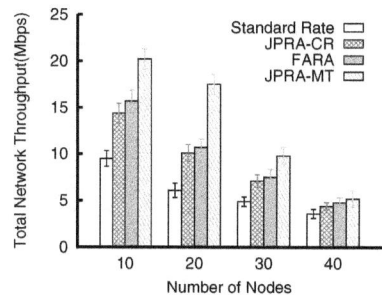

Figure 11: Network throughput with 1472 bytes packet size .

to achieve the preset target delivery ratio. These gains are important, since potentially JPRA-CR can reduce interference; moreover, it can lower energy costs since it requires less transmission power to achieve a certain performance.

Overheads: A receiver needs to send feedback information to the transmitter to perform rate and power reallocation. WARP's CSMA/CA MAC layer uses standard 802.11 DATA and ACK packet formats, but allows us to piggyback rate and power information on the ACK packets (as with 802.11n ACK packets [3]). Bit-rate information is encoded using 6 bits (up to 64 values). Each sub-carrier power value is a signed real value between 0 and 2.0, which can be represented using 7 bits. When channel remains stable for multiple packets we just send a normal ACK (without any additional information) which tells the receiver to use power and rate values from the previous instance. For 48 data sub-carriers to encode bit-rate and power information we require 48*6 bits and 48*7 bits, respectively. In a worst case scenario, JPRA-MT incurs a 5.2% overhead and JPRA-CR incurs a 2.5% overhead compared to Standard Rate adaptation for a payload size of 1472 bytes. Note that the corresponding overhead increase with FARA is 2.4%. In our testbed the average overhead that we observed for JPRA-MT and JPRA-CR was 4.1% and 1.40%, respectively. The average overhead of FARA was 1.82%, which is more than JPRA-CR since JPRA-CR is more stable and requires less feedback. We wish to point out here that our throughput results, already account for this overhead i.e., the gains are in spite of the overhead .

5.3 Simulations

In order to evaluate our scheme in larger settings we perform simulations in NS-3. In order to have a realistic OFDM PHY layer model, we use a modified version of NS-3 which includes a detailed physical layer model called PhySim-Wifi[14]. PhySim-WiFi includes a physical layer implementation of the OFDM PHY specification for the 5 GHz band as well as a wireless channel emulation. Specifically in our simulations we use NS-3.9 with the PhySimWiFi 1.1 module. Key Physical and MAC layer parameters are given in Table 3 . We further modified PhySim-WiFi in order to implement JPRA-CR, JPRA-MT and FARA. Standard rate adaptation is already implemented in NS-3 but we modified it to use EVM instead of SNR as rate adaptation metric. We implemented these schemes in a similar fashion as we implemented them on our testbed. We repeat the calibration phase in the simulator in order to calculate per sub-carrier EVM, EVM to sub-carrier Power mapping and EVM to rate thresholds. EVM to sub-carrier power mapping follows a similar third degree polynomial curve that we observed on our testbed.

To create frequency selectivity, the sub-carrier gains $|h_n^2(t)|$ are generated based on path loss and fading. The path loss model is

Carrier Frequency	5320MHz
RF Bandwidth	20MHz
Number Of Sub-carriers	64
Modulation Schemes	BPSK, QPSK, 16QAM, 64QAM
Payload Length	512/1472 bytes
Tx Power	20dBm
MAC Protocol	CSMA/CA

Table 3: MAC/PHY parameters used in simulation

characterized by a standard model $|h_{PL}|^2 = K * 1/d^\alpha$ parameterized by $K = 40.14 dB$ and the path loss exponent $\alpha = 3.5$. The fading component is implemented based on [20] and parameterized with the values provided in [24] (Table 1 of the reference). The resulting fading gains feature correlation in time and in frequency. The environment is further characterized by an RMS delay spread of 25ns.

Simulation setup: Our simulation topology consists of 40 nodes, arranged in a 8×5 grid. The inter-node distance in the grid is 20m. We consider randomly selected transceiver pairs (also referred to as traffic pairs) to send data packets. We consider 4 sets of experiments while varying the number of active links. In particular, we experiment with 5, 10, 15 and 20 traffic pairs, active simultaneously. Since the topology is fixed, a higher number of traffic pairs implies higher traffic intensity. We use CBR (constant bit rate) to generate packets of size 512 bytes and 1472 bytes. Simulation time for each run is 120 seconds, and our results are the average of 10 of such simulation runs.

Simulation results: Figures 10 and 11 present the results for network throughput with different node densities with each scheme. With JPRA-CR, the total network throughput is 71% higher as compared with Standard rate adaptation, when node density is 20. JPRA-MT improves the total network throughput by upto 185% in comparison to Standard rate and also registers a gain of 64% over FARA. Similar to our experimental results FARA outperformed JPRA-CR by 9% in simulations; again, this is due to per sub-carrier rate adaptation which leads to higher throughput. As the number of competing links increase, the gain for each scheme increases up to a certain traffic intensity. After this *sweet point* the gains are diminished since interference and contention become the dominant impairments.

With increased interference we observe an increasing number of packet collisions. Interference reduces the opportunity to gain from power redistribution since it affects all the sub-carriers.

233

6. CONCLUSIONS

In this paper, we make a case of adaptively distributing uneven power levels to OFDMA sub-carriers to cope with frequency selectivity in fading. To validate our thesis that this will provide significant throughput benefits, we design JPRA, for jointly selecting the transmission power and bit-rate for the sub-carriers that compose a channel. We design two versions of JPRA: (a) JPRA-CR, a simpler scheme, which while less effective in improving throughput, provides increased stability in the rate in use, simplicity in transceiver design and potential power savings and (b) JPRA-MT which provides significant throughput gains albeit with lower rate stability. We implement both variants on our WARP radio testbed and also perform simulations to showcase its benefits in larger settings. We show that JPRA-MT can provide upto a 35 % increase in throughput compared to the state of the art, FARA scheme that only performs per sub-carrier rate adaptation.

7. ACKNOWLEDGEMENTS

We thank the MobiHoc'12 reviewers and our shepherd, Dr. Jerry Chiang, for feedback on this paper. This work was supported in part by Cisco University grant and Multi-University Research Initiative (MURI) grant W911NF-07-1-0318.

8. REFERENCES

[1] Onoe rate control.

[2] Wireless lan medium access control (mac) and physical layer (phy) specifications: High-speed physical layer in the 5 ghz band. *Part 11, Standard ed.,IEEE802.11 Working Group* (1999).

[3] *Part 11 : Wireless LAN Medium Access Control (MAC) and Physical Layer (PHY) Specifications.* IEEE Computer Society, 2009.

[4] BALAS, E., AND ZEMEL, E. An algorithm for large zero-one knapsack problems. *Operations Research* (1980).

[5] BHARTIA, A., CHEN, Y.-C., RALLAPALLI, S., AND QIU, L. Harnessing frequency diversity in wi-fi networks. In *Proceedings of the 17th annual international conference on Mobile computing and networking* (2011), MobiCom.

[6] BICKET, J. C. Bit-rate selection in wireless networks. Tech. rep., Master's thesis, MIT, 2005.

[7] DUDZINSKI, K., AND WALUKIEWICZ, S. Exact methods for the knapsack-problem and its generalizations. *European Journal Of Operational Research* (1987).

[8] HOLLAND, G., VAIDYA, N., AND BAHL, P. A rate-adaptive mac protocol for multi-hop wireless networks. In *Proceedings of the 7th annual international conference on Mobile computing and networking* (2001), MobiCom.

[9] ISHIKAWA, H., FUJII, M., ITAMI, M., AND ITOH, K. Bi-directional ofdm transmission using adaptive modulation that spreads data symbols. In *Power Line Communications and Its Applications, IEEE International Symposium on* (2006).

[10] JUDD, G., WANG, X., AND STEENKISTE, P. Efficient channel-aware rate adaptation in dynamic environments. In *Proceedings of the 6th international conference on Mobile systems, applications, and services* (2008), MobiSys.

[11] KELLERER, H., PFERSCHY, U., AND PISINGER, D. *Knapsack problems.* Springer, 2004.

[12] KITTIPIYAKUL, S., AND JAVIDI, T. Subcarrier allocation in ofdma systems: beyond water-filling. In *Signals, Systems and Computers. Conference Record of the Thirty-Eighth Asilomar Conference on* (2004).

[13] MCKINLEY, M., REMLEY, K., MYSLINSKI, M., KENNEY, J., SCHREURS, D., AND NAUWELAERS, B. Evm calculation for broadband modulated signals. *64th ARFTG Conf. Dig* (2004).

[14] MITTAG, J., PAPANASTASIOU, S., HARTENSTEIN, H., AND STROM, E. G. Enabling Accurate Cross-Layer PHY/MAC/NET Simulation Studies of Vehicular Communication Networks. *Proceedings of The IEEE* (2011).

[15] PISINGER, D. A minimal algorithm for the multiple-choice knapsack problem. *European Journal of Operational Research* (1994).

[16] RAHUL, H., EDALAT, F., KATABI, D., AND SODINI, C. G. Frequency-aware rate adaptation and mac protocols. In *Proceedings of the 15th annual international conference on Mobile computing and networking* (2009), MobiCom.

[17] SADEGHI, B., KANODIA, V., SABHARWAL, A., AND KNIGHTLY, E. Opportunistic media access for multirate ad hoc networks. In *Proceedings of the 8th annual international conference on Mobile computing and networking* (2002), MobiCom.

[18] SEN, S., SANTHAPURI, N., CHOUDHURY, R. R., AND NELAKUDITI, S. Accurate: Constellation based rate estimation in wireless networks. In *NSDI* (2010).

[19] STEEL, R., AND TORRIE, J. *Principles and procedures of statistics.* 1960.

[20] WANG, C.-X., PATZOLD, M., AND YAO, Q. Stochastic Modeling and Simulation of Frequency-Correlated Wideband Fading Channels. *IEEE Transactions on Vehicular Technology* (2007).

[21] WILLINK, T. J., AND WITTKE, P. H. Optimization and performance evaluation of multicarrier transmission. *IEEE Transactions on Information Theory* (1997).

[22] WONG, C. Y., CHENG, R. S., D., K. B. L. R., MURCH, R. D., MEMBER, S., AND MEMBER, S. Multiuser ofdm with adaptive subcarrier, bit, and power allocation. *IEEE Journal on Selected Areas of Communications* (1999).

[23] WONG, S. H. Y., YANG, H., LU, S., AND BHARGHAVAN, V. Robust rate adaptation for 802.11 wireless networks. In *Proceedings of the 12th annual international conference on Mobile computing and networking* (2006), MobiCom.

[24] ZHAO, X., KIVINEN, J., AND VAINNIKAINEN, P. Tapped delay line channel models at 5.3 GHz in indoor environments. *Composites Part A-applied Science and Manufacturing* (2000).

A Distributed Resource Management Framework for Interference Mitigation in OFDMA Femtocell Networks

Jongwon Yoon [*]
University of Wisconsin,
Madison, WI
yoonj@cs.wisc.edu

Mustafa Y. Arslan
University of California,
Riverside, CA
marslan@cs.ucr.edu

Karthikeyan Sundaresan
NEC Labs America Inc.
Princeton, NJ
karthiks@nec-labs.com

Srikanth V. Krishnamurthy
University of California,
Riverside, CA
krish@cs.ucr.edu

Suman Banerjee
University of Wisconsin,
Madison, WI
suman@cs.wisc.edu

ABSTRACT

Next generation wireless networks (i.e., WiMAX, LTE) provide higher bandwidth and spectrum efficiency leveraging smaller (femto) cells with orthogonal frequency division multiple access (OFDMA). The uncoordinated, dense deployments of femtocells however, pose several unique challenges relating to interference and resource management in these networks. Towards addressing these challenges, we propose RADION, a *distributed* resource management framework that effectively manages interference across femtocells. RADION's core building blocks enable femtocells to opportunistically find the available resources in a completely distributed and efficient manner. Further, RADION's modular nature paves the way for different resource management solutions to be incorporated in the framework. We implement RADION on a real WiMAX femtocell testbed deployed in a typical indoor setting. We extensively evaluate two solutions integrated with RADION, both via prototype implementation and simulations and quantify their performance in terms of quick and efficient self-organization.

Categories and Subject Descriptors

C.2.1 [**Network Architecture and Design**]: Wireless Communication

General Terms

Design, Experimentation, Management, Performance

Keywords

Resource Management, Distributed, OFDMA, Femtocell

---

[*]Jongwon Yoon and Mustafa Arslan interned in the Mobile Communications and Networking department at NEC Laboratories America Inc, Princeton during this work.

1. INTRODUCTION

Recent industry reports predict that mobile data usage will grow by a factor of 26 by 2015 with around 40% of the mobile data originating indoors (i.e., homes and enterprises) [1]. The consequent demand for higher capacity has led to smaller cells, called femtocells, that operate using the same OFDMA-based technology as the macro cells [2]. Femtocells offer improved coverage and a cheap and energy-efficient deployment option for indoor usage [3, 4]. They provide numerous advantages to both customers and service providers. For example, customer equipment (e.g., 4G-capable smart phone, laptop) can save energy on the uplink (while enjoying high throughput), since it does not have to transmit to a distant macro base station. For service providers, femtocells offer (i) a cost-effective way of providing coverage, (ii) increased system capacity via spatial reuse and (iii) reduced operational expenses and subscriber churn. Since users can seamlessly migrate between femtocells and macrocells on a licensed spectrum, femtocells offer an attractive alternative to the unlicensed and often congested WiFi systems.

Unlike the planned deployment of macrocell base stations by the cellular operators, femtocells are deployed in an unplanned fashion within buildings. Hence, as femtocell deployments continue to grow, interference will inevitably be a performance limiting factor in a manner similar to that experienced by residential WiFi networks (e.g. [5]). There are two potential sources of such interference for femtocells: (i) interference between femtocells and the macrocell, and (ii) interference between femtocells. This paper focuses on the problem of interference mitigation between multiple femtocells (in range) that stems from their unplanned deployment [1].

Inapplicability of solutions used in macrocells or in WiFi networks: The problem of interference mitigation through efficient resource management is not new and has been extensively studied in different settings. For instance, this problem arises even in the design of macrocell topologies by cellular operators. However, the planned nature of such macrocell deployments leads to very efficient frequency planning and other coordination-based solutions (e.g., fractional frequency re-use), that cannot be applied to our unplanned femtocell scenarios. Efficient design strategies of WiFi networks have also needed to address the problem for unlicensed spectrum. The basic toolkit for WiFi protocols to address interference problems include random access channel contention strategies

---

[1]Other work has focused on the macro-femto interference problem [6], and is beyond the scope of this paper.

that *de-synchronize* competing transmitters in time through methods of carrier sensing and back-off. Unfortunately, femtocells cannot leverage such strategies since their MAC-PHY protocols are required to follow the same *synchronous* channel access methods as their macrocell counterparts. More specifically, standard OFDMA femtocells do not employ carrier sensing based deferral. Hence, they cannot sense the spectrum occupied by other cells to tune themselves onto orthogonal frequencies to resolve interference. Finally, unlike in WiFi, a single OFDMA frame carries data for multiple clients, and our resource management strategies are constrained by these requirements.

RADION – Distributed femtocell resource management: We propose RADION, a *framework* for distributed management of time-frequency resources in OFDMA-based femtocell networks. This framework is designed for the scenario where nearby femtocells cannot explicitly coordinate or interact with each other, and hence is suitable for unplanned residential deployments. Our specific solution requires each femto base station (BS) to intelligently probe availability of resources and use them in an opportunistic and distributed manner. We contrast this design to a centralized resource management solution, called FERMI, that was recently proposed for femtocells [7]. More specifically, FERMI focused on femtocell deployments in enterprise settings, where multiple femtocells were assumed to *cooperate* towards a network-wide objective. In particular, the design in FERMI required the centralized collection of the global network view, constructed through explicit information exchange between the various femtocells. Such cooperation is not realistic for residential settings. Hence the solution in RADION is stylized such that each residential femto BS will try to optimize for its own local objective, without explicitly exchanging information with neighboring femtocell BSs.

Several challenges arise in designing a distributed resource management framework for OFDMA femtocells: (**i**) OFDMA schedules multiple clients in the same frame. The clients experience different levels of interference and hence, a resource management framework has to account for the characteristics of each client. Specifically, clients with strong interference need to operate on orthogonal resources (i.e., frequency isolation), while clients with weak interference can operate on all frequency resources (i.e., reuse). For the efficient use of resources in OFDMA frames, we need to first differentiate the clients. (**ii**) Given that multiple clients share frame resources in OFDMA networks, the next challenge is "how to accommodate multiple clients of various classes in the same frame?". The frame structure has to be carefully managed for various clients considering their interference levels and demands. In particular, each cell needs to determine how much resources can be reused without causing interference to the neighboring cells and how much frequency resources need to be isolated to mitigate the interference from other cells. The frame structure can impact the network wide resource reuse as well, where multiple contention domains are involved. (**iii**) The resources allocated to one femtocell directly impact the resources for the interfering cells. We need to determine the time and frequency resources of operation for the clients in each cell while accounting for the resources used by neighboring cells. Resource allocation for each femtocell should be adaptive to the network changes, but without explicit coordination due to lack of any central component in the system.

In RADION, we address the above challenges through the specific design of the following three building blocks:

- Client Categorization: Active probing is used by the BS to categorize clients into two classes, those that require resource isolation to mitigate the interference and those that can reuse the spectrum, based on their interference levels. Client cat-

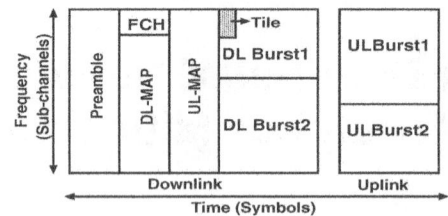

Figure 1: Illustration of the WiMAX frame structure.

egorization is a critical module for efficient use of frame resources given that multiple clients have to share the resources in each OFDMA frame.

- Resource Decoupling: Once the clients are categorized, RADION is capable of scheduling multiple clients in the same frame by employing a three-zone frame structure. A zone is a dedicated area of the frame in which data to clients can be scheduled. This new frame structure enables frame resource reuse, and hence promotes more efficient use of spectrum. It also decouples resources across contention domains to promote better network-wide reuse.

- Two-phase Adaptation and Allocation: The resources used by each BS are determined by a fast and iterative mechanism that is completely distributed. RADION employs a two-phase adaptation for determining jointly the time and frequency resources for each femtocell. Further, it allows for individual execution at each BS without coordination.

Contributions: We design and implement RADION on a WiMAX testbed, consisting of three femtocells in an unplanned indoor environment. To the best of our knowledge, this is the first implementation based design and evaluation of a self-organizing, distributed framework for OFDMA femtocells. We evaluate the various components of RADION to highlight the functionality and accuracy of each. RADION maintains standards compatibility; it is immediately deployable with commercial OFDMA clients (e.g. WiMAX clients). Further, RADION's components and modules are applicable to the other OFDMA technologies, LTE and LTE-A. We wish to point out that RADION is modular. In particular, depending on the context and the objectives, different resource allocation solutions can be easily incorporated within the RADION framework.

The rest of the paper is organized as follows. In §2, we provide background on WiMAX and OFDMA and discuss related work. §3 discusses the challenges in managing resources in OFDMA femtocell networks and briefly describes how RADION addresses them. In §4, the functional blocks in RADION are described. We evaluate RADION using our testbed and simulations in §5. §6 concludes the paper.

2. BACKGROUND AND RELATED WORK

WiMAX Preliminaries: RADION is implemented on a WiMAX femtocell testbed. In WiMAX, the spectrum is divided into multiple sub-carriers and several sub-carriers are grouped to form a sub-channel. There are two modes of grouping. In *distributed grouping*, sub-carriers are picked as per a pre-determined permutation. This allows a user to see uniform gain and interference across different sub-channels. In *contiguous grouping*, a contiguous set of sub-carriers is grouped to form a sub-channel. Distributed grouping, the default mode in 802.16e standard, is considered in our work.

WiMAX has a two-dimensional frame (see Fig.1) that carries data across time (symbols) and frequency (sub-channels). A combination of a symbol and a sub-channel constitutes a *tile*, which is the basic unit of resource allocation. Data to multiple mobile stations (MSs) are scheduled as rectangular bursts of tiles in a frame and frames are sent out periodically (every 5ms). Mainly, frames consist of two parts; downlink (from BS to the MS) and uplink (from the MS to the BS). Downlink frames contain the preamble, control and data bursts. While the preamble allows a MS to associate with the BS, the control consists of FCH (frame control header) and MAP. The DL-MAP indicates where each burst is placed in the frame, which MS it is sent to, and what modulation and coding scheme (MCS) should be used to decode it.

Related Work: While OFDMA standards (e.g., WiMAX, LTE) have been drafted recently, related research has existed for some time [8]. Efforts that focus on the macro-femtocells interference [9] and the interference to cell-edge users in OFDMA macrocells exist. The *localized* (cell edge) interference and *planned* cell layouts have aided various interference management solutions [10]. These also include fractional frequency partitioning (FFR) approaches [11], where the spectrum is partitioned into pre-determined static sets. Unlike macrocells, femtocells are deployed in an unplanned manner without coordinated operations and are hence, vulnerable to interference. This necessitates novel interference mitigation solutions. There have been recent studies [9, 12] in this direction but are restricted to theory with several simplifying assumptions that restrict their scope and deployment.

Recently in [7], FERMI, a *centralized* resource management solution for enterprise femtocell deployments was proposed. However, RADION is a completely *distributed* solution targeting unplanned deployment settings (such as residences), where cooperation among femtocells (implicitly assumed in [7]) is not realistic. The concepts of zoning and client categorization were also used in FERMI; however, access to global knowledge at a central controller made it easy to design these functions. RADION allows each femtocell to intelligently determine its zones in a conservative manner. Unlike in FERMI, to account for inaccuracies in distributed zone determination, a transition zone is introduced (details later). This in turn makes the client categorization in RADION also different from that in FERMI; clients that access the medium in the transition zone have to be chosen with care. Due to lack of coordination, accurate client categorization is all the more critical for efficient use of resources.

Diverging from the cellular context, recent efforts show the benefits of OFDMA in WiFi by building systems that enable dynamic spectrum fragmentation [13] and adaptive channel width [14]. In WiFi, APs use one among multiple 20 MHz channels and several conventional distributed channel selection algorithms [5] can be used to configure APs on different channels to avoid interference. However, in femtocells, the entire spectral chunk (say 20MHz) is available to all the cells. They need to operate on mutually orthogonal subsets of frame resources (sub-channels and time symbols) to avoid interference. Thus, resource allocation has to adapt to network dynamics (such as traffic, load etc.). Here, we seek to distributively and quickly determine the resources for each cell. Further, resource allocation constitutes just one component in our broader goal of designing a framework for distributed resource management for femtocells.

3. RESOURCE MANAGEMENT CHALLENGES

The problem of resource management is for each femtocell to distributively determine the frame resources (tiles) that it can use to schedule its clients. Since the resource allocation decisions of one cell impact multiple other cells, efficient mechanisms are needed to quickly converge to a network-wide resource allocation. We first

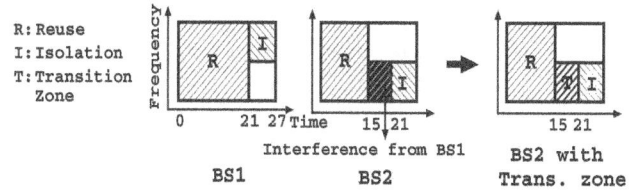

Figure 2: Introducing transition zone.

discuss the challenges in achieving this objective and briefly describe how RADION addresses each of them.

Client Categorization: OFDMA schedules multiple clients in a frame. Different clients may experience different levels of interference from neighboring femtocells; clients subject to strong interference need to operate on orthogonal sub-channels partitioned across cells (i.e., frequency isolation), while clients with weak interference can use the entire spectrum (i.e., reuse) and still tolerate interference via link (rate) adaptation. Differentiating the clients is key in realizing good spectral efficiencies. A client *categorization* is included within RADION to accurately differentiate between such clients (which can reuse the spectrum *reuse clients*) and those that need spectral isolation (*isolation clients*). Since sensing the medium is not possible with standard OFDMA femtocells, RADION uses an intelligent active probing technique to achieve client categorization with high accuracy. Note that the accuracy of categorization has a direct impact on how efficiently the tiles are utilized.

Resource Decoupling among the Clients: Once the clients are categorized, next question is "how to accommodate multiple clients of different categories (*isolation* and *reuse* clients) in the same frame ?". Frame resources have to be carefully assigned to both the *isolation* and *reuse* clients; else, it leads to under-utilization of network resources. Each frame can be segmented into two zones, the reuse and the isolation zone, to accommodate the two types of clients (BS1 in Fig.2) ;clients in the reuse zone will receive data encoded on the entire spectrum, while clients in the isolation zone will receive on a subset of frequencies (determined by an allocation algorithm). The static FFR approaches, using pre-determined sizes of reuse and isolation zone across cells, will however not work in femtocell deployments where interference is pervasive (i.e., not localized). Further, the load of a femtocell as well as interference from other BSs also need to be taken into account in adapting the zone sizes of a particular cell. RADION uses a novel *three-zone* frame structure to address these issues as will be described in §4.2; in a nutshell the additional third zone is a *transition zone* that prevents resource coupling across cells (discussed next).

Resource Decoupling across the Femtocells: When two interfering femtocells have different reuse zone sizes (based on their loads), the larger reuse zone will interfere with the isolation zone of the other cell (BS2 in Fig.2). Having a common reuse zone for the two cells is essential to avoid this. However, irrespective of whether the maximum or the minimum of the reuse zone sizes is chosen as this "common" zone size, it is easy to see that there is under-utilization in one of the cells (either part of the reuse zone or part of the isolation zone is not utilized). More importantly, this coupling can propagate across cells resulting in network-wide under-utilization. RADION's *transition zone* intelligently localizes this resource coupling and prevents such propagation.

Resource Allocation: Each femtocell has to determine its zone sizes and resource usage in a completely distributed manner. RADION uses an iterative, joint time (zone sizes) and frequency (sub-channels in the isolation zone) resource allocation algorithm that converges to efficient allocations and adapts to network dynamics quickly and efficiently.

(a) Measurement zones

(b) Categorization results

(c) Illustration of three-zone structure

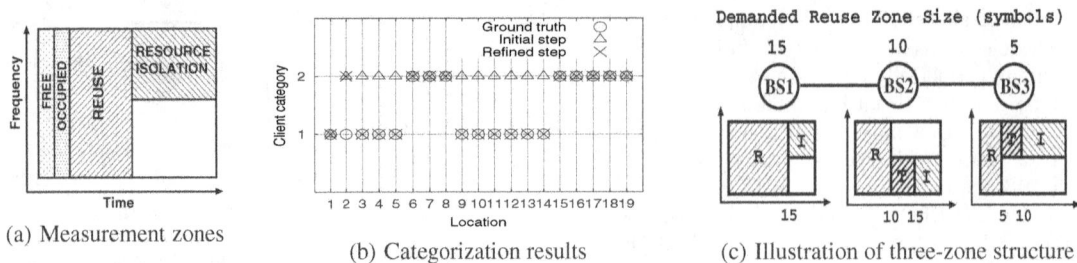

Figure 3: Free and Occupied zones are used for client categorization (a). Refinement step yields accurate categorization results (b).

4. RADION AND ITS COMPONENTS

We next describe RADION in detail by elaborating on the functionalities of its building blocks. A detailed description of our prototype implementation and the testbed used for experiments are given in §5.

4.1 Client Categorization

Initial Step: The first component in RADION categorizes clients into two classes; the first needs just link adaptation (class 1) and the second needs resource isolation (class 2). This is achieved through active probing and Burst Delivery Ratio (BDR) estimation. We define the BDR to be the ratio of successfully delivered bursts to the total number of transmitted bursts. To facilitate client categorization, the frame structure contains two measurement zones of equal size (2 symbols) – *free* and *occupied* zones as depicted in Fig. 3(a). Each BS performs the following two steps towards categorizing a client leveraging these two zones: **(i)** Schedule the client's data in the *occupied* zone and schedule it in the *free* zone probabilistically. It is possible that more than one BS would simultaneously schedule its client in the *free* zone. In order to avoid this, a random access mechanism with probability $\frac{1}{n}$ is used where n is the number of interfering BSs. Keep track of the resulting BDR in both zones over K frames. **(ii)** Determine the normalized throughput per tile in the two zones T_{occ} and T_{free} corresponding to their BDRs. If $T_{free} \geq (1 + \alpha)T_{occ}$ then the client is categorized as class 2; otherwise class 1. Through exhaustive measurements, we set $K = 25$ frames with $\alpha = 0.25$ for high accuracy (>90%) of categorization.

Refinement Step: In the above step, while clients in class 1 are identified accurately, not all clients categorized as class 2 may require resource isolation. With resource isolation, a BS allocates only a subset of resources to a client. For clients with low-moderate interference, link adaptation to cope with interference may be a better option than sacrificing resources through isolation. Thus, to further refine the categorization of class 2 clients, we factor in the loss of resources due to isolation. This was missing in the step **(ii)**, since equal resources were used in the *occupied* and *free* zones. However, the amount of isolated resources available to a cell depends on the resource allocation algorithm. If the resources assigned to the isolation zone is a fraction f of that of the reuse zone, the BS refines the status of a client in class 2 by scheduling the client on resources in the *isolation* zone and determining its normalized per tile throughput in this zone, T_{isol}. It retains the client in class 2 only if $f \cdot T_{isol} \geq (1 + \beta)T_{occ}$; the client is reverted to class 1 otherwise. Here, β (0.05, experimentally obtained) is used to avoid oscillations in categorization. RADION further sub-classifies clients in class 2 as those that benefit significantly from resource isolation (class 2h: $\frac{f \cdot T_{isol}}{T_{occ}} \geq (1 + \alpha)$) and those that benefit marginally from it (class 2l: $(1 + \beta) \leq \frac{f \cdot T_{isol}}{T_{occ}} < (1 + \alpha)$). The benefits of such a sub-classification will be discussed in §4.2.

Evaluation: We consider two cells (1 and 2); clients 1 and 2 belong to the two cells respectively. We generate multiple topologies with different levels of interference by varying the location of client 1 in the presence of interfering cell 2. The isolation zone of each cell has the same number of symbols as the occupied zone but operates on an orthogonal half the sub-channels. First, for every location of client 1, its ground truth is determined by scheduling data to the client in the *occupied* and *isolation* zones and determining it as class 2 if $0.5 \cdot T_{isol} \geq T_{occ}$. Then our three step categorization algorithm is executed. The categorization results of the initial and the refinement steps are shown in Fig.3(b). It is seen that the initial step wrongly categorizes clients in ten locations, however, the refinement step corrects most of them. The only erroneous classification (location 2) was due to a change in channel conditions during the process. The clients who need to be in class 2 are correctly classified even with the initial step, while refinement only adds more clients from class 2 to class 1.

(a) Using two-zone structure

(b) Using three-zone structure

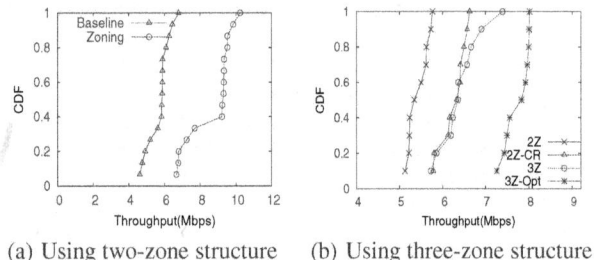

Figure 4: Throughput benefits of multi-zone structure.

4.2 Resource Decoupling

Two-zone Structure: To schedule clients of both classes in the same frame, we use two variable size data *zones* (reuse and isolation zones) as in Fig.3(a). The reuse zone operates on all sub-channels and schedules class 1 clients, while the isolation zone schedules class 2 clients on only a contiguous subset of *sub-channels*. To understand the benefits of the two-zone structure, we conduct experiments with two BSs. Each BS has two clients (one in each class), and interferes with the class 2 client of the other BS. The baseline scheme operates the two BSs on two orthogonal sets of sub-channels, with each BS scheduling both its clients within its own subset. This is compared against a two-zone scheme where a BS schedules its class 1 client on all sub-channels, while the isolation zone uses the other half of the frame for its class 2 client on half the sub-channels. We generate various interference topologies and the CDF of the net throughput is plotted for the two schemes in Fig.4(a). We see that with the two zones, class 1 clients can be scheduled in tandem to reuse sub-channel resources effectively, yielding over a 35% throughput gain over the conservative isolation scheme.

Drawbacks of Two-zone Structure: A two-zone structure enables resource reuse, but is insufficient in a multi-cell context. Different cells will have different reuse zone sizes based on the load generated by the class 1 clients. If two interfering cells were to

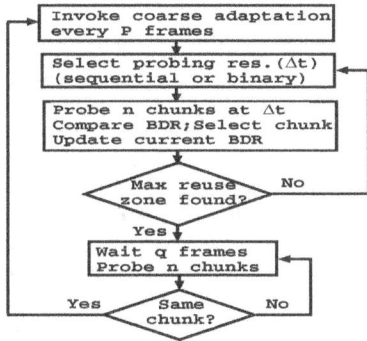

Figure 5: Flow chart of RADION's two-phase adaptation.

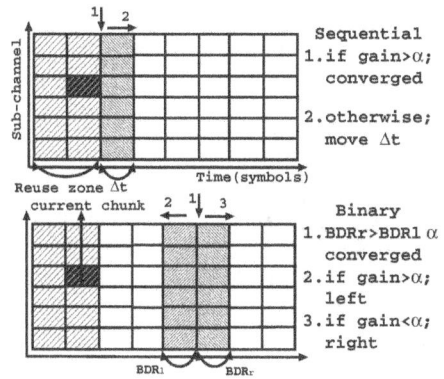

Figure 6: Details of two probing methods.

operate independently, the larger reuse zone will overlap with the isolation zone of the other cell and hence, interfere with the isolation clients of that cell (Fig.2). Consequently, it is important to use a common reuse zone between interfering cells. For the example topology in Fig.3(c), the result in Fig.4(b) indicates that the throughput of BS2 (with the smaller reuse zone) is degraded when its isolation zone starts right after its reuse zone without accounting for the larger reuse zone of BS1 (2Z curve). This is in comparison to even a simple scheme (2Z-CR curve) that starts the isolation zone of BS2 only after the end of the reuse zone of BS1, leaving the region between the reuse zones of the two BSs unused in BS2.

Need for Common Reuse Zone: RADION uses the maximum of the reuse zone sizes within the interference neighborhood as the common reuse zone for two reasons: (i) since the deployment is uncoordinated and non-cooperative, there is no incentive for a cell to decrease its reuse zone size; (ii) given the absence of sensing by the clients, active probing by the BS to determine resource availability can be employed to *only* determine the reuse zone of a cell with a larger zone size effectively (elaborated in §4.3).

Three-zone Structure: RADION uses a three-zone transmission structure as shown in Fig.3(c). While a BS's *reuse zone* remains the same (for class 1 clients), its *isolation zone* only begins from the end of the common (maximum) reuse zone in its neighborhood. The region between its reuse zone and common reuse zone is the *transition zone*. RADION intelligently picks class 2 clients (using the sub-classification in §4.1) to be scheduled in the transition zone on the same subset of sub-channels as the isolation zone. Specifically, class $2l$ clients are scheduled in the transition zone, while class $2h$ clients operate in the isolation zone. There are two benefits to such an approach. First, the transition zone allows selected class 2 clients to reuse resources without incurring significant interference. Second, operating the transition zone on the same subset of sub-channels as the isolation zone prevents the common reuse zone from propagating to the entire network (across multiple contention domains), thereby eliminating under-utilization due to resource coupling.

Evaluation: The benefits of RADION's three-zone structure are evident in Fig.4(b). While both 3Z and 3Z-Opt employ a three-zone structure, 3Z-Opt employs intelligent scheduling of class 2 clients in the transition and isolation zones, 3Z randomly schedules class 2 clients in the two zones. It is clear that the three zone structure with sub-classification yields a 35% throughput improvement.

4.3 Distributed Allocation Framework

We now describe the distributed resource allocation process. The goal of each cell is to determine the size of the reuse (s_r) and transition (s_t) zones as well as the specific contiguous set of sub-channels ($\mathcal{C}$) for operation in the isolation zone. Each BS first classifies its clients into classes 1, $2l$ and $2h$. The preamble used by each BS

is chosen from a standardized set (of orthogonal sequences); using this, the clients measure the signal strength to various BSs and hence lock on to a BS with strong signal strength. The same process is used by each client in class $2h$ to determine its set of strong interferers (received signal strength over a threshold, S_{th}). Using feedback from these clients, the BS determines the super-set of strong interferers. The cardinality of this set including itself (n) determines the fair share allocation of sub-channels ($m = \frac{N}{n}$, where N is total number of sub-channels) for the clients in the isolation zone (class $2h$)[2]. Then the BS determines the desired size of its reuse zone (s_r). This is proportional to the relative traffic load from the clients in the two classes. Next, the BS determines the common reuse zone (in time, s_t) in its interference neighborhood as well as the specific set of m sub-channels (in frequency, $\mathcal{C}$) for operation in the isolation zone (details to follow). After determining the resource allocation parameters, clients in classes 1, $2l$ and $2h$ are scheduled in the reuse, transition and isolation zones respectively. Determination of the resource allocation parameters in RADION is accomplished with the following joint time-frequency *probe-adapt* mechanism.

4.3.1 Two-phase Adaptation

RADION employs a combination of both coarse and fine time scale adaptations (Fig.5). Every BS picks a period P of coarse adaptation (order of several seconds; thousands of frames). P is picked from a set of large prime numbers to reduce the frequency of overlap of adaptation (and probing) periods across cells. The goal of coarse adaptation is to track coarse network dynamics such as (de)activation of cells/clients, load changes, etc., that happen at the granularity of several seconds. Every P frames, each BS triggers a series of fine time scale adaptations automatically. Once triggered, the goal of fine adaptation is to quickly converge to the right set of parameters (s_t, $\mathcal{C}$), for a given set of network conditions. During fine adaptation, the BS performs a *probe-and-adapt* procedure every q frames till convergence, where q is randomly selected from $[1, 0.1P]$ and operates at the granularity of hundreds of milliseconds. The randomness of q minimizes probing collisions.

Employing coarse adaptation in isolation will result in long recovery times in the event of probing collisions across femtocells. This leads to large periods of degraded performance. On the other hand, employing fine adaptation in isolation will require continuous probing to track network dynamics, thereby resulting in large overhead. RADION strikes a good balance between coarse and fine adaptations; fine adaptation is suspended after quick convergence to an efficient resource allocation and invoked again only in the next

---

[2]The solution can be extended to weighted allocations, albeit at the cost of information exchange between femtocells.

Algorithm 1 RADION: Distributed Resource Allocation Framework

1: $m = \frac{N}{n}$, $c_i \in \mathcal{C}$, $s_t \leftarrow s_r$, b_i: BDR, e_i: counter, $\forall e_i = 0$
2: Probe c_c, update b_c, $s_t \mathrel{+}= \Delta t$, f_c: current frame
3: **while** $(f_c \bmod P) \equiv 0$
4:      **for** $i = 1 : n$
5:          Probe c_i, update b_i /*probe n frequency chunks*/
6:      $b_u : u = \max_{i:c_i \in \mathcal{C}}(b_i)$ /*find the max BDR*/
7:      **if** $(b_u > b_c \cdot \alpha) \;\|\; (b_c > \beta)$
8:          Select c_f : call Algo.2 or 3
9:          $c_c \leftarrow c_f$, update b_c (we found s_t)
10:         Pick $q \in [1, 0.1P]$, wait q frames /*fine adaptation*/
11:         **for** $i = 1 : n$
12:            Probe c_i, update b_i
13:         Select c_f : call Algo.2 or 3
14:         **if** $c_c \equiv c_f$
15:            $\exists c_i \in \mathcal{C} \setminus \{c_f\}$, s.t $b_i > \beta$, $e_i\mathrel{+}\mathrel{+}$
16:            $\exists e_i \geq 2$, pick one of the c_i, $e_i = 0$; otherwise $c_i = \emptyset$
17:            $c_c \leftarrow c_c \cup c_i$, goto step 2 /*frequency converged*/
18:         **else**
19:            $c_c \leftarrow c_f$, goto step 10 /*re-do fine adaptation*/
20:      **else**
21:         $s_t \mathrel{+}= \Delta t$, $c_c \leftarrow c_u$, update b_c, goto step 4

Algorithm 2 Gibbs Sampler: Frequency Resource Selection

1: Temperature parameter: $T = 0.05$
2: $\forall c_i \in \mathcal{C}$ compute the probability:
$\quad \pi(c_i) = (e^{\frac{b_i - 1}{T}}) / (\sum_{i=1}^{n} e^{\frac{b_i - 1}{T}})$
3: Sample a random variable $rand$ with law π
4: Select c_f according to $rand$

Algorithm 3 Greedy: Frequency Resource Selection

1: index: $i = \arg\max_{i:c_i \in \mathcal{C}}(b_i)$, $c_f \leftarrow c_i$

---

coarse adaptation period. Thus, the BS spends a large fraction of its P frames operating on an efficient allocation with its probing and adaptation mechanism constituting only a small portion of it. Further, even during the probing-adaptation procedure, data scheduling to clients is not interrupted and is seamlessly incorporated into it.

4.3.2 Probe and Adapt

The goal of fine adaptation is to ensure quick convergence in the determination of resource parameters in both time (common reuse zone) and frequency (sub-channels in isolation zone) domains. This is achieved with a joint time-frequency adaptation algorithm. Each BS probes a vertical strip of resources (a resource region) in the frame, of size $\Delta t \times N$ (i.e., encompassing all sub-channels in time Δt), where Δt is the granularity of probing in the time domain (few symbols). The frequency domain is further probed in *chunks* of Δf contiguous sub-channels ($\Delta f = m$). Δt and Δf can be varied to tradeoff fine grained allocation and convergence time.

Joint Probing in Time and Frequency: When coarse adaptation is triggered, the BS probes resource regions after its own reuse zone to determine the common reuse zone in its neighborhood (step 2 in Algo.1). The intuition is that since the interfering cell with the largest reuse zone will use all sub-channels till its reuse zone, when frequency chunks are probed within the largest reuse zone, they will exhibit similar (degraded) BDRs, while when probed beyond the largest reuse zone, there will be at least one frequency chunk, whose BDR exceeds those of the other chunks by α (see inference in §4.1). This observation is used by every BS to determine the common reuse zone. Specifically, to probe in a vertical resource region, the BS transmits data to a client in each of n randomly

chosen frequency chunks of size m. Since P varies across cells, and the frequency chunk to be probed is chosen at random, probing conflicts across resource regions are avoided. Each chunk is probed for twenty-five frames and the BDR on each of these chunks is estimated (step 4,5); the maximum BDR (across chunks) is compared to the client's current recorded BDR (step 6,7).

Convergence in Time: We consider two approaches to probing in the time domain: sequential and binary search (Fig.6). In sequential probing (outlined in the pseudo-code), the vertical strip to be probed is advanced sequentially by Δt till a gain exceeding α or a high value ($> \beta = 0.8$) of BDR (for BS constituting the maximum reuse zone) is seen compared to the current BDR (time convergence) (step7). Otherwise, the current BDR is updated based on the maximum BDR with the recent probing (step 21). In binary search, two adjacent vertical strips are probed and the BDR with the left and right strips are compared. If $BDR_{right} > BDR_{left} \cdot \alpha$ then size of common reuse zone has been detected (time convergence). Otherwise the maximum value of the BDR (across frequency chunks) is compared with the current BDR to determine if the direction of adaptation should be to the left ($> \alpha$) or right ($\leq \alpha$), and the current BDR is updated only when the region probed is within the maximum reuse zone. If there are multiple clients in class $2h$, they are probed together in each of the chunks and decisions are made with respect to each client. Since different clients may receive interference from different cells, the common reuse zone varies with respect to clients. Time domain probing continues till the common reuse zone for each client in class $2h$ is determined, with the largest common reuse zone determining the termination of the transition zone for the cell.

Convergence in Frequency: Once the common reuse zone is detected, the BS simultaneously has the BDR information on n frequency chunks, with multiple frequency chunks potentially available for operation. We consider two approaches for the selection of a frequency chunk (step 8,9): greedy and Gibbs sampler (Algo.2 and 3). While the greedy scheme is deterministic and picks the chunk yielding the highest BDR, the Gibbs sampler is probabilistic and favors chunks with higher BDR [5]. It has a temperature parameter T, which can be varied with time to provide an annealed version that converges to stable states of low potential (low interference and high BDR). While convergence in the time domain (common reuse zone) can be achieved with high accuracy, frequency domain convergence is sensitive to frequency selectivity and channel errors. Hence, the frequency chunk selected is confirmed by another iteration of fine adaptation, which probes the frequency chunks alone (i.e., no increment of Δt) after q frames (step 10-13). If the same frequency chunk yields the highest BDR, then there is convergence in the frequency domain and the chosen time and frequency parameters are employed for operation till the next coarse adaptation (step 14-17). Otherwise, probing in the frequency domain is repeated every q frames till contention (interference) is alleviated (step 18,19). Thus, by probing vertical strips of frequency chunks, RADION determines both the common reuse zone (s_t) and the set of sub-channels ($\mathcal{C}$ in isolation zone) simultaneously; this leads to quick convergence.

4.3.3 Handling Network Dynamics

Client (dis)-associations impact the traffic load of a cell and consequently, the resource allocations in the isolation zone. Ideally, every cell has to share the frequency resources in the isolation zone of a frame in the contention domains that it belongs to (cliques in the interference/conflict graph), with its ideal share being determined by the size of the largest contention domain that it belongs to. When a new cell is introduced or an existing cell leaves (or traffic ceases)

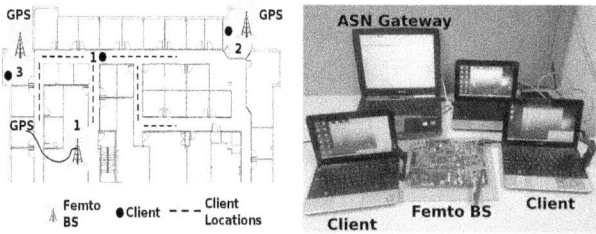

Figure 7: Deployment and picture of our WiMAX testbed.

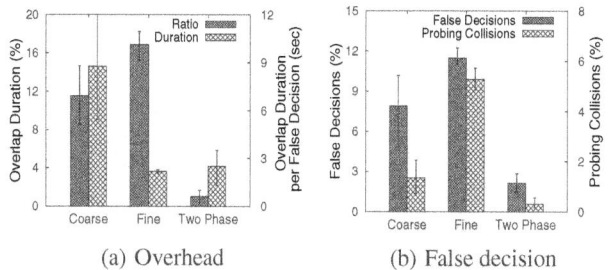

(a) Overhead   (b) False decision

Figure 8: Two-phase adaptation outperforms other schemes.

the contention domain, the existing share of frequency resources decreases or increases, respectively. However, this change has to be detected by each cell in a completely distributed manner in RADION; this allows cells to contract and expand their sub-channel allocations in the isolation zone. Such a feature is also useful in improving the resource utilization in the network. Note that, since a cell (A) does not have information on its contention domains (requires global knowledge), it computes its fair share (say x, $x \leq$ ideal share) based on its interfering neighbors. However, if one of its neighboring cells (B) belongs to a larger contention domain (hence has a lower share $< x$), then some resources (unused by B) will be under-utilized in cell A's contention domain. The ability to probe and expand resource usage will avoid such under-utilization that is a by-product of distributed operations (smaller granularity of chunk sizes ($< \Delta f$) lowers such under-utilization).

RADION allows a cell to adapt its sub-channel usage as follows. Although a cell selects one of the frequency chunks for operation upon convergence, it keeps track of the BDR in other frequency chunks and hence the potential set of chunks unused by its neighboring cells. It continues to monitor such unused chunks for an additional period of P frames (coarse adaptation period), giving its neighbors enough time to detect and use their fair share. If some of these chunks still continue to be available, then the BS decides to expand its resource usage by adding one of the unused chunks to its allocation (step 16). Adding one chunk at a time, allows other cells in its contention domain to also share the unused resources in a fair manner. This expansion of resource usage will address cases when cells switch off or cease to carry traffic. However, if after expansion, the ceased traffic in a cell restarts or a new cell enters the contention domain, this will be detected in the form of degraded BDRs on the frequency chunks or as a new interferer sensed by its clients. In either case, the BS will contract to its conservative share of sub-channels (in the isolation zone) computed based on its updated set of interfering neighbors and re-run its adaptation algorithm. Any resulting under-utilization in its contention domains will be addressed subsequently through its resource expansion mechanism.

5. SYSTEM EVALUATION

5.1 Testbed and Prototype Implementation

Testbed: Our testbed consists of three femtocells deployed in an indoor environment. The testbed components are shown in Fig.7. We use PicoChip's femto BSs that run WiMAX (802.16e). Our clients are laptops with commercial USB WiMAX cards [15]. All the three cells operate on a 8.75 MHz bandwidth with a carrier frequency of 2.59 GHz. For this frequency, an experimental license has been obtained. We achieve synchronization across femtocells via external GPS modules [16]. Given the cost of procuring programmable WiMAX femtocells, our testbed is restricted to three femtocells. However, we augment our evaluations with simulations, where we study scalability under dense deployments with large number of BSs. Unlike WiFi, femtocells perform syn-

chronous frame transmissions without carrier sensing. Hence, it makes more sense to generate different interference topologies by varying the locations of clients. This provides a finer control on the inter-BS interference magnitude as opposed to changing the locations of the BSs, and also covers a wide range of scenarios that include both line-of-sight (LOS) and non-LOS links. We also consider ideal link adaptation where we sequentially run the experiment over all MCS levels and record the one delivering the highest throughput for the given topology.

Implementation: RADION is implemented on the PicoChip platform [17], which provides a *base reference* implementation of the 802.16e standard. We significantly extend and modify the MAC scheduler ($\approx$2000 lines C code) to realize the components in RADION. Our key modifications include: **(1)** Client Categorization: We introduce two measurement zones (i.e., free and occupied zones) for client categorization and schedule data bursts in each measurement zone. **(2)** Resource Decoupling: Each BS tracks its clients' categories to schedule them in the appropriate zone. **(3)** Probe-and-Adapt Module: The two-phase adaptation algorithm is implemented on each BS; adaptations are triggered at frame boundaries. Further, RADION's modular nature allows us to incorporate the two variants of channel selection as outlined in §4.

5.2 Prototype Evaluations

Having evaluated the first two components of RADION in §4, here we focus on evaluating the two-phase adaptation algorithm. To understand the efficiency of algorithm we first evaluate the adaptation process in time and frequency domains in isolation, followed by their joint evaluation. We conclude the evaluation of RADION's adaptiveness to network dynamics. We create multiple clique topologies where clients in class $2h$ are within the transmission range of other BSs. The BSs operate on frames with 30 sub-channels and 22 time symbols for data. Each experiment is run for at least 10 minutes and is repeated multiple times to generate confidence results.

5.2.1 Frequency Domain Convergence

Here, we fix the reuse zone of all the three cells to be the same, thereby eliminating the need for determining a common reuse zone. Hence, the focus is only on frequency domain convergence in the isolation zone for class $2h$ clients; every cell has to identify and operate on a contiguous set of 10 sub-channels each (3 contending cells). We evaluate RADION's two-phase (coarse + fine) adaptation against the coarse and fine time scale adaptations in isolation. The coarse adaptation period is chosen to be a prime number of frames in [1000,6000] for each BS and is fixed for subsequent adaptations. Similarly, the period of fine adaptation is chosen randomly in [1,600]. Further, selection of sub-channels after probing follows the greedy approach (Gibbs sampling considered later).

To understand the impact of false decisions on system (throughput) performance, we consider the following metrics: **(i)** *fractional overlap duration*: the fraction of time that cells operate on over-

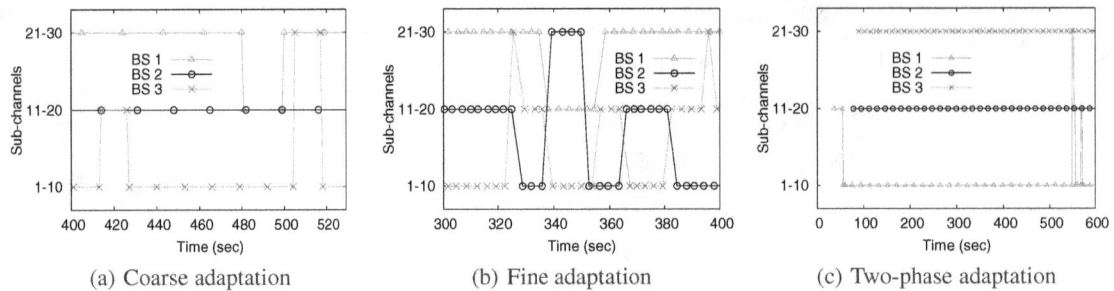

(a) Coarse adaptation (b) Fine adaptation (c) Two-phase adaptation

Figure 9: Three convergence patterns.

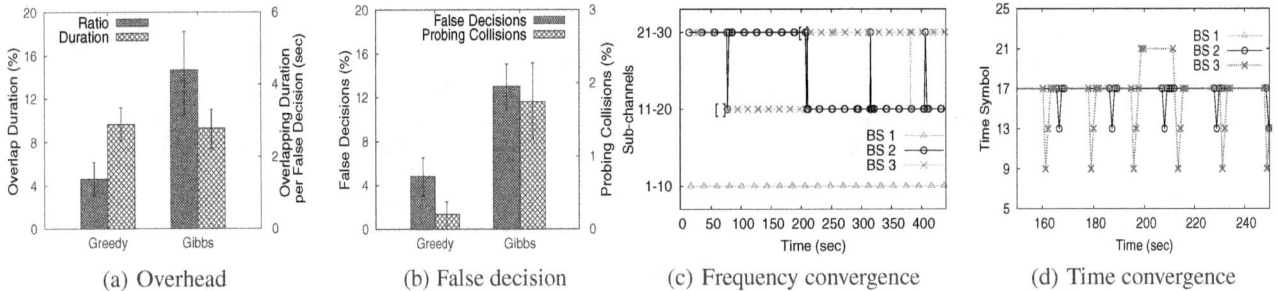

(a) Overhead (b) False decision (c) Frequency convergence (d) Time convergence

Figure 10: Greedy selection outperforms Gibbs sampler (a and b). Convergence patterns of greedy selection (c and d).

lapping resources (leading to collisions); **(ii)** *overlap duration per false decision*: ratio of the net duration of resource overlap to the total number of false decisions (resource overlap ≡ collisions); **(iii)** *fractional false decisions*: ratio of net false decisions to total number of decisions; **(iv)** *fractional probing collisions*: ratio of net probing collisions to total number of decisions. A probing collision occurs if two or more cells probe at the same time and choose the same resources for isolated operations resulting in a false decision. However, a false decision can also occur due to inaccurate BDR estimates and/or asymmetric interference patterns. We omit the throughput results and rather focus on the convergence to orthogonal resources among the BSs. Throughput degradation only occurs when the BSs operate on the same resource therefore, the *fractional overlap duration* reflects the throughput result indirectly.

The average results of multiple runs of frequency convergence are in Figs.8(a) and 8(b). Coarse adaptation incurs fewer false decisions compared to fine adaptation, but the time to recovery (overlap duration) is large per false decision. In contrast, while the overlap duration per false decision is small for fine adaptation, probing collisions and hence, false decisions increase with the number of active probes. RADION's two-phase adaptation strikes a good balance between coarse and fine adaptations, resulting only in 2% of the frames colliding and hence outperforms others in all metrics.

To illustrate, we present one particular example of sub-channel convergence as a function of time in Fig.9. We found similar patterns for all other runs. While the cells maintain orthogonality of sub-channels for most part of the experiment, several differences are observed between the schemes. With fine adaptation (Fig.9(b)), at 352 seconds, BS2 switches from sub-channels 21-30 to 1-10, which is already occupied by BS3. This is a false decision due to an inaccurate probing estimate. After BS2 switches to sub-channels 1-10, BS3 switches to sub-channels 11-20 when its next adaptation is invoked (after q frames). Thus, each cell reacts to others' decisions to avoid interference and maintains orthogonal sub-channel usage. While the number of such switches is small with coarse adaptation (Fig.9(a)), recovery from a false decision is also slow. While recovery is fast with fine adaptation, multiple switches result in both increased probing overhead and false decisions. The

two-phase adaptation (Fig.9(c)) exhibits the best of both adaptations; it incurs fewer switches while also providing fast recovery from false decisions.

5.2.2 Time Domain Convergence

Here, the three cells are pre-assigned orthogonal sets of sub-channels in the isolation zone. We focus on their convergence to the common reuse zone. Recall that RADION probes multiple frequency chunks and compares their BDRs to determine convergence to this common zone. We study both convergence and the impact of the number of probing frames used, for estimating BDR. Sequential search is used for time domain adaptation. The three cells are set different reuse zone demands of 9, 13 and 17 symbols, respectively. BS3, having the maximum reuse zone, will quickly converge to the common reuse zone, while the other two cells will require adaptation. For a given number of probing frames (per chunk), we run the adaptation experiment over 100 times and determine the fraction of cases where the common reuse zone is accurately determined by cells 1 and 2. We repeat this experiment by varying the number of probing frames. We observe that 25 probing frames are sufficient to correctly determine the common reuse zone in over 94.5% of the cases (more frames only provided marginal improvements).

5.2.3 Joint Time-Frequency Convergence

With three BSs (1,2,3) in a clique topology with reuse zone demands set at 17, 13 and 9 symbols respectively, we run the joint time-frequency adaptation process of RADION at each BS. From the results of multiple runs in Fig.10(a) and 10(b), we see that the greedy sub-channel selection yields quick convergence with a low false error rate. The Gibbs sampler incurs a higher fraction of collided frames compared to the greedy. Since both the schemes employ the two-phase adaptation, the average collision duration per false decision is similar with the two schemes (Fig.10(a)). While one might expect the probabilistic nature of Gibbs sampling to yield better convergence [5], this is only true if inferences are based on sensing as opposed to probing; the probabilistic selection increases the number of probing collisions and hence false decisions (Fig.10(b)). Although the greedy approach is deterministic, diver-

(a) Clique topology

(b) Three-links chain topology

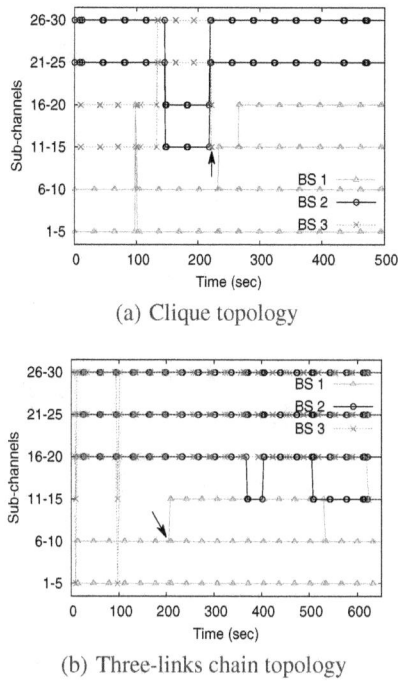

Figure 11: Convergence of RADION with network dynamics.

sity in sub-channel gain across BSs (and their clients) implicitly results in cells picking different frequency resources for their operation. Since both the schemes employ the same time-domain adaptation procedure, their convergence error in the time domain remains similar and less than 5%.

The time-frequency convergence patterns for the greedy scheme are shown in Figs.10(c) and 10(d). Here we present one example of the multiple runs but we confirm very similar pattens for all runs. The frequency adaptations are directly indicated. False convergences in the time domain are indicated by brackets to capture the duration for which the cell operates with a wrong common reuse zone. We see that the number of probing collisions and the resulting quick switch in sub-channel resources to maintain orthogonality in the frequency domain. We omit the pattern of Gibbs sampling for the sake of brevity; however, we highlight that it shows many more frequent switches as the results (Fig.10(a) and 10(b)) show. The time convergence pattern for the greedy approach is expanded in Fig.10(d). We see that BS3 computes the common reuse zone falsely as 21 symbols (instead of 17) at 200 secs; this is corrected in the next coarse adaptation period. Till the reuse zone is corrected, BS3's false decision does not cause interference to BS1's and BS2's class $2h$ clients since it continues to use its isolated resources in its transition zone. However, BS3's class $2h$ clients now incur unfairness as their operational resources are reduced by four symbols. Since the number of false decisions is very small, we let these correct themselves in the next coarse adaptation period.

In summary, we find that the joint time-frequency adaptation process in RADION yields quick and accurate convergence at each femtocell to the common reuse zone as well as to provide orthogonal sub-channels in the isolation zone.

5.2.4 Network Dynamics: Single Contention Domain

We now evaluate RADION's ability to adapt to network changes. First, we consider a single contention domain where all BSs cause interference to each other. In the clique topology of three cells, RADION's adaptation algorithm allows each BS to detect its fair share and converge to orthogonal sets of 10 sub-channels in the isolation

zone. The frequency convergence pattern is shown in Fig.11(a), where frequency resources are probed at the granularity of five sub-channel chunks. RADION's resource expansion mechanism allows each BS to expand its frequency resources in the isolation zone if it probes empty resources for more than a coarse adaptation period. When BS3 is switched off at about 210 seconds (indicated by the arrow), its sub-channels 11-20 become free. BS1 probes the availability of chunks 11-15 and 16-20 but decides to expand its resources only to 1-15, to allow fair access to other cells in the contention domain. However, since the chunk 16-20 remains available in the next adaptation period as well, BS1 continues to expand its resources to 1-20. BS2 fails to grab the chunk 16-20 due to an inaccurate BDR estimate. Thus, while RADION paves the way for a distributed fair sharing of unused resources, distributed operations without information sharing prevents it from controlling the utilization-fairness tradeoff in the best way possible.

5.2.5 Network Dynamics: Multiple Contention Domains

Here we evaluate the adaptiveness in a multiple contention domain by creating chain topology BS3-BS1-BS2; the ideal fair shares of all BSs are 15 sub-channels each. However, since BS1 does not have global information, it computes its fair share to be 10 sub-channels (based on two interfering neighbors) without realizing that its neighbors belong to different contention domains. BS2 and BS3 compute their fair shares as 15 sub-channels each. RADION's resource expansion can help BS1 salvage under-utilized resources. The convergence is shown in Fig.11(b). Initially, while BS2 and BS3 converge to operate on sub-channels from 16-30, BS1 converges to sub-channels 1-10. RADION's resource expansion/contraction feature is enabled after 200 seconds. At this point, BS1 probes the available chunk from 11-15 and expands its allocation to 1-15; this remains stable till 370 secs. At 370 secs, BS2 expands its allocation to 11-30 due to an inaccurate probing, which is immediately rectified in the next adaptation. This again happens at around 510 secs. However, this time, before BS2 rectifies its decision, BS1 responds to interference by contracting its allocation back to its initial fair share of 1-10. This in turn prompts BS3 to probe and expand its allocation to 11-30 (similar to BS2).

The above experiment demonstrates that when a new BS joins the contention domain, it is detected by other cells in the domain; these BSs update (contract) their fair share and run the adaptation process to determine their isolated resources. Stated otherwise, it succinctly captures both the expansion and contraction features incorporated into RADION to track network changes. It also indicates the transition between a fair allocation and one with high utilization. However, to finely control such transitions, information exchange across multiple cells is required, which may not be feasible in residential environments. RADION's *best-effort* utilization-fairness tradeoffs are particularly suited for such environments.

5.3 Evaluation through Simulations

Simulation Set-up: To understand RADION's effectiveness in dense deployments, we resort to simulations. We simulate a single contention domain and increase the number of BSs to stress test RADION's convergence in the frequency domain. We consider two versions of coarse adaptation: (a) a BS picks a fixed prime number of frames P from [8000, 16000], and (b) P is varied (randomly) across adaptation periods. A larger range is chosen for P to avoid excessive probing collisions. For fine adaptation, q is chosen from [$n \times s$, 600], where n is the number of BSs and s is the number of probes sent on each chunk (set to 10 frames). The variable range is based on the observation that smaller ranges are unfair when a large number of BSs simultaneously probe and adapt. The maximum

	5 BS Max Q: 600 - 1600 - 2600 - 3600	7 BS Max Q: 600 - 1600 - 2600 - 3600	10 BS Max Q: 600 - 1600 - 2600 - 3600	12 BS Max Q: 600 - 1600 - 2600 - 3600
Collision Time Fraction	Greedy: 0.049 - 0.029 - 0.021 - 0.016 Gibbs: 0.131 - 0.094 - 0.073 - 0.062	Greedy: 0.21 - 0.113 - 0.076 - 0.057 Gibbs: 0.309 - 0.248 - 0.204 - 0.17	Greedy: 0.648 - 0.46 - 0.326 - 0.234 Gibbs: 0.619 - 0.524 - 0.46 - 0.403	Greedy: 0.847 - 0.72 - 0.585 - 0.459 Gibbs: 0.759 - 0.685 - 0.615 - 0.567
False Decision Fraction	Greedy: 0.015 - 0.009 - 0.006 - 0.004 Gibbs: 0.04 - 0.03 - 0.023 - 0.019	Greedy: 0.05 - 0.026 - 0.017 - 0.012 Gibbs: 0.075 - 0.058 - 0.048 - 0.039	Greedy: 0.15 - 0.094 - 0.062 - 0.042 Gibbs: 0.129 - 0.102 - 0.085 - 0.073	Greedy: 0.215 - 0.154 - 0.112 - 0.082 Gibbs: 0.157 - 0.128 - 0.108 - 0.095
Collision per False Decision	Greedy: 439 - 657 - 960 - 1217 Gibbs: 429 - 639 - 885 - 1124	Greedy: 408 - 629 - 890 - 1139 Gibbs: 402 - 629 - 857 - 1095	Greedy: 310 - 521 - 751 - 1000 Gibbs: 342 - 547 - 772 - 997	Greedy: 241 - 425 - 632 - 856 Gibbs: 296 - 485 - 691 - 908

Figure 12: Results for the fine adaptation process. Fine adaptation is simulated for different period ranges.

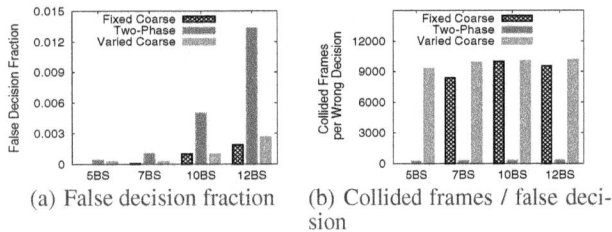

(a) False decision fraction

(b) Collided frames / false decision

Figure 13: Microscopic results for greedy selection.

(a) Greedy selection

(b) Gibbs selection

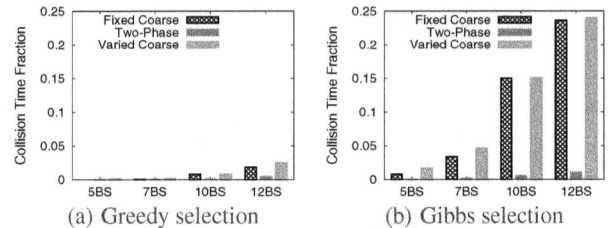

Figure 14: Performance of two-phase adaptation.

range of q is also varied among 600, 1600, 2600 and 3600 frames, to study its impact.

Results: The performance of RADION with an increased number of femtocells is presented in Figs.12, 13 and 14. These results clearly corroborate our findings from the prototype evaluation, thereby demonstrating RADION's scalability in dense deployments. While fewer false decisions are incurred with coarse adaptations (Fig.13(a)), the collision duration per false decision is also longer (Fig.13(b)). The contrary is true with fine adaptation (smaller collision duration per false decision but a higher number of false decisions) as seen in Fig.12. RADION combines the best of coarse and fine adaptation and provides a significant reduction in the time spent in collisions (Fig.14). As with prototype results, the greedy approach outperforms Gibbs sampling; increased probing collisions and hence, false decisions are seen with Gibbs sampling. The table in Fig.12 shows the impact of q's range on fine adaptation. Increasing the range of q provides more time for estimating BDR on the current frequency chunk. Hence, false decisions due to inaccurate BDR estimates and thus, collision durations are reduced. However, adapting q's range is a double-edged sword. For small ranges, its performance is dominated by false decisions, while at large ranges it is dominated by collision duration per false decision. Hence, adapting q's range alone is not sufficient; the two-phase process as in RADION is needed.

6. CONCLUSIONS

We design and implement RADION, arguably the first self organizing resource management framework for OFDMA femtocell networks. RADION consists of three key building blocks i.e., *client categorization, resource decoupling* and *two-phase adaptation and allocation*. RADION allows appropriately chosen clients to opportunistically reuse the spectrum while isolating resources for the other clients in a distributed way. We implement RADION using a WiMAX testbed to show its quick convergence to an efficient resource allocation in real settings. We also demonstrate the scalability and efficacy of RADION in larger scale settings with simulations. We only consider downlink performance, however, a similar approach can be applied to the uplink. As part of future work, we plan to investigate the impact of power control at the BSs.

Acknowledgments

We thank Sampath Rangarajan and Rajesh Mahindra from NEC Labs for their invaluable support and suggestions. J. Yoon and S. Banerjee have been supported in part by the following grants of the US National Science Foundation: CNS-1040648, CNS-0916955, CNS-0855201, CNS-0747177, CNS-1064944, and CNS-1059306. M. Arslan and S. V. Krishnamurthy have been supported by the Multi University Research Initiative grant W911NF-07-1-0318.

7. REFERENCES

[1] "Cisco Visual Networking Index: Global Mobile Data Traffic Forecast Update, 2010-2015", Feb 2011.

[2] R. Van Nee and R. Prasad, "OFDM for Wireless Multimedia Communications", *Artech House*, 2000.

[3] H. Ekstrom, A. Furuskar, J. Karlsson, M. Meyer, S. Parkvall, J. Torsner, and M. Wahlqvist, "Technical Solutions for the 3G Long Term Evolution", *IEEE Comm. Mag.*, vol. 44, pp. 38-45, 2006.

[4] S. Yeh, S. Talwar, S. Lee, and H. Kim, "WiMAX Femtocells: A Perspective on Network Architecture, Capacity, and Coverage", *IEEE Comm. Mag.*, vol. 46, pp. 58-65, 2008.

[5] B. Kauffmann, F. Baccelli, A. Chaintreau, V. Mhatre, K. Papagiannaki, and C. Diot, "Measurement-Based Self Organization of Interfering 802.11 Wireless Access Networks", In *INFOCOM*, 2007.

[6] V. Chandrasekhar and J. Andrews, "Uplink Capacity and Interference Avoidance for Two-Tier Femtocell Networks", *IEEE Trans. on Wireless Communications*, 2007.

[7] M. Y. Arslan, J. Yoon, K. Sundaresan, S. V. Krishnamurthy, and S. Banerjee, "FERMI: A FEmtocell Resource Management System for Interference Mitigation in OFDMA Networks", *MobiCom*, 2011.

[8] 3GPP, "Technical Specification Group Radio Access Networks; 3G Home NodeB Study Item Technical Report (release 8)", *TR 25.820 V1.0.0 (2007-11)*, Nov 2007.

[9] J. Yun and K. G. Shin, "CTRL: A Self-Organizing Femtocell Management Architecture for Co-Channel Deployment", In *MobiCom*, 2010.

[10] D. Lopez-Perez, G. Roche, A. Valcarce, A. Juttner, and J. Zhang, "Interference Avoidance and Dynamic Frequency Planning for WiMAX Femtocells Networks", In *IEEE ICCS*, 2008.

[11] R. Chang, Z. Tao, J. Zhang, and C. Kuo, "Dynamic Fractional Frequency Reuse (FFR) in Multi-cell OFDMA Networks", In *ICC*, 2009.

[12] K. Sundaresan and S. Rangarajan, "Efficient Resource Management in OFDMA Femto Cells", In *ACM MOBIHOC*, May 2009.

[13] L. Yang, W. Hou, Z. Zhang, B. Zhao, and H. Zheng, "Jello: Dynamic Spectrum Sharing in Digital Homes", In *IEEE INFOCOM*, 2010.

[14] T. Moscibroda, R. Chandra, Y. Wu, S. Sengupta, P. Bahl, and Y. Yuan, "Load-Aware Spectrum Distribution in Wireless LANs", In *ICNP*, 2008.

[15] Accton, http://www.accton.com/.

[16] TeraSync, http://www.terasync.net/.

[17] PicoChip, http://www.picochip.com/.

Towards Intelligent Antenna Selection in IEEE 802.15.4 Wireless Sensor Networks *

Mubashir Husain Rehmani, Stéphane Lohier, Abderrezak Rachedi
Université Paris-Est Marne-la-Vallée, LIGM, PasNet, France
rehmani@univ-mlv.fr, lohier@univ-mlv.fr, rachedi@univ-mlv.fr

Thierry Alves, Benoit Poussot
Université Paris-Est Marne-la-Vallée, ESYCOM, France
talves@univ-mlv.fr, benoit.poussot@univ-mlv.fr

ABSTRACT

We plan to design and implement software defined intelligent antenna switching capability to wireless sensor nodes based on link quality metric, such as Received Signal Strength Indicator (RSSI). In this paper, as a first step, we discuss the preliminary results of our newly designed radio module (Inverted-F Antenna) for 2.4 GHz bandwidth wireless sensor networks. In this perspective, we consider the TelosB motes and compare the performance of the built-in TelosB antenna with our proposed antenna. Experimental results confirm the effectiveness of the proposed radio module i.e., 5% to 12% gain over the built-in radio module of the TelosB motes.

Categories and Subject Descriptors

C.2.1 [**Computer-Communication Networks**]: Network Architecture and Design—*Wireless communication*

General Terms

Algorithms, Design, Measurement, Performance

Keywords

Wireless sensor networks, IEEE 802.15.4, ZigBee, RSSI, LQI

1. INTRODUCTION

With the advancement in technology and the availability of cheaper devices, wireless sensor nodes become more and more abundant, resulting in the creation of massively deployed Wireless Sensor Networks (WSNs). The application of these WSNs ranges from environmental monitoring like wildlife tracking, habitat monitoring, forest fire detection to military applications. However, the performance of these WSNs depends upon the quality of the wireless link, the built-in antenna available on the sensor device, and the antenna diversity. In WSNs, there are several application scenarios where a clear line-of-sight (LOS) between the sender

*This work is part of the project entitled "Wireless Sensor Networks for vehicle and pedestrian traffic assessment applications", funded by the university of Paris Est under the PPS (Projet Pluriannuel Structurant) program.

and receiver is not present. This results in fading of the signal and causes multipath propagation. Antenna diversity is a way in which two antennas are attached with the sensor node to improve the quality and reliability of the wireless link [1].

In IEEE 802.15.4, the quality of the link is measured by two metrics: Link Quality Indicator (LQI) and Received Signal Strength Indicator (RSSI). These two metrics are offered by the IEEE 802.15.4 physical layer, which can then be used by the routing layer to select good quality routes. For instance, the impact of LQI-Based routing metrics on the performance of a One-to-One routing protocol for IEEE 802.15.4 Multihop Networks has been studied by the authors in [2]. The authors in [3] provided a detailed study on the comparison of LQI and RSSI metrics. However, aforementioned works [2, 3] did not consider the antenna diversity.

Based on two link quality metrics (RSSI and LQI), our ultimate goal is to design and implement software defined intelligent antenna switching capability to wireless sensor nodes. More precisely, we want to attach an external antenna with the sensor nodes besides the built-in antenna to achieve antenna diversity. Then, based on wireless link condition, the sensor node switch to the appropriate antenna for communication.

As a first step, we design a new radio module, an Inverted-F Antenna for 2.4 GHz bandwidth wireless sensor networks. We then discuss the preliminary results of our antenna by considering TelosB motes and compare the performance of the built-in TelosB antenna with our proposed antenna. Experimental results confirm the effectiveness of the proposed radio module over the built-in radio module of the TelosB motes.

2. INVERTED-F ANTENNA (IFA) DESIGN

We now describe the design and connection procedure of our antenna with the TelosB mote, which is widely used in WSNs research. According to the TelosB datasheet [4], we have implemented an SubMiniature version A (SMA) connector in order to connect our small external antenna. This antenna is called as Inverted-F Antenna (IFA). Compared to the embedded antenna, this one is different in the sense that it couples to its one ground plane. This particularity makes the antenna working on Ultra Wide Band (UWB) from 2070 to 3140 MHz for a -10 dB matching bandwidth. The current excited by the IFA on its ground plane make

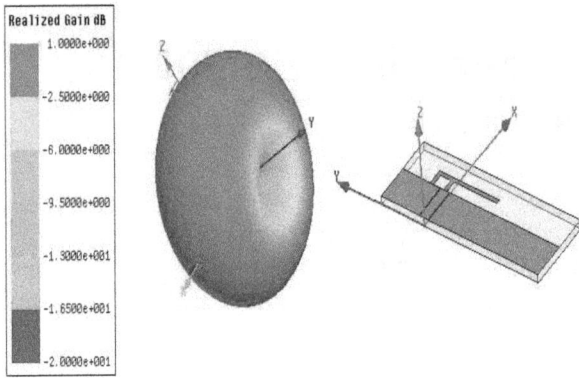

Figure 1: Realized Gain (dB) and Design of Inverted-F Antenna for 2.4 GHz WSNs.

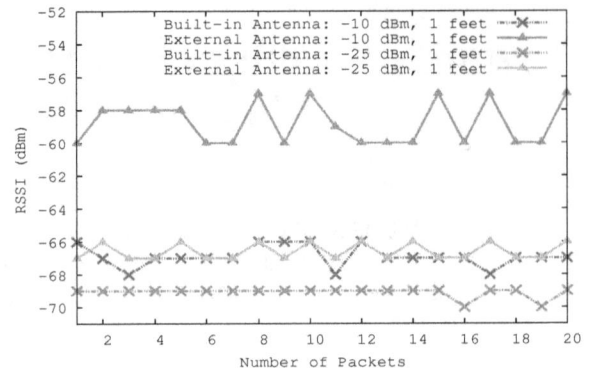

Figure 2: Number of Packets and RSSI for -10 dBm and -25 dBm Power.

the radiation pattern to have a null on the Y-axis. The gain is almost constant is the XZ plane and is around 1 dBi (cf. Fig. 1 for the design and radiation pattern). This antenna has two advantages :

-Firstly, the matching bandwidth is much more important from the embedded antenna. Making it a little less sensitive to proximity effects.

-Secondly, as the gain is almost constant in the ZX plane, the TelosB can be used in horizontal position, limiting troubles from the USB connector.

3. PERFORMANCE ANALYSIS

We consider Telosb motes [5], which are equipped with Chipcon CC2420 radio module [6]. The transceiver of TelosB motes operate in 2.4 GHz band. We use RSSI as a link quality metric to study the performance of our newly designed antenna. The RSSI provides the signal strength at the receiver (in dBm) [3]. The CC2420 calculates the RSSI over 8 symbol periods and stores the result in its RSSI.RSSI_VAL register. Chipcon specifies the following formula to compute the received signal power (P) in dBm: P = RSSI VAL + RSSI OFFSET, where RSSI OFFSET is about -45. We refer to this power, P (in dBm) as RSSI throughout this paper.

We consider two types of TelosB motes in our experiment: (1) sink node, which is connected with the serial port of the Laptop, and (2) the sending mote, which sends packet with a delay of 500 ms. Total 20 packets were sent by the sending mote. As soon as the sink node receives packet by the sending mote, it saves the RSSI value of each corresponding packet. The sink node and the sender node is placed 1 feet apart. We carried out our experiment at 2 different power levels: -10 dBm and -25 dBm. We perform our experiments in RF Anechoic Chamber to avoid the effect of reflection and interference. To change the transmission power, we change the default transmission power of CC2420 i.e., 0 dBm (CC2420_DEF_RFPOWER = 31), to -10 dBm and -25 dBm, by changing the register values to 11 and 3, respectively.

Fig. 2 compares the RSSI values of built-in antenna and external IFA antenna for each received packet with -10 dBm and -25 dBm power levels. The IFA antenna provides higher power levels i.e., RSSI values at the receiver. More precisely, when power level is -10 dBm, the external IFA antenna pro-

vide -58 dBm power compared to -66 dBm power for the built-in antenna, i.e., the gain of 12%. And when the power level is -25 dBm, the external IFA antenna provide -66 dBm power compared to -69 dBm power for the built-in antenna, i.e., the gain of around 5%. In summary, results in Fig. 2 confirm that the newly designed IFA antenna can provide good link reliability, suitable for wireless radio communication in IEEE 802.15.4 based WSNs.

4. CONCLUSION AND FUTURE WORK

In this paper, we discussed our preliminary results of the newly deigned radio module (Inverted F Antenna) for 2.4 GHz wireless sensor networks. Experimental results confirmed the effectiveness of our proposed radio module over the built-in TelosB radio module. This is in fact an ongoing work and as plan of our future work, we intend to integrate software defined switching capability to TelosB motes to change the antenna decision adaptively. This antenna switching decision will be based on link quality metrics, such as RSSI and LQI. We also plan to study different routing metrics with antenna diversity under multi-hop network configuration. The energy consumption of IFA antenna is also an important aspect and we also plan to study it.

5. REFERENCES

[1] K. Sasloglou, I.A. Glover, K. Kae-Hsiang, I. Andonovic. Wireless sensor network for animal monitoring using both antenna and base-station diversity. IEEE ICCS 2008

[2] Carles Gomez, Antoni Boix, and Josep Paradells. Impact of lqi-based routing metrics on the performance of a one-to-one routing protocol for IEEE 802.15.4 multihop networks. *EURASIP Journal on Wireless Communications and Networking, 2010:6:1-6:20, July 2010.*

[3] Kannan Srinivasan and Philip Levis. RSSI is under appreciated. *In Proceedings of EmNets, 2006.*

[4] Moteiv Corporation, telos datasheet http://www2.ece.ohio-state.edu/ bibyk/ee582/telosMote.pdf

[5] MEMSIC Inc, http://www.memsic.com, 2011.

[6] Chipcon cc2420 datasheet, http://www.chipcon.com, 2011.

Localization Using Bluetooth Device Names

Troy A. Johnson
Department of Computer Science
Central Michigan University
Mount Pleasant, MI, 48859
johns4ta@cmich.edu

Patrick Seeling[*]
Department of Computer Science
Central Michigan University
Mount Pleasant, MI, 48859
pseeling@acm.org

ABSTRACT

In this work, we present a scheme based on Bluetooth friendly device names to enable power-optimized ad-hoc localization of mobile devices. Eliminating the service discovery and connection (including potential pairing) phases in Bluetooth allows for speedier and more power-efficient conveying of location information using friendly device names. Furthermore, we observe that using the signal strength commonly provided in reference APIs of mobile OSs, client distances can be calculated with high accuracy and without additional power penalties.

Categories and Subject Descriptors

C.2.1 [**Network Architecture and Design**]: Wireless communication

General Terms

Algorithms, Experimentation, Performance

Keywords

Bluetooth, localization

1. INTRODUCTION

While fine grained outdoor localization can typically be achieved using the Global Positioning or similar systems, fine grained indoor and cooperative localization techniques are manifold and subject of ongoing research efforts. Approaches to indoor positioning have gathered industry interests and are currently in implementation as solutions by major companies, such as Google or Nokia. Typical approaches include wireless network location finger-printing or Bluetooth beacons, utilizing the pervasiveness of Bluetooth in short-range communication scenarios. In [1], the authors use mobile devices as location beacons, which are processed by an enterprise infrastructure. An detailed approach using Bluetooth beacons with services as well as fingerprinting and device names is described in [3]. In our prior works, we provided a general evaluation for localization in the context of power savings, see [5].

---

[*]Please direct correspondence to P. Seeling.

The approach we follow here is a combination of the one presented in [3] and the one introduced in [2], where Bluetooth friendly device name *tags* were used in interaction design on a university campus.

2. DESIGN

For localization from information of multiple neighboring nodes (i.e., without user intervention), a mobile device performs the inquiry phase (yielding neighboring device identifiers) once, followed by a name resolution phase for each of the n responding devices as implemented in the mobile OS.

Utilizing the Bluetooth friendly device names to convey the actual location information itself eliminates (i) the service discovery, connection and exchange periods required if the functionality were to be implemented as a Bluetooth service as well as (ii) the typically enforced device pairing, which is undesired for background services.

An additional benefit of using friendly device names is that most smart phone operating systems perform the Bluetooth name resolution automatically as part of a (combined) device discovery phase and provide the received signal strength from the hardware interface as well; this allows for finer grained estimation of the distance to the nodes responding to the initiated discovery without additional impacts on the power consumed.

Using up to 48 bytes (characters) from the friendly device name field (typically max. 255), we propose to use a `:geo:` tag followed by comma-separated latitude and longitude values (both signed and with leading zeros) and an accuracy estimator (in m) followed by an epoch time stamp of the location used in the name. Fixed markers (beacons) can use zero as time stamp. A mobile device labeled `phone` would hence appear as
`phone:geo:+043.586541,-084.775811,3,1331327432` in the output of a device inquiry.

3. EVALUATION

The initial design is implemented as a mobile application that utilizes Bluetooth friendly device name resolution. For evaluation purposes, we implemented the system using BlackBerry 9800 smart phones as (static) Bluetooth beacons and a client-side application using the Android operating system on a Nexus S smart phone.

Using a LOS configuration outdoors, we evaluate the accuracy of our approach using 6 different configurations of 3 beacons (approx. 20m distance) and one mobile device as illustrated in Fig. 1. We illustrate the resulting accuracy estimations in Fig. 2, which are based on simple RSSI-distance

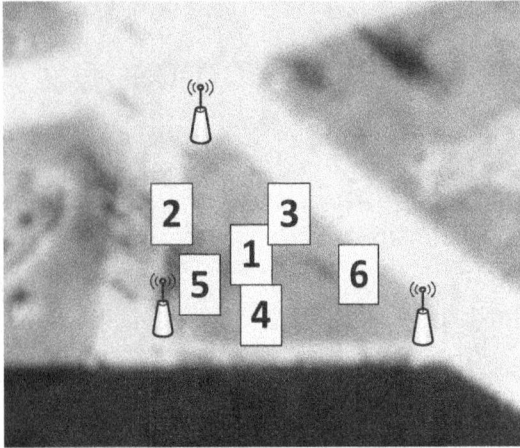

Figure 1: Courtyard beacon-based LOS evaluation scenarios.

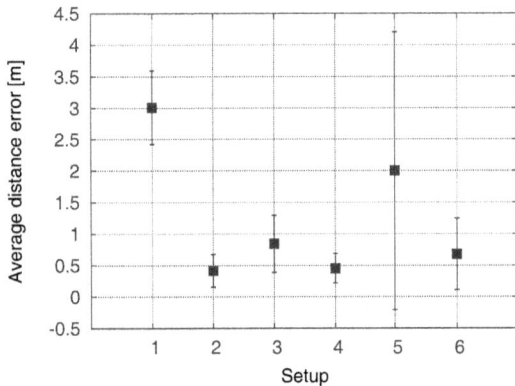

Figure 2: Accuracy for Bluetooth beacon-based location estimation.

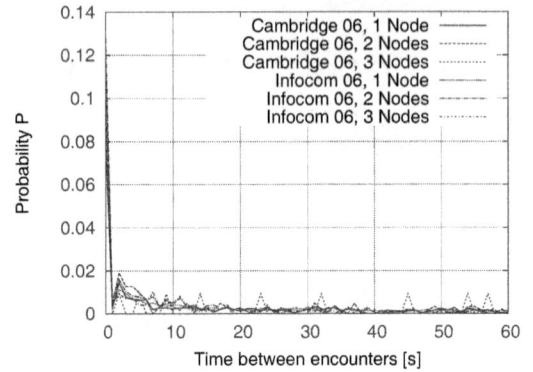

Figure 3: Probability for nodal encounter with requirements for localization.

estimations without optimization. We overall find that for 43 measurements, an approximate accuracy of 1.83m with a standard deviation of 1.41m can be achieved. Assuming the default values for Bluetooth, the total resolution time for b beacons can be described as $10.24 + b \cdot 5.12$s, which results in localization costs of $L = 7.03$ [Wsm] for the 3 beacon scenario here, with more details in [5].

A first look at a mobile cooperation extension to the fixed beacon approach is to utilize mobility traces from [4]. If an assumed 3 nodes would be required to remain in contact during the entire period of 25.6s, for the Infocom 2006 trace with 78 nodes, only an average $P = 0.014$ of nodal encounters would be suitable to determine a position while mobile ($SD = 0.008$). Reducing the accuracy and required scanning time to the two/one node equivalents, we derive $P = 0.027$ ($SD = 0.014$) and $P = 0.09$ ($SD = 0.0357$), respectively. Similarly, when evaluating the time required to meet another set of nodes, we observe for these traces and the Cambridge variant with 36 nodes in Fig. 3 that most encounters are in close proximity time-wise, independent of the number nodes required or scenario. In addition to the heavy tail behavior, we derive that meeting times are somewhat clustered and finding other nodes to derive information from might be challenging.

4. DISCUSSION

We presented a Bluetooth device name based approach to sharing location information as extension of [3], without lookup tables or service connections, by suggesting the use of a `:geo:` tag in the friendly device name. A first look at a mobile cooperative extension indicates potential challenges, which we strive to investigate in our ongoing works.

We are implementing a cooperative extension, which allows mobile devices to determine their location based on beacons as well as other mobile devices and dynamically share their location. In this cooperative context, we will evaluate security and time constraints in addition to power consumption and accuracy.

5. REFERENCES

[1] G. Anastasi, R. Bandelloni, M. Conti, F. Delmastro, E. Gregori, and G. Mainetto. Experimenting an indoor bluetooth-based positioning service. In *Proc. IEEE Int. Conf. on Distr. Comp. Sys.*, pages 480–483, Providence, RI, USA, 2003.

[2] N. Davies, A. Friday, P. Newman, S. Rutlidge, and O. Storz. Using bluetooth device names to support interaction in smart environments. In *Proc. ACM MobiSys*, pages 151–164. ACM, 2009.

[3] A. Huang and L. Rudolph. A privacy conscious bluetooth infrastructure for location aware computing. In *Proc. of SMA 2005 Symposium*, Singapore, Jan. 2005.

[4] A. Mei and J. Stefa. Swim: A simple model to generate small mobile worlds. In *INFOCOM 2009, IEEE*, pages 2106–2113, Apr. 2009.

[5] P. Seeling. Power consumption evaluation for cooperative localization services. In *Proc. of the Int. Joint Conf. on Comp., Inf., and Sys. Sciences, and Eng. (CISSE)*, pages 1–5, Online, Dec. 2011.

Channel Width Assignment using Relative Backlog: Extending Back-pressure to Physical Layer

Parth H. Pathak, Sankalp Nimbhorkar, Rudra Dutta
Computer Science Department, North Carolina State University, Raleigh, NC, 27695
phpathak@ncsu.edu, sunimbho@ncsu.edu, dutta@csc.ncsu.edu

ABSTRACT

With recent advances in Software-defined Radios (SDRs), it has indeed became feasible to dynamically adapt the channel widths at smaller time scales. Even though the advantages of varying channel width (e.g. higher link throughput with higher width) have been explored before, as with most of the physical layer settings (rate, transmission power etc.), naively configuring channel widths of links can in fact have negative impact on wireless network performance. In this paper, we design a cross-layer channel width assignment scheme that adapts the width according to the backlog of link-layer queues. We leverage the benefits of varying channel widths while adhering to the invariants of back-pressure utility maximization framework. The presented scheme not only guarantees improved throughput and network utilization but also ensures bounded buffer occupancy and fairness.

Categories and Subject Descriptors

C.2.1 [**Computer-Communication Networks**]: Network Architecture and Design

General Terms

Design, Algorithm, Performance

Keywords

Utility maximization, Back-pressure, Channel width adaptation, IEEE 802.11

1. INTRODUCTION

Even though the advantages of variable channel width assignment have been explored before [2], a few fundamental questions surrounding the problem have largely remained unanswered. Some of the questions are -

(1) Do link-level benefits of higher throughput using increased channel width come at other network-wide performance penalties?

(2) How can we assign channel width to links such that flows can achieve maximum throughput while maintaining fairness among the flows?

(3) Since variable channel width can introduce significant variations among achievable per-link throughput, how can we design a channel width assignment scheme that can guarantee stability of queues and bounded buffer occupancy?

This paper attempts to answer the questions by utilizing back-pressure principles of utility maximization framework

for the purpose of variable channel width assignment. We show that substantial gains in throughput and network utilization can be achieved along with improved fairness when back-pressure framework is used as a directive for channel width assignment. The central idea is to assign higher channel widths to queues with larger backlog (relative to its interference neighborhood).

1.1 The Back-pressure Framework

We assume a network in which transmission time is divided into slots. For a network graph $G = (V, E)$, let $f \in F$ denote a flow from $s(f)$ to $d(f)$. Let $l(u, v) \in E$ denote a link between node u and v, and $\gamma_{l(u,v)}$ be the transmission rate of link $l(u, v)$. Transmission rates of all links in E are presented by $\Gamma = \{\gamma_{l(u,v)}, l(u, v) \in E\}$. Let χ be the set of all possible combinations of rates at which links can operate. Every node $u \in V$ maintains a separate queue (Per Destination Queue - PDQ) for destinations of all flows in F. Packets received by u for a flow f destined to $d(f)$ are queued in $Q_u^{d(f)}$. Let $|Q_u^{d(f)}(t)|$ denote the size of the PDQ maintained at node u for destination $d(f)$ at time t. Every node shares its PDQ length information with all its neighbors at the beginning of time slot t. Using this, a node u calculates $D_{l(u,v)}^{d(f)}(t) = |Q_u^{d(f)}(t)| - |Q_v^{d(f)}(t)|$ for all $l(u, v) \in E$ and all $f \in F$. Now, for every link $l(u, v) \in E$, let $\Delta_{l(u,v)}(t) = \max_{f \in F}\left(D_{l(u,v)}^{d(f)}(t)\right)$. Back-pressure scheduling [3] suggests that Γ at time t should be chosen such that -

$$\Gamma(t) = \max_{\Gamma \in \chi} \sum_{l(u,v) \in E} \left(\gamma_{l(u,v)} \Delta_{l(u,v)}(t)\right) \qquad (1)$$

It was proved in [3] that a routing/scheduling policy that can achieve a solution of Equ.1 is throughput optimal. Unfortunately, the above mentioned problem is proven to be NP-hard in wireless case due to interference constraints.

2. CHANNEL WIDTH ASSIGNMENT

It can be observed that any approximation of solution to Equ.1 should allow PDQs with higher backlog to transmit at a proportionally higher transmission rates. This higher transmission rate can be achieved by assigning the endpoints of PDQs (radio interfaces) wider channel widths.

Let $P(t)$ be the set of all PDQs in the network at time t. For any PDQ p_i with backlog $b(p_i)$, let $N(p_i)$ be the set of links whose PDQs interfere with the link of p_i. Let $max(p_i)$ (and $min(p_i)$) be the backlog of PDQ with maximum (minimum) backlog in $N(p_i)$. Algorithm 1 presents two *centralized* and *greedy* schemes of channel width assignment. The first scheme (Algo. 1 - Case 1) targets a *continu-*

ous assignment in which a PDQ can be assigned any width (in Hz) from total spectrum width of D Hz (60 MHz here as in most 802.11 radios). It determines the relative backlog of a PDQ as compared to its neighborhood PDQs in order to proportionally assign a channel width to it. In the case of Algo. 1 - Case 2, only a few pre-selected channel widths can be used by each PDQs. As in 802.11 standards, we choose 4 discrete levels of 5, 10, 20 and 40 MHz.

Algorithm 1 Channel Width Allocation

CASE 1: CONTINUOUS

At the beginning of each time slot t

Sort $P(t)$ in descending order of backlog

for $i = 1 \rightarrow |P(t)|$ **do**

channel width $cw(p_i) = \frac{(b(p_i) - min(p_i)) \cdot D}{max(p_i) - min(p_i)}$

if $cw(p_i)$ width not available in $N(p_i)$ **then**

assign maximum available width in $N(p_i)$ to p_i

end if

end for

CASE 2: DISCRETE

widthArray[4] = {5, 10, 20, 40}

At the beginning of each time slot t

Sort $P(t)$ in descending order of backlog

for $i = 1 \rightarrow |P(t)|$ **do**

$l = \frac{(b(p_i) - min(p_i)) \cdot 4}{max(p_i) - min(p_i)}$

$cw(p_i) = widthArray[l]$

while $cw(p_i)$ width not available in $N(p_i)$ **do**

$l = l - 1; cw(p_i) = widthArray[l]$

end while

end for

The scheme also reduces the transmission power of radios operating at a lower channel width in such a way that communication range of every node remains constant. Also, note that the number of radios necessary in order for both channel width allocation strategies to work can be very high (e.g. in worst case, D radios of continuous and $D/5$ radios discrete schemes). This along with centralized nature of scheme makes it difficult to implement in practice. Nevertheless, the scheme works well as a proof-of-concept and sets important performance benchmarks for our ongoing distributed protocol design.

3. SIMULATION RESULTS

We choose a 7×7 grid with 10 randomly chosen source-destination pairs. Each source node generates 1500 bytes packets using poisson random process with mean of 1350 packets/second. Bi-directional protocol interference model is used for determining the link interference relationships. Also, utility based source node injection rate control [1] is employed. The results of simulation and observations are shown in Figs. 1 and 2. We compare the schemes with two other schemes. First, a *fixed width channel assignment* is used which utilizes back-pressure scheduling but 60 MHz spectrum is divided into 3 channels of 20 MHz fixed widths. Second, a *random scheme* in which the same 3 channels are used but instead of scheduling the links using back-pressure policy, links are randomly chosen for scheduling.

4. DISTRIBUTED PROTOCOL DESIGN

We are currently developing a back-pressure based distributed channel width assignment protocol on CSMA/CA

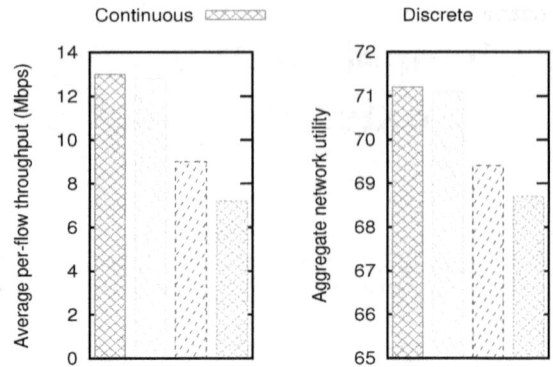

Figure 1: (a) **Average per-flow throughput** (x_f) **increases significantly (by factor of 1.4) in variable channel width case, (b) Network Utility** ($\Sigma_{f \in F} log(x_f)$) **also increases**

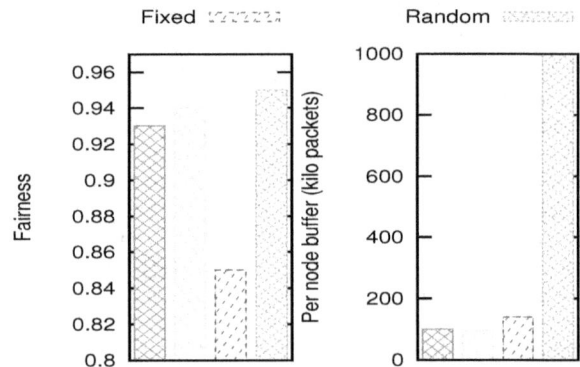

Figure 2: (a) **Jain's fairness index** ($\frac{(\Sigma_{i=1}^{|F|} x_i)^2}{\Sigma_{i=1}^{|F|} x_i^2}$) **increases because backlogged queues are served at a faster rate in variable channel width cases, (b) Note that buffer occupancy reduces by 25% as compared to random scheme. Even with comparison to fixed-width case, buffer occupancy reduces while increased fairness, throughput and utility in case of varying channel widths**

MAC that can estimate the benchmark set by centralized scheme presented here. Apart from distributed selection of channel mid-point and width, and transmission synchronization, the problem becomes further more challenging due to the fact that in real-world scenarios each node possesses only a few limited number of radios.

5. ACKNOWLEDGMENTS

This work is supported by the U.S. Army Research Office (ARO) under grant W911NF-08-1-0105 managed by NCSU Secure Open Systems Initiative (SOSI). The contents of this paper do not necessarily reflect the position or the policies of the U.S. Government.

6. REFERENCES

[1] U. Akyol, M. Andrews, P. Gupta, J. Hobby, I. Saniee, and A. Stolyar. Joint scheduling and congestion control in mobile ad-hoc networks. In *IEEE INFOCOM 2008*.

[2] R. Chandra, R. Mahajan, T. Mosci., R. Raghavendra, and P. Bahl. A case for adapting channel width in wireless networks. ACM SIGCOMM 2008.

[3] L. Tassiulas and A. Ephremides. Stability properties of constrained queueing systems and scheduling policies for maximum throughput in multihop radio networks. *Automatic Control, IEEE Transactions on*, 1992.

Reliable Link Lifetime-Based Cluster-Head Election in Wireless Ad Hoc Networks

Dongsheng Chen, Alireza Babaei and Prathima Agrawal
Dept. of Electrical and Computer Engineering, Auburn University
{dzc0017, ababaei, agrawpr}@auburn.edu

Categories and Subject Descriptors

C.2.1 [**Computer-Communication Design**]: Network Architecture and Design- *Wireless Communications*

General Terms

Algorithms, Design, Performance, Reliability

Keywords

Ad hoc network, Cluster-head election, Cluster update, Stability

1. INTRODUCTION

In mobile ad hoc networks (MANETs), the mobility of nodes results in dynamic change of network topology which incurs frequent communication interruptions, larger delay and lower performance. To alleviate this problem and improve the scalability, introducing a hierarchical structure has proven to be very effective [1]. Clustering algorithms proposed in the literature can be classified as identifier-based, connectivity-based, weight-combination, and energy-aware [2]. In the identifier-based approach, each node is assigned a unique ID which is broadcasted to its neighbors using HELLO messages. In the LID algorithm [3], a node with minimum ID is selected as the cluster-head (CH). LID may result in low network lifetime as the number of cluster-heads may become unnecessarily large. To reduce the number of clusters, HD algorithm is proposed in [3] in which the node with largest degree (largest number of neighbors) is elected as the CH. The HD algorithm, however, does not consider the impact of mobility on the stability of the clustering structure. In [4], WCA chooses a CH by striking a balance between few factors including: sum of the relative distances of nodes to their neighbors, degree difference, mobility and power left. It needs to assign the appropriate weight to each of these factors before selecting or updating the CH. In [5] a probabilistic approach is proposed to elect CH and achieve load balancing without considering the relative stability of the cluster structure. While taking many factors affecting the CH election into consideration, the proposed algorithm in this paper employs no weights combination unlike the previous proposed approaches. It utilizes only the one-hop neighbor information from their HELLO messages. The required information is available if nodes are equipped with GPS.

2. MODELS

Two models for link lifetime and maximum cluster-head update intervals are established. These models will be used for parameters design in our proposed algorithm.

Link Lifetime Model: We estimate the link lifetime using the information extracted from the HELLO messages periodically broadcast by the 1-hop neighbors of each node. Here we take the example of link $(0,1)$ in Fig. 1, to estimate the link lifetime. In this figure, v_1 and v_2 are the current velocities of nodes 0 and 1; (x_1, y_1) and (x_2, y_2) are current updated locations of nodes 0 and 1, respectively. $\vec{v}_1$ and $\vec{v}_2$ are the velocity vectors, θ_1 and θ_2 are the velocity directions of nodes 0 and 1; $d_{link(0,1)}$ is the estimated length of movement of node 1 while in the transmission range of node 0 (i.e., while there is a link between them) and $t_{link(0,1)}$ is the link lifetime.

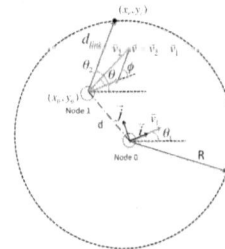

Figure 1. Example of calculating link lifetime between node 0 and

We analytically derive $d_{link(0,1)}$ and using that, the link lifetime, i.e., $t_{link(0,1)} = d_{link(0,1)} / v_{(0,1)}$, where $v_{(0,1)}$ is the relative velocity of node 1 with respect to node 0. Thus, each node can update the link lifetime with neighbors in its neighbor list after receiving theirs' HELLO message.

We assume: 1) nodes are uniformly distributed in the networks, 2) both nodes' speed magnitude and angle are uniform random variables and are uncorrelated; i.e. v is uniformly distributed in $(0, a]$ (a is maximum speed), and a is uniformly distributed in $(-\pi, \pi]$; 3) a link can be established between two nodes if they are within transmission range of each other.

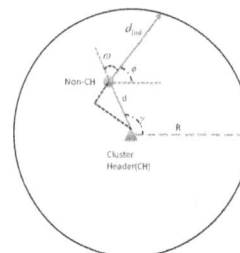

Figure 2. Expected link lifetime model of cluster member

In Fig. 2, d and γ are used to describe the location of a typical node. Also, we have $t_{link} = d_{link} / v$. We obtain the expected link lifetime: $\bar{\tau} = E\{t_{link}\}$ analytically:

$$\bar{\tau} = E(t_{link}) = \iiint_{v\,\varphi\,d\,\gamma} t_{link} \cdot f_{v,\varphi}(v,\varphi) \cdot f_d(d) \cdot f_\gamma(\gamma) \cdot dd \cdot d\gamma \cdot d\varphi \cdot dv$$

$$= \frac{8R}{3\pi} \cdot \frac{1}{2\pi a^2} \cdot \iint_{v_1\,\varphi} (\ln \left| \frac{a + \sqrt{a^2 - v_1^2 \sin^2 \varphi}}{v_1 + v_1 \cos\varphi} \right| \cdot d\varphi \cdot dv_1$$

Fig. 3 shows that the expected link lifetime decreases with increasing node velocity; In Fig. 4, the expected link lifetime $\bar{\tau}$ is linearly proportional to transmission range R which can also be seen from the obtained analytical result.

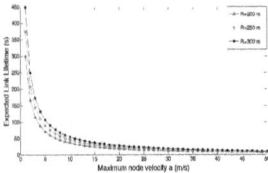

Figure 3. Expected link lifetime with varied a

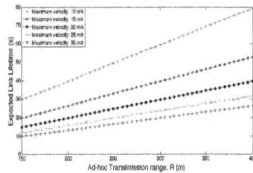

Figure 4. Expected link lifetime with varied R (m)

3. CLUSTER HEAD ELECTION

Double M Principles: A good cluster must 1) include as many neighbors as possible. 2) Members do not move in/out frequently between different clusters. In other words, it is desirable that a cluster-head has the maximum number of one-hop neighbors within its transmission range and minimum rate of nodes moving out of its transmission range.

Area-Based Approximation of Double M Principle: We approximate the requirements of the double M principle. Considering node i, we use the area S_i, shown in Fig. 5, to estimate the qualification of node i in being a CH. In this figure, the vertical axis shows the number of neighbors within transmission range of a node as time goes on. If the area S_i is large, it implies that first: the number of neighbors of node i when conducting the cluster formation/ reformation is relatively large; and second: This number is decreased with a lower rate which means that the number of neighbors remains relatively large as time goes on. As shown in Fig. 5, $S_i > S_j$ which means that node i satisfies the requirements of the double M principle better than node j and therefore is more qualified in being a CH.

4. SIMULATION RESULTS

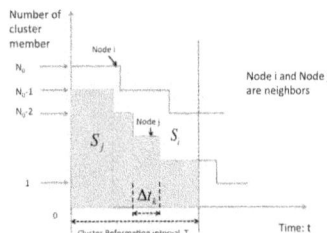

Figure 1. Area-based cluster-head election method

In this section, we define three metrics: cluster head update rate, member exchange rate, and average number of cluster head in network. To analyze performance of RLCH, we compare it with LID, HD, WCA. Each group of simulation was conducted 15 times and result was averaged. In figures 6 and 7, we study the CH update rate. We can observe that our proposed RLCH can achieve better performance compared to the other three algorithms. Figures 8-11 show that RLCH has a smaller member exchange rate and number of cluster head compared with other three algorithms in most of cases. Since WCA and RLCH have a close cluster-head update rate in high mobility or small R; the member exchange rate of RLCH is close to that of WCA.

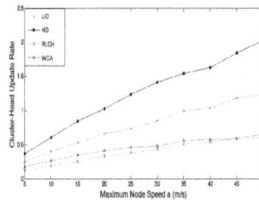

Figure 6. Cluster-head update rate with varied maximum node speed

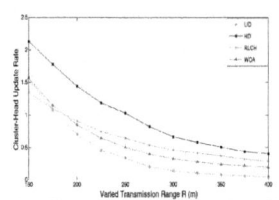

Figure 7. Cluster-head update rate with varied transmission

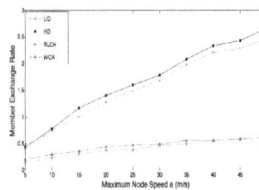

Figure 8. Member exchange rate with varied maximum node

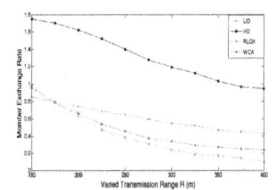

Figure 9. Member exchange rate vs. varied transmission

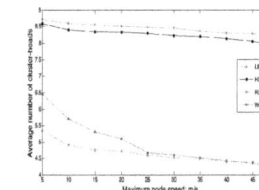

Figure 10. Average number of cluster-head vs. varied maximum node speed

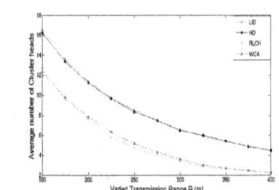

Figure 11. Average number of cluster head vs. varied transmission range

5. REFERENCES

[1] Zhao, S., and Raychaudhuri, D. 2009. Scalability and Performance Evaluation of Hierarchical Hybrid Wireless Networks. IEEE/ACM Transactions on Networking (Oct, 2009), vol.17, No.5.

[2] Agarwal, R. and Motwani, M. "Survey of clustering algorithm for MANET", International Journal on Compute Science and Engineering, vol.1(2) (2009), 98-104.

[3] Gerla, M. and Tsai, J.T. 1995. Multiuser, Mobile, Multimedia Radio Networks. Wireless Networks, vol. 1 (Oct. 1995).

[4] Chatterjee, M., Das, S.K., and Turgut, D. 2000. An On-Demand Weighted Clustering Algorithm (WCA) for Ad hoc Networks. In proceeding of IEEE Globecom'00, 1697-701.

[5] Kawadia, V and Kumar. 2003. P.R, Power Control and Clustering in Ad Hoc Networks. INFOCOM, pp. 459-469.

Price-Reward for Data Relaying and Handover Management in Wireless Networks

Muhammad Shoaib Saleem
Institut Télécom - Télécom SudParis
9, Rue Charles Fourier
Évry, France
shoaib.saleem@it-sudparis.eu

Éric Renault
Institut Télécom - Télécom SudParis
9, Rue Charles Fourier
Évry, France
eric.renault@it-sudparis.eu

ABSTRACT

We propose *Price-Reward for Data Relaying and Handover in Wireless Networks*, a handover and data relay management algorithm for wireless networks. The algorithm works in dynamic environment in which mobile nodes move randomly. The mathematical model presented explains how cooperative diversity helps to augment network coverage plus revenue and ensures connection reliability during mobility and handover situations.

Categories and Subject Descriptors

C.2.1 [**Network Architecture and Design**]: Wireless communication

General Terms

Algorithms

Keywords

Wireless Networks, Stackelberg Model, Price-Reward

1. INTRODUCTION

The use of the Internet in wireless networks always demand for better QoS during mobility. A wide range of proposal suggest either to adopt high performance protocols or replace the contemporary Internet architecture with a new one that incorporates all the desired features required today. Seamless mobility in wireless networks is an issue addressed widely. We, in [3], proposed an mobile node architecture for the Future Internet called *Network of Information Mobile Node (NetInf Mobile Node)*. It possesses features like seamless handover and data relaying during mobility to sustain the QoS during an ongoing session.

However, in order to analyze its performance, the two possible methods are the analytical and the simulation approaches. In an analytical approach, a mathematical model is developed with the help of some available tools. For example, in this paper we are using *Game Theory*. In an environment, where nodes do not have complete information about their network and are moving randomly, the *learning schemes* in [4] can aide mobile nodes to update their strategies and utility functions. The cooperative diversity

of the network is ensured by using *Stackelberg model* approach where an Access Point *(AP)* introduces the *price (λ)-reward (μ)* parameter. The prime objective of the game presented in this work is that all the players maximize their utilities or payoff. The algorithm which is developed using these techniques is named *Price-Reward for Data Relaying and Handover in Wireless Networks.*

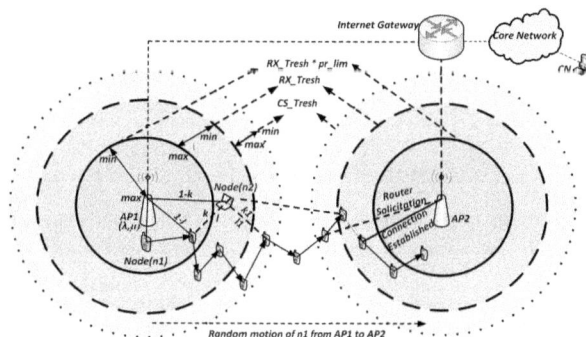

Figure 1: Data Relay and Handover

2. THE *ALGORITHM*

The working principle of our algorithm is explained with the help of a scenario illustrated in the Figure 1. The system has two wireless Access Points *AP1* and *AP2*, two mobile nodes, *n1* and *n2* and a Corresponding Node *CN*. We assume that *n1* is in session with *CN*. The area covered by an *AP* is divided into three levels as defined in [1]. The three levels are *(RX_Tresh * pr_lim)*, *(RX_Tresh)* and *(CS_Tresh)* with minimum and maximum threshold values as shown.

The algorithm performs its functions in three phases after being initialized by the network when the *(λ,μ)* values are advertised. The first phase involves the *(RX_Tresh)* level where after some time *t*, *n1* starts moving away from *AP1*. As a result the *RSS (Received Signal Strength)* value decreases in *(RX_Tresh*pr_lim)* level where $(0 \leq pr\_lim \leq 1)$ and defines the minimum and maximum *RSS* values. This triggers the data relaying process. Node *n1* allocates *l* bits of its data stream to be relayed by a neighboring node. It broadcasts packets carrying data forwarding request. Mobile Node *n2*, which has already updated itself with *(λ,μ)*, acknowledges *n1* request by proposing *k* fraction of its channel to relay *n1's l* bits of data stream. *n1* upon receiving the reply, informs *AP1* and directs *l* bits of data stream towards *n2*. The direct link between *n1* and *AP1* carries *(1-l)* bits of data and *n2* uses *(1-k)* fraction of its channel to stream its own data.

The second phase starts when the *RSS* becomes minimum in *(RX_Tresh)* level and the probability of packet loss between *n1* and *AP1* link increases. In such circumstances, *n1* finds it hard to traffic its data via direct link with *AP1*. At this moment, the network announces new (λ,μ) values with increased incentive (μ) for relaying nodes (*n2* in our case). Following that, *n1* diverts most of its data stream, say l_1 | $l_1 > l$ to the link between *n1* and *n2*. *n2* updates (λ,μ) information and allocates fraction of its channel, $k_1|k_1 > k$, for *n1*'s l_1 data stream. Once *n1* enters *(CS_Tresh)* level, *n2* starts relaying most of the *n1* data stream.

The final phase starts when the *RSS* value in *(CS_Tresh)* level gets equal to zero. The link between *n1* and *AP1* break-offs. This critical phase lasts when eventually *N1* detects *(CS_Tresh)* level of *AP2*. In *(RX_Tresh)$_{min}$* of *AP2*, *n1* sends a router solicitation message to *AP2*. *AP2* checks the *n1* authenticity with the help of Internet gateway and sends back router advertisement. While accepting the request of stronger connection, *n1* requests *AP1* via *n2* to inform Internet gateway to detour *CN* traffic towards *AP2*. Once the connection between *n1* and *AP2* establishes, *n1* terminates its link with *n2*. The acknowledgement of route diversion is received via *AP2*.

3. THE SYSTEM MODEL

Let $\mathcal{N}$ be the set of mobile nodes in a wireless network. From [4] we adopt a *Combined fully Distributed Payoff and Strategy Reinforcement Learning* (CODIPAS-RL) scheme for mobile nodes to define a strategy that can help to maximize their utility functions. Each mobile node has a set of finite action. For example, we consider a scenario of two mobile nodes (*n1,n2*) with $\mathcal{A}_1$ and $\mathcal{A}_2$ be the sets of finite number of actions in each respectively. Based on each action, a mobile node updates its strategy and payoff functions. Let a_{n_i} be an action chosen by any node $n_i \mid i \in \mathcal{N}$. The learning rates $r_{n_i,t}$ and $q_{n_i,t}$ are assigned for the strategy update($x_{n_i,t+1}$) and estimated payoff ($\hat{u}_{n_i,t+1}(a_{n_i})$) functions respectively as shown in eq.(1). $\mathbb{1}_{a_{n_i,t}=a_{n_i}}$ is an indicator function for the chosen action a_{n_i} at time t and $U_{n_i,t}$ is the observed utility function for any node n_i.

$$x_{n_i,t+1} = x_{n_i,t}(a_{n_i}) + r_{n_i,t}U_{n_i,t}(\mathbb{1}_{a_{n_i,t}=a_{n_i}} - x_{n_i,t}(a_{n_i}))$$
$$\hat{u}_{n_i,t+1}(a_{n_i}) = \hat{u}_{n_i,t}(a_{n_i}) + q_{n_i,t}\mathbb{1}_{a_{n_i,t}=a_{n_i}}(U_{n_i,t} - \hat{u_{n_i,t}})$$
$$(1)$$

We consider a network with finite number of states for a node n_i represented by the set $\mathcal{S}_i$ in which $s_{n_i,j}$ is a state for any node n_i where $j = (1,2,....M)$. Using $U_{n_i,t}$, the strategy selected at time t with a_{n_i} as the chosen action, is $x_{n_i,t}(a_{n_i})$ and the strategy function $x_{n_i,t+1}$ is updated using eq.(1). The estimated utility $\hat{u}_{n_i,t+1}(a_{n_i})$ function is also updated through the same procedure.

Eq.(2) & eq.(3) represents *n1* & *n2* utility functions where P_{n1} & P_{n2} are their powers respectively. $f(.)$ is the efficiency function from [2] and γ is the SNR. For three different paths we have γ_{n1n2}, γ_{n1a1} and γ_{n2a1}. *n1 and n2* optimize their utilities using eq.(2) & eq.(3).

$$Node1 : maxU_{n1}(P_{n1},l) \qquad (2)$$
$$\equiv max\Big(\frac{1}{P_{n1}} - \lambda\Big)\big[(1-l)f(\gamma_{n1a1}(P_{n1})) + lf(\gamma_{n1n2}(P_{n1}))\big]$$
$$s.t. \begin{cases} 0 \leq P_{n1} \leq P^{max}, 0 \leq l \leq 1, \\ lf(\gamma_{n1n2}(P_{n1})) \leq kf(\gamma_{n2a1}(P_{n2})). \end{cases}$$

$$Node2 : maxU_{n2}(P_{n2},k) \qquad (3)$$
$$\equiv max\Big[(1-k)\Big(\frac{1}{P_{n1}} - \lambda\Big) + k(\mu-\lambda)\Big]f(\gamma_{n2a1}(P_{n2}))$$
$$s.t. \begin{cases} 0 \leq P_{n2} \leq P^{max}, 0 \leq k \leq 1, \\ lf(\gamma_{n1n2}(P_{n1})) \geq kf(\gamma_{n2a1}(P_{n2})). \end{cases}$$

For every (λ,μ) pair the two nodes compete to maximize their $U_{n_i,t}$ using eq.(2) & eq.(3) for different strategies in eq.(1). Once the point of equilibrium is achieved, the network utilizes the equilibrium values (P_{n1}^*,l^*) and (P_{n2}^*,k^*) of *n1, n2* and calculates its net-revenue R_{net} using eq.(4) for different values of (λ,μ).

$$R_{net}[(P_{n1}^*,l^*),(P_{n2}^*,k^*)] = max_{(\lambda \geq =0,\mu \geq =0)}R \qquad (4)$$

where R is,

$$R = \sum \lambda(Throughput_{allnodes}) - \sum \mu(Throughput_{relayingnodes})$$

4. CONCLUSION AND FUTURE WORK

We proposed a data relaying and handover management algorithm for wireless networks. The mathematical model discussed explains the strategy selection and utility maximization methods. We are currently using NS-2 patch developed by NIST [1] (for heterogeneous wireless networks) for implementation and performance evaluation for our algorithm. The robustness of the algorithm will be proved by showing how quickly it can converge to an equilibrium state when a mobility or a handover scenario is taken into account.

5. REFERENCES

[1] "The Network Simulator NS-2 - NIST add-on - Mac 802.11", National Institute of Standards and Technology (NIST), January 2007.

[2] D. Goodman and N. Mandayam. Power control for wireless data. *IEEE International Workshop on Mobile Multimedia Communications, (MoMuC '99)*, 1999.

[3] M. Saleem, E. Renault, and D. Zeghlache. NetInf Mobile Node Architecture and Mobility Management Based on Lisp Mobile Node. *IEEE Consumer Communications and Networking Conference (CCNC)*, 2011.

[4] Q. Zhu, H. Tembine, and T. Basar. Distributed Strategic Learning with Application to Network Security. *American Control Conference (ACC)*, 2011.

JIM-Beam: Using Spatial Randomness to Build Jamming-Resilient Wireless Flooding Networks

Jerry T. Chiang
Advanced Digital
Sciences Center
Singapore
jerry.chiang@adsc.com.sg

Dongho Kim
University of Illinois at
Urbana–Champaign
Urbana, IL, USA
dkim99@illinois.edu

Yih-Chun Hu
University of Illinois at
Urbana–Champaign
Urbana, IL, USA
yihchun@illinois.edu

ABSTRACT

Since a transmitter can only be at one location at a time, a jammer must jam in a narrowband fashion in the spatial domain. We propose JIM-Beam, a narrowband jamming-resilient flooding protocol that randomizes the orientation of a node's directional antenna over time. We use ns-2 simulations to show that JIM-Beam provides improvements in packet delivery ratio over flooding naïvely and flooding using the uncoordinated frequency hopping protocol.

Categories and Subject Descriptors

C.2.1 [**Network Architecture and Design**]: Wireless communication

Keywords

availability, flooding network

1. INTRODUCTION

Wireless network protocols, such as routing or key assignment, often use *flooding* first to ensure packet delivery to all reachable network nodes. Jamming-resilient flooding is thus crucial in providing long-term network performance.

Strasser et al. propose the Uncoordinated Frequency Hopping (UFH) scheme in which a transmitter and a receiver each selects a random channel to transmit and to listen, respectively [4]. When the transmitter and the receiver *rendezvous*, the receiver is able to receive the transmission. Fig. 1(a) illustrates the UFH scheme.

We borrow the *randomization* concept from UFH and propose a flooding protocol that uses directional antennas: Each network node reorients its directional antenna toward different and randomly selected directions at different times. A node can reorient its directional antenna either mechanically (like a radar) or electrically (e.g. by array reformation or antenna reconfiguration [1]). Fig. 1(b) illustrates our protocol, which we name *JIM-Beam*: Jamming and Interference Mitigation using Beam antennas.

JIM-Beam provides one advantage over UFH: Since each adversarial transmitter can only be in a single place at any instance in time, it can only attack in a *narrowband* fashion in the spatial domain. We assume that no adversary can react to future events or authenticate forged messages.

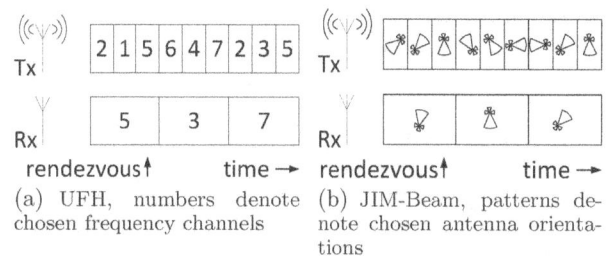

(a) UFH, numbers denote chosen frequency channels

(b) JIM-Beam, patterns denote chosen antenna orientations

Figure 1: Illustration of UFH and JIM-Beam

2. RELATED WORK

Several previous studies propose using directional antennas to reduce interference. Sani et al. proposed adaptively changing the directional antenna orientation in order to minimize the received interference [3]. Noubir used sectored antennas, and studied the theoretic minimum number of jammers to reduce the network connectivity index to 0 [2].

Strasser et al. proposed the UFH scheme [4]. JIM-Beam borrows the randomization concept from UFH and randomizes the orientation of each node's directional antenna.

3. JIM-BEAM FLOODING PROTOCOL

In the JIM-Beam protocol, each node is equipped with a directional antenna. If a node receives a packet it has not seen before, the node rebroadcasts the packet B times so other nodes can receive the packet and pass it on. Before a node wants to transmit a packet, either to initiate a new flood or to forward a packet, the sender backs off by choosing a backoff counter BC randomly, BC $\xleftarrow{r}$ $[0, \text{BC}_{max}]$ (where BC_{max} is a system parameter), and transmitting using the directional antenna after BC time slots. JIM-Beam does not require a wideband spectrum to provide jamming resiliency.

4. EVALUATION

4.1 Methodology

We implement JIM-Beam, UFH, and naïve flooding in ns-2 to compare these protocols. We augment the original ns-2 with a directional antenna patch[1]. We let each node reorient its directional antenna every one second. We equip each simulated node with a home-made cantenna, which is built using a Pringles can such that the can is able to rotate

---

[1]source code at: http://cbg.me/2009/02/adding-directional-antenna-and-multiple-interface-support-to-ns-233/

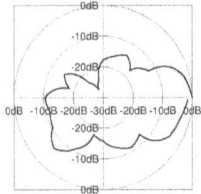

Figure 2: Radiation pattern of cantenna

(a) Field dimension = 500×500 m (b) Field dimension = 1500×1500 m (c) Field dimension = 4000×4000 m

Figure 3: Packet delivery ratio versus the number of jammers

about the line feed. Fig. 2 shows the radiation pattern of our cantenna, which has maximum directivity of 7.244. Our JIM-Beam implementation uses only one frequency channel.

To implement UFH, we augment the ns-2 packet structure with a "channel number" field, and let each node randomly choose one of eight frequency channels every one second. A UFH receiver then discards any incoming packet not bearing the channel number matching that of the receiver.

We assume each jammer uses an omni-directional antenna and has the same transmission power as a benign node. We further assume each jammer spreads his power evenly across all available frequency bands; this represents a best case scenario since a reactive jammer can always degrade the network performance more.

We place 100 benign nodes uniformly randomly on the field, and similarly place one to nine additional jammers on the field. The source node broadcasts a 256-byte packet. We simulate three field dimensions: 500×500 m, 1500×1500 m, and 4000×4000 m. Since the number of benign nodes are fixed, the results capture the performance of the flooding protocols with respect to node density.

When a node receives a new data packet, the node relays that packet with parameters $B = 50$ and $\text{BC}_{\max} = 2000$ sslot, where sslot = 2 ms is a *small slot*. We simulate each scenario 250 times, each 100 simulated seconds in duration, and measure the packet delivery ratio (PDR), i.e. the fraction of benign nodes that receive the broadcast packet.

4.2 Simulation Results

Fig. 3 shows the PDR versus the number of jammers, and compares the results between: 1. JIM-Beam; 2. UFH; and 3. naïve flooding. We plot a different curve for each scheme; and a different plot for each field size. With all protocols, intuitively, the PDR decreases in the number of jammers.

The results also show that the performance of the protocols depends on the density of the network. In networks with high and medium node density, JIM-Beam offers better but similar performance as UFH. In sparse networks, JIM-Beam significantly outperforms UFH using the same transmission power since the directional antenna extends a JIM-Beam node's transmission range.

5. JAMMING RESILIENCE OF JIM-BEAM

For ease of analysis and only for this section, we analyze the security of JIM-Beam assuming each receiver is equipped with a *sectored antenna* with solid beam-width Ω. Let each packet be t in duration and with 0 error correcting capability.

If the jammer wishes to corrupt a data packet with probability p, the jammer needs to make at least $p - (2\Omega)/360°$

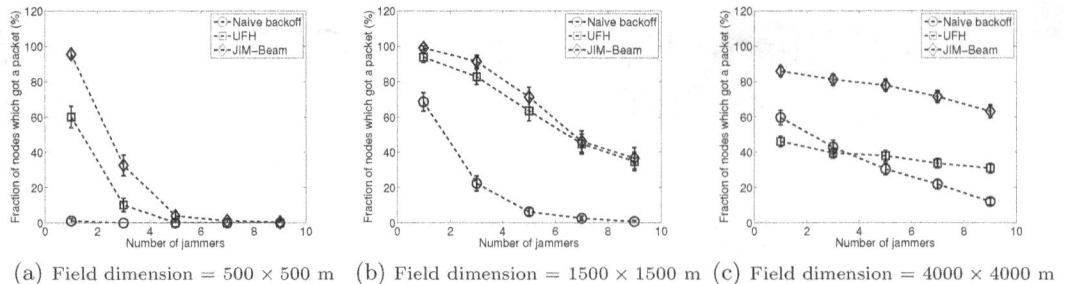

of a revolution around the receiver within time t, usually resulting in an unrealistically high centripetal force.

6. CONCLUSION

We propose a flooding protocol that uses random antenna orientation to mitigate the jamming attack. One particular benefit of using directional antennas to reject interference is that jammers cannot perform the equivalent of wideband jamming, since the location of each malicious node is a single-source of interference. We also show that the jamming resilience of JIM-Beam comes from physical, and not computational, restrictions.

We use ns-2 simulations to evaluate the effectiveness of the JIM-Beam protocol. Our results show that JIM-Beam can significantly outperform flooding naïvely or flooding using the UFH protocol.

7. ACKNOWLEDGMENTS

This material is based upon work partially supported by USARO under Contract No. W-911-NF-0710287 and by NSF under Contract No. NSF CNS-0953600. This study is also partially supported by the research grant for the Human Sixth Sense Program at the Advanced Digital Sciences Center from Singapore's Agency for Science, Technology and Research (A*STAR). We would also like to thank Dr. J. T. Bernhard for the very helpful discussions.

8. REFERENCES

[1] J. Boerman and J. Bernhard. Performance study of pattern reconfigurable antennas in MIMO communication systems. *IEEE Transactions on Antennas and Propagation*, 56(1):231 –236, Jan. 2008.

[2] G. Noubir. On connectivity in ad hoc networks under jamming using directional antennas and mobility. In *Wired/Wireless Internet Communications*, volume 2957 of *LNCS*, pages 521–532. 2004.

[3] A. A. Sani, L. Zhong, and A. Sabharwal. Directional antenna diversity for mobile devices: characterizations and solutions. In *Proceedings of the sixteenth annual international conference on Mobile computing and networking (MobiCom '10)*, pages 221–232, 2010.

[4] M. Strasser, C. Pöpper, S. Capkun, and M. Cagalj. Jamming-resistant key establishment using uncoordinated frequency hopping. In *Proceedings of the 2008 IEEE Symposium on Security and Privacy (SP 2008)*, pages 64–78, May 2008.

Extracting Jamming Signals to Locate Radio Interferers and Jammers

Zhenhua Liu
Department of CSE
University of South Carolina
Columbia, SC 29208
liuz@cse.sc.edu

Hongbo Liu
Department of ECE
Stevens Institute of
Technology
Hoboken, NJ 07030
hliu3@stevens.edu

Wenyuan Xu
Department of CSE
University of South Carolina
Columbia, SC 29208
wyxu@cse.sc.edu

Yingying Chen
Department of ECE
Stevens Institute of
Technology
Hoboken, NJ 07030
yingying.chen@stevens.edu

Categories and Subject Descriptors

C.2.0 [**Computer-Communication Networks**]: General—
Security and protection, Data communications

Keywords

Jamming, Radio interference, Localization

1. INTRODUCTION

We study the problem of localizing a constant jammer that continually emits radio signals, regardless of whether the channel is idle or not. Current jammer-localization methods mostly rely on parameters derived from the affected network topology, such as packet delivery ratios (PDR) [3] and nodes' hearing ranges [2]. Using these parameters makes it difficult to accurately localize jammer's position. We seek to localize jammer by using directly the strength of jamming signals (JSS).

Localizing a jammer utilizing JSS is challenging. First, the jamming signals are embedded in the regular network traffic. The commonly used received signal strength (RSS) measurement associated with a packet does not corresponding to JSS. Second, to improve the accuracy of wireless device localization, the existing RSS-based localization algorithms often rely on obtaining a site survey of radio RSS fingerprints during the training phase. Obtaining a JSS site survey is infeasible for jamming, since a jammer does not cooperate with localization algorithms and the jammer's transmission power is unknown.

We formulated the jammer localization problem as an error minimizing problem and used a simulated annealing algorithm for finding the best solution. Our experiments showed that our algorithm outperforms existing jammer localization approaches by 71.82%.

2. LOCALIZATION FORMULATION

Our jammer localization approach works as follows. Given a set of JSS measurements, for every estimated location, we are able to provide a quantitative evaluation feedback in-

MobiHoc'12, June 11–14, 2012, Hilton Head, South Carolina, USA.
ACM 978-1-4503-1281-3/12/06.

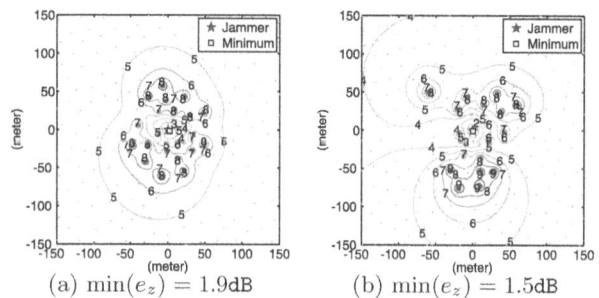

Figure 1: Illustration of error contours for e_z in a network of 200 nodes.

(a) $\min(e_z) = 1.9\text{dB}$ (b) $\min(e_z) = 1.5\text{dB}$

dicating how close it is to the true jammer's location. Although unable to adjust the estimation directly, it is possible, from a few candidate locations, to select one that is closest to the true jammer's location with high probability, making searching for the best estimate feasible. Our jammer localization approach can be divided into the two steps. (a) *Signal-Strength Collection.* The jammer localization algorithm relies on boundary nodes for sampling and collecting JSS for jammer localization. The boundary nodes are those nodes affected by jamming, and they can no longer receive packets from some of their neighbors but a subset of them. All boundary nodes obtain JSS measurements locally. (b) *Best-Estimate Searching.* Based on the obtained JSS, a designated node will search for the best estimate of jammer's position with the help of the evaluation feedback metric. There are several challenges associated with this search-based algorithm:

1. What metric is appropriate to quantify the accuracy of jammer localization?

2. How do we obtain strength of jamming signals, which may be embedded in regular transmission?

3. How do we efficiently search for the best estimate?

2.1 Localization Evaluation Metric

We quantify the evaluation feedback metric e_z as the estimated standard deviation of the random attenuation X_σ, assuming the jammer was indeed located at the estimated location.

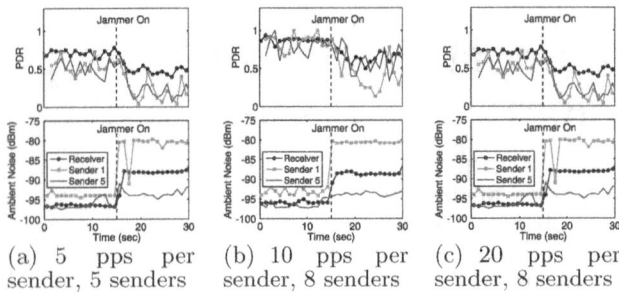

(a) 5 pps per sender, 5 senders   (b) 10 pps per sender, 8 senders   (c) 20 pps per sender, 8 senders

Figure 2: Time series plots of estimated ANF when the jammer was turned on in three traffic cases.

Algorithm 1 Acquiring the Ambient Noise Floor (ANF). ANF approximates the strength of jamming signals.

1: **procedure** MEASUREJAMMINGRSS
2:    $\mathbf{s} = \{s_1, s_2, ..., s_n\}$ = MeasureRSS()
3:    **if** var($\mathbf{s}$) < varianceThresh **then**
4:      $\mathbf{s}_a = \mathbf{s}$
5:    **else**
6:      $upperBound = \alpha[\max(\mathbf{s}) - \min(\mathbf{s})]$ ▷ $\alpha \in [0,1]$
7:      $\mathbf{s}_a = \{s_i | s_i < upperBound, s_i \in \mathbf{s}\}$
8:    **end if**
9:    **return** mean($\mathbf{s}_a$)
10: **end procedure**

Figure 1(a) and Figure 1(b) both show that e_z reaches its minimum near the jammer's true location, which suggests that e_z indicates how close an estimation is to the true jammer's location.

Assume a jammer J located at (x_J, y_J) starts to transmit at the power level of P_J, and m nodes located at $\{(x_i, y_i)\}_{i \in [1,m]}$ become boundary nodes. We can formulate the jammer localization problem as an optimization problem,

PROBLEM 1.

$$\underset{\mathbf{z}}{minimize} \quad e_z(\mathbf{z}, \mathbf{p})$$
$$subject\ to \quad \mathbf{p} = \{P_{r_1}, \ldots, P_{r_m}\};$$

where $\mathbf{z}$ are the unknown variable vector of jammer, e.g., $\mathbf{z} = [x_J, y_J, P_J + K]$, where K is an antenna-related constant; and $\{P_{r_i}\}_{i \in [1,m]}$ are the JSS measured at the boundary nodes $\{1, \ldots, m\}$.

2.2 Measuring Jamming Signals

While it is difficult, if ever possible, to extract signal components contributed by jammer or collision sources, we discover that it is feasible to derive the JSS based on periodic measurement of ambient noise, i.e., ambient noise floor (ANF).

To derive JSS, our scheme involves sampling ambient noise values regardless of whether the channel is idle or busy. In particular, each node will sample n measurements of ambient noise at a constant rate, and denote them as $\mathbf{s} = [s_1, s_2, \ldots, s_n]$. The measurement set $\mathbf{s}$ can be divided into two subsets: (1) $\mathbf{s}_a$: the ambient noise floor set measured when only the jammer is active; (2) $\mathbf{s}_c$: the *combined* ambient noise set measured when a jammer and one or more senders are active. The JSS is approximately the average of ANFs, i.e., mean($\mathbf{s}_a$). To extract $\mathbf{s}_a$ from all measurements, we designed Algorithm 1.

To verify our scheme, we conducted experiments involving one receiver and eight senders, implemented on MicaZ

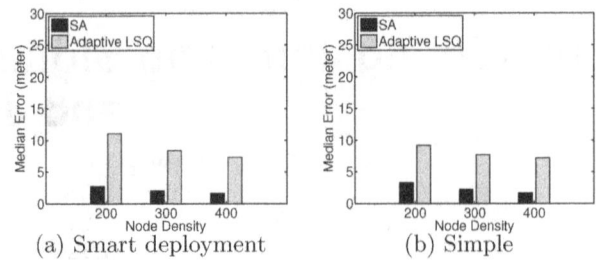

(a) Smart deployment   (b) Simple

Figure 3: Impact of node density on median localization errors.

nodes. The results depicted in Figure 2 show that our scheme is able to adaptively calculate the ANF as the jammer became active at the 15th second. Encouragingly, prior to jamming, even in the high collision case where the average PDRs are less than 50% as shown in Figure 2(c), the receivers were able to derive a stable ANF.

2.3 Finding the Best Estimation

To search for the best estimation, we propose to use a simulated annealing algorithm(SA) [1]. An SA searches for the optimal solutions by modeling the physical process of heating a material and then controlled lowering the temperature to decrease defects. At each iteration, the simulated annealing algorithm compares the current solution with a randomly-generated new solution, replaces the current solution based on a probability governed by both the new object function value and temperature. By accepting 'worse' solutions occasionally, the algorithm can avoid being trapped in local minima, and is able to explore solutions globally.

3. PERFORMANCE VALIDATION

We studied the SA-based jamming localization algorithm, and compared it to the prior work by Liu *et al.* [2], i.e. the Adaptive LSQ algorithm. Figure 3 depicts the median localization errors for both algorithms when $\{200, 300, 400\}$ nodes were deployed in 300-by-300 meter networks and a jammer was located at the center. The experimental network deployments include two types: the smart deployment with nodes following a uniform coverage and the simple deployment with a random coverage.

Firstly, we observed that SA consistently outperformed Adaptive LSQ algorithm in all the node densities and deployment setups. The median errors for SA that ranges from 1.6 to 3.3 meters, are much smaller than the errors of Adaptive LSQ that range from 7 to 11 meters. Secondly, as the network node density increases, the accuracy of all algorithms improves: median errors reduced from 2.7 to 1.6 meters for SA in a smart deployment and from 7.2 to 11 meters for Adaptive-LSQ in a smart deployment. This is because, given a jammer, a higher node density results in a larger number of boundary nodes, which in turn improves the accuracy of our algorithm.

4. REFERENCES

[1] P. V. Laarhoven and E. Aarts. *Simulated Annealing: Theory and Applications*. Springer, 1987.
[2] Z. Liu, H. Liu, W. Xu, and Y. Chen. Exploiting jamming-caused neighbor changes for jammer localization. *IEEE Transactions on Parallel and Distributed Systems*, 23(3), 2012.
[3] K. Pelechrinis, I. Koutsopoulos, I. Broustis, and S. V. Krishnamurthy. Lightweight jammer localization in wireless networks: System design and implementation. In *Proceedings of the IEEE GLOBECOM*, 2009.

A Novel Misbehavior Evaluation with Dempster-Shafer Theory in Wireless Sensor Networks

Muhammad R. Ahmed
Faculty of information Sciences and Engineering
University of Canberra
ACT, Australia
+61262012329

muhammad.ahmed@canberra.edu.au

Xu Huang
Faculty of information Sciences and Engineering
University of Canberra
ACT, Australia
+61262012430

xu.huang@canberra.edu.au

Dharmendra Sharma
Faculty of information Sciences and Engineering
University of Canberra
ACT, Australia
+61262012153

dharmendra.sharma@canberra.edu.au

ABSTRACT

Misbehavior evaluation in Wireless Sensor Network (WSN) has attracted significant attention to provide the complete security in the network. Many schemes have been proposed to identify misbehaving nodes in WSN. Most of these mechanisms relay on either predefined threshold or a set a training dataset. But it is not realistic to set a threshold or collecting training dataset of an attack ahead of time. In this paper, we have proposed a misbehaviour evaluation technique based on the dempster-shafer theory by combining the observation result by multiple neighbor nodes.

Categories and Subject Descriptors

H.4.3 [**Communications Applications**]: Language Contructs and Features –

Keywords: Wireless Sensor Network, misbehavior, security, dempster-shafer theory.

1. INTRODUCTION

Wireless sensor networks are a new technology for collecting data with autonomous sensors. Recently, this technology became more popular because of its application and cost. It consists of large number of low cost, low power and multifunctional sensors embedded with short range wireless communication capability. Sink in which all data is transmitted in an autonomous way has high capacity of storage and analysis power. [1] The application of WSN includes battlefield surveillance, border monitoring, habitat monitoring, intelligent agriculture, home automation, etc.

In order to ensure its functionality of WSN especially in malicious environments, misbehavior evaluation or security mechanisms are essential. The protection system of a WSN usually relies on the following two criterias: (i) authentication and secure protocols and (ii) intrusion and attack (misbehavior) detection. Misbehavior in WSN can take upon different forms: packet dropping, modification of data etc. in the network the compromised entity or misbehaved node of WSN acts as a legitimate node. in order to function WSN misbehavior evaluation is significant.

So far, not much attention has been given to save and secure the network by misbehavior evaluation of the node. In this paper, we have proposed Dempster–Shafer theory (DST) based misbehavior evaluation mechanism with neighbor nodes parameters observation as DST has the feature of dealing with uncertainty.

2. RELATED WORK

To do misbehavior evaluation in wireless sensors networks several works has been done in the past but efficient methods to deal with uncertainty like DST was not given significant attention. For detection of abnormal behavior of the nodes Staddon et al [2] proposed to trace the failed nodes in sensor networks at the base station assuming directional routing tree. This work can identify the failed node but directional. Watchdog like technique was proposed by Marti et al [3], this technique can detect the packet dropping attack by letting nodes listen to the next hope nodes broadcasting transmission. In this multiple watchdogs work collaboratively. So, reputation system is necessary. Zhang et al [4] proposed a scheme which is the first work on intrusion detection in wireless ad hoc networks. A new architecture but provides protection from attack on ad hoc routing.

3. METHODOLOGY

3.1 Concept

The main concept behind the node misbehavior evaluation in WSN is portrayed in Figure 1. Evidence or belief function allows one to represent and fuse information evaluation provided by more or less reliable and conflicting sources on the same hypothesis.

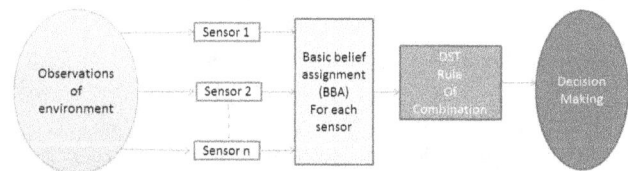

Figure 1. Concept of Node Misbehavior Detection.

3.2 Dempster-Shafer Framework

In DST, probability is replaced by an uncertainty interval bounded by belief and plausibility. Belief is the lower bound of the interval and represents supporting evidence. Plausibility is the upper bound of the interval and represents the non-refuting evidence. In this reasoning system, all possible mutually exclusive hypothesis (or events) of the same kind are enumerated in the frame of discernment Θ. A basic belief assignment (BBA) or mass function is a function $m: 2^{\Theta} \rightarrow [0, 1]$, and it satisfies two following conditions:

$$m(\phi) = 0 \qquad (1)$$

$$\sum_{A \in 2^{\Theta}} m_i(A) = 1 \qquad (2)$$

where ϕ is the empty set and a basic belief assignment (BBA) that satisfies the condition $m(\phi) = 0$ is called normal. The subsets A of Θ with nonzero masses are called the focal elements of m and the $m(A)$ indicates the degree of belief that is assigned to the exact set of A and not to any of its sub- sets. There are also two other definitions in the theory of evidence. They are belief and *plausibility* functions associated with a BBA and are defined respectively, as follow:

$$Bel(A) = \sum_{B \subseteq A} m_i(B) \qquad (3)$$

$$Pl(A) = \sum_{B \subseteq A} m_i(B) \qquad (4)$$

Bel(A) represents the total amount of probability that is allo- cated to A, while Pl(A) can be interpreted as the maximum amount of support that could be given to A. Note that, the three functions m, Bel and Pl are in one-to-one correspondence and by knowing one of them, the other two functions could be derived.

For each possible proposition Dempster-Shafer theory gives a rule of combining sensors observation. Let m_1 and m_2 be two BBA's by two independent items of evidence. These pieces of evidences can be combined using Dempster rule of combination.

$$(m)(X) \equiv [m_1 \oplus m_2] = \frac{\sum_{A \cap B = X} m_1(A) m_2(B)}{1 - \sum_{A \cap B = \phi} m_1(A) m_2(B)} \qquad (5)$$

More than two belief function can be combined with pairwise in First Page Copyright Notice

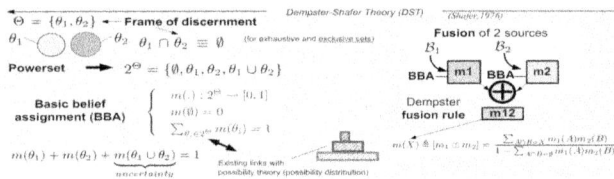

Figure 2. Dempster Shafer Theory [5]

3.3 Evaluation in WSN

In temperature collection WSN we consider the normal temperature range is $T = 8$ to 10 degree centigrade based on the Gaussian distributing with 1 sigma based on the approach taken by holder *el at* [6] , and $\sim T$ means the temperature is out of range and consider the node M is misbehaved. So, the frame of discernment consists of two probabilities concerning the attacker node A: $\Theta = \{T, \sim T\}$. Hence, for Ω the power set has three focal elements: hypothesis $H = \{\sim T\}$, $H = \{T\}$ and universe hypothesis $U = \Theta$ meaning node M is either misbehaved or a good node. We consider that neighbor node N_1 is a trusted node with the probability β. Based on the node N_1 information if node M is misbehaved, the basic probability assignment will be as follows.

$$m_1(H) = 0$$

$$m_1(\sim H) = \beta$$

$$m_1(U) = 1 - \beta \qquad (6)$$

4. EXAMPLE

In the paper we have given some mathematical calculation and results for the combined degree of belief that the node M is misbehaved

Table 1: Combine degree of belief calculation

Trust probability of the neighbor node			Combined degree of Belief
N_1	N_2	N_3	
0.9	0.8	0.2	0.975
0.2	0.2	0.9	0.878
0.8	0.8	0.8	0.828

From the table 1 we can see that the calculation is done by assigning the different trust probability to the neighbor and combine degree of belief is 0.975, 0.878, 0.828 respectively. From the high belief is it concluded that the node is behaving abnormally.

5. CONCLUSTION

In this paper a misbehavior evaluation framework in wireless sensor network is proposed with Dempster-Shafer theory of evidence combination method. The mathematical calculation shows that the result depends on the neighbor nodes reliability. Moreover, the conflict increases with the number of sources they had developed.

6. REFERENCES

[1] Huang X, Ahmed M, Sharma D, 2012 "Timing Control for Protecting from Internal Attacks in Wireless Sensor Networks", *IEEE, ICOIN 2012 conference* February (2012)

[2] Staddon J., Balfanz D., and Durfee G. 2002, "Efficient tracing of failed nodes in sensor networks," in *WSNA 2002, pp. 122-130, Atlanta*, USA (2002)

[3] Marti S., Giuli T.J., Lai K., Baker M. 2000, "Mitigating Routing Misbehavior in Mobile Ad Hoc Networks," *ACM MOBICOM 2000*, pp. 255-265, Boston, USA, August (2000)

[4] Zhang Y., Lee W. 2000, "Intrusion Detection in Wireless Ad-hoc Networks," *ACM MOBICOM 2000*, pp. 275-283, Boston, USA, August (2000).

[5] Tacnet J., Dezert. J, Hubert. M, 2000 "AHP and Uncertainty Theories for Decision Making using the ER-MCDA Methodology", *International Symposium on Analytic Hierarchy/Network Process, Sorrento (Italy)*, 15-18 june, 2011.

[6] Holder C., Boyles R., Robinson P., Raman S., and Fishel G., 2006 "Calculating a daily Normal temperature range that reflects daily temperature variability", *American Meteorological Society*, June (2006)

Your Smartphone Can Watch the Road and You: Mobile Assistant for Inattentive Drivers

[Extended Abstract]

Sanjeev Singh, Srihari Nelakuditi, and Yang Tong
University of South Carolina
{singhsk,srihari,tongy}@cse.sc.edu

Romit Roy Choudhury
Duke University
romit.rc@duke.edu

ABSTRACT

Motor vehicle accidents are one of the leading causes of death. While lane departure warning, blind spot warning, and driver attention monitoring systems for avoiding collisions have been in development for quite sometime, to date mostly luxury cars are only equipped with these safety features. As a cheaper and ubiquitous alternative, we explore how a smartphone can assist an inattentive driver by leveraging its front and back cameras apart from other sensors. The challenge, however, is given the resource constraints of a smartphone, how quickly and accurately can it detect an unintended maneuver and alert the driver. In this paper, we describe our on-going attempt to address this challenge.

Categories and Subject Descriptors

I.4.8 [**Image Processing and Computer Vision**]: Scene Analysis—Sensor Fusion; K.4.1 [**Computers and Society**]: Public Policy Issues—Human Safety

Keywords

Smartphone, Driver Inattention, Accident Avoidance

1. INTRODUCTION AND MOTIVATION

According to National Highway Traffic Safety Administration (NHTSA), more than 6 million police-reported motor vehicle crashes occurred in the United States in 2007 [1]. Around 41000 people lost their lives and another 2.5 million people were injured in these crashes. The core cause of these crashes is that drivers fail to pay attention to the road and the traffic either because they are drunk, distracted, or drowsy. Apart from advocating defensive driving, developing and deploying technologies to detect and alert inattentive drivers of unintentional actions is essential to avoid vehicle accidents and save human lives.

There has been active research towards developing systems that make driving safer [2, 3]. These include lane departure warning, blind spot warning, and driver attention monitoring systems. While these systems are quite valuable in enhancing the safety, they are pricey too. Therefore these safety features are commonly fitted only in luxury vehicles such as Lexus and Cadillac. While third party equipment is available from manufacturers like Iteris [3], the cost of

MobiHoc'12, June 11–14, 2012, Hilton Head Island, SC, USA.
ACM 978-1-4503-1281-3/12/06.

custom hardware and inconvenience of its installation can discourage drivers from adopting these systems.

Towards developing an affordable alternative for bringing safety features to economy cars such as Corolla and Pontiac, we propose to leverage smartphones that are always present with people. These smartphones are armed with front and back cameras apart from other sensors. The driver's smartphone may be placed above the dashboard or on the windshield such that it can watch the driver and the road too. The back camera can see the lane markings on the road ahead and detect lane crossings. The front camera can observe the driver's face for inattention and also cover the blindspot on the driver's side. The microphone can help infer driver's intention by identifying the sound of lane change signal. By combining the capabilities of these sensors and employing image processing algorithms, a smartphone may be able to approximate the safety features of luxury cars. In other words, our core conjecture can be posed rhetorically as, *can a smartphone turn a Pontiac into a Cadillac?'*

There are several challenges in realizing a smartphone based system for alerting inattentive drivers. First, currently only one of the two cameras can be active at a time. It is not possible to continually access the front and back camera views. Instead, the application needs to switch between them which results in increased latency and decreased frame rate. Fig. 1 shows the frame rate achieved with varying time between camera switches on a Samsung Galaxy Note. With the march of technology, it is hoped that soon both cameras can be used at the same time. Second, given the computational resource constraints of a smartphone, standard computer vision and image processing algorithms are not directly applicable. Finally and more importantly, the application should detect an unintended action and alert the driver quickly and accurately. This paper makes a case that these challenges can be addressed adequately.

We argue that even with any shortcomings, smartphone based approach is appealing. As smartphones become more advanced, the proposed system gets better over time and makes safety features affordable and available to all drivers.

2. RELATED WORK

Luxury cars from Mercedes-Benz and BMW have lane departure warning system to alert the driver with a vibrating steering [4]. The systems used by many auto manufacturers are based on core technology from Mobileye [2]. Iteris [3] is another company which offers similar systems. Both Mobileye and Iteris use radars as well as cameras for this purpose. Recently, there has been active work on using smartphones

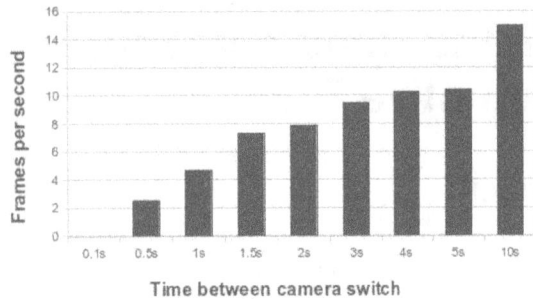

Figure 1: The resulting frame rate on a Samsung Galaxy Note with varying time between camera switches. Frequent switching reduces frame rate.

to assist drivers. SignalGuru [5] advises the driver to maintain a certain speed while approaching a signal for fuel efficiency. iOnRoad [6] is an app that warns drivers when they get too close to a vehicle. To the best of our knowledge, currently there is no smartphone based system that monitors both the driver and the road to ensure safety.

3. OUR PRELIMINARY IMPLEMENTATION

An obvious first step in using the proposed system is to mount a phone on the windshield or above the dashboard of a car. Fig. 2 shows a reasonable setting where a phone is fixed to the windshield above the drivers gaze on the road.

Figure 2: From left to right: Smartphone mounted on windshield; Back and front camera views.

Real-time performance for image processing is challenging on the phone, whose computational resources are limited compared to the type of desktop platform usually used in computer vision. The methods are chosen keeping in mind that there should be less computational overhead, since we do not have that luxury in a smart phone. OpenCV [7], is used on a Samsung Galaxy Note, running GingerBread, with 1.4 GHz processor, 8 MP back camera and 2 MP front camera. We observed a lag of around 30 ms while switching from the back to front camera, and a lag of around 120 ms while switching from the front to the back camera.

Lane Change Detection: Canny edge detection is run on back camera frames and then Hough transform is used to find lanes. The lines with slope outside a predefined range are eliminated. The resulting lane markers are shown in Fig. 3. The closest pair of markers with opposing slopes are then identified as the driving lane. A region is fixed between the two lanes which covers around 80% of the lane width. If a lane marker enters that region, it means a lane change is going to occur. Depending on which lane marker enters the region, the direction of the lane change can also be found.

Vehicle Detection: By training a Haar classifier over positive and negative samples, we can detect vehicles as in Fig. 4.

Figure 3: Lane tracking.

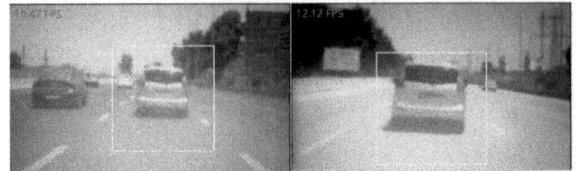

Figure 4: Vehicle tracking.

4. ON-GOING AND FUTURE WORK

Driver Attention Detection: Eye gaze detection may not be possible with the current smartphone cameras. However, the driver's head pose can be detected using smartphones. There are some well known ways to detect the head pose, but we have to choose a lightweight method. A geometric method would be appropriate, which detects the pose based on the relative configuration of eyes, nose, and the mouth. As an alternative, we can train classifiers for different kinds of head poses, and use it for detecting driver's attention.

Lange Change Signal Detection: The periodic nature of the lane change signal sound makes its detection feasible. But at high decibel levels of music and traffic noise, it may not be quite audible. However, since the frequency of the signal sound is low, we can use a low pass filter and then perform signal detection, yielding high detection accuracy.

Blind Spot Detection: The front camera covers the driver side windows. It can track vehicles in the driver's side blind spot, and alert the driver if she attempts a lane change. Since the front end of each class of vehicles look similar, we can train a classifier that can detect vehicles. We can also use optical flow to detect moving objects in the blind spot.

We plan to implement and integrate the above methods into a reliable warning system. Our goal is to push the smartphone based system far enough that it is considered an acceptable alternative to safety features in luxury vehicles.

5. REFERENCES
[1] "Traffic safety facts,"
    http://www-fars.nhtsa.dot.gov/Main/index.aspx.
[2] "Mobileye applications," http://www.mobileye.com/
    en/manufacturer-products/applications.
[3] "Autovue lane departure warning,"
    http://www.iteris.com/upload/datasheets/LDW\
    _Final_web.pdf.
[4] "Toyota safety," http:
    //www.toyota.com/esq/pdf/08_SMS_C.Tinto.pdf.
[5] E. Koukoumidis, L.S. Peh, and M.R. Martonosi,
    "Signalguru: Leveraging mobile phones for collaborative
    traffic signal schedule advisory," in ACM MobiSys, 2011.
[6] "ionroad," http://www.ionroad.com/aboutior.
[7] G. Bradski, "The OpenCV Library," Dr. Dobb's
    Journal of Software Tools, 2000.

A Distributed Channel Selection Scheme for Multi-Channel Wireless Sensor Networks

Amitangshu Pal, and Asis Nasipuri
Electrical & Computer Engineering, The University of North Carolina at Charlotte
Charlotte, NC 28223-0001, USA
apal@uncc.edu, anasipur@uncc.edu

ABSTRACT

We consider the channel assignment problem in multi-channel wireless sensor networks for maximizing the network lifetime. We assume a data collection traffic pattern where all nodes forward data periodically to a sink and propose a distributed channel selection scheme that tries to maximize the lifetime of the nodes by controlling the energy consumption from overhearing. Some initial experimental results are included to show the effectiveness of the proposed scheme.

Categories and Subject Descriptors

C.2.2 [**Network Protocols**]: Routing protocols

Keywords

Wireless sensor networks, multi-channel routing, distributed algorithms

1. INTRODUCTION

Single channel sensor networks using contention based MAC protocols suffer from the overhearing problem where nodes waste energy by receiving packets intended for other nodes. One way to reduce this is to coordinate sleep cycles of neighboring nodes, which can be a complex problem for large-scale networks without a network-wide time synchronization mechanism. In this work we consider usage of multiple channels with dynamic channel selection to control the overhearing in the network. We consider *data collecting* wireless sensor networks where all nodes sense some paramemers and periodically forward them to the sink. We assume a multi-channel transmission model where nodes can choose their own channels for reception, which they monitor by default, and any node wishing to transmit to another node needs to temporarily switch to the channel of the receiver for transmission. This leads to a multi-channel tree rooted at the sink, where individual links can be on different channels as determined by the receive channel of the corresponding receiver (Figure 1(a)).

2. DESIGN OF OUR SCHEME

We define *receiver channel* as the channel on which a node receives packets. On the other hand *transmit channel* is the channel on which a node transmits, which is the receiver channel of its intended destination. Since nodes listen to their receiver channel by default, overhearing is limited to neighboring transmissions on a node's receiver channel only. In our scheme, nodes select their receiver channels to enable distribution of traffic over multiple orthogonal channels. While transmit channels are chosen dynamically to *prolong the lifetime of the neighboring node with the worst battery health*, by remaining on the same receiver channel, frequent switching is reduced. Note that channel selection is tied to parent selection, which leads to route determination. Hence the proposed approach leads to a joint channel selection and routing in the WSN.

With these objectives, we propose a channel selection scheme that runs in two stages. In the first stage all the nodes are on the default channel and runs the *Collection Tree Protocol (CTP)* [1]. In this stage nodes choose their channels collaboratively but do not switch. In the second stage, they switch to their respective receiver channels. We assume that all nodes broadcast periodic beacon messages, which include their hop-count, their receiver channel, and a parameter indicating their battery health. The battery health parameter is calculated based on the state-of-charge of its battery and usage, which is explained later.

First stage: Nodes that are immediate neighbors of the sink are termed as first level nodes (hop-count = 1). Neighbors of the first level nodes that have hop-count = 2 are second level nodes and so on. Each first level node chooses a random backoff, and selects the least used channel in its neighborhood when the backoff timer expires. This channel becomes its receiver channel. All the first level nodes then send their hop-count, health and chosen receiver channel in the beacon messages.

Then all second level nodes go on random backoff, and choose the least used channel in their neighborhood, when their timer expire. If there are more than one channel that are least used, the tie is broken as follows: for any channel c, each second level node calculates $\mathcal{H}_c = \min\{H_i\} \ \forall \ i \in S_c$ where S_c is the set of neighbors that are in receiver channel c and H_i is the health of node i. Then a node chooses the receiver-channel j such that $\mathcal{H}_j = \max\{\mathcal{H}_c\} \ \forall \ c$. All second level nodes then send their chosen channel, hop-count and health parameter through beacon messages. This process is repeated in successive levels. After a certain time interval τ, all nodes switch to their receiver channels and the second stage begins. In the first stage all nodes store their parents as well as the parent's receiver channel. Parents are the nodes whose beacons are received by the test node and whose hop-counts are less than that of test node.

MobiHoc'12, June 11–14, 2012, Hilton Head Island, SC, USA.
ACM 978-1-4503-1281-3/12/06.

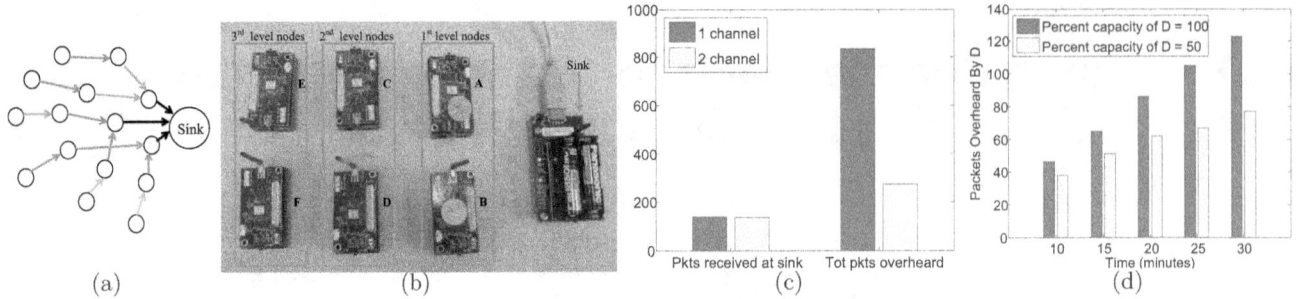

Figure 1: a) A multi-channel tree for WSN. (b) Experimental setup.(c) Comparison of packets received and total overhead with single channel and double channel. (d) Effectiveness of transmit channel selection.

Second stage: In the second stage the nodes remain in their receiver channel. In this stage nodes perform parent selection, and consequently, their transmit channel, dynamically. The first level nodes, while transmitting DATA packets switch to the channel of the sink (default channel). All nodes other than first level nodes, while transmitting the DATA packets, choose a channel c with a probability of $\frac{\mathcal{H}_c}{\mathcal{H}}$ where $\mathcal{H} = \sum \mathcal{H}_i \ \forall$ channel i in the node's neighbor. This ensures that the receiver channel of the node with the worst health is chosen with the smallest probability. Thus overhearing is minimized for the neighboring node with worst battery health. Then it chooses the parent among all its parents on c with a probability proportional to their health. Beacons are generally transmitted alternatively in different channels, so that neighbors that are on different channel get the new health periodically.

In a real network, nodes may join the the network at any time. Thus when a node joins the network, it first stays in the default channel for τ time (first stage). If it does not receive any message from any neighbor within τ, it chooses a channel randomly and stays in that channel as its receiver channel and goes in the second stage.

Battery health calculation: Based on the experimentally validated model [2], we represent the estimated average current consumption in a node by

$$\mathcal{I} = \frac{I_{Rt}T_{Rt}}{T_{rui}} + \frac{I_{Dt}T_{Dt}}{T_D} + O\left(I_{Rr}T_{Rr} + I_{Dr}T_{Dr}\right)$$
$$+ \ F.I_{Dt}T_{Dt} + \frac{I_s T_s}{T_D} + 8I_P T_P \qquad (1)$$

where I_x and T_x represent the current drawn and the duration, respectively, of the event x; and T_{RUI} and T_D represent the route update (beacon update) and data intervals, respectively. Transmission/reception of route uptate packets is denoted by R_t/R_r, data transmit/receive is denoted by D_t/D_r and processing and sensing are denoted as P and S, respectively. O and F are the overhearing and forwarding rate respectively. With this, the *health* of a mote can be calculated as $H \propto \frac{B}{\mathcal{I}}$ where B is the capacity of the battery.

We consider $MICAz$ nodes, which operate in a voltage range of 2.7V to 3.3V. The actual battery voltage is related to the ADC reading as follows: $V_{bat} = \frac{1.223 \times 1024}{VoltageADC}$. Thus we assume that when battrey voltage is greater than or equal to 3V (VoltageADC = 417 from Mica voltage sensor), the capacity is 100% and when it goes lower than 2.6V (VoltageADC = 482) the capacity is 0%. Between these two levels it is assumed that the capacity varies linearly with the

voltage, approximately. Thus we come up with an approximation expression for the capacity $B = \frac{482 - VoltageADC}{0.65}$. Even if the battery decays are not linear, we assume this linearity for simplicity. The health of a node is calculated using equation(1) in a periodic interval.

3. EXPERIMENTAL RESULTS

We implement our proposed scheme in TinyOS using Mica motes, and use an experimental setup as shown in Figure 1(b). The beacon interval, DATA interval and τ are chosen to be 30, 36 and 180 seconds, respectively.

We perform several experiments. First, we compare the performance of our channel selection scheme using two channels with the single channel case. The results are shown in Figure 1(c), where we run the experiment for 15 minutes. From Figure 1(c) we observe that the total packets received by the sink is almost similar for both cases (implying that packet delivery performance is not affected by channel switching), whereas the overhearing effect is drastically reduced in case of two channels.

To show the effectiveness of choosing transmit channels dynamically, we made the battery capacity of all nodes to be 100 in one case; in the second case, we change the capacity of node D to be 50. The number of packets overheard in node D (Figure 1(d)) is much lower for the second case as the proposed channel selection scheme reduces the transmissions in D's receiver channel.

4. CONCLUSIONS

This paper proposes a channel selection scheme for wireless sensor networks. Through experimental study, we show the effectiveness of our proposed scheme. But the experimental setup is fairly small; thus our future work aims to implement our routing scheme with large number of nodes.

5. ACKNOWLEDGEMENT

This work was supported by NSF grant CNS-1117790.

6. REFERENCES

[1] O. Gnawali, R. Fonseca, K. amieson, D. Moss, and P. Levis. Collection tree protocol. In *SenSys*, pages 1 14, 2009.

[2] A. Nasipuri, R. Cox, . Conrad, L. V. der el, B. Rodriguez, and R. McKosky. Design considerations for a large-scale wireless sensor network for substation monitoring. In *LCN*, pages 866 873, 2010.

Can Smartphone Sensors Enhance Kinect Experience?

[Extended Abstract]

Rufeng Meng, Jason Isenhower, Chuan Qin, Srihari Nelakuditi
Computer Science and Engineering, University of South Carolina, Columbia, SC 29208
{mengr, isenhowj, qinc, srihari}@cse.sc.edu

ABSTRACT

Kinect has become quite popular for gaming as it tracks players' natural gestures without a controller like other gaming systems. But it has some inherent limitations such as occlusion problem and fails to track a player accurately if there is an obstacle. In this work, we propose to leverage smartphone to supplement Kinect. We explore how to fuse information from Kinect's tracking with the smartphone's sensor readings to improve Kinect gaming experience.

Categories and Subject Descriptors

I.4.8 [**Image Processing and Computer Vision**]: Scene Analysis; K.8.0 [**Personal Computing**]: General

Keywords

Kinect, Smartphone, Games, Sensor Fusion

1. INTRODUCTION AND MOTIVATION

Kinect is one of the most successful systems that promote the use of natural and intuitive gestures from the users. Kinect employs cameras and depth sensor (see Fig. 1) to track the user's body in real time. It uses an infrared sensor to project light patterns on users in front of it. The light pattern is distorted by the users due to the different distance from the users to Kinect. The depth sensor uses the distorted light pattern to calculate the depth (i.e. the distance from Kinect to user) based on some reference frame corresponding to the projected light pattern [1].

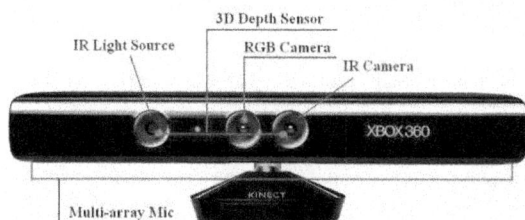

Figure 1: Kinect Sensors

Although Kinect is very intuitive, it has some inherent limitations. One of the major problems Kinect suffers from

is occlusion. When some joints of the body are hidden by any obstacle, Kinect loses its tracking and tries to infer the coordinates of the hidden joints. But in many instances, its inference can be quite inaccurate. Kinect also has a relatively small view range (horizontal view: 57 degrees, vertical view: 43 degrees) which limits its ability to simultaneously track multiple users (currently Kinect can only actively track 2 users and detect 4 additional users at the same time). Kinect is designed for relatively short distance usage (less than 5 meters), which limits its applicability in wider spaces. Although Kinect has other limitations and drawbacks, we plan to focus on these three aspects.

We propose to enhance the capabilities of Kinect by leveraging smartphones. Nowadays, many people have smartphones and it is reasonable to expect that they carry them (in pocket or on belt) even when they are playing games. Almost all the smartphones have accelerometer, gyroscope, microphone, etc. When the player is moving, the smartphone sensors can produce information related to the movement. For example, when the player is kicking or rotating, the accelerometer and gyroscope can potentially measure the real-time acceleration and rotation speed of the player's action. Although Kinect can track the player's movement, it may not know how hard a player is doing a kicking action or the real-time speed of a rotation of the body. The smartphone can help to improve the accuracy and sensitivity of Kinect and potentially mitigate its limitations.

We plan to fuse the data from Kinect and smartphone's sensors to enhance the capability and user experience of Kinect system for both the developers and players.

2. RELATED WORK

Some work in combining Kinect with smartphones has been done in literature. The authors of [2] built an interactive augmented reality system, which uses Kinect to capture the motion, depth map and real images. They also use smartphone, which is attached to a cap, to measure the heading, pitch and roll of the head from the phone's accelerometer and compass. [3] presents a solution which utilizes a Kinect to sense gesture input and uses Android smartphones to interact with the display wall. In the system shown in [4], Kinect is used to track the user behavior, and the acceleration information from mobile phone is utilized to complement the coarse contextual data of Kinect.

3. OUR APPROACH

Kinect continuously tracks and records the video image, skeletal joint coordinates, and depth information for the users in its viewing range. This information is transmitted to a Xbox in real-time. Fig. 2 illustrates the proposed smartphone-enhanced Kinect system. The user carries smartphone in pocket when (s)he plays games or uses Kinect applications. In our design, the smartphone also connects with the Xbox. The sensor readings from the smartphone are also sent to the Xbox in real-time. The Xbox fuses the data from Kinect and smartphone sensors to better track the users.

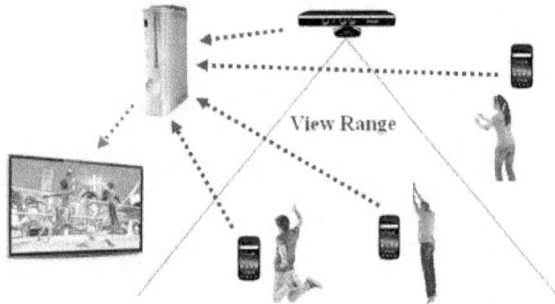

Figure 2: Smartphone-enhanced Kinect System

We propose to enhance Kinect system in three aspects with the aid of smartphones: broaden its tracking range, improve its accuracy, increase its hearing distance.

Current Kinect systems can only actively track 2 users and detect 4 additional users simultaneously. When a user goes out of its view, Kinect will stop tracking the user and the character corresponding to the user in the game might be destroyed. With the aid of a smartphone, even when the user runs out of Kinect's view, Xbox can still detect that the user is still playing by analyzing the pattern of the smartphone's sensor readings. Moreover, since Kinect has tracked this user in the past, Xbox can remember some characteristics (e.g. height, shoulder width) of the user. This allows Xbox to construct or maintain the character in the game corresponding to the user who is outside Kinect's view. Smartphone's sensors could tell whether the user is spinning, jumping, etc., enabling the Xbox to render the character's movements. In this way, the user could still be in the game even if she has moved out from Kinect's view temporarily. In other words, smartphone broadens Kinect's tracking range which makes it possible to detect and track more users simultaneously.

Kinect suffers from occlusion problem. When some joints of the user are hidden, Kinect loses its tracking and tries to infer the coordinates for these hidden joints. But in many instances, this inference is not accurate, sometimes it is totally incorrect. We do a simple experiment to study this problem. Figure 3 shows the user doing a spin in front of the Kinect. A smartphone is in the user's pocket and the built-in gyroscope is recording the user's rotation. The two graphs in the bottom of Figure 3 show the rotation result calculated from Kinect's skeletal joint coordinates (we use the two joints on left and right shoulders to calculate the rotation of the user) and smartphone's gyroscope readings. From the graph, it is evident that there are some moments when the user's joints are unseen, Kinect fails to track and gives inaccurate joint coordinates, which leads to the rugged circle. On the other hand, smartphone's gyroscope can record the user's rotation in any direction and measure the rotation accurately, which leads to the perfect circle at the bottom right of Figure 3.

Each smartphone has an embedded microphone. When

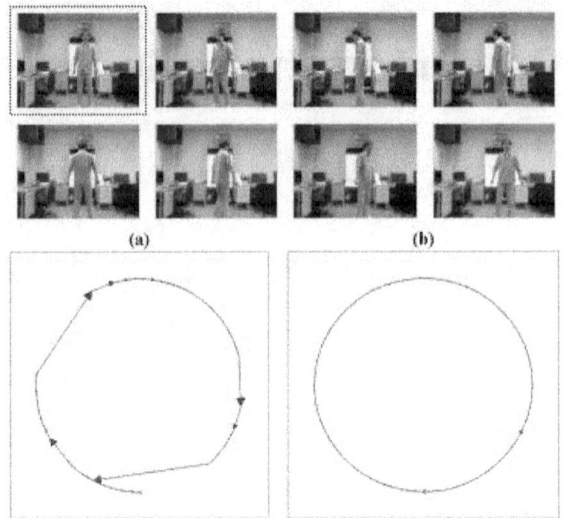

Figure 3: User spins (shown in the top). Related rotation retrieved from Kinect (bottom left) and smartphone's gyroscope (bottom right).

the smartphone is carried by the user, its microphone is much closer to the user compared to Kinect's microphones. So the smartphone could collect the voice sample and transmit to Xbox to fuse with the voice recorded by Kinect. In this way, we can improve Kinect's hearing distance.

4. ONGOING WORK

We are currently analyzing and fusing the sensor data from both Kinect and smartphone. We plan to develop a prototype and do more experiments to see in which aspects and to what extent the smartphone sensors can help enhance the capabilities of Kinect. Depending on how a user carries the smartphone, it may only measure the movement of one particular part of the body (one joint in Kinect's viewpoint). Different games and applications require the movement of different parts of the user's body. Therefore, we need to investigate which categories of games and applications are likely to benefit most from smartphone-enhanced Kinect system. Our long-term goal is to create a set of enhanced Kinect API based on current Kinect SDK for developers, which allows subtler and smoother control over the characters in games as well as actions in applications.

5. REFERENCES

[1] John MacCormick, "How does the kinect work?," http://users.dickinson.edu/~jmac/selected-talks/kinect.pdf.

[2] L. Vera, J. Gimeno, I. Coma, and M. Fernández, "Augmented mirror: interactive augmented reality system based on kinect," *Human-Computer Interaction–INTERACT 2011*, pp. 483–486, 2011.

[3] V. Kulkarni, H. Wu, and Y. Jin, "Bayscope: Evaluating interaction techniques for large wall-sized displays," in *CHI 2012*, Austin, TX, USA, May 2012, ACM.

[4] L. Norrie and R. Murray-Smith, "Virtual sensors: rapid prototyping of ubiquitous interaction with a mobile phone and a kinect," in *ACM Mobile HCI*, 2011.

Authors Index

Abu-Affash, Karim 65
Addepalli, Sateesh 225
Agrawal, Prathima 251
Ahmed, Muhammad R. 259
Alves, Thierry 245
Ammar, Mostafa H. 145
Arslan, Mustafa Y. 215, 235
Babaei, Alireza 251
Banerjee, Suman 235
Carzaniga, Antonio 155
Che, Xin 75
Chen, Dongsheng 251
Chen, Shigang 95
Chen, Xu 205
Chen, Yingying 257
Chiang, Jerry T. 255
Diot, Christophe 165
Dutta, Rudra 249
Efrat, Alon 65
Eriksson-Bique, Sylvester David....................65
Eshghi, Soheil 175
Han, Bo 5
Han, Junze 135
Han, Kai 85
Hu, Yih-Chun 255
Huang, Jianwei 205
Huang, Xu 259
Isenhower, Jason 265
Johnson, Troy A. 247
Ju, Xi 75
Jung, Taeho 135
Khalil, Issa M. 115
Khazaei, Koorosh 155
Khouzani, MHR. 175
Khreishah, Abdallah 115
Kim, Dongho 255
Knightly, Edward W. 1
Krishnamurthy, Srikanth V...........215, 225, 235
Kruegel, Christopher 195
Kuhn, Fabian 155
Kumar, V.S. Anil. 35
Lakafosis, Vasileios 145
Li, Jason Hongjun.....................................105
Li, Tao 95
Li, Xiang-Yang 135
Li, Xinfeng 25
Liu, Hongbo 257
Liu, Xiaohui 75
Liu, Yang 85
Liu, Zhenhua 185, 257
Lohier, Stéphane 245
Lu, Songwu 185
Luan, Wentao 185
Luo, Jun 85

Luo, Wen 95
Ma, Chao 135
Mao, Xufei 135
Meng, Rufeng 265
Miller, Kyle 55
Nasipuri, Asis 263
Nelakuditi, Srihari 261, 265
Nimbhorkar, Sankalp 249
Pal, Amitangshu 263
Pathak, Parth H. 249
Pei, Guanhong 35
Pelechrinis, Konstantinos 225
Pietiläinen, Anna-Kaisa 165
Polishchuk, Valentin 65
Poussot, Benoit 245
Qiao, Yan 95
Qin, Chuan 265
Rachedi, Abderrezak 245
Ramasubramanian, Srinivasan 65
Rangarajan, Sampath 215
Rehmani, Mubashir Husain. 245
Renault, Éric 253
Roy Choudhury, Romit 261
Sagduyu, Yalin Evren. 105
Saleem, Muhammad Shoaib 253
Sankararaman, Swaminathan 65
Sanne, Atresh 55
Sarkar, Saswati 175
Seeling, Patrick 247
Segal, Michael 65
Shahbazi, Moloud 225
Sharma, Dharmendra 259
Shi, Cong 145
Shroff, Ness B. 125, 175
Singh, Sanjeev 261
Singh, Shailendra 225
Srinivasan, Aravind 5
Srinivasan, Kannan 55
Sundaresan, Karthikeyan 215, 225, 235
Symington, Andrew 15
Tang, Shaojie 45, 135
Teng, Jin 25
Tong, Yan 261
Trigoni, Niki 15
Vaidya, Nitin 3
Venkatesh, Santosh S. 175
Vishwanath, Sriram 55
Wan, Peng-Jun 45
Wan, Zhiguo 45
Wang, Xinbing 185
Wang, Zhu 45
Wu, Jie 115
Xiang, Liu 85
Xiang, Qiao 75

Xu, Boliu 135
Xu, Wenyuan 257
Xu, Xiaohua 45
Xuan, Dong 25
Yang, Lei 105, 195
Yang, Yang 125
Yoon, Jongwon 235
Zegura, Ellen W.. 145

Zhang, Boying 25
Zhang, Hongwei 75
Zhang, Zengbin 195
Zhao, Ben Y.. 195
Zheng, Haitao 195
Zheng, Yuan F.. 25
Zhu, Junda 25

www.ingramcontent.com/pod-product-compliance
Lightning Source LLC
Chambersburg PA
CBHW061351210326

41598CB00035B/5950